HUMAN PERFORMANCE PHYSIOLOGY AND ENVIRONMENTAL MEDICINE AT TERRESTRIAL EXTREMES

Edited by
KENT B. PANDOLF, Ph.D.
MICHAEL N. SAWKA, Ph.D.
RICHARD R. GONZALEZ, Ph.D.
U.S. Army Research Institute of Environmental Medicine
Natick, Massachusetts

WCB Brown & Benchmark

Disclaimer and Distribution Statements

The views, opinions, and/or findings contained in this book are those of the authors and should not be construed as an official Department of the Army position, policy, or decision, unless so designated by other official documentation.

Citations of commercial organizations and trade names in this book do not constitute an official Department of the Army endorsement or approval of the products or services of these organizations.

Approved for public release; distribution is unlimited.

Library of Congress Cataloging in Publication Data:

PANDOLF, KENT B. 1945 -

HUMAN PERFORMANCE PHYSIOLOGY AND ENVIRONMENTAL MEDICINE AT TERRESTRIAL EXTREMES

Cover Design: Gary Schmitt

Copy Editors: Michele Rose

Lynn Hendershot

Library of Congress Catalog Card number: 86-73129

ISBN: 0-697-14823-8

Printed in the United States of America by Brown & Benchmark, 2460 Kerper Boulevard, Dubuque, IA 52001.

10 9 8 7 6 5 4 3

About The Editors

Kent B. Pandolf received his B.S. degree from Boston University in 1967 and his M.A. degree in 1968, M.P.H. degree in 1970, and Ph.D. degree in physiology from the University of Pittsburgh in 1972. After a post-doctoral research fellowship at the John B. Pierce Foundation Laboratory and Yale University School of Medicine, he joined the staff at the U.S. Army Research Institute of Environmental Medicine in 1973. He is currently the Director of the Military Ergonomics Division at this Institute and holds the rank of Adjunct Professor of Health Sciences at Boston University and Adjunct Professor of Environmental Medicine at Springfield College.

Michael N. Sawka received his B.S. degree in 1973 and M.S. degree in 1974 from East Stroudsburg University, and his Ph.D. degree in exercise physiology from Southern Illinois University at Carbondale in 1978. After a post-doctoral fellowship and subsequently becoming Assistant Professor at Wright State University School of Medicine from 1978 to 1980, he joined the U.S. Army Research Institute of Environmental Medicine. He currently is Chief, Physiology Branch, Military Ergonomics Division. His primary research interests are integrative physiology as well as exercise and environmental physiology.

Richard R. Gonzalez received his B.S. degree from the University of Texas in 1961, his M.S. degree from the University of San Francisco in 1966 and his Ph.D. degree in physiology and biophysics from the University of California, Davis in 1970. From 1970 to 1983 he was involved with research on thermoregulatory modeling, mechanisms of heat transfer, and physical factors in the environment at the John B. Pierce Foundation Laboratory and Yale University School of Medicine. He joined the U.S. Army Research Institute of Environmental Medicine in 1983 and is Chief, Biophysics Branch, Military Ergonomics Division. He is currently Adjunct Professor of Environmental Science and Physiology at the Harvard Medical School.

Foreword

In 1986, the U.S. Army Research Institute of Environmental Medicine (USARIEM) commemorated its 25th anniversary. The Institute was officially activated as an installation of the U.S. Army Medical Research and Development Command on July 1, 1961 with the mission "to conduct basic and applied research to determine how heat, cold, high terrestrial altitude, and work affect the soldier's life processes, his performance and his health." In reality, the Institute is much older than that because its history dates back to laboratories established during World War II. It was founded as a composite of elements of the Army Medical Research Laboratory located at Fort Knox, KY, and of the Environmental Protection Research Division of the Quartermaster Research and Development Command at Natick, Massachusetts. Thus, the Institute was originally a merger of a medical and a quartermaster unit. The Institute currently has a staff of 151 personnel including 18 officers, 56 enlisted personnel, and 77 civilians, including 43 scientists at the doctoral level and another 24 with masters' degrees. Scientists and support personnel are grouped organizationally into eight Divisions, of which four (Altitude Research Division, Cold Research Division, Heat Research Division, and Military Ergonomics Division) contributed to the preparation of this book.

Our Institute's scientific staff has access to many unique and highly specialized chambers that are capable of supporting animal and human experimentation in a wide variety of simulated environmental extremes. Two altitude chambers with an air lock can simulate altitudes up to 7,620 meters, temperatures from −22° C to 50° C, and humidity from 20% to 80% over extended periods of time. The Institute also maintains a field laboratory on the summit of Pike's Peak, Colorado at an altitude of approximately 4,300 meters. Additionally, our Institute houses 14 environmental chambers which are collectively capable of providing controlled temperatures ranging from −40° Cto 60° C. Facilities are also available for water immersion research in a thermally-controlled pool which is 3 meters × 3 meters × 4.2 meters (deep) holding 36,000 liters of water and capable of controlling water temperature between 5° C and 45° C.

Many of the scientists at this Institute felt that a book should be prepared which provides the student and professional with the most recent information on the physiological and medical responses to the environmental extremes of heat, cold, altitude, diving and poor air quality. This book is dedicated to this task and to further commemorate our Institute's 25th anniversary.

Guest Referee Editors

The Editors gratefully acknowledge the assistance of the following Guest Referee Editors in the review of these chapters.

Thomas K. Akers, Ph.D.

Larry G. Berglund, Ph.D.

Jack A. Boulant, Ph.D.

Ralph P. Francesconi, Ph.D.

A. Pharo Gagge, Ph.D.

Carl V. Gisolfi, Ph.D.

Robert F. Grover, M.D., Ph.D.

Allan J. Hamilton, M.D.

Margaret A. Kolka, Ph.D.

Kenneth K. Kraning, II, Sc.D.

Stephen R. Muza, Ph.D.

Lou A. Stephenson, Ph.D.

John R. Sutton, M.D.

C. Bruce Wenger, M.D., Ph.D.

Andrew J. Young, Ph.D.

Contributors

Captain Lawrence E. Armstrong, Ph.D.
Research Physiologist
Heat Research Division
U.S. Army Research Institute of Environmental Medicine
Natick, Massachusetts

Allen Cymerman, Ph.D.
Director, Altitude Research Division
U.S. Army Research Institute of Environmental Medicine
Natick, Massachusetts

Charles S. Fulco, M.A.T.
Research Physiologist
Altitude Research Division
U.S. Army Research Institute of Environmental Medicine
Natick, Massachusetts

Richard R. Gonzalez, Ph.D.
Chief, Biophysics Branch
Military Ergonomics Division
U.S. Army Research Institute of Environmental Medicine
Natick, Massachusetts

Murray P. Hamlet, D.V.M.
Director, Cold Research Division
U.S. Army Research Institute of Environmental Medicine
Natick, Massachusetts

Roger W. Hubbard, Ph.D.
Director, Heat Research Division
U.S. Army Research Institute of Environmental Medicine
Natick, Massachusetts

Margaret A. Kolka, Ph.D.
Research Physiologist
Physiology Branch, Military Ergonomics Division
U.S. Army Research Institute of Environmental Medicine
Natick, Massachusetts

Mark K. Malconian, D.O.[1]
12 Mere Lane
Kennebunk, Maine

William D. McArdle, Ph.D.
Professor
Laboratory of Applied Physiology
Department of Health and Physical Education
Queens College of City University of New York
Flushing, New York

Stephen R. Muza, Ph.D.[2]
Technical Director
Pulmonary Function Laboratory
Temple University Hospital
Philadelphia, Pennsylvania

Kent B. Pandolf, Ph.D.
Director, Military Ergonomics Division
U.S. Army Research Institute of Environmental Medicine
Natick, Massachusetts

Major Paul B. Rock, D.O., Ph.D.[1]
Department of Medicine
Fitzsimmons Army Medical Center
Aurora, Colorado

William R. Santee, Ph.D.
Research Physical Scientist
Biophysics Branch, Military Ergonomics Division
U.S. Army Research Institute of Environmental Medicine
Natick, Massachusetts

Michael N. Sawka, Ph.D.
Chief, Physiology Branch
Military Ergonomics Division
U.S. Army Research Institute of Environmental Medicine
Natick, Massachusetts

---

[1] Former Major, U.S. Army Medical Corps, stationed at the U.S. Army Research Institute of Environmental Medicine, Natick, MA.
[2] Former Captain, U.S. Army Medical Service Corps, stationed at the U.S. Army Research Institute of Environmental Medicine, Natick, MA.

Lou A. Stephenson, Ph.D.
Research Physiologist
Physiology Branch, Military Ergonomics Division
U.S. Army Research Institute of Environmental Medicine
Natick, Massachusetts

Michael N. Toner, Ph.D.[3]
Associate Professor
Laboratory of Applied Physiology
Department of Health and Physical Education
Queens College of City University of New York
Flushing, New York

C. Bruce Wenger, M.D., Ph.D.
Research Pharmacologist
Physiology Branch, Military Ergonomics Division
U.S. Army Research Institute of Environmental Medicine
Natick, Massachusetts

Andrew J. Young, Ph.D.
Research Physiologist
Physiology Branch, Military Ergonomics Division
U.S. Army Research Institute of Environmental Medicine
Natick, Massachusetts

Captain Patricia M. Young, Ph.D.
Research Biochemist
Altitude Research Division
U.S. Army Research Institute of Environmental Medicine
Natick, Massachusetts

---

[3] Captain, U.S. Army Reserve, stationed at the U.S. Army Research Institute of Environmental Medicine, Natick, MA.

Contents

1

Characteristics of the Thermal Environment

WILLIAM R. SANTEE, Ph.D.

RICHARD R. GONZALEZ, Ph.D.

OUTLINE

THE THERMAL ENVIRONMENT . . . PATHWAYS OF HEAT EXCHANGE

Broadly defined, the environment consists of everything external to an organism's surface. Not all environmental parameters have a significant impact on physiology or human performance. In terms of the impact of the physical environment on an organism's physiology, the primary effect of the environment is that it determines the potential for heat exchange between a person and the environment. The actual heat exchange is determined by the interaction of the environment, certain physical characteristics of the organism such as posture and surface characteristics and internal physiology. The environmental parameters which determine the potential for heat exchange constitute the "thermal environment." Although the references to the thermal environment are common in the literature, a formal definition is rarely presented. Tracy, *et al.* (68) defined the thermal environment as "a biophysical aggregate of air temperature, wind speed, relative humidity, and radiation."

The overall impact of the environment on an individual during the course of a single day is difficult to establish, because people do not usually occupy a single, simple, homogeneous environment. In the course of a day, we move between environments—from room to room and indoors to outside. Even in a single outdoor location, environmental conditions change; daylight to night, calm to windy, warm to cold.

A triathlon event (62) is an example of how complex the interaction between individuals and their environment can become. An athlete in a triathlon typically starts with a morning swim. In cold water, heat loss is rapid, but the response is partially offset by the intense muscular activity. The participants emerge from the water, and begin the bicycle stage, perhaps along a winding coastal highway, alternating exposure to sun and shade, wind and calm as well as the air flow created by the speed of their own motion. As the morning progresses, the sun rises higher in the sky, increasing both air temperatures and solar radiation. As the ground heats, wind movement, directed as up and down slope breezes and more general winds change, in strength and direction. Along the coast the air tends to be humid, limiting the cooling value of body sweat. At the end of the cycle ride the participants begin a marathon, generally running on more level terrain under the afternoon sun as air temperatures reach their daily maximum. In the warmer air, away from the ocean moisture, sweat evaporates more readily, allowing more heat to be lost. For the slower participants, the sunlight wanes at sunset and they will continue their run into the night, either benefiting from the absence of direct sunlight and cooler air temperatures, or becoming chilled due to the combination of colder air and radiant heat loss.

From the preceding account, it should be apparent that, quantitatively, the environment involves a measurement of a number of variables. In the hypothetical triathlon, the participant's environment changed with time and place as the race progressed from morning to night and out of the ocean, along the coast and out across the flats. Our environment, therefore, has both a temporal aspect related to daily and annual cycles and a spatial aspect. Even a laboratory physiologist must take environmental variation into account if an attempt is made to select laboratory conditions which approximate a "representative" environment.

In order to characterize the environment, it is first necessary to understand the basic pathways of heat exchange between individuals and their environment. Once we have established how the environment impacts on an individual, then the question becomes which environmental parameters do we need to measure and what meteorological instruments can be used to collect the necessary data. In collecting any data, environmental or physiological, it is necessary to consider the cost and other limited resources, such as time or technical expertise relative to the utility of that information. In this chapter we will present a variety of methods.

Energy Balance Equation

The basis for measuring the effects of the environment on an organism is derived from the First Law of Thermodynamics used in the heat balance equation:

$$S = M - (\pm W_k) \pm E \pm R \pm C \pm K, \quad [W \cdot m^{-2}] \tag{1}$$

Internal heat production is represented by metabolism (M). Heat exchange occurs by evaporation (E), conductance (K), convection (C) or radiation (R). The remaining energy is either work (W_k, where + is positive work representing energy leaving the system and − is negative or eccentric work) or heat storage (S).

The heat exchange pathways, evaporation, conductance, convection and radiation can be subdivided into two categories, mechanisms of dry or sensible heat exchange and insensible or wet heat exchange. The potential for dry heat exchange is determined primarily by the environment whereas evaporative heat exchange is strongly influenced by sweating rate, a physiological function, as well as the limits imposed by the external environment.

Sensible or Dry Heat Exchange

Dry heat exchange (or non-evaporative heat exchange) does not involve the evaporation of water from the body surface and can be determined directly as a function of the measurable (sensible) difference in temperatures between the organism and its environment. Dry heat exchange is particularly useful when the ability to sweat is inhibited, such as by individuals wearing impermeable clothing.

When evaporative heat loss cannot occur, the potential for dry heat exchange represents the variability in the thermal environment.

Convection. Convection is heat exchange between a surface and a fluid, normally air or water. The rate of convective heat exchange is dependent on the density of the fluid, the temperature gradient between the surface and the fluid, the area of exposure, and the flow rate and turbulence of the fluid. Equation 2 is an equation for convective heat exchange:

$$C = h_c (T_s - T_a), \quad [W \cdot m^{-2}] \tag{2}$$

The variables are the temperatures of the surface and the fluid ($T_s - T_a$ in K), and the convective heat transfer coefficient (h_c in $W \cdot m^{-2} \cdot K^{-1}$). Determination of h_c is not an easy step in calculating convective heat loss as it is a complex function of fluid density, flow and shape or posture. There are two types of convection, free and forced convection.

"Free" convection is primarily a function of fluid density and is important only in fluids that are still or moving at very low velocities or flow rates. Free convection results from density gradients forming in the fluid surrounding the subject, due to the warming/cooling of fluids which are in close proximity to the body surface. Free convection creates a stable, static gradient of fluid surrounding the surface and, in effect, alters the rate of heat exchange between the surface and the environment.

"Forced" convection is a function of fluid velocity in addition to fluid properties and is important at higher wind speeds. Fluid movement disrupts the static layering responsible for free convection. The flow characteristics of the fluid are important because a smooth, turbulent free (laminar) flow creates a series of layers of fluid of increasing velocity above the surface. The higher the velocity of the fluid, the thinner the layers. The thickness and velocity present within these layers affect the rate of heat transfer between the surface and the surrounding fluid. The faster that fluid flows over the surface, the less time there is for heat transfer to a given volume of fluid. As a consequence, the slower moving fluid layer which is closest to the surface is directly mixed with the temperature of the surface (14,17,47). Turbulence in the fluid flow disrupts the layers or lamination, bringing warmer/colder fluid into proximity to the surface. Turbulence can also be caused by surface texture or roughness and irregular surfaces (14,30,49).

The rate of heat exchange by convection is given by the heat transfer coefficient (h_c). To calculate h_c, a series of empirically derived dimensionless variables are used: the Nusselt number (Nu), the Grashof number (Gr), the Prandlt number (Pr), and the Reynolds number (Re). The Nusselt number is related to h_c by the equation:

$$h_c = Nuk_f/L, \quad [W \cdot m^{-2} \cdot K^{-1}] \tag{3}$$

Where L is the characteristic dimension or shape factor and k_f is the thermal conductivity of the fluid: for air (14,49) it is a function of air temperature (°C):

$$k_a = 2.41 \cdot 10^{-2} + 7.8 \cdot 10^{-5} \cdot T_a, \quad [W \cdot m^{-1} \cdot {}^{\circ}C^{-1}] \tag{4}$$

The Grashof and Prandlt non-dimensional (N.D.) numbers are used to calculate the Nusselt number for free convection. In air (T_a 10–50° C [55]), the Prandlt number is a constant (0.72). The Grashof number can be calculated according to the following equation:

$$Gr = a \cdot g \cdot L^3 (T_s - T_a)/\nu^2, \quad [N.D.] \tag{5}$$

The variables are the coefficient for the thermal expansion of air (a), the acceleration due to gravity (g) and the kinematic viscosity of air (ν). Temperatures may be expressed either °C or K in equation 4, provided both T_s and T_a agree in scale. The latter term can be calculated from T_a (°C):

$$\nu = 1.33 \cdot 10^{-5} + 9.0 \cdot 10^{-8} \cdot T_a, \quad [m^2 \cdot s^{-1}] \tag{6}$$

The equation for the Nusselt number for free convection is:

$$Nu = cGr^a Pr, \quad [m \cdot s^{-1}] \tag{7}$$

But, substituting a constant value (c′) for cPr in air, the equation becomes:

$$Nu = c'Gr^a, \quad [N.D.] \tag{8}$$

A value of 0.50 for c′ was derived from Campbell (14) for cylinders, spheres and flat surfaces.

For forced convection, the Nusselt number is calculated from the Prandlt and Reynolds numbers:

$$Nu = cPr^aRe^b, \quad [N.D.] \tag{9}$$

The constants, a, b and c, are dependent on shape factors. As noted, the Prandlt number can be treated so, just as in the case of free convection, the equation can be simplified by substituting a constant value for Pr^a:

$$Nu = c''Re^b, \quad [N.D.] \tag{10}$$

Nishi *et al.* (1970) cites "compromise" values of 0.33 and 0.55 for c″ and b, respectively, for a cylinder.

The Reynolds number can be calculated from the equation:

$$Re = v\nu/L, \quad [N.D.] \tag{11}$$

Where v is the wind speed. The characteristic dimension (L) is determined by the shape and actual dimensions of the subject. For a cylinder, the usual model for a standing human, L is the diameter.

The original relationships were derived in wind tunnels (laminar flow). In natural environments, however, air flow near the surface often is turbulent. Also, the geometry of human and animal forms is complex compared to the models used to derive the original relationships. Mitchell (49) presents alternative equations for calculating the characteristic dimension and Nusselt number for small animals and humans in outdoor environments.

Nishi and Gagge (55) proposed the use of an adjusted heat transfer coeffficient, $\bar{h}_c$, to adjust for the activity of the subject. The sublimination rate of napthalene spheres, fixed just above the body surface, were used to estimate the convective heat transfer rate. For a subject resting on a stationary cycle they calculated h_c at 3.1, but for a subject cyclying at 60 rpm, $\bar{h}_c$ was 6.0 (wind speed $\sim 0.15\ m \cdot s^{-1}$).

Conduction. Conduction (K) is heat exchange between two solid surfaces in direct contact. The rate of conductance is dependent on the temperature difference between the two surfaces, the thermal conductivity of the surface materials (k_c). The distance through which the heat is conducted is l.

$$K = (k_c/l)(T_a - T_s), \quad [W \cdot m^{-2}] \tag{12}$$

For a standing human with adequately insulated feet, conductance is of very limited importance in calculating the total heat exchange. Standing bare-footed on ice or holding a cold-soak wrench will cause significant local heat loss or even cold injury, but generally we avoid long-term or extreme exposure of the extremities to conductive heat loss. Lying uninsulated on ground that is significantly hotter or colder than the body will result in significant heat exchange. The daily activity of greatest continuous duration is usually sleep. Participants in winter camping are well aware of the importance of insulation from the ground. More than one winter camper has awakened in the morning to discover they are lying in a depression created by body heat melting the packed snow beneath the tent floor.

Radiation. The radiation term in the energy balance equation is a complex variable representing the net effective radiation balance of an individual. Six radiation terms determine the radiation exchange between an object and its environment. There are five incoming radiation variables; three solar terms and two thermal or "heat" terms. The sixth term is radiation emitted or radiated by the subject out into the environment.

In most biometeorological studies, solar radiation refers to radiation in the 400 to 750 nm wavelengths emitted and received directly from the sun. Thermal radiation is in the near infra-red range from 800 to $8.0 \cdot 10^4$ nm. Solar radiation may be received directly from the sun as "direct" solar radiation; scattered, sky or "diffuse" solar radiation; and solar radiation "reflected" from the ground (Fig. 1-1). Incoming thermal radiation consists of "sky" and "ground" thermal radiation. In SI units, all radiation intensities are expressed in $W \cdot m^{-2}$ and the radiation terms for the net radiation equation are calculated by multiplying the radiation intensity by the correct surface area. In the case of direct solar radiation, the calculation of the correct surface area presents a rather interesting problem.

The amount of direct solar radiation received depends on the intensity of the solar radiation and the posture and position of the subject relative to the direct parallel beams of solar radiation. Although direct solar radiation falls across the entire body surface area exposed to direct sunlight, the amount of direct solar radiation received is equal to the full intensity of the sunlight measured (or corrected) normal to the solar beam times the cross-section area of an object normal to the solar rays (A_p). This relationship is known

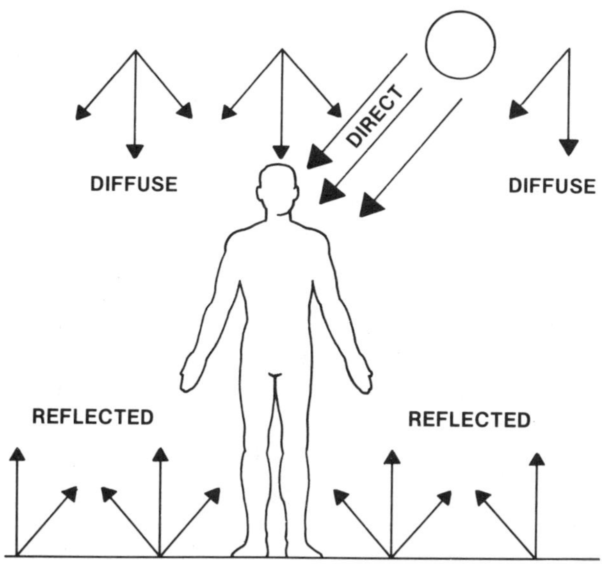

Figure 1-1. *The impact of direct, diffuse and reflected solar radiation are dependent on the orientation of the individual relative to the radiation source.*

as Lambert's cosine law (2,14,65). To clarify further, assume that a subject is represented by an upright cylinder (Fig. 1-2).

If the sun's rays came straight down from overhead, the full intensity of the radiation would fall only on the top of the cylinder, an area equal to the circular top of the cylinder (38,66,69). Now assume that the sun's rays are coming directly from the horizon. Although the radiation is spread over the entire half of the cylinder, the total amount of solar radiation received is equal to the full strength of the direct rays times the area of a rectangle with height equal to the height of the cylinder and width equal to the diameter of the cylinder. If the source of the solar beam moves from directly overhead to the horizon, the cross-sectional area of the cylinder normal to the solar rays (A_p) goes from a minimum circular area to a max-

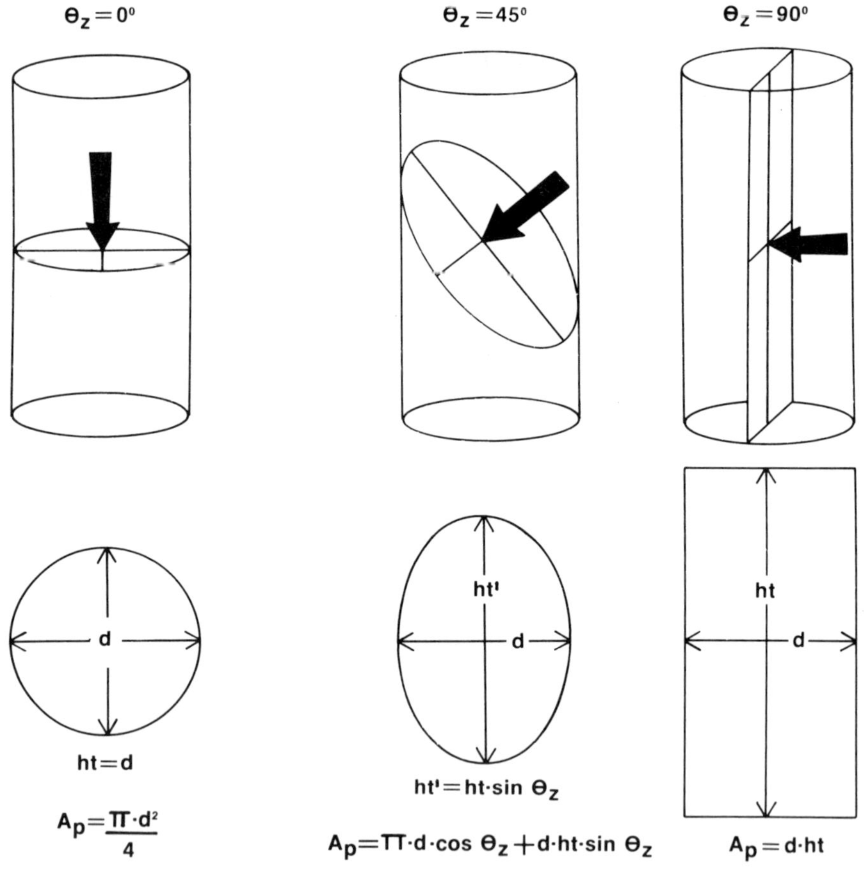

Figure 1-2. *The position of the sun relative to an object determines the cross-sectional area which receives the full intensity of direct solar radiation.*

imum rectangular area. In between, A_p is an oval, with a short axis equal to the diameter of the cylinder and the longer axis lengthening from the diameter to the height of the cylinder, as the solar source changes from vertical to horizontal relative to the subject cylinder. The longer axis height (ht′) can be calculated by determining the angle between the sun and the horizon (solar elevation θ_s), or deviation from vertical (the zenith angle θ_z) and multiplying by the cosine or sine respectively of the solar or zenith angle. Underwood *et al.* (69) presents the following equation to calculate the cross-sectional or projected area (A_p) of a cylinder normal to the beam:

$$A_p = \pi \cdot r^2 \cdot \cos\theta_z + 2 \cdot r \cdot ht' \cdot \sin\theta_z, \quad [m^2] \tag{13}$$

Calculation of the zenith angle from the time of day, site coordinates, daily solar declination, and other information is described in appendix A of this chapter.

Human shapes are a little more complex than simple cylinders and a more pragmatic method for determining the correct cross-section angle for a given zenith angle has been utilized. The basic method is to use a device that holds a camera at an equal distance from the subject while traversing an arc simulating the actual solar angle. A reference sphere which always presents the same cross-sectional area to the camera may also be photographed (10). By using a planimeter, or simply cutting out and weighing the prints, the surface area of the model normal to light entering the camera aperture can be determined. The clothed surface area of an individual can be calculated from the DuBois nomogram for nude surface area (A_D) and proportional values for each zenith angle can be determined (20).

$$A_D = 0.202\, wt^{0.425}\, ht^{0.725}, \quad [m^2] \tag{14}$$

Height (ht) is in m and weight is in kg. Those derived proportional values can be applied to the calculated surface area of any subject in the same posture and orientation relative to the sun.

Underwood *et al.* (69) present a more specific equation for determining A_p, the body area normal to the direct solar beam for an "average man":

$$A_p = 0.043 \sin\theta_s + 2.997 \cos\theta_s \sqrt{(0.02133\cos^2\phi + 0.0091\sin^2\phi)}, \quad [m^2] \tag{15}$$

The solar elevation angle (θ_s) is the complimentary angle of θ_z. The azimuth angle (ϕ) is the orientation of the object relative to the sun. An individual facing directly towards or away from the sun has a ϕ of 0° and sideways (shoulder towards the sun) has a ϕ of 90°.

Determining the amount of incoming diffuse solar, reflected solar, ground thermal, and sky thermal radiation is not as difficult.

All these radiation sources are treated as isotrophic, that is the radiation is emitted in all directions from an area source. For diffuse or scattered solar radiation, the solar radiation strikes all surface areas exposed to the sky with equal intensity over their entire area. For a subject cylinder, the diffuse radiation would fall with equal intensity on all of the top and side surfaces of the cylinder. Only the bottom area of the cylinder is sheltered from the diffuse radiation. Non-solar thermal radiation from the sky utilizes the same surfaces as diffuse solar radiation. Solar and thermal radiation reflected or emitted from the ground strikes only the sides of the cylinder in our model. For a different shaped model, such as sphere, some of the radiation will fall above the equator of the model.

One other physical characteristic of the subject determines how much incoming radiation is received. All light striking the surface of an object is reflected, absorbed, or transmitted. The solar load of an object is a product of the total solar irradiance (direct, diffuse, and ground reflected) and absorptivity (α); the percentage of light absorbed by the object. For human subjects, radiation that is transmitted through the clothing surface is generally absorbed by the body tissues, so "absorptivity" is calculated by subtracting the proportion of light reflected from 1.

All surfaces with a temperature above absolute zero emit thermal radiation. The Stefan-Boltzmann law defines the emission (R_b) of a perfect blackbody as:

$$R_b = \sigma T_s^4, \quad [W \cdot m^{-2}] \tag{16}$$

Sigma (σ) is the Stefan-Boltzmann constant ($5.67 \times 10^{-8}\ W \cdot m^{-2} \cdot {}^\circ K^{-4}$) and T_s is the absolute surface temperature of the blackbody in degrees Kelvin (°K). Emissivity (ϵ) is a proportional term for the amount of thermal radiation, relative to a blackbody, that a surface emits. By multiplying the calculated R_b of a perfect blackbody by the emissivity (ϵ), the radiation emitted by an object can be determined. A perfect blackbody is defined as a body that completely absorbs all radiation. According to Kirchoff's law emissivity equals absorptivity at a given wave length. Except for polished or highly reflective surfaces, the emissivity of most surfaces is 0.95 or greater in the IR or thermal wavelengths (2,14).

The term for R_b in equation 16 is expressed as a flux, a rate of energy flow per unit area ($W \cdot m^{-2}$). Radiative heat exchange does not occur evenly on the body surface. For that reason, some biologists prefer to express heat exchange as the total *rate* per animal (W). Human physiologists prefer to express heat exchange as fluxes. For a human physiologist, it would be desirable to express the total radiation exchanged with the environment as a single term related to surface area. Mean radiant temperature ($\bar{T}_r$) is defined as "the temperature of a uniform black enclosure in which man would exchange the same radiant heat as in his nonuniform environment" (31). Expressing the radiation term as a pseudo-temperature allows the investigator to use radiation in calculations in units of temperature rather than radiation ($W \cdot m^{-2}$). When expressed as $\bar{T}_r$, the net radiation received is treated as if it were equally distributed on all exposed surfaces.

$$R_{net} = h_r (\bar{T}_r - T_s), \quad [W \cdot m^{-2}] \tag{17}$$

Mean radiant temperature can be calculated directly from the temperature of a Vernon globe thermometer (70). That method is presented and discussed in the instrument section of this chapter. The radiant heat transfer coefficient (h_r) can be calculated as a linear relation (5,31):

$$h_r = 4\, A_r \sigma \epsilon [(\bar{T}_r + T_a)/2]^3, \quad [W \cdot m^{-2} \cdot K^{-1}] \tag{18}$$

Gagge (26) uses the term Effective Radiant Flux (ERF) for the net radiation exchanged between a *man-shaped object* with a surface temperature equal to air temperature and the environment.

$$ERF = \sigma\, \alpha\, A_r/A_D\, (\bar{T}_r^4 - T_a^4), \quad [W \cdot m^{-2}] \tag{19}$$

Where A_r is the effective radiating surface area of the body. For all practical purposes, A_r is equal to the exposed surface area of the subject or object. In the case of the cylinder model, only the cylinder base, in direct contact with the surface, would not be considered an emitting surface. For a standing human, this would be an area equal to the foot surface area and the areas "shaded" by other body parts; such as the inner surfaces of legs, arms, and fingers.

There are several approaches to calculating the radiant heat load. $\bar{T}_r$ obscures the importance of solar exposure by expressing radiation as a uniformly distributed flux. Another approach is to calculate the total (whole animal) radiative heat exchange rate from the six separate thermal and solar radiation terms. The latter approach is desirable if the various radiation terms can be quantified because it allows the contribution of different radiation sources to be considered independently. The incoming solar load ($\dot{Q}_s$) can be calculated by summing the three solar parameters:

$$\dot{Q}_s = A_p \alpha I_D + A_1 \alpha I_d + A_2 \alpha I_r, \quad [W] \tag{20}$$

A_1 and A_2 are the surface areas exposed to radiation from the sky and ground, respectively. Both areas may be estimated as equal to or less than the effective radiating surface area (A_r) (1,11). Fanger (22) gives the ratio of A_r/A_D as 0.73 for a standing man and 0.70 for a sitting man. Berglund *et al.* (7) use $0.7 \cdot A_D$ for A_1 and $A_r/2$ for A_2 for a cycling subject. The incoming thermal radiation ($\dot{Q}_{ti}$) can be calculated by summing the incoming ground and sky thermal loads:

$$\dot{Q}_{ti} = A_1 \epsilon R_{sky} + A_2 \epsilon R_{gr}, \quad [W] \tag{21}$$

The net radiation balance is calculated by summing the incoming radiation and subtracting the outgoing or emitted radiation term. Emitted radiation is calculated by entering the surface temperature, emissivity, and effective radiating surface area (A_r) of the subject into the Stefan-Boltzmann equation:

$$\dot{Q}_e = A_r \cdot \sigma \cdot \epsilon \cdot T_s^4, \quad [W] \tag{22}$$

The net radiation absorbed (Qabs) is the sum of the absorbed solar and thermal radiation:

$$\dot{Q}_{abs} = \dot{Q}_s + \dot{Q}_{ti}, \quad [W] \tag{23}$$

The ERF can be calculated from the total (whole animal) exchange rate (Q_t) by dividing by A_r.

$$\dot{Q}_t = \dot{Q}_{abs} - \dot{Q}_e, \quad [W] \tag{24}$$

$$ERF = \dot{Q}_t / A_r, \quad [W \cdot m^{-2}] \tag{25}$$

Insensible or Evaporative Heat Exchange

Insensible heat exchange is the result of the evaporation or condensation of water on the body surface. Insensible, "wet" or evaporative heat exchange is normally a one-way heat flow from an individual body surface to the environment. The basic principle involved is that the phase change from liquid to water vapor requires 2.45 J/kg, the latent heat of vaporization for water (14,54). Because the heat of vaporization is "absorbed" without changing the measured temperature of the water, the heat exchange is considered "insensible." A third term for insensible or evaporative heat exchange is "moist" heat exchange, because unlike convective, conductive or radiative heat exchange, water is required. The physical determinant of the evaporative potential is the water concentration gradient between the body surface and the environment. The physiological limits are the sweating rate and the level of individual hydration.

The basic equation for determining evaporative heat exchange (E) is:

$$E = \omega h_m (P_{s,sk} - P_w), \quad [W \cdot m^{-2} \cdot Torr^{-1}] \tag{26}$$

The two water vapor pressures are saturated pressure at skin temperature ($P_{s,sk}$) and saturated vapor pressure at the ambient dew-point temperature (P_w). The difference between the two water vapor pressures is used by physiologists as a convenient analog to the actual water concentration gradient. ω is the wetted skin surface area fraction (24,31,39) and h_m is the mass transfer coefficient.

A common simplification is to use the Lewis number (L_R), 2.2 K/Torr, to relate the mass heat transfer coefficient to the convective heat transfer coefficient (60). The Lewis number defines the relationship between thermal diffusivity and the diffusion coefficent for water vapor in air (57,60). The Lewis relationship (60) relates h_m to h_c by the equation:

$$h_m = 2.2 \cdot h_c, \quad [W \cdot m^{-2} \cdot Torr^{-1}] \tag{27}$$

Relatively little attention is focused on air moisture below 0° C. Intuitively we recognize that colder air has very little moisture capacity and will be quickly saturated by a very low level of absolute moisture. However, if the temperature of very cold air is raised, the moisture or water capacity of the air can be increased significantly. If for example, saturated arctic air is inhaled, the air is rapidly warmed as it travels through the nasal and oral cavities into contact with the moist lung tissue. As the cold saturated air is warmed it essentially becomes very "dry" in terms of moisture capacity. An analogous situation arises if air is warmed as it comes in contact with the body surface. As a consequence, despite the saturated or near saturated state of cold air, dehydration is a common problem during activities under arctic conditions.

Another aspect of the effect of temperature on the water carrying capacity of air is that as warm air moves out from the skin through clothing in extreme cold, it is cooled and water condenses in the clothing; reducing the insulation afforded by the clothing. Sweat can also be absorbed directly from the skin into the clothing. By adjusting the insulation to prevent heat building and sweating, and by ventilating the clothing to pass moist air directly into the outside environment instead of through the clothing, the loss of insulation due to moisture can be avoided. This will be considered in Chapter 2.

Heat Storage

Heat storage (S) is not an "environmental stress" term in the heat balance equation. What the storage term actually represents is whether an individual is able to maintain a thermal equilibrium with the environment. It is the net result of the interaction of environmental conditions—certain physical parameters, (such as clothing insulation or absorptivity), and physiology (especially hydration levels). If the individual is not in equilibrium, the sign of the storage term will indicate whether there is a net loss to the environment (negative) or heat gain (positive). A positive flow occurs when an individual is unable to transfer excess body heat to the environment. The basic principle that relates body temperature to heat storage parallels insensible loss of heat via the latent heat of vaporization. The accumulation or loss of sufficient heat (the specific heat (c_p) of body tissue is about 3.5 kJ/(kg·K) (7,16,52) will result in a net change in body temperature. A net negative heat flow will result in hypothermia and a positive flow in hyperthermia.

What is important to remember about the heat storage term is that as body temperature ($\bar{T}_b$) changes, the thermal gradient between an individual and the environment also changes and as a result, net heat exchange is altered in direct relation to the new gradient. As body temperatures change in response to storage in a stable environment, the storage term tends towards zero or equilibrium. The question is whether equilibrium will be reached before body core temperatures reach a critical tolerance level.

Other Environmental Factors

Altitude. The primary effects of the reduced pressure, due to increasing elevation, are physiological. The most important physiological parameter is the reduction in oxygen concentration with increasing altitude. In terms of the thermal environment, higher altitudes receive more solar radiation because the air mass that the incoming radiation travels through, before reaching the earth's surface, is reduced. Reflected radiation levels may also be increased because of the high albedo of rock and ice or snow (14,64). As air density is reduced, this results in a reduction in the rate of convective heat transfer. The convective heat transfer coefficient is a power function of the barometric pressure associated with higher elevations (28,56):

$$h_c' = h_c\,(P_b/760)^{0.55}, \quad [W \cdot m^{-2} \cdot K^{-1}] \tag{29}$$

Where P_b is the barometric pressure in Torr (760 Torr = 1 atmosphere; 1 atm = 1.01325 × 10^5 Pascals, Pa). An increase in evaporative heat transfer can also be related to the change in the thermal diffusivity (D). The mass transfer coefficient, h_m, can also be adjusted for the effects of pressure (23,49):

$$h_m' = h_m(760/P_b)^{0.45}, \quad [W \cdot m^{-2} \cdot Torr^{-1}] \tag{30}$$

The most interesting aspect of altitude is that the Lewis relationship between convective and evaporative heat exchange becomes uncoupled as pressure is significantly reduced (Gonzalez, unpublished).

Precipitation. Relatively little research has been done on the effects of precipitation, especially rainfall, on heat exchange. This may be due to the fact that the normal response to significant rainfall is to retreat to shelter or put on protective clothing. Rain has several effects. First, rain water is often colder than air temperature. Second, the thermal conductivity of water is greater than air. Third, the insulation provided by clothing is reduced by wetting (9). Marathon runners may actually appreciate a light rain because the increased heat loss on a cool day prevents heat storage.

Being soaked in a rain is equivalent to completely saturating one's skin surface with sweat; but while it is raining, the air is completely moisture saturated so no evaporation should occur with no forced convection if the air is close to skin temperature. If the air temperature is colder than skin, air entering the clothing can be warmed and the moisture capacity thereby increased, allowing some evaporation from the skin. As the air is recooled, the water capacity is again reduced and condensation will occur. If the air is rapidly removed from the clothing before it is cooled to air temperature, recondensation will not occur inside the clothing.

Air Pollution. Conventional weather forecasting rather than micrometeorology is the best predictor of the concentration of air pollution. The site, concentration and toxicity of air pollutants is determined by the location of the site relative to local and regional pollution sources, the direction and speed of upper air movements, terrain features, air temperature and solar radiation, pollutant chemistry, precipitation and particle or aerosol size. The impact of a concentrated pollutant plume on an elevated terrain feature may result in very high local pollution levels under certain meteorological conditions. Different chemical pollutants mix and react in the atmosphere, dependent in part on the presence of solar radiation, air temperature, and humidity. Air temperature and particle size also affect the buoyancy of some air-borne pollutants. Precipitation may scavenge or scrub out pollutants, rendering the expression "pure as the driven snow" a quaint anachronism in many places.

Underwater Diving Environment. Divers encounter a relatively simplified thermal environment. The two primary environmental parameters are water temperature and pressure. Water temperature is frequently stratified, but relatively stable compared to the fluctuations in air temperature. Unless a current is present, most heat exchange is by free convection. Because of the higher thermal conductivity of water, a relatively small difference in temperature between the body surface and the surrounding water will result in very rapid heat exchange. An ambient temperature that would not be life threatening in air can rapidly result in hypothermia and death in water. Pressure is directly related to water depth, so an accurate depth gauge should suffice to predict pressure. As in the case of altitude, the primary consequences of higher pressure are physiological rather than thermal—once corrections for increased fluid density are made. The physiology of diving is discussed in greater detail in a later chapter of this volume.

ENVIRONMENTAL DETERMINANTS OF HEAT EXCHANGE POTENTIAL AND INSTRUMENTS FOR MEASUREMENT

The pathways for heat exchange are evaporation, convection, conductance, and radiation, but the primary meteorological parameters that we measure to calculate or estimate the potential effects of the environment are temperature, wind speed, radiation, and humidity. At any given time, we occupy only a single point in our total potential environment. The term "nannoclimate" (67) is used to indicate the meteorological conditions that immediately impact on an organism in a given time and place; basically an instantaneous climate. The "microclimate" of a species is the aggregate of nannoclimates occupied by a species in a given environment. Among human physiologists the term "microclimate" refers to conditions between the skin surface and the clothing of a subject. Consequently, it would be more appropriate to refer to the site "micrometeorology" rather than the microclimate when referring to human physiology.

The roadblock to characterizing the environment is generally insufficient knowledge regarding which instruments to use. In an academic research setting, consultation with meteorology faculty is recommended. Bakken (4) offers a list of sources for portable, remote instrumentation.

Temperature

Technically temperature is the "mean kinetic energy of the molecules of a substance," or an equivalent definition. The "zeroth law" of thermodynamics, however, focuses on the functional consequence of temperature differences (63). The zeroth law states that "if two objects are brought into contact through a diathermal wall, and the objects are the same temperature, no net heat flow will occur." Temperature is, therefore, a measurement of the potential for heat exchange between two objects or substances. Measurement of air and ground temperature are, therefore, important environmental parameters which determine the potential for heat exchange via convection and conduction between an individual and its environment.

Location of Measurements. In common with several other meteorological parameters, temperature demonstrates a vertical profile or gradient in many environments. During the later daylight hours of the summer, the ground is heated by solar radiation and warmer than the air above. The presence of snow or ice would reduce the effects of solar radiation on air temperatures near the ground. In the desert regions, the difference between ground and air temperatures can be considerable, with observed air temperatures at 2 m of 43° C and ground temperatures of 68° C (v. Palm Springs,CA [59]). A standing individual may experience a considerable temperature profile in the desert standing on a hot surface and experiencing a vertical air temperature profile from his feet to his head. For example, in deep desert canyons in southern Utah, the difference in temperature between boulders, in and out of direct sunlight, may be used to provide natural cooling for hikers that overheated in the late morning as they hiked along the canyon bottom in full sunlight. By lying down on rock still cold from the previous night, heat is conducted from their bodies into the rock surface, effectively dumping their excess heat. A certain amount of caution is necessary when this method of cooling is employed because rattlesnakes and other animals also utilize the cooler microenvironment of the shaded rocks. The position where temperature is measured is determined, according to the basic criteria given by Platt and Griffth (57), by which site is most representative of the conditions experienced by the subject. Typically, if a single temperature is taken it is at approximately head level for a standing man, between 1.7 and 2 m. The Wet Bulb Globe Temperature measurements are taken at 4 ft (1.2 m). Ideally, of course, multiple temperatures, including ground and a uniform series of air temperatures, should be measured. Santee (61) utilized a series of measurements including 10 cm below surface, ground, 0.5, 1.0, 1.5, 2.0 and 3.0 m (Fig. 1-3). The first value is not of interest to most investigators and the 3 m value is primarily used in calculating a temperature profile.

Thermometers. A thermometer is essentially any "instrument" that is used to measure temperature. A thermometer works by relating a change in an intrinsic property of the sensor element to the mean kinetic energy of the substance being measured. For practical measurements within the human tolerance range there are directly read liquid and metal thermometers and electronic analog thermometers, which are frequently automatically recorded.

There are three basic types of electronic "thermometers": thermocouples, thermistors, and thermopiles (5,57). When incorporated with the appropriate electronics they are generally faster and more accurate than directly read thermometers. Electronic thermometers

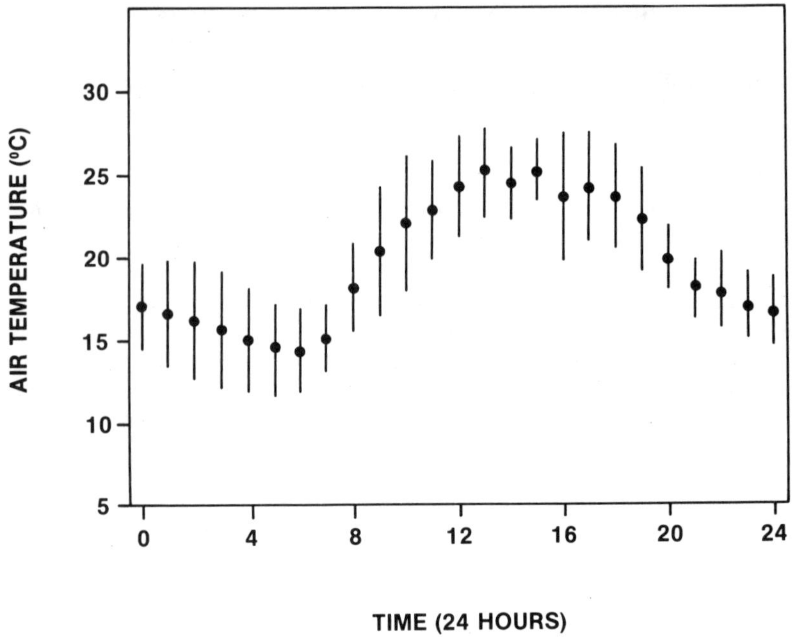

Figure 1-3. *Hourly mean air temperature (°C) and standard deviations observed at 2 m during June, 1982 in west-central Indiana (Santee, 1985).*

are incorporated into the majority of automated temperature collection systems. Their cost and the bulk/weight of the necessary electronics are the negative aspects of these instruments.

Wind

Measurement Sites. In natural environments air movement, like temperature, frequently varies along a vertical profile (30). Air movement is classified as laminar or turbulent. Laminar flow is linear in parallel currents; whereas, turbulent flow has eddies or other cross currents which are difficult to measure and enhance convective heat transfer by disrupting the boundary layer and thermal gradient characteristic of laminar flow and by increasing the surface area. Near the ground, friction or drag at the ground-air interface slows the wind speed and creates turbulence. Uneven surface features or vegetation can enhance the turbulence at ground level. Because of the vertical distribution of wind speeds it is extremely desirable to use multiple measurement points along a vertical profile. With two known points, wind speeds along a vertical profile can be calculated with reasonable accuracy (14,57), but as in the case with temperature a direct measurement is more desirable than a calculated value. A second consideration regards fetch, the unobstructed distance the wind travels before reaching the subject or measurement point. Building, vegetation, and other obstacles create complex air movements which make the analysis of air flow much more difficult. The general rule for a wind speed measurement sites is that the fetch should be ten times the height of the obstruction (57). For example, if a shrub is 2 m high, the anemometer should be 20 m from the shrub. Anemometers are often not ideally sited according to that criteria, but investigators should be especially concerned about buildings, walls, and vegetation that create abrupt microscale variations in wind speed (Fig. 1-4).

Hand-held instruments. The simplest method for approximating the wind speed is the "Beaufort scale," based on the observable effects of wind in the outdoor environment. Such observations are clearly not too precise, but careful observation may be no worse than a single measurement made with a hand-held instrument at a poorly selected or uncharacteristic site. The presence of the observer operating a hand-held anemometer or wind gauge creates an obstruction that is undesirable, and generally hand-held observations are frequently assigned a status just slightly more desirable than a

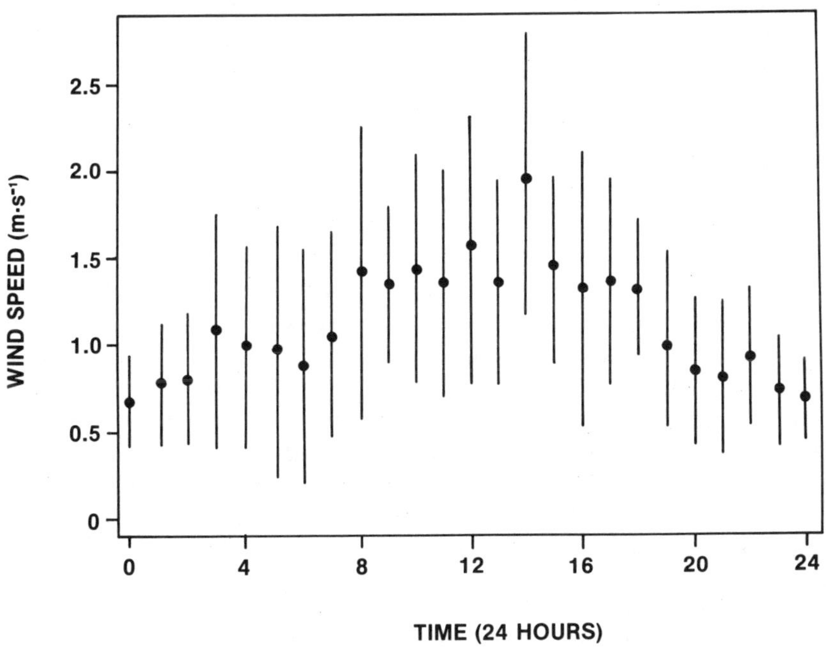

Figure 1-4. *Mean wind speed ($m \cdot s^{-1}$) and standard deviation observed at 3 m during June, 1982 in west-central Indiana (Santee, 1985).*

guess. Hand-held instruments may be mechanical or electronic. An example of a simple hand-held wind gauge is based on the same principle as chimneys, known as the Bernoulli effect. As wind blows across the mouth of a narrow tube, pressure within the tube is reduced and light-weight objects are drawn upward by a combination of higher pressure below, and lower pressure above. In this simple gauge, a small ball inside the tube is drawn up the tube and its position relative to a scale on the side of the tube indicates the wind speed. Proper orientation of the tube and the observer is necessary. The device is very inexpensive and adequate for gross wind measurements. The U.S. Forest Service includes this type of wind gauge in its "fire kit," a simple belt mounted packet of meteorological instruments used to measure meteorological parameters at a fire site.

Mechanical Anemometers. More complex mechanical anemometers consist of a propeller or set of cups that are rotated by the force of the wind. The two disadvantages of these devices are the inertia due to the mass of the cups or propeller and frictional resistance of the bearings which limits the measurement of low wind speeds and the fact that all measurements are for air movement in a single plane. The threshold of an accurate cup anemometer is 0.2 to 0.3 $m \cdot s^{-1}$, due to the inertia of the cups that must be overcome before the instrument will rotate, creating a reading.

Electrical Anemometers. Mechanical anemometers usually operate with some type of electronic pulse counter that "counts" the number of rotations. In the context of this chapter, an "electronic anemometer" is an instrument that operates directly on the principle of convective heat exchange. In principle, such devices are quite simple. In a "hot wire anemometer," a thin wire is heated and maintained through electronic circuitry at a high, constant temperature through use of a wheatstone bridge. The power required to maintain the constant temperature of the wire is directly proportional to the convective heat exchange between the wire and the surrounding air and the "cooling power" of the air flow can be calculated from the power required and the air temperature. An electronic anemometer, therefore, actually measures convection rather than linear wind speed, which is actually the rationale for measuring wind speed to calculate the potential for convective heat exchange.

In addition to hot wire anemometers, spherical or globe heated anemometers which operate on the same principle also exist. The advantage of the latter is that orientation relative to the air stream is not as important, whereas a hot wire anemometer must be carefully positioned relative to the air flow. In general, heated electronic anemometers are indoor laboratory instruments that are accurate, expensive and rather delicate. The potential to measure low wind speeds, turbulent as well as laminar air flow, and accuracy are very attractive features of heated electronic anemometers. However, sensitive electronic anemometers for outdoor use are normally custom-built by knowledgeable individuals and require careful calibration.

In actual field use, the less complex, more rugged, and less expensive cup anemometers are more practical than electronic anemometers. The inertia factor can be compensated by adding a small constant "correction value," usually close to the threshold value, to the instrument reading and ignoring the free convective heat exchange.

Radiation

For many physiological studies, the investigator may be satisfied by deriving a value for the mean radiant temperature ($\bar{T}_r$) from a black globe thermometer (discussed in the section on combination instruments) as a descriptor of the radiation conditions during the study. For outdoor studies during daylight hours, unless conditions are uniformly clear or heavily overcast, more precise measurement of solar radiation is desirable. During indoor studies where radiation is unevenly distributed, such as an office with a heat radiator or a large window, direct measurement of radiation is also important (47).

UV Radiation. In terms of heat exchange, the most important wave lengths for radiation are in the visible (400 to 750 nm) and the "thermal" or near infra-red (800 to 8.0×10^4 nm) wave lengths. Short wave radiation is biologically disruptive because of the higher energy levels (Planck's law), but due to the screening effect of atmospheric ozone, the actual radiation levels (2% of total) attributed to UV radiation on the earth's surface has been considered negligible. As a consequence, the fact that the pyranometers and radiometers, commonly used to measure radiation, do not measure UV wave lengths was not considered a significant deficiency. Concern about the possible increase in ultra-violet (UV) radiation resurfaced when studies of the ozone levels over Antarctica revealed the appearance of a "hole" (15,43). Because of the biological importance of UV radiation and the possible increase in UV radiation reaching the earth's surface, investigators may soon have to consider the consequences of not monitoring UV radiation.

Direct, Diffuse Sky and Reflected Solar Radiation. The optimum instrumentation for measuring the direct, diffuse and reflected components of solar radiation would be to use three instruments to measure each parameter individually. A more efficient use of instrumentation is to measure global and diffuse radiation. Global radiation is the combined value for diffuse and direct solar radiation impacting a horizontal surface (Fig. 1-5).

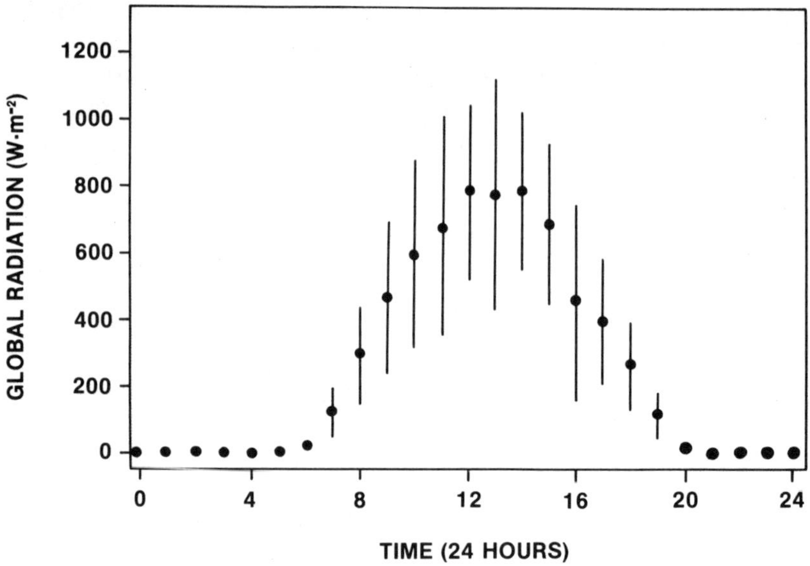

Figure 1-5. *Hourly mean global radiation ($W \cdot m^{-2}$) and standard deviations observed during June, 1982 in west-central Indiana (Santee, 1985).*

Diffuse radiation can be measured through use of a shade ring or shadowband. The shadowband consists of a platform to hold a pyranometer and an adjustable band of metal that is adjusted to block direct solar radiation from striking the pyranometer. The shaded pyranometer, with a small correction due to the width of the shadowband, directly measures diffuse solar radiation. When the diffuse radiation value is subtracted from the global radiation value, the remainder is the direct solar component impacting a horizontal surface. By dividing that value by the cosine of the zenith angle (Lambert's law), the full intensity of direct solar radiation can be calculated. Reflected solar radiation can be estimated by multiplying the global radiation value by the albedo (reflectivity) of the ground surface which can be found in most meteorology texts for common surface types. Reflected solar radiation can also be measured directly with an inverted pyranometer (4,57).

Radiometers. Radiometers are instruments that measure radiation. The most common instruments in current use are net radiometers based on electric thermopiles protected by a transparent dome of plastic or glass. Glass domes screen out long-wave radiation ($>3 \times 10^3$ nm). Net radiometers use clear plastic screens which allow the penetration of light from 300 to $6.0 \cdot 10^4$ nm. Net radiometers measure both incoming ground and sky radiation and electronically "subtract" the difference to determine the net value (4,23,48). In calculating the radiation balance of an object, the sky and ground radiation values, not the net product, are the desired values. The usual solution to this problem is to use a pair of net radiometers. One is modified with either a special dome or cup, painted white on the outside and black on the inside, and fitted with a thermocouple temperature sensor (40,41). The modification is a simple project consisting of locating a properly sized metal can or cup, painting the radiation guard, attaching a thermocouple to the inside, and taping around the bottom hemisphere. The radiometer is furnished with a dome that may be modified and directly installed in place of the regular dome. Conversion cups are also commercially available.

The radiometer fitted with the radiation guard measures the total incoming sky radiation (i.e., direct and diffuse solar radiation plus sky thermal). The radiation term recorded from the modified net radiometer is the product of the incoming sky radiation minus the radiation received from the black inside surface of the radiation shield. By entering the temperature of the shield and assuming an emissivity of 0.95 for the inside surface, the Stefan-Boltzmann equation can be used to calculate the radiation received as total incoming ground radiation. By adding that calculated value to the radiometer reading, a term for the total sky radiation can be obtained. By subtracting the global radiation, which is the combined incoming solar term, the actual incoming thermal sky radiation can be determined. By subtracting the net radiation term from the unmodified radiometer from the uncorrected total sky radiation term, a total ground radiation term can be calculated. The thermal ground radiation can be determined by subtracting the reflected solar radiation term from the total ground radiation term.

Pyranometers. Pyranometers are radiometers that measure shortwave or visible radiation (4,48,57). As noted above, the glass globes or domes filter out thermal and UV radiation. Colored filters may be ordered which measure only a narrow band of the visible spectrum. If fitted with a quartz dome, pyranometers can measure UV radiation. The better quality pyranometers use a thermopile as the radiation sensor, but less expensive silicon cell detectors should be adequate for most field studies.

Direct solar radiation can be measured with a pyrheliometer which tracks the sun's movement across the sky, measuring solar radiation normal to the sun's rays. The expense of a pyrheliometer would not usually be justified for normal field work. The use of a shadowband to measure diffuse radiation and then determine direct solar radiation should provide adequate data.

Pyranometers are mounted on level stands above the ground. For measuring global and diffuse radiation, the only requirement is that the instruments not be shaded by the surrounding vegetation or other surface features. If an inverted pyranometer is used to measure reflected solar radiation, the instrument is usually mounted on a ring at the end of an extension arm. The height of this pyranometer should be 1–1.5 m above a level surface representative of the study site if possible.

Because of the cost and relative complexity of measuring incoming solar radiation, a considerable effort has been made to calculate direct and diffuse solar radiation values (12,14,21,34,40,70). On clear days, in simple terrain, calculated direct and diffuse solar radiation values are acceptable for general physiological studies. The cost of operating a single, silicon cell pyranometer is not excessive and is particularly useful on heavily overcast days when virtually all solar radiation is diffuse. For partial cloud cover, and especially transitory sky cover, the relationship between I_D and I_d is difficult to define. Commercially manufactured shadowbands are equivalent in cost to a top quality pyranometer. Design and construction of a shadowband is sufficiently complex that it cannot be recommended as a routine laboratory or instrument shop project. In theory it is possible to temporarily block direct solar radiation to obtain a quick estimate of diffuse radiation from a pyranometer, but the practice cannot be recommended at this time. The best solution for utilizing a single pyranometer to obtain values for both I_D and I_d is to relate diffuse radiation to calculated direct solar radiation and observations of cloud or sky cover. However, at this time no simple, reliable relationship has been determined (4,13,29).

Air Moisture or Humidity

Absolute, Relative Humidity, Vapor Pressure and Dew-Point Temperature. Many individuals have difficulty with the concept of air moisture. Part of this is perception. We can appreciate the effects of wind and radiation by sensing the difference between shade and full sunlight or the cooling effects of a breeze, but humidity is so pervasive that we cannot readily escape it without totally leaving the environment.

Humidity is an attribute of the physical environment generally recognized by its role affecting climate, weather, and thermal comfort. In addition, besides its effect on the natural environment, humidity is important in simulated environments and life support systems (19,28,38).

As with the other physical variables—air temperature, barometric pressure, and wind velocity—humidity has long been documented and transcribed as a regular factor in history which spans weather prediction. For example, the meteorologist gives attention to boundary layer properties on the earth's atmosphere that include humidity turbulence. Increasing interest is prevalent nowadays in quantification of the water vapor gradients important to the humidity effects on weather patterns, air quality, and overall energy transfer. In the artificial simulation of the environment, such as in spacecraft or submersible life support systems, humidity is a vital part of safety and comfort (19,28).

Such interest and needs have provided the impetus for more accurate humidity measuring devices. The following is a brief review of the essential elements of humidity and its analysis. In its simplest concept, humidity is the amount of water, in the vapor phase, present in a gaseous atmosphere. Typically, the quantity of water vapor in a gas can be expressed in multiple ways: wet bulb temperature, mixing ratio (i.e., pounds per cubic foot), dew point temperature, parts per million, and partial pressure.

One of the properties often overlooked is the concept of Dalton's Law in considering humidity effects. Dalton was the first to recognize that total pressure (P_m), exerted by a mixture of ideal gases or vapors, is the sum of the pressures of each gas if it were to occupy the same volume solely (28,29). The pressure which each gas component of such a constituent gas mixture (i.e., "air") exerts is its partial pressure. Thus, air must be considered a mixture of the gases oxygen, nitrogen, water vapor, and inert gases. All measurements of humidity can be, therefore, reduced to the effects of water vapor pressure, as we shall see constantly in this volume, and all definitions encountered vaguely as "humidity effects" can be (and should be) expressed in terms of its apparent vapor pressure. Thus, relative humidity is the ratio of the actual vapor pressure in the mixture to the saturation vapor pressure with respect to water at the prevailing dry bulb (ambient) temperature (28).

Added to this is the concept of dew point. *Dew point temperature* is the saturation temperature to which air (or any gas) must be cooled at constant pressure so that it will be saturated with respect to water. *Frost point* is the definitive saturation temperature to which the air (or any gas) has to be cooled so that it will be saturated with respect to ice (18,29).

Since the definition applicable to relative humidity includes saturation vapor pressure with respect to "dew" or water, generally super-cooled dew point tables are used below 0° C (32° F) instead of frost point tables. Little, if any, reliable experimental data exist for super-cooled water below 0° C, so typical dew point tables are extrapolated from 0° C (32° F) to −51° C (−60° F). Thus, relative humidity is a meaningless concept when discussing equivalent dew points below −51° C (−60° F).

The concept of %rh is usually defined with respect to water for the following reasons: a) many humidity sensors, which are responsive to relative humidity, indicate %rh with respect to water; b) clouds, usually at temperatures below 0° C (32° F), consist of supercooled water; c) the atmosphere describing "air quality" is often supersaturated with respect to ice at temperatures below 0° C and this fact prevents relative humidity from exceeding 100%. Thus, it has become a convenient empirical index.

One index of ambient humidity, which is also widely used to gauge air quality, is the parts per millon (PPM) index. PPM by volume is the ratio of the amount of water vapor to the amount of dry gas on a volume basis. For example, PPMv for an air system at total pressure (P_b) of 760 Torr and −57.2° C frost point ($P_w = 1.26 \times 10^{-2}$ Torr) would be:

$$\begin{aligned} PPM_v &= [P_w/(P_b - P_w)] \times 10^6, \quad \text{[ppm by volume]} \\ &= [(1.26 \times 10^{-2})/(760 - 1.26 \times 10^{-2})] \times 10^6 \\ &= 16.6 \text{ ppm by volume} \end{aligned} \tag{31}$$

PPM by weight, on the other hand, is defined as the ratio of the amount of water vapor to the amount of dry carrier gas on a gravimetric basis. It is determined by the product of the molecular weight ratio of water vapor to carrier gas. The PPM is obtained by volume. For the example above, if the carrier gas is hydrogen and the total system is at 760 Torr, PPMw would be:

$$\begin{aligned} PPMw &= PPMv \cdot \frac{\text{Mol. weight of } H_2O}{\text{Mol. weight of hydrogen } (H_2)} \\ &= 16.6 \cdot 18/2 = 153 \text{ [PPM by weight]} \end{aligned} \tag{32}$$

One property which should be recognized is that as the total pressure of a gas sample changes, all the partial pressures included in the total pressure also alter in the same ratio. This is especially important in considering any analysis of dew point properties in hyperbaric or hypobaric pressures. For our example, the frost point of −57.2° C at a total pressure of 760 Torr would alter dramatically at a total pressure of 4380 Torr (5.76 atmospheres, ATA) as follows:

$$\begin{aligned} \text{vapor pressure at new dew point} &= P_w|4380/760, \quad \text{[Torr]} \\ &= 1.26 \times 10^{-2} \text{ Torr} \times 5.76 \\ &= 7.26 \times 10^{-2} \text{ Torr} \end{aligned} \tag{33}$$

This would bring the frost point to roughly −42.5° C.

From a physiological perspective, the importance of air moisture is related to the capacity of the surrounding air to absorb water from the body surface. Measurements of absolute and specific humidity measure the actual quantity of water present in the atmosphere, expressing the values as mass of water per volume or air mass, respectively. Relative humidity, dewpoint temperature, and water vapor pressure are more directly related to the evaporative potential of the environment, which is a function of the combination of absolute water and temperature. The key to determining what the ambient potential for evaporation is to determine the difference between the actual air moisture content and the saturation value.

Psychrometrics. One of the easiest ways to visualize the concept of humidity is by the use of a psychrometric chart (Fig. 1-6).

Typical psychrometric charts show ambient or dry bulb temperatures along the bottom horizontal axis. Saturation temperatures (dew point and wet bulb) are expressed by a curved line on the left side. Along the vertical line on the right, ambient water vapor pressure (Torr or kPa) is depicted and parallel to the vertical axis absolute moisture content (as gH_2O/g dry air or grains of H_2O/pound of dry air) is often drawn. Moisture content is linear but the ambient water vapor scale is not. Lower percentages of relative humidity, less than 100% saturation, are represented as diverging curved lines below the saturation curve. Dew point temperatures will always be indicated on such a chart by the intersection of the horizontal (dry bulb) lines with the 100% rh curve. Wet bulb temperatures will be indicated by the intersection of a 45-degree sloping line with the 100% rh curve. Psychrometric charts, such as the one described, are useful in describing dependent variables (vapor pressure, dew point temperatures) as a function of a temperature index such as operative temperature or effective temperature (27,28). Such charts allow the graphical display of the heat balance equation (27,31) and other heat stress indices in a psychrometric format. One problem occurs in that description of enthalpic lines (depicting true heat content, $J.g^{-1}$ dry air) and wet bulb temperature lines are confounded. This is because wet bulb lines are often used as representive of both properties. An "enthalpic deviation" line is used to correct for enthalpy from the wet bulb line; often these values are not clear-cut. Another approach, first developed by Mollier (50) and more recently used by Gagge *et al.* (27), unifies heat strain lines of a new effective temperature. In a Mollier chart, by depicting dry bulb temperature on a vertical left axis and ambient water vapor pressure on a bottom horizontal axis and humidity ratio on a top horizontal axis, enthalpic lines can be drawn more exactly without application of a correction factor added to wet bulb temperature lines.

Wood (74) circumvented many of the problems, inherent in

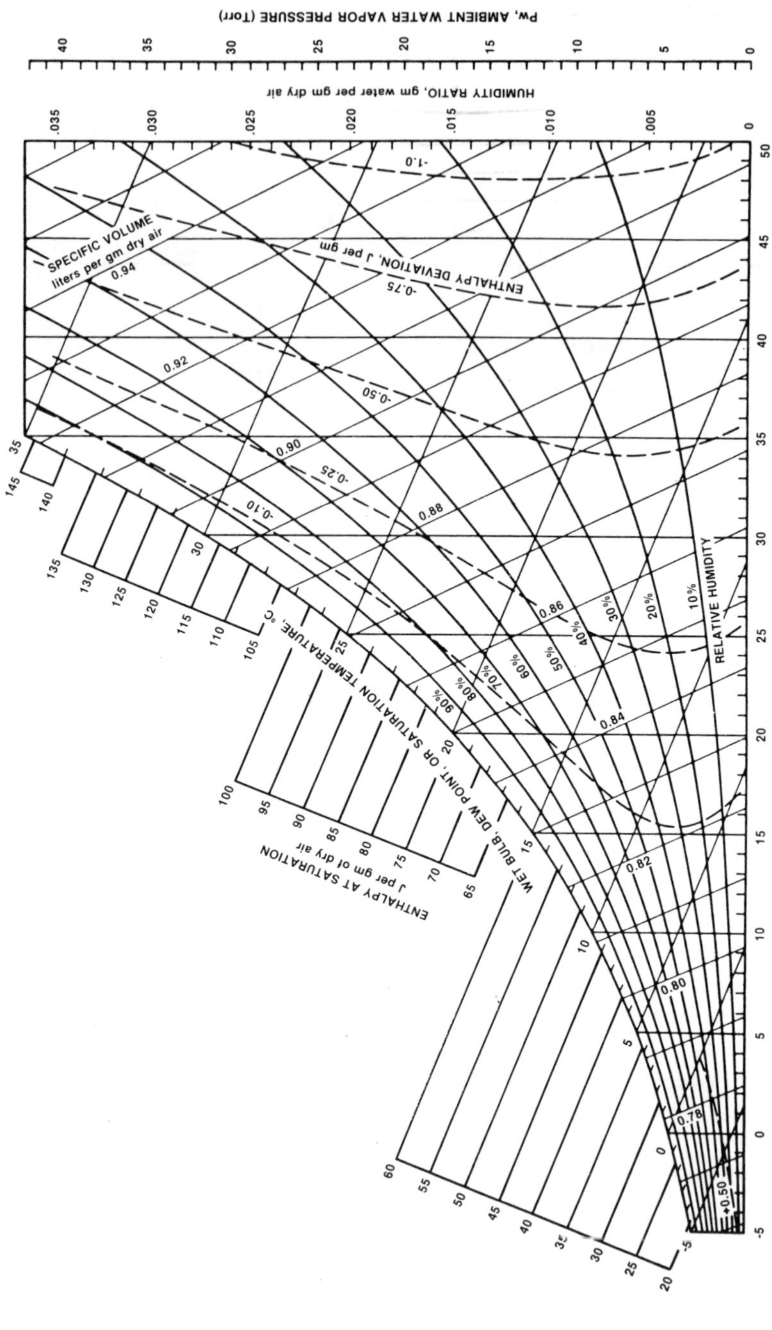

Figure 1-6. *Psychometric chart (adapted from Chambers, 1970).*

conventional psychrometric charts, by construction of a psychrometric chart with dew point temperature (y-axis) and dry bulb temperature (x-axis) coordinates. He also showed that the original Antoine equation was an accurate approximation of saturated vapor pressure of water from 0° C to 60° C. This equation is used quite frequently in physiological and clothing research to obtain saturated vapor pressures (P,t). The equation is as follows, where:

$$P,t = \exp^{(16.6536 - 4030.183/(t+235))}, \quad [\text{kPa}] \tag{35}$$

$$= \exp^{(18.6686 - 4030.183/(t+235))}, \quad [\text{Torr}] \tag{36}$$

With t representing the given temperature (°C), which must be used to obtain P,t.:

The classical psychometric chart as described holds for variables defined at constant barometric pressure (sea level). Construction of psychrometric charts for various other total pressures (altitudes) is, however, possible (18,34,44). Characteristically, the effect of lowered barometric pressure increases %rh, ambient water vapor pressure, and humidity ratio for a given dry bulb temperature (18,45). Enthalpy lines are not appreciably affected and the slope of the wet bulb lines are not changed. However, although the saturation curve and each given %rh lines are displaced to the left with lowered Pb. The converse effect on %rh lines occurs with hyperbaric environments.

Humidity Measuring Devices. There exist a multitude of humidity measuring instruments which have been categorized extensively according to accuracy and sensitivity (72). A hierarchy of those frequently used in the laboratory is presented in Appendix B.

Pressure and/or Altitude

Barometric pressure is often monitored, even in standard temperature and pressure controlled laboratory studies, to adjust physiological measurements for variations from standard temperature and pressure (STP) conditions. As noted in the discussion of altitude effects, both convective and evaporative heat exchange rates are altered by changes in pressure. However, pressure is usually systematically monitored as an environmental parameter only when the study occurs at sufficient elevation to affect oxygen levels (2500 m).

The simplest method for obtaining a gross estimate of the effects of altitude on pressure is to obtain a topographic map which will allow a reasonably accurate determination of the altitude of the study site. In the United States, state indexes to topographic maps may be obtained from the U.S. Geological Survey (USGS), Washington, D.C. 20242. The state indexes may list local map sources in addition to the regional USGS offices. In addition, map collections may be located at universities or colleges, either in the general library collection or in the geology or geography department. Standard tables can be used to adjust STP values according to the change in altitude.

If a more precise measurement of barometric pressure is desired, a hand-held barometer, either mechanical or electronically operated, can be purchased. Barometers measure changes in atmospheric pressure due to both altitude and high and low pressure weather systems. Barometers are classified as mercury or aneroid barometers, depending on whether the pressure sensitive element is mercury or a diaphragm, spring or other non-liquid sensor. An altimeter interprets pressure differences as elevation rather than force.

Generally, the hand-held barometers will suffice if only a few reference readings are collected. If a continuous record is desired an electronic, automatically recorded analog barometer is the best solution.

Combination Instruments

Direct measurement of a single environmental parameter is generally the preferred method; however, precise environmental monitoring is expensive and time consuming. Industrial and military safety personnel often are less interested in the fundamental theory of heat exchange since they require easy, quick and accurate accessment of the potential risks at a minimal investment in time and money. A minimum number of simple, inexpensive, easy to operate and maintain instruments also serve as an attractive solution to environmental monitoring when the researcher faces a reduced budget. A few such instruments are described here.

The best example of a simple combination instrument in widespread use is the Vernon or black globe thermometer. A sphere always presents the same cross-sectional area (A_p) normal to the direct solar radiation, regardless of the sun's position. Consequently, the impact of direct solar radiation on a sphere is dependent on only I_D. A sphere also presents a constant, uniform surface area to diffuse and reflected solar and incoming thermal radiation. As indicated in the section on radiation instrumentation, the mean radiant temperature can be calculated from the black globe temperature (T_g, °C), air temperature (°C), and the wind speed (48,75):

$$\bar{T}_r = T_g^4 + 0.247 \cdot 10^{-8} \cdot (\sqrt{}(v) \cdot (T_g - T_a)), \quad [°C] \qquad (38)$$

The equation for T_r also indicates the primary disadvantage of the globe thermometer. The observed measurement is the combined effect of radiation and wind speed. An error in the measurement of wind speed will result in an error in the calculation of $\bar{T}_r$.

Another instrument that simultaneously measures the effects of several environmental parameters is the "naturally aspirated" wet bulb thermometer. A hand-held psychrometer is spun to obtain maximum or, at least, uniform air flow; but a natural wet bulb simply has a saturated wick over the bulb. As a consequence, wind speed, air temperature, and evaporation are all variables which contribute to the observed value. Again, there is a problem of separating the effects of wind from humidity. Both the globe thermometer and the natural wet bulb are used in the calculation of the Wet Bulb Globe Temperature (WBGT) index (78).

Environmental Data Loggers

The best way to obtain an accurate, continuous record of environmental conditions is to collect the data from meteorological instruments on a data logger or data acquisition system (4). For physiological studies it may be desirable to record physiological responses and environmental parameters in the same data file. All of the microenvironmental parameters can be measured by meteorological instruments which generate similar voltage output signals.

The simplest data collection devices are field portable data loggers which are completely battery-powered. These basic data loggers may have built-in thermocouple reference thermometers, but are limited in the number of total input channels. The addition of multiplexers to increase the number of input channels and a tape printer will enhance the performance of the logger, but at the cost of increased weight, bulk, and expense.

Some instruments, for example a hot wire anemometer, require an initial excitation or signal conditioner to operate the sensor. A D-cell battery-powered data logger has limited capabilities for generating an excitation signal. Other data loggers have lead-acid batteries or can be converted to operate off AC power sources. Batteries discharge rapidly when air temperatures drop below 0° C. An alkaline battery-powered data logger was successfully operated through an Indiana winter in an unheated shed. It was done by placing a 100 W light bulb about 8 cm above the battery pack to warm the batteries (Santee, unpublished). Larger portable data loggers may operate off a lead-acid battery pack or AC current and the number of input channels can be increased significantly. Other data logging systems are available for fixed sites.

Climatic Data

An important source of basic information regarding a site's environment is the compiled climatic data. However, excessive reliance on local climatic summaries, even during the planning stages, can present problems. The most serious shortcoming of climatic tables is the presentation of unsupported mean temperatures. In the fall, under the clear skies associated with high pressure systems, it is not uncommon to see a daily minimum-maximum temperature range of 10 to 26° C in central Texas, even though the mean daily temperature is a pleasantly cool 16° C.

An important question regarding local climatic data is how representative the data collection site is of the local climate. In one case, the National Weather Service (NWS) forced the relocation of the local meteorological station. The town in which the station was located was an important tourist area by the shores of a desert reservoir. Data were collected on the well-watered lawn of the local fire station. When NWS personnel requested that the meteorological instruments be located in more representative desert conditions, local interests objected, fearing that high temperatures would be reported regionally and nationally as the daily extreme, thereby discouraging potential visitors. International Falls, Minnesota (and by association the entire state of Minnesota) has acquired a reputation for extremely cold winters by frequently having the coldest officially reported temperature in the contiguous United States, even though much lower temperatures occur at other locations. In Utah, a local television weatherman established a remote weather station in a small, high elevation depression where cold air accumulated and began to report phenomenally low temperatures. The relevance of extreme high and low temperatures is reduced when the size of the local population is taken into consideration.

More than one study has attempted to utilize data collected at a fixed meteorological station located several kilometers from the actual study site. In many locations, this may result in serious errors (4). As noted in the preceding paragraphs, both temperature and wind speed vary considerably in the vertical profile. Both parameters are strongly influenced by local topography, terrain features, and vegetation. Moisture is strongly influenced by vegetation, air movement, bodies of water, etc. Reflected solar and thermal ground radiation are strongly dependent on the nature of the surface. Direct and scattered solar radiation measurement may be useful on clear or overcast days, provided sky cover is equivalent at the study and instrument sites. Scattered cloudiness or rapidly changing cloud cover reduce the value of remote radiation measurements.

METHODS FOR QUANTIFYING THE THERMAL ENVIRONMENT

Below an elevation of 2500 m, only those physical parameters which affect heat exchange, the parameters which define the thermal environment, have a significant impact on human physiology. Heat exchange between a subject and the environment can follow four basic pathways. Dry or sensible heat exchange can occur by convection, conductance, or radiation. Evaporation, the fourth pathway, is termed wet or insensible heat exchange.

Heat exchange may occur simultaneously along all four pathways. From the heat balance equation, it is clear that the actual heat exchange is the sum of the individual terms. It would be desirable for the purpose of the statistical analysis of data to express the net result of the interaction of the different heat exchange pathways as a net potential for heat exchange.

In response to the desire for a single temperature or index that will sum up the thermal environmental conditions (the weather report for a layman), several different indices have been developed. The simplest to understand is the "wind-chill" index (49,63,70). We all recognize intuitively that the presence or absence of wind affects how warm or cold we feel. The wind-chill index is a method for quantitatively expressing the combined effects of wind and air temperature as an "equivalent" temperature. For example, a combination of an air temperature of −34.4° C and 2.2 $m \cdot s^{-1}$ wind speed represents the same potential for heat loss from exposed skin as still air at −38° C. The practice of media "weather" reporters to cite only wind-chill, to describe outdoor climatic conditions, is a disservice to the public. Although wind-chill is a useful method for emphasizing the importance and potential danger of convective heat loss to laymen, the insulation provided by clothing and any form of wind protection (either wind-proof clothing or physical barriers [cf. Chapter 2]) quickly reduces the utility of wind-chill. There is a considerable difference between a wind-chill of −40° C produced by an air temperature of −17.8° C (0° F) and a wind speed of 10.0 $m \cdot s^{-1}$ (22.5 mph) and an actual air temperature of −40° C (or °F) in still air when you go to start your car in the morning.

To a large degree the perception of the environment is determined by culture and personal environment. A great number of scientists that deal in human factors research are employed at urban centers where the normal extremes of environmental stress are experienced only during transitory movements from one sheltered environment to another. The majority of scientists, as well as the general population of urban centers, are engaged in sedentary indoor activities. Exposure to extreme environments tend to occur in the context of brief exposures during recreation activities such as skiing, mountain climbing, or driving across the desert. Consequently, for most of us, our anthropmorphic perception of severe environments is that of a tourist. For more primitive cultures, outdoor workers and the military, environmental exposure is not as selective or elective. An Inuit seal hunter may be forced to squat motionless for hours on an icefield waiting for his prey. Military personnel may be facing an enemy force in a dense jungle or a high mountain pass, unable to retreat or advance. The thermal environment experienced by these individuals cannot be adequately described by a standard environment appropriate for office workers or joggers. Media meteorologists offer the public a forecast that will allow them to plan activities to avoid extreme outdoor exposure or to adjust their clothing. The indoor work or home environment may be quite different. For a physiologist attempting to anticipate or even replicate environmental stress, there is a need for a more sophisticated "forecast."

Ecologists frequently have to describe microenvironments that are quite different from normal human environments. Porter and Gates (58) introduced the idea of a climate space (Fig. 1-7) which uses radiation, wind speed and air temperature as three axes to plot

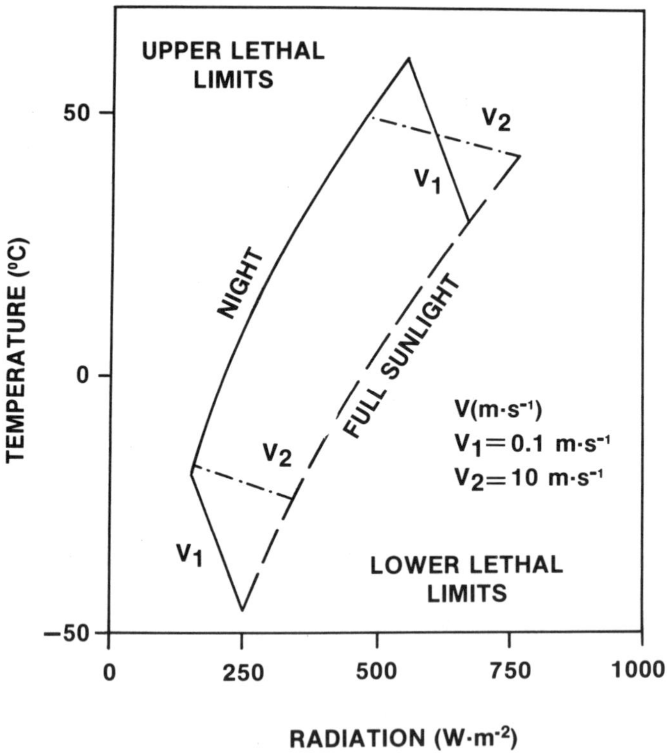

Figure 1-7. *Simplified climate space diagram (redrawn with permission, Porter and Gates, 1969).*

the limits of a species' thermal environment. The radiation limits are set by the maximum (solar and thermal) and minimum (nighttime, clear sky) possible radiation. The upper and lower temperature limits are the upper and lower critical temperatures of the species. Within the boundary limits each point represents a unique combination of radiation and air temperature. The effects of wind speed are demonstrated in a two-dimensional diagram by drawing sloped lines that indicate equivalent temperatures at a specified wind speed. The climate space diagram especially defines different combinations of temperature, radiation, and wind as points within the overall potential environment; the parameters that define the potential for dry heat exchange. The introduction of the climate space concept was followed by a series of attempts by physiological ecologists to develop equations for an "equivalent black-body temperature (T_e);" a single "temperature" that was equivalent to the simultaneous effects of wind, radiation, and temperature on the subject species (51).

During the vigorous debate over the best formulation for expressing T_e (3), ecologists found that the idea of expressing the net potential for heat exchange, as an equivalent temperature, was not the exclusive providence of field biologists. One of the earliest and most successful "equivalent temperatures" (6) was operative temperature (T_o), which was introduced by Gagge (25) as a method of expressing all the parameters of dry heat as a single variable for human subjects. Operative temperature was defined (31,73) as "the temperature of an imaginary isothermal 'black' enclosure in which a man would exchange the same heat by radiation, convection, and conduction from his skin surface at temperature (T_s) as he would in his actual non-uniform environment." (Fig. 1-8)

$$M - E = (h_c + h_r) \cdot (T_s - T_o), \quad [W \cdot m^{-2}] \tag{39}$$

In addition to metabolism and evaporation the other variable is surface temperature (T_s). Operative temperature can be calculated from either of the following equivalent equations:

$$T_o = (h_c T_a + h_r T_r)/(h_c + h_r), \quad [°C] \tag{40}$$

$$T_o = T_a + H_r/(h_c + h_r), \quad [°C] \tag{41}$$

$$H_r = h_r (\bar{T}_r - T_a), \quad [°C] \tag{42}$$

H_r is the effective radiant field. Bakken *et al.* (5) presents alternate equations based on whole body net radiation, which avoid calculation of T_r.

Operative temperature does not account for evaporative heat exchange, but as noted earlier in saturated air or when protective

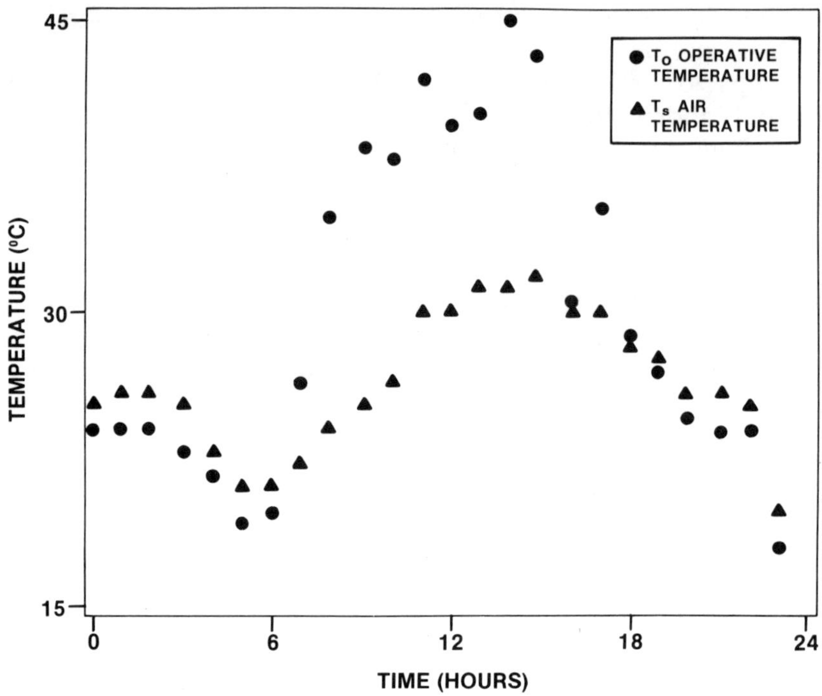

Figure 1-8. *Air (T_s) and operative (T_o) temperature (°C) for June 12, 1982 in west-central Indiana.*

clothing eliminates evaporative exchange, dry heat exchange represents all of the variability in the potential environmental stress. One shortcoming of operative temperature is that while it is a good predictor of dry heat exchange potential, it is not a wholly adequate index or reference for comparison between environments because a change in the wind speed alters the convective heat transfer coefficient (h_c). Standard operative temperature (T_{so}) eliminates that problem by adjusting the operative temperature to a standard convective condition (usually still air [25]).

Building from a base of operative temperature, a series of other equivalent temperatures were formulated, including humid operative temperature (T_{oh}) and standard humid operative temperature (T_{soh}). Humid operative temperature is essentially a term for expressing how the combined parameters for dry and evaporative heat exchange "operate" on a human subject. Standard humid operative temperature is analogous to T_{so} with the assumption of a standard condition for humidity (saturated).

Operative temperature and humid operative temperature are referred to as "rational" temperatures because the equations are derived from actual heat transfer theory and, therefore, have a sounder biophysical basis (6,31). Other indices of thermal stress, such as the WBGT index (78), are derived empirically from observations of meteorological measurements and human performance. WBGT is an accessible, easily understood and easily measured index. For a scientist, a major shortcoming of this index is that it cannot be used to quantify or identify the contributions of different environmental parameters to the overall heat stress. WBGT and similar empirical indices can only relate an observed value to an anticipated level of environmental stress for a given set of conditions. If significantly different clothing is worn, which alters the potential for heat exchange, the empirical relationships that are the basis for the WBGT index, are altered.

REFERENCES

1. Arens, E., R. Gonzalez and L. Berglund. Thermal comfort under an extended range of environmental conditions. *ASHRAE Trans.* 92(1A):410–419, 1986.
2. ASHRAE. *ASHRAE Handbook: 1985 Fundamentals.* Atlanta, GA: American Society of Heating, Refrigerating and Air-conditioning Engineers, Inc., 1985.
3. Bakken, G.S. How many equivalent black-body temperatures are there? *Thermal Biol.* 6:59–60, 1981.
4. Bakken, G.S. and T.H. Kunz. Microclimate methods. In: T.H. Kunz (ed). *Ecological and Behavioral Methods for the Study of Bats.* Washington, D.C.: Smithsonian, 1987.
5. Bakken, G.S., W.R. Santee and D.J. Erskine. Operative and standard operative temperature: tools for thermal energetic studies. *Amer. Zool.* 25:933–943, 1985.
6. Belding, H.S. The search for a universal heat stress index. In: Hardy, J.D., A.P. Gagge and J.A.J. Stolwijk (eds.). *Physiological and Behavioral Temperature Regulation.* Springfield, IL: C. Thomas, pp. 193–202, 1970.
7. Berglund, L., D. Fashena, XH. Su and A. Gwosdow. Absorbed solar radiation from measured sweat rate. *Proceedings of the Thirteenth Annual Northeast Bioengineering Conference.* New York: Institute Electrical Electronics Engineers, pp. 507–510, 1987.
8. Bligh, J. and K.G. Johnson. Glossary of terms for thermal physiology. *J. Appl. Physiol.* 35:941–961, 1973.
9. Breckenridge, J.R. Use of a wetted cover to reduce heat stress in impermeable clothing. *Technical Report* T7/80. Natick, MA: US Army Research Institute of Environmental Medicine, 1980.
10. Breckenridge, J.R. Effective area of clothed man for solar radiation. *Technical Report* EP-157. Natick, MA: Quartermaster Res. Engineering Command, 1961.
11. Breckenridge, J.R. and R.F. Goldman. Solar heat load in man. *J. Appl. Physiol.* 31:659–663, 1971.
12. Bruno, R. A correction procedure for separating direct and diffuse insolation on a horizontal surface. *Solar Energy* 20:97–100, 1978.
13. Bristow, K.L. and G.S. Campbell. On the relationship between incoming solar radiation and daily maximum and minimum temperature. *Agricult. Meteorol.* 31:159–166, 1985.
14. Campbell, G.S. *An Introduction to Environmental Biophysics.* New York: Springer-Verlag, 1977.
15. Cicerone, R.J. Changes in stratospheric ozone. *Science* 237:35–42, 1987.
16. Clark, J.A., A.J. McArthur, J.L. Monteith and A.E. Wheldon. The physics of microclimate. In K. Cena and J.A. Clark (eds.). *Bioengineering, Thermal Physiology and Comfort.* New York: Elsevier, pp. 13–27, 1981.
17. Clark, R.P. Human skin temperature and convective heat loss. In K. Cena and J.A. Clark (eds.). *Bioengineering, Thermal Physiology and Comfort.* New York: Elsevier, pp. 57–76, 1981.
18. Chambers, A.B. A psychrometric chart for physiological research. *J. Appl. Physiol.* 29:406–412, 1970.
19. Drake, R.M., J.E. Funk and J.B. Moegling. Sensible heat transfer in the Gemini and Apollo space suits. In *Thermal Problems in Biotechnology.* New York: American Society of Mechanical Engineers, pp. 96–112, 1968.

20. Dubois, D. and E.F. Dubois. A formula to estimate the approximate surface area if height and weight be known. *Arch. Internal Med.* 17:863–871, 1916.
21. Erbs, D.G., S.A. Klein and S.A. Duffie. Estimation of the diffuse radiation fraction form hourly, daily and monthly average global radiation. *Solar Energy* 28:293–302, 1982.
22. Fanger, P.O. *Thermal Comfort.* New York: McGraw-Hill, 1970.
23. Fritschen, L.J. Minerature net radiometer improvements. *J. Appl. Meteorol.* 4:528–532, 1965.
24. Gagge, A.P. A new physiological variable associated with sensible and insensible perspiration. *Amer. J. Physiol.* 120:277–287, 1937.
25. Gagge, A.P. Standard operative temperature. A generalized temperature scale applicable to direct and partitional calorimetry. *Amer. J. Physiol.* 131:93–103, 1940.
26. Gagge, A.P. Effective radiant flux, an independent variable that describes thermal radiation of man. In Hardy,J.D., A.P. Gagge and J.A.J. Stolwijk (eds.). *Physiological and Behavioral Temperature Regulation.* Springfield, IL: C. Thomas, pp. 34–45, 1970.
27. Gagge, A.P. Thermal sensation and comfort in dry humid environments. In P.O. Fanger (ed.) *CLIMA 2000, Volume 4: Indoor Climate.* Copenhagen: VVS Kongress-VVS Messe, pp. 77–82, 1985.
28. Gagge, A.P. and Y. Nishi. Heat exchange between human skin surface and thermal environment. In D.H.Lee (ed.), *Handbook of Physiology: Section 9. Reaction to Environmental Agents.* Rockville, MD: American Physiological Society, pp. 69–92, 1983.
29. Galanis, N. and R. Chatigny. A critical review of the ASHRAE solar radiation model. *ASHRAE Trans.* 92(1B):18–26, 1986.
30. Gieger, R.G. *The Climate Near the Ground.* Cambridge, MA: Harvard University Press, 1965.
31. Gonzalez, R.R., Y. Nishi and A.P. Gagge. Experimental evaluation of standard effective temperature; a new biometeorological index of man's thermal discomfort. *Intern. J. Biometeorol.* 18:1–15, 1974.
32. Graichen, H., R. Rascati and R.R. Gonzalez. Automatic dew-point temperature sensor. *J. Appl. Physiol.* 52:1658–1660, 1982.
33. Greenspan, L. A pneumatic bridge hygrometer for use as a working humidity standard. In A. Wexler and W.A. Wildhack (eds.). *Humidity and Moisture: Measurement and Control in Science and Industry. Fundamentals and Standards.* New York: Reinhold, pp. 433–443, 1965.
34. Haines, R.W. How to construct high altitude psychrometric charts. *Heating, Piping and Air Conditioning* Oct:114–116, 1961.
35. Harrington, J.B. Solar radiation in a clear-cut strip: a computer algorithm. *Agricult. Meteorol.* 33:23–39, 1984.
36. Harrison, L.P. Fundamental concepts and definitions relating to humidity. In A. Wexler and W.A. Wildhack (eds.). *Humidity and Moisture: Measurement and Control in Science and Industry. Fundamentals and Standards.* New York: Reinhold, pp. 3–69, 1965.
37. Harrison, L.P. Some fundamental considerations regarding psychrometry. In A. Wexler & W.A. Wildhack (eds.). *Humidity and Moisture: Measurement and Control in Science and Industry. Fundamentals and Standards.* New York: Reinhold, pp. 71–103, 1965.
38. Heath, J.E. Reptilian thermoregulation: evaluation of field studies. *Science* 146:783–784, 1964.
39. Ideriah, F.J.K. A model for calculating direct and diffuse solar radiation. *Solar Energy* 26:447–452, 1981.
40. Idso, S.B. Transformation of a net radiometer into a hemispherical radiometer. *Agricult. Meteorol.* 9:109–121, 1971.
41. Idso, S.B. Simplifications in the transformation of net radiometers into hemispheric radiometers. *Agricult. Meteorol.* 10:473–476, 1972.
42. Kaufman, W.C. and D.J. Bothe. Wind chill reconsidered, Siple revisited. *Aviat. Space Environ.* 57:23–26, 1986.
43. Kerr, R.A. Has stratospheric ozone started to disappear? *Science* 237:131–132, 1987.
44. Kusuda, T. Calculation of the temperature of a flat-plate wet surface under adiabatic conditions with respect to the Lewis relation. In R.E. Ruskins (ed.) *Humidity and Moisture: Measurement and control in Science and Industry. Principles and Methods of Measuring Humidity in Gases.* New York: Reinhold, pp. 16–32, 1965.
45. Kusuda, T. Algorithms for psychrometric calculations. National Bureau of Standards Rpt. No. 9818:1–22, 1969.
46. Leone, O.J. Improvements in dew-point measurements of gases by use of Peltier devices. In R.E. Ruskins (ed.). *Humidity and Moisture: Measurement and control in Science and Industry. Principles and Methods of Measuring Humidity in Gases.* New York: Reinhold, pp. 635–642, 1965.
47. McIntyre, D.A. Radiant heat from lights, and its effect on thermal comfort. *Lighting Res. Technol.* 8:121–128, 1976.
48. McIntyre, D.A. *Indoor Climate.* London: Applied Science, 1980.
49. Mitchell, J.W. Heat transfer from spheres and other animal forms. *Biophysics* 16:561–569, 1976.

50. Mollier, R. Ein neues Digramm fur Damfluftgemische. *Z. Vergleich. Duetch Ingr.* 67:869–872, 1923.
51. Morhardt, S.S. and D.M. Gates. Energy exchange analysis of the Belding ground squirrel and its habitat. *Ecol. Monogr.* 44:17–44, 1974.
52. Nautical Almanac Office. *The Air Almanac.* Washington, D.C.: US Government Printing Office, 1970.
53. Nautical Almanac Office. *Ephemeris of the Sun, Polaris and other Selected Stars: Seventieth Edition.* Washington, D.C.: US Government Printing Office, 1979.
54. Nishi, Y. Measurement of thermal balance of man. In K. Cena and J.A. Clark (eds.). *Bioengineering,Thermal Physiology and Comfort.* New York: Elsevier, pp. 29–39, 1981.
55. Nishi, Y. and A.P. Gagge. Direct evaluation of convective heat transfer coefficient by napthalene sublimination. *J. Appl. Physiol.* 29:830–838, 1970.
56. Nishi, Y. and A.P. Gagge. Effective temperature scale useful for hypo- and hyperbaric environments. *Aviat. Space Environ. Med.* 48:97–107, 1977.
57. Platt, R.B. and J. Griffiths. *Environmental Measurement and Interpretation.* New York: Reinhold, 1966.
58. Porter, W.P. and D.M. Gates. Thermodynamic equilibrium of animals with the environment. *Ecol. Monogr.* 39:227–244, 1969.
59. Porter, W.P., J.W. Mitchell, W.A. Beckman and C.B. DeWitt. Behavioral implications of mechanistic ecology. *Oecologia* (Berlin) 13:1–54, 1973.
60. Rapp, G.M. Convective mass transfer and the coefficient of evaporative heat loss from human skin. In: J.D. Hardy, A.P. Gagge and J.A.J. Stolwijk, (eds.). *Physiological and Behavioral Temperature Regulation.* Springfield, IL: C. Thomas, pp. 55–80, 1970.
61. Santee, W.R. *The Thermal Environment and Behavior of Breeding Male Redwinged Blackbirds (Agelaius phoeniceus),* Thesis, Indiana State University, 1985.
62. Scott, D. *Dave Scott's triathlon training.* New York: Simon & Schuster, 1986.
63. Sears, F.W., M.W. Zemansky and H.D. Young. *University Physics. Sixth Edition.* Reading, MA: Addison-Wesley, 1982.
64. Siples, P.A. and C.F. Passel. Measurements of dry atmospheric cooling in subfreezing temperatures. *Proc. Amer. Phil. Soc.* 89:177–199, 1945.
65. Strahler, A.N. *Physical Geography: Third edition.* New York: Wiley, 1969.
66. Swan, H. *Thermoregulation and bioenergetics: Patterns for survival.* New York, NY: Elsevier, 1974.
67. Tracy, C.R. Minimum size of mammalian homeotherms: role of thermal environment. *Science* 198:1034–1035, 1977.
68. Tracy, C.R. and K.A. Christian. Ecological relations among space, time, and thermal niche axes. *Ecology* 67:609–615, 1986.
69. Underwood, C.R. and E.J. Wand. The solar radiation area of man. *Ergonomics* 9:155–168, 1966.
70. Vernon, H.M. The measurement of radiant heat in relation to human comfort. *J. Ind. Hyg.* 14:95–111, 1932.
71. Weiss, A. and J.M. Norman. Partitioning solar radiation into direct and diffuse, visible and near- infra-red components. *Agricult. Meteorol.* 34:205–213, 1985.
72. Wexler, A. and R. Hyland. The NBS standard hygrometer. In A. Wexler and W.A. Wildhack (eds.). *Humidity and Moisture: Measurement and Control in Science and Industry. Fundamentals and Standards.* New York: Reinhold, pp. 389--432, 1965.
73. Winslow, C.E.A., L.P. Herrington and A.P. Gagge. Physiological reactions of the human body to varying environmnetal temperatures. *Amer. J. Physiol.* 120:1–22, 1937.
74. Wood, L.A. The use of dew-point temperature in humidity calculations. *J. Res. Nat. Bureau Standards.* 74:117–122, 1970.
75. Woodcock, A.H. and J.R. Breckenridge. *Technical Report* BP-7 Theory of the globe thermometer. Natick, MA: Quartermaster Research and Engineering Center, 1957.
76. Woodcock, A.H. and J.R. Breckenridge. *ASME Paper* 57-SA-64 Analysis of energy exchange between man and his environment. New York: American Society of Mechanical Engineers, 1957.
77. Wylie, R.G. The properties of water-salt systems in relation to humidity. In A. Wexler and W.A. Wildhack (eds.). *Humidity and Moisture: Measurement and Control in Science and Industry. Fundamentals and Standards.* New York: Reinhold, pp. 507–517, 1965.
78. Yaglou, C.C. and D. Minard. Control of heat casualties at military training camps. *Arch. Ind. Health* 16:302–316, 1957.

Appendix A . . . Calculation of the Zenith Angle

The zenith angle (θ_z) can be calculated from the following parameters: site latitude (L) and longitude (l), time (civil or clock time, t_{cl}), and solar declination angle (δ).

The hour angle converts clock time (t_{cl}) into a solar angle (h).

$$ch = \Delta l \cdot (1440/360), \quad [\text{min}] \tag{43}$$

$$t_{sun} = (t_{cl} \cdot 60) - ch, \quad [\text{min}] \tag{44}$$

The sun is directly over the standard meridan (l_s) at 12:00 noon, plus or minus a correction factor (the equation of time). Hour angle correction (ch) means that at a site of longitude 87° 13.5′W, the sun is directly overhead on the average day (mean solar time) at 12:50 EST because the site is west of the standard time meridian (75° W). The investigator should also make an allowance for the sampling interval when selecting t_{cl}.

$$\begin{aligned} I_s &= 75^\circ \text{ W} \\ l &= 87.25 \\ \Delta l &= -12.25 \\ ch &= ((-12.25) \cdot 4) \\ &= -49^\circ \end{aligned}$$

If data are collected in 15 minute blocks, data recorded at 12:00 spans the period from 11:45 to 12:00. The mid-point of the sampling period 11:52:30 (12:00—(15/2)), may be a more appropriate clock time.

$$h = ((t_{sun} - 720)/1440) \cdot 2 \cdot pi + \Delta E, \quad [\text{radians}] \tag{45}$$

The correction factor from the equation of time, ΔE, is an adjustment that relates apparent solar time to mean solar time. This correction is essentially an adjustment due to the fact that the earth's orbit and rotation are not uniform. The equation of time correction can be found in the *ASHRAE Fundamentals Handbook* (1985), *The Ephemeris* or the *Air Almanac*. It may also be calculated (2,65).

The following equation is presented for either the cos θ_z or sin θ_s, the compliment of the solar elevation angle (2,14).

$$\cos \theta_z = \sin L \cdot \sin \delta + \cos L \cdot \cos \delta \cdot \cos h, \quad [\text{N.D.}] \tag{46}$$

The declination angle (D) is also published in *The Ephemeris* or equivalent reference. It also may be approximated with the appropriate equation (2,14).

Appendix B . . . Humidity Measuring Devices

The multitude of humidity measuring instruments have been categorized according to accuracy and sensitivity (36,37) in the following way with highest precision being the NBS standard (72):

Hierarchy of Humidity Instruments

TYPE 1:	NBS Standard		
TYPE 2:	a. Optical chilled mirror dew point	b. Double pressure generator	
TYPE 3:	a. Wet Bulb psychrometer	b. Saturated salt solutions	c. Pneumatic bridge
TYPE 4:	a. Saturated heated lithium chloride	b1. Piezoelectric b2. Electrolytic b3. Dunmore Elements	c1. Mechanical c2. Specotroscopic

Type 1 is the most precise gravimetric hygrometer developed by the National Bureau of Standards (NBS, Washington, D.C.) (70). In this instrument, a test sample flows from a humidity generator through a drying tube and a precision volume measuring column in a temperature controlled bath. Thus, the required measurements of temperature, weight, pressure, and volume are known. One disadvantage is in its operation time which may be up to 30 h at low humidity ranges. NBS generally uses a two pressure humidity generator as a definitive source reference hygrometric (type 2). In this device the gas sample is saturated at constant temperature, followed by an expansion to a lower pressure which may be at the same or a higher or lower temperature. By Dalton's law (ratio of partial pressures vary directly with total pressure) %rh is determined accurately by the ratio of the test chamber pressure (P_c) to the saturation pressure (P_{st}):

$$\%\text{ rh at } P_{st} = P_c/P_{st} \times 100, \quad [\%] \qquad (37)$$

As is apparent, the relative humidity is not dependent arbitrarily on the measurement of water content of a testing sample, but on measurement of pressures only in an isothermal condition. This system, although accurate, is not wholly practical for continuous monitoring. For this reason the chilled mirror optical dew point calibration techniques have been developed. At saturation temperature, or the dew point, an air-water mixture is saturated with respect to water or the ice point. Theoretically, if the mirror is completely clear, the rate of water molecules condensing on the mirror surface is equal to the rate of those being evaporated from the surface and entering the atmosphere. Thus, at equilibrium, the water vapor partial pressure of the condensate equals the water vapor partial pressure of the gas atmosphere (ambient water vapor pressure, P_w). The necessary requirements are that the mirror surface must be cooled at the precise saturation temperature. Cooling of the mirror surface is usually done by continually flowing refrigerant liquids, acetone, liquid CO_2, or other cooling fluids.

Recently, precise and miniaturized dew point sensors (32) have been developed using the thermoelectrical heating pump technique, first developed by Peltier (46). These are becoming useful in clothing water vapor analysis and in methods which ascertain the sweating rate of small areas of the skin, mentioned in several chapters in this book. In this method, the dew-point is measured by a thermocouple or thermal module operable by a servo-loop or microprocessor device, which detects and signals changes in the temperature of a sensing plate. The latter is alternately cooled and heated to cause moisture in the ambient gas, close to a surface, to condense between a pair of electrodes supported on either side of the sensing plate. The sensing plate is composed of a high thermally conductive material. As condensate forms on the plate, the change in resistance (analogous to change in reflectance of a chilled mirror) is detected. This type of hygrometer is extremely valuable because of its small size, accuracy, and almost continuous direct absolute humidity reading property.

The thermodynamic wet bulb temperature, most frequently used in laboratory or field experiments, is an example of Type 3 hierarchy humidity devices. The wet bulb temperature is either obtained by covering a glass thermometer, thermocouple, or thermistor with a moistened wick aspirating the system at a constant flow ($>3\ m \cdot s^{-1}$), and then noting the lowest cooling depression. In theory, the wet bulb temperature of moist air at constant pressure, temperature, and humidity is the temperature at which water evaporates into the air to saturate it without gain or loss of heat (i.e., adiabatically). The method is fairly accurate; however, errors in measurement exist due to placement of the sensing device such as a thermocouple into the interface of the wick, purity of water, how clean the wick is, air flow, radiation effects, density, viscosity, and thermal conductivity of the gas. Many of these properties are associated with the barometric pressure, temperature, and type of gas so wet bulb psychrometry is not wholly accurate for use in either hyperbaric or hypobaric environments.

Another example of a Type 3 humidity device includes the pneumatic bridge hygrometer (33). In this system, two sets of optimum flow nozzles are connected in series in an arrangement analogous to the Wheatstone bridge. Mass flow rate of a gas through the downstream nozzle is arranged to be proportional to the gas upstream pressure. Any change in mass flow induced by a desiccant placed between the two nozzle arrangements is introduced as a variable which affects the pressure in a test branch. A differential manometer is used to compare the test branch and a reference branch. Although a bit cumbersome, the change in water vapor concentration produced by the desiccator in the test branch of the pneumatic bridge, in comparison to the absolute value of pressure in the reference, produces a very accurate reading.

Less accurate, but easily determined Type-3 systems include use of saturated and unsaturated aqueous salt solutions (77). Experimentally a given salt: H_2O solution determines, by its ambient temperature, the equilibrium water vapor pressure. Thus, water vapor pressure versus temperature curves of many salt solutions, such as lithium chloride, ammonium chloride, calcium chloride, etc. may be used. In these solutions the dissolving salt, along with water, depresses the solution. In supersaturated solutions where a solid phase of salt and water vapor coexist in isothermal expansion, the vapor pressure depression reaches a limit. Therefore, the temperature at which equilibrium (by mass balance) is reached will signify the vapor pressure of the saturated solution very accurately. This is somewhat analogous to the chilled mirror dew point hygrometric concept already discussed.

Type-4 humidity devices are widely used because of their simple operation, but have shortfalls compared to the devices previously discussed. The family of heated lithium chloride sensors, although widely used in air quality or meteorological measurements, are affected by direct water exposure. These sensors have wide drift and often exhibit errors in calibration properties as a result of changing solution concentration and contamination. Also, response time of such sensors is slow; for example, a lithium chloride sensor may need almost five minutes to record 67% step change.

A whole series of resistive and electrolytic hygrometer and spectroscopic (infrared/ultraviolet) are being introduced. The spectroscopic hygrometers measure the humidity of a gas sample by detection of the energy absorption in the water vapor bands. A typical system requires, as an energy source, a thermal radiation energy detector and an optical system to discriminate the wavelengths and a measuring device to quantitate the decreases in radiation caused by water vapor present in the optical path. The response time is very fast; typically, 90% of the final reading occurs within five seconds. These instruments, however, are still relegated to use of separate calibration by use of the higher order methods already described.

2

Biophysics of Heat Transfer and Clothing Considerations

RICHARD R. GONZALEZ, Ph.D.

OUTLINE

INTRODUCTION

This chapter continues with an account on the interaction of the homeotherm and its surroundings, primarily the thermal environment. Chapter 1 was directed towards the thermal exchange properties in unclothed persons. This chapter considers the human as a clothed homeotherm responsive not only to its internal body temperature control characteristics, but affected by the properties inherent in the clothing layer. Unlike animals with pelts (skin with fur), clothing in humans is a part of the environmental barrier, influencing the physical processes of energy exchange between the body and the ambient. The skin (and the pelt in other animals) is the integral node which allows this immediate interaction; clothing, alternatively, is a passive instrument in the regulation of thermal energy processes between the body and the environment.

The aspect of thermophysics, which is incorporated in this chapter, is relatively new, developing during or soon after World War II as physicists entered into the domain of physiological sciences as apparent in Newburgh's classic book [105,106,134]. Never really defined, thermal biophysics (or rather environmental biophysics) is more nearly a branch of environmental, systemic physiology. It embraces fundamental laws and principles governing energy exchanges between a system (the homeotherm usually) and its ambient. It also studies the energy transformations and functions of the organism's systems in physical terms. The systems considered in humans include alterations in the internal body and surface temperature, heat flow and distribution of heat which affects the boundary layer of the surface, and the thermal resistance of the clothing which acts in series with the layer [72,73,83,114,137,138].

Also considered in the above systems studied in thermal biophysics are the characteristics of modes of heat transfer: radiation and convection, conduction, and evaporation. It is useful to distinguish these terms. *Heat transfer* deals with an analysis of the rate of heat transfer in a system governed by the laws of thermodynamics; energy changes by *heat flow* are not directly measurable but heat flow is always related to temperature which is a measured quantity.[1] Whenever there is a temperature difference or gradient (denoted in this chapter as K, kelvin or °C deg C), heat flows from the region of high temperature to that of low temperature. By knowing the temperature distribution, *heat flux* is assessed as the amount of heat transfer per unit area per unit time.

The organism can be in a physical or functional state that is almost steady (that is, non-varying with time) or varies from instant to instant (transient) as a response to a forcing (constraining) factor.

[1] A fascinating account is given by King [83] who notes that . . . "no thermal measurements were possible until after the invention of the thermometer in 1593 by Galileo (although Drebel of Holland also made one). These instruments were actually thermobaroscopes since they were open to the environment and responded to barometric pressure. . . ."

Multiple variables and functions are necessary for a complete quantitative description of the state of the system. Chapter 1 pointed out that a triathlon's performance depends on a number of variables. Those that are independent are variables of state and those that are quantitatively measurable as a function of the latter are dependent variables called state functions. In describing human heat balance with the environment (formally introduced in Chapter 1), nine independent variables are considered: metabolic energy production, M (an endogenous source); the positive (or negative, eccentric) work accomplished, W; the dry bulb temperature T_a; the dew point temperature T_{dp}; the mean radiant temperature $\bar{T}_r$; a linear radiation transfer coefficient h_r; a convective heat transfer coefficient h_c, which includes the effect of air movement; the thermal insulation of the clothing around the skin surface I_{cl} and the time of exposure (t) in the environment.

The dependent variables are those affected in the heat balance equation covered thoroughly in Chapter 1. These encompass evaporative heat loss from the skin surface, E_{sk}, the average skin temperature $\bar{T}_{sk}$, the skin wettedness w, and the rate of change in mean body temperature as a weighted value of internal core temperature and average skin temperature [27,28,29,50,53].

Historical Perspective

An integral part of clothing biophysics is related to its history. For this reason a hint of its developmental roots is relevant. More extensive sketches are found elsewhere for the interested reader [29,39,64,105,143]. Historical accounts of primitive man show a clear link of clothing properties and their protection against the thermal environment. In the early writings of Buddha [64], for example, clothing protection is directly implied by his account of proper dress to ward off "hostile environments" rather than as an entreatment of clothing as a decorative tool. The fact that the efficacy of clothing offers a resistance to the loss of heat due to the fiber's trapping of stagnant air was realized in early times. Renbourn states [119] that . . . "As early as 1620, Francis Bacon suspected that the warmth of wool, skin, and feathers and the like was due to the confinement and separation of air . . ." Also, Robert Hook as early as 1664 pictured the use of artificial fibers as thermal insulators in place of natural fibers [64]. Before the 1940's, little attention was directed towards quantification of clothing properties other than sparse accounts from the Harvard Fatigue Laboratory, the John B. Pierce Foundation Laboratory, the Russell Sage Institute, and various other research institutions in the United Kingdom, particularly the Shirley Institute [1,47,51,71,72,77,134]. A fundamental (and now universal) unit which appeared early in 1941 expressed the insulating effects of clothing in terms of the clo unit. Gagge, Burton, and Bazett [55] first described this unit. Their aim was to provide a standardized and practical measure which expressed thermal activity and insulation in both metric and English nomenclature. To review briefly, if thermal equilibrium occurs without the necessity of major physiological adjustments, the three properties important in energy balance at an optimal skin temperature are heat production of the body (M), the insulating value of the clothing, and the environmental temperature. The use of practical units for thermal activity and insulation provide a uniform system that describe comfortable conditions in relation to the heat exchange of humans with the environment. Gagge *et al.* [55] defined resting metabolic activity as 50 kcal per hour per square meter of the surface area of the individual. This unit is *one met.* The unit for thermal insulation of clothing is the insulation necessary to maintain, in comfort, a sitting (resting) subject in a normally ventilated room (air movement, V = 10 cm/s) at a temperature of 21° C and a humidity of less then 50%. This unit is *one clo.* Since thermal insulation is the resistance offered to flow of heat, its measure is done by considering the ratio of the difference in temperature between two surfaces to the flow of heat per unit surface area. Gagge *et al.* [55] estimated the heat transmitted through the clothing as 38 kcal/($m^2 \cdot h$), (76% of 50). The total insulation, which is the sum of the insulation of the clothing, I_{cl}, and that of the ambient air, I_a, (albeit in the original definition this property was not considered in the unit of clo) is given by the equation:

$$I_T = (I_{cl} + I_a) = \{33 - 21°\ C\}/38 = 0.32,\ \text{clo units} \qquad \text{(Eq. 2-1)}$$

where, 33° C signifies the "comfortable average skin temperature." From previous work by Winslow *et al.* at the John B. Pierce Foundation Laboratory [133,134], the values for the insulation of air (I_a) went into the definition of clo [Gagge, personal communication, 1987]. Burton [28] initially calculated these values by the following empirical equation:

$$I_a = 1/[0.61 \cdot (T_a/298)^3 + 0.19 \cdot V \cdot (298/T_a)],\ \text{clo} \qquad \text{(Eq. 2-2)}$$

where, T_a is in °C and air velocity (V) is in cm/s. I_a is thus equivalent to 0.14°C/($kcal/m^{-2} \cdot h^{-1}$) in the definition above. The insulation of intrinsic clothing, which is equal by definition to one clo becomes:

$$I_{cl} = 0.32 - 0.14 = 0.18°\ C/(\text{kcal} \cdot \text{per sq m per h})\ \text{or} \qquad \text{(Eq. 2-3)}$$
$$= 0.155\ m^2 \cdot °C \cdot W^{-1}$$

Burton calculated the equation for T_a ranges of 25 to −40° C at various air movements and found that I_a varied fewer than 0.1 clo (+ or $-0.016\ m^2$ °C per watt), so he suggested elimination of the temperature factor in its calculation. The equation was modified by $I_a = 1/[0.61 + 1.87(P_b/760 \cdot V)^{0.5}]$ for altitude and V = m/s [6,29] (Figure 2-1).

Further developments [24,26,62] to Burton's equation show that the boundary air layer expressed in clo units is [24]

$$I_a = 6.46/(h_r + h_c),\ \text{clo} \qquad \text{(Eq. 2-4)}$$

Where 6.46 converts °C · m^2/W clo units. This analogy applies, as we will point out later, very usefully in clothing studies which use manikin and human data as a cooling efficiency coefficient, in which

$$F_{cl} = I_a/(f_{cl} \cdot I_T),\quad \text{[N.D., non-dimensional]} \qquad \text{(Eq. 2-4')}$$

In the above equations the radiation coefficient, h_r, is a function of surface and air temperatures, h_c is the respective convective heat transfer coefficient (a function of V), f_{cl} is a factor which alters the boundary layer insulation as clothing surface area increases, roughly 15% for each clo unit [25,43,56,58,113].

Figure 2-2 shows the total insulation (I_T) as the insulation

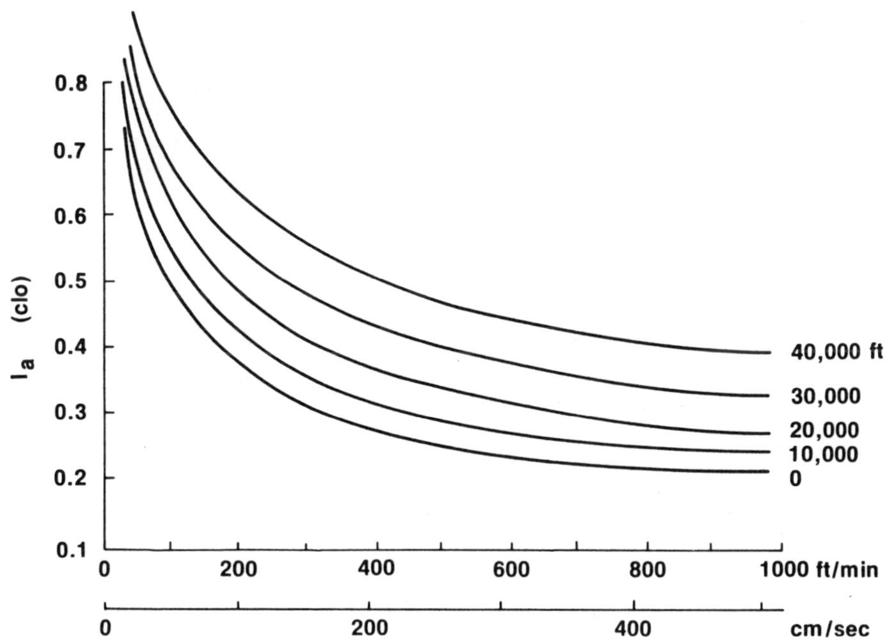

Figure 2-1. *Relationship of insulation of the boundary air layer (I_a), in clo units, to air velocity at different altitudes. [Redrawn from Reference 6].*

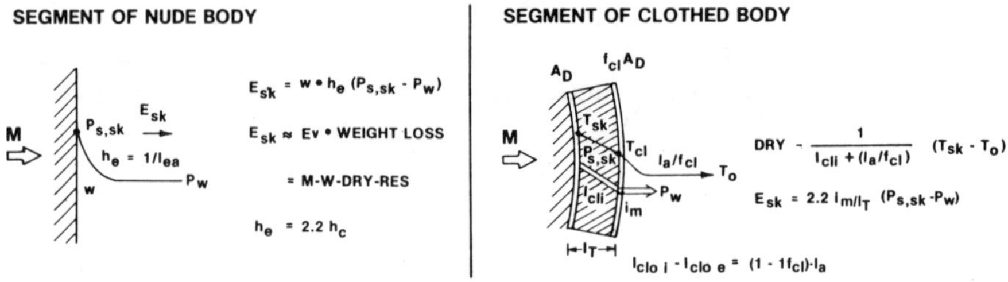

Figure 2-2. *Sensible and insensible clothing parameters active in the skin surface to ambient interface in comparison to the nude skin. Heat generated internally is dissipated on the skin surface and flows through a variable resistance affected by different intrinsic, and evaporative clothing coefficients.*

[24,26,78,80,96,112] from the skin surface to the environment, which includes the property of the increased surface area (f_{cl}) and the added resistance at the surface of the clothed body (I_a).

$$I_T = 6.46\,(\bar{T}_{sk} - T_o)/H_s \qquad \text{(Eq. 2-5)}$$

Where I_T = total insulation, clo units
H_s = dry heat loss per m^2 skin area, $W \cdot m^{-2}$
T_{sk} = mean skin temperature, °C
T_o = operative temperature, °C
$6.46 = 1\ \text{clo}/(0.1547\ m^2 \cdot K \cdot W^{-1})$

The *effective clothing insulation* (I_{cle}) is the insulation [110] from the skin to the clothing surface, which excludes any added effect of the increased surface area of the clothed body (f_{cl}) and uses the human body surface (i.e., DuBois) as its standard surface.

$$I_{cle} = I_T - I_a = 6.46\,(\bar{T}_{sk} - T_o)/H_s - I_a,\ \text{clo} \qquad \text{(Eq. 2-6)}$$

Where I_{cle} = effective clothing insulation, clo units
I_a = resistance at the surface of the clothed body, clo

The *intrinsic clothing insulation* (I_{cl}) is the insulation [78,96,112] from the skin to the clothing surface:

$$I_{cl} = I_T - I_a/f_{cl} = 6.46(\bar{T}_{sk} - T_o)/H_s - I_a/f_{cl} \qquad \text{(Eq. 2-7)}$$

Where I_{cl} = intrinsic clothing insulation, clo units
f_{cl} = clothing area factor (typically for each clo unit increases by 20% ±5%).

The following gives a composite of the relation between I_{cl}, I_{cle} and I_T:

$$I_{cl} = I_{cle} + I_a(1 - 1/f_{cl}) = I_T - I_a/f_{cl} \qquad \text{(Eq. 2-8)}$$

Insulation values for various common clothing ensembles should specify whether they refer to the total, the effective, or the intrinsic clothing insulation. I_{cl} is usually measured with a heated copper manikin, while f_{cl} is measured with a photographic method [43,96,112]; I_{cle} is often measured directly on humans by direct or partitional calorimetry [65,75,101,110]. One clo unit is equivalent roughly to the insulative value of a normal business suit (the fashion and weight in 1941) at 21° C. Peirce and Rees [cited in Reference 47] of the Shirley Institute introduced the "tog" unit around 1946, a smaller unit of resistance roughly equivalent to light summer clothing [1 tog = 0.645 clo or 0.1° C/($m^2 \cdot W$)]. Other units have been developed in conformity with the initial clo definition used in describing thermal resistances of animal pelts (such as r_T, in which 1 clo corresponds to about 200 s/m, or 2 s/cm) [33,100].

Some Methods for Clothing Evaluation

Scientists around the late 1940s almost extensively used heated flat plate or guarded ring cylinders [8,47,105] for their thermal insulation measurements. In the flat plate apparatus, a central heated plate is thermostatically controlled at a set temperature and it is guarded beneath and at the sides by other heated elements maintained at the same temperature. The given fabric is laid smoothly over the measuring area and guard ring and the rate of heat flow through the fabric into the air is analyzed by the energy required to maintain the source at constant temperature.

In order to distinguish more closely between insulating values of various whole ensembles developed for use during exposure to environmental extremes, researchers needed faster access to data. Electrically-heated manikins became essential. This was especially evident from an offshoot of a physiological study by Belding in 1943 [Breckenridge, personal communication] at the Harvard Fatigue Laboratory on electrically heated flying suits and in another study by him on heated casualty blankets with built-in thermostats.

One copper manikin developed at this time allowed alterations in the surface temperatures of the hands and feet without alterations in the rest of the body. In parallel time periods, but independent of each other, first generation manikins appeared in many other laboratories. An early manikin [134] constructed for the J.B. Pierce Laboratory Co. had 0.8 mm copper sheeting molded over a papier mache structure that had a surface area of 1.8 sq. m. Power was provided to the manikin by 16 variable wattage light bulbs enclosed in copper mesh cages. Each cage was located at various segments of the body which simulated the temperature distribution of a "comfortably cool" person (in agreement with the classic clo definition), while nude and at rest in $T_a = 30° C$. Heat input in this manikin was variable up to 4 mets.

The Quartermaster Climatic Research Laboratory [QM] (a predecessor to USARIEM) around this time period was also experimenting with prototype manikins. One model was constructed of stovepipe and sheet metal. It lacked arms and a head and had a robust torso with a central electrical heater and a fan to circulate air within its shell. The earliest rigorous copper manikins used by the QM was patterned after the Belding manikin [45]. By 1944, sizing standards for humans were obtained. Three plastic, jointed manikins were constructed for the Army Air Forces based on the anthropomorphic characteristics of some 2961 persons [40].

The concepts extended to clothing insulation assessed the dry heat exchange properties adequately in heated copper manikins using the clo value and many of the other laboratories between 1949–1969 built their own [85,94,99]. These were single circuit models that controlled a constant average skin surface temperature in compliance with the original definition of clo, or many had specialized circuits for evaluation of variable heat flow through different areas of the skin surface [41,85,94,99,123]. Currently, a robotic model that simulates sweating is being developed [13].

Information obtained with heated copper manikins (relative to the dry thermal insulation of clothing) supplemented and eventually reduced to a lower plane data that were acquired with flat plates. Studies with flat plates still remain valuable because insulative layers of samples are homogeneous in their thermal layers so a rule of thumb of 4 clo per inch (0.16 clo/mm) estimates I_T adequately [47,80]. However, this property does not hold on highly curved surfaces [129]. Layers of insulation which are parallel on a flat surface are not on curved surfaces. As apparent from manikin studies, either the thermal conductivity is reduced (by an added trapped air space) or the external heat transfer coefficient is altered by air movement. The geometrical configuration of the human surface area also affects such convective and radiative properties which cannot be wholly simulated by flat plates or cylinders.

One missing property in clothing assessment was addressed to the insensible component present in humans which is impeded by clothing and is activated by thermoregulatory sweating. In 1937, Gagge [50] introduced a skin wettedness component (w) which is important to clothing evaluation, because it describes the ratio of evaporative heat loss of the total skin surface area wet with sweat (E_{sk}) to the maximum possible limited by the environment (E_{max}). The latter property is strongly affected by the evaporative heat transfer coefficient (h_e') and the water vapor transfer gradient. As water vapor from thermoregulatory sweating accumulates within clothing, this raises the local humidity, and therefore, slows down the rate of sweat evaporation and removal of latent heat possible.

Limited studies up to 1949 addressed the quantification of this vapor impedance factor, but separation of $w \cdot h_e'$ was not possible [73]. One of these studies produced an artificial "sweating" apparatus made up of a cylinder of wet blotting paper internally heated [47]. In 1955, Whelan *et al.* [132] developed a relationship between clothing and impedance to water diffusion. However, the impedance to moisture diffusion compared masses of still air, rather than transient properties which are affected by convection, conduction and thermal radiation. How scientists approached many of these problems will be covered in the next sections.

Belding *et al.* first employed in 1945, the above equations in defining human, *in situ*, clothing insulation [8]. By using Burton's I_a—the insulation value of air, they determined I_T as:

$$\begin{aligned} I_T &= (I_{cl} + I_a), \text{ clo} \\ &= \{6.46(T_{sk} - T_a)A_d\}/[M - 0.68 \cdot E \\ &\quad + 0.97\dot{m}_b(0.67\Delta T_{re} + 0.33\Delta \bar{T}_{sk})] \end{aligned} \quad \text{(Eq. 2-9)}$$

Where 6.46 = the reciprocal of the clo value, which converts total resistance W/(m^2 °C) into clo units.

M = metabolic heat production, W

A_D = Dubois surface area, m^2

E = evaporative heat loss from successive weighing of the clothed subjects (g/h) × 0.68 W h/g)

$\dot{m}_b$ = weight of the unclothed subject, kg

ΔT_{re} = rate of rectal temperature change, °C/h

$\Delta \bar{T}_{sk}$ = rate of fall of average skin temperature *excluding* hands, feet, and head, °C/h.

Belding [6,8,11,12] found that under equivalent environmental conditions (i.e., analogous T_a and rh) the most exact physiological studies still produced anywhere from 5 to 15% errors in thermal insulation values. Alternatively, manikin values in a less complex form give no greater than 2% errors in the hands of experienced persons [15,24,26,62,75].

A multitude of factors affect the clothing insulative value; a few that impact directly on physiological responses to the thermal environment include:

1. the fabric's thermal conduction and extent of trapped air layers;
2. the fabric's dispersion over the skin surface area, which extends the total surface open to the environment;

3. variations in skin temperature distribution and heat flow at various sites;
4. variations in the clothing surface covering the skin: none on the face and hands, presence of arteriovenous anastomosis (AVA's) in the extremities, and vasodilatory activity in the face [29,39,42,44,49,91,105,129,130].

Other concerns present with clothing resistance determinations on humans are those associated with time factors and measurement limitations. The clothing insulation values determined on human subjects must be taken without the influence of insensible perspiration becoming a variable and in an environment where thermal equilibrium is possible. In the transient phases of an experiment, for example, absolute clothing insulation is highly variable and dependent on heat capacity of the clothing which is a function of the pre-conditioning temperatures [4,39,44,136,140]. If heat content of an ensemble measured during the transient period is larger than that present at steady-state, initial clothing insulation measurements are excessively high (i.e., dry heat loss is overestimated). Therefore, a limitation of using humans for routine determinations of thermal insulation is centered on the calculation of heat content (or heat debt) especially when the body is losing excessive heat.

Heat and Mass Transfer Analogues

Heat loss is achieved through four modes: conduction, convection, radiation, and evaporation. Conduction (K_s) occurs when heat from within the body flows through the skin and into cooler objects in direct contact with it. This is done as fast-moving molecules hit slower molecules and transfer energy to them, thereby producing heat: no transfer of material occurs, just a net transfer of energy. Convection (C) is defined as the exchange of heat between hot and cold objects by physical movement in a liquid or gas. This is readily apparent with elevations in air movement across the body. What is not always realized is that the body surface is surrounded by a layer of rising heated air, the *natural convection boundary layer,* which occurs because body surface temperature is normally higher than the ambient air. As the air layer surrounding the body heats up, it rises as a result of its buoyancy. In still air this layer serves as a micro-environment [37] which is modified by local temperature and humidity and which allows the transportation of small particles such as dust, pollen, and bacteria. Modifications to the boundary layer also occur by normal movement of the individual and presence of clothing. A similar micro-environment is present when the body rests immersed in another fluid such as water [135] as we will show later.

Mass transfer is a complex of several occurrences. There occurs properties of mass transfer that are associated with convection by which mass is transported macroscopically (in bulk) from one place to another in the flow system [83,114]. Alternatively, when a mixture of gases or liquids occurs in the same area so that a concentration gradient appears in the same system, mass transfer results in a microscopic level as a result of diffusion from the higher to the lowest concentration. Mass diffusion also occurs from a temperature gradient in the system, a property called thermal diffusion. Conversely, a concentration gradient of mixed fluids can produce a temperature gradient and thereby results in heat transfer. The latter effects are termed coupled phenomena which often occur in semi-impermeable or fully impermeable clothing [90,140].

Heat Transfer Laws. In considering heat loss from the skin to the environment which is our focus in this chapter, a description is done by use of the Newton's law of cooling. The rate of temperature loss based on Newton's law varies in proportion to the temperature difference between the cooling body and the environment [83,84,114]. Kleiber [84], however, argued that in homeotherms—by definition of their mechanism—there is no cooling since body temperature remains nearly constant. He reasoned that since homeotherms keep changes in body temperatures {dT} nearly constant, the proper format is that of Fourier's law of heat flow which describes sensible heat flux between the body and its environment as:

$$dQ/dt = A_D \cdot \{k/L\} \cdot (T_c - \bar{T}_{sk}), \text{ Watt} \qquad \text{(Eq. 2-10)}$$

Where Q is heat (Joule/s = W), t is unit time, A_D is the surface area (m^2); k is the thermal conductivity of the surface layer ($W \cdot m^{-1} \cdot K^{-1}$); L is its thickness (m), and T_c is core temperature (K) and $\bar{T}_{sk}$ is average surface (skin) temperature (K).

If a resistance to heat flow, r_s, is defined as 1/(k/L) or r_s = (L/k); i.e., [m · K/W] · m, Eq. 2-10 (as an analogue to Ohm's law) becomes:

$$dQ/dt = A_D(r_s)^{-1} \cdot \{T_c - \bar{T}_{sk}\}, \text{ Watt} \qquad \text{(Eq. 2-11)}$$

in which the rate of heat loss is inversely proportional to the resistance against heat flow (i.e., insulation) and directly proportional to surface area and temperature difference between the core and skin surface [32,73]. The association between the two laws is apparent by introduction of the heat capacity of the body, cb, (=0.965 W · h/(K · kg) and mass, m_b (kg):

where, $dQ = c_b\, dT$ or $dT = dQ/c_b$ and

$$dT/dt = A_D/\{c_b \cdot m_b \cdot r_s\}\,(T_c - T_a), \text{ °C/h} \qquad \text{(Eq. 2-12)}$$

Organisms continuously lose water by diffusion through the skin into the air [20,53,67]. The mass diffusion in a binary mixture such as air-water, in which the temperature gradient does not produce excessive thermal diffusive flow, also occurs through layers of porous clothing in a one dimensional mode. Molecular conduction and convection are also important processes occurring in heat transfer through clothing materials, particularily non-porous garments [20,90,114]. In these cases conduction is:

$$F_x = -k_s \, dT/dx \qquad \text{(Eq. 2-13)}$$

where, F_x is heat flux, $W \cdot m^{-2}$,
k_s is thermal conductivity of the fluid (air-water), $(W \cdot m^{-1} \cdot K^{-1})$,
dT/dx is the temperature gradient in the x direction, K/m.

Convective flux (Joule/$(s \cdot m^2)$ or $W \cdot m^{-2}$) in the direction of x through clothing material is determined by:

$$F_x = \rho \cdot c_a \cdot V_x \cdot T \qquad \text{(Eq. 2-14)}$$

where, ρ is the air density, $g \cdot cm^{-3}$, or (kg/m^3)
c_a is the specific heat, $W \cdot g^{-1} \cdot K^{-1}$, or $J/(kg \cdot K)$, and
V_x is the air movement in m/s.

Water transport through clothing also incorporates components of molecular diffusion and convective mass transport. The diffusion of water vapor in air is characterized by Fick's first law [114], similar to Fourier's law for heat conduction previously discussed, in which mass flux of a component per unit area is proportional to the concentration gradient [83,100,114,122].

$$\dot{m} = -D \, dC_v/dx, \quad kg \cdot m^{-2} \cdot h^{-1} \qquad \text{(Eq. 2-15)}$$

Where the mass flux (density) of water vapor, $\dot{m}$, in a direction in space, x, (meter) is dC_v/dx.
C_v has units of kg/m^3.
D is the diffusivity for water vapor in air, m^2/s, or m^2/h; this is independent of C_v but it is a function of temperature and pressure. The product of $\dot{m}$ and the latent heat of vaporization (670 W h/kg) determines the flux density of latent heat (J_v) in W/m^2 [122].

For predicting moisture loss through clothing, the integration $\int$ of dx/D gives a resistance to diffusion (r_v), m/h or m/s.

$$\dot{m} = -\Delta C_v/r_v, \quad kg \cdot m^{-2} \cdot h^{-1} \qquad \text{(Eq. 2-15')}$$

An analogous development for sensible heat flux was introduced by Cena and Clark [33] as an alternative thermal resistance (r_T). When normalized to a standard temperature, pressure environment (so the volumetric specific heat of air is a constant), the clo unit is roughly equivalent to r_T of 200 s/m, or 2 s/cm. An I_T of 5 clo would have a resistance in this form equal to 10 s/cm.

Often C_v is expressed by the ideal gas law [114] as:

$$C_v = P/[R_w(T + 273)], \quad kg \cdot m^{-3} \qquad \text{(Eq. 2-15'')}$$

Where R_w is the gas constant for water vapor, 3.46 Torr $\cdot$ $m^3/(kg \cdot K)$ and (T + 273) is in deg Kelvin.

P expresses the partial pressure in the gaseous surface system from a high concentration ($P_{s,sk}$) to the lower vapor pressure (P_w).

Mass flux by convection in the x-direction through a layer of clothing is expressed by:

$$F_x = V_x \cdot C_v, \quad W \cdot m^{-2} \qquad \text{(Eq. 2-16)}$$

As in heat transfer analysis, all the above fluxes can be incorporated into a mass conservation equation [114] which is solved to determine the space distribution of the fluid within the area concerned [44,114].

As pointed out before, heat and water vapor conservation equations (i.e., continuity) are uniquely tied together [90,114,125]. The coupling presents problems in determination of permeation efficiency with semi-permeable materials. This is because as water vapor from thermoregulatory sweating enters the skin-air gap and the intrinsic clothing layer, residual heat flux occurs. The boundary condition for highest water vapor concentration is assumed as that saturated vapor existing at the skin surface ($P_{s,sk}$). Also the specific heat of air is a function of the vapor concentration which is itself affected by the skin-air temperature in the gap [83].

ENERGY BALANCE AND INSULATIVE PROPERTIES

Metabolic Heat Production

As explained by others in the next chapters, the conversion of metabolic energy into mechanical work varies in efficiency with type of muscle and the speed of its contraction. Efficiency is least at high and low speeds. A typical efficiency is about 10 to 20%. Most of the energy liberated is in the form of heat. This heat is primarily dissipated at the skin surface and some loss occurs via the lungs [58]. The ability to dissipate the heat the body produces is proportional to the surface area (the Dubois surface area in humans). Kleiber's paper [84] points out that Fourier's law represents heat flow proportional to the surface area, but Newton's law characterizes rate of body cooling inversely symmetrical to the cube root of body mass. "That is because S (the surface area, A_D) is proportional to weight to the 2/3 power and the (body) heat capacity is proportional to weight itself . . . thus, surface/heat capacity or $w^{2/3}/wc_b = w^{-1/3}$."

This leads to an interesting analogy with respect to heat loss. In the cold, animals who reduce surface area would have the edge (in terms of natural selection) over those that cannot alter their surface area to mass ratio sufficiently [63].

A general plot of the avenues of heat exchange and metabolic heat production as a function of ambient or operative temperature (introduced in Chapter 1) often helps define environmental limits. A thermoneutral zone is described in which metabolism is at a minimum. No excessive physiological effort is necessary in this zone by sweating, vasoconstriction or vasodilation. As shown in Figure 2-3, in unclothed humans, this zone is relatively small; clothed, the zone increases by a wider span which is governed by size and insulative value of the clothing. Typically, the larger the homeotherm, the

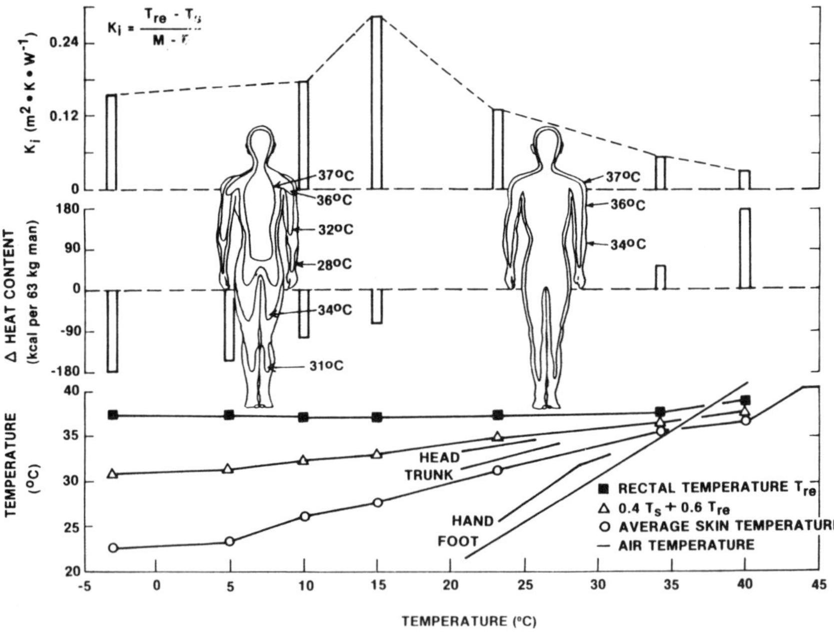

Figure 2-3. *The response of nude man to environmental temperature, tissue insulation (K_i), change in heat content, and various body temperatures. The manikin superimposed shows isotherms of body temperature distribution in an ambient of about 7° C and 27° C. Tissue resistance is lowest in warm zones but increases to a maximum in the cold, then decreases with variable heat flux incurred by shivering. A mean body temperature is calculated from the ratio of skin and rectal temperature.* [*Adapted from Reference 32*].

greater is the metabolic energy per surface area and greater is the thickness of insulation [7,27,32,72].

Homeotherms must balance any heat gain through metabolism by a subsequent heat loss through skin and respiratory passages. Body temperature regulation also adjusts vasomotor action. This alters conditions of heat exchange similar to alterations apparent in changes in resistance to clothing. The skin of the human body is supplied by a rich network of blood vessels and also responds to adrenergic (acral regions) drives (see Chapter 4) as well as to thermal changes. With cold stress, blood flow to the skin is greatly reduced, but transient heat flow to extremities may occur by the mechanism of cold-induced vasodilation (CIVD) [29,32,129,130] which is modified to a great extent by body heat content. Heat flux through the cold skin must pass variable insulation barriers that are often higher than the fixed resistance typically seen at full dilation (1 to 1.5 cm of fatty tissue) [27,29,32,64]. Variations in cutaneous blood flow distribution thereby allow skin to alter its resistance in order to attentuate thermal gains derived from metabolic heat production as well as from heat gains from the environment. In the cold, the outer layers of the body (described as a 'shell') offer a fixed resistance through fat, muscle layer, and skin thickness (Figure 2-3) [3,6,32].

In cold weather conditions the body heat alone is not able to maintain the proper heat balance. Appropriate clothing, therefore, as an added factor to skin resistance becomes a vital property in maintaining heat balance not only for comfort, but also for health [63]. One method the body uses in circumstances where the outside weather conditions are too cold, or the clothing inadequate for maintaining a heat balance for the total body, is to sequester the warm blood in toward the vital organs and, in effect, limit the body heat flux to such extremities as hands, feet, and ears. Extremity protection is a vital determinant for cold weather clothing ensembles. During extreme conditions hypothermia can occur, and the body's core temperature can be lowered excessively to the point of death. Clothing must also, therefore, prevent excessive heat loss from the body as a whole. An estimate of the interaction of clothing and exercise level during constant exposures to various ambient temperatures is given in Figure 2-4.

Since clothing adds another layer to human skin, as such it imposes a barrier to heat transfer by all three avenues [7,27,32,53,58]. Interference with air motion across the skin decreases transfer by convection and the consequent potential to lose heat by evaporation. The decrease in the boundary layer immediately surrounding the shell becomes an impediment to heat loss when air tempera-

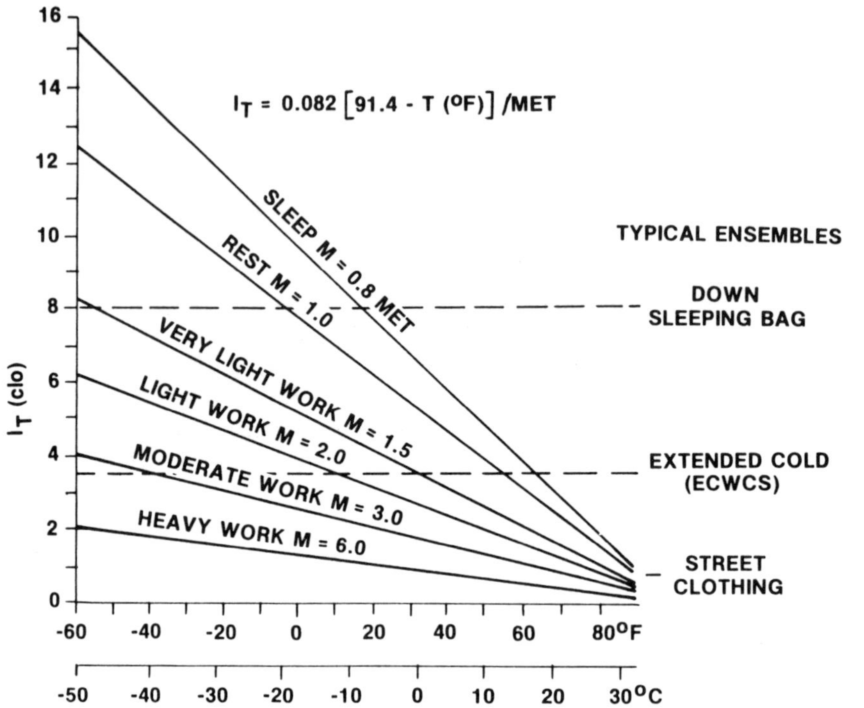

Figure 2-4. *Total insulation (I_T, clo) of clothing plus air necessary for comfort at various metabolic rates [1 met = 100 watts] (Adapted from Reference 29).*

tures are lower than skin temperature and is a protective measure when air temperature is higher than skin temperature.[2] The properties of clothing affect the evaporative potential primarily when ambient water vapor pressure is high [7,53,64], particularly while exercising in wet tropics (dewpoint >15° C) or if the clothing type is especially impermeable. A "psychrometric range of a clothing" is described based on optimum resistance values of sensible and insensible heat loss [98,99].

Sensible Heat Loss

When humans are clothed they do not exchange heat directly with the environment, except at exposed skin sites. Any sensible heat (that which is associated with a temperature difference because of dry (non-evaporative) radiative plus convective heat exchange) must arise from internal heat transported to the skin, from there to

[2] i.e., the familiar sauna effect which when the boundary skin-air layer is disrupted, creates a "burning" sensation.

each successive layer of clothing then to the environment (Figure 2-2).

Dry heat exchange between the clothed skin and the environment is best described by operative temperature (T_o, see Chapter 1) which includes resident heat transfer coefficients verified in vapor permeable material (h_{cl}, h_c, and h_r) [28,56]. As pointed out before (see Eqs. 2-4 and 2-4′), Burton first proposed that the efficiency factor of clothing (F_{cl}) could be expressed as a ratio of thermal resistance between the clothing surface and the air to the resistance between the skin surface and air ($I_a + I_{cl}$) modified by f_{cl} [24,25,26,43,58]. This factor is composed of the ratio of surface area of the clothing layer to the Dubois [58] skin surface area (A_{cl}/A_D). The exchange is expressed as:

$$(R + C) = DRY = h^* \cdot (\bar{T}_{sk} - T_o), \quad W \cdot m^{-2} \qquad \text{(Eq. 2-17)}$$

where the combined coefficient (h^*)[3]

$$h^* = \left\{\frac{1}{f_{cl}(h_c + h_r)} + \frac{1}{h_{cl}}\right\}, \; W \cdot m^{-2} \cdot K^{-1}$$

Where h_{cl} depends on intrinsic clothing conductance [46,56,113].

Insensible Heat Loss

In 1961, Woodcock [137] introduced a new parameter, denoted as an index of the permeability of clothing to water vapor based on his studies with wetted, unheated cylinders. He coined the parameter "water vapor permeability index" or i_m. The original measurement was with a wet cylinder in an uncontrolled, but precisely measured insulated room, from which the i_m was calculated using a similar analogy for evaporative heat transfer that was done previously for sensible heat flow. This allowed the formulation of a transfer characteristic as the ratio of the thermal resistance of the clothing layer, plus the overlying air layer (i.e., the I_T) to the resistance to evaporative heat loss per unit of vapor difference across clothing, plus the air layer (I_T/R_e). When applying the respective resistances in terms of a wet bulb psychrometer in which total sensible heat loss is zero:

$$I'/E' = \{T_a - T_{wb}\}/\{P_{dp} - P_a\}, \quad °C/Torr \qquad \text{(Eq. 2-18)}$$

Where I' = the value of I_T applied to a wet-bulb thermometer, (°C)
E' = the value of evaporative resistance (R_e) applied to a wet-bulb thermometer, (Torr)

---

[3] h* describes effective combined coefficients comprised of one or more variables; h without the asterisk is the simple thermal coefficient described by the thin film of fluid juxtaposed to a source or sink of thermal energy.

P_{dp} = the saturated vapor pressure of water at the wet-bulb temperature, T_{wb}.

The reciprocal of the ratio I′/E′ is also referred to as the psychrometric constant (2.2° C/Torr) which when used as a description for complete evaporation at the skin to air boundary layer incorporates the Lewis relation coefficient (LR). The ratio of $(I_T/R_e)/(I'/E')$ is the i_m of the original Woodcock definition. A recent review of applied heat and mass transfer analysis of porous clothing material points out that i_m is actually a modifier of the LR coefficient; it combines transfer coefficients of h_{cl}, h_r, h_c, and f_{cl} [46]. By definition, i_m is a non-dimensional constant which has the value of 1.0 for skin-air layers and from 0 to 1.0 for additional layers depending on extent of "moisture impedence." Figure 2-5 describes the upper limits based on the original analysis of i_m in terms of environmental zones described by low permeability and theoretical highest permeability (i_m = 1).

In the mid-1960s many of Woodcock's suppositions were modified for usage in heated copper manikins. Breckenridge, for example, devised a cotton "skin" which fitted tightly around the com-

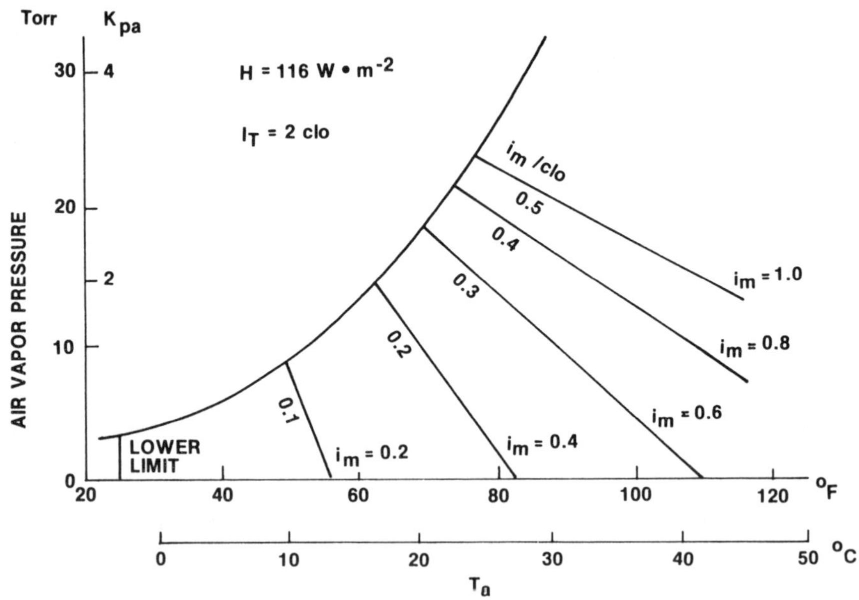

Figure 2-5. *Characteristic upper limits of ambient conditions on a psychrometric chart (air vapor, pressure in Torr and kilopascal) in terms of Woodcock's {i_m} permeation constant. H is net heat flux on the skin surface. Typical porous garments have an i_m of 0.4–0.5. [Adapted from Reference 137].*

plete surface area of a manikin. In terms of i_m and the ratio i_m/I_T (Figure 2-5), the total evaporative heat loss of the skin was now measurable on a human-sized model if the cotton skin remained completely wet and thereby 100% skin wettedness occurred. The evaporation from such a manikin's surface (E_{sk}) is described as:

$$E_{sk} = \{2.2 \cdot i_m/I_T\} (P_{s,sk} - P_w), \quad W \cdot m^{-2} \qquad \text{(Eq. 2-19)}$$

Where the term in brackets is the effective evaporative heat transfer coefficient.

Another approach useful to quantify evaporative heat exchange is by the relation [53,108,113].

$$E_{sk} = w \cdot h_e \cdot F^*_{pcl} (P_{s,sk} - P_w), \quad W \cdot m^{-2} \qquad \text{(Eq. 2-20)}$$

where, w (skin wettedness) is the fraction of the body surface wet with sweat [50,65] and F^*_{pcl} incorporates permeation coefficients effective for moisture loss through porous, thin garments. The evaluation of evaporative resistance is possible now by use of sensitive humidity meters or techniques [49,68,69,76] which allow measurement of dew point close to the skin surface and, therefore, the local skin wettedness at a point site. Figure 2-6 is an example of one such evaluation at various regional skin sites under clothing (I_T of 0.8 clo) in which E_{sk} was determined by use of Equation 2-15 and its rationale.

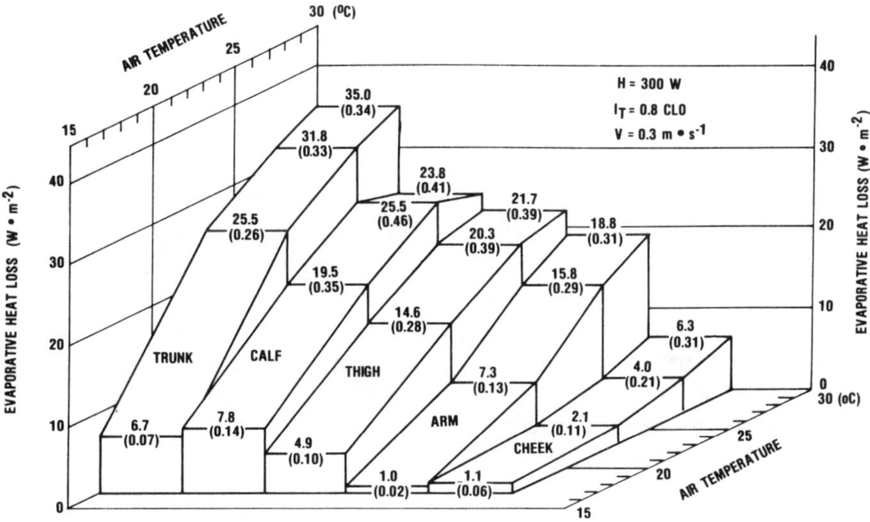

Figure 2-6. *Local skin wettedness at 5 skin sites (n) under a garment and skin evaporation by weight loss determined during exercise at various air temperatures (T_a = 16 to 30° C).*

THE BIOPHYSICS OF CLOTHING MATERIALS

Fiber Conduction and Physical Properties

A brief review of the intrinsic properties of clothing materials helps in understanding how sensible and latent heat flux occurs through the whole clothed body. Fiber properties are described by at least seven factors: fiber diameter, crimp and elastic moduli, thermal characteristic and bulk density, moisture and wicking tendencies (water transmission from the skin to the exterior in garments), air permeability, and element of contact [14,39,64].

Fiber composition is often classified according to surface area. For example, average merino wool has a total surface area of around 800 ft^2/lb [14]; for an average suit, this is an area of some 2500 ft^2. Other smaller-diameter fibers (cotton and silk) may reach surface areas of 3000 ft^2/lb. As such, the surface area allows closer contact between the skin and the ambient air. The effectiveness of thermal insulation is a property of the air trapped in the fabric. When the thermal conductivity (k_f) of a wool fabric is plotted against its bulk density, a curvilinear rise is seen, with k_f flat until it reaches a bulk density of 0.5 $g \cdot cm^{-3}$; below this bulk density, the thermal conductivity approaches that of air, 0.026 $W \cdot m^{-1} \cdot K^{-1}$ [14,31,33,44,114].

Since early times, it has been recognized that the insulating property of fabrics can be altered by extending the annular spacing between skin and the garment. Wetting of the fabric is not a factor in the alteration of this property, the reason being that there is still freedom of circulation of warm air within the space. The effect is noticeable at low wind speeds and in still air. With high wind speeds impinging on the protective layer, all materials tend toward an equal thermal resistance and the critical property is the volume of air enclosed rather than the type of fabric [14,24,39,47].

The extent of contact between a garment and the skin surface is a more important property than its fiber composition. Since bulk density is the more important criterion in thermal resistance, manufactures have attempted to alter air permeability in discrete garments. Although covered as a separate topic later, in brief, air permeability is altered by allowing diffusion of fresh air through the clothing (zippers) or other natural clothing vents (bellows property), particularly in low-permeable garments [62,91,92]. Knitted fabrics have a high degree of permeability to air, while twill weave fabrics exhibit low permeability.

The effect of increased contact of a garment with the skin also increases chilling or clamminess [18,47,69,97]. Continuous-filament fibers create this effect primarily by contact. On the other hand, fibers with natural elasticity or those made with high synthetic elasticity show little contact with the skin. Actual thermal characteristics (vapor flow, thermal conductivity) are affected by the volume of trapped air [14,39,47,64].

Wool logically was used in garment construction because it is an excellent insulator due to its resilience, by which the fiber springs back to maintain the original volume of trapped air for an extended period of time [14,33,105]. The honeycombed tubular structure and scales present in wool fibers allow them to retain their resilience, which is an advantage when wet since the thermal insulating properties remain stable. Wool fibers can also transmit moisture (wick) from the inner to the exterior boundary layer. Cotton, however, is far less resilient than wool; it wicks rapidly but absorbs moisture as well [14,44,119,138,140]. Cross-linking of cotton fibers, a process that reduces its hydroscopic properties, often allows substantial wicking with less postexercise evaporation [48,97].

Silk fibers also have high absorbency, but they are more expensive compared to similarly absorbent synthetic fibers. Of the synthetic fibers, acrylic and polyester are the most widely used. Acrylic is a resilient fiber and traps air much like wool. It is not absorbent but does wick moisture to some extent. Polyester does possess limited absorbency and resiliency, but generally, by itself, has poor wicking properties. New organic additives to the fiber help the wicking action because the hollow fiber is given an inner hydrophobic property but an outer hydrophilic surface. Garments made from such treated polyester can dispel moisture by a "spreading" motion (by diffusive flux) from the inner fiber to the exterior surface [64].

In essence, newer treated synthetic fibers used in clothing manufacture are treated so that they allow the fiber's properties to enhance the transport of insensible perspiration away from the skin surface. Skin wettedness at a point site should therefore be reduced; however, this has not been proven [64]. Treated synthetic fibers dispel moisture by capillary transfer of moisture in a cascade fashion from one fiber adjacent to another—by hydrophobic transfer of water molecules (since the fibers have variable or little moisture absorption), or by a spreading action. The carrying capacity of water vapor transfer ($\dot{m}_{max}$) through fabrics is limited and governed by the intensity of exercise [44,64,65,76,90,92,103].

Polypropylene is a synthetic fiber with an organic base with "wicking" properties. It has low thermal conductivity, low moisture regain and weight (0.9 g/cm^2). These properties seem ideal fiber properties for underclothing during exercise and cold wear except that undergarments made with this fiber trap oils from sweat, exhibit low melting points, and the organic compound wears out [64].

Other synthetic fibers use polytetrafluoroethylene (PTFE), either laminated onto woven or knit fabrics or applied along with polyurethane as a porous coating to common textiles (cotton, nylon). The theory of operation is that such garments allow water vapor generated from sweating to pass through micropores (diameters of 5×10^{-7} to 3×10^{-6} m). Since latent heat loss by water vapor generates molecules of water averaging 4×10^{-10} m in diameter [64] but drizzling rain has a diameter of roughly 5×10^{-4} m, by simple membrane physics these garments supposedly provide "breathability" but also protection from rain.

Along with fiber characteristics, fabric construction is a factor in the ability of a garment to conduct heat and moisture [31]. Such fabrics may be a woven, knitted, or nonwoven material. Woven fabrics have interlacing yarns that intersect at 90° angles. If the interlacing is tight, not much stagnant air can be trapped, so the thermal insulation is minimal. Tightly woven fabrics with synthetic hydrophobic fibers are effective wind breakers. On the other hand, knits can be distinguished by their interlocking loops and thus are bulkier than wovens. The looping characteristics allow the stagnant air to be trapped and, thus, also serve to insulate better than woven material. Thermal underwear, sweaters, ski pants, etc. which allow movement, are typical knits.

Fiber pile clothing, favored by backpackers, is a knit with a polyester pile placed close to the body layer and a knit backing close to the outer surface. The latest fabric construction incorporates nonwoven fabrics in which layers of fiber material are bonded or matted together chemically (melt- blown) or by pressure [64]. Familiar nonwoven fabrics are felt products. More recent ones include thick mats, incorporating polyesters or acrylic fibers with outer woven shells. These nonwoven fabrics retain a degree of resilience when wet and trap air effectively. Thus, they have sufficient thermal insulation. One nonwoven batting made of polypropylene fibers has the trade name Thinsulate [74]. This fabric is made up of olefins spun into a microfiber construction that effectively trap air molecules on the surface of the fibers and, therefore, raise the amount of air per given volume. It is claimed to have almost twice the insulative value of a similar thickness of wool or down-filled material [74].

In the past, high moisture regain was a desirable property in the dissipation of vapor [4,119]. High-regain fibers protected the wearer from sudden alterations in air temperature and humidity. For example, vapor transients produced during movement from a comfortable, dry environment to the outdoors, (where the air temperature is much lower but high in relative humidity) would not be felt sharply when wearing a high-regain garment such as wool. The degree of buffering to a temperature change, exhibited by a particular textile is dependent to an extent on the regain-relative humidity relationship [14,44]. The higher the regain for a given relative humidity change, the greater the water absorption [47,90]. However, with the advent of synthetics with wicking or high moisture permeation properties, the smaller is the value of high-regain fibers. Complications of water vapor dissipation exist in garments with these properties [44,65,76,90].

A recent study in which moisture permeation was ascertained by dew point sensors underneath clothing [65] during light exercise (31% $\dot{V}O_2max$, 350 W) shows this complexity. A PTFE ensemble with high clothing insulation did display almost 45% more permeation than an impermeable garment and only 26% less than a completely porous track suit during experiments of fan-directed wind speeds of up to 3 $m \cdot s^{-1}$. However, these results were from observations in which the rate of weight loss was observed for up to 2 h at a T_a of 25° C and low dew point temperatures. They should not be applied to transient conditions or higher work rates in which sweating is exacerbated. Other studies also show [44,76,90] that semipermeable garments allow latent heat transfer by moisture permeation to an extent not greater than 82 $W \cdot m^{-2}$ of regulatory sweating or, for steady, light activity (175 $W \cdot m^{-2}$), about 46% of M [64]. Limitations are set by the absolute surface area of microporous openings available compared to the high volumes of sweat secreted during high-intensity exercise. Subsequent condensation of vapor underneath the outer layer, when ambient temperatures drop below about 15° C, also restrict the latent heat loss through semipermeable clothing [44,90].

Aluminized suits (with reflective fibers) are also an attractive item in ensemble construction especially during exercise outdoors [22,28,39]. However, such low emissivity fabrics generally have poor moisture permeation properties [22,26]. However, one recent study by Berglund *et al.* [16]. will be discussed later that discounts this effect in skin tight body suits.

Encumbrance due to thick clothing also often serves to produce a decrement in performance [11,12,70]. Much of the decrement is associated with added weight. For example, Teitlebaum and Goldman [127] observed that persons doing treadmill exercise (1.6 $m \cdot s^{-1}$) in an arctic uniform weighing 6000 g displayed a metabolic rate 80 W higher than when the same weight was carried in a belt.

Long-Wave and Short-Wave Radiation

All materials are absorbers and emitters of thermal radiation. The second law of thermodynamics requires that, for a given wavelength, each substance's emissivity and absorptance be the same. Emissivity is defined as the ratio, for a given wavelength interval, of the energy emitted per unit area divided by the energy emitted by a black-body of the same area. For real materials, the actual absorptance is a fraction of the wavelength. A so-called black-body is defined as one that absorbs all wavelengths. Such a cavity, held at a given temperature, will generate radiant energy according to a function of wavelength (λ) [73]. Infrared radiation (beyond the red end of the visible spectrum) includes wavelengths from 0.4 μm to about 150 μm (far infrared) [124].

Thermal radiation is the "radiant energy emitted by a medium that is due solely to the temperature of the medium." [124]. The temperature of the medium governs the actual emission of thermal radiation. Of concern to our topic on clothing properties and thermal exchanges is the two distinct bands: solar (shortwave) radiation occurring in the range of 0.3 μm and longer waves occurring at about 0.7 to 150 μm.

Human skin has variable thickness over different parts of the body and is therefore heterogenous to incident radiation. In the spectral region from 0.3 μm to 3 μm there is a maximum reflectivity (Figure 2-7b) for white and black skin at about 0.8 to 1.2 μm, roughly the λm of maximum radiation for most infrared heaters. From 3 μm to 20 μm, reflectance becomes independent of skin pigmentation and is quite low.

Skin pigmentation is an important variable at short wavelengths near 0.4 to 0.7 μm; a darkly pigmented person will become markedly more heated by direct radiation >2500° K than would one of lighter skin.

The wearing of clothing, however, minimizes the above differences as shown in the above figure [1,6,7]. By decreasing emissivity of clothing, a reduction in radiative heat loss adds an additional protection against heat stress. Emissivity (ϵ_s) of human skin varies with spectral region, but skin emits in the infrared spectrum (5 μm to 20 μm). It is the emissivity value in this range that is important in human heat exchange.

In the presence of clear days with desert heat, clothing also decreases heat transfer by thermal radiation but affects the evaporative process. Typical decreases in the amount of radiant heat on the skin can be about 30–40% with ordinary working clothes near industrial furnaces [7]. Early work by Adolph *et al.* [1,7,105] showed that solar load in semi-nude men walking in the desert was about 233 watts. When clothing of light color was worn, demand on evaporative (sweating) heat transfer was lessened by some 116 watts (roughly 50% of solar load) which drops the effective air and globe temper-

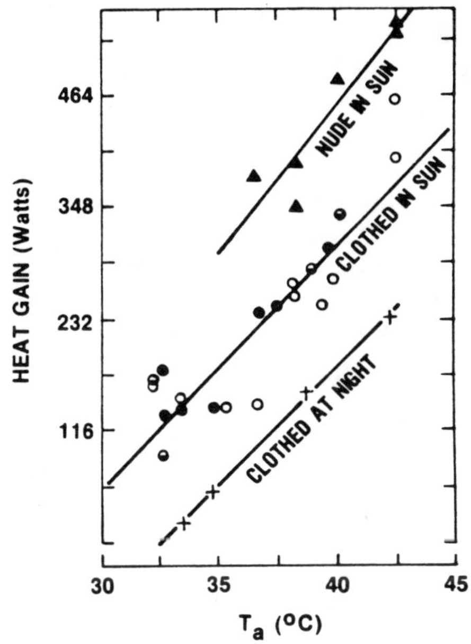

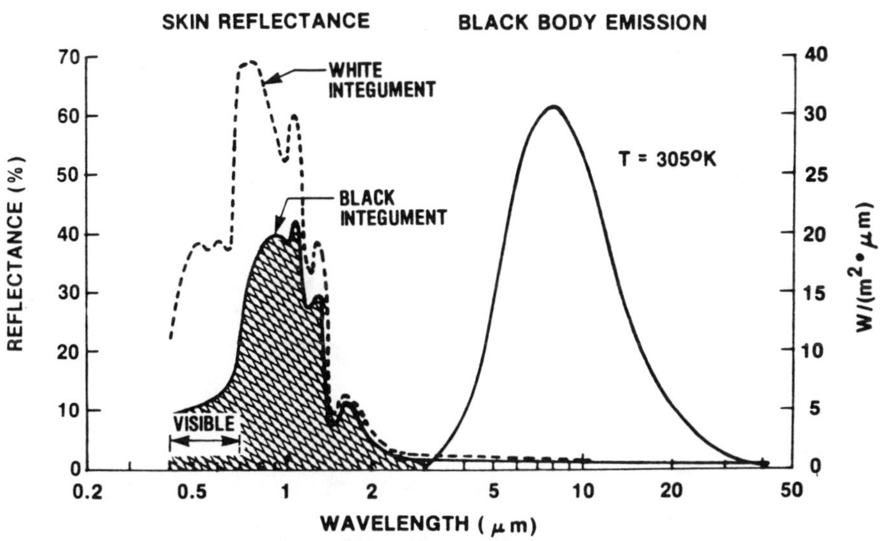

Figure 2-7a&b. *(A):— Heat gain of a man wearing clothes at various times of the day as a function of air temperature; (B)—reflectance (%) for black and white skin and emissive power of a black body (W/(m^2 · μm)) plotted as a function of wavelength. [Adapted from References 1 and 73].*

atures by 4 to 5° C [6,64]. Logically, the thick, loose fitting, light colored clothing used in the hot-dry zones of the Middle East by natives offers a guard for thermal extremes spanning excessive heat by solar gain in the day and rapid cold spells at night [6]. In coastal areas of such zones, ventilation of clothing by activity allows some evaporation, but in humid zones where air temperature stays about 32–33° C with humidities in the 90% rh range, clothing is more of an impediment since the effective ambient water vapor pressure gradient is so small [26,62,64,136].

The two ways of measuring emissivity of skin, analysis of reflecting power or analysis of absorbing power in the above ranges, are technically difficult. Hardy [73] compared the emitting power of the skin by analyzing skin temperature accurately with an infrared radiometer and compared this with the temperature of a "perfect black-body" using the Stefan-Boltzmann law. Hardy found $\epsilon_s = 0.98 \pm 1\%$, a value constant for $>.4\ \mu m$; ϵ_s is, therefore, ~ 1.0 [73,101].

To an extent, color of clothing affects its reflective properties to solar radiation as apparent in Figures 2-7b and 2-8.

Breckenridge and Pratt showed little effectiveness of color on heat exchange [21]. But Rees [120] showed that the effectiveness of

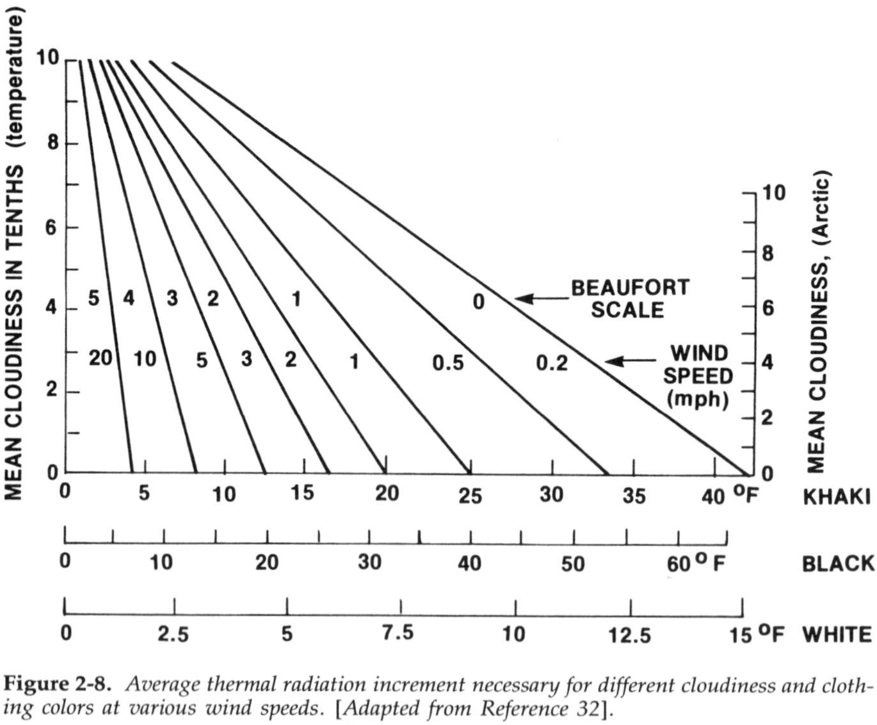

Figure 2-8. *Average thermal radiation increment necessary for different cloudiness and clothing colors at various wind speeds. [Adapted from Reference 32].*

dyed fabrics as reflectors of solar radiation (wavelengths 0.7 to 0.9 μm) is color dependent. For similar samples (cotton poplin), treated with various shades of black, blue, khaki, white, or coated with aluminum paint, highest reflection coefficients were with white and aluminum and black cloth the most absorptive. However, the interesting property was the mix in the black reflection coefficients dependent on the dye used. Reflection coefficients in the black samples ranged from 6% to 37%.

Berglund *et al.* [16] reported recently that color and weave also influence absorbed radiation. Subjects exercising in skin-tight, form-fitted garments with varied wicking material increased rate of weight loss differentially, as a direct result of color in respect to incident direct radiation. Rate of weight loss (by thermoregulatory sweating) for a given increase of radiation was highest with blue knits followed by black, white, nude (shorts), and a silver body suit (90% polypropylene: 10% Lycra). More importantly, extrapolation of the results in terms of extent of hypohydration based on loss of body water show that the high reflectance of the silver suit was water sparing compared to nude skin for up to 5 h in a 27° C environment with solar intensity of about 847 $W \cdot m^{-2}$.

A good deal of information has been provided on radiative heat exchange between the clothed human body and its surroundings [21,25,43,52,54,57,58,66]. For the most part, this exchange can be expressed by the Stefan-Boltzmann law [Chapter 1] and is included in the heat balance equation (insensible heat loss and storage are not considered) in the sections covered previously.

The linear radiative coefficient [57] described for the clothed case is essentially:

$$h_r = 4 \cdot \sigma \cdot \epsilon \cdot (A_r/A_D)\, f_{cl}\, \{[(T_{cl} + T_o)/2] + 273\}^3 \quad W \cdot m^{-2} \cdot K^{-1} \qquad \text{(Eq. 2-21)}$$

Sensible heat loss becomes:

$$R = 4 \cdot \sigma \cdot \epsilon \cdot (A_r/A_D)\, f_{cl}\, \{[(T_{cl} + T_o)/2] + 273\}^3 \cdot (T_{cl} - T_r) + C, \; W \cdot m^{-2} \qquad \text{(Eq. 2-22)}$$

In Equations (2-21 and 2-22), σ is the Stefan-Boltzmann constant and includes the emissivities of human skin ($\epsilon = 0.98$) and clothing ($\epsilon = 0.95$) [43,53,73].

The term (A_r/A_D) is the ratio of a 4π radiating area of the body surface "open" to thermal radiation over its Dubois total surface area (A_D, m^2). A_r is always less than A_D because extremities or other surfaces radiate to each other, rather than to exterior surfaces. Other variables in the assumption of A_r depend on posture (sitting, standing, or crouching) and clothing folds. Several researchers have provided useful methods to establish A_r in respect to A_D [57,70,112] with variations from $0.65 \cdot A_D$ for sitting to $0.85 \cdot A_D$ for a "spread eagle" position. A value of $0.72 \cdot A_D$ is usual for a standing clothed person. The factor (f_{cl}) expresses the increase in A_D due to clothing and can range from 1.0 for 0 clo to as high as 1.5 for 3 clo [58].

The effective radiant field (ERF) describing the net radiant exchange with all sources that radiate at temperatures different from the isotropic ambient T_a is an important concept to clothing interaction with the environment [57]. Effective radiant field is defined as the net radiant heat exchange between all enclosing surfaces and directional heat sources and sinks by an occupant whose surface is hypothetically at the ambient temperature T_a. In other words, ERF is the net radiant energy exchanged by the clothed or unclothed occupant with only those surfaces and sources whose temperatures differ from the ambient air (T_a). $(ERF)_n$ is a function of the radiating temperature of the source n in absolute temperature, the absorptance α of the exposed body or clothing surface for the emission of αK^4 by the source, and an appropriate view factor (F_{h-n}) that describes the fraction of a person's total body surface irradiated by the source [52,58,112].

Any radiant source n contributing to the total ERF is [57,58]:

$$(ERF)_n = f_{cl}\,(A_r/A_D)F_{h-n}\alpha_n[(T_n + 273)^4 - (T_a + 273)^4],\ W \cdot m^{-2} \quad \text{(Eq.2-23)}$$

The view factor F_{h-n}, refers to the body area A_D itself rather than the radiating area of the body surface ($A_r f_{cl}$) corrected for clothing. Only for the simplest geometrical arrangements can values of F_{h-n} be easily evaluated [57,112].

In every case during thermal equilibrium, $(ERF)_n$ will include the energy absorbed by the body system via the outer surface of the body; it is solely a function of T_r and T_a and is not a function of the temperature of the assymetric.

ERF, therefore, allows one to assess the "radiation-needed" zone when outdoor air temperature drops below 20° C in the clothed condition. The radiation is based on a direct, easily-quantified concept taking into account the possible solar radiation load necessary to maintain an optimum comfort zone.

Convection

Convection is a more complex process than conduction because it entails the mass movement of the medium which is transferring thermal energy from one site to another [83,114,122]. A simple example is the connection of a source and sink of thermal energy with a tube of liquid or gas. At the warm end, the fluid rises to give place to a cooler fluid. At the other end of the tube, the warmed fluid flows along the top from the warm to cold end where it deposits the thermal energy to the sink. Energy is conveyed from source to sink continuously unless blocked by a resistor.

At the source or sink of thermal energy, the air flow is low enough to be streamlined. Outside this film, where speeds are higher, the fluid becomes turbulent. As pointed out in Chapter l, boundary layer flow is described as laminar or turbulent according to the ratio of buoyancy forces to viscous forces in the layer (Grashof number, Reference 114). Buoyancy forces dominate if the Grashof number is greater than 10^{10} and the layer becomes turbulent. For example, in a naked standing person at T_a of 15° C, air flow will be laminar up to the plane of center of the abdomen, have a transition region of 50–60 centimeters, and have turbulent flow about the upper torso and head [37]. The conventional non-dimensionless numbers (Chapter 1) are used to establish convective exchange primarily for uncovered or semi-covered cylinders. Rationally derived equations extended for human shapes [32,73,104,113,115,116] have been formulated, but there is little information on clothed humans other than by calorimetry.

For forced convection the Nusselt number is a function of the Reynolds (Re) number and Prandtl number (Pr) and is described by the relation

$$(\mathrm{Nu}) \sim (\mathrm{Re})^{0.55} \cdot (\mathrm{Pr})^{0.33}, \quad [\mathrm{N.D.}] \qquad \text{(Eq. 2-24)}$$

Where Re $= \rho \cdot V \cdot x/\mu$ or $V \cdot d/\nu$ in which ρ is the mass density ($kg \cdot m^{-3}$), V is the velocity in m/h, μ the absolute viscosity in kg/(h · m), x is the diameter (m) or distance, and ν is the kinematic viscosity in $m^2 \cdot h^{-1}$.

Gagge and Nishi [58] point out the following: (a) for free convection, h_c varies with the temperature gradient ΔT, the gravitational constant G (in the Grashof number) and density; (b) for forced convection, h_c varies with density and air motion and is independent of ΔT; (c) for both free and forced convection, h_c varies inversely with some function of the diameter (x) or length (1) of the heated object. In a space shuttle, free convection does not exist, as G = 0, and all convection becomes forced. Barometric pressure (P_b) affects density and thus h_c. The h_c for a large object would be lower than one for a small object in any given environment.

In the transitional region between free and forced convection, there is no simple relationship as described above. For the resting human body with a probable ΔT of 5° C from surface to ambient air, the transition occurs at 0.2 − .3 $m \cdot s^{-1}$ (40–60 fpm) ambient air movement. Below this level of air movement h_c may be considered as a constant [117].

As pointed out previously, the convective heat transfer coefficient h_c, defined as the ratio of the total heat exchange by convection from body surface to the temperature gradient from the body surface to the ambient air temperature is difficult to evaluate accurately in man. Optimum calorimetric measurements with the heat balance equation [38] are possible if the total dry heat exchange (R + C) can be accurately measured from the observed M_{sk} and E_{sk} and under conditions of thermal equilibrium when S is probably zero. Calorimetric measurements of the DRY heat exchange under these conditions result always in an accurate value for the combined heat transfer coefficient (h). The value of h_c is now found by subtracting from h a calculated value for h_r [38,51,104,133,134,139].

The physical factors associated with h_c are not easily measurable. Conditions of free and forced convection may exist simultaneously, convection is also affected by the motion of the ambient air, and by activity level. The ambient air movement (V) caused by either turbulent or laminar flow, may affect h_c differently [34,37,114,116,117].

The naphthalene method, rationalized for humans by Nishi and Gagge [10], gave the first observed measurements of h_c on humans during exercise, treadmill walking and free walking without the use of calorimetry. Many of the properties applied earlier to human and thermal manikins are now applicable to other models.

In one recent study [34] with an articulated, moveable manikin—the sectional heat transfer coefficients were mapped-out while the manikin simulated walking. The manikin was not internally heated, according to a procedure reminiscent of Woodcock's [137] original studies of unheated flat plates outlined previously. The study addressed the effect of the walking motion, per se, on convective heat exchange. A naphthalene sublimation plate technique was developed closely analogous to the Nishi and Gagge napthalene sphere studies [107]. However, in the manikin study napthalene disks, molded to follow the contours of the copper manikin, were affixed to various body segments in a flush configuration rather than outside the air boundary layer. The manikin then simulated walking at four different gaits (0.2 to 0.9 $m \cdot s^{-1}$) under constant air temperature (30° C) at different ambient wind speeds (0.4 to 0.7 $m \cdot s^{-1}$).

The amount of naphthalene weight loss, via its sublimation, was translated to the effective convective heat transfer coefficient (h_c, $W \cdot m^{-2} \cdot K^{-1}$) based on the Chilton-Colburn analogy between heat and mass transfer [36]. It was found that walking, or the motion of the swinging limbs of the manikin during walking, generally decreased (rather than exacerbated as has been thought) the forced h_c as measured on the arms and legs (by some 5–7%, $p < 0.05$). The implications of this research are that during running the major bulk of the increase in heat loss in humans is a function of the effective evaporative heat transfer coefficient (h_{e*}). Thus, in the relationship of the permeation efficiency factor applied to clothing properties:

$$i_m = h_e^*/LR, \quad \text{N.D.} \qquad \text{(Eq. 2-25)}$$

Where a critical component is h_e^* which is affected by the evaporative heat resistances of the air and clothing layers.

Clothing Ventilation

Body motion alters effective heat exchange by convection both outside the boundary layer and within the clothing worn [24,62]. This reduces the effective and intrinsic thermal insulation [98,99]. Thus the rate of convective heat loss and evaporative heat loss are both augmented. Belding [11,12] showed that a 50% reduction occurred in the intrinsic clo value of an arctic ensemble, primarily by the "stirring" and exchange of air within and below the clothing by the body movement. Protective clothing is designed with pathways for facile dissipation of moisture and sensible heat as pointed out before. Cold weather multipurpose garments require provisions for both. In cold environments, the clothing must be dry and warm for long periods of resting activity (Figures 2-2 and 2-4), but protective against wind which reduces the total clothing insulation (Figure 2-8). The garment must allow means for dispelling heat during short bursts of exercise [24,62,91].

Givoni and Goldman [60] described the effect of movement of the wearer to pump additional air during exercise. This was related obliquely to "effective" values of clo and i_m/I_T in which:

$$V_{eff} = V_a + 0.004\,(M - 105), \quad m \cdot s^{-1} \qquad \text{(Eq. 2-26)}$$

Where M = watts and V_a = wind speed, $m \cdot s^{-1}$

Breckenridge [24] showed rationally that these series of coefficients affect uniquely the Woodcock i_m/I_T relationship extended for human and manikin evaluations that become dependent on the effective wind speed owing to pumping. Consider the effect of a 10 $m \cdot s^{-1}$ wind on the i_m/I_T relationship in a typical manikin evaluation of common combat fatiques (with armor and helmet). Dry insulation (I_T) of 1.88 clo at V = 0.3 $m \cdot s^{-1}$ is modified by the effect of wind:

$$I_T = a\,V_{eff}^B, \text{ clo} \qquad \text{(Eq. 2-26')}$$

Where a = intercept of V normalized at 1 $m \cdot s^{-1}$

$$B = [\ln I_T, V_F - \ln I_T, V_i]/(\ln V_F - \ln V_i)$$

Where the natural log of I_T,V_F is the clo value observed at the high wind speed (0.81 clo) and I_T,V_i is the clo value observed at 0.3 $m \cdot s^{-1}$ (1.88 clo).

The coefficient for B above is first determined by inputing experimental values to obtain B = −0.24. The coefficient of a is determined as $1.88 = a\,(0.3)^{-.24}$ and for 1 $m \cdot s^{-1}$, a is 1.4. The effect of wind on I_T for this test is therefore:

$$I_T = 1.4\,(V_{eff})^{-0.24}, \text{ clo} \qquad \text{(Eq. 2-26'')}$$

Similar evaporative coefficients are derived for this one evaluation as:

$$i_m = c\,V^d_{eff}, \text{ [N.D.]}$$
$$= 0.406\,V^{0.147}_{eff} \qquad \text{(Eq. 2-27)}$$

The composite effect of wind on the i_m/I_T of this typical example is:

$$i_m/I_T = \frac{0.406V^{.147}_{eff}}{1.4\,V_{eff} \cdot V^{-0.24}_{eff}}$$
$$= .29\,V^{0.387}_{eff} \qquad \text{(Eq. 2-28)}$$

Such coefficients can be incorporated into useful models [115] that will predict probable heat casualty and extension of exercise time in various thermal environments.

Trace gas methods are used to quantify the ventilation properties [92]. In general, the improvement of ventilation by body movement in the heat during heavy exercise is minimal compared to the excessive body moisture required to become dissipated. In cold environments, however, with wind speeds greater than 2 to 5 $m \cdot s^{-1}$, the effect of wind penetration on the garment far outweighs the balance of internal air circulation by body movement. Latent heat cannot be dissipated and in very cold climates, condensation of moisture occurs [24,91,136,140]. A remedial option is by openings in the garment to dispel excessive metabolic heat [91].

CLOTHING RESISTANCE PROPERTIES IN DIVERSE ENVIRONMENTS

Hypobaric Environments

The effects of diverse environments: hypobaric (altitude) hyperbaric and aquatic, are uniquely tied to both the systemic physiological changes and skin-clothing environmental interface. Readers are referred to extensive reviews on each topic [5,58,81,86,118,131,142] and specific chapters on the physiological effects.

As shown in Figure 2-1, altitude (hypobaric stress) alters the insulation of the air layer (I_a) as affected by wind speed. The lower the wind speed and the lower the atmospheric pressure, the greater is the value of the I_a or lower the heat conductance. To maintain a given metabolic rate without increasing sensible heat flux when wind speed is elevated also requires an appropriate increase in the total clothing insulation. Effect of altitude at 15,000 feet (428 Torr) in a 440 fpm (2.2 $m \cdot s^{-1}$) only increases I_a from 0.3 to 0.4 clo [6,28,29,120]. However, little data exist on the effect of such wind penetration or compression of the intrinsic clothing layer. The effect of any increased thermal resistance is of little consequence in the extremities. Recently our laboratory evaluated specialized, new mountaineering boots. These boots have rigid plastic with a closed (gas-filled) cellular foam designed to insulate effectively in sea-level cold environments. Thermal insulation data were obtained at simulated high altitude and sea-level on a regionally-heated copper foot [22,121] that measures I_T at 29 sections. Hypobaric exposure [from sea-level to 30,000 ft, $T_a = -6.5$ to $-8.0°$ C] confirmed that I_T increases (by 11%). This was due, in part, to the hypobaric pressure effects on the wool sock (intrinsic and air layer) but primarily because the closed cell foam expanded by 21%. The paradoxical effect of increased insulation is of little benefit to hikers at high altitude with such inserts in a rigid boot. The volumetric expansion results in a net fluid force exerted as a vector towards the soft tissue of the individual's feet. The net effect is like a ligature which may result in frostbite if limited blood flow is present in the toes as well [111].

Gagge and Nishi [58] showed that the heat and mass transfer properties that govern exchange at the skin-environment interface have certain relationships affected by barometric pressure: (a) the heat transfer coefficient (h_c) in still air, which is dependent on metabolic activity or forced convection, varies as a factor of (P_b/760); (b) the effective Lewis relation coefficient, based on the ratio of the evaporative heat transfer coefficient (h_e^*) to h_c^* is also elevated with hypobaric environments. This is due to the effect of P_b on density and mass diffusivity of the fluid on the surface properties; (c) these latter properties also change the extent of convective heat exchange (C) and the thermal efficiency factor of clothing as noted before.

The effect of hypobaric environments on the Lewis constant is $2.2 \times (760/P_b)^{.45}$, °C/Torr, which increases the maximum evaporative power of the environment. Figure 2-9 is from a recent study in which clothed ($I_T = 0.5$) subjects exercised at their respective $\dot{V}O_2$ peak in a given environment (Tglobe = T_o = 20° C) *at three simulated altitudes* (P_b = 760 Torr, 554 Torr and 428 Torr). In general, based on partitional calorimetry, the drop in h_c is definitely affected by P_b

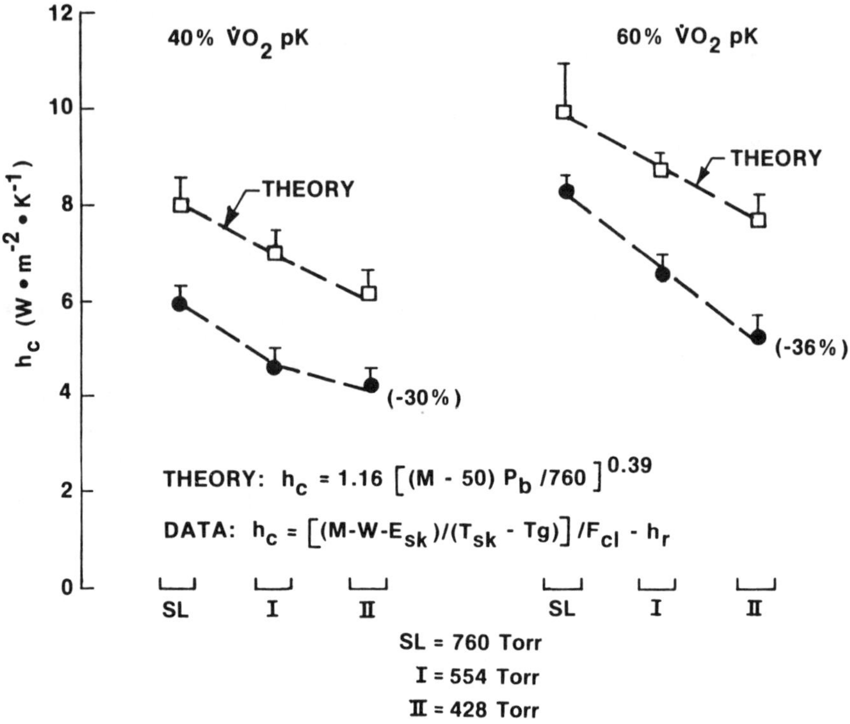

Figure 2-9. *Convective heat transfer coefficient calculated from a theoretical equation based on absolute metabolism and from partitional calorimetry during hypobaric exposure up to 15,000 ft. at T_a = 20° C.*

and elevated at each higher exercise intensity. The theoretical equation [shown in Figure 2-9] based on absolute metabolism, however, overpredicted h_c based on the calorimetric results.

Figure 2-10 shows that chest skin wettedness (by direct dew point sensors), under a layer of porous clothing (I_{cl} = 0.4), increases significantly during rest in hypobaric environments (T_a of 30° C) both in comparison to sea level and cooler environments. Characteristically, as skin sensible heat loss drops with hypobaric environments, skin diffusion increases possibly from augmented interstitial fluid pressure to the skin. These results can have impact on climbers in high terrestrial areas particularly with an augmented intensity of solar radiation at high altitude [6,53].

Hyperbaric Environments

Although specific physiological consequences are covered in Chapter 15, special problems exist with hyperbaric exposure

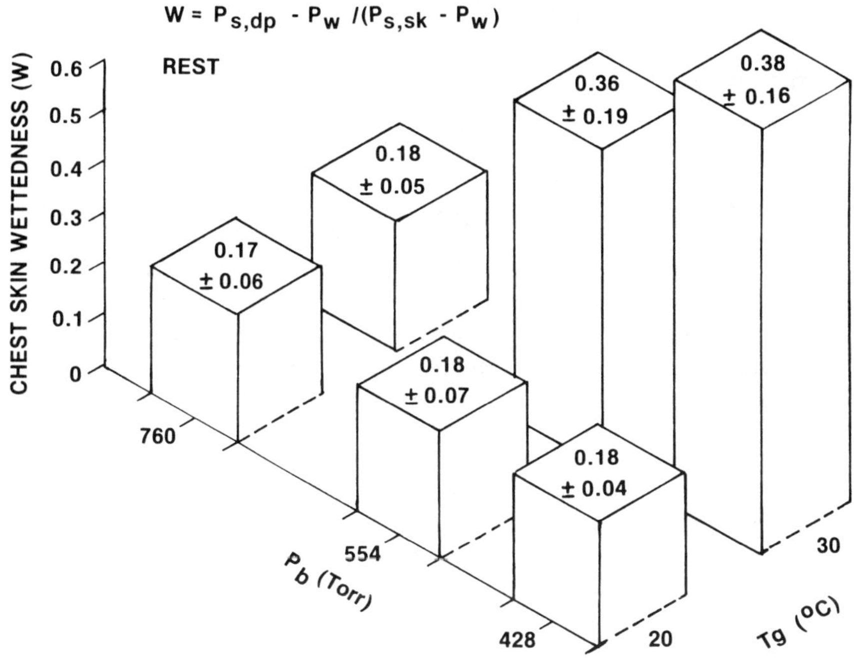

Figure 2-10. *Chest skin wettedness determined by a dew point sensor during hypobaric exposure at rest.*

[5,118,142] in clothed individuals. Only brief accounts can be given here.

Several basic properties affecting the heat exchange occur when a comfortably clothed person at sea level (1 ATA) is exposed to deep mines or pressurized tunnelling cabins where a gas mixture of helium and oxygen (Heliox) or other gas mixtures (nitrogen, etc.) are used as the ambient environment. As explained in other sections, convection in the air layer is governed by the Nusselt number which in laminar, free convection is associated with the Grashof number and Prandtl number [114]. The only variable strongly dependent on P_b is density. When air velocity is constant, h_c increases as a function of $P_b{}^{.52}$ to $P_b{}^{.62}$ [142]. Rate of heat transfer by thermal radiation is not affected by atmospheric pressure. The mass transfer coefficient, however, increases with increasing pressure as h_c increases.

Sensible heat loss increases with increasing barometric pressure, but with increasing ambient temperature (R + C) may be a gain on the clothed individual. Since air velocity decreases with increasing pressure, the rate of transfer by C alone increases less rapidly than the increase in P_b. Effectiveness of sweating and its evaporation, especially in hot chambers, diminishes with the increase in atmospheric pressure and the moisture accumulates on the skin surface increasing its film density.

For cold stress in hyperbaric environments, the gas property having the greatest influence on thermal insulation is thermal conductivity which is independent of atmospheric pressure. The effect of increased sensible heat loss reduces overall thermal insulation and heat transfer is elevated from such exposed skin sites as the face and hands. Increasing insulation by clothing to compensate for these effects results in encumbrance and inhibition of evaporation especially during moderate activities [58,142]. The insulative value of a thin garment gives enough protection against cold environments with increasing atmospheric pressures but such a garment creates an effective barrier for insensible heat loss.

Gagge and Nishi [58] showed rationally that in 30 ATA, in the unclothed state (0.06 clo) the evaporative heat transfer coefficient decreases to about 20–30% of the sea level value. In the clothed state, the effective evaporative coefficient which includes the clothing conductance factors is about 8–10% of the sea level value. Thus, in a hyperbaric environment a normal porous garment with a 0.5 clo sea-level value or i_m/I_T of 0.45 has the same net effect to a wearer's comfort as a vapor impermeable garment ($i_m/I_T = 0.12$).

Aquatic Environments

Physiological responses during water immersion are covered in Chapter 9 and in several reports [81,86,102]. Generally, physiological responses that affect thermal exchange during water immersion are buffered by the specific insulation of the tissues.

Heat exchange during partial or complete water immersion follows principles covered in earlier sections. In complete water immersion, the radiative and evaporative components are absent although during exercise, the forcing drive of increased internal body temperature certainly stimulates sweat glands; their secretion and mass flux is mixed with the boundary layer flow resident with the free or turbulent convection of ambient water flow. No data exist addressing these properties.

A major obstacle to an adequate description of thermal exchange in water immersion is the determination of convective heat loss. Heat loss by this component is many times greater than in air because of the higher thermal capacity of water. Theoretically-determined values [114,116,117,135] are subject to assumptions of body shape and water motion. Heat transfer by conventional analyses of the surface-fluid assumptions indicate that the sensitivity is based on the skin to water temperature differences which offer little thermal resistance [128]. For example, consider the $h_{c,a}$ of a 0.02 m outer diameter horizontal cylinder in atmospheric air (1 ATA). With free convection at $\Delta T = 25$ K, the combined coefficient ($h_c + h_r$) is about 8 $W \cdot m^{-2} \cdot K^{-1}$. In water, $h_{c,w}$ is about 741 $W \cdot m^{-2} \cdot K^{-1}$, 93 times greater. A nude human in comparable circumstances would have an $h_{c,a}$ of about 3 $W \cdot m^{-2} \cdot K^{-1}$, but estimates of $h_{c,w}$ could occur from 100 to 230 $W \cdot m^{-2} \cdot K^{-1}$, very predictable with theory, or some 77 times greater depending on calorimetry, heat flow transducer errors, etc. [102,135].

With partial immersion outdoors R from solar load (i.e., head and neck exposed), respiratory evaporative, and convective (E_{res} + C_{res}) heat loss must be taken into account along with the net heat flux arriving at the skin surface due to the metabolic drive [$H_{sk} = M - (E_{res} + C_{res})$] in any heat balance.

For the nude completely immersed body, the heat balance is: [58]

$$\pm S = H_{sk} - h_{c,w}(\bar{T}_{sk} - T_w), \quad W \cdot m^{-2} \qquad \text{(Eq. 2-29)}$$

The estimate of heat storage (S) in warm water (+S) or heat debt (−S), due to cold water immersion lower than a comfortable $\bar{T}_{sk}$, is affected by metabolic drive, specific resistance of tissues, etc. S becomes intimately affected by the percentage of body fat and the convective heat transfer coefficient from the exposed body surface to the water [126,128]. This coefficient has been estimated by heat flow discs (which are subject to calibration errors) in a precise swimming flume at water velocities from still to 0.97 $m \cdot s^{-1}$ impinging on a swimming person. The $h_{c,w}$ values varied for the unclothed subjects from a value of 230 $W \cdot m^{-2} \cdot K^{-1}$ at rest in still water, 460 at rest in moving water, and 580 $W \cdot m^{-2} \cdot K^{-1}$ while actively swimming [102].

In complete water immersion when the body is covered with various protective clothing garments, the effective heat transfer coefficient in Equation 2-29 becomes h_w^*, modified by a thermal efficiency factor [$F_{cl,w} \cdot h_{c,w}$], as shown by Gagge and Nishi [58]. The factor $F_{cl,w}$ is dependent on the ratio of the boundary-water resistance to the total insulation of the suit ($I_w/I_{T,w}$), in clo units.

The total clothing insulation ($I_{T,w}$) is the sum of the skin-surface insulation of water I_w, in clo units, and the intrinsic insulation of the garment in water. Calculations are most accurately done with heated manikins where total H_{sk} is a known variable. Since no radiative effects occur, $(I_w)^{-1}/0.1547$ can be used to estimate $h_{c,w}$. For example, a skin-free boundary water insulation of about 0.028 clo is 230 $W \cdot m^{-2} \cdot K^{-1}$.

Some reasonable values of total insulation during water immersion are obtained by physiological means through indirect calorimetry. By this classical method, total insulation and tissue insulation (a sum of the insulation of fat, muscle, and skin layers) was estimated in studies on Korean women breath-hold divers who now use wet suits and dry suits [62].

Tissue insulation is first found by estimation of the deep core to skin temperature gradient. Total insulation from the "shell" (Figure 2-2) to water (T_w) is:

$$I_T = (T_c - T_w)/H_s, \quad m^2 \cdot K \cdot W^{-1} \qquad \text{(Eq. 2-30)}$$

where, T_c is deep core temperature (°C), T_w is water temperature (°C) and H_s is in $W \cdot m^{-2}$.

The challenge is in the development of suitable clothing for water immersion that allows the extension of the critical thermoneutral zone by substitution of adequate insulation which humans lack. This focus has produced immersion suits that are "wet," with variable thicknesses of foam materials or suits that are "dry," therefore operable in air, but often necessary during involuntary immersion [88]. Dry suits often become too easily compressed in water and lose their air measured thermal resistance. On the other hand, serious hyperthermia occurs above 15° C water temperature [19] unless auxillary aids to dissipate heat are designed into the suit (by zippers, open hood vents, etc.). Bogart *et al.* [19] showed, by use of a manikin, a technique for the estimation of upper environmental limits and minimum water temperatures for protective suits. The analysis is done by manipulation of the basic sensible heat transfer equation during separate air and water immersion.

Cena and Clark [33] point out that in various animal pelts there are examples of wet and dry suits. In seals and polar bears, for example, the uncompressed water layer of 2–10 mm (about 0.32 to 1.6 clo) juxtaposed to the skin acts as an effective thermal insulator permitting deep dives in these animals. Alternatively, the penguin's feather pelt acts as a suitable "dry" suit with a very efficient air-insulator.

The significance of the estimation of wet-suit and cotton suit insulation, used by contemporary Korean breath-hold divers, is associated with the outcome that, frank cold acclimation (which was prevalent in semi-nude divers) is not wholly present anymore. In the past, these divers certainly relied on intrinsic tissue insulation which widened their thermoneutral zone during cold-water immersion. Classical cold-acclimation has been replaced by modern convenience.

HUMAN THERMAL COMFORT/DISCOMFORT AND MODELS OF ENERGY EXCHANGE

Behavioral Thermoregulation

Roughly 80% of the healthy population is sedentary and of these almost 90% of the persons are clothed with an intrinsic clo value of from 0.3 to 4 clo [Kraning, personal communication, 1987]. For this reason some discussion of thermal comfort attributes is a necessary ingredient to the role of clothing. Extensive research produced a description of a zone of environmental temperatures in which the body is in a state of physiological thermal neutrality. The resting clothed person maintains thermal steady-state (i.e., heat balance) such that internal body and skin temperatures are constant without excessive physiological regulatory activity such as sweating, vasoconstriction or vasodilation of skin blood vessels. This is also the zone of "neutral" temperature sensation.

Fanger [43] equates thermoneutrality with preferred comfort state or preferred temperature. For individuals clothed at an intrinsic value of 0.6 clo, air velocity near still conditions (V at 0.10 to 0.15 $m \cdot s^{-1}$), 50% rh and sedentary activity, the thermoneutral (comfort) temperature is invariable regardless of age, sex or season. Fanger demonstrated that other preferred comfort temperatures during thermal balance may be ascertained for any metabolic heat production ($\geq$ 3 met), for various clothing insulations (I_{cl}), dry bulb and mean radiant temperature, and ambient vapor pressures as well as various coefficients of heat and mass transfer.

During thermal neutrality there occurs: a) a linear relationship between skin temperature ($\bar{T}_{sk}$) and metabolic rate (M) and b) a linear function between evaporation of sweat and (M). Fanger suggested a predicted mean sensation (PMV) for an individual from the thermal load on the body; a thermal load is presumed to occur whenever there is a difference between heat production and the heat loss to the actual environment for a person supposedly kept near heat balance (i.e., comfort) at the mean skin temperature and sweat secretion determined for a specified activity level. Thus, a change in heat loss is the prominent effect which occurs if a person (having a given metabolic rate, clothing insulation value, etc.) is displaced from a comfortable environment to a warmer or colder environment. The implication that thermal comfort is equivalent to a sensation of neutral temperature is not wholly accurate or adequate. Other variables such as circadian rhythm (Chapter 7), fatigue, endocrine (humoral) factors and sensitivity to pain play a significant role.

In humans, the origin of dissatisfaction with the thermal environment can include psychological, physiological or physical factors. Winslow *et al.* [133] related unpleasant cold conditions in sedentary subjects to skin temperatures below 33.5–34° C and unpleasant warm conditions to accumulation of sweat on the skin surface (skin wettedness). Chatonnet and Cabanac [35] showed that the judgment of thermal displeasure was relative to the state of hyper- or hypothermia of the body. Gagge *et al.* [59] showed that any sudden thermal transient towards a comfortable environment caused an immediate sensation of comfort, although at the moment of the initial change the actual skin temperature and rate of sweating could be at levels not associated optimally with a feeling of comfort.

Benzinger [15] suggested that internal body temperature is the important variable affecting thermal comfort. His subjects, resting in a water bath at 38.5° C, felt comfortable until tympanic temperature began to rise, at which time thermal discomfort increased. The inference was that the sensation of overheating was a function of deep body temperature with minimal effect from elevated skin temperature. However, in a study by Berglund and Gonzalez [17] in which $\bar{T}_{sk} = T_a = 35°$ C and skin wettedness was close to 100%, a spray of water caused immediate shivering and cold discomfort with no drop in esophageal temperature. Allnutt and Allan [2] elevated internal body temperature to 38° C but cooled the skin to 30° C. Under these conditions subjects judged themselves as comfortable. Gagge and Hardy [57] showed, by using infrared radiation, that onset of thermal sweating at T_a from 20 to 31° C was immediate and accompanied a sensation of warmth, induced by radiant heat prior to any increase in internal (tympanic) body temperature. These studies indicate that $\bar{T}_{sk}$ and internal body temperature (as an integrated drive) are both important parameters in judgments of the state of thermal comfort.

Marks and Gonzalez [95] explored the way pleasantness and unpleasantness of a thermal stimulation of the blackened forehead depend on both the magnitude of the intensity and extent of surface area of radiant stimulation. While air temperatures were held constant at 5, 10, 15, 20 and 30° C, irradiance was varied randomly from no stimulus to as high as 800 $mW \cdot cm^{-2}$ impinging on the forehead.

Figure 2-11 shows results averaged from two clothed subjects in this study. When radiant intensity increases, the estimation of that stimulus first becomes gradually pleasant then markedly unpleasant; the shape of the curve is dependent upon the air temperature. A 2° C increase in forehead skin temperature, produced by the thermal radiation, feels pleasant in the cold but unpleasant in warm air. The hedonic value and the intensity of sensation depends

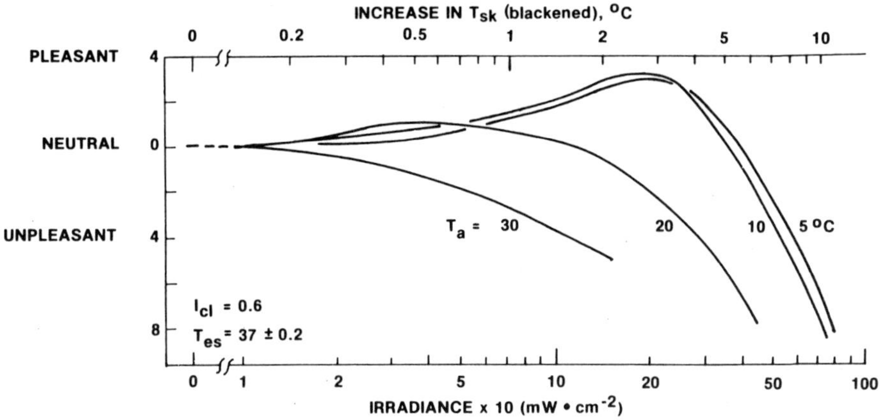

Figure 2-11. *Relationship between psychophysical hedonic estimates of radiant stimuli to the forehead and thermal irradiance at 4 constant air temperatures and associated forehead skin temperature.* [*Redrawn from Reference 95*].

on size and duration of stimulation. These are indications of spatial and temporal summation.

In these experiments, internal body temperature (T_{es}) in the clothed subjects remained close to normal levels (37–37.2° C). Therefore, in the absence of any strong internal drive resulting from hyper- or hypothermia, the hedonic value is strongly governed by superficial skin temperatures. Deviations in internal temperature can promote, and often dominate, overall sensations of thermal affect, a phenomenon coined "alliesthesia." [30]

The sense of warm discomfort, especially while clothed and sedentary and during thermal radiation, is affected by two other principal physical properties of the skin. These properties are its temperature ($\bar{T}_{sk}$) and wettedness (w) [18,50,56,59,66,69], both intimately involved in heat exchange with the environment. The following processes are likely involved: (a) the heat transported by blood (SkBF or K_s) flowing from the body core to the periphery, (b) the sensible heat exchange between the skin surface and the environment by radiation and convection, (c) the combined effect of skin and internal body temperature on the sweating drive necessary for regulation of body temperature, and (d) the ability of the sweat to evaporate and thus provide cooling of the body.

Gwosdow *et al.* [69] recently showed that an element in the increased skin wettedness affecting thermal discomfort with clothing is related to the frictional forces between cloth and skin (Figure 2-12).

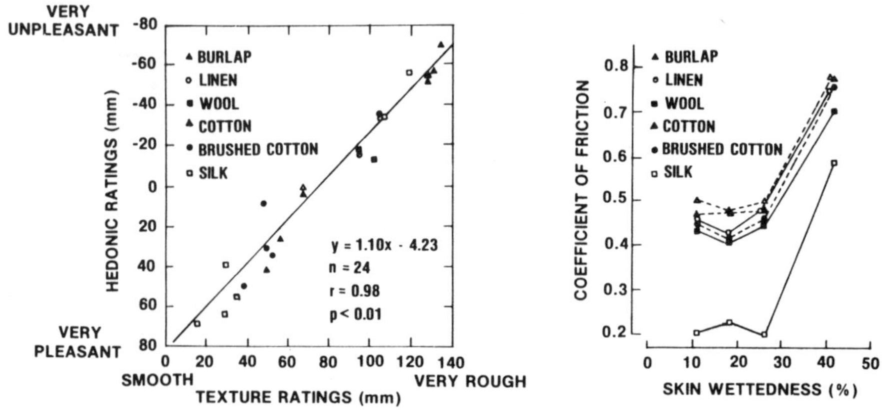

Figure 2-12. *Relationship between hedonic estimates and texture ratings of 6 fabrics (left panel) and coefficient of friction plotted as a function of local skin wettedness calculated by a dew point sensor. [Redrawn from Reference 69].*

The affective perception to fabrics deemed as pleasant was to those fabrics exhibiting the lowest coefficient of friction (how the cloth moved through the skin).

Model-Clothing Interaction

The information gained by the use of models is an invaluable part of the focus required in ascertaining the interactions of human-environment-clothing-and task performance as presented in the scheme in Figure 2-13.

As implied by this servo-mechanistic scheme, a critical element

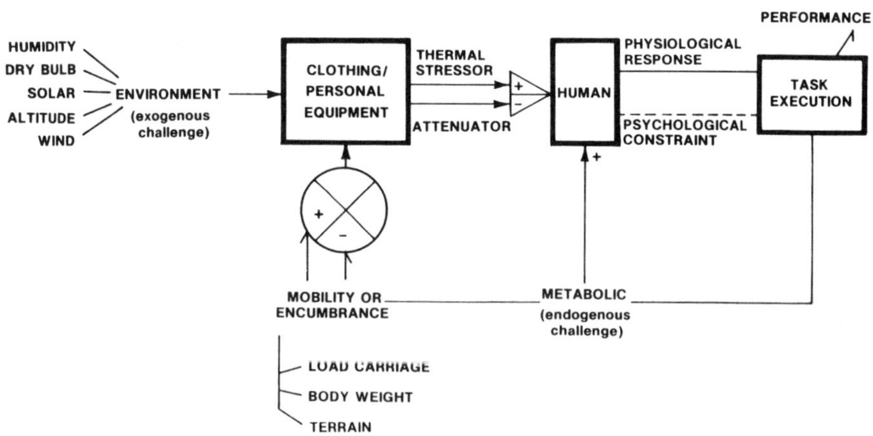

Figure 2-13. *Servo-control scheme of the human-clothing-environmental interaction.*

to the research into military clothing and task performance is directed towards protective measures or attentuation against internal or environmental extremes which cannot be compensated by normal heat exchange processes [7,9,10,26,41,60,61,62,70,82,88,89,105,128]. This objective is in direct contrast to that of civilian clothing research which focuses largely on establishing comfort requirements (or zones). Interestingly, both objectives trace their roots to early quantification gained by the experimentation with thermal manikins and electrical models which are still actively used today.

Development of Operational and Rational Models of Heat Strain

By far the greatest impact of the fundamental studies involved with human research, thermal manikins, and biophysical devices directly related to clothing properties has been the blend of the thoughts of researchers active in such studies. One of the earliest attempts to model human performance was by Belding and Hatch in their Heat Stress Index (HSI) which is analogous with elements of Gagge's skin wettedness properties [9]. On the other hand, the initial coefficients derived from the Woodcock and Breckenridge research [24,137,138,139,140,141] were used by Givoni and Goldman in 1972 to develop an operational model [60]. The latter model also was used to predict deep body temperatures and heart rate responses to a wide environmental range where individuals would reach thermal equilibrium or become a heat casualty dependent on clothing, work level and various other factors such as load carriage, terrain coefficients, solar coefficients, etc. [115].

The different research focuses produced two disparate approaches to thermal modelling which, curiously, use the same coefficients and factors derived by either manikins or human clothing studies. One approach is directed towards heat stress and strain risk analysis which is a more empirical attribute that necessitates definitions to operational activities (i.e., prediction of heat or cold casualties, hypovolemic levels where status of water requirements is a critical property). Recent developments to this approach have been made by the formulation of a predictive, operationally-derived model. This model utilizes and extends many of the approaches formulated by Woodcock and Breckenridge as presented in the original Givoni-Goldman equations. An example of the utility of the USARIEM operational model features [115] is in its prediction of limits to work and water requirements for individuals in chemical protective clothing, while heat acclimatized or non-acclimated. Consider the following hypothetical, but not uncommon, input parameters that are used in such a model:

1. Clothing System: Impermeable Protective (clo = 2)
2. Metabolic Work Rate: 175 Watts
3. Level of Heat Casualties <5%
4. Heat Acclimation Status NONE AND FULLY
5. Wind Speed Calm [roughly 5 $m \cdot s^{-1}$]
6. Solar Load None
7. Humidity 20%
8. Air Temperature 40 and 49° C

The results to the iteration are:

		T_a = 40° C/20% rh		T_a = 49° C/20% rh	
		non-accl	fully-accl	non-accl	accl
Work/Rest cycle (min)	______	12/48	NO LIMIT	NONE	NO LIM
Maximum one time (min)	______	110	NO LIMIT	72	NO LIM
Water Requirements					
at Work	______	1 liter/h		1.5 liter/h	
at Rest	______	0.7		1.1	

A useful parameter in this model (evident by #3 above) is the use of threshold first developed by Goldman [61] as indicative of the incidence or probability of heat casualties as a function of rectal temperature (or indirectly heat storage, in J/g). The following were developed as thresholds:

EXTENT OF HEAT CASUALTY	*HEAT STORAGE (J/g)
UNCOMFORTABLE	4.78
25% CHANCE	7.17
50% CHANCE	9.56
75% CHANCE	11.95
DEFINITE	>12

*[e.g., kcal × 4186/70 kg = J/g for standard man]

To have an individual slightly uncomfortable, but in a tolerable level, heat storage should be less than 4.8 J/g (<80 kcal). If a standard man's (A_D = 1.8 sq.m) metabolic heat production is at 145 $W \cdot m^{-2}$ (2.5 met), rate of heat dissipation mostly by sweat evaporation, must be greater than 169 W or 4 g/min.

Another strategy used in clothing models is the rational approach which derives its foundation from homeostatic analysis employing the body heat balance equation [10,46,51,54,56,59,66]. A wealth of research has produced sound indices such as Operative

Temperature and Standard Effective Temperature. [46,56,114]. A "psychrometric range of a clothing system" concept is another useful alternative [99].

A rational environmental index, practical for clothing interaction with the environment, is a way of describing the combined effects of physiological and thermal exchanges. It is almost an intractable ambition of many scientists who deal with thermal biophysics. An effective temperature (ET*), in contrast to old ET which was based on sensory responses [77], is a rational index of the thermal stress and consequent physiological or thermal responses. The term "rational" refers to the fact that the index integrates almost all of the problems encountered in the thermal environment and predicts multiple interaction. Such indices are derived in a logical manner from the principles of heat transfer physics and physiological control theory and validation. These indices more readily substitute for vague environmental impact of T_a, relative humidity, wind speed, or empirical indices such as Windchill, Beufort scale (as evident in Figure 2-8), or WBGT which will not be addressed in this chapter [6,32,77,89,139].

Heat and water vapor transport through clothing in rational indices currently are treated in the form of heat exchange from the skin by radiation and conduction and evaporation as simultaneous, but separate, phenomena as fixed resistors in normal, porous garments. Some criticism has been addressed to this fact especially with activity in the cold which involves more complex transport processes [90,93]. However, for 90% of the world population problems, thermal interaction with the environment covers responses to low activity, indoor occupancy which requires such rational indices [43,56]. The other 10% of the problems can be adequately handled by models which do involve sophisticated treatment [44,93,125].

Rational indices are often written in terms of *Standard Temperatures* that allow comparison of two or more diverse environments with different thermal characteristics (i.e., hyperbaric with Heliox versus normal sea level). Surprisingly, little utilization has been made to the advantages of a rational index; many scientists, engineers, Army personnel and managers preferring description in vague ambient temperature form. Only one engineering society has adopted a Standard Effective Temperature.

A Standard Temperature is defined as the temperature of a specified hypothetical environment in which a body would experience a similar thermal exchange as in the actual environment (i.e., "feel" or respond to the same sharpness). The Standard Environment is chosen as one that is typical of a person's everyday environment; it is specified by the following characteristics [66]:

[a] Sea Level;
[b] Uniform temperature [T_a = mean radiant temperature (T_r)];
[c] Relatively still air movement ($V = 0.1 - 0.18\ m \cdot s^{-1}$) which corresponds to a convective heat transfer coefficient (h_c,ST) of $2.9\ W \cdot m^{-2} \cdot K^{-1}$);
[d] Sedentary activity (1 met);
[e] Clothing insulation of 0.6 clo (intrinsic);

The following three rational temperature indices have been developed [51,54,59,109]: operative temperature, humid operative temperature, and a recent variation of ET* [46]. Operative temperature has been covered extensively in this chapter and Chapter 1; it combines the effects of sensible heat transfer from the skin and accounts for this property. Standard Operative temperature is the same as Operative except, in the imaginary isothermal environment, the person is exposed to the specified standard conditions summarised above. Humid Operative temperative combines the effects of sensible *and* insensible heat transfer from the skin by convection, radiation, and total evaporation. Standard Humid Operative temperature is defined the same way as for Humid Operative temperature except the isothermal environment has the added condition of 100% rh. All the latter have improvements by more exact rational quantification of mass transfer coefficients [46]. Each one could be easily modified to incorporate dependent, simultaneous heat and mass transfer in multi-layer garments by fundamentals discussed before.

SUMMARY

The gist of the problems covered in this chapter is the idea clothing allows adaptation or modifications to a thermal environment in respect to the work requirements (Figure 2-13). This is a very complex design property left to man which Nature does implicitly.

As in all mechanistic models, developments take place slowly, because appropriate coefficients have to be verified with the responses they intend to predict by physiological experiments. Currently, ET* and humid operative temperature, for example, have been improved to predict the enthalpy of the thermal environment perceived by humans more rigorously [46]. The new ET* no doubt will certainly be developed for prediction of more complex properties of clothing interaction.

REFERENCES

1. Adolph, E.F. and Associates. *Physiology of Man in the Desert.* Interscience Publication Inc. New York, NY:1947.
2. Allnutt, M.F., and J.R. Allan. The effects of core temperature elevation and thermal sensation of performance. *Ergonomics.* 16:189–197, 1973.
3. Aschoff, J., and R. Wever. Kern und schale im warmehauschalt des menshen. *Naturwissenchaften.* 45:477–485, 1958.
4. Behmann, F.W. The influence of climatic and textile factors on the heat loss in drying of moist clothing. In: *Biometeorology.* New York, NY: Pergamon Press, 1962.
5. Behnke, A.R. Temperature and comfort in pressurized environments. In: *Human Factors 1970,* Bulletin SF-70-3. New York, NY: ASHRAE, 1970.
6. Belding, H.S. Protection against dry cold. In: L.H. Newburgh (ed.). *Physiology of Heat Regulation and Science of Clothing.* Philadelphia, PA: W.B. Saunders Co., pp. 351–367, 1949.
7. Belding, H.S. Resistance to heat in man and other homeothermic animals. In: A.H. Rose (ed.). *Thermobiology* New York, NY: Academic Press, pp. 479–510, 1967.
8. Belding, H.S., R.C. Darling, D.R. Griffin, S. Robinson and E.S. Turrell. Evaluation of thermal insulation provided by clothing. In: L.H. Newburgh and M. Harris (eds.). *Clothing Test Methods,* Subcommittee on clothing. Washington, DC: Nat. Res. Council. p. 9, 1945.
9. Belding, H.S., and T.F. Hatch. Index for evaluating heat stress in terms of resulting physiological strain. *Heating, Piping, and Air Conditioning.* 27:129–136, 1955.
10. Belding, H.S., and E. Kamon. Evaporative coefficients for prediction of safe limits in prolonged exposures to work under hot conditions. *Fed. Proc.* 32:1598–1601, 1973.
11. Belding, H.S., H.D. Russell, R.C. Darling and G.E. Folk. Thermal responses and efficiency of sweating when men are dressed in Arctic clothing and exposed to extreme cold. *Am. J. Physiol.* 149:204–222, 1947.
12. Belding, H.S., H.D. Russell, R.C. Darling and G.E. Folk. Analysis of factors concerned in maintaining energy balance for dressed men in extreme cold; effects of activity on the protective value and comfort of an arctic uniform. *Am. J. Physiol.* 149:223–239, 1947.
13. Bennett, D.W., and J.A. Yount. Design, construction, and testing of an anthropomorphic robot. A.N.S. Topical. Robots XI. April. pp. 1–6, 1987.
14. Bennett, R.D. Physical and chemical properties of clothing. *Chemistry in Canada.* Chemical Institute (Canada). January pp. 33–36, 1952.
15. Benzinger, T.H., and C. Kitzinger. Direct calorimetry by means of the gradient principle. *Rev. Sci. Instrum.* 20:849–860, 1949.
16. Berglund, L.G., D. Fashena, X. Su and A. Gwosdow. Absorbed solar radiation from measured sweat rate. *Proc. 13th Ann. Northeast Bioenerg. Conf.* New York, NY: IEEE, pp. 507–510, 1987.
17. Berglund, L.G. and R.R. Gonzalez. Mass transfer characteristics of the sweating human. *Special Technical Publication ASME.* 10–15 December, pp. 173–180, 1978.
18. Berglund, L.G. and R.R. Gonzalez. Evaporation of sweat from sedentary man in humid environments. *J. Appl. Physiol.* 42:767–772, 1977.
19. Bogart, J.E., J.R. Breckenridge and R.F. Goldman. Thermal protection of commercial dry suit diving systems. *USARIEM Technical Report T1/81.* Natick, MA., 1981.
20. Brebner, D.F., D. McK. Kerslake and J.L. Waddel. The diffusion of water vapor through human skin. *J. Physiol.* (London) 132:225–231, 1955.
21. Breckenridge, J.R., and R.L. Pratt. Effect of clothing color on solar heat load. *Technical Report EP-155,* Natick, MA., 1961.
22. Breckenridge, J.R. Insulating effectiveness of metallized reflective layers in cold weather clothing systems. *USARIEM Technical Report T2/78.* Natick, MA., 1978.
23. Breckenridge, J.R. Effect of wet insulation in vapor barrier cold-weather boots. *Textile Res. J.* 37:809–811, 1967.
24. Breckenridge, J.R. Chapter 11. Effects of body motion on convective and evaporative heat exchanges through clothing. In: N.R.S. Hollies and R.F. Goldman (eds.). *Clothing Comfort: Interaction of Thermal, Ventilation, Construction, and Assessment Factors.* Ann Arbor, MI: The Fiber Society, Inc., 1977, pp 153–166.
25. Breckenridge, J.R., and R.F. Goldman. Human solar heat load. *ASHRAE Trans.* 78:110–119, 1972.
26. Breckenridge, J.R. and R.F. Goldman. Effect of clothing on bodily resistance against meteorological stimuli. In: J.F. Tromp (ed.). *Progress in Human Biometeorology.* Amsterdam: Sweits and Zeitlinger, Inc, 1977, pp. 194–208.
27. Burton, A.C. Human calorimetry: II. The average temperature of the tissues of the body. *J. Nutr.* 9:261–268, 1935.
28. Burton, A.C. Transfer of heat to the ambient air and the thermal insulation of the ambient

air. In: L.H. Newburgh and M. Harris (eds.). *Clothing Test Methods.* Washington, D.C.: National Res. Council, 1945, p. 37.
29. Burton, A.C., and O.G. Edholm. *Man in a Cold Environment: Physiology and Pathological Effects of Exposure to Low Temperatures.* London: E.A. Arnold, 1955.
30. Cabanac, M. Plaisir ou deplaisir de la sensation thermique et homeothermie. *Physiol. Behav.* 4:359–369, 1969.
31. Cain, B., and B. Farnworth. Two new techniques for determining the thermal radiative properties of thin fabrics. *J. Thermal Insulation.* 9:301–322, 1986.
32. Carlson, L.D. and A.C.L. Hsieh. *Control of Energy Exchange.* New York, NY: Macmillan and Co., pp. 151, 1970.
33. Cena, K. and J.A. Clark (eds.). *Bioengineering, Thermal Physiology and Comfort.* Elsevier, New York. 1981, pp. 289.
34. Chang, S.Kw., E. Arens and R.R. Gonzalez. Determination of the effect of walking on the forced convective heat transfer coefficient using an articulated manikin. *ASHRAE Trans.* (in press).
35. Chatonnet, J., and M. Cabanac. The perception of thermal comfort. *Int. J. Biometeorol.* 9:183–197, 1965.
36. Chilton, T.H. and A.P. Colburn. Mass-transfer (absorption) coefficients: Prediction from data on heat transfer and fluid friction. *Indust. Eng. Chem.* 26:1183–1187, 1934.
37. Clark, R.P. Human skin temperature and convective heat loss. In: K. Cena and J.A. Clark (eds.). *Bioengineering, Thermal Physiology and Comfort.* New York, NY: Elsevier, 1981, pp. 57–76.
38. Colin, J., and Y. Houdas. Experimental determination of coefficient of heat exchange by convection of the human body. *J. Appl. Physiol.* 22:31–38, 1967.
39. Crow, R.M. Heat and moisture transfer in clothing systems. part I. Transfer through materials, a literature review. *Defence Res. Establishment Ottawa* Technical Note:74-27, pp. 1–18, 1974.
40. Damon, A., F.E. Randall, R.S. Benton and A.C. Brues. The importance of human sizing standards. *Air Surgeon's Bulletin.* 1:8–9, 1944.
41. Drake, R.M., J.E. Funk and J.B. Moegling. Sensible heat transfer in the Gemini and Apollo space suits. In: J.A. Chato (ed). *Thermal Problems in Biotechnology* New York, NY: Amer. Soc. Mech. Engr., 1968, pp 96–112.
42. Elnas, S., and I. Holmer. Thermal insulation of handwear measured with an electrically-heated hand model. In: K. Amundin, C. Brunis, and A. Bran Persson (eds.). *Proc. of the Intl. Conf. on Protective Clothing Systems* Sweden: Nat. Def. Res. Inst., pp. 243–250, 1983.
43. Fanger, P.O. Thermal Comfort: Analysis and Applications in Environmental Engineering. Copenhagen: Danish Technical Press, 1970.
44. Farnworth, B. A numerical model of the combined diffusion of heat and water vapor through clothing. *Textile Res. Inst. J.* 11:653–655, 1986.
45. Fitzgerald, J.E. A study of the copper man. Phase I: Physical characteristics; thermometry; air clo evaluations. *Climatic Res. Lab. Report.* Lawrence, Ma. 28 Aug., 19 pp, 1946.
46. Fobelets, A., and A.P. Gagge. Rationalization of the effective temperature, ET*, as a measure of the enthalpy of the human indoor environment. *ASHRAE Trans.*, (in press).
47. Fourt, L., and M. Harris. Physical properties of clothing in fabrics. In: L.H. Newburgh (ed.). *Physiology of Heat Regulation and Science of Clothing.* Philadelphia, PA: W.B. Saunders Co., 1949.
48. Frick, J.G. Jr., R.M. Dullman and R.M. Reinhart. Modification of crosslinking effects in cotton fabrics by swelling and methylation. *J. Appl. Polym. Sci* 9:2311–2318, 1965.
49. Fujitsuka C. and K. Ohara. Studies on water vapor pressure gradient from external air through clothing to the skin in relation to external humidity and clothing conditions. *J. Human Ergol.* 6:75–85, 1977.
50. Gagge, A.P. A new physiological variable associated with sensible and insensible perspiration. *Am. J. Physiol.* 120:277–287, 1937.
51. Gagge, A.P. Standard operative temperature, a generalized temperature scale, applicable to direct and partitional calorimetry. *Am. J. Physiol.* 131:93–105, 1940.
52. Gagge, A.P. Effective radiant flux: an independent variable that describes thermal radiation in man. In: J.D. Hardy, J. Stolwijk and Gagge (eds.). *Physiological and Behavioral Temperature Regulation.* Springfield, IL: C.C. Thomas, 1970, p. 34.
53. Gagge, A.P. Partitional Calorimetry in the Desert. In: M.K. Yousef, S.M. Horvath, and R.W. Bullard (eds.). *Physiological Adaptations: Desert and Mountain.* New York, NY: Academic Press, pp. 23–51, 1972.
54. Gagge, A.P. Rational temperature indices of man's thermal environment and their use with a 2-node model of his temperature regulation. *Fed. Proc.* 32:1572–1582, 1973.
55. Gagge, A.P., A.C. Burton, and H.C. Bazett. A practical system of units for the description of the heat exchange of man with his environment. *Science.* 94:428–429, 1941.

56. Gagge, A.P., A.P. Fobelets, and L.G. Berglund. A standard predictive index of human response to the thermal environment. *ASHRAE Trans.* 92:1–22, 1986.
57. Gagge, A.P., and J.D. Hardy. Thermal radiation exchange of the human by partitional calorimetry. *Am. J. Physiol.* 23:248–258, 1967.
58. Gagge, A.P. and Y. Nishi. Heat exchange between the human skin surface and thermal environment. In: D. Lee (ed.). *Handbook of Physiology: Reactions to Environmental Agents.* Bethesda, MD.: American Physiological Society, pp. 69–92, 1983.
59. Gagge, A.P., J.A.J. Stolwijk, and Y. Nishi. An effective temperature scale based on a simple model of human physiological regulatory response. *ASHRAE Trans.* 77:247–262, 1971.
60. Givoni, B., and R.F. Goldman. Predicting rectal temperature response to work, environment, and clothing. *J. Appl. Physiol.* 32:812–821, 1972.
61. Goldman, R.F. Tolerance time for work in the heat when wearing CBR protective clothing. *Mil. Medicine.* 128:776–786, 1963.
62. Goldman, R.F. Clothing design for comfort and work performance in extreme thermal environment. *Trans. N.Y. Acad. Sci.* 36:531–544, 1974.
63. Gonzalez, R.R. Work in the North: physiological aspects. *Arct. Med. Res.* 44:7–17, 1986.
64. Gonzalez, R.R. Biophysical and physiological integration of proper clothing for exercise. In: K.B. Pandolf (ed.). *Exercise and Sport Sciences Reviews.* New York, NY: Macmillan Co., 1987, pp 261–295.
65. Gonzalez, R.R. and K. Cena. Evaluation of vapor permeation through garments during exercise. *J. Appl. Physiol.* 52:928–935, 1985.
66. Gonzalez, R.R., and A.P. Gagge. Magnitude estimates of thermal discomfort during transients of humidity and operative temperature and their relation to the new ASHRAE effective temperature (ET*). *ASHRAE Trans.* 79:88–96, 1973.
67. Goodman, A.B., and A.V. Wolf. Insensible water loss from the human skin as a function of ambient vapor concentration. *J. Appl. Physiol.* 26:203–207, 1969.
68. Graichen, H., R. Rascati and R.R. Gonzalez. Automatic dew point temperature sensor. *J. Appl. Physiol.* 52:1658–1660, 1982.
69. Gwosdow, A.R., J.C. Stevens, L.G. Berglund, and J.A.J. Stolwijk. Skin friction and fabric sensations in neutral and warm environments. *Textile Res. J.* 9:574–580, 1986.
70. Haisman, M.F., and R.F. Goldman. Physiological evaluations of armored vest in hot-wet and hot-dry climates. *Ergonomics.* 17:1–12, 1974.
71. Hardy, J.D., and E.F. Dubois. The technique of measuring radiation and convection. *J. Nutr.* 15:461–475, 1938.
72. Hardy, J.D., A.T. Milhorat, and E.F. Dubois. Basal metabolism and heat loss of young women at temperatures from 22 degrees to 35 degrees Celsius. *J. Nutr.* 21:383–404, 1941.
73. Hardy, J.D., Heat transfer. In: L.H. Newburgh (ed.). *Physiology of Heat Regulation and The Science of Clothing.* Philadelphia, PA: W.B. Saunders, Co., 1949, pp. 78–109.
74. Hauser, E.R., D.D. Keane and G.D. Nelson. Thermographic evaluation and heat flow analysis of clothing insulation and garment design. *Internal Tech Report. 3M Co. St. Paul, Minn. IRIE 78,* C9-119, 1978.
75. Holmer, I. Heat exchange and thermal insulation compared in woolen and nylon garments during wear trials. *Textile Res. Inst. J.* 9:511–518, 1985.
76. Holmer, I. and S. Elnas. Physiological evaluation of resistance to evaporative heat transfer by clothing. *Ergonomics.* 24:63–74, 1981.
77. Houghten, F.C., and C.P. Yaglou. Determining lines of equal comfort. *ASHVE Trans.* 29:163–176 and 361–384, 1923.
78. International Standard ISO 7730-1984. *Estimation of the thermal resistance of clothing ensembles.* Annex C., 1984.
79. Kang, D.H., Y.S. Park, Y.D. Park, I.S. Lee, D.S. Yeon, S.H. Lee, S.Y. Hong, D.W. Rennie, and S.K. Hong. Energetics of wet-suit diving in Korean women breath-hold divers. *J. Appl. Physiol.* 54:1702-1707, 1983.
80. Kaufman, W.C., D. Bothe and S.D. Meyer. Thermal insulating capabilities of outdoor clothing materials. *Science* 215:690–691, 1982.
81. Keatinge, W.R. *Survival in Cold Water.* Oxford and Edinburgh: Blackwell Scientific Publications, 1969.
82. Kerslake, S.M. A heated manikin for studies of air ventilated clothing. *FPRC Memo 214, Royal Air Force, Institute of Aviation Medicine, Farnborough-Hants,* 1964.
83. King, A.L. *Thermophysics.* San Francisco: W.H. Freeman, 1962.
84. Kleiber, M. Trophic responses to cold. *Fed. Proc.* 22:772–774, 1963.
85. Korsgaard, V., and Th. Lund Madsen. A thermal mannequin for the determination of the thermic effects of the indoor climate on human beings. *Tech Rept 9. Tech Univ. of Denmark,* 20 pp, 1967.
86. Lambertsen, C.J. (ed.). *Underwater Physiology.* Proceedings on the Fourth Symposium on

Underwater Physiology. New York, NY: Academic Press, 1971.
87. Lewis, W.K. The evaporative of a liquid into a gas. *ASME Trans.* 4:325–335, 1922.
88. Light, I.M., A. Avery, A.M. Grieve. Immersion suit insulation: the effect of dampening on survival estimates. *Aviat. Space Environ. Med.* 58:964–969, 1987.
89. Lind, A.R. Prediction of safe limits for prolonged occupational exposure to heat. *Fed. Proc.* 32:1602–1606, 1973.
90. Lotens, W.A., and E.J.G. van de Linde. Insufficiency of current clothing descriptions. In: R. Henane (ed.). *Intl. Conference of Biophysical and Physiological Evaluation of Protective Clothing.* Lyon, France: Inst. des Santes Armees, 1983, pp. 1–13.
91. Lotens, W.A. Clothing design for work in the cold. *Arct. Med. Res.* 46:3–12, 1987.
92. Lotens, W.A., and G. Havenith. Ventilation of rainwear determined by a trace gas method. *Institute for Perception Technical Rept 6.1.5.* The Netherlands, 1986.
93. Lotens, W.A., J. Varkevisser, and K. Kwant. Hematran, a computer model for the calculation of heat and mass transfer in multiple-layer assemblies. *Institute for Perception Technical Rept B87-63.* The Netherlands, 1987.
94. Madsen, Th. Lund. Description of a thermal mannequin for measuring the thermal insulation capacity of the clothing. *Tech. Univ. of Denmark Report 48.*, 16 pp., 1976.
95. Marks, L.E., and R.R. Gonzalez. Skin temperature modifies the pleasantness of thermal stimuli. *Nature* 247:473–475, 1974.
96. McCullough, E.A., and B.W. Jones. A comprehensive data base for estimating clothing insulatin. *Tech Rept 84-01. Manhattan, Kansas: Inst. for Env. Res.*, Kansas State Univ., 1984.
97. Meechels, J.H., R.M. Demeler and E. Kachel. Moisture transfer through chemically treated cotton fabrics. *Textile Res. J.* 36:375–384, 1966.
98. Meechels, J.H., and K.H. Umbach. Thermophysiological properties of clothing systems. *Milliand Testilberichte* (English Edition) December, pp. 75–81, 1977.
99. Meechels, J.H., and K.H. Umbach. The psychrometric range of clothing systems. In: N.R.S. Hollies and R.F. Goldman (eds.). *Clothing Comfort: Interaction of Thermal, Ventilation, Construction, and Assessment Factors.* Ann Arbor, MI: The Fiber Society, Inc., 1977, pp. 133–151.
100. Monteith, J.L., and G.S. Campbell. Diffusion of water vapour through integuments-potential confusion. *J. Thermal Biol.* 5:7–9, 1980.
101. Mitchell, D. and A.J. Van Rensburg. Assessment of clothing insulation: The problem and its solution using direct calorimetry on exercising men. *Arch. Sci. Physiol.* 27:149–162, 1973.
102. Nadel, E.R., I. Holmer, U. Bergh, P.-O. Åstrand, and J.A.J. Stolwijk. Energy exchanges of swimming man. *J. Appl. Physiol.* 36:465–471, 1974.
103. Nagata, H. Evaporative heat loss and clothing. *J. Human Ergol.* 7:169–175, 1978.
104. Nelson, N., L.W. Eichna, S.M. Horvath, W.B. Shelley, and T.F. Hatch. Thermal exchange of man at high temperature. *Am. J. Physiol.* 151:626–652, 1947.
105. Newburgh, L.H. (ed.). *Physiology of Heat Regulation and the Science of Clothing.* San Francisco, CA: W.B. Saunders Co., 1949.
106. Newburgh, L.H., and M. Harris (eds.). *Clothing Test Methods.* CAM Report No. 390. Ann Arbor, MI: Edwards Bros., Inc., 1945.
107. Nishi, Y., and A.P. Gagge. Direct evaluation of convective heat transfer coefficient by naphtalene sublimation. *J. Appl. Physiol.* 29:830–838, 1970.
108. Nishi, Y., and A.P. Gagge. Moisture permeation of clothing—a factor governing thermal equilibrium and comfort. *ASHRAE Trans.* 76:137–145, 1970.
109. Nishi, Y., and A.P. Gagge. Humid operative temperature. A biophysical index of thermal sensation and discomfort. *J. Physiol.* (Paris) 63:365–368, 1971).
110. Nishi, Y., R.R. Gonzalez and A.P. Gagge. Direct measurement of clothing heat transfer properties during sensible and insensible heat exchange with the thermal environment. *ASHRAE Trans.* 81:183–199, 1975.
111. Oakley, E.H. The design and function of military footwear: a review following experiences in the South Atlantic. *Ergonomics* 27:631–637, 1984.
112. Olesen, B.W. and R. Nielson. Thermal insulation of clothing measured on a movable thermal manikin and on human subjects. *Tech. Rept. #7206/00/914 Tech. Univ. Denmark,* 1983, pp. 107.
113. Oohori, T., A.P. Gagge, and L.G. Berglund. Comparison of current two-parameter indices of vapor permeation of clothing—as factors governing thermal equilibrium and human comfort. *ASHRAE Trans.* Part I, 90:1–16, 1984.
114. Ozisik, M.N. *Heat Transfer: A Basic Approach.* New York, NY: McGraw-Hill, 1985.
115. Pandolf, K.B., L.A. Stroschein, L.L. Drolet, R.R. Gonzalez, and M.N. Sawka. Prediction modeling of physiological responses and human performance in the heat. *Comput. Biol. Med.* 6:319–329, 1986.
116. Rapp, G.M. Convective mass transfer and the coefficient of evaporative heat loss from human skin. In: J.D. Hardy, A.P. Gagge, J.A.J. Stolwijk, (eds.). *Physiological and Behavioral Temperature Regulation* Springfield, IL: C.C. Thomas, 1970.

117. Rapp, G.M. Convective heat transfer for nude man, cylinders and spheres at low air velocities. *ASHRAE Trans.* 79:75–87, 1973.
118. Raymond, L.W. The thermal environment for undersea habitats. *Human Factors 1970,* Bulletin SF-70-3. New York, NY: ASHRAE, 1970.
119. Renbourn, E.T. The science of clothing hygiene: past and present. *J. Textile Inst.* 51:469–485, 1960.
120. Rees, W.H. Physical factors determining the comfort performance of textiles. *Presentation of the III Shirley Seminar.* Manchester, U.K., 15 June, 1971.
121. Santee, W.R., and T.L. Endrusick. Biophysical evaluation of footwear for cold-weather climates. *Aviat. Space, Environ. Med.* (in press).
122. Sibbons, J.L.H. Coefficients of evaporative heat transfer. In: J.D. Hardy, A.P. Gagge, and J.A.J. Stolwijk (eds.). *Physiological and Behavioral Temperature Regulation.* Springfield, IL: C.C. Thomas, 1970, pp. 108–138.
123. Soper, D.J.G. A description of the RCAF copper man—with some notes on its use. *RCAF Report IAM57-4, Inst. Aviat. Med. Royal Canadian Air Force,* Toronto, 1957.
124. Sparrow, E.M., and R.D. Cess. *Radiation Heat Transfer.* Belmont, CA: Wadsworth Publ. Co, 1966.
125. Stewart, J.M., and R.F. Goldman. Development and evaluation of heat transfer equations for a model of clothed man. *The South Afr. Mech. Engr.* 28:174–178, 1978.
126. Toner, M.M., W.L. Holden, M.E. Foley, J.E. Bogart, and K.B. Pandolf. Influence of clothing and body-fat insulations on thermal adjustments to cold-water stress. *Aviat. Space Environ. Med.* (in press).
127. Teitlebaum, A., and R.F. Goldman. Increased energy cost with multiple clothing layers. *J. Appl. Physiol.* 32:743–748, 1972.
128. Tikuisis, P., and R.R. Gonzalez. Rational considerations for modelling human thermoregulation for cold water immersion. *ASHRAE Trans.* (in press).
129. Van Dilla, M., R. Day, and P. Siple. Special problems of hands. In: L.H. Newburgh (ed.). *Physiology of Heat Regulation and The Science of Clothing.* Philadelphia, PA: W.B. Saunders Co., 1949, pp. 374–386.
130. Vanggaard, L. Physiological reactions to wet-cold. *Aviat. Space Environ. Med.* 46:33–36, 1975.
131. Webb, P. Body heat loss in undersea gaseous environments. *Aerospace Med.* 41:1282–1287, 1970.
132. Whelan, M.E., L.E. MacHattie, A.C. Goodings, and L.H. Turl. The diffusion of water vapor through laminae with particular reference to textile fibers. *Textile Res. J.* 25:197–207, 1955.
133. Winslow, C-E.A., L.P. Herrington, and A.P. Gagge. Physiological reactions of the human body to varying environmental temperatures. *Amer. J. Physiol.* 120:1–22, 1937.
134. Winslow, C-E.A., and L.P. Herrington. *Temperature and Human Life.* Princeton, NJ: Princeton Univ. Press, 1949.
135. Witherspoon, J.M., R.F. Goldman, and J.R. Breckenridge. Heat transfer coefficients of humans in cold water. *J. Physiol.* (Paris) 63:459–462, 1971.
136. Woodcock, A.H. Wet-cold II. A theoretical interpretation of the sensation of damp cold experienced by cold man. *Tech Rept #64-06-001, Quartermaster Climatic Res. Lab. Lawrence, MA.* 40, pp. 1953.
137. Woodcock, A.H. Moisture permeability index, a new index for describing evaporative heat transfer through fabric systems. *Quartermaster Res. and Eng. Command Tech. Rept.,* EP-149:1-12, 1961.
138. Woodcock, A.H. Moisture transfer in textile systems, Part I. *Text. Res. J.* 32:628–633, 1962.
139. Woodcock, A.H., and J.R. Breckenridge. A model description of thermal exchange for the nude man in hot environments. *Ergonomics* 8:223–235, 1965.
140. Woodcock, A.H., and T.E. Dee. Wet-cold I. Effect of moisture on transfer of heat through insulating materials. *Tech Rept. #7-64-06-01. Quartermaster Climatic Res. Lab., Lawrence, MA.* 18 pp. 1950.
141. Woodcock, A.H., H.L. Thwaites and J.R. Breckenridge. Clothed Man. *Mech. Engr.* Aug., pp. 71–74, 1959.
142. Wissler, E.H. An analysis of heat stress in hyperbaric environments. In: C.E. Johnson, M.L. Nuckols, and P.A. Clow (eds.). *Hyperbaric Diving Systems and Thermal Protection.* New York, NY: Amer. Soc. Mech. Engrs., 1978, pp 53–74.
143. Wulsin, F.R. Adaptation to climate among non-European peoples. In: L.H. Newburgh (ed.). *Physiology of Heat Regulation and the Science of Clothing.* Philadelphia, PA: W.B. Saunders Co., 1949, pp. 3–69.

3

Physiological Responses to Acute Exercise-Heat Stress

MICHAEL N. SAWKA, Ph.D.

C. BRUCE WENGER, M.D., Ph.D.

OUTLINE

INTRODUCTION

In this chapter, we discuss the normal thermoregulatory, circulatory and metabolic responses of humans performing exercise in the heat. Muscular exercise increases metabolic rate above resting levels, thus producing a considerable amount of heat that must be dissipated to the environment to protect the health of the individual. In comparison with the broad temperature range for terrestrial climates (-88° to 58° C), the human body temperature is usually regulated within a narrow range (35° to 41° C) to preserve normal physiological function. The human possesses an elaborate set of mechanisms which regulates body temperature and includes two control systems: behavioral temperature regulation and physiological temperature regulation. Behavioral temperature regulation operates largely through conscious, willed behavior to employ any means available. Physiological temperature regulation operates through the autonomic nervous system, and includes control of: a) rate of metabolic heat production, b) heat flow via the blood from the core to the skin, and c) sweating. Even without heat stress, a relatively high cardiac output is required to maintain the high level of metabolism during exercise. When heat stress is combined with exercise, the cardiovascular system may be pushed close to its limit to simultaneously support the competing metabolic and thermoregulatory demands for blood flow. The following pages will consider human temperature regulation, control of heat dissipating responses and the problems associated with maintaining adequate cardiac output during exercise-heat stress.

THERMOREGULATORY CONTROL

Behavioral Temperature Regulation

Selection of a suitable micro-environment is the most primitive thermoregulatory response, and is seen in all vertebrates. Even fish and reptiles, which are poikilothermic ("cold-blooded"), will, given the opportunity, control their body temperatures by moving to a warmer or cooler micro-environment. Human beings have a wider range of options and by choice of clothing, which impedes heat loss from the body and thus increases thermal insulation, and by using shelter, ventilation and air conditioning, they can live for extended periods in the most extreme climates on earth. Thermal sensation and thermal discomfort presumably represent the direct motivation for behavioral thermoregulation, and in human subjects they can be measured by psychophysical means. Some workers (e.g., 31) have considered both physiological and behavioral thermoregulatory responses to be controlled in the same way. However, during transients in environmental temperature, thermal sensation and thermal discomfort appear to change much more quickly than do either core and skin temperatures (64, Chapter 2) or physiological thermoregulatory responses (34). Thermal sensation and discomfort thus seem to anticipate changes in the body's thermal state, perhaps by using nervous signals related to rate of change of skin temperature (see below). The anticipatory nature of thermal sensation and discomfort may be beneficial, since behavioral thermoregulation is accomplished by discrete responses, whereas physiological thermoregulation is accomplished by responses which occur continuously, and whose intensity is easily adjusted, as long as the thermal stress is present.

In human beings the drives for behavioral thermoregulation are often overridden by other factors. Thus, concern for personal appearance may cause people to dress inappropriately for their activity and thermal environment; and well-motivated individuals, such as athletes and military recruits, may exercise in hot environments to the point of endangering themselves.

Physiological Regulation

Among the most familiar examples of non-living regulatory systems are refrigerators and air conditioning systems for buildings, which are controlled by thermostats and operate so as to maintain the inside temperature at a relatively constant level. Although more sophisticated systems do exist, a typical refrigeration or air conditioning system can operate at only two levels of cooling, on and off, and is not capable of a graded response. When the inside temperature rises above the desired level, the thermostat turns on the compressor in the refrigeration unit; and when the temperature falls below that level, the thermostat turns the compressor off. (In practice the response of such a system lags somewhat, so that it is not constantly switching on or off in response to minute temperature changes.) Such a system is called an on-off system. The variable which a control system acts to maintain within narrow limits (in this example, temperature) is often called the regulated variable, and the quantities which it controls in order to accomplish this (in this example, rate of heat removal averaged over several on-off cycles) are called controlled variables.

Rather than operating only at one of two levels, on and off, most physiological control systems can produce a graded response according to the disturbance in the regulated variable. In many instances, changes in the controlled variables are proportional to displacements of the regulated variable from some threshold value, and such control systems are called proportional control systems. Figure 3-1 shows how reflex control of three heat-dissipating responses, sweating, skin blood flow, and forearm venous volume, depends on body core temperature (represented by esophageal temperature) and mean skin temperature. In Figure 3-1, each response has an esophageal temperature threshold. This threshold is the esophageal temperature at which the response starts to increase, and the threshold depends on the mean skin temperature. Note that the temperature controller is of the proportional type, and that these thermoregulatory responses depend on both core and mean skin temperatures: at any given skin temperature, each response is proportional to esophageal temperature, and increasing the skin temperature lowers the threshold level of esophageal temperature and increases the response at any given core temperature. (In addition to the role of skin temperature in the reflex control of thermoregulatory responses, skin temperature also has a local effect that modifies the local response to the reflex signal. This will be discussed later in the chapter.) Although the most accurate mathematical representation of the reflex control of the thermoregulatory effectors may well be a complicated function of tissue temperatures at a number of sites, and contain non-linear elements, it is convenient and useful to represent the control in terms of a linear combination of core temperature (T_c) and mean skin temperature ($\bar{T}_{sk}$), and to represent the control of each thermoregulatory effector response by an equation of the form:

$$R - R_o = a_1 \cdot T_c + a_2 \cdot \bar{T}_{sk} + b \qquad \text{(Eq. 3-1)}$$

where R is a thermoregulatory response;

R_o is a basal value of R;

and a_1, a_2, and b are constants.

In terms of Figure 3-1, a_1 is the slope of a particular $R:T_c$ relation and a_2 is the effect of a unit change in $\bar{T}_{sk}$ on the $R:T_c$ relation. The ratio $a_1:a_2$ is about 9:1 for all of the heat dissipating responses: sweating (96), vasodilation as represented by forearm blood flow (172), and release of vasoconstriction as represented by finger blood flow (171). (There are differences in nervous control between finger and forearm blood flow which are discussed in Control of Skin Blood Flow, below.)

The coefficients a_1 and a_2 represent the sensitivity of R to changes in T_c and $\bar{T}_{sk}$, respectively, and the ratio of 9:1 means that a change of 1° C in core temperature elicits about nine times as great a change in thermoregulatory response as a 1° C change in mean skin temperature. However, since skin temperature typically varies over a much wider range than does core temperature, the importance of

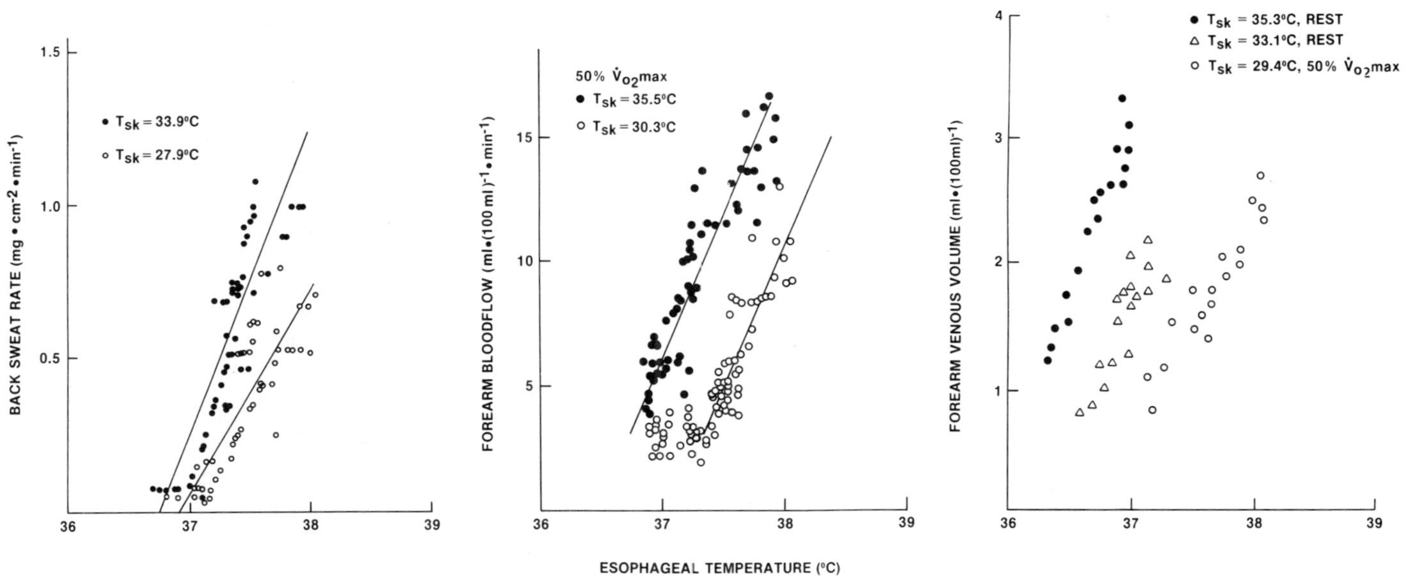

Figure 3-1. *The relations of back sweating rate (left), forearm blood flow (center), and forearm venous volume (right) to esophageal and mean skin temperatures (T_{es} and $\bar{T}_{sk}$). Sweating data are from four subjects performing cycle exercise at an O_2 consumption rate of 1.6 ℓ/min. Measurements of forearm blood flow and forearm venous volume are from one subject each. Forearm temperature was kept at 36.8° C during measurements of blood flow, and at 35.1° C during measurements of forearm volume to eliminate a difference in local temperature between experiments at different ambient temperatures. Local temperature was not controlled independently during measurements of sweating, so that the difference between conditions includes a small effect of local skin temperature, appearing as a difference in slope. Left panel drawn from data of Reference 139; center panel modified from Reference 172; right panel modified from Reference 170.*

skin temperature in thermoregulatory control is much greater than what the 9:1 ratio might suggest. The sensitivity of these thermoregulatory responses to core temperature allows the thermoregulatory system to keep core temperature relatively constant. Their sensitivity to mean skin temperature allows the system to respond appropriately to changes in environmental temperature over a wide range with little change in body core temperature. For example, the skin temperature of someone who enters a hot environment will rise and elicit sweating before any possible change in core temperature. On the other hand, an increase in heat production within the body, as occurs during exercise, has relatively little effect on skin temperature and elicits the appropriate heat-dissipating responses mostly in response to a rise in core temperature.

Once these thermoregulatory effector responses are sufficient to dissipate heat at the rate at which it is being produced, thermal balance is restored—i.e., the storage term in the heat balance equation (Chapter 1) is zero—so that core temperature stops rising and reaches a steady-state level. In the language of control theory, the rise in core temperature which elicits heat-dissipating responses sufficient to re-establish thermal balance is called a "load error." Even in a cool environment, core temperature must rise during exercise in order to increase skin blood flow to carry heat at a greater rate to the skin, whether or not sweating also is required. In a cool environment, the rate of skin blood flow occurring at a given level of core temperature will be less than in a warm environment, because of the lower mean skin temperature; but the cooler skin will also remove more heat from each liter of skin blood flow (see Heat Transfer From Skin Blood Flow, below), thus allowing the necessary core-to-skin heat transfer to be achieved with a lower rate of skin blood flow than in a warm environment.

During exercise the increase in core temperature is proportional to the metabolic rate, and nearly independent of environmental temperature over a fairly wide range (Fig. 3-6). This was first observed by Marius Nielsen in 1938, in experiments in which subjects exercised for one hour in ambient temperatures from 5 to 36° C. He showed that with increases in ambient temperature, reductions in heat loss by radiation and convection (dry or "sensible" heat loss) were exactly balanced by increases in evaporative cooling ("insensible" heat loss) (107) (Fig. 3-5). Nielsen concluded that core temperature increased during exercise as the consequence of "a change in the regulatory adjustment of body temperature", and not of an insufficient ability to dissipate heat, as had previously been thought (108). His conclusion has often been understood to mean that the body's "thermostat" is re-set at a higher level during exercise, or equivalently that the core temperature thresholds for the thermoregulatory effector responses are elevated, as they are in fever (53,94). However, it is unnecessary to invoke any such threshold changes, since even without them an increase in core temperature is necessary to evoke heat-dissipating responses during exercise, as we have already shown. (We will comment further on the difference between the rise in core temperature with exercise and that with fever after developing the concept of thermoregulatory "set point".)

Besides fever, heat acclimation (Chapter 4), phase of the menstrual cycle, and time of day (Chapter 7) change the core temperature thresholds for the thermoregulatory responses. Since each of these factors has been shown to shift thresholds for several different thermoregulatory responses by the same amount (see References 56,75 for further discussion) it is useful to introduce the concept of a thermoregulatory "set point", which serves as a reference temperature in the control of all the thermoregulatory responses, and which is changed by these factors. A change in set point, then, produces a corresponding change in the threshold for each thermoregulatory response. Figure 3-2 shows schematically the difference between the rise in core temperature with exercise and that with fever. Conceptually, the difference between fever and exercise in Figure 3-2 is that the former involves an increase in thermoregulatory set point, and the latter does not (149). The observations on which this conceptual difference is based are these: First, although heat production sometimes increases substantially (through shivering) early during fever when core temperature is rising, it does not need to stay high to maintain the fever, and in fact it returns nearly to pre-febrile levels once the fever is established. During exercise, on the other hand, an increase in heat production not only causes the elevation in core temperature, but is necessary to sustain it. Second, for almost the whole time that core temperature is rising during fever, rate of heat loss is usually lower (and is never higher) than it was before the fever began. During exercise, however, the heat dissipating responses, and thus rate of heat loss, start to increase early and continue increasing, in response to rising core temperature.

The function of the human thermoregulatory system is shown schematically in Figure 3-3. Certain factors that act peripherally on the sweat glands and blood vessels, and are associated with local temperature and heat acclimation, are omitted here, but will be discussed later in this chapter and in Chapter 4. The scheme presented in Figure 3-3 presumes that there are thermal receptors in the skin and body core which send information about their temperatures to some central integrator, which generates a "thermal command signal" that participates in the control of sweating, vasodilation, and

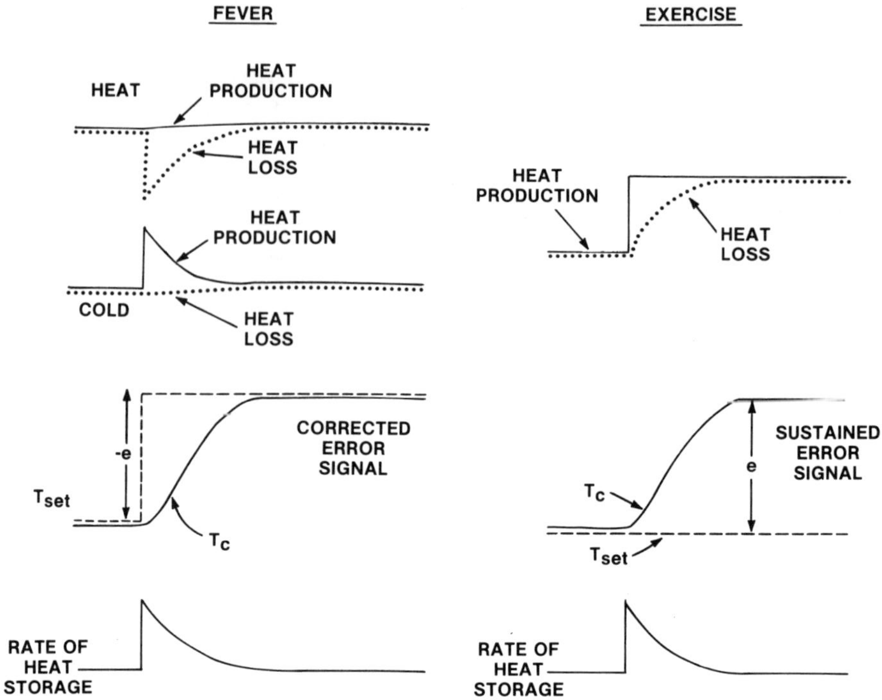

Figure 3-2. *Differences between the elevation of core temperature in fever and the elevation of core temperature in exercise. In fever, the primary event is a rise in set point temperature (T_{set}), which initially causes a negative load error or error signal ($-e$). Heat-dissipating responses are inhibited and heat production is stimulated until T_c rises enough to correct the error signal and establish a new thermal balance in which heat production and heat loss are near (or slightly above) their values before the fever. During exercise T_c rises because of the increase in heat production. T_{set} is unchanged, and heat-dissipating responses are elicited as T_c rises. T_c continues to rise until it is high enough that the error signal (e) elicits heat loss responses at a rate sufficient to match heat production, and produce a new thermal balance. Figure reproduced from Reference 149, with permission.*

vasoconstriction. The notion of such a thermal command signal is supported by the evidence for a common thermoregulatory set point, mentioned above, and by evidence that core and skin temperature are combined in the same way in the control of sweating, vasodilation, and vasoconstriction (56). Division of the thermal receptors in the body into those in skin and those in core is convenient and useful in practice, but it is somewhat artificial since, as discussed later in the chapter, neither the skin nor the core has a uniform temperature most of the time. The thermal receptors in the core (or at least those that participate in the control of thermoregulatory responses) are very unevenly distributed, and in laboratory mammals are concentrated especially in the hypothalamus, which also is where much of the integration occurs (15). Changes of only a few tenths of 1° C in the temperature of the anterior preoptic area of the hypothalamus elicit changes in the thermoregulatory effector responses, and this area contains many neurons which increase their firing rate in response either to warming or to cooling. It is not known how much thermal receptors inside the core but outside the hypothalamus contribute to thermoregulation, but such thermal receptors have been reported in a number of other sites, including the hearts of rabbits (41), pulmonary vessels of sheep (14), and spinal cords of dogs (159). It is presumed that the human hypothalamus also has a high concentration of thermal receptors, although there is no direct evidence on this point. There is evidence, however, for a high concentration of thermal receptors somewhere in the distribution of the carotid arteries, since infusing hot saline solution into the carotid artery causes more vasodilation in the hand than do infusions into other arteries (32).

Since skin temperature generally is not uniform, the contribution of skin temperature to thermoregulatory control is usually expressed in terms of an appropriately weighted mean skin temperature. Neurophysiological studies show three types of thermal receptors in skin: a) One type responds to heating with a transient burst of activity and an increase of static activity (63,69); b) one type responds to cooling with a transient burst of activity and an increase of static activity (63,69); and c) one type responds to several stimuli, including warming, with a continuous discharge (63). The transient discharges that occur at the beginning of heating or cooling give the central integrator information about changes in skin temperature as they are occurring, and this feature may account for the sensitivity of sweating not only to skin temperature, but also to the rate at which skin temperature is changing (96).

In summary, if some disturbance, such as an increase in metabolic heat production or environmental temperature, upsets the thermal balance and causes additional heat to be stored in the body, temperatures in the core or skin (or both) will increase, and changes in these temperatures will be detected by the thermal receptors. In response to information from these receptors, the thermoregulatory controller in the central nervous system will call for responses that increase heat loss. Unless the heat stress exceeds the capacity of the thermoregulatory system, these responses will increase until they are sufficient to restore heat balance and prevent further increases in body temperatures. A parallel argument applies if the disturbance causes a net heat loss from the body, except that besides decreasing heat loss, the body can also increase heat production to restore heat balance (Chapter 9).

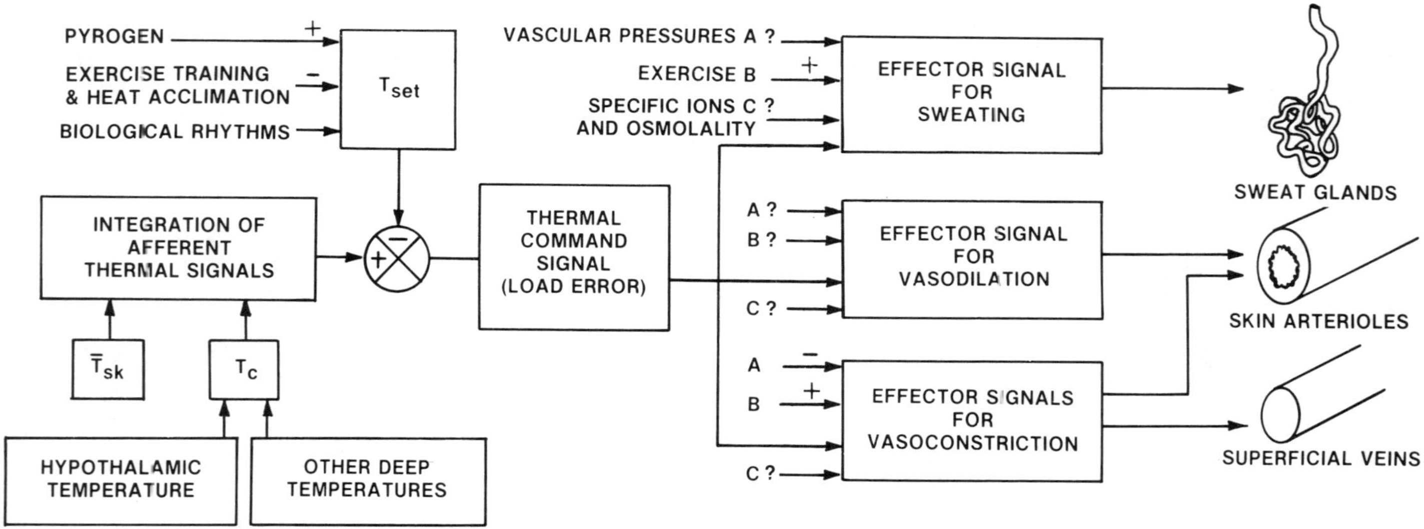

Figure 3-3. *Schematic model of the control of human thermoregulatory effector responses.*

The thermoregulatory system both controls responses that affect body temperatures and receives information about body temperatures. It therefore is said to receive "feedback" about the effect of its actions on body temperatures; and since the system responds to temperature changes in such away as to *oppose* further temperature changes in the same direction, it is said to be a "negative feedback" system. Negative feedback is not a characteristic only of the thermoregulatory system; it is an essential element of any regulatory system, in the sense in which the concept of regulation has been developed in this chapter.

BODY TEMPERATURES

Core Temperature

Fundamental to the study of human temperature regulation is the accurate measurement of deep body or core temperature. Core temperature is measured during physical exercise, to either estimate average internal temperature or provide information concerning the core temperature input to the thermoregulatory controller. It might seem that this basic measurement should by now, more than a century since the pioneering work of Claude Bernard (10), be well defined and broadly accepted. Instead, there is still debate among thermal physiologists concerning the best measurement site and the correct interpretation of the different indices of core temperature (16).

The idea that a single temperature measurement can provide an index of the average internal temperature may not be reasonable. By 1950, it was clear to several investigators that the notion of a single measurement site representing the average internal temperature was fictitious (8,9,71,89). In fact, Mead and Bonmarito (89) stated that "it is apparent that no single regional temperature could indicate, except by chance, the average of all internal temperatures." The idea that a single temperature measurement can provide a close approximation of the core temperature input to the thermoregulatory controller has several problems, both theoretical and practical. First, the thermoregulatory controller receives temperature input from many core sites (56). Second, although the hypothalamic temperature is believed to be a major input into the thermoregulatory controller, neither brain temperature (175) nor hypothalamic temperature has been well quantitated in humans.

The temperature within a given body region depends upon: (a) the metabolic rate of the surrounding tissues, (b) the source and magnitude of blood flow, and (c) the temperature gradients between contiguous body regions. Considerable temperature gradients exist between and within different orifices, body cavities and blood vessels (45,71,90,147). For resting humans, about 70% of the metabolic heat is produced by internal organs and viscera within the body core (4,151). During muscular exercise, however, up to about 90% of the metabolic heat can be produced by the skeletal muscles. Because of the different source of metabolic heat during exercise as compared to rest, temperatures measured during exercise within a given body region may change relative to other body regions (3,90,129,130). For example, during rest in a comfortable environment, skeletal muscle temperature is lower than core temperature; however during exercise, the temperature within the active skeletal muscle exceeds core temperature (3,129,130), while the temperature within the inactive skeletal muscles usually does not increase (3). Blood perfusing the active skeletal muscles is warmed, and the warmer blood carries heat to other body regions and consequently core temperature is elevated (3,90).

The requirements for an ideal site to measure core temperature are: (a) the measurement site should be convenient, (b) the measurement should not be biased by environmental conditions, and (c) the measured changes should quantitatively reflect small changes in arterial blood temperature (29,147). In addition, the response time for temperature changes is an important consideration when selecting a measurement site for certain experimental situations, especially those involving exercise. A subject's core temperature is often measured at the esophagus, rectum, mouth and tympanum/auditory meatus. The relative advantages and disadvantages of these and other measurement sites are well understood (3,30,33,54,85,130). During the subsequent paragraphs these core temperature measurement sites will be systematically evaluated for their compliance with the stated requirements for an ideal site.

Esophageal Temperature. Esophageal temperature is obtained by inserting a catheter, containing a thermocouple or thermistor, through the nasal passage and into the throat and then swallowing it. Since some individuals find this procedure to be uncomfortable, a topical anesthetic, such as lidocaine, is often used on the catheter and/or in the throat to alleviate some discomfort. It is important to use a fairly stiff plastic catheter which will not easily curl back on itself when advanced within the esophagus. At the level of the left atrium, the heart and esophagus are in contact and are fairly isothermal (16). If the catheter is passed too far, such as into the stomach, the temperature and response times are altered; likewise, if the catheter is too high, the pulmonary ventilation will cool or heat (dependent upon ambient conditions) the temperature sensor (16,33,73,174). The proper catheter length can be predetermined from anatomical measurements and may be verified by a radiogram, an acoustical signal, or by monitoring an electrocardiogram from the catheter tip (18). In our laboratory, an anatomical measurement of one-fourth the subject's height (R.R. Gonzalez, unpublished anatomical measurements of cadavers) is used as the catheter distance from the nostril to the heart level within the esophagus. A final methodological consideration is the avoidance of swallowing during measurement, as the passage of saliva will spuriously lower the esophageal temperature values. Swallowing can be avoided by having the subject spit or by using a suction device to remove saliva from the mouth (16).

Most thermal physiologists agree that esophageal temperature is the best non-invasive index of core temperature for humans. Several investigators have found that esophageal temperature responds rapidly to changes in blood temperature elicited by extracorporeal circulation (93,147), and by body cooling during anesthesia (30). For example, Shiraki *et al.* (147) found that esophageal temperature quantitatively reflected pulmonary artery (mixed venous blood) temperature with an average difference of 0.1° C and a lag time of about one minute during their hyperthermia experiments. Therefore, esophageal temperature provides a good index of blood temperature. Several investigators have simultaneously measured esophageal temperature and the temperature of active skeletal muscles during exercise (3,128,129,130). Each study demonstrated that during exercise esophageal temperature and active muscle temperatures achieved steady-state values in 15–20 minutes. The rapid response time for esophageal temperature is due to the low heat capacity of the esophagus and its proximity to the heart; therefore, it can be rapidly warmed or cooled.

Rectal Temperature. Rectal temperature is widely employed by physiologists because it is a comfortable and non-invasive measurement site. Rectal temperature is obtained by inserting a temperature sensor a minimum of five centimeters past the anal sphincter. Temperature measurements are uniform within the rectum from 5–27 centimeters past the anal sphincter (1,9,76,106). One early study (89) reported that rectal temperature values varied by as much as 0.8° C depending upon the insertion depth; this variability apparently resulted from the use of a very thin flexible catheter which caused great variability for temperature sensor placement within the pelvis. The use of a more rigid rectal catheter seems to solve this problem (1,9,76,89,106). Another potential problem is that a properly placed temperature sensor may later slip to less than five centimeters beyond the anal sphincter during muscular exercise. This potential problem can be solved by attaching a bulb to the rectal catheter at the desired insertion depth, that will abut the anal sphincter.

Rectal temperature values are generally higher than values measured in arterial blood (7,45) and other core temperature sites. During exercise it takes approximately 25–40 minutes to achieve steady-state rectal temperature values (3,60,90,106,128,129). These steady-state values are generally ~0.2° C higher than simultaneously measured esophageal temperature values (33,106,129,130) and they are independent of the environmental temperature (60,99,152). As a result, the steady-state rectal temperature provides a good index to assess body heat storage (130,154). The main problem with rectal temperature is that it is slow to respond to changes in blood (90) and other core temperatures. Figure 3-4 provides a representative record of rectal and esophageal temperature responses to two exercise bouts spaced by a rest period. Note that the slow response results in rectal temperature being less than esophageal values during periods of rapid heat storage, and greater than esophageal values during periods of rapid heat loss. The reason for the slow response of rectal temperature to thermal transients is probably a low rate of blood flow to the rectum compared to other measurement sites (7,93). The slow response time makes rectal temperature a poor core temperature index for estimating the input to the thermoregulatory controller (44,54,129).

Oral Temperature. Physicians routinely measure oral temperature to determine the presence of fever. For this, a thermometer is placed sublingually and allowed to equilibrate prior to recording of the temperature. The equilibration time necessary to obtain a stable oral temperature value ranges from 3 to 10 minutes and depends upon the ambient temperature (85,154). During the equilibration and measurement period the subject must breathe through the nose, as ventilation from mouth breathing can artificially lower oral temper-

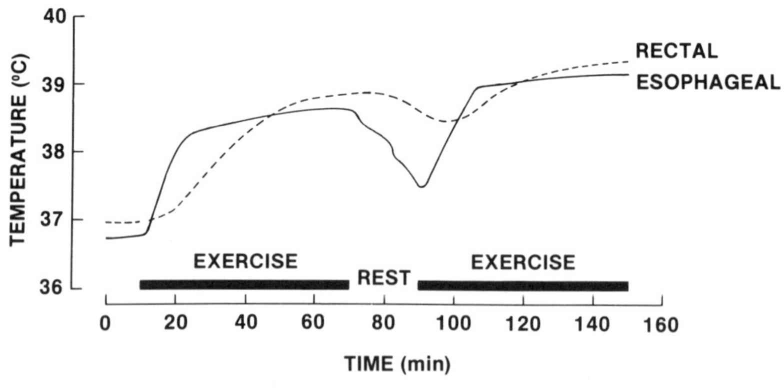

Figure 3-4. *Rectal and esophageal temperature responses to rest and exercise in the heat.*

ature values in cold environments and artificially elevate oral temperature values in hot environments. Some investigators have sealed their subjects' mouths to insure nasal breathing while obtaining oral temperature measurements (33,54). Finally, both swallowing and the ingestion of fluid must be avoided as they will bias oral temperature values (33,153).

The tongue is perfused by the lingual branch of the external carotid artery and has a high blood flow per gram of tissue (68). This enables the tongue to be an effective heat exchanger and allows sublingual temperatures to approximate blood temperature (72); however, changes in head and face skin temperatures will bias oral temperature (88,148). For example, McCaffrey *et al.* (88) measured oral temperature while cooling and heating areas of the head and neck. The oral temperature values followed the changes in cutaneous temperature of the head. Also, in cold environments, cheek cooling will affect the superficial segment of the parotid salivary duct and cool the saliva reaching the mouth, thus biasing the temperature measurement (148). Therefore, unlike esophageal and rectal temperature values, oral temperature values are not always independent of the environmental temperature.

Under favorable conditions, oral temperature values are very similar to esophageal temperature values (33,43,54,85). During steady-state light intensity exercise, which can be performed with nasal breathing, oral temperature values have been found to correlate well ($r = 0.83$–0.92) with simultaneously measured esophageal and rectal temperature measurements (85,153). During steady-state moderate intensity exercise, which can be performed with nasal breathing, oral temperature values may be lower than esophageal and rectal temperature values, perhaps due to a sympathetically mediated vasoconstriction of the tongue vasculature (85,154). Another possibility is that additional cooling of the upper respiratory passages may occur at greater exercise intensities (154).

Tympanic/Auditory Meatus Temperature. Tympanic temperature is obtained by inserting a small temperature sensor into the ear canal and advancing it until it rests against the tympanic membrane. Proper placement is determined by the subject's hearing a sound when the temperature sensor touches the tympanic membrane. Some subjects find this contact to be uncomfortable (16). In addition, there are reports of the temperature sensor's perforating the tympanic membrane (38,155,163). Because of the potential discomfort and trauma as well as placement problems associated with tympanic measurements, some investigators have chosen instead to measure the temperature of the external auditory meatus. For this measurement, a temperature sensor is placed in an ear plug and inserted into the external auditory meatus. Placement of the temperature sensor is important since there is a substantial (~0.5° C) temperature gradient along the wall of the meatus (29).

Tympanic/auditory meatus temperature measurements do not provide a reliable index of the level of core temperature during either rest or exercise (29,60,86,87,88,99). Depending upon the environmental conditions, tympanic/auditory meatus temperature values can be lower or higher than simultaneously measured steady-state rectal (60,99) and esophageal (86,87,88) temperature values. In addition, local head heating and air flow to the face will bias the temperature of the external meatus (86,87,88). Therefore, tympanic/auditory meatus temperature is not independent of the environmental conditions. Since the tympanic/auditory meatus temperature values are biased by facial skin temperature, the measured temperature is the sum of a variable weighting (depending upon the environmental condition) of skin and core temperature (60,88,99). However, if ambient temperature is constant and neither esophageal or oral measurements can be obtained, the rapid change in tympanic/auditory meatus temperature can be of some value in estimating the core temperature input to the thermoregulatory controller during exercise (113).

Skin Temperature

Skin temperature is measured for the purposes of: (a) calculating the mean body temperature for heat storage determinations, (b) calculating sensible (radiative and convective) heat exchange and skin conductance, and (c) integrating into an index of the skin temperature input to the thermoregulatory controller. Although the skin surface is easily accessed (unlike the core), measurement problems often occur because the skin represents the boundary between two media, tissue and the ambient air. As a result, changes in skin temperature might result from physiological adjustments (cutaneous blood flow, sweat evaporation) or alterations in the environment (air motion, temperature, radiation). Generally, skin temperatures are measured from temperature sensors in contact with the skin's surface or from non-contact infrared methods. For the former method, care needs to be taken to insure that the temperature sensor remains in good thermal contact with the skin, since otherwise the measurement will be biased by the ambient temperature.

Although a single skin temperature measurement can be useful for biophysical calculations, thermal physiologists are more often interested in the average or mean skin temperature (65,177). The mean skin temperature ($\bar{T}_{sk}$) represents the sum of weighted individual skin temperatures. Generally, the weighting is based on the percentage of body surface area that is represented by the body region from where the temperature is measured (91). For example, Hardy and DuBois (65) divided the body into 12 regions while Winslow and colleagues (177) divided the body into 15 regions for skin temperature measurements. Numerous investigators have attempted to minimize the number of measurement sites necessary to obtain a valid estimate of mean skin temperature (115,158). Mitchell and Wyndham (91) compared nine of these shortened equations which are used to estimate mean skin temperature. They recommended that when it is difficult to measure a large number of sites, the equation developed by Ramanthan (115) be used, where:

$$\bar{T}_{sk}, {}^{\circ}C = 0.3\,(\text{chest} + \text{upper arm temperatures}) + 0.2\,(\text{thigh} + \text{calf temperatures}) \quad \text{(Eq. 3-2)}$$

Mean skin temperature values that are based on regional weighting according to the percentage of body surface area are particularly useful for the calculation of mean body temperature and heat storage.

Nadel and colleagues (101) have developed a novel approach to determine the regional weightings for the mean skin temperature calculation. Their approach was to base the regional weighting on the skin's thermal sensitivity and not on its percentage of body surface area. These investigators reasoned that thermal receptors are not evenly distributed over the skin's surface, and warming of a body region having the greatest number of thermal sensors would have the greatest influence on the thermoregulatory effector responses. Nadel and colleagues had subjects rest in a range from warm to hot environments while they applied thermal radiation to selected skin areas and measured the change in sweating rate at the thigh. Based upon the thermal sensitivity data (change in thigh sweating rate), which was adjusted for regional surface area, they developed the following equation:

$$\begin{aligned}\bar{T}_{sk}, {}^{\circ}C = {} & 0.21\,(\text{face temperature}) \\ & + 0.21\,(\text{mean of chest and back temperatures}) \\ & + 0.17\,(\text{abdomen temperature}) \\ & + 0.15\,(\text{thigh temperature}) \\ & + 0.08\,(\text{calf temperature}) \\ & + 0.12\,(\text{upper arm temperature}) \\ & + 0.06\,(\text{forearm temperature})\end{aligned} \quad \text{(Eq. 3-3)}$$

Subsequently, Libert *et al.* (81) conducted a similar study in which they heated larger skin areas by smaller temperature increments but for greater durations than in the experiments of Nadel and colleagues. Libert *et al.* (81) concluded that the weighting factors based on thermal sensitivity were not different from the weightings based on percent of body surface area. It seems that more research is needed concerning the use of a thermal sensitivity approach for calculation of the mean skin temperature. Both of these studies were performed on semi-nude and resting subjects (81,101), and the results obtained may not be directly applicable to clothed or exercising subjects. Finally, the thermal sensitivity approach may eventually become the preferable index for estimating the skin temperature input to the thermoregulatory controller.

CORE TEMPERATURE RESPONSES TO EXERCISE

Environment and Exercise Intensity

As depicted in Figure 3-4, during dynamic exercise core temperature initially increases rapidly and subsequently increases at a reduced rate until heat loss equals heat production and essentially steady-state values are achieved. The elevation of core temperature represents the storage of metabolic heat which is a by-product of skeletal muscle contraction. At the initiation of exercise, the metabolic rate increases immediately; however, the thermoregulatory effector responses for heat dissipation respond more slowly. The thermoregulatory effector responses which enable sensible (radiative and convective) and insensible (evaporative) heat loss to occur, increase inproportion to the rate of heat production (103). Eventually, these heat loss mechanisms increase sufficiently to balance metabolic heat production allowing a steady-state core temperature to be achieved.

During muscular exercise, the magnitude of core temperature elevation at steady-state is largely independent on the environmental condition and is proportional to the metabolic rate (58,82,107, 108,152). These concepts were first presented by Nielsen (107) who had three subjects perform exercise at several intensities (up to ~3.0 $\ell\ 0_2 \cdot min^{-1}$) in a broad range (5° to 36° C with low humidity) of ambient temperatures. Figure 3-5 presents the heat exchange data for one subject during an hour of cycle exercise at a power output of 147 watts and at an aerobic metabolic rate of approximately 650 watts. The difference between metabolic rate and total heat loss represents the energy used for mechanical work and the heat storage. Note that the total heat loss and, therefore, the heat storage and elevation of core temperature were constant for each environment. The relative contributions of sensible and insensible heat exchange to the total heat loss, however, varied with the environmental conditions. In the 10° C environment, the large skin-to-ambient temperature gradient facilitated sensible heat exchange which accounted for ~70% of the total heat loss. As the ambient temperature increased, this gradient for sensible heat exchange diminished and there was a greater reliance upon insensible heat exchange. When the ambient temperature is equal to skin temperature, insensible heat exchange will account for essentially all of the heat loss. In addition, when the ambient temperature exceeds the skin temperature, there is a sensible heat gain to the body.

Nielsen's finding that the magnitude of core temperature elevation is independent of the environmental conditions, is inconsistent with the personal experience of most athletes. For example, a runner will certainly experience greater hyperthermia if he/she competes in a 40° C than in a 20° C environment (119). Lind (82) has shown that the magnitude of core temperature elevation during exercise is independent of the environment only within a range of conditions or a "prescriptive zone." Figure 3-6 presents a subject's steady-state core temperature responses during exercise performed at three metabolic intensities while in a broad range of environmental conditions. The environmental conditions are represented by the "old" effective temperature, which is an index that combines the effects of dry bulb temperature, humidity and air motion. Note that during exercise, the greater the metabolic rate, the lower is the upper limit of the prescriptive zone. In addition, Lind found that

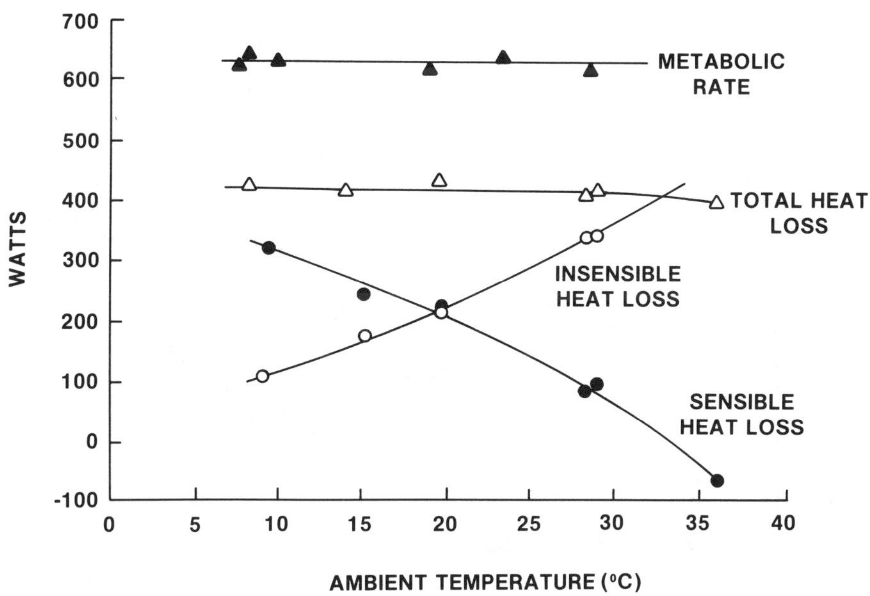

Figure 3-5. *The heat exchange data averaged over one hour for one subject performing constant intensity exercise in a variety of ambient temperatures. The difference between metabolic rate and total heat loss is the sum of mechanical power (147 W) and mean rate of heat storage. Redrawn from Reference 107.*

even within the prescriptive zone there was a small but significant positive relationship between the steady-state core temperature and the old effective temperature. It seems fair to conclude that throughout a wide range of environmental conditions, the magnitude of core temperature elevation during exercise is largely but not exclusively independent of the environment. During exercise with a substantial combined metabolic and environmental heat stress, the thermoregulatory load error is great enough to result in further elevated steady-state core temperature levels. In addition, in humid environments the evaporative capacity of the environment may be insufficient to dissipate the individual's metabolic heat from exercise and result in further elevated core temperature levels (58).

As stated, within the prescriptive zone the magnitude of core temperature elevation during exercise is proportional to the metabolic rate (107,130,152). For any individual, the greater the metabolic rate, the higher the individual's steady-state core temperature during exercise (129). The relationship between metabolic rate and core temperature is good for a given individual, but does not always hold well for comparisons between different individuals. Åstrand (6) first reported that the use of relative intensity (percent of maximal oxygen uptake), rather than actual metabolic rate (absolute intensity),

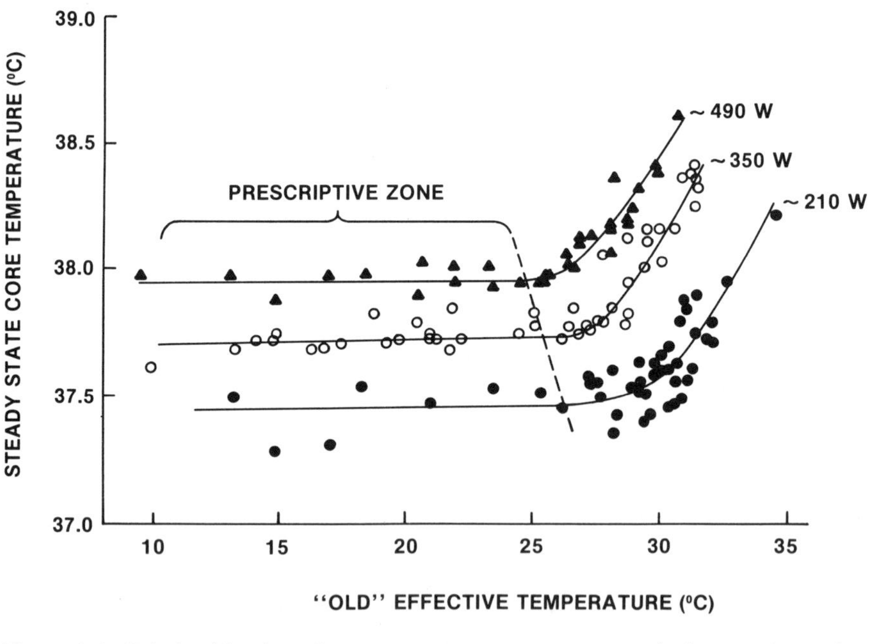

Figure 3-6. *Relationship of steady-state core temperature responses during exercise at three metabolic rates to the environmental conditions. Redrawn from Reference 82.*

removes most of the inter-subject variability for the core temperature elevation during exercise. Figure 3-7 shows the relationship of steady-state core temperature to both absolute and relative intensities in a moderate environment (130).

Davies and colleagues (35,36,37) attempted to clarify the relationship between relative intensity and the steady-state core temperature response to exercise. Their investigations were performed in environments with dry bulb temperatures ranging from 5° to 25° C (relative humidity is <50%) and with the subjects performing exercise at relative intensities ranging from ~20% to ~90% of their maximal oxygen uptake. Figure 3-8 presents data redrawn from their studies; note that for a group of laboratory and field experiments (Figure 3-8A) there was a curvilinear relationship between steady-state core temperature and relative intensity (37), described by the quadratic equation:

$$T_c, °C = 37.25 - 0.00264\ (\%\dot{V}O_2max) + 0.00037\ (\%\dot{V}O_2max)^2.$$

Figure 3-8B presents two subjects' steady-state core temperature values during exercise, at 65% and 85% of their maximal oxygen uptake, in relation to the ambient dry bulb temperature (35). Core temperature was found to be independent of dry bulb temperature from 5 to 20° C at the 65% relative intensity; however, core temperature was influenced by dry bulb temperature at the 85% relative intensity. These data are consistent with those of Lind (82), showing that the prescriptive zone is smaller at higher metabolic rates.

The preceding investigations have shown that in humans, core temperature changes are related to the relative exercise intensity. It

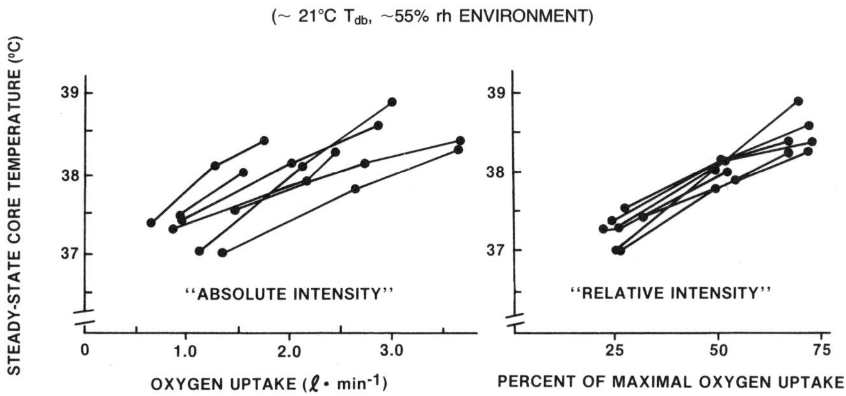

Figure 3-7. *Relationship of steady-state core temperature responses during exercise to the absolute and relative metabolic rates. Redrawn from Reference 130.*

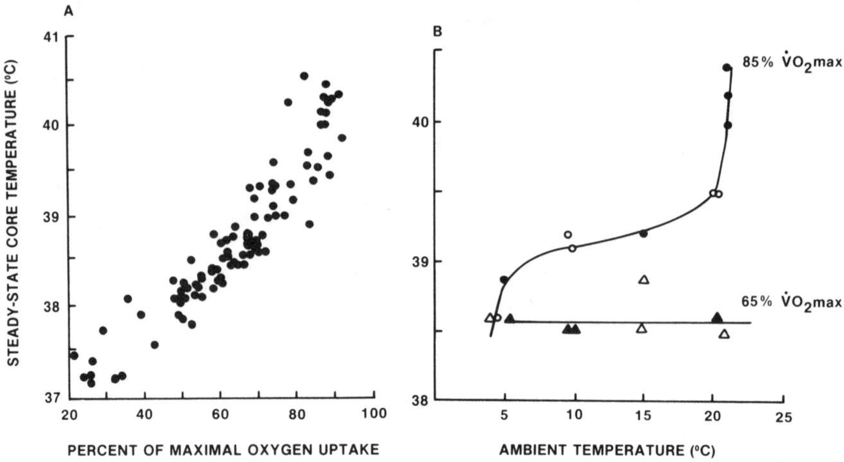

Figure 3-8. *Relationship of steady-state core temperature responses during exercise to the relative exercise intensity and the ambient temperature. Panel 3-8A is redrawn from Reference 37 and panel 3-8B is redrawn from Reference 35.*

would, therefore, seem logical to expect that any condition lowering maximal oxygen uptake (and thus increasing relative intensity) would also elicit an elevated core temperature response at a given absolute (thus higher relative) intensity. This hypothesis has been tested by several investigators with differing results (61,79,105,127). Greenleaf *et al.* (61) lowered three subjects' maximal oxygen uptake by simulating altitudes of 2,000 m (−11%) and 4,000 m (−32%) in a hypobaric chamber. Their subjects' core temperature responses during cycle exercise at a given metabolic rate were similar whether exposed to sea level or to simulated altitude. However, their subjects' resting core temperature values decreased with simulated altitude, and there may have been a greater increase in core temperature from rest to exercise at the 4,000 m than sea level experiments. Rowell *et al.* (127) lowered six subjects' maximal oxygen uptake (−27%) by having them breathe hypoxic (11–12% O_2) gases during cycle exercise in a comfortable environment. They found similar steady-state core temperatures during exercise at a given metabolic rate whether the subjects were breathing normoxic or hypoxic air. The core temperature at rest, however, was only measured with the subjects' breathing normoxic air. In contrast, Nielsen (105) found that lowering of maximal oxygen uptake (−15 to −28%) by carbon monoxide poisoning resulted in a higher core temperature during cycle exercise at a given metabolic rate, but the relation to relative intensity was essentially unchanged. Finally, Kolka *et al.* (79) had eight subjects perform cycle exercise in a hypobaric chamber at sea level, and then at simulated altitudes of 2,600m and 4,600m. Their subjects' maximal oxygen uptakes were reduced from sea level values by −8% (2,600 m) and −19% (4,600 m). They found that the increase in core temperature from rest to exercise was dependent upon the relative intensity.

It would also seem logical to expect that any condition increasing maximal oxygen uptake (thus lowering relative intensity) might elicit a lowered core temperature at a given metabolic rate. Two studies have used autologous erythrocyte infusion to increase subjects' maximal oxygen uptake (comfortable environment) and subsequently measured their core temperature responses to exercise in the heat (138,140). The first study (138) used subjects who were unacclimated to the heat, and found that steady-state core temperatures at a given metabolic rate were not significantly reduced after erythrocyte infusion. However, the calculated heat storage values were generally lower during the post- than during the pre-infusion heat stress tests. The second study (140) used heat-acclimated subjects, and found lowered steady-state core temperature responses at a given metabolic rate during the post- compared to the pre-infusion heat stress tests.

In conclusion, during muscular exercise the core temperature response is proportional to the metabolic rate. If the metabolic rate is expressed as a percentage of the individual's maximal oxygen uptake, then much of the inter-subject variability disappears for the magnitude of steady-state core temperature values. Acute changes in maximal oxygen uptake, however, may or may not modify the steady-state core temperature response at a given metabolic rate. There is some evidence, however, that the increase in core temperature from rest to exercise might be related to the relative intensity during simulated altitude experiments. The inconsistent results between investigations probably reflect differences in subject status (e.g., fitness, acclimation state) and/or differences in the methods employed to acutely alter maximal oxygen uptake.

Exercise Type

All of the investigations discussed so far have employed leg (lower body) exercise (treadmill or cycle). There is debate as to whether upper body exercise (e.g., arm-crank) results in different core temperature values than those elicited by lower body exercise at the same metabolic rate. Maximal effort arm-crank exercise elicits an oxygen uptake that is approximately 70% of that obtained during maximal effort cycle exercise. It is uncertain whether the core temperature response during arm-crank exercise would be determined by the relative intensity of the upper or lower body muscle mass. If these responses are determined by the relative intensity (with respect to the musculature employed) then arm-crank exercise would be expected to elicit a higher core temperature for a given metabolic rate than would lower body exercise.

In 1947, Asmussen and Nielsen (5) studied two subjects' core temperature responses to arm-crank and cycle exercise at the same metabolic rates. They found that arm-crank exercise elicited lower rectal temperature values than cycle exercise. In addition, this difference widened as metabolic rate increased. Since the authors were concerned that rectal temperature may have been spuriously high during cycle exercise because of the warm venous blood returning from the leg muscles, they conducted additional experiments in which they monitored stomach temperature. In agreement with their rectal temperature data, the stomach temperature values were consistently lower during arm-crank than cycle ergometer exercise. In 1968, Nielsen (104) examined the core temperature responses of two subjects during arm-crank and cycle ergometer exercise. Rectal temperature values (mean of values obtained at four depths ranging from 12–27 cm past the anal sphincter) were found to be consistently lower during arm-crank than cycle exercise at a given metabolic rate. In contrast, esophageal temperature values appeared to be similar for the two exercise types at a given metabolic rate (Figure 3-9).

In 1984, Sawka *et al.* (142) measured nine subjects' rectal temperature responses during arm-crank and cycle exercise performed at similar absolute and relative intensities. During the experiments matched for absolute intensity, the subjects' steady-state rectal temperature values were the same for both exercise types. On the other hand, during the experiments matched for relative intensity, the subjects' rectal temperature values were lower during arm-crank than cycle exercise. Subsequently, in a different group of experiments, these investigators (139) found that their subjects' steady-state esophageal temperature values were the same for arm-crank and cycle exercise at the same metabolic rate.

It seems clear that during upper body exercise the magnitude of core temperature elevation is determined by the metabolic rate and not the relative exercise intensity. Several early studies (5,104) reported that rectal temperature values were lower during arm-crank than cycle exercise at a given metabolic rate. Those findings suggest that rectal measurements may not provide the best index of core temperature during upper body exercise. Later studies, however, reported no difference for rectal (142) or esophageal temperature (104,139) responses between the two exercise types at a given met-

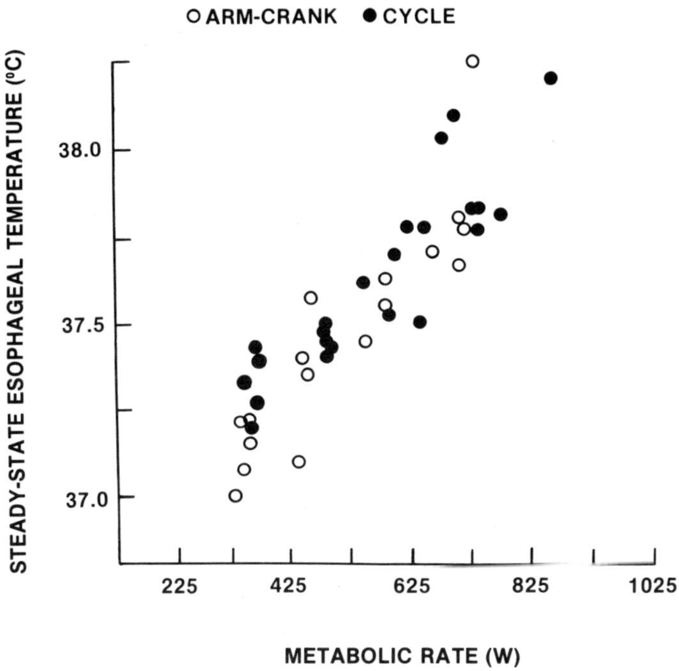

Figure 3-9. *Relationship of steady-state esophageal temperature responses to arm-crank and cycle exercise at a given absolute metabolic rate. Redrawn from Reference 104.*

abolic rate. The apparent discrepancy concerning the differences in rectal temperature responses may reflect methodological differences between the studies.

METABOLISM

Metabolic Rate

As defined by the heat balance equation (Chapter 1), body heat storage is equal to the difference between metabolic heat production and the heat loss to the environment. Physical exercise can increase total body metabolism by 5–15 times the resting rate in order to provide energy for skeletal muscle contraction. Depending upon the exercise task, between 70 and 100% of the metabolic rate is released as heat and needs to be dissipated in order to restore body heat balance.

The effects of acute heat stress on man's ability to achieve maximal aerobic metabolic rates during exercise have been well studied. Most investigators find that maximal oxygen uptake is reduced in a hot as compared to a comfortable environment (77,126,128,143,145), but some investigators report no differences (124,176). For example, in one study (143) maximal oxygen uptake was $0.2\ \ell \cdot min^{-1}$ lower in a 49° C, as compared to a 21° C environment. The physiological mechanism responsible for the decreased maximal oxygen uptake is probably the diversion of blood to the cutaneous vasculature which could: (a) reduce the portion of the cardiac output perfusing the contracting musculature and/or, (b) cause a peripheral pooling of blood, reduce the central blood volume and thus reduce cardiac filling pressure and cardiac output as discussed later.

Acute heat stress increases the metabolic rate for resting humans (27,28,40) and anesthetized dogs (144). Does heat stress alter an individual's metabolic requirements for performing a given submaximal exercise task? The answer to this question not only affects the calculation of the heat balance equation, but might also have implications for the nutritional requirements of individuals exposed to hot environments. This latter concern stimulated Consolazio and colleagues (27,28) and later others to address the question of metabolic requirements for humans performing exercise in hot environments. Many investigators have found that to perform a given submaximal exercise task the metabolic rate is greater in a hot than comfortable environment (27,28,40,48). Some investigators have reported the opposite results (20,112,176,179). The subjects' state of heat acclimation does not account for whether they demonstrate an increased or decreased metabolic rate during submaximal exercise in the heat. We believe that this discrepancy can be explained and we will attempt to do so in the following paragraphs. All of these investigators, except one (40), only calculated the aerobic metabolic rate during submaximal exercise.

Dimri *et al.* (40) had six subjects perform cycle exercise at three power output levels in each of three environments. Figure 3-10 presents their subjects' total metabolic rate (bottom) and the percentage of this metabolic rate which was contributed by aerobic and anaerobic metabolic pathways. The anaerobic metabolism was calculated by measuring the post-exercise oxygen uptake that was in excess of resting baseline levels. Note that to perform exercise at a given power output, the total metabolic rate increased with the elevated ambient temperature. More importantly, the percentage of the total metabolic rate contributed by anaerobic metabolism also increased with the ambient temperature. The increased anaerobic metabolic rate exceeded the increase of total metabolic rate during exercise at the elevated ambient temperatures. Therefore, if only the aerobic metabolic rate had been quantified, Dimri *et al.* (40) would have reported a decreased metabolic rate in the heat for performing exercise at a given power output. The investigations which reported a lower metabolic rate during exercise in the heat also reported increased plasma or muscle lactate levels (112,176,179) or an increased respiratory exchange ratio (20) which also suggests an increased anaerobic metabolism. Likewise, other investigators have reported that plasma lactate levels are greater during submaximal exercise in a hot as compared to a comfortable environment (39,40,48,95,120).

Skeletal Muscle Metabolism

Two investigations have examined the effects of environmental heat stress on skeletal muscle metabolism during exercise (48,179). Fink *et al.* (48) had six subjects perform 45 minutes of cycle exercise (70 to 85% of $\dot{V}O_2max$) in a cold (9° C) and a hot (41° C) environment. They found greater plasma lactate levels and increased muscle glycogen utilization during exercise in the heat. Also, muscle triglyceride utilization was reduced during exercise in the heat as compared to the cold. In addition, serum glucose concentration increased and serum triglyceride concentration decreased during ex-

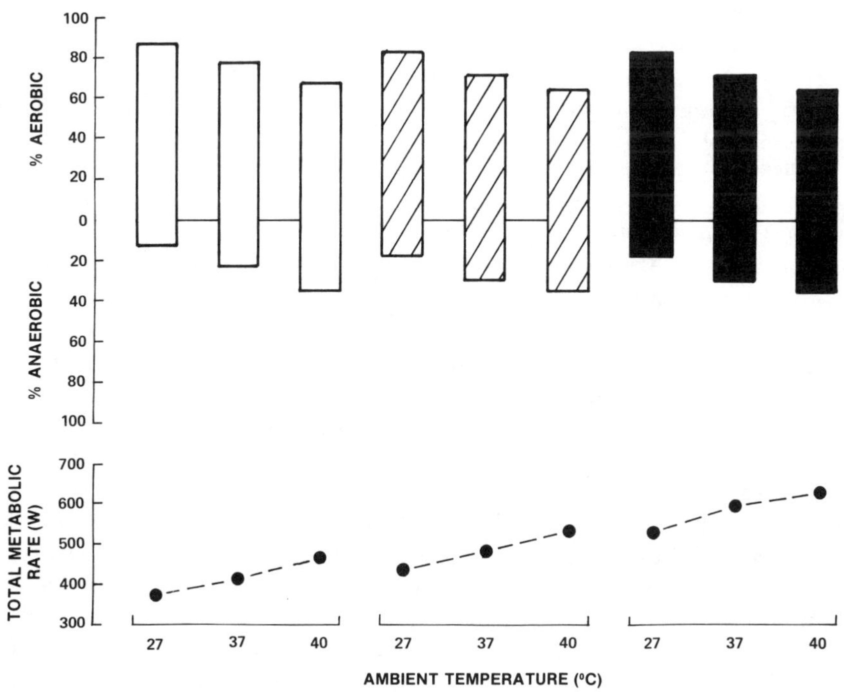

Figure 3-10. *The total metabolic rate and percentage contribution of aerobic and anaerobic metabolism during exercise at different ambient temperatures. Drawn from data presented in Reference 40.*

ercise in the heat, compared to the opposite responses during exercise in the cold. During exercise in the heat, the increased muscle glycogen utilization was attributed to an increased anaerobic glycolysis resulting from local muscle hypoxia, which was caused by a reduced muscle blood flow. Since these investigators (48) did not perform the control experiments in a comfortable environment, the differences reported could be due partially to the effects of the cold exposure.

Young *et al.* (179) had 13 subjects perform 30 minutes of cycle exercise (70% of $\dot{V}O_2max$) in a comfortable (20° C) and a hot (49° C) environment. They found that skeletal muscle and plasma lactate concentrations were greater during exercise in the heat. Interestingly, they also found a relationship ($r = 0.70$) between the subjects' percent fast twitch fibers and the increase of muscle lactate in the hot environment (Figure 3-11). These investigators found no difference in muscle glycogen utilization between the two experimental conditions. Young *et al.* (179) speculated that during exercise in the heat, an alternative glycolytic substrate might have been utilized, such as blood borne glucose. Rowell *et al.* (125) have shown a dramatic increase in hepatic glucose release into the blood during exercise in a hot compared to a comfortable environment. Such an increased release of hepatic glucose could account for the elevated serum glucose concentration reported in the hot environment by Fink *et al.* (48).

The data from Dimri *et al.* (40) and Young *et al.* (179) both support the concept of increased anaerobic metabolism during submaximal exercise in the heat. Much of the other support for this concept is based on the findings that, during submaximal exercise, the plasma lactate accumulation is greater in a hot than in a comfortable environment (39,40,48,95,120,176,179). However, any inference about metabolic effects at the muscle level from changes in plasma lactate is open to debate. Plasma lactate concentration reflects the balance between muscular production, efflux into the blood, and removal from the blood. Rowell *et al.* (125) have shown that during exercise in the heat the splanchnic vasoconstriction reduced hepatic removal of plasma lactate. Therefore, the greater blood lactate accumulation during submaximal exercise in the heat can be attributed, at least in part, to a redistribution of blood flow away from the splanchnic tissues.

The question remains as to what physiological mechanism(s) might be responsible for an increased anaerobic metabolism during exercise in the heat. As will be discussed, one possibility is the redistribution of blood to the cutaneous veins for heat dissipation; this may result in reduced perfusion of the active skeletal muscles and

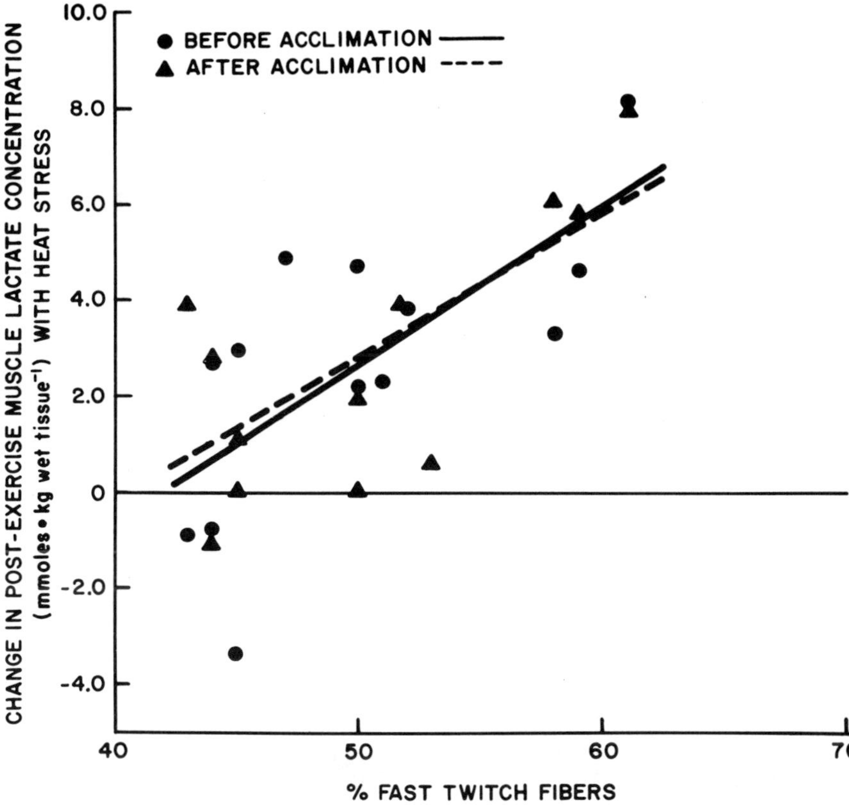

Figure 3-11. *The relationship between the subject's percentage of fast-twitch muscle fibers and the magnitude of muscle lactate increment in a hot compared to comfortable environment during exercise (179).*

thus local tissue hypoxia (176). Hypoxia would be expected to cause a shift from aerobic to anaerobic pathways in order to provide energy for the skeletal muscle contraction. Another possible explanation for increased anaerobic metabolism, is that acute heat exposure might increase the recruitment of fast-twitch motor units. Fast-twitch skeletal muscles derive a greater percentage of their total energy expenditure from anaerobic pathways than do slow-twitch fibers, regardless of their level of perfusion (141). In addition, fast-twitch fibers expend a greater amount of energy to develop the same amount of tension than do slow-twitch fibers (141,166). A final possibility is that an elevated temperature may reduce the efficiency of skeletal muscle (19) and, therefore, increase the energy cost of contraction.

In conclusion, acute heat stress has marked effects on metabolism. During maximal exercise, an individual's maximal oxygen uptake is reduced by heat stress. During submaximal exercise, the total energy expenditure is probably greater in a hot than in a comfortable environment. The greater total energy expenditure may simply reflect the individual's discomfort in the heat, thus making more extraneous movements during exposure. On the other hand, biochemical and physiological mechanisms may also contribute to the greater total energy cost. Finally, it seems clear that there is a shift from aerobic to anaerobic metabolic pathways during exercise in the heat.

EVAPORATIVE HEAT LOSS

Sweat Gland and Evaporation

Figure 3-5 illustrates that when the ambient temperature increases, there is a greater dependence upon insensible (evaporative) heat loss to defend core temperature during exercise. For humans, unlike most animals, respiratory evaporative cooling is small compared to total skin evaporative cooling (92). This provides the advantage of having a greater surface area available for evaporation. The eccrine glands secrete sweat onto the skin surface which causes evaporative cooling when it is converted from liquid to water vapor. The rate of evaporation is dependent upon the gradient between the skin and ambient water vapor pressure, and the evaporative heat transfer coefficient (see Chapter 1); and the wider the gradient, the greater the rate of evaporation for a given mass transfer coefficient. When a gram of sweat is vaporized at 30° C, 2.43 kilojoules (kJ) of heat (the latent heat of evaporation) is absorbed in the process (167). Wenger (167) has shown that ambient vapor pressure and solute content do not alter the latent heat of evaporation.

During cycle ergometer exercise it is not unusual for an individual to have a metabolic rate of 800 W ($\dot{V}O_2 \sim 2.3\ \ell \cdot min^{-1}$). If the exercise performed is 20% efficient then the remaining 80% of the metabolic rate is converted to heat in the body, so that 640 W (i.e., $0.64\ kJ \cdot s^{-1}$ or $38.4\ kJ \cdot min^{-1}$) needs to be dissipated to avoid body heat storage. The specific heat of body tissue is $\sim 3.5\ kJ \cdot kg^{-1} \cdot {}^{\circ}C^{-1}$ (see Chapter 1) so that a 70 kg man has a heat capacity of $245\ kJ \cdot {}^{\circ}C^{-1}$. If such a man performed exercise in an environment that enabled only insensible heat exchange, and he did not sweat, the body temperature would increase by ~1.0° C every 6.4 min ($245\ kJ \cdot {}^{\circ}C^{-1} \div 38.4\ kJ \cdot min^{-1}$). Since the latent heat of evaporation for sweat is 2.43 $kJ \cdot g^{-1}$, this man would need to evaporate ~16 g of sweat per min ($38.4\ kJ \cdot min^{-1} \div 2.43\ kJ \cdot g^{-1}$) in order to achieve a steady-state body temperature. This amount of sweat output is not unreasonable, as values in excess of 30 g per min have been reported during exercise in the heat (146).

Thermoregulatory sweat in humans is secreted by about 1.6–4 million eccrine glands (80,131). The number of sweat glands per unit of skin surface area varies considerably between body regions (80). Eccrine sweat glands are most numerous on the sole of the foot and least numerous on the back (131). The fine structure of the eccrine sweat gland consists of the secretory coil, duct and skin pore. The secretory coil is composed of clear cells as well as dark mucoid and myoepithelial cells. The clear cells contain many mitochondria and are believed to be responsible for the secretion of sweat and electrolytes (131). The clear cells are probably dependent upon glucose as a substrate, but it is unclear whether the source is local glycogen stores or blood borne glucose (131). Sweating can cause glycogen depletion in the clear cells, but this apparently does not occur in heat acclimated individuals (131). The myoepithelial cells provide structural support for the secretory epithelium, but do not contribute directly to the expulsion of sweat as once believed (133).

The amount of sweat secreted by the eccrine gland is dependent upon the structure and function of the stimulated gland, as well as the sudomotor signal from the central nervous system. Sato and Sato (137) have recently excised human eccrine sweat glands and performed in vitro analyses of their structure and function. They found that individuals who were self-categorized as copious sweaters had larger eccrine glands. In addition, as the sweat gland size increased there were: (a) a greater maximal sweating rate per gland, (b) a greater sweating rate per unit tubular length or unit volume of secretory coil, and (c) a greater cholinergic sensitivity of the sweat gland. These findings emphasize the importance of eccrine gland size and cholinergic sensitivity on the thermoregulatory sweating response.

Neural Control of the Sweat Gland

The sudomotor signal descends through the brain stem and spinal tracts to exit into the paravertebral ganglionic chain. The post-ganglionic sympathetic fibers which innervate the eccrine gland are nonmyelinated class C fibers that are primarily cholinergic. The eccrine sweat glands respond primarily to thermal stress through sympathetic cholinergic stimulation. However, it appears that circulating catecholamines, in particular epinephrine, facilitate thermoregulatory sweating as there are α and β adrenergic receptors associated with eccrine sweat glands (2,114,134,135,136). The relative effects on sweat secretion are 4:1:2 for cholinergic, α adrenergic and β adrenergic receptor stimulation, respectively (131). Robertshaw has reported in Macaca speciosa that epinephrine released from the adrenal medulla will enhance the sweating rate during exercise in the heat (118). Finally, vasoactive intestinal polypeptide (VIP) may be released at the eccrine sweat gland (84). Lundberg *et al.* (84) have shown that VIP and acetylcholine are concomitantly released from the same post-ganglionic fibers innervating exocrine glands of the cat. Although both transmitters caused exocrine secretions, the VIP primarily acts by dilation of local vasculature to facilitate sweat gland function. For many years it has been known that eccrine sweat secretion and cutaneous vasodilation are associated (83). The co-release of acetylcholine and VIP at the eccrine gland might explain this relationship (78,84).

Figure 3-12 represents the pattern of eccrine secretion, as measured by dew point hygrometry, during exercise in a comfortable environment. Note that sweat is normally secreted in a cyclic pattern, which may reflect the pattern of action potentials transmitted via the sudomotor neurons and subsequent neuroglandular release of acetylcholine (46,47). Thermoregulatory sweating can begin within a few seconds (162) to minutes (59) after starting muscular exercise. This variability for the onset time of thermoregulatory sweating is influenced by skin temperature, previous sweating and other factors. The increase in thermoregulatory sweating closely parallels the increase in body temperature (100,129). As the sweating rate increases toward maximum levels, first there is a recruitment of sweat glands and then an increased sweat secretion per gland (70,116,132). Therefore, the sweat secretion for a given region of skin is dependent upon both the density of sweat glands as well as the sweat secretion per gland (70,100,116,132). Nadel *et al.* (100) have also demonstrated that individual regions of skin have different sweating responses for a given core temperature. Figure 3-13 presents the regional local sweating responses plotted against core temperature. Note that for a given core temperature the back and chest have the greatest sweating rates. Conversely, the limbs will have relatively high sweating rates only after a substantial elevation in core temperature.

Skin Temperature and Ambient Water Vapor Pressure

For a given sudomotor signal to the eccrine gland, local skin temperature and skin wettedness influence the amount of sweat secreted. Bullard and co-workers (24) were among the first to systematically evaluate the relationship between local skin temperature and thermoregulatory sweating. They found that for an active eccrine gland, an elevated local skin temperature induced a greater sweat-

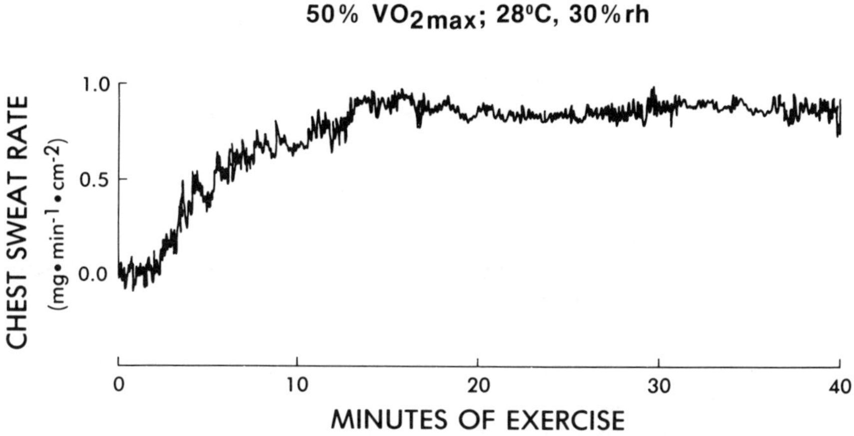

Figure 3-12. *The pattern of eccrine sweating, as measured by dew point hygrometry, during exercise in the heat.*

ing rate. These results have since been verified and expanded upon by many other investigators (46,47,96,109,110).

Nadel *et al.* (96) conducted a study on the importance of skin temperature in the regulation of sweating. These investigators independently varied the mean skin temperature, core temperature, and local skin temperature, while measuring thigh sweating rate during rest and post-exercise. They concluded that: (a) at a constant skin temperature sweating rate was proportional to core tempera-

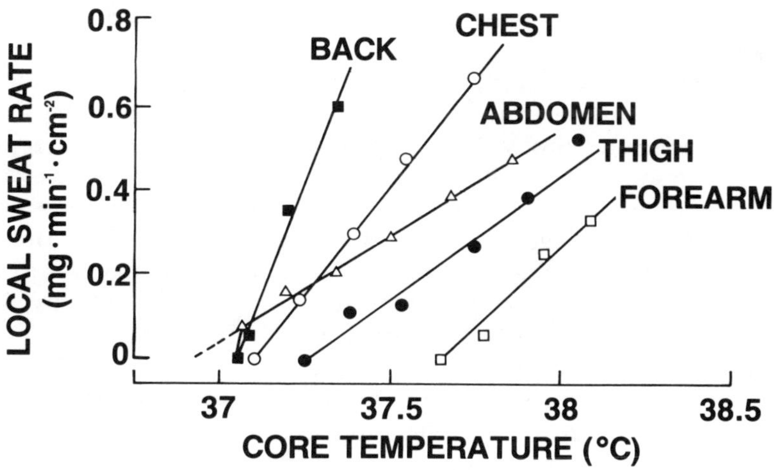

Figure 3-13. *The regional sweating rate responses plotted against core temperature. Redrawn from Reference 100.*

ture, (b) at a constant core temperature sweating rate was proportional to the mean skin temperature, and (c) at a given combination of core and mean skin temperatures local sweating was dependent on the local skin temperature. It was proposed that the local skin temperature acted as a multiplier to the central control signal, represented by a linear additive model of core and mean skin temperatures, to determine local sweating rate.

The physiological mechanism by which elevated local skin temperatures enhances the sweating response is unclear. However, several factors may contribute to enhanced sweating. Some investigators believe that local skin heating results in a greater release of neurotransmitter substance for a given sudomotor signal arriving at the eccrine sweat gland (24,46,47,96,109,110). The greater neurotransmitter release would, in turn, stimulate greater sweat production and release. In addition, Ogawa and colleagues (109,110) have provided evidence that local heating increases the eccrine glands responsiveness to a given amount of neurotransmitter substance. It is unknown if this increased glandular responsiveness is receptor mediated or reflects increased cellular metabolism within the secretory coil.

Remember, the rate of sweat evaporation depends upon air movement and the water vapor pressure gradient between the skin and the environment, so that sweat tends to collect on the skin in still or moist air. Wetting the skin surface causes a reduction in sweat secretion (26,59,102), an effect called hidromeiosis. The mechanism(s) responsible for hidromeiosis is/are unknown, although several hypotheses have been proposed. Peiss *et al.* (111) suggest that wetting the skin might cause the stratum corneum to swell and cause mechanical obstruction of the sweat duct. This idea was systematically tested by Brown and Sargent (23) who supported the concept that hydration of the stratum corneum was responsible for hidromeiosis. Nadel and Stolwijk (102) proposed that sweat may follow an osmotic gradient (from the solutes left behind by evaporated sweat) from the proximal duct to the skin surface; and have hypothesized that excessive skin wettedness might reduce, via dilution, the osmotic gradient along the sweat duct, thus reducing sweating rate. Candas *et al.* (26) suggested that hidromeiosis may result, by negative feedback, from locally wet skin which reduces local sweat output. All of the above hypotheses, with the exception of the first, are tentative and more research is merited concerning skin wettedness and the mechanisms of sweat suppression. Interestingly, if the environmental evaporative capacity is increased and subsequently sweat on the skin evaporates, the sweating rate will be potentiated (102,156). For example, Nadel and Stolwijk (102) demonstrated that individuals exposed to increased air motion, had an elevated sweating rate at a given core temperature during exercise in a warm-dry environment.

Several investigators (160,178) have proposed that the eccrine sweat glands may fatigue during prolonged periods of high sweat output; the notion being that the fatigued eccrine gland is unable to respond to a constant or even increasing sudomotor signal (23). The evidence supporting the idea of sweat gland fatigue is that: (a) during prolonged (three-five hours) periods of heat exposure, both the steady-state (55,159,178) and maximal (178) sweating rates are reduced and, (b) these "fatigued" eccrine glands demonstrate reduced responsiveness to the subdermal injection of a sudorific agonist (159). Examination of these studies, however, can lead one to conclude that hidromeiosis and/or dehydration (Chapter 6) may account for much of the so-called glandular fatigue.

Brown and Sargent (23) conducted a series of experiments to determine if hidromeiosis could be responsible for the proposed eccrine gland fatigue. They found that during prolonged (four-eight hours) heat exposure, the decline in sweat output could be reversed if the subject were simply moved from a moist heat to a dry heat. Brown and Sargent (23) then subdermally injected a sudorific agonist into the skin and measured the local sweating responses both before and after a four-to six-hour walk in the heat. These injections were made in areas of intact skin and in areas of skin that were stripped of the stratum corneum. They found that total body sweating rate decreased after several hours and continued to decrease throughout the duration of the exercise-heat exposure. During this period the local sweating rates decreased in the intact skin, but remained more nearly constant for the stripped skin. Likewise, injection of the sudorific agonist demonstrated a reduced sweat responsiveness in the intact skin but no change in sweat responsiveness in the stripped skin. They concluded first that excessively wetted skin results in swelling of the stratum corneum which obstructs the sweat duct and causes hidromeiosis, and second that hidromeiosis accounts for the so-called sweat gland fatigue in their experiments.

In conclusion, during exercise in hot environments, the evaporation of eccrine sweat is an important avenue of heat exchange for humans. The eccrine sweat glands are primarily controlled by the thermoregulatory center(s) via sympathetic cholinergic nerves, but local temperature modifies sweat secretion. Core temperature provides the primary input to the thermoregulatory controller for sweating; however, skin temperature also provides a smaller input. In addition, wetting of the skin surface causes a reduction in sweat secretion, an effect called hidromeiosis.

SKIN CIRCULATION AND HEAT TRANSFER

Control of Skin Blood Flow

Skin blood flow carries heat by convection between the deep body tissues and the skin. When core and skin temperatures are low enough that sweating does not occur, raising skin blood flow brings skin temperature nearer to blood temperature, and lowering skin blood flow brings skin temperature nearer to ambient temperature. In such a situation, the body is able to control sensible (convective and radiative) heat loss by varying skin blood flow, and thus skin temperature. In conditions in which sweating occurs, skin blood flow continues to increase with increases in core or skin temperature. However, in these conditions the tendency of skin blood flow to warm the skin is approximately balanced by the tendency of sweating to cool the skin. Therefore, there is usually little change in skin temperature and sensible heat exchange after sweating has begun, and skin blood flow serves primarily to deliver to the skin the heat that is being removed by sweat evaporation. Skin blood flow and sweating thus work in tandem to dissipate heat under such conditions.

Skin circulation is affected by temperature in two ways: local skin temperature affects the vascular smooth muscle directly, as discussed later in this section; and temperatures in the core and of the skin elsewhere on the body affect skin blood flow by reflexes operating through the sympathetic nervous system (Fig. 3-1). Blood flow in much of the human skin is under a dual vasomotor control (51,122). During cold exposure (and also with certain non-thermal reflexes—see Figure 3-3), skin blood flow is reduced through the action of vasoconstrictor fibers. These fibers presumably are adrenergic, since their action is blocked by bretylium (11). In the hands, feet, lips, ears, and nose, these fibers seem to be the predominant vasomotor innervation, and the vasodilation that occurs in these regions during heat exposure is largely the result of withdrawing vasoconstrictor activity. Over most of the skin area, however, vasoconstrictor activity is already minimal under conditions of thermal comfort, and vasodilation during heat exposure depends on intact sympathetic innervation, since it can be prevented or reversed by regional nerve block (12,51). Since it depends on the action of neural signals, such vasodilation is sometimes referred to as active vasodilation. Both active vasoconstriction and active vasodilation play a major part in controlling skin blood flow of the upper arm, forearm, thigh, and calf (12). However, active vasoconstriction is believed not to have a major role in controlling skin blood flow on the trunk or on most of the head (13,52).

Although active vasodilation occurs in limited skin areas (mostly paws and tail) in many animal species, human beings seem to be unique in the intensity of the response and the amount of skin involved (122). The vasoactive agonist responsible for active cutaneous vasodilation in man has not yet been identified, but some investigators have suggested that the vasodilation is mediated through the action of sweat glands. It has been proposed that sweat glands, when stimulated to secrete sweat, also release bradykinin, a powerful vasodilator, into the interstitial space of the skin (51), where it acts on the arterioles of the skin. This hypothesis now seems doubtful, and Rowell (122) has summarized the arguments against it. Nevertheless, active vasodilation may be linked to sweat gland activity through some other mechanism, such as release of vasoactive intestinal polypeptide (VIP), since VIP-like immunoreactivity has been reported in the terminals of the cholinergic nerves supplying human sweat glands (161). Evidence suggesting a relationship of active vasodilation to sweating includes observations on some patients with anhidrotic ectodermal dysplasia, a rare hereditary condition in which sweat glands are absent. Although these patients had normal vasoconstrictor activity in their forearms, and their forearm skin vasodilated normally in response to local heating, no active vasodilation occurred in response to whole-body heating (17). Since patients with anhidrotic ectodermal dysplasia usually have normal autonomic function, it is tempting to conclude that active vasodilation is secondary to some action of the sweat glands. Other explanations are possible, however, such as that in this disorder both the sweat glands and the nerve endings responsible for active vasodilation failed to develop during embryogenesis.

Besides contributing to the reflex control of skin blood flow, skin temperature also affects skin blood flow through direct actions on the blood vessels themselves (21,22), as shown in Figure 3-14. Local temperature acts on skin blood vessels in at least two ways. First, local cooling potentiates, and heating weakens, the contractile response of vascular smooth muscle to norepinephrine and other constrictor agonists, apparently by changing the affinity of α_2 adrenergic receptors for these agonists (75). Second, in human forearm skin (and presumably in other skin regions where active vasodilation occurs) local heating causes vasodilation, and local cooling causes vasoconstriction, even in the absence of nervous signals (173). Through this direct vasodilator effect, local heating increases skin blood flow so much that most of the heat delivered to the skin is carried away by the blood and little is conducted to the deeper tissues. This response reinforces thermoregulatory vasomotor responses, and also tends to protect the skin from heat injury.

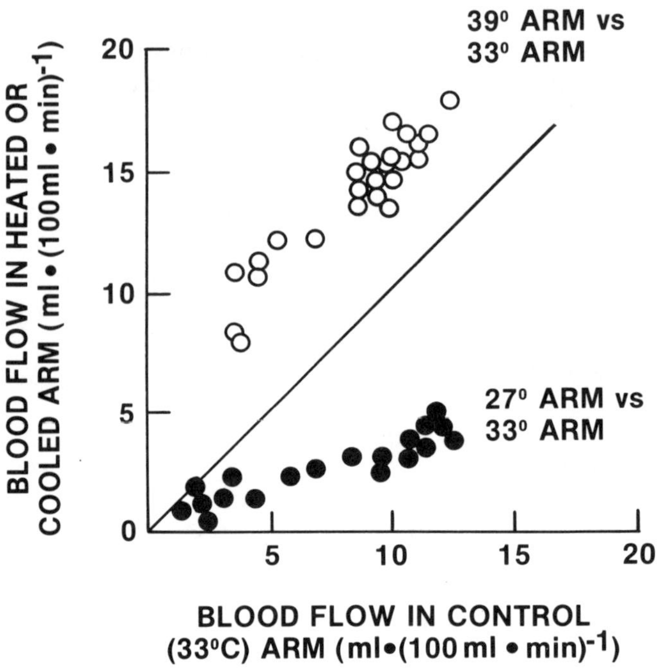

Figure 3-14. *Effect of local temperature on forearm blood flow during cycle exercise. The skin of the experimental forearm was either heated to 39° C (open circles) or cooled to 27° C (filled circles), and the skin of the control forearm was maintained always at 33° C. Blood flow in the experimental forearm is plotted against blood flow measured simultaneously in the control forearm. If there were no effect of local temperature, plotted points would lie on the line of identity. Redrawn from Reference 169.*

Heat Transfer from Skin Blood Flow

Blood has a volume specific heat of 3.85 $kJ \cdot \ell^{-1} \cdot °C^{-1}$ (0.92 $kcal \cdot \ell^{-1} \cdot °C^{-1}$) so that if a liter of blood at, say, 37° flows through the skin and returns to the body core 1° C cooler, the body core loses 3.85 kJ of heat. The muscle blood flow necessary to sustain aerobic metabolism depends on the O_2 content of the arterial blood, which is seldom much above 200 ml O_2 per liter. The energy released when 200 ml O_2 is consumed depends somewhat on the substrate being oxidized, but is about 4.2 kJ. Thus, even if the muscle uses all of the oxygen in the arterial blood, it needs to be only 1.09° C warmer than the arterial blood in order for the blood to remove the heat that the muscle is producing. Therefore, the blood flow needed to sustain aerobic metabolism provides amply for removing heat from exercising muscle.

The rate of core-to-skin heat transfer by skin blood flow depends on the rate of skin blood flow and on the temperature difference between blood leaving the core on its way to the skin and blood returning to the core from the skin. In the most efficient case, this temperature difference is equal to the difference between core and skin temperature (T_c and T_{sk}), and core-to-skin heat flow (HF) is:

$$HF = (SkBF \cdot 3.85kJ \cdot \ell^{-1} \cdot °C^{-1} + K_o) \cdot (T_c - T_{sk}) \qquad \text{(Eq. 3-4)}$$

where SkBF is skin blood flow. SkBF must be expressed in $\ell \cdot s^{-1}$, rather than the more usual $\ell \cdot min^{-1}$, to compute HF in W; and K_o is the thermal conductance of the tissues when skin blood flow is at a minimum. K_o depends on body surface area and thickness of subcutaneous fat, but typically is 10–18$W \cdot °C^{-1}$ (66).

In practice, especially in a cool subject, heat transfer is less efficient. As the warm blood leaves the body core via the arteries, it flows through superficial tissues that are cooler than the core but warmer than the skin, and as it flows through these tissues, the blood gives up some of its heat to them. The blood that enters the skin is thus cooler than the core, so that heat loss from the blood to the skin is less than that given by Equation 3-4. Some of the heat that the blood has lost on its way to the skin eventually finds its way to the skin by conduction, but some of this heat is returned to the cooled venous blood as the blood passes through the same layers of tissue on its way back to the core through which it passed on its way outward. One may think of this heat as having taken a "short circuit" from the arteries to the veins, without having passed through the skin. This process, the back-and-forth exchange of heat between the blood and the tissues through which it passes on the way between core and skin, thus has the effect of reducing convective heat transfer by the blood, or alternatively of requiring more blood flow to cause a given heat transfer than would be required in the most efficient case assumed in Equation 3-4.

The difference between actual core-to-skin heat transfer and that predicted by Equation 3-4 is subject to physiological control in the limbs. To explain this, it is first necessary to describe the dual venous drainage of the limbs, consisting of deep veins, which ordinarily drain blood mainly from the muscles; and superficial veins, which lie in the skin and subcutaneous tissue and ordinarily drain blood mainly from the skin. However, the deep and superficial veins are connected by many penetrating veins, so that potentially blood from anywhere in the limbs can return to the heart through either deep or superficial veins. The deep veins have a relatively poor sympathetic innervation (165), but the superficial veins have a rich sympathetic innervation (165), constricting when the skin or body core is cooled and dilating when the skin or core is warmed (164,170). In a cool subject, therefore, most blood from the limbs returns to the core via the deep veins. Since these veins and the major limb arteries lie adjacent to each other for a considerable part of their length, some of the heat from the arterial blood flow is conducted to the cooler venous blood and returned to the body core, and so is conserved. Such heat exchange between parallel streams moving in opposite directions is called "counter-current" heat exchange. In a warm subject, however, counter-current heat exchange is no longer advantageous, and it is considerably diminished by dilation of the large superficial veins, which allows much of the venous blood to return to the core along paths far removed from the major limb arteries. Dilation of the large superficial veins thus increases core-to-skin heat transfer by minimizing counter-current heat exchange. In addition, dilation of the smaller superficial veins may further improve heat transfer by increasing the time that the blood remains in the skin, as suggested by evidence that the degree of venous filling affects heat transfer to the skin (57).

Venous occlusion plethysmography is the most quantitative noninvasive method for measuring blood flow. However, it can be used only on the extremities. Furthermore, the part of the body being studied must be elevated above the heart to provide venous drainage, so that the use of this technique is limited to the hand and forearm in most situations. Since muscle blood flow is generally not influenced by thermal reflexes (42), changes in hand or forearm blood flow due to thermal reflexes can be taken to represent primarily changes in skin blood flow. Measurements of heat flow, either through the skin as a whole or through a restricted area, are sometimes divided by the temperature difference between core and skin to give a thermal conductance. Even in the absence of blood flow, there will be some heat flow due to the thermal conductivity of the tissues themselves (Equation 3-4), but after the contribution of this heat flow to thermal conductance is subtracted, the remainder can be used as an index of skin blood flow. However, such calculations only give values for the *minimum* blood flow rate which could produce the observed heat flow. For reasons discussed earlier, the actual blood flow may be, and probably usually is, greater; but accurate data on the relation between heat flow and blood flow are available only for the extremities and only in a few conditions. Furthermore, if this method is used to provide an index of blood flow to the entire skin, potentially serious errors may be introduced by regional variation in skin temperature. Given the lack of a better practical index, forearm blood flow is often used as an index of whole-body skin blood flow, since both types of vasomotor control, constrictor and dilator, are represented in forearm skin. However, very little information is available concerning how good an index forearm blood flow actually is under different conditions.

CARDIOVASCULAR CONSEQUENCES OF THERMOREGULATORY INCREASES IN SKIN BLOOD FLOW

Thermoregulatory Skin Blood Flow Responses to Exercise-Heat Stress

Although skin temperature is higher the warmer the environment, core temperature is relatively unaffected by environmental temperature over a fairly wide range, as discussed above. Thus in a warm environment the core-to-skin thermal gradient is relatively narrow, and skin blood flow increases in response to the high skin temperature, so as to achieve core-to-skin heat transfer sufficient for thermal balance (Equation 3-4). During exercise, metabolic rate and heat production may be ten or more times their levels at rest. A healthy but not athletic, young 70-kg man may have a metabolic rate at rest of 80 W and a $\dot{V}O_2max$ of 3.5 $\ell \cdot min^{-1}$, which corresponds to a metabolic rate of 1215 W. Even though core temperature rises during exercise and thus widens the core-to-skin temperature gradient somewhat, this small effect on core-to-skin heat transfer is not enough to match the increase in metabolic heat production. Besides widening the core-to-skin temperature gradient, the rising core temperature elicits reflex increases in skin blood flow, which eventually produce a rate of core-to-skin heat transfer sufficient to re-establish heat balance.

During exercise in the heat the narrow core-to-skin temperature gradient and the high rate of metabolic heat production that needs to be dissipated combine to require high levels of thermoregulatory skin blood flow. It is not possible to measure whole-body skin blood flow directly, but we can make an approximate estimate of skin blood flow during exercise from Equation 3-4. Let us consider again the man performing cycle exercise at a metabolic rate of 800 W and 20% efficiency, for a rate of heat production of 640 W. If at steady state his mean skin temperature is 35.5° C and core temperature is 39° C, and if we ignore respiratory heat loss and assume an intermediate value of $14W \cdot °C^{-1}$ for K_o, Equation 3-4 becomes $640\ W = (SkBF \cdot 3.85kJ \cdot \ell^{-1} \cdot °C^{-1} + 14W \cdot °C^{-1}) \cdot (3.5°\ C)$. Since $1\ W = 0.001\ kJ \cdot s^{-1}$, solving for SkBF we have $SkBF = 0.044\ell \cdot s^{-1} = 2.64\ \ell \cdot min^{-1}$. This is a crude estimate of skin blood flow, and should be taken only as an approximate figure. Although the man in this example is under considerable thermal stress, his estimated skin blood flow is well below the maximum obtainable. Rowell estimated the maximum skin blood flow obtainable by the combined effects, local and reflex, of heating the skin and body core. He computed skin blood flow of heated resting subjects as the sum of increases in cardiac output and decreases in splanchnic, renal, and muscle blood flow during heating (121). At the end of heating, core temperature was 39.1° C and still rising at an undiminished rate, skin temperature was 40.5° C, and computed skin blood flow was 7.8 $\ell \cdot min^{-1}$ and had nearly stopped rising, and so was probably near the maximum obtainable by skin and whole-body heating. Such a high skin blood flow may be incompatible with the muscle blood flow required during exercise, and at any event is unrepresentatively high for exercise because of the effects of the high skin temperature.

Circulatory Strain During Exercise-Heat Stress

It is evident from the foregoing examples that during exercise-heat stress thermoregulatory skin blood flow, though not precisely known, may be substantial. The higher skin blood flow will generally, but not always, result in a higher cardiac output, and one might expect the increased work of the heart in pumping this blood to be the major source of cardiovascular strain associated with heat stress. The work of the heart in providing the skin blood flow necessary for thermoregulation in the heat does, in fact, impose a substantial cardiac strain on patients with severe cardiac disease (25). In healthy subjects, however, the cardiovascular strain associated with heat stress results mostly from reduced cardiac filling and stroke volume (Fig. 3-15), which require a higher heart rate to maintain cardiac output. This occurs because the venous bed of the skin is large and compliant and, moreover, dilates reflexly during heat stress (122; Fig. 3-1). Therefore, as skin blood flow increases, the blood vessels of the skin become engorged and rather large volumes of blood pool in the skin, thus displacing blood from the thorax and reducing central blood volume and cardiac filling (122,123; Fig. 3-16). Since about 70% of the blood volume in an upright human being is below heart level (122), it is in the upright posture that the cardiovascular effects of this blood pooling are greatest. Thus, for example, stroke volume during exercise is lower in the upright than in the supine posture, and this difference between postures is greatest in the heat (117).

In addition, exercise and heat stress affect plasma volume in two ways. First, they cause fluid movements between plasma and tissues (67; Chapter 6). These fluid movements can occur rather quickly (Fig. 3-15), well before any substantial losses of fluid have occurred by sweating. The overall magnitude and direction of these

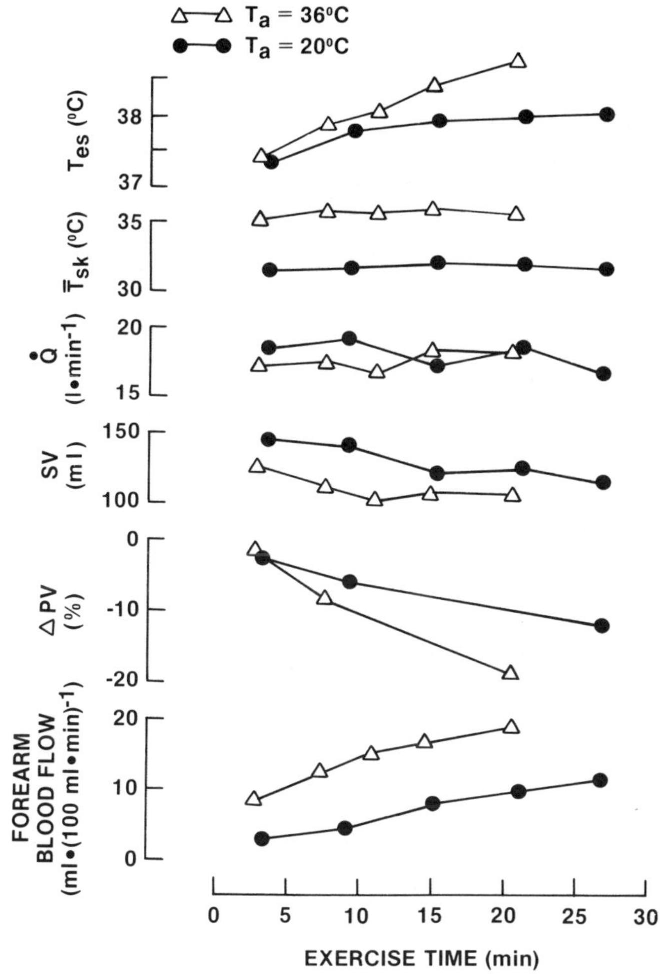

Figure 3-15. *Thermal and circulatory responses of one subject during cycle exercise at 70% $\dot{V}O_2max$ in ambient temperatures of 20 and 36° C, showing (from top) esophageal and mean skin temperatures, cardiac output, stroke volume, percent change in plasma volume, and forearm blood flow. Drawn from data presented in Reference 97.*

fluid movements depends on a number of factors, such as temperature, exercise type and intensity, hydration level, and status of heat acclimation, so that the overall effect may be either to increase or to decrease plasma volume, and thus to alleviate or to aggravate the circulatory effects of pooling of blood in the skin. Second, much fluid is lost by sweating during exercise and heat stress. Since the main solute in sweat is sodium chloride, a disproportionately large

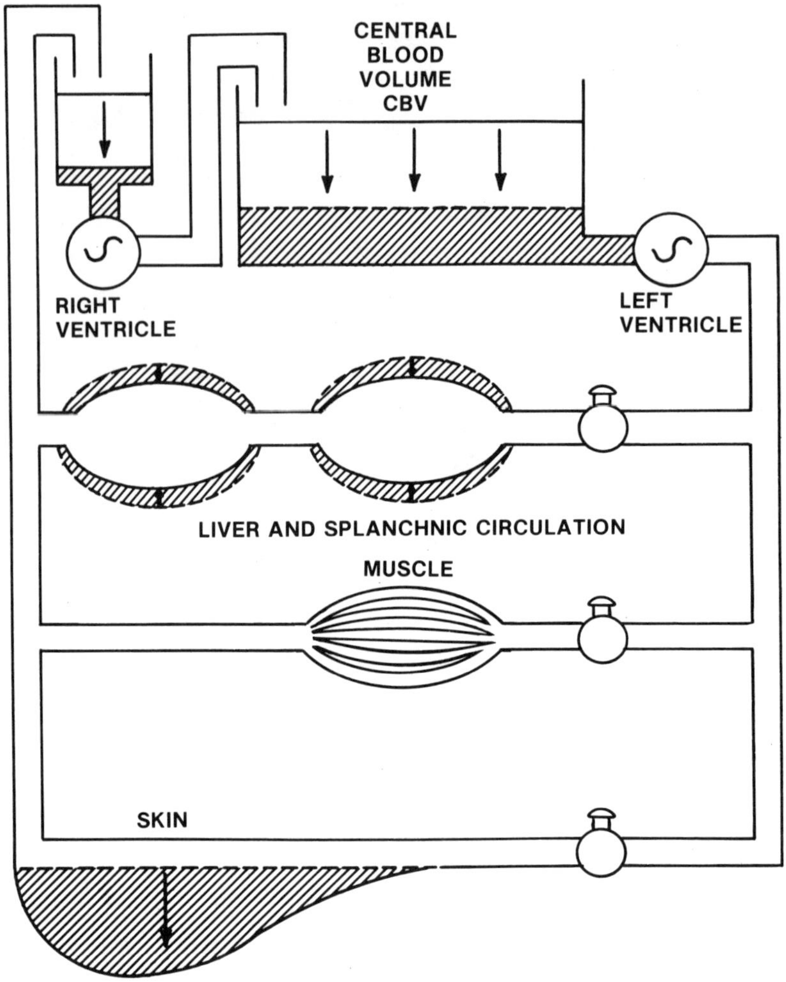

Figure 3-16. *Schematic diagram of the effects of skin vasodilation on the thoracic reservoirs from which the ventricles are filled, and also effects of compensatory vasomotor adjustments in the splanchnic circulation. Redrawn from References 122,123.*

fraction of the water in the sweat will be lost at the expense of extracellular fluid, including plasma, to the extent that the body's sodium ion content is reduced. The disproportionate loss from extracellular fluid will be minimized if the sweat is dilute, as is the case with subjects well acclimatized to heat. If the water and salt lost by sweating are not replaced, plasma and extracellular fluid volumes will be progressively reduced during exercise-heat stress.

Compensatory Responses to Maintain Cardiovascular Homeostasis During Exercise-Heat Stress

There are several reflex adjustments that compensate for peripheral pooling of blood and possible decreases in plasma volume, and help to maintain cardiac filling, cardiac output, and arterial pressure during exercise-heat stress. Splanchnic and renal blood flows are reduced during exercise (121). The reduction in splanchnic blood flow is in proportion to relative exercise intensity (i.e., as a percent of $\dot{V}O_2$max), and presumably the same is true of renal blood flow (121). These blood flows also undergo a graded and progressive reduction in subjects heated at rest; and in the splanchnic bed, at least, the vasoconstrictor effects of temperature and of exercise appear to be additive, so that at any exercise intensity the reduction in splanchnic blood flow is greater at a higher skin temperature (121). Reduction of renal and splanchnic blood flow allows a corresponding diversion of cardiac output to skin and exercising muscle. In addition, the splanchnic vascular bed and the hepatic vascular bed into which it drains via the hepatic portal vein are very compliant, so that a reduction in splanchnic blood flow is followed by a reduction in the amount of blood pooled in these beds (Fig. 3-16). A substantial volume of blood can thus be mobilized from these beds to help maintain cardiac filling during exercise and heat stress.

Since skin blood flow can reach high levels during exercise-heat stress, and since high skin blood flow may produce substantial peripheral pooling of blood, one might ask whether exercise affects skin blood flow in the same way that it affects renal and splanchnic blood flow. Since exercise quickly warms the body, and since thermal factors play such a large role in control of skin blood flow, especially during exercise in the heat, it is useful to re-state the question so as to separate thermal from non-thermal influences on skin blood flow. We shall do this by asking whether exercise affects the relation of skin blood flow to core and skin temperatures and, if so, whether the effect of exercise is proportionate to relative exercise intensity (percent $\dot{V}O_2$max). Under conditions of high cardiovascular strain, such as during very intense exercise or at high skin temperatures, skin blood flow at a given core and skin temperature is known to be reduced during exercise (121,123). However, at moderate skin temperatures, the relation of forearm blood flow to core and skin temperatures is independent of exercise intensity over a range from mild to fairly intense exercise (74,172; see Reference 168 for further discussion). A more recent study has confirmed these findings, and extended them by showing that at a sufficiently high exercise intensity, skin blood flow is lower than would be expected from the levels of core and skin temperature, and this difference between actual and expected blood flows becomes greater as exercise intensity increases further (157). Thus in the service of cardiovascular homeostasis, the body begins to reduce splanchnic and renal blood flows at fairly low levels of cardiovascular strain, while sparing thermoregulatory skin blood flow, which begins to be compromised only at fairly high levels of cardiovascular strain.

Skin blood flow also is affected, especially during exercise in the heat, by other reflexes which function to maintain cardiac output and blood pressure, and which are elicited in situations in which cardiac filling is compromised. Thus during cycle exercise forearm vasodilation begins at a higher core temperature in the upright than in the supine posture, and this postural difference is greater at higher ambient temperatures, in which there presumably is more peripheral pooling (117). Likewise, during cycle exercise in the heat, forearm vasodilation begins at a higher core temperature if blood volume has been reduced before the start of exercise by a 4-day course of diuretics (98) or by acute blood withdrawal (49). However, the core temperature threshold for forearm vasodilation seems not to be affected by acute expansion of blood volume before the start of exercise (49). In some experiments, a break has been observed in the slope of the relation of forearm blood flow to core temperature (49,97,98), so that the slope of the relation is much lower beyond the break (Fig. 3-17). Moreover, such breaks seem to occur at a lower forearm blood flow under conditions of greater cardiovascular strain (e.g., during exercise with reduced blood volume, 98). These breaks are not intrinsic to the thermal control of skin blood flow, since they can be removed by having the subject exercise in the supine position (Fig. 3-17), which favors return of blood to the heart, and cardiac filling. These breaks may represent the recruitment of a compensatory vasomotor adjustment to limit cutaneous vascular pooling.

Lower body negative pressure (LBNP) is an experimental technique in which the lower half of the body is enclosed in a rigid box to which negative pressure, down to about 90 Torr below atmospheric, is applied to cause pooling of blood in the lower part of the body. Application of LBNP reduces central blood volume and central venous pressure, thus lowering the pressures sensed by the cardiopulmonary baroreceptors. With levels of LBNP not far below atmospheric, there is little effect on arterial pressure, but mean arterial pressure is reduced with higher levels of LBNP. Forearm blood flow is reduced during application of LBNP, even at levels low enough that arterial pressure is not affected, so that the response of forearm blood flow represents vasoconstriction in response to the reduction in central venous pressure (122). If forearm blood flow has previ-

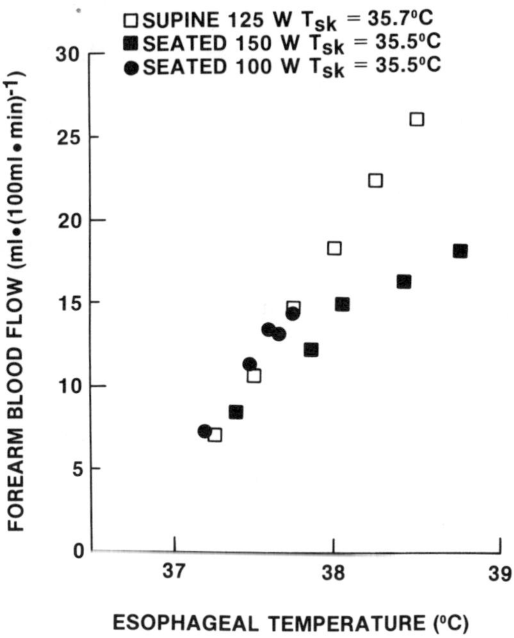

Figure 3-17. *Relation of forearm blood flow to esophageal temperature during exercise in supine and seated positions in an ambient temperature of 36° C. Data points are averaged from duplicate experiments on one subject. Modified from Reference 97.*

ously been raised by heating the subject, the decrease in blood flow in response to LBNP can be much greater than in a subject at a comfortable temperature (122). Intravascular pressures were not measured in the studies described in the previous paragraph (49,97,98,117), and impairment of cardiac filling was inferred from reductions in stroke volume. However, the changes in the control of forearm blood flow that were observed in conditions in which stroke volume was reduced were probably mediated by the cardiopulmonary baroreceptors, and probably represent the same reflex as that elicited by applying LBNP.

At a given level of skin blood flow, the volume of blood that pools in the cutaneous vessels depends on the compliance of the cutaneous veins, and thus can be reduced by constriction of these veins. With light to moderate exercise, the cutaneous veins constrict at the beginning of exercise, but relax within a few minutes (122), perhaps in response to the increase in core temperature. However, cutaneous venous volume seems to be more sensitive to exercise than is skin blood flow (170), and the venoconstriction during more intense exercise is sustained (50,62).

All of the adjustments discussed above help to maintain cardiac filling and cardiac output during exercise in the heat. In addition, sympathetic stimulation of the heart increases the rate and vigor of cardiac contraction, and thus contributes further to maintaining cardiac output under these conditions. Most of this cardiac output goes to supply exercising muscle, but it is not known whether or not blood flow to exercising muscle is affected by heat stress, and quantitative measurements of blood flow through exercising muscle are not available. Rowell (122) has summarized some of the evidence bearing indirectly on this point, but this evidence is insufficient to make a clear case either way. Some other recent evidence, though also somewhat indirect, suggests that heat stress may, in fact, compromise muscle blood flow, since the initial rise in blood lactate during exercise is more rapid during exercise at an ambient temperature of 35° C than at 25° C (95).

CONCLUSION

Human body core temperature is regulated by two distinct control systems, behavioral and physiological. Behavioral thermoregulation involves the conscious, willed use of whatever means are available to minimize thermal discomfort. Behavioral thermoregulatory responses characteristically are discrete rather than continuous. Control of thermal discomfort is not precisely characterized, but discomfort appears to be proportional to core and skin temperatures, and also to include elements that lead or anticipate thermal changes. Physiological thermoregulation employs primarily unconscious responses that are controlled through the autonomic nervous system in a proportional control fashion according to the levels of core and skin temperatures. Since temperature generally is not uniform throughout the body core and since there are thermosensitive elements at a number of sites in the body core, there is no one site whose temperature represents "true" core temperature. Furthermore, although the hypothalamus has an especially high density of thermosensitive elements, it is not known how core temperatures at various core sites are combined in the control of the thermoregulatory effector responses. Nevertheless, in practice changes in esophageal temperature, which follows central blood temperature rather closely, can account fairly accurately for changes in thermoregulatory effector responses.

The physiological responses controlling body heat loss are skin blood flow and sweating. During exercise, the increase in heat production causes core temperature to rise until it reaches a new steady state where heat loss responses have increased enough to bring heat production and heat loss into balance. The core temperature elevation at steady-state is proportional to the metabolic rate and is largely independent of the environmental conditions. Because of the proportional control nature of the thermoregulatory system, this core temperature increase is a "load error" and not the result of a change in the set-point temperature. Depending on the environmental temperature the relative contributions of sensible (radiative and convective) and insensible (evaporative) heat exchange to the total heat loss will vary, so that the hotter the environment the greater the dependence on insensible heat loss, and thus on sweating. The input of skin temperature to the thermoregulatory system allows the heat loss responses to be adjusted according to environmental temperature with little change in core temperature

During exercise in the heat, the primary cardiovascular problem is to provide simultaneously sufficient blood flow to exercising skeletal muscle to support its metabolism, and sufficient skin blood flow to dissipate the heat released from the exercising muscle. In hot environments, the core-to-skin temperature gradient is reduced, so that skin blood flow needs to be relatively high to achieve heat transfer sufficient for thermal balance. This high skin blood flow causes pooling of blood in the compliant skin veins, especially below heart level. In addition, sweat secretion can result in a net loss of body water, and thus reduce plasma and blood volume. Thus, both through pooling of blood in the skin and through reduced plasma volume, heat stress can reduce cardiac filling. Compensatory responses include reductions in splanchnic and renal blood flow, which mobilize blood from the splanchnic vascular bed, and also allow cardiac output to be diverted from the abdominal viscera to skin and exercising muscle; increased cardiac contractility, which helps to defend stroke volume in the face of impaired cardiac filling; and increased heart rate to compensate for decreased stroke volume. If these compensatory responses are insufficient, skin and muscle blood flow will be impaired. The impairment of skin blood flow will reduce the effectiveness of thermoregulation, possibly leading to dangerous hyperthermia; and the impairment of muscle blood flow will reduce exercise performance, and may lead to exhaustion, during exercise in the heat.

REFERENCES

1. Abrams, R.M. and J.P. Royston. Some properties of rectum and vagina as sites for basal body temperature measurements. *Fertility and Sterility* 35:313–316, 1981.
2. Allan, J.A. and I.C. Roddie. The role of circulating catecholamines on sweat production in man. *J. Physiol., London* 227:801–814, 1972.
3. Aikas, E., M.J. Karvonen, P. Piironen and R. Ruosteenoja. Intramuscular, rectal and oesophageal temperature during exercise. *Acta Physiol. Scand.* 54:366–370, 1962.

4. Aschoff, J. and R. Wever. Kern und Schale in Wärmehaushalt des Menschen. *Naturwissenschaften* 45:477–485, 1958.
5. Asmussen, E. and M. Nielsen. The regulation of the body-temperature during work performed with the arms and with the legs. *Acta Physiol. Scand.* 14:373–382, 1947.
6. Åstrand, I. Aerobic work capacity in men and women. *Acta Physiol. Scand.* 49:(Suppl.169):64–73, 1960.
7. Aulick, L.H., S. Robinson and S. Tzankoff. Arm and leg intravascular temperature of men during submaximal exercise. *J. Appl. Physiol.* 51:1092–1097, 1981.
8. Bazett, H.C., L. Love, M. Newton, L. Eisenberg, R. Day and R. Forster II. Temperature changes in blood flowing in arteries and veins in man. *J. Appl. Physiol.* 1:3–19, 1950.
9. Benedict, F.G. and E.P. Slack. A comparative study of temperature fluctuations in different parts of the body. Washington D.C.: Carnegie Institute of Washington, 1911.
10. Bernard, C.*Lecons sur la Chaleur Animale, sur les Effets de la Chaleur et sur la Fievre.* Paris: Libraire J.-B. Bailliere et fils, 1876.
11. Blair, D.A., W.E. Glover, B.S.L. Kidd and I.C. Roddie. The effect of bretylium tosylate (Darenthin) on certain vascular reflexes in human limbs. *J. Physiol., London* 149:74P–75P, 1959, Abstract.
12. Blair, D.A., W.E. Glover and I.C. Roddie. Vasomotor fibers to skin in the upper arm, calf and thigh. *J. Physiol., London* 153:232–238, 1960.
13. Blair, D.A., W.E. Glover and I.C. Roddie. Cutaneous vasomotor nerves to the head and trunk. *J. Appl. Physiol.* 16:119–122, 1961.
14. Bligh, J. The receptors concerned in the respiratory response to humidity in sheep at high ambient temperature. *J. Physiol., London* 168:747–763, 1963.
15. Boulant, J.A. Hypothalamic control of thermoregulation. In: P.J. Morgane and J. Panksepp (Eds.) *Handbook of the Hypothalamus Vol. 3, Part A. Behavioral Studies of the Hypothalamus.* New York: Marcel Dekker, 1980, pp. 1–82.
16. Brengelmann, G.L. Dilemma of body temperature measurement. In: K. Shiraki and M.K. Yousef (Eds.) *Man in a Stressful Environment; Thermal and Work Physiology.* Springfield, IL: C.C. Thomas, 1987, pp. 5–22.
17. Brengelmann, G.L., P.R. Freund, L.B. Rowell, J.E. Olerud and K.K. Kraning. Absence of active cutaneous vasodilation associated with congenital absence of sweat glands in humans. *Am. J. Physiol.* 240:H571–H575, 1981.
18. Brengelmann, G.L., J.M. Johnson and P. Hong. Electrocardiographic verification of esophageal temperature probe position. *J. Appl. Physiol.* 47:638–642, 1979.
19. Brooks, G.A., K.J. Hittelman, J.A. Faulkner and R.E. Beyer. Temperature, skeletal muscle mitochondrial functions and oxygen debt. *Am. J. Physiol.* 220:1053–1059, 1971.
20. Brouha, L., P.E. Smith, Jr., R. De Lanne and M.E. Maxfield. Physiological reactions of men and women during muscular activity and recovery in various environments. *J. Appl. Physiol.* 16:133–140, 1960.
21. Brown, G.M., D. Hatcher and J. Page. Temperature and blood flow in the forearm of the Eskimo. *J. Appl. Physiol.* 5:410–420, 1953.
22. Brown, G.M. and J. Page. The effect of chronic exposure to cold on temperature and blood flow of the hand. *J. Appl. Physiol.* 5:221–227, 1952.
23. Brown, W.K. and F. Sargent. Hidromeiosis. *Arch. Environ. Health.* 11:442–453, 1965.
24. Bullard, R.W., M.R. Banerjee, F. Chen, R. Elizondo and B.A. MacIntyre. Skin temperature and thermoregulatory sweating: a control systems approach. In: J.D. Hardy, A.P. Gagge and J.A.J. Stolwijk (Eds.) *Physiological and Behavioral Temperature Regulation.* Springfield, IL: C.C. Thomas, 1970, pp. 597–610.
25. Burch, G.E. and N.P. DePasquale. *Hot Climates, Man and His Heart.* Springfield, IL: C.C. Thomas, 1962.
26. Candas, V., J.P. Libert and J.J. Vogt. Sweating and sweat decline of resting men in hot humid environments. *Eur. J. Appl. Physiol.* 50:223–234, 1983.
27. Consolazio, C.F., L.O. Matoush, R.A. Nelson, J.A. Torres and G.J. Isaac. Environmental temperature and energy expenditures. *J. Appl. Physiol.* 18:65–68, 1963.
28. Consolazio, C.F., R. Shapiro, J.E. Masterson and P.S.L. McKinzie. Energy requirements of men in extreme heat. *J. Nutrition* 73:126–134, 1961.
29. Cooper, K.E., W.I. Cranston and S. Snell. Temperature in the external auditory meatus as an index of central temperature changes. *J. Appl. Physiol.* 19:1032–1035, 1964.
30. Cooper, K.E. and J.R. Kenyon. A comparison of temperature measured in the rectum, oesophagus and on the surface of the aorta during hypothermia in man. *Brit. J. Surg.* 44:616–619, 1957.
31. Corbit, J.D. Behavioral regulation of body temperature. In: J.D. Hardy, A.P. Gagge, and J.A.J. Stolwijk (Eds.) *Physiological and Behavioral Temperature Regulation.* Springfield, IL: C.C. Thomas, 1970, pp. 777–801.
32. Cranston, W. I. Temperature regulation. *Brit. Med. J.* 2:69–75, 1966.

33. Cranston, W.I., J. Gerbrandy and E.S. Snell. Oral, rectal and oesophageal temperatures and some factors affecting them in man. *J. Physiol., London* 126:347–358, 1954.
34. Cunningham, D.J., J.A.J. Stolwijk and C.B. Wenger. Comparative thermoregulatory responses of resting men and women. *J. Appl. Physiol.* 45:908–915, 1978.
35. Davies, C.T.M. Influence of skin temperature on sweating and aerobic performance during severe work. *J. Appl. Physiol.* 47:770–777, 1979.
36. Davies, C.T.M. Thermoregulation during exercise in relation to sex and age. *Eur. J. Appl. Physiol.* 42:71–79, 1979.
37. Davies, C.T.M., J.R. Brotherhood and E. Zeidifard. Temperature regulation during severe exercise with some observations on effects of skin wetting. *J. Appl. Physiol.* 41:772–776, 1976.
38. Dickey, W.T., E.W. Ahlgren and C.R. Stephen. Body temperature monitoring via the tympanic membrane. *Surgery* 67:981–984, 1970.
39. Dill, D.B., H.T. Edwards, P.S. Bauer and E.J. Levenson. Physical performance in relation to external temperature. *Arbeitsphysiologie* 3:508–518, 1930/31.
40. Dimri, G.P., M.S. Malhotra, J. Sen Gupta, T.S. Kumar and B.S. Aora. Alterations in aerobic-anaerobic proportions of metabolism during work in heat. *Eur. J. Appl. Physiol.* 45:43–50, 1980.
41. Downey, J.A., R.F. Mottram and G.W. Pickering. The location by regional cooling of central temperature receptors in the conscious rabbit. *J. Physiol., London* 170:415–441, 1964.
42. Edholm, O.G., R.H. Fox and R.K. MacPherson. The effect of body heating on the circulation in skin and muscle. *J. Physiol., London* 134:612–619, 1956.
43. Edwards, R.J., A.J. Belyavin and M.H. Harrison. Core temperature measurement in man. *Aviat. Space Environ. Med.* 49:1289–1294, 1978.
44. Eichna, L.W. Thermal gradients in man: Comparison of temperatures in the femoral artery and femoral vein with rectal temperatures. *Archiv. Physical Med.* 30:584–592, 1949.
45. Eichna, L.W., A.R. Berger, B. Rader and W.H. Becker. Comparison of intracardiac and intravascular temperature in man. *J. Clin. Invest.* 30:353–359, 1951.
46. Elizondo, R. Local control of eccrine sweat gland function. *Fed. Proc.* 32:1583–1587, 1973.
47. Elizondo, R.S. and G.S. Johnson. Peripheral effector mechanisms of temperature regulation. In: Z. Szelenyi and M. Szekely (Eds.) *Advances in Physiological Sciences. Contributions to Thermal Physiology*. New York: Pergamon, 1980, pp. 397–408.
48. Fink, W.J., D.L. Costill and W.J. Van Handel. Leg muscle metabolism during exercise in the heat and cold. *Eur. J. Appl. Physiol.* 34:183–190, 1975.
49. Fortney, S.M., E.R. Nadel, C.B. Wenger and J.R. Bove. Effect of acute alterations of blood volume on circulatory performance in humans. *J. Appl. Physiol.* 50:292–298, 1981.
50. Fortney, S.M., C.B. Wenger, J.R. Bove and E.R. Nadel. Effect of blood volume on forearm venous and cardiac stroke volume during exercise. *J. Appl. Physiol.* 55:884–890, 1983.
51. Fox, R.H. and O.G. Edholm. Nervous control of cutaneous circulation. *Brit. Med. Bull.* 19:110–114, 1963.
52. Fox, R.H., R. Goldsmith, and D.J. Kidd. Cutaneous vasomotor control in the human head, neck and upper chest. *J. Physiol., London* 161:298–312, 1962.
53. Fox, R.H. and R.K. Macpherson. The regulation of body temperature during fever. *J. Physiol., London* 125:21P–22P, 1954, Abstract.
54. Gerbrandy, J., E.S. Snell and W.I. Cranston. Oral, rectal, and oesophageal temperatures in relation to central temperature control in man. *Clin. Sci.* 13:615–624, 1954.
55. Gerking, S.D. and S. Robinson. Decline in the rates of sweating of men working in severe heat. *Am. J. Physiol.* 147:370–378, 1946.
56. Gisolfi, C.V. and C.B. Wenger. Temperature regulation during exercise: old concepts, new ideas. In: R.L. Terjung (Ed.) *Exercise and Sport Sciences Reviews*, Vol 12. Lexington, MA: Collamore Press, 1984, pp. 339–372.
57. Goetz, R. H. Effect of changes in posture on peripheral circulation, with special reference to skin temperature readings and the plethysmogram. *Circulation* 1:56–75, 1950.
58. Gonzalez, R.R., L.G. Berglund and A.P. Gagge. Indices of thermoregulatory strain for moderate exercise in the heat. *J. Appl. Physiol.* 44:889–899, 1978.
59. Gonzalez, R.R., K.B. Pandolf and A.P. Gagge. Heat acclimation and decline in sweating during humidity transients. *J. Appl. Physiol.* 36:419–425, 1974.
60. Greenleaf, J.E. and B.L. Castle. External auditory canal temperature as an estimate of core temperature. *J. Appl. Physiol.* 32:194–198, 1972.
61. Greenleaf, J.E., C.J. Greenleaf, D.H. Card and B. Saltin. Exercise-temperature regulation in man during acute exposure to simulated altitude. *J. Appl. Physiol.* 26:290–296, 1969.
62. Hanke, D.M., M. Schlepper, K. Westermann and E. Witzleb. Venentonus, Haut-und Muskeldurchblutung an Unterarm und Hand bei Beinarbeit. *Pflügers Arch.* 309:115–127, 1969.
63. Hardy, J.D. Physiology of temperature regulation. *Physiol. Rev.* 41:521–606, 1961.
64. Hardy, J.D. Thermal comfort: Skin temperature and physiological thermoregulation. In:

J.D. Hardy, A.P. Gagge, and J.A.J. Stolwijk (Eds.) *Physiological and Behavioral Temperature Regulation*. Springfield, IL: C.C. Thomas, 1970, pp. 856–873.
65. Hardy, J.D. and E.F. DuBois. The technique of measuring radiation and convection. *J. Nutr.* 15:461–475, 1938.
66. Hardy, J.D. and E.F. DuBois. Differences between men and women in their response to heat and cold. *Proc. Natl. Acad. Sci. USA* 26:389–398, 1940.
67. Harrison, M.H. Effect of thermal stress and exercise on blood volume in humans. *Physiol. Rev.* 65:149–209, 1985.
68. Hellekant, G. Circulation of the tongue. In: N. Emmelin and Y. Zetterman (Eds.) *Oral Physiology*. London: Pergamon Press, 1972, pp. 127–137.
69. Hensel, H. Temperature receptors in the skin. In: J.D. Hardy, A.P. Gagge and J.A.J. Stolwijk (Eds.) *Physiological and Behavioral Temperature Regulation*. Springfield, IL: C.C. Thomas, 1970, pp. 442–453.
70. Hertzman, A.B., W.C. Randall, C.N. Peiss and R. Seckendorf. Regional rates of evaporation from the skin at various environmental temperatures. *J. Appl. Physiol.* 5:153–161, 1952.
71. Horvath, S.M., A. Rubin and E.L. Foltz. Thermal gradients in the vascular system. *Am. J. Physiol.* 161:316–322, 1950.
72. Ilsley, A.H., A.J. Rutten and W.B. Runciman. An evaluation of body temperature measurement. *Aneasth. Intensive Care* 11:31–39, 1983.
73. Jaeger, J.J., E.C. Deal, D.E. Roberts, R.H. Ingram and E.R. McFadden. Cold air inhalation and esophageal temperature in exercising humans. *Med. Sci. Sports Exerc.* 12:365–369, 1980.
74. Johnson, J.M. Responses of forearm blood flow to graded leg exercise in man. *J. Appl. Physiol.* 46:457–462, 1979.
75. Johnson, J.M., G.L. Brengelmann, J.R.S. Hales, P.M. Vanhoutte and C.B. Wenger. Regulation of the cutaneous circulation. *Fed. Proc.* 45:2841–2850, 1986.
76. Karlberg, P. The significance of depth of insertion of the thermometer for recording rectal temperature. *Acta Paediatrica* 38:359–366, 1949.
77. Klausen, K., D.B. Dill, E.E. Phillips and D. McGregor. Metabolic reactions to work in the desert. *J. Appl. Physiol.* 22:292–296, 1967.
78. Kolka, M.A. and L.A. Stephenson. Cutaneous blood flow and local sweating after systemic atropine administration. *Pflügers Arch.* 410:524–529, 1987.
79. Kolka, M.A., L.A. Stephenson, P.B. Rock and R.R. Gonzalez. Local sweating and cutaneous blood flow during exercise in hypobaric environments. *J. Appl. Physiol.* 62:2224–2229, 1987.
80. Kuno, Y. *Human Perspiration*. Springfield, IL.: C.C. Thomas, 1956.
81. Libert, J.P., V. Candas, J.C. Sagot, J.P. Meyer, J.J. Vogt and T. Ogawa. Contribution of skin thermal sensitivities of large body areas to sweating response. *Jap. J. Physiol.* 34:75–88, 1984.
82. Lind, A.R. A physiological criterion for setting thermal environmental limits for everyday work. *J. Appl. Physiol.* 18:51–56, 1963.
83. Love, A.H.G. and R.G. Shanks. The relationship between the onset of sweating and vasodilatation in the forearm during body heating. *J. Physiol., London* 162:121–128, 1962.
84. Lundberg, J.M., A. Anggard, J. Fahrenkrug, T. Hokfelt and V. Mutt. Vasoactive intestinal polypeptide in cholinergic neurons of exocrine glands: functional significance of coexisting transmitters for vasodilation and secretion. *Proc. Natl. Acad. Sci. USA.* 77:1651–1655, 1980.
85. Mairiaux, P., J. Sagot and V. Candas. Oral temperature as an index of core temperature during heat transients. *Eur. J. Appl. Physiol.* 50:331–341, 1983.
86. Marcus, P. Some effects of cooling and heating areas of the head and neck on body temperature measurements at the ear. *Aerospace Med.* 44:397–402, 1973.
87. Marcus, P. Some effects of radiant heating of the head on body temperature measurement at the ear. *Aerospace Med.* 44:403–406, 1973.
88. McCaffrey, T.V., R.D. McCook and R.D. Wurster. Effect of head skin temperature on tympanic and oral temperature in man. *J. Appl. Physiol.* 39:114–118, 1975.
89. Mead, J. and C.L. Bonmarito. Reliability of rectal temperatures as an index of internal body temperature. *J. Appl. Physiol.* 2:97–109, 1949.
90. Melette, H.C. Skin, rectal and intravascular temperature adjustments in exercise. *Am. J. Physiol.* 163:734, 1950, Abstract.
91. Mitchell, D. and C.H. Wyndham. Comparison of weighting formulas for calculating mean skin temperature. *J. Appl. Physiol.* 26:616–622, 1969.
92. Mitchell, J.W., E.R. Nadel and J.A.J. Stolwijk. Respiratory weight losses during exercise. *J. Appl. Physiol.* 32:474–476, 1972.
93. Molnar, G.W. and R.C. Read. Studies during open-heart surgery on the special characteristics of rectal temperature. *J. Appl. Physiol.* 36:333–336, 1974.
94. Nadel, E.R. A brief overview. In: E.R. Nadel (Ed.) *Problems with Temperature Regulation during Exercise*. New York: Academic Press, 1977, pp. 1–10.

95. Nadel, E.R. Effects of temperature on muscle metabolism. In: *Biochemistry of Exercise* H.G. Knuttgen, J.A. Vogel and J. Poortmans (Eds.) Champaign, IL: Human Kinetics Publishers, 1983, pp. 134–143.
96. Nadel, E.R., R.W. Bullard and J.A.J. Stolwijk. Importance of skin temperature in the regulation of sweating. *J. Appl. Physiol.* 31:80–87, 1971.
97. Nadel, E.R., E. Cafarelli, M.F. Roberts and C.B. Wenger. Circulatory regulation during exercise in different ambient temperatures. *J. Appl. Physiol.* 46:430–437, 1979.
98. Nadel, E.R., S.M. Fortney and C.B. Wenger. Effect of hydration state on circulatory and thermal regulations. *J. Appl. Physiol.* 49:715–721, 1980.
99. Nadel, E.R. and S.M. Horvath. Comparison of tympanic membrane and deep body temperatures in man. *Life Sci.* 9:869–875, 1970.
100. Nadel, E.R., J.W. Mitchell, B. Saltin and J.A.J. Stolwijk. Peripheral modifications to the central drive for sweating. *J. Appl. Physiol.* 31:828–833, 1971.
101. Nadel, E.R., J.W. Mitchell and J.A.J. Stolwijk. Differential thermal sensitivity in the human skin. *Pflügers Arch.* 340:71–76, 1973.
102. Nadel, E.R. and J.A.J. Stolwijk. Effect of skin wettedness on sweat gland response. *J. Appl. Physiol.* 35:689–694, 1973.
103. Nielsen, B. Regulation of body temperature and heat dissipation at different levels of energy and heat production in man. *Acta Physiol. Scand.* 68:215–227, 1966.
104. Nielsen, B. Thermoregulatory responses to arm work, leg work and intermittent leg work. *Acta Physiol. Scand.* 72:25–32, 1968.
105. Nielsen, B. Thermoregulation during work in carbon monoxide poisoning. *Acta Physiol. Scand.* 82:98–106, 1971.
106. Nielsen B. and M. Nielsen. Body temperature during work at different environmental temperatures. *Acta Physiol. Scand.* 56:120–129, 1962.
107. Nielsen, M. Die Regulation der Körpertemperatur bei Muskelarbeit. *Skand. Arch. Physiol.* 79:193–230, 1938.
108. Nielsen, M. Heat production and body temperature during rest and work. In: J.D. Hardy, A.P. Gagge, and J.A.J. Stolwijk (Eds.) *Physiological and Behavioral Temperature Regulation.* Springfield, IL: C.C. Thomas, 1970, pp. 205–214.
109. Ogawa, T. Local effect of skin temperature on threshold concentration of sudorific agents. *J. Appl. Physiol.* 28:18–22, 1970.
110. Ogawa, T. and M. Asayama. Quantitative analysis of the local effect of skin temperature on sweating. *Jap. J. Physiol.* 36:417–422, 1986.
111. Peiss, C.N., W.C. Randall and A.B. Hertzman. Hydration of the skin and its effect on sweating and evaporative water loss. *J. Invest. Dermatol.* 26:459–470, 1956.
112. Petersen, E.S. and H. Vejby-Christensen. Effect of body temperature on steady-state ventilation and metabolism in exercise. *Acta Physiol. Scand.* 89:342–351, 1973.
113. Piironen, P. Sinusoidal signals in the analysis of heat transfer in the body. In: J.D. Hardy, A.P. Gagge and J.A.J. Stolwijk (Eds.) *Physiological and Behavioral Temperature Regulation.* Springfield, IL: C.C. Thomas, 1970, pp. 358–366.
114. Quinton, P.M. Sweating and its disorders. *Ann. Rev. Med.* 34:429–452, 1983.
115. Ramanathan, N.L. A new weighting system for mean surface temperature of the human body. *J. Appl. Physiol.* 19:531–533, 1964.
116. Randall, W.C. Quantification and regional distribution of sweat glands in man. *J. Clin. Invest.* 25:761–767, 1946.
117. Roberts, M.F. and C.B. Wenger. Control of skin blood flow during exercise by thermal reflexes and baroreflexes. *J. Appl. Physiol.* 48:717–723, 1980.
118. Robertshaw, D. Catecholamines and control of sweat gland. In: H. Blaschko, G. Sayers and A.D. Smith (eds.), *Handbook of Physiology. Endocrinology.* Bethesda, MD: American Physiological Society, Sec. 7, Vol. 6, 1975, pp. 591–603.
119. Robinson, S. Temperature regulation in exercise. *Pediatrics.* 32:691–702, 1963.
120. Robinson, S., D.B. Dill, J.W. Wilson and M. Nielsen. Adaptations of white men and negroes to prolonged work in humid heat. *Am. J. Trop. Med.* 21:261–287, 1941.
121. Rowell, L.B. Human cardiovascular adjustments to exercise and thermal stress. *Physiol. Rev.* 54:75–159, 1974.
122. Rowell, L.B. Cardiovascular adjustments to thermal stress. In: J.T. Shepherd and F.M. Abboud (Eds.) *Handbook of Physiology. The Cardiovascular System. Peripheral Circulation and Organ Blood Flow.* Bethesda, MD: American Physiological Society, sect. 2, Vol. 3, 1983, pp. 967–1023.
123. Rowell, L.B. Cardiovascular aspects of human thermoregulation. *Circ. Res.* 52:367–379, 1983.
124. Rowell, L.B., J.R. Blackmon, R.H. Martin, J.A. Mazzarella and R.A. Bruce. Hepatic clearance of indocyanine green in man under thermal and exercise stresses. *J. Appl. Physiol.* 20:384–394, 1965.
125. Rowell, L.B., G.L. Brengelmann, J.B. Blackmon, R.D. Twiss and F. Kusumi. Splanchnic

blood flow and metabolism in heat-stressed man. *J. Appl. Physiol.* 24:475–484, 1968.
126. Rowell, L.B., G.L. Brengelmann, J.A. Murray, K.K. Kraning and F. Kusumi. Human metabolic responses to hyperthermia during mild to maximal exercise. *J. Appl. Physiol.* 26:395–402, 1969.
127. Rowell, L.B., P.R. Freund and G.L. Brengelmann. Cutaneous vascular response to exercise and acute hypoxia. *J. Appl. Physiol.* 53:920–924, 1982.
128. Saltin, B., A.P. Gagge, U. Bergh and J.A.J. Stolwijk. Body temperatures and sweating during exhaustive exercise. *J. Appl. Physiol.* 32:635–643, 1972.
129. Saltin, B., A.P. Gagge and J.A.J. Stolwijk. Body temperatures and sweating during thermal transients caused by exercise. *J. Appl. Physiol.* 28:318–327, 1970.
130. Saltin, B. and L. Hermansen. Esophageal, rectal and muscle temperature during exercise. *J. Appl. Physiol.* 21:1757–1762, 1966.
131. Sato, K. The physiology, pharmacology and biochemistry of the eccrine sweat gland. *Rev. Physiol. Biochem. Pharmacol.* 79:51–131, 1977.
132. Sato, K. and R.L. Dobson. Regional and individual variations in the function of the human eccrine sweat gland. *J. Invest. Derm.* 54:443–449, 1970.
133. Sato, K., A. Nishiyama and M. Kobayashi. Mechanical properties and functions of the myoepithelium in the eccrine sweat gland. *Am. J. Physiol.* 237:C177–C184, 1979.
134. Sato, K. and F. Sato. Cyclic AMP accumulation in the beta adrenergic mechanism of eccrine sweat secretion. *Pflügers Arch.* 390:49–53, 1981.
135. Sato, K. and F. Sato. Pharmacologic responsiveness of isolated single eccrine sweat glands. *Am. J. Physiol.* 240:R44–R51, 1981.
136. Sato, K. and F. Sato. Role of calcium in cholinergic and adrenergic mechanisms of eccrine sweat secretion. *Am. J. Physiol.* 241:C113–C120, 1981.
137. Sato, K. and F. Sato. Individual variations in structure and function of human eccrine sweat gland. *Am. J. Physiol.* 245:R203–R208, 1983.
138. Sawka, M.N., R.C. Dennis, R.R. Gonzalez, A.J. Young, S.R. Muza, J.W. Martin, C.B. Wenger, R.P. Francesconi, K.B. Pandolf and C.R. Valeri. Influence of polycythemia on blood volume and thermoregulation during exercise-heat stress. *J. Appl. Physiol.* 62:912–918, 1987.
139. Sawka, M.N., R.R. Gonzalez, L.L. Drolet and K.B. Pandolf. Heat exchange during upper- and lower-body exercise. *J. Appl. Physiol.* 57:1050–1054, 1984.
140. Sawka, M.N., R.R. Gonzalez, A.J. Young, S.R. Muza, K.B. Pandolf, W.A. Latzka, C.R. Valeri and R.C. Dennis. Effects of acute polycythemia and hydration on blood volume and exercise-heat performance. *Physiologist* 30:205, 1987, Abstract.
141. Sawka, M.N., J.S. Petrofsky and C.A. Phillips. Energy cost of submaximal isometric contractions in cat fast and slow twitch muscles. *Pflügers Arch.* 390:164–168, 1981.
142. Sawka, M.N., N.A. Pimental and K.B. Pandolf. Thermoregulatory responses to upper body exercise. *Eur. J. Appl. Physiol.* 52:230–234, 1984.
143. Sawka, M.N., A.J. Young, B.S. Cadarette, L. Levine and K.B. Pandolf. Influence of heat stress and acclimation on maximal aerobic power. *Eur. J. Appl. Physiol.* 53:294–298, 1985.
144. Schumacker, P.T., J. Rowland, S. Saltz, D. Nelson and L.H. Wood. Effects of hyperthermia and hypothermia on oxygen extraction by tissues during hypovolemia. *J. Appl. Physiol.* 63:1246–1252, 1987.
145. Sen Gupta, J., P. Dimri and M.S. Malhotra. Metabolic responses of Indians during submaximal and maximal work in dry and humid heat. *Ergonomics* 20:33–40, 1977.
146. Shapiro, Y., K.B. Pandolf and R.F. Goldman. Predicting sweat loss responses to exercise, environment and clothing. *Eur. J. Appl. Physiol.* 48:83–96, 1982.
147. Shiraki, K., N. Konda and S. Sagawa. Esophageal and tympanic temperature responses to core blood temperature changes during hyperthermia. *J. Appl. Physiol.* 61:98–102, 1986.
148. Sloan, R.E. and W.R. Keatinge. Depression of sublingual temperature by cold saliva. *Brit. Med. J.* 1:718–720, 1975.
149. Stitt, J.T. Fever versus hyperthermia. *Fed. Proc.* 38:39–43, 1979.
150. Stolwijk, J.A.J. and J.D. Hardy. Partitional calorimetric studies of responses of man to thermal transients. *J. Appl. Physiol.* 21:967–977, 1966.
151. Stolwijk, J.A.J. and J.D. Hardy. Temperature regulation in man—a theoretical study. *Pflügers Arch.* 291:129–162, 1966.
152. Stolwijk, J.A.J., B. Saltin and A.P. Gagge. Physiological factors associated with sweating during exercise. *Aerospace Med.* 39:1101–1105, 1968.
153. Strydom, N.B., J.F. Morrison, J. Booyens and J. Peter. Comparison of oral and rectal temperatures during work in heat. *J. Appl. Physiol.* 8:406–408, 1956.
154. Strydom, N.B., C.H. Wyndham, C.G. Williams, J.F. Morrison, G.A.G. Bredell and A. Joffe. Oral/rectal temperature difference during work and heat stress. *J. Appl. Physiol.* 20:283–287, 1965.
155. Tabor, M.W., D.M. Blaho and W.R. Schriver. Tympanic membrane perforation: complication of tympanic thermometry during general anesthesia. *Oral Surg.* 51:581–583, 1981.

156. Taylor, C.L. and K. Buettner. Influence of evaporative forces upon skin temperature dependency of human perspiration. *J. Appl. Physiol.* 6:113–123, 1953.
157. Taylor, W.F., J.M. Johnson, W.A. Kosiba and C. M. Kwan. Graded cutaneous vascular responses to dynamic leg exercise. *J. Appl. Physiol.* (In press) 1988.
158. Teichner, W.H. Assessment of mean body surface temperature. *J. Appl. Physiol.* 12:169–176, 1958.
159. Thauer, R. Thermosensitivity of the spinal cord. In: J.D. Hardy, A.P. Gagge, and J.A.J. Stolwijk (Eds.) *Physiological and Behavioral Temperature Regulation.* Springfield, IL: C.C. Thomas, 1970, pp. 472–492.
160. Thaysen, J.H. and I.L. Schwartz. Fatigue of the sweat glands. *J. Clin. Invest.* 34:1719–1725, 1955.
161. Vaalasti, A., H. Tainio and L. Rechardt. Vasoactive intestinal polypeptide (VIP)-like immunoreactivity in the nerves of human axillary sweat glands. *J. Invest. Dermatol.* 85:246–248, 1985.
162. Van Beaumont, W. and R.W. Bullard. Sweating: its rapid response to muscular work. *Science* 141:643–646, 1963.
163. Wallace, C.T., W.E. Marks, W.Y. Adkins and J.E. Mahaffey. Perforation of the tympanic membrane, a complication of tympanic thermometry during anaesthesia. *Anesthesiology* 41:290–291, 1974.
164. Webb-Peploe, M.M. and J.T. Shepherd. Response of dogs' cutaneous veins to local and central temperature changes. *Circ. Res.* 23:693–699, 1968.
165. Webb-Peploe, M.M. and J.T. Shepherd. Response of large hindlimb veins of the dog to sympathetic nerve stimulation. *Am. J. Physiol.* 215:299–307, 1968.
166. Wendt, I.R. and C.L. Gibbs. Energy production of rat extensor digitorum longus muscle. *Am. J. Physiol.* 224:1081–1086, 1973.
167. Wenger, C.B. Heat of evaporation of sweat: thermodynamic considerations. *J. Appl. Physiol.* 32:456–459, 1972.
168. Wenger, C.B. Non-thermal factors are important in the control of skin blood flow during exercise only under high physiological strain. *Yale J. Biol. Med.* 59:307–319, 1986.
169. Wenger, C.B., R.B. Bailey, M.F. Roberts and E.R. Nadel. Interaction of local and reflex thermal effects in control of forearm blood flow. *J. Appl. Physiol.* 58:251–257, 1985.
170. Wenger, C.B. and M.F. Roberts. Control of forearm venous volume during exercise and body heating. *J. Appl. Physiol.* 48:114–119, 1980.
171. Wenger, C.B., M.F. Roberts, E.R. Nadel and J.A.J. Stolwijk. Thermoregulatory control of finger blood flow. *J. Appl. Physiol.* 38:1078–1082, 1975.
172. Wenger, C.B., M.F. Roberts, J.A.J. Stolwijk and E.R. Nadel. Forearm blood flow during body temperature transients produced by leg exercise. *J. Appl. Physiol.* 38:58–63, 1975.
173. Wenger, C.B., L.A. Stephenson and M.A. Durkin. Effect of nerve block on response of forearm blood flow to local temperature. *J. Appl. Physiol.* 61:227–232, 1986.
174. Whitby, J.D. and L.J. Dunkin. Temperature difference in the oesophagus. Preliminary study. *Brit. J. Anaesth.* 40:991–995, 1968.
175. Whitby, J.D. and L.J. Dunkin. Cerebral, oesophageal and nasopharyngeal temperatures. *Brit. J. Anaesth.* 43:673–676, 1971.
176. Williams, C.G., G.A.G. Bredell, C.H. Wyndham, N.B. Strydom, J.F. Morrison, J. Peter, P.W. Fleming and J.S. Ward. Circulatory and metabolic reactions to work in heat. *J. Appl. Physiol.* 17:625–638, 1962.
177. Winslow, C.-E.A., L.P. Herrington and A.P. Gagge. A new method of partitional calorimetry. *Am. J. Physiol.* 116:641–655, 1936.
178. Wyndham, C.H., N.B. Strydom, J.F. Morrison, C.G. Williams, G.A.G. Bredell and J. Peter. Fatigue of the sweat gland response. *J. Appl. Physiol.* 21:107–110, 1966.
179. Young, A.J., M.N. Sawka, L. Levine, B.S. Cadarette and K.B. Pandolf. Skeletal muscle metabolism during exercise is influenced by heat acclimation. *J. Appl. Physiol.* 59:1929–1935, 1985.

4

Human Heat Acclimatization

C. Bruce Wenger, M.D., Ph.D.

OUTLINE

INTRODUCTION

Overview

Reliance on sweating rather than panting for evaporative heat loss potentially allows humans to use almost the whole skin surface, rather than the much smaller surface of the upper respiratory passages, for evaporation. This advantage has its price, however, in the form both of a burden on maintenance of fluid and electrolyte balance, and of increased difficulty in meeting the circulatory requirements of exercise. In the first instance, electrolytes secreted in the sweat are lost to the body, while electrolytes in the saliva of a panting animal can be swallowed and reused; and especially in a humid environment, sweat may be secreted more rapidly than it can be evaporated, so that the excess drips from the skin and is wasted. In the second instance, if the evaporation of sweat from the skin is to be effective in removing heat produced by the viscera and exercising skeletal muscle, skin blood flow must increase greatly in order to carry this heat to the skin. The increase in skin blood flow causes pooling of blood in the cutaneous vascular bed, thus reducing thoracic blood volume and cardiac filling and stroke volume (Chapter 3). As a result, heat stress reduces the ability of humans to perform sustained exercise.

Other manifestations of heat stress, which depend on the degree of physiological strain produced, include rapid pulse, narrow pulse pressure, headache, nausea and vomiting, dizziness, cramps, shortness of breath, dependent edema, flushing of the face and neck, and orthostatic hypotension (11,36,84). Indeed, a feeling of lassitude and increased fatigue during exercise are common experiences at the beginning of a spell of hot weather. However, if the weather continues hot for a few days, the lassitude abates and the ability for sustained exercise returns, as one becomes "acclimatized" to heat. This acclimatization is the result of a number of physiological adaptations, some clearly defined, others less well understood, which serve both to improve the body's thermoregulatory function and to alleviate the circulatory strain and salt loss occurring during heat stress. This chapter will describe the process of acclimatization and discuss some of what is known about the underlying mechanisms.

The terms "acclimation" and "acclimatization" are often used interchangeably to refer to any adaptive changes which occur when an individual undergoes prolonged or repeated exposure to a stressful environment, and which reduce the physiological strain produced by such an environment. However, this chapter will follow the recommendation of the International Union of Physiological Sciences, and call such changes "acclimation" if produced in a controlled laboratory setting, and "acclimatization" if produced by a change in the natural environment, whether by a change of season or place of residence. "Acclimatization" will also be used as an inclusive term instead of the awkward phrase "acclimatization and/or acclimation."

Acclimation to heat is produced by repeated exposure to heat sufficient to raise internal body temperature (most effectively accomplished with exercise) and provoke moderate to profuse sweating. Moderate exercise in the heat for an hour a day is usually sufficient to produce an effect in unacclimatized subjects, and changes in the responses to heat exposure begin to appear during the first few days. Acclimatization to heat includes two parts: First, certain cardiovascular changes occur that progressively reduce the heart rate and cardiovascular strain associated with a given level of exercise in the heat. Second, sweating during a heat exposure begins earlier, a given elevation in core temperature elicits a higher sweat rate, and higher sweat rates can be sustained. The combined results of these physiological changes can be quite dramatic.

Figure 4-1 illustrates these effects in three healthy young men, acclimated by daily treadmill walks in dry heat for 10 days. On the first day, heart rate and rectal temperature during exercise rose to levels much higher than those reached during exercise at the same intensity in an ambient temperature of 25° C (cool control). After 10 days, however, final heart rate and rectal temperature during exercise fell by more than 40 beats $\cdot$ min^{-1} and 1° C respectively, to values little higher than in cool control conditions. In addition, sweat production rose 10%, skin temperature fell about 1.5° C, and the metabolic cost of treadmill walking fell 4%.

During prolonged heat exposure, especially with high humidity, sweat rate declines (Chapter 3). This effect is often referred to as inhibition or "fatigue" of the sweat glands, though much of the effect may be due to hidromeiosis (Chapter 3). Heat acclimatization, especially to humid heat, increases the ability of the sweat glands to resist this inhibition, and thus to sustain the necessary sweat rate during prolonged heat stress. Figure 4-2 shows average responses of 10 students at the beginning and end of a program of acclimation to humid heat, during which they performed bench-stepping exercise for four hours a day. When the subjects were unacclimated, sweat rate declined after the first hour in the heat; but once acclimated, the subjects could maintain high sweat rates throughout the four hours. As a consequence, heart rate and rectal temperature rose substantially less during the second through fourth hours after acclimation.

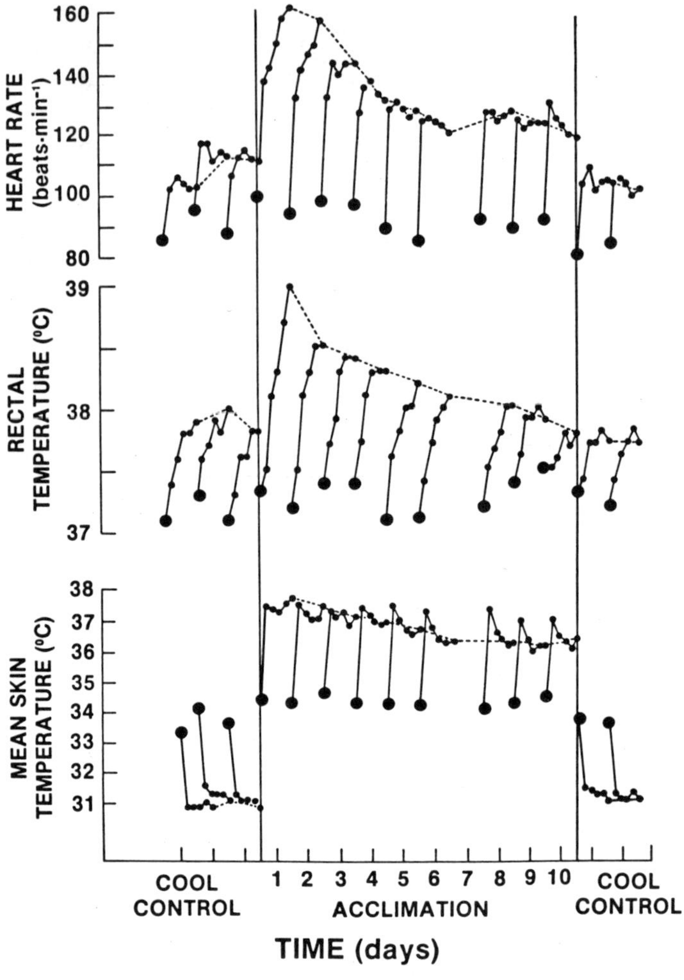

Figure 4-1. *Change in the responses of heart rate, rectal temperature, and mean skin temperature during exercise in a 10-day program of acclimation to dry heat, together with responses during exercise in a cool environment before and after acclimation. Each day's exercise consisted of five 10-min periods of treadmill exercise, separated by 2-min rest periods. Large circles show values before the start of the first exercise period each day, small circles show values at the ends of successive exercise periods, and dotted lines connect final values each day. Redrawn from Reference 37.*

In conclusion, the traditional hallmarks of heat acclimation are lower thermal and cardiovascular strain, manifested primarily as lower core temperature and heart rate and secondarily as alleviation of the symptoms of heat strain (11), and increased sweat production during a given level of heat stress or exercise in the heat. In addition,

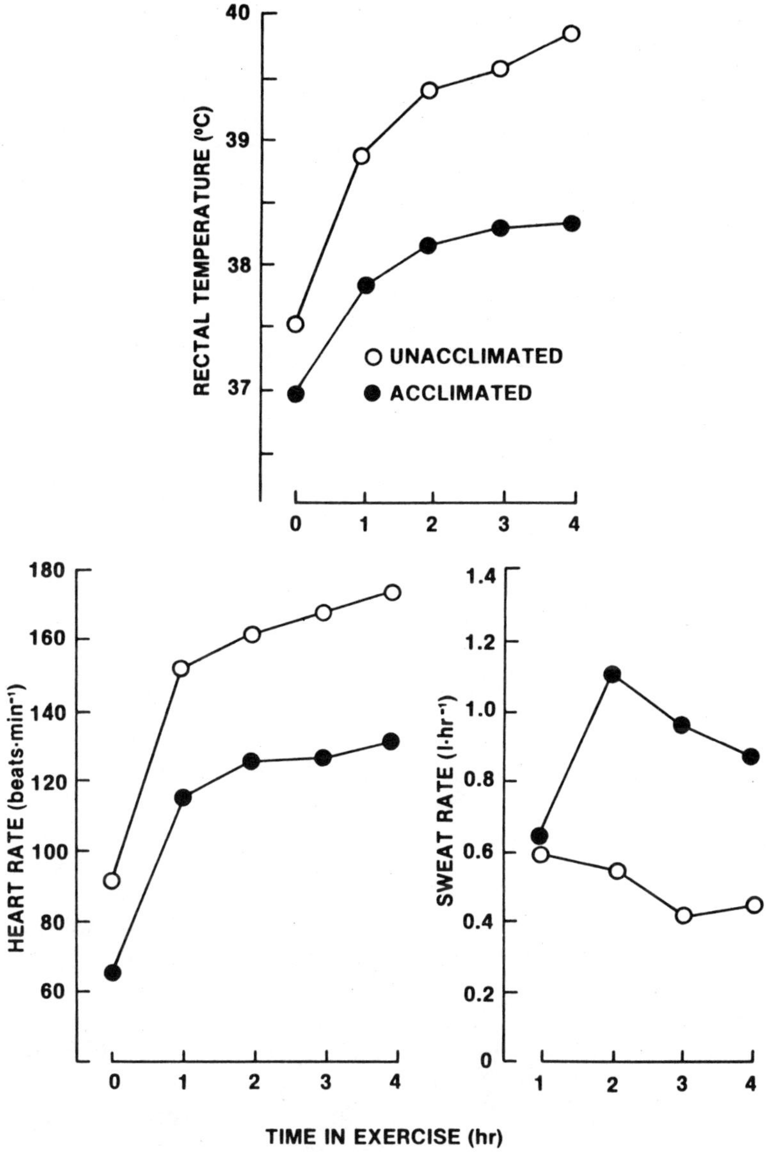

Figure 4-2. *Responses of medical students during 4 hours of exercise in humid heat, before and after a program of acclimation to humid heat, during which subjects performed bench-stepping for 4 hours a day. Redrawn from Reference 140.*

sweat usually becomes more dilute, and acclimated men are generally more comfortable and can work longer in the heat.

Differences Between Dry and Humid Heat: Biophysical and Physiological Significance

Hot weather is less uncomfortable, and exercise more easily borne, if humidity is low than if it is high. This familiar experience reflects the biophysical differences between hot dry and hot humid environments and their physiological consequences. Because of these biophysical differences, there are certain physiological responses that confer much greater benefit in humid heat than in dry heat; and one might expect acclimatization to humid heat to produce somewhat different physiological adaptations from acclimatization to dry heat. There is evidence suggesting that this is so, but the literature directly comparing acclimation to humid heat and acclimation to dry heat is rather meager. There are a few reports that acclimation to dry heat confers a substantial advantage in humid heat also (11,36), but there seems to be only one report in which the responses of subjects acclimated to humid heat and those acclimated to dry heat were compared by testing both groups of subjects in the same conditions before and after acclimation (43). This report is discussed later in the section Inhibition of Sweat Gland Activity, Including Hidromeiosis.

The rate of sweat evaporation is proportional to the wetted skin area and also to the difference between the saturation vapor pressure of water at skin temperature and the ambient water vapor pressure (Chapters 1,2). To produce a high rate of evaporative cooling in a humid environment, then, it is necessary to make up for the effect of the high ambient water vapor pressure by maintaining either a higher vapor pressure at the skin (which requires a higher skin temperature) or a larger wetted skin area, as compared to what would be necessary in a dry environment. Unless core temperature is allowed to rise along with skin temperatures, with no reduction in core-to-skin temperature gradient, higher skin temperature must be achieved by increasing core-to-skin thermal conductance, which presumably requires a higher skin blood flow. One difference that might be expected between acclimatization to humid heat and acclimatization to dry heat is for the former to involve greater circulatory adaptations, to support higher skin blood flow with minimal circulatory strain. The wetted skin area may be increased either by increasing sweating selectively in those areas that sweated least before acclimatization, or by more intense stimulation of sweating generally, so that even some skin areas that tend to sweat less profusely sweat enough to be completely wetted. In the latter case, those areas that sweat more profusely would produce more sweat than can be evaporated, so that the excess would drip from the body and be wasted. Therefore, one might also expect human heat acclimatization to be accompanied by selective increases in regional sweating, so as to make more efficient use of the skin surface for evaporation and at the same time to minimize waste of sweat by dripping.

Acquisition and Loss of Heat Acclimatization

Conditions Necessary to Produce Acclimatization. Regular heavy exercise in the heat is the most effective stimulus for developing heat acclimatization. However, heat exposure during moderate exercise, or even during rest, confers definite acclimation, though to a lesser degree (11,36,84). This occurs both during the change from winter to summer, and during experimental heat exposure. Rectal temperature (60,70) and heart rate (70,118) during a standard environmental heat stress or exercise in the heat are lower in summer than in winter, sweat is secreted more profusely and is more dilute (60), and heart rate is lower and systolic pressure is better maintained during head-up tilt (118). Nine to twelve days with passive heat exposure of resting subjects increases the sweat rate elicited at any given level of core temperature (56,58) and reduces the heart rate and skin and core temperatures and increases the sweat rate reached during exercise in the heat (43). Probably because subjects are already partly acclimatized, experimental heat acclimation occurs more quickly in warm weather (36). Up to a point, at least, the degree of acclimation acquired is proportional to the the daily heat stress. In a study in which different groups of subjects were acclimated for 12 days by exposure at rest to different intensities of heat and for different periods from 30 to 120 min per day, the percent increase in sweat loss over the course of acclimation was proportional to the amount of sweat secreted during acclimation (45). Likewise, daily 100-min exercise bouts in the heat conferred more acclimation after nine days than did 50-min bouts; however, adding a second exercise bout each day provided no further benefit (80). Acclimation to heat and to cold can coexist (33), and can even be induced concurrently, with heat exposures and cold exposures on mornings and afternoons of the same day (50).

In experimental heat acclimation by daily exercise in a hot environment, most of the improvement in heart rate, core and skin temperatures, and sweat rate is acquired in the first week, although there is no sharp end to the improvement (3). Heart rate shows the most rapid reduction (84,98,136), most of which occurs in four to five days (84). After seven days the reduction in heart rate is nearly complete, and most of the improvement in core and skin temperature also has occurred (62,98). The increase in sweating (62,84) and in the ease of walking in the heat (62) may be more prolonged, taking up to a month.

Relation of Acclimation to Physical Fitness and Aerobic Training. Aerobic training, even in a temperate environment, elevates core temperature and elicits sweating, and thus can improve thermoregulatory function and heat tolerance. For example, 10 days of cycle exercise at 70 to 80% of maximal oxygen uptake ($\dot{V}O_2max$) increased the slope of the relation of sweat rate to core temperature (89), and three months of interval training, alternating cycle exercise at 25 and 110% $\dot{V}O_2max$, lowered the core temperature threshold for sweating by 0.5° C (57). Good aerobic condition hastens acquisition of acclimation (98), and the speed with which core temperature during exercise declines over the course of acclimation is directly related to pre-acclimation $\dot{V}O_2max$ (90). Conversely, the majority of southern African mining recruits that cannot acclimate to heat have $\dot{V}O_2max$ below 2.5 $\ell \cdot min^{-1}$ (67), and patients with chronic congestive heart failure (who presumably have very low $\dot{V}O_2max$) cannot acclimate effectively (17); however, age itself appears to present no obstacle to acclimation of fit and healthy subjects (94). On the other hand, continued aerobic training at moderate levels appears to confer no further advantage in heat tolerance beyond that associated with good physical condition (11,36), and acclimation reduces the correlation between aerobic fitness and thermoregulatory fitness (19). Furthermore, Strydom and Williams concluded that aerobic training in cool conditions is not an adequate substitute for heat acclimation (125). They based this conclusion on observations made on subjects who were tested during 4 hours of exercise in the heat, then trained in bench stepping at a $\dot{V}O_2$ of 1.5 to 1.6 $\ell \cdot min^{-1}$ for 4 hours per day for 12 days at 21° C ambient temperature, and then again tested in the heat. Although subjects' physical condition improved substantially, their physiological responses in the heat were improved only during the first 2 hours of exercise.

The question of whether or not aerobic training induces heat acclimation has been the subject of much controversy. Gisolfi and Cohen (48) have reviewed the literature pertinent to this controversy, and note that the points at issue include the duration and intensity of the aerobic training programs, and the duration and conditions of the tests used to assess heat tolerance. They conclude that two weeks of intense training in a temperate environment cannot substitute for exercise in the heat to produce heat acclimation, but that 8 to 11 weeks of interval training (i.e., training in which periods of intense effort alternate with periods of mild effort) in a temperate environment can produce substantial improvement in heat tolerance, and that endurance runners can maintain thermal balance for up to 4 hours of mild exercise in dry or humid heat (48), in contrast to the observations of Strydom and Williams (125) cited above.

Another aspect of the controversy seems to involve a theoretical question of what constitutes full heat acclimation. Highly-trained distance runners who were already quite tolerant to exercise in conditions of moderate heat stress were reported to acclimate further by exercising in more severe dry heat (2,91). Since these athletes increased their heat tolerance further by exercise in intense dry heat, Wyndham (135) argued that they had not been acclimated by aerobic training alone. In making that argument, Wyndham seems to assume implicitly that one either is or is not acclimated, i.e., that acclimation is an absolute state rather than a matter of degree, even though others (see Ref. 6) recognize degrees of acclimation. Most of Wyndham's research has been concerned with physiological responses to hot humid environments, in which the rate at which sweat can be evaporated is much lower than what is possible in a dry environment, and also lower than the rate at which an acclimatized subject can sweat. In humid heat, therefore, the biophysics of heat exchange sets limits to the severity of exercise-heat stress to which subjects can acclimate, and depending on the wet-bulb temperature, these limits occur at low to moderate levels of $\dot{V}O_2$ (139). If subjects can be acclimated fairly readily to these limiting conditions, it may not be unreasonable to think of acclimation as an absolute state in hot humid conditions. However, it probably is unreasonable to think of acclimation as an absolute state in hot dry environments. In fact, Wyndham's laboratory has not determined whether or not their subjects, after acclimation to humid heat, are capable of further increasing their tolerance to exercise in hot dry environments.

Loss of Acclimation. The benefits of acclimation are lessened or undone by sleep loss, intercurrent infection, and alcohol abuse (6,84), salt depletion (84), and dehydration (84,104). Moreover, heat acclimation is transient, and gradually disappears if not maintained by repeated heat exposure, although partial losses due to a few days' lapse are easily made up (6). It is generally agreed that the improvement in heart rate, which develops more rapidly, is also lost more rapidly than are the thermoregulatory improvements (80,90,99,131). However, there is a great variation in the literature concerning the speed with which acclimation is lost, with the time taken for nearly complete loss varying from 17 days for women acclimated for 7 days (21) to 34 days for soldiers acclimated for 14 days (1). Soldiers acclimated for 9 days in another study (90) may have retained their acclimation even longer, although they were followed only for 18 days after the end of acclimation. In all tests during this time they retained at least 71% of their improvement in heart rate and 82% of their improvement in rectal temperature during exercise in the heat. As this comparison may suggest, physically fit subjects retain heat acclimation longer (11,90,98). Warm weather also may favor retention of acclimation (36), but there is conflicting evidence on this point (131), and intermittent cold exposure is reported not to hasten the loss of heat acclimation (122).

Comparison of Acclimation with Acclimatization, Including Adaptations of Peoples Indigenous to Tropical Climates

Acclimatization in a natural environment differs from experimental acclimation in that the heat stress in the latter is more precisely defined and usually more severe, but most commonly is intermittent (although in some studies subjects stayed in a climatic chamber throughout acclimation—see Refs. 9, 11), and takes up to a few weeks rather than months or years. Warm summers, as discussed above, should be considered as inducing some natural acclimatization.

In general, Europeans living in tropical climates have, as a result of their natural acclimatization, the same sorts of responses during exercise in the heat (i.e., higher sweat rate and lower core and skin temperatures and heart rates than unacclimated controls) as are acquired in a program of artificial acclimation (54). However, although better acclimatized than their temperate counterparts, such naturally acclimatized persons need to undergo more severe exercise-heat stress in order to equal artificially acclimated subjects (75,86). The literature is not unanimous, however, in reporting that residence in the tropics lowers core temperature. There are two reports that Europeans and resident Indians in Pondichery in southern India had rectal temperature *higher* by about 0.6° C, and also a wider core-to-skin thermal gradient and thus lower core-to-skin conductance (87,92), and lower hand blood flow (87), both at rest and after exercise (33° C db, 24.6–25.5° C wb), than Europeans in Paris. Although these responses would tend to minimize cardiovascular strain, these reports (87,92) are rather puzzling, not only because of the way that these Europeans acclimatized to heat, but also because in a standard heat stress test, Indians newly arrived in Britain had *lower* core temperature than unacclimatized Europeans, though they sweated the same (34).

Newcomers to the tropics are continually bathed in sweat, but sweat less after several years (72). Natives of the tropics sweat and drip less than newcomers during exercise (38), and at rest they are relatively dry to the touch (38,72). In standard heat tests, Japanese immigrants in the Philippines began to sweat at a lower level of heat stress than did Filipinos or tropical-born Japanese, and the response of those born or acclimated long-term in the tropics appears to be a means to avoid excessive sweating (72). Although they sweat less profusely, people born in tropical Asia, both Japanese and members of the indigenous populations, have more functioning sweat glands than do Japanese in Japan (72). Subjects raised in temperate climates appear to have some sweat glands that are fully developed histologically but do not secrete, and perhaps such glands would secrete, had these subjects been born and raised in the tropics (24).

There is some evidence that peoples indigenous to tropical climates, and also their descendants living in more temperate climates, are better adapted than Europeans to humid heat. In interpreting the following studies, however, it should be remembered that the subject populations were not matched for such factors as lean body mass and aerobic fitness.

In a study of American soldiers (5), black soldiers had slightly but significantly lower heart rates, rectal temperatures, and sweat rates than white soldiers during marches in hot humid weather. These differences between black and white soldiers seemed unrelated to where the soldiers had lived before, since southern white soldiers were no more heat tolerant than northerners. The differences disappeared in hot dry weather, if the soldiers marched in the shade or were fully clothed, so as to eliminate differences in absorption of solar radiation by the skin.

In a study of Mississippi sharecroppers conducted from June to August (95), white sharecroppers had somewhat higher final rectal and skin temperatures and heart rates than black sharecroppers after a standard 2-hour treadmill walk ($5.6\ km \cdot h^{-1}$, 8.6% grade) in humid heat. The difference owed, at least partly, to lower efficiency (i.e., higher oxygen uptake in relation to mechanical work) of the whites. The whites also sweated more, both absolutely and (by 14%) in relation to their heat dissipation. Since the blacks had mean skin temperatures 0.54° C lower than the whites, differences in sensible heat exchange are in the wrong direction to explain differences in sweating. The blacks thus used their sweat more efficiently, presumably wasting less by dripping. In another study of Mississippi sharecroppers (41), the whites had, normalized for surface area, the same plasma volume as the blacks ($1.89\ \ell \cdot m^{-2}$), but a smaller interstitial fluid volume as computed from thiocyanate space (4.33 vs. $5.08\ \ell \cdot m^{-2}$). The white laboratory workers conducting this study increased their plasma volume and decreased their interstitial fluid volume on going to Mississippi, so as to reach similar values to the white sharecroppers' (41). It is unfortunate that these fluid volumes were measured only at rest: it would be interesting to know whether the blacks' larger interstitial fluid volume allowed them to maintain a larger plasma volume during exercise.

Some of the above differences between black and white Americans find parallels in data reported from Africa. At high wet bulb temperatures, acclimated African laborers sweated much less, and had somewhat lower heart rates than reported for acclimated Europeans at similar levels of heat stress assessed as predicted 4-hour sweat rate (137). In another study, unacclimatized whites were less heat-tolerant (as indicated by symptoms of heat strain) than unacclimatized Bantu during 4 hours' exercise in humid heat, and had higher rectal temperatures even though they sweated significantly more (140). The whites also had a greater metabolic rate during a given exercise task (140). After acclimation, differences in heat tolerance and rectal temperature disappeared, and differences in metabolic rate were less, but persisted. However, after the first hour the whites still sweated more, and the difference in sweat rate increased with time (140).

Fully acclimated Europeans and Nigerians have similar rectal temperatures and sweat rates early during exercise under severe conditions, but the Nigerians can continue longer and are less likely to collapse (75). The sweat of Nigerians fully acclimated to humid heat is more dilute than that of similarly acclimated Europeans. However, that report makes no mention either of controlling salt intake, or of the relation of sweat salt concentration to sweat rate, even though sweat rate affects salt concentration in sweat (Fig. 4-4). During exercise in humid heat, sweat rate of acclimated Nigerians (76) or Bantu (140) decreases more quickly than that of the Europeans. Furthermore, Europeans living in Nigeria but not fully acclimated, thermoregulate as well as similarly acclimatized Nigerians during moderate heat stress. However, in more severe heat stress they thermoregulate less well, even though they sweat more than their Nigerian counterparts.

East Indians appear to adopt a somewhat different strategy from either Europeans or blacks. In a standard heat stress test, unacclimated Indians in Britain had the same core temperature as unacclimated Europeans, but sweated less. After 10 days' acclimation of both groups, the Indians sweated as much as unacclimated Europeans, while Europeans increased their sweat rate, and decreased their core temperature enough that it was lower than that of the Indians (34). Thus, like blacks, the Indians used their sweat more efficiently than Europeans, but did not defend their core temperature as well as Europeans when both groups were fully acclimated.

In summary, short-term residence in the tropics confers certain adaptations to heat that are similar in nature, though less in degree, to those produced by an acclimation program employing regular heavy exercise in the heat; and in this respect it resembles other sorts of sustained or repeated mild-to-moderate exercise-heat stress. Long-term residence induces in addition changes in the pattern of sweating that minimize excess sweating in hot humid conditions. Comparisons of different population groups are somewhat risky, since apart from the obvious factors of race and—in some cases—history of place of residence, other factors that might have affected subjects' responses usually have not been controlled. However, studies conducted both in Africa and in the United States indicate that in hot humid conditions blacks waste less sweat than whites, experience less circulatory strain, and consume slightly less oxygen than whites during similar exercise tasks.

CHANGES IN THERMOREGULATORY RESPONSES

With acclimatization to heat, the body's ability to meet the demands of heat stress increases. All of the physiological responses for heat dissipation are greater at any given core and skin temperatures, so that heat balance during exercise-heat stress is achieved at lower core and skin temperatures. This information by itself is insufficient to predict the effect of acclimation on circulatory responses during thermal steady state. However, the biophysics of heat exchange require somewhat more evaporative cooling after acclimation, because of the cooler skin and resulting change in sensible heat exchange. Sweat rate therefore must be greater after acclimation, since the responses that reduce waste of sweat by dripping develop far too slowly to occur during a program of acclimation in the laboratory.

One of the most dramatic effects of heat acclimatization is the strikingly improved feeling of comfort and well being that takes place after a few days. This improvement must owe almost entirely to the lowering of core temperature and heart rate, since at the same core temperature resting subjects are *more* uncomfortable after acclimation than before (45). The greater discomfort at a given core temperature after acclimation may owe partly to an effect of heart rate, which in resting subjects is also higher at a given core temperature after acclimation.

Sweating

Acclimatization greatly increases the sweating capacity, and peak sweat rate increases from up to 1.5 liters of sweat per hour in a healthy man unacclimatized to heat, to as much as 2–3 liters per hour in men well acclimatized to heat (35,74). Such high sweat rates cannot be sustained for long, however, and maximum daily sweat production is usually taken to be 10–15 liters (71). Furthermore, especially with acclimatization to humid heat, the increase in total sweating capacity is accompanied by increased ability to sustain sweating during prolonged heat exposure, as mentioned above; and by increased local sweating capacity in those skin areas that sweated the least before acclimation, thus helping to maximize the wetted skin area during humid heat exposure. These changes combine to produce large increases in sweat rate in humid heat after acclimatization. The rate of evaporative cooling increases much less, however, since much of the additional sweat produced is wasted by dripping. In dry heat, by contrast, nearly all of the sweat produced is evaporated, and acclimation causes only a modest increase in sweat production, perhaps only enough to account for the lesser heat storage and to compensate for the alteration in sensible heat exchange (radiation and convection) associated with lower skin temperature (24).

The sweat glands are innervated with sympathetic nerves that secrete acetylcholine (Chapter 3) and thus are stimulated by cholinomimetic drugs such as acetylcholine and methacholine, a synthetic analog of acetylcholine that resists hydrolysis by cholinesterases; and they are inhibited by anticholinergic drugs such as atropine. These drugs have been used to study sweating in some of the studies discussed below.

Relation to Core Temperature. Sweating begins at lower core and skin temperatures after heat acclimation (24,46,58,89,93,134); and the plateau in the relation of sweat rate to core temperature occurs at a higher sweat rate (134) (Fig. 4-3). Although one report (89) stated that exercise training increases the slope of the relation of sweat rate to core temperature, while acclimation lowers the threshold, this distinction does not always hold: In several other studies exercise training changed the threshold (57,93), and acclimation changed the slope (56,93,134).

After acclimation a greater sweating response is elicited not only by a standard thermal stress, but also by a standard electrical stimulus (20) or by pharmacological stimulation with standard doses of methacholine (25,69). Therefore the changes in sweating that occur with acclimation must owe at least partly to changes in the sweat glands themselves, and cannot be ascribed wholly to a change in the sympathetic signal. Histological studies indicate that there are morphologically normal sweat glands that do not secrete, but these do not seem to be recruited by acclimation, so that the increase in sweat rate is accomplished by an increase in the activity of the al-

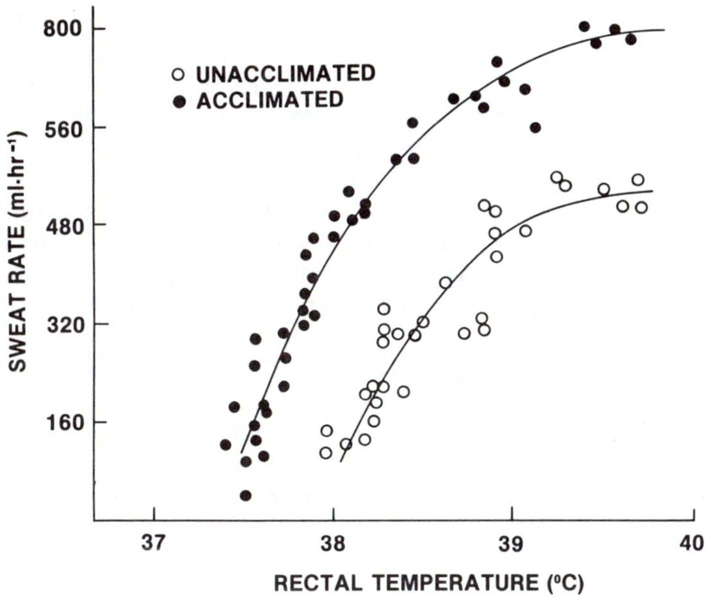

Figure 4-3. *The relation between sweat rate, averaged over four hours, during bench stepping 58 min of each hour in humid heat, and rectal temperature in unacclimated and acclimated Bantu miners. Acclimation shifted the relation leftward about 0.6° C, increased the slope of the steep portion, and raised the plateau of sweat rate. Redrawn from Reference 134.*

ready-functional individual glands (24). This increase in sweat rate seems to take the form of increases both in the sensitivity to sympathetic stimulation, accounting for the increased sensitivity of the sweating response to changes in core temperature (and also to electrical or pharmacological stimulation), and in each gland's sweating capacity. This increase in the sensitivity of the acclimated sweat glands is probably brought about largely by the increased sweat secretion by the glands during the acclimation process; for this reason it is sometimes described as "training" of the sweat glands. The evidence for this conclusion is as follows: the effect of acclimation on the sweat glands is suppressed in any skin region where sweating was suppressed during acclimation, whether by local cooling (20,44) or by soaking the skin in 35° C water to inhibit sweating before immersing the subject in a 39° C bath to induce acclimation (15). In addition, the effect is enhanced on forearms where the sweating response during acclimation was augmented by local heating to 43° C (44). Furthermore, repeated daily intradermal injections of methacholine increased the local sweating response elicited by methacholine injections—just as acclimation did, although the effect of acclimation was larger (25).

However, Chen and Elizondo have disputed the concept of sweat-gland "training." Their arguments are that in their experience, repeated stimulation of the sweat glands with electrical current or iontophoresis of cholinergic agents, and heating the skin to 40° C during acclimation (they consider heating to 43° C to be an unphysiological way to elicit sweating, since such sweating is not blocked by atropine) all failed to mimic the effect of acclimation on sweating (20). In fact, repeated pharmacological stimulation reduced the local sweating response, an effect that they ascribed to desensitization of the neuroglandular junction (20). Cold applications on the arm proximal to the test site reduced sweating at the test site by 80%, presumably by impairing transmission of nervous impulses. Since such cold applications during acclimation did not prevent the normal augmentation of sweating at the test site, Chen and Elizondo proposed that some trophic factor released during whole-body heating brings about the sweating changes, and that its action is suppressed locally by cooling the glands (20).

In its simplest form, the notion of sweat-gland training implies a) that the increase in sensitivity of the sweat glands after training depends on the amount of sweating elicited by the training stimulus and the scheduling of the application of this stimulus; and b) that the response to training is relatively unaffected by the nature of the training stimulus, whether whole-body heating, with or without local manipulations to modify the sweating response, or electrical or pharmacological stimulation. The chief difficulty in resolving the conflicting views concerning sweat-gland training is that on the basis of the published reports it is impossible a) to compare the different kinds of training stimuli used according to the amount of sweating that they elicited when they were applied; and thus b) to determine whether the effect on sweat-gland sensitivity depended on the kind of stimulus used or on the amount of sweating that it elicited each time it was applied. Thus, for example, it is not possible to tell whether repeated electrical stimulation of sweating is incapable of increasing sweat-gland sensitivity, or whether the amount of sweating that it elicited when applied by Chen and Elizondo (20) was too small to produce a training effect.

Composition of Sweat and its Effect on Salt Balance. Since sweat is hypotonic to extracellular fluid, secreting sweat glands must do osmotic work, reabsorbing sodium by active transport. As sweat rate increases, the rate at which sweat glands reabsorb sodium does not increase in the same proportion, so that sodium concentration in sweat increases with sweat rate. Sodium concentrations in sweat can vary enormously, from less than 5 up to 60 meq $\cdot \ell^{-1}$, depending on sweat rate, state of acclimatization, and the action of mineralocorticoids (97). Heat acclimatization seems to increase the sweat glands' capacity for sodium reabsorption, and both the salt concentration of sweat at any given sweat rate, and the slope of the relation of concentration to sweat rate are reduced by acclimation (4) (Fig. 4-4).

Persons who are well acclimatized to heat can often secrete very dilute sweat, with sodium concentration less than 5 meq $\cdot \ell^{-1}$. However, those who are less well acclimatized may lose large amounts of salt in their sweat and become substantially salt depleted. This salt-sparing effect of acclimatization depends on the adrenal cortex, and is not seen in patients with Addison's disease (97). Aldosterone secretion is increased by sodium depletion and by heat stress (68); and aldosterone seems to be the hormone primarily responsible for this salt-sparing effect. Aldosterone lowers sodium concentration in thermal sweat and sweat induced by methacholine, but this effect is delayed compared to the renal effect (27), and is not evident for the first 12 h after administration (22,23). Aldosterone also seems to reduce sweat rate slightly (27), and may be partly responsible for the reduction in sweating that accompanies dehydration (78). Furthermore, spironolactone, a competitive antagonist of aldosterone, increases sodium concentration in thermal sweat of acclimating subjects (22).

However, it is not clear whether aldosterone (along with other endogenous steroids with lesser mineralocorticoid actions) is the exclusive cause of this salt sparing in sweat, or whether in addition

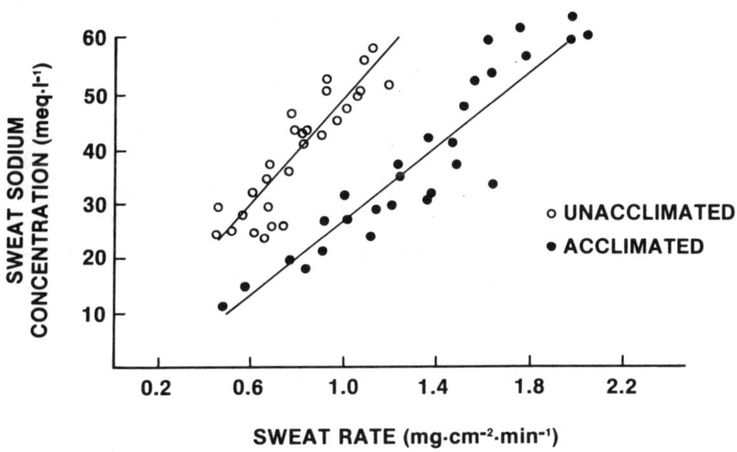

Figure 4-4. *The relation between sweat sodium concentration and sweat rate on the back of one subject before and after acclimation by daily immersion in hot water, 1 hour a day, for three weeks. Redrawn from Reference 4.*

the presence of aldosterone is required in order to permit other factors associated with acclimatization to exert a salt-conserving effect on the sweat glands. McCance (81) found that severe experimental salt depletion reduced sodium and chloride concentrations in sweat substantially, but not nearly so much as has been reported in some studies of heat acclimation. Kirby and Convertino (64) report that over the course of acclimation sweat sodium concentration decreased, so that in spite of a 12% increase in sweat rate, sweat sodium losses decreased 59%. Plasma aldosterone concentration during exercise also decreased, and the ratio of sodium reabsorbed (from an isotonic sweat precursor) to plasma aldosterone concentration thus approximately tripled. They proposed that acclimation increased sweat gland responsiveness to aldosterone. Another possible interpretation, since exercise and heat stress elicit aldosterone secretion (see below) and since the action of aldosterone on the sweat glands is delayed, is that this action on the sweat glands is cumulative, and not due solely to the present aldosterone concentration.

Distribution of Sweating. In humid heat, a greater portion of the sweat production is on the limbs after acclimation than before (44,59,73,117). Since the mean sweating intensity (i.e., sweat rate per unit area) is much lower on limbs than on the trunk before acclimation (Chapter 3), acclimation tends to make the sweating intensity over the skin surface more uniform. In a humid environment this is beneficial, since by making greater use of the skin of the limbs for evaporation of sweat, the body is able to evaporate a greater portion of any additional sweat secreted. Sweat glands on the arm are smaller (and also more densely distributed) than those on the chest, and thus may have more potential for increasing their output (24).

Inhibition of Sweat Gland Activity, Including Hidromeiosis. Sweat rate declines during prolonged intense heat exposure. Much of the decline is due to hidromeiosis, associated with wetness of the skin (Chapter 3), especially in humid heat. However, since other factors, such as dehydration, also may cause sweating to decrease (Chapters 3 and 6), it is often unclear what caused sweat rate to decline during a particular study.

As noted above, the decline of sweat rate due to sweat-gland inhibition in humid conditions occurs sooner in those who are unacclimatized (24,45, Fig. 4-2). Most reports of the effects of acclimation on this phenomenon have been rather descriptive. However, Fox *et al.* in a well-designed study compared the effects of acclimation to dry and to humid heat on this inhibition of sweating (43). They acclimated resting subjects with controlled hyperthermia, maintaining core temperature near 38.2° C for 2 hours a day for 12 days, using dry heat for one group and moist heat for the other group. Both groups had their left arms in plastic bags for sweat collection. These bags created a warm, humid microclimate. Both groups increased sweat rate over the course of acclimation, but on each day the dry group sweated an average of 300 $ml \cdot h^{-1} \cdot m^{-2}$ more, probably because of a lesser hidromeiotic effect.

After acclimation both groups showed similar decreases in heart rate and core and skin temperatures and similar increases in sweating during a standard exercise test in moderately humid heat. In a 2-hour controlled hyperthermia test while they rested in very humid heat, both groups had about the same whole-body sweat rates. All arms which had experienced humid heat during acclimation (i.e., both arms of the "humid" group and the left arms of the "dry" group) had similar increases in their sweat production during this test, and the increases for these arms were proportionately two to three times as great as the increases in whole-body sweat rate, consistent with what was described in Distribution of Sweating, above. Also during the controlled hyperthermia test in humid heat, sweat rates of these arms declined more slowly than before acclimation. During the same test the right arms of the "dry" group, which had not experienced humid heat during acclimation, also had a higher initial sweat rate than before acclimation, but thereafter their sweat rate declined as fast as before acclimation, so that during the two hours of the test these arms secreted substantially less sweat than did either the contralateral arms or the arms of the "humid" group.

These results indicate that in this study (43), most of the improvement in the ability to maintain high sweat rates in high humidity after acclimation probably owed to a diminution of hidromeiosis, and depended on local factors associated with local skin wetting during acclimation. In addition, acclimation rapidly reduces the amount of glycogen depletion in the sweat glands that is caused by acute heat stress (24,25). This effect of acclimation may also play a role in enabling the sweat glands to sustain high rates of secretion, if glycogen depletion is a factor contributing to sweat gland fatigue.

Skin Circulation

Blood Flow. In a dry environment, one might expect the increase in sweating to cool the skin more than the core, thereby widening the core-to-skin thermal gradient for heat loss and allowing for a lower skin blood flow, and so reduce the cardiovascular strain. Such a widening of the core-to-skin thermal gradient has been reported in some studies of acclimation (e.g., 37,87,133), especially in dry heat, but not in others (e.g., 98). There are certain possible mechanisms that might lessen cardiovascular strain without a change in core-to-skin thermal gradient, such as more efficient use of skin blood flow for cooling, either by longer residence time in skin, selective routing through vessels best suited to heat transfer, or preferential vasodilation in cooler skin areas. However, none of these has been shown to occur. Since heart rate was found to be markedly lower at any level of hand or forearm blood flow after heat acclimation (46), changes in skin blood flow may not fully explain the reduction of cardiovascular strain after acclimation. Unfortunately, it is not known how well the two most widely measured indices of skin blood flow, hand blood flow and forearm blood flow, represent skin blood flow in the body as a whole (Chapter 3); and the lack of suitable means to estimate skin blood flow quantitatively anywhere other than on the extremities is a serious methodological problem in this area.

Vasodilation occurs in the forearm at a lower core temperature after acclimation than before (46,93, Fig. 4-5). The same is true wherever else on the body the effect of acclimation on skin blood flow has been studied: in the hand, though less consistently there (46); and on the chest (46,51) and ear (46) as estimated from changes in thermal conductance. These changes parallel changes in the core-temperature threshold for sweating (46). In addition, acclimation increases the maximum plateau level of chest conductance reached at high core temperature (51), but maximum blood flow or conductance after acclimation seems not to have been measured elsewhere. However, this observation suggests that acclimation may produce changes in the skin blood vessels, somewhat analogous to changes in the sweat glands. Such changes either might allow the skin arterioles to dilate more, or might allow more efficient use of skin blood flow for heat transfer, as speculated upon earlier.

Although skin blood flow at a given core and skin temperature is higher after acclimation, this information does not tell us about the actual levels of skin blood flow reached with a given combination of exercise and environmental conditions, since core and skin temperatures are lower. Hand blood flow is reported to show no consistent change (55) or to be lower (87) after acclimation; and forearm blood flow has been reported to be lower (55,133) or unchanged (132) after acclimation. However, although Wood and Bass found no overall change in forearm blood flow between the beginning and end of acclimation, they did find it to reach a minimum on the third and fourth days of acclimation, which coincided with the disappearance of symptoms of heat strain (132). It may be that in their study an immediate cause of the disappearance of symptoms was a decrease in skin blood flow, which must then have been superseded by other adjustments, since evidence of acclimation persisted.

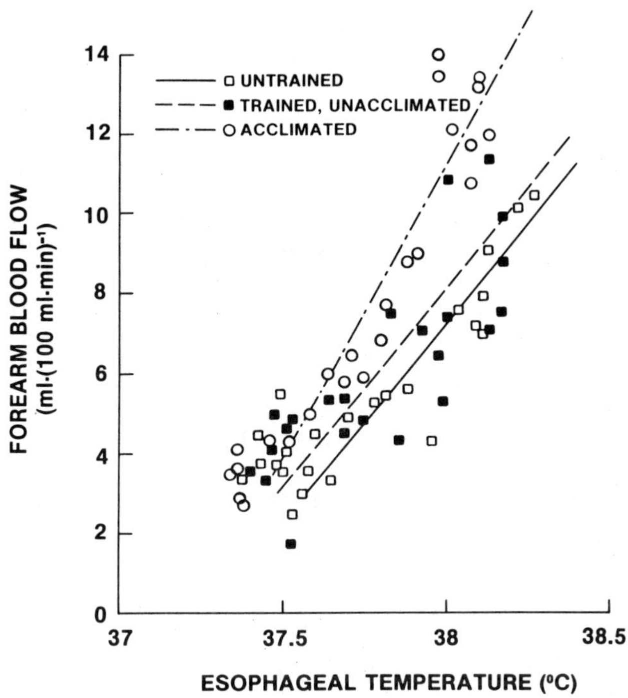

Figure 4-5. *Relation of forearm blood flow to esophageal temperature of one subject during 15 min exercise at 25° C before a program of exercise training on a cycle ergometer 1 hour a day for 10 days, after the end of the training program, and again after a 10-day program of acclimation to humid heat, 70 min (two 30-min periods of cycle exercise separated by 10 min rest) a day for 10 days. Changes in slope of the relation at different stages were inconsistent among the eight subjects in this study. However, the esophageal-temperature threshold for vasodilation was significantly ($P < 0.01$) lowered by training (mean change = 0.20° C) and again by acclimation (mean change = 0.26° C). Redrawn from Reference 93.*

Moreover, the variability among these reports of the effect of acclimation on skin blood flow may indicate that the results depend on the type of thermal stress employed.

Venous Tone and Volume. In a cool environment, acclimation lowers the core-temperature threshold for venodilation (Fig. 4-6) in the same way that it lowers thresholds for vasodilation and sweating. Cutaneous venomotor control seems to be much more susceptible to non-thermal influences associated with exercise than is the control of skin blood flow (Chapter 3), but nothing is known about how acclimation changes cutaneous venomotor responses to exercise. Thus, what we know about the effect of acclimation on venomotor control does not tell us about the effect of acclimation on cutaneous venous tone or peripheral pooling of blood during ex-

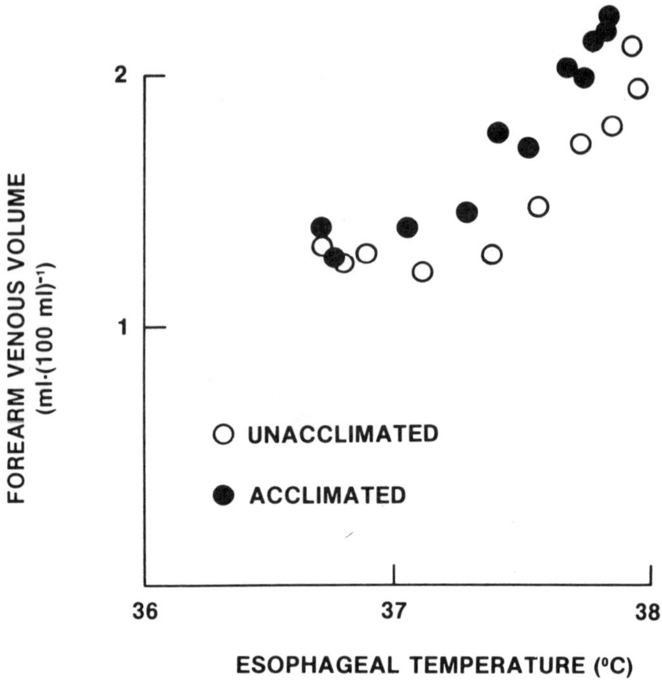

Figure 4-6. *Relation of forearm venous volume, at a congesting pressure of 30 torr, during 15 min exercise in 15° C ambient temperature, before and after a program of acclimation to humid heat, 1 hour a day for 10 days. Drawn from unpublished data of C. B. Wenger and M. F. Roberts.*

ercise-heat stress, just as was the case with skin blood flow, above. Wood and Bass, in the same acclimation study mentioned earlier (132), measured venodilation as the change in arm volume upon increasing venous congesting pressure from 1 to 31 torr. They found the least venodilation on the third and fourth days of acclimation, the same time as when forearm blood flow was minimum and the symptoms of heat strain disappeared. Like the transient decrease in forearm blood flow, this diminution of venodilation did not persist. Since venomotor activity is only one of several factors governing the peripheral pooling of blood, the amount of blood contained in veins that are not deliberately congested might be more pertinent to explaining other cardiovascular events occurring with acclimation. Studies of this have yielded conflicting results. Scott *et al.*, using infrared photographs, reported that acclimation produced a slight but somewhat indefinite increase in the size of uncongested superficial forearm veins (109). Whitney, on the other hand, reported that the "forearm venous reservoir" was smaller after acclimation (130).

He estimated the size of the "forearm venous reservoir" from the increase in uncongested forearm volume occurring in the first 20 min of heat exposure, before the start of exercise. However, he found no simple relation of this volume change to venous tone and distensibility, and venous capacity at 50 torr collecting pressure was unchanged by acclimation (130).

Thermoregulatory Set Point

Fox *et al.* reported that in tests on subjects heated at rest, acclimation lowers the core-temperature thresholds for sweating and vasodilation in such a way that the increases in arm and chest blood flow accompany the onset of sweating, both before and after acclimation (46). On the other hand, Roberts *et al.*, testing subjects who were heated by leg exercise, found that exercise training and heat acclimation both produced larger shifts in the core-temperature threshold for forearm vasodilation than for sweating (93). The difference likely owes to the use by Roberts *et al.* of exercise, since especially with inexperienced subjects, exercise is likely to introduce non-thermal influences to the control of skin blood flow, which might abate with training or familiarization. Therefore it is likely that, as proposed in a recent review (49), acclimation involves a lowering of the thermoregulatory "setpoint". The observation that acclimation lowers the threshold for venodilation in a cool environment (Fig. 6) in the same way suggests that cutaneous venomotor reflexes also participate in such a lowering of thermoregulatory "set point".

CHANGES THAT ARE NOT SPECIFICALLY THERMOREGULATORY

Circulatory Changes

Heart Rate and Stroke Volume. On the first day of exercise in the heat, heart rate reaches much higher levels than in temperate conditions, and stroke volume is lower. Thereafter, heart rate begins to decrease and, usually, stroke volume starts to increase as early as the second day of heat acclimation (e.g., 136). These changes are rapid at first, but continue more slowly for about a week. The time course of the decrease in heart rate is roughly similar to that of the increase in plasma volume, and the two changes are significantly correlated (115), though the change in heart rate appears to lead the change in plasma volume somewhat (Fig. 4-7). Note that increases in stroke volume are not reported consistently. For example, two studies (102,136) report increased stroke volume with little change in cardiac output as heart rate fell; but another study (133) reports a decrease in cardiac output, associated with a decrease in "surface

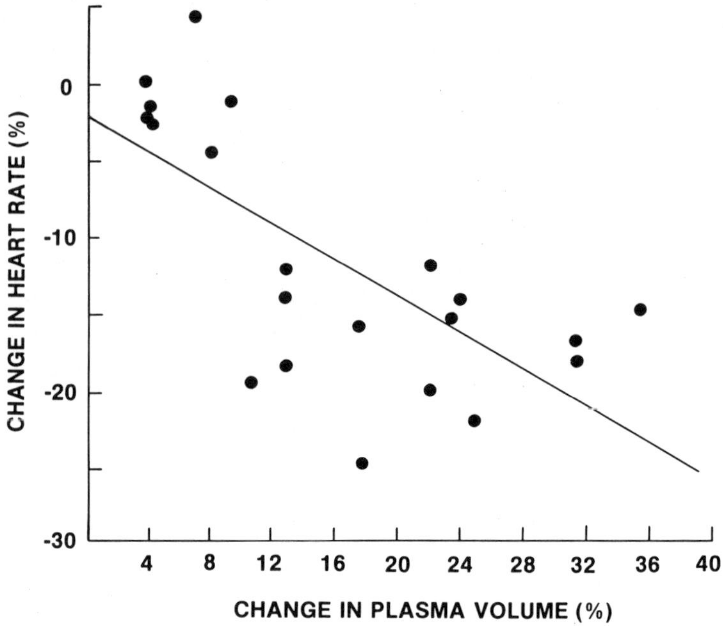

Figure 4-7. *The relation between the percent change in heart rate at the end of exercise to percent change in plasma volume during 10 days of acclimation to humid heat, in which subjects peformed 4 hours of cycle exercise a day. Changes in heart rate and plasma volume are with respect to values on the first day of acclimation.*

blood flow'' (estimated calorimetrically) as heart rate fell, and little change in stroke volume; and still another study (138) reports a mixed pattern, with two subjects showing a steady increase in stroke volume, one a transient increase, reversing after the sixth day, and one showing no increase. The reason for these differences is not clear: Reference 102 describes dry heat acclimation, and References 133, 136, and 138 all describe humid heat acclimation. Acclimation does not change systolic blood pressure at rest, but diastolic pressure and heart rate fall, probably because of higher stroke volume and lower peripheral resistance (108).

Three explanations have been proposed for the decrease in heart rate over the course of heat acclimation. One is the increase in plasma volume. In spite of the fact that plasma volume at rest tends to return toward control values after about 10 days of heat acclimation (see below) without any decline in the ability to exercise in the heat (8), the increase in plasma volume at rest during the first week or so of acclimation probably plays an early role in reducing heart rate and circulatory strain. Furthermore, as discussed later, Senay has hypothesized that even after plasma volume at rest returns toward normal, plasma volume during exercise in the heat is still larger than before acclimation, because of an enhanced hemodilution response during exercise in the heat (112).

Another proposed explanation for the decrease in heart rate is increased venous tone, since venoconstriction can mobilize up to 25% of the blood volume (8). However, Wood and Bass found a significant reduction in forearm venous volume only on the third and fourth days of acclimation (132). Thus, although peripheral venomotor adjustments may play a transitional role in heat acclimation, there is no evidence of their making a persistent contribution to the reduction in circulatory strain. It is possible that venous beds elsewhere in the body participate in the circulatory changes that occur with acclimation, but this matter has not been investigated. A reduction in peripheral pooling of blood would improve cardiac filling and alleviate circulatory strain whether the pooling was reduced by venoconstriction or by other means. Although the progressive lowering of uncongested forearm venous volume that Whitney observed during heat acclimation (130) probably does not represent venomotor changes, as noted above, it may indicate a reduction in peripheral pooling of blood secondary to a lower skin blood flow. In his study, and in situations in which there is evidence for reductions in skin blood flow after acclimation (e.g., 37,55,87,133), secondary decreases in peripheral pooling may play a role in improving cardiovascular function. However, since in many studies of acclimation a sustained decrease is observed neither in skin blood flow nor in core-to-skin thermal conductance (see above), such circulatory changes do not provide a general explanation for the circulatory improvements in acclimation. Either an increase in plasma volume during exercise or a decrease in peripheral pooling should be expected to act by raising central (thoracic) blood volume and central venous pressure. There have been no measurements of changes in central venous pressure with heat acclimation (101), but central blood volume was not consistently or significantly changed by 12 days of heat acclimation, even though stroke volume increased (102).

A third proposed explanation is that the fall in heart rate is secondary to the fall in core and skin temperatures with heat acclimation (102). Since heart rate at a given core temperature is the same or higher after acclimation than before (45,46), lower body temperatures are a necessary condition for lower heart rate with a given exercise-heat stress after acclimation. However, besides lowering core and skin temperatures during exercise-heat stress, acclimation lowers the core temperature thresholds for vasodilation and venodilation, so that lower core and skin temperatures by themselves are not a sufficient explanation for lower heart rate after acclimation. Furthermore, the explanation based on lower core and skin temperatures does not include an explicit mechanism. One may propose that acclimation reduces heart rate by reducing skin blood flow or peripheral pooling of blood at thermal steady state during exercise heat stress. However, the available evidence gives little support to such a mechanism, as discussed above. In addition, it is difficult to see how this explanation is sufficient to explain a reduction in heart rate with no comparable reduction in either metabolic rate (although there is evidence for a very small reduction in metabolic rate, as discussed later) or core-to-skin thermal conductance. A final reservation with respect to this explanation is that exercise heart rate usually decreases more rapidly than exercise core temperature over the course of acclimation (see above, and Reference 135), so that exercise core temperature is still decreasing when exercise heart rate has essentially reached its new level.

It may be an over-simplification to expect a single explanation to account for the improvement in cardiovascular function in the heat after heat acclimation. There may be several mechanisms that participate, and their relative contributions may vary, both over the course of one subject's acclimation and also among subjects, according to their constitution and thermal history. In fact, Senay *et al.* noted in one report that each subject adapted to heat stress in his own way (115). However, the evidence presently available does not seem sufficient to establish either one explanation or a coherent group of explanations capable of explaining the cardiovascular improvements resulting from heat acclimation.

Plasma Volume and Composition. An early report describing effects on blood volume of seasonal changes is that of Bazett *et al.* (10). They found no consistent seasonal effect on hematocrit, but both plasma volume and red cell volume were higher in summer than winter. Their data are highly variable, with seasonal changes in blood volume as high as 30–40% in some cases. Yoshimura (141) reported an increase in plasma volume in summer and decrease in winter, with a mean seasonal difference of 8.6%. Yoshimura (141) also reported increases in total body water (TBW) and in volume of extracellular fluid (ECF) in summer. In contrast to the work of Bazett *et al.* (10), Yoshimura found lower hematocrit and plasma protein concentration as plasma volume expanded, so that red cell and plasma protein masses were relatively constant. Although plasma sodium and chloride concentrations decreased, their total content in plasma increased in summer (141).

In general, plasma volume expands during heat acclimation as protein is added to the vascular space, but the time course and amount of expansion are variable (101). Harrison *et al.* (53) acclimated six subjects for 11 days with cycle exercise in the heat. Blood volume underwent an increase of 8.7% complete by the fourth day and persisting for the entire 11 days, through a plasma volume increase of 14.7%. Since there was little change in plasma protein concentration, total circulating mass of plasma protein increased. At the end of acclimation, the loss of blood volume during cycle exercise was greater than at the beginning, by 2% of initial blood volume, and for this reason Harrison *et al.* doubt that the hemodilution plays any major role in the cardiovascular adaptation occurring with acclimation (53). This conclusion is not entirely convincing, however, since blood volume during exercise was still almost 7% greater at the end of heat acclimation than at the beginning. (As Chapter 6 discusses, different leg exercise modes differ markedly in their effects on plasma volume, so that the above experience with cycle exercise is not representative of other exercise modes.)

Expansion of plasma volume at rest begins early and rapidly during heat acclimation, and then continues more slowly for about a week. Senay *et al.*, (115) reported plasma volume to increase 24% after ten days of cycle exercise, 40–50% $\dot{V}O_2max$, for 4 hours a day in humid heat. Collins and Weiner, in their review (27), give a range of 12 to 27% for the increase in blood volume and 6 to 16% for the increase in ECF. The few studies of acclimation going beyond 7 to 10 days report a return of resting plasma volume toward control level: Bass *et al.* studied men exercising and living in heat for the duration of their study, and found resting plasma volume by dilution of Evans' blue up 27% after 5 days, but only 14.5% at 14 days (9). Corresponding values for ECF by dilution of thiocyanate were 16% and 14%; and for TBW by dilution of antipyrine, 2.4% and 1.7% (9). Wyndham *et al.* reported resting plasma volume by dilution of radioiodinated serum albumin up 10.4% at one week, but only 4.8% at three weeks; corresponding values for ECF by dilution of ^{82}Br were 3% and 1.4%; and for TBW by dilution of tritiated water, 7.4% and 5.7% (136).

From the data cited in the previous paragraph it is evident that there is much variability in the ECF and TBW changes that accompany resting plasma volume changes during heat acclimation, but in proportional terms they tend to be substantially less than corresponding changes in plasma volume. The events causing these changes in plasma volume are matters of conjecture. From the fact that the increases in resting plasma volume are proportionately larger than those in ECF, Wyndham (135) infers that salt retention and expansion of ECF are not the main factors in the expansion of plasma volume. Some of the gain in plasma volume may, in fact, be at the expense of the interstitial space (e.g., 41), and may be associated with a translocation of protein and fluid from the interstitial space (see below). Senay proposed that a change in composition of interstitial space, perhaps in the sol/gel ratio, may make more protein available to be moved (110). Water balance also seems to be an important factor in this translocation of protein. Isotonic expansion of the plasma volume of dogs increases plasma protein content, perhaps by helping to "flush" protein from the interstitial space. Furthermore, Senay (111) reported that during heat exposure at rest, there was greater and more rapid hemodilution after acclimation provided that water balance was maintained; but that with dehydration, the advantage of acclimatization was lost.

As two studies cited earlier show, if a program of acclimation is continued for two weeks or more, the early increase in plasma volume at rest is followed by a return toward pre-acclimation levels (9,136). Results consistent with this finding are found in other reports. Bass *et al.* (7) found plasma volume by dilution of Evans' blue to be 20 ml *lower* after 3 weeks acclimation, but made no measurements at intermediate times. Fortney and Senay (42) reported plasma volume by the carboxyhemoglobin method up 6.4% after three weeks. This was substantially lower than is generally reported earlier during acclimation, but in this study plasma volume had already risen 7.4% during a program of exercise training (42). However, the eventual return of blood volume at rest toward control levels, which occurs during acclimation lasting more than two weeks or so, is not associated with any impairment of ability to work in heat (8). This finding suggests that the increases in plasma volume in the first few days permit the early cardiovascular changes, and then give way to other adjustments that occur later (101). It also raises the question of what other adjustments do come into play at this stage to maintain the ability to exercise in the heat. However, as a qualifying note it should be remembered that except for Reference 42 all the longer-term studies of changes in plasma volume or blood volume have relied on dilution of labels attached to plasma proteins. Since acute changes in hematocrit and plasma protein concentration during exercise in the heat are often incommensurate with each other, Wyndham believes that plasma proteins may move across the capillary membrane so rapidly that determinations of plasma volume by dilution of labels attached to protein are open to doubt (135).

In the early phase of acclimation when plasma volume is rising, increases in plasma volume are significantly correlated with increases in stroke volume and decreases in heart rate (115), as noted earlier. In one study of a 3-week course of acclimation mentioned above, plasma volume at rest decreased from 13.4% above control at the end of the first week to 9.5% above control at the end of the third week. This change in plasma volume was accompanied by a decrease in stroke volume at the end of 1 hour of exercise in the heat from 10.4% to 4.7% above the corresponding value on the first day in the heat (136). This rough relationship between plasma volume and stroke volume may be fortuitous, since there may not be a dependable relation between plasma volume at rest and after 1 hour of exercise in the heat; and the few other reports that present measurements of plasma volume after two weeks or more of acclimation do not include corresponding measurements of circulatory responses. However, it is possible that after several weeks of acclimation the body is able to use skin blood flow more efficiently for heat dissipation, as speculated earlier, and thus is able to support a given intensity of exercise in the heat with a lower cardiac output.

Another adjustment that may supersede the early increase in plasma volume at rest is a change in body fluid distribution at the beginning of exercise, as Senay has hypothesized (112). His concept is that expansion of plasma volume at rest is a short-term response, to be replaced after a while by an increased ability to hemodilute rapidly and early during exercise in the heat, by flushing protein into the vascular space and keeping it there. Thus even after several weeks of acclimation, when plasma volume at rest has returned nearly to pre-acclimation levels, plasma volume during exercise is still high, in his concept, because of the greater hemodilution during exercise. (Although Senay does not mention this point directly, his concept seems to require, as plasma volume at rest returns to pre-acclimation levels, that the hemodilution during exercise increase further, in order to reach the same plasma volume during exercise as occurred at an earlier point in acclimation.) The evidence for this mechanism is that exercise training reduces the tendency toward hemoconcentration or increases the tendency toward hemodilution, and that acclimation further increases the tendency toward hemodilution during exercise in the heat. (Most of the studies that he cites employed bench-stepping exercise.) A collateral argument in support of this proposed mechanism is that among mining recruits in South Africa, heat-intolerant individuals are distinguished by a reduced ability to hemodilute during exercise in humid heat, and also by reduced ability to maintain whatever hemodilution does occur (113). To provide at least a partial explanation of how training and acclimation exert these effects on the fluid movements, Senay notes that as exercise intensity increases, hemoconcentration starts to occur at about the lactate turn point. Therefore, since training increases both $\dot{V}O_2max$ and the percent of $\dot{V}O_2max$ at which the lactate turn point occurs, it will presumably increase the tendency toward hemodilution at any absolute exercise intensity. Since acclimation reduces lactate accumulation (142), it is likely to favor hemodilution in a similar way. This concept is plausible; and it has the attraction that, along with the short-term expansion of plasma volume at rest, it offers a generally applicable explanation of the reduction of circulatory strain by acclimation. Its chief weakness is that it seems to rest on extrapolation from fluid shifts observed during the phase of acclimation when plasma volume at rest is elevated to fluid shifts that are hypothesized to occur during a later phase of acclimation, when plasma volume at rest has returned to preacclimation levels. Measurements of circulatory responses, plasma volume, and hemodilution are needed at a number of different stages of an acclimation program lasting several weeks, in order to put this concept on firmer ground. Finally the reader should be reminded that fluid shifts are very sensitive to exercise type (Chapter 6), so that it is somewhat risky to draw conclusions about exercise in general from observations made mostly on one exercise type.

Other Circulatory Changes. Another possibility for reducing circulatory strain is some sort of cardiac adaptation that would allow a given stroke volume to be achieved at a lower central venous pressure. Horowitz *et al.* (61) reported that 2 months of heat acclimation increased the compliance and reduced ventricular stiffness of isolated rat hearts, so that there was a greater diastolic volume at any end-diastolic pressure. This change would, of course, tend to increase stroke volume. Schaible and Scheuer (106) reported that isolated hearts of rats that had undergone physical training by swimming showed greater pump performance, developed higher ventricular pressure, and increased ventricular pressure more rapidly during systole, and relaxed more quickly in diastole than did control hearts. This was true even though heart masses were not significantly greater. In addition aldosterone, which is proposed to have a role in heat acclimation (see below), is reported to have an inotropic effect in anesthetized dogs (79) and on isolated cat hearts (127); and injected daily into rats for three days before they were killed, it tended to have an inotropic effect on the heart muscle preparations, though this effect fell short of statistical significance (128).

Laboratory heat acclimation (120), and also natural winter-to-summer acclimatization in Beersheva, Israel (118) are reported to increase tilt table tolerance; and in laboratory acclimation, at least, the degree of improvement is proportional to that in exercise heart rate and rectal temperature (120). Acclimation also reduces or abolishes the drop in cardiac output that occurs in the heat on standing (109). Ladell (77) reports that Nigerians, who are tropical Africans, have better tilt-table tolerance than resident Europeans with the same degree of acclimatization of thermoregulatory responses, so there may be long-term adaptations that are not reflected in sweat rate and core temperature. [Southern African Bantu, however, seem not to share this advantage (77).] Mack *et al.* (85) report that the slope of the relation of forearm vascular resistance to central venous pressure is less steep in aerobically trained subjects. Thus, if acclimation is like training in this respect, then in acclimated subjects either central venous pressure or stroke volume is defended in other ways than through peripheral vascular resistance. Acclimation does not lessen the decrement in $\dot{V}O_2max$ caused by 5% thermal dehydration (18), or by heat (105). Also, although aerobic training may improve the work performance of hypohydrated subjects in a temperate environment, acclimation gives no further advantage (18).

Nothing is known about how heat acclimation changes the distribution of cardiac output, but one might expect the smaller increase in core temperature and heart rate to be associated with reduced vasoconstriction in viscera and kidneys during exercise (100,101). The fall in urinary excretion of norepinephrine, suggesting reduced sympathetic nerve activity, and in plasma renin activity (PRA) after acclimation suggests that this may in fact occur (100,101).

Endocrine and Metabolic Changes

Hormonal and Electrolyte Changes. At the start of acclimatization, especially if subjects' sodium intake is restricted, fasting blood sugar and urinary nitrogen loss may rise (123), the latter reflecting a negative nitrogen balance in the body; and if sweating is profuse there may be a net loss of sodium. Sodium retention first by kidneys and then by sweat glands increases, and urinary excretion of 17-hydroxy corticoids and 17-ketosteroids may increase for a few days (123). Within a few days, nitrogen balance and sodium balance are restored, and fasting blood sugar returns to normal. Once sodium balance is restored, the kidneys may stop retaining sodium, though the sweat glands continue to do so. This sequence of events is initiated by an increase in the secretion of adrenocorticotrophic hormone (ACTH) in response to the circulatory effects of the exercise-heat stress, and also, if sweating is profuse, of sodium depletion. ACTH elicits secretion of all the adrenal corticosteroids, including cortisol, adrenal androgens, and aldosterone. Cortisol causes gluconeogenesis and a net catabolism of protein, which increase the fasting blood sugar and urinary nitrogen loss; and metabolites of cortisol and androgens are excreted in the urine as 17-hydroxy corticoids and 17-keto steroids. Both cortisol and aldosterone contribute to the sodium retention, first by the kidneys within a few hours, and after a delay of 1 to 2 days, by the sweat glands. Secretion of aldosterone is elicited not only by ACTH, but also by elevated plasma potassium concentration and by angiotensin II. The role of angiotensin II is especially important during acclimatization, since it is an element of the renin-angiotensin system, which is activated by renal sympathetic stimulation, catecholamines, and reductions in renal blood flow; and it helps to maintain secretion of aldosterone later on, after ACTH secretion has returned to normal. All of these influences are likely to be present during exercise-heat stress, and the rise in plasma renin activity correlates with final heart rate (40). Heat exposure by itself increases plasma aldosterone concentration (68) and aldosterone excretion in the urine (26), and the data of Knochel *et al.* (66) indicate that aldosterone secretion is elevated by exercise not only in the heat, but also in cool conditions.

After a period of time, ACTH secretion returns to normal as aldosterone by itself becomes adequate to re-establish and maintain sodium balance. Once sodium retention by the sweat glands is by itself sufficient to maintain sodium balance, the kidneys "escape" from the salt-conserving effect of aldosterone and adjust their excretion of sodium so as to maintain sodium balance, even as sweat sodium concentration remains low. The ongoing need for an acclimated subject to sustain salt-conserving responses is revealed by the observations that the pattern just described can be reproduced in an acclimatized subject by suddenly reducing sodium intake (28), or by administering spironolactone (29), which blocks the action of aldosterone.

It is of historical interest that Conn at one time developed the concept that the pituitary-adrenal axis, acting via ACTH, was fundamentally associated with the mechanism of heat acclimatization (28). However, as later work cast doubt on the importance of this mechanism, he modified his position to give the primary role to aldosterone (29), and his name is associated with much of the classic work on the role of this hormone. He showed that acclimation can be temporarily disrupted by pharmacological manipulations. This can be accomplished by administering spironolactone, as described above, and also by giving a 7-day course of desoxycorticosterone (a synthetic steroid which, like aldosterone, causes the kidneys to conserve sodium) to inhibit aldosterone production, and then abruptly discontinuing the administration of desoxycorticosterone (29). In either case, there is a transitory (about two days) loss of acclimatization, including higher core temperature and heart rate, weight loss, decreased exercise-heat tolerance, nausea, faintness, and other manifestations of heat strain during exercise-heat stress. At the same time there is a transitory increase in loss of sodium, first in urine, then sweat, until acclimatization is re-established (29).

Up to this point, the role of aldosterone in heat acclimatization might seem to be rather straight-forward. However some authors report that the decrease in sodium loss in sweat depends on the sodium balance and is not a necessary part of the acclimation process, since it may not occur if salt intake is maintained at a high level (e.g. 27,121,129). Likewise, giving supplementary NaCl sufficient to maintain sodium balance during acclimation prevents the increase in urinary excretion of tetrahydroaldosterone that occurs if a sodium deficit is allowed to develop (121). On the other hand Davies *et al.* reported that although supplementary NaCl (7.5 g per day) during acclimation reduced plasma renin activity and aldosterone concentration compared with values that occurred during acclimation without supplementary NaCl, it did not affect the reduction in sweat sodium concentration that occurred with acclimation (32). Furthermore, plasma aldosterone concentrations during rest at room temperature (13), during passive heat exposure (13), or at the end of exercise in the heat (13,32) are reported not to be significantly different after acclimation. Similarly, Finberg and Berlyne (39) report that the increased sodium retention after acclimation is not associated with increased plasma aldosterone concentration. They report further that plasma renin activity and aldosterone concentration increase less during exercise after acclimation than before, and less in summer than in winter (39). From observations like the foregoing Bonner *et al.* (13) concluded that the adrenal cortex is not an important factor in maintaining acclimation after sodium balance is achieved. However, Kosunen *et al.* report that although plasma aldosterone peaks about 0.5 h after a 20-min sauna, and after 2 h is no longer significantly above control levels, there are an increase in plasma sodium concentration and a hemodilution lasting over the next two days (68). These results indicate either that the effects of aldosterone outlast the aldosterone itself by two days or so, or that heat stress also elicits sodium retention by another mechanism that Kosunen *et al.* did not identify. If the former possibility is the correct one, it may be that the sodium conservation associated with acclimation owes to a cumulative effect of the transient increases in aldosterone concentration occurring with each spell of the exercise-heat stress that produces and maintains the state of acclimation. If this is so, it follows that the increase in sodium conservation might develop over the course of some days of acclimation without any need for successive spells of exercise-heat stress to elicit higher levels of plasma aldosterone concentration than they did at the start of acclimation.

Results of acclimation programs in which subjects were given mineralocorticoids also do not yield any clear conclusions about their role in the acclimation process. Administering 5 or 10 mg of desoxycorticosterone acetate (DCA) per day on the last 2 days before acclimation and the first day of acclimation to acclimating men who are being maintained on a low-salt diet raises plasma sodium concentration and lowers sweat sodium and chloride concentrations on the first 2 days. However this does not affect sweat rate, heart rate, core or skin temperature, or plasma volume at rest or the change in plasma volume on the first day in the heat (96). Braun *et al.* studied the effects of daily administration of 1 mg of *d*-aldosterone for the three days before the start, and for the first six days of a program of acclimation to dry heat, in which subjects attempted 90 min of treadmill exercise each day (14). Compared with responses of subjects receiving no drug, aldosterone significantly increased endurance time on the first 3 days of acclimation, significantly lowered heart rate on the first 4 days, and significantly lowered final rectal temperature during exercise on the first 5 days. However, the authors concluded that aldosterone did not shorten the time required for acclimation, which they determined to be 8 days with or without aldosterone (14). Differences between this study and the study mentioned previously (96) probably owe not only to the number of days on which the drug was given, but also to the choice of drug and dose. Since the mineralocorticoid action of aldosterone is 30 times as great as that of desoxycorticosterone, the mineralocorticoid action of 1 mg per day of aldosterone in the latter study (14) is 3 or 6 times as great as that of 5 or 10 mg per day of desoxycorticosterone in the former study (96). Likewise, Bass and Henschel (8) and Collins and Weiner (27) observe that DCA mimics the fluid and electrolyte changes of acclimation, but neither produces "preacclimation" nor hastens acclimation. Furthermore, increased aldosterone secretion appears not to be necessary to acclimation if sodium intake is adequate to maintain sodium balance (22,27). The foregoing studies show that administered mineralocorticoids can reproduce certain features of heat acclimation. However, since none of them examined the effect of administered mineralocorticoid on endogenous aldosterone production or on plasma aldosterone concentrations, their implications for the role of endogenous aldosterone in heat acclimation are unclear.

Certain additional cautions in interpreting the foregoing studies are the following: In certain earlier studies (e.g., 26,96,121,123) urinary excretion of aldosterone was used as an index of aldosterone secretion. However, since the liver accounts for most metabolism of aldosterone (107), changes in urinary excretion may not represent changes in secretion. Also, in addition to affecting renin secretion, exercise-heat stress, by decreasing hepatic blood flow (100), may reduce hepatic clearance of renin and steroids, prolonging their action.

Acclimatized men in whom sodium conservation by the sweat glands has been developed to its maximal extent can sweat 4 to 9 liters per day and stay in salt balance on 5 grams or less of NaCl per day (28,30). It should be pointed out, however, that in these subjects sodium-conserving capacity of the sweat glands was developed by a gradual reduction of dietary sodium while they continued daily exercise in the heat. With a daily exchange of NaCl of about 10 g, most Caucasians who are not undergoing exercise-heat stress will be in salt balance and have high sweat sodium concentrations (78). If they are suddenly faced with a heavy requirement for sweating, they may undergo a substantial net loss of sodium before their mechanisms for sodium conservation become fully active. Therefore usual recommendations for salt consumption for men performing heavy exercise in the heat are that acclimatized men in a relatively steady state of salt balance need no sodium supplement beyond what is contained in a normal diet, but that unacclimatized men should have their diet supplemented by 10 grams of salt per day unless water is in short supply (78).

It is sometimes asserted that high salt intake is disadvantageous in subjects undergoing exercise-heat stress. Dasler *et al.* (31) compared the effects on acclimating subjects of different salt intakes from 15 to 30 grams per day, and reported that excess sodium intake in that range impairs heat acclimatization. However, since their subjects' daily energy expenditure during exercise in the heat was 1250 $kJ \cdot m^{-2}$ of body surface, it is likely that even the least salt intake was well in excess of sweat losses. It is therefore not possible to draw from their report any conclusions about the use of salt supplements which approximately replace salt losses.

Macfarlane and Robinson (82) reported that in summer, a standard 4-hour heat exposure produces a small increase in plasma osmolality and a large rise in antidiuretic activity; but in winter, no change in plasma osmolality and a smaller rise in antidiuretic activity. Leithead has suggested that antidiuretic hormone may play a role in limiting sweating in dehydrated subjects (78).

In unacclimated subjects, plasma potassium concentration increases during exercise in the heat. Acclimation reduces or reverses this increase (47,114); and the time of the reversal of the trend, from an increase to a decrease, roughly coincides with the tendency to hemodilution (112). This tendency to hemodilution figures prominently in one hypothesis proposed to explain the long-term improvement in cardiovascular function following acclimation (112), as discussed earlier.

Changes in Metabolic Rate and Substrate. There are a number of reports of lower basal or resting metabolic rate accompanying heat acclimatization. Basal metabolic rate of British soldiers stationed in Singapore was reported to fall an average 5 to 6% during the first year, and to change no further thereafter (83). Many soldiers showed no change (83), but in some persons basal metabolic rate may fall up to 10% with acclimatization (124). Collins and Weiner in their review (27) note that there is controversy, but conclude that the effect is real but small (about 5% on average), and variable among subjects. This lowering of metabolic rate may be secondary to altered thyroid function, since exposure to moderate heat appears to reduce thyroid function (27).

Heat acclimation has little effect on $\dot{V}O_2max$, and, in fact, does not ameliorate the decrease in $\dot{V}O_2max$ occurring at very high environmental temperatures (105). It does provide slight reduction in the metabolic rate elicited by walking (3–5%) (103) and cycling (2%) (142). Greater reductions in the metabolic rate elicited by bench-stepping have been reported (e.g., 119,126), but that also seems to decrease with practice. Lactate accumulation and consumption of muscle glycogen during exercise are greater in the heat than in a cool environment. Acclimatization reduces lactate accumulation in plasma and muscle both in warm and cool environments (142). However, there are conflicting data concerning the effect of acclimation on muscle glycogen utilization: Young *et al.* (142) report no effect, but King *et al.* (63) and Kirwan *et al.* (65) report lower glycogen usage after acclimation, all studying cycle exercise. Kirwan *et al.* (65) also report more use of free fatty acids, and a trend to higher leg muscle blood flow. Consistent with this report of greater use of free fatty acids, King *et al.* (63) found a trend, not statistically significant, to lower respiratory quotient.

Effects on Heat Disorders

Heat stroke is the most severe of the recognized heat disorders, and, unless treated promptly and vigorously, is associated with a high mortality. Heat stroke may be divided into two forms, depending on the pathogenesis. In the classical form, the primary pathogenic factor is environmental heat stress which overwhelms an impaired thermoregulatory system, whereas in exertional heat stroke the primary factor is metabolic heat production. Victims of the exertional form thus tend to be younger and more physically fit (typically, soldiers—especially recruits—and athletes) than victims of the classical form. Among recruits, heat stroke occurs typically in the first few weeks of basic training, and thus before they are fully acclimatized. Much of the protective effect of acclimatization obviously owes to the thermoregulatory improvement. Beyond that, however, multiple hemorrhages are almost a constant finding in fatal heat stroke at autopsy, and fibrinolysis has been implicated as a factor in heat stroke mortality (116). Environmental heat stress increases the fibrinolytic activity of whole blood, and the finding that this effect is diminished by heat acclimation (12) may help to explain the protective effect of acclimation against heat injury.

Heat syncope is a temporary circulatory failure due to pooling of blood in the peripheral veins and the consequent decrease in diastolic filling of the heart. The primary cause of the peripheral pooling is a large increase in thermoregulatory skin blood flow, but an inadequate baroreflex response is probably an important factor contributing to the syncope. Heat syncope usually occurs in individuals who are standing with little activity. Heat acclimation rapidly reduces the susceptibility to heat syncope (Chapter 8). Since naturally acclimatized Nigerians are reported to be virtually immune, while Europeans with a similar degree of acclimatization of their thermoregulatory responses are not (77), adaptations other than strictly thermoregulatory ones seem to be involved in protecting against this disorder.

Heat cramps is an acute disorder consisting of brief, recurrent, and often agonizing cramps in skeletal muscles, typically those which have recently been engaged in intense exercise. It is mentioned here because of the traditional view that patients are characteristically physically fit men, well acclimatized to heat. (Chapter 8, however, presents evidence against this view.) They typically have been drinking adequate amounts of water but not replacing salt lost in the sweat, and so they are usually hyponatremic. Although the hyponatremia is thought to be involved in the pathogenesis of the cramps, hyponatremia is rather common, but heat cramps are an unusual accompaniment. The fact of their losing enough salt to become hyponatremic suggests that, although physically fit and able to exercise enough to sweat profusely, they are not well enough acclimatized for their sweat glands to conserve salt.

Some apparently healthy individuals cannot acclimate successfully to heat and so are heat-intolerant. In one study of acclimation (58), two subjects did not increase their sweat rate or begin to sweat earlier, and their sweat rates declined during every heat exposure, whereas the other subjects, who acclimated normally, showed declines in sweat rate only from the third heat exposure onward, when sweat rates had increased above those on the first day. These two heat-intolerant subjects also complained of symptoms of heat stress throughout the study. A small proportion of Bantu mining recruits do not acclimate to heat, and show persistently high rectal temperature in the standard 8-day program of exercise in humid heat used to acclimate the recruits. These heat intolerant recruits, on the average, tend to be smaller and older than those that do acclimate, but there is a great deal of overlap in these respects. One striking differential factor is that mean $\dot{V}O_2max$ in the intolerant group was 2.4 $\ell \cdot min^{-1}$, as compared to 3.1 $\ell \cdot min^{-1}$ in the tolerant group; and all of the heat tolerant recruits, but only 37% of the intolerant ones, had $\dot{V}O_2max$ above 2.5 $\ell \cdot min^{-1}$ (67). In addition, the intolerant subjects hemodiluted less during bench stepping in humid heat, and maintained the initial hemodilution less well (113).

One aspect of the effects of heat stress is strikingly demonstrated in epidemiological studies of individual heat waves. Bridger and Helfand analyzed the July, 1966 heat wave in eastern Missouri and central Illinois, and related the death rates of various age groups to the reported daily temperatures (16). This heat wave came on suddenly and was the first truly hot spell of the year, and the authors proposed that for those reasons it may have had an especially severe effect on mortality (16). Few of the excess deaths during this heat wave were due to recognized heat disorders, and most were ascribed to aggravation of other disease processes by heat stress. The authors' proposal thus suggests the importance of natural acclimatization in protecting patients suffering from such other diseases.

CONCLUSIONS

Nearly complete acclimation to a given level of exercise-heat stress can be acquired in 7 to 10 days of daily exercise in the heat. In addition, a certain degree of acclimatization is acquired during passive heat exposure without exercise, or by living in a hot environment without performing hard physical exercise. Acclimation and acclimatization involve both thermoregulatory improvements and cardiovascular improvements that seem not to be directly related to the thermoregulatory changes. Since core-temperature thresholds for sweating, increases in skin blood flow, and venodilation all shift toward lower core temperature, and since moreover the changes in sweating threshold often occur without any change in the slope of the relation between sweat rate and core temperature, part of the change in thermoregulatory control probably consists of a lowering of the thermoregulatory "setpoint" in the central nervous system. In addition there are peripheral changes, at least in the control of sweating. These include increased sensitivity of the sweat glands to nervous and pharmacological stimulation, increased sweating capacity, increased ability to reabsorb salt and thus secrete dilute sweat, and improved ability to sustain high sweat rates, especially in humid conditions. These changes are brought about by "training" of the sweat glands, including especially sweating in humid conditions, and by the action of adrenal cortical mineralocorticoids. It is not known whether or not there are peripheral changes affecting the skin blood vessels. However, the observation that maximal thermal conductance on the chest is increased after acclimation indicates that such changes may occur.

Besides the thermoregulatory changes there are changes that improve cardiovascular function in the heat. Although in some cases the core-to-skin thermal gradient widens with acclimation, and presumably allows the subject to achieve thermal balance with a lower skin blood flow and so with less secondary peripheral pooling of blood, this is not a consistent finding. It is therefore unlikely that thermoregulatory changes are sufficient to explain the improvement in cardiovascular function. A number of other responses during acclimation have been reported or proposed that would tend to alleviate cardiovascular strain, among them increased plasma volume at rest, decreased skin blood flow, and decreased cutaneous venous capacity. However, it is unlikely that any of these can provide a general explanation for the improved cardiovascular function brought about by acclimation, since they either appear to be transient, or have not been reported consistently. Other possible mechanisms are greater hemodilution during exercise-heat stress after acclimation, and cardiac changes involving either increased contractility or increased diastolic compliance. However, these mechanisms are hypothetical at this point, and more experimental evidence is required to establish any of them. The number of these potentially beneficial responses, and the transient or inconsistent nature of most of them, suggest two conclusions: First, there may be several phases of acclimation, with the cardiovascular improvement in each phase being due to a different mechanism or mix of mechanisms. Second, each mechanism may play a greater or lesser role in a particular case, depending on the nature of the heat stress, the type of exercise, and the subject's constitution and thermal history. In fact, the authors of one study of acclimation (115) commented that each subject adapted to the stress in his own way.

There are also certain hormonal and metabolic changes with heat acclimation. One of these is a change in sweat gland function which allows the glands to secrete very dilute sweat. When sweating is profuse and prolonged, this change is essential to maintain sodium balance. Adrenal cortical steroids, chiefly aldosterone, are necessary to this change, and play at least a permissive role. Increased aldosterone secretion, or else a cumulative effect of aldosterone secreted in response to the acclimating stress, may have other beneficial effects besides that on sodium balance, but this is uncertain. In addition there is a small and inconsistent reduction in basal metabolic rate, which may be associated with a reduction in thyroid function, and a small reduction in the metabolic rate elicited by submaximal exercise.

REFERENCES

1. Adam, J.M., R.H. Fox, G. Grimby, D. J. Kidd and H. S. Wolff. Acclimatization to heat and its rate of decay in man. *J. Physiol., London* 152: 26P–27P, 1960, Abstract.
2. Adams, W. C., R. H. Fox, A. J. Fry and I. C. MacDonald. Thermoregulation during marathon running in cool, moderate, and hot environments. *J. Appl. Physiol.* 38: 1030–1037, 1975.
3. Adolph, E. F. Life in Deserts. In: E. F. Adolph *et al.* (eds.) *Physiology of Man in the Desert.* New York, NY: Interscience, 1947, pp. 326–341.
4. Allan, J. R., and C. G. Wilson. Influence of acclimatization on sweat sodium concentration. *J. Appl. Physiol.* 30: 708–712, 1971.
5. Baker, P. T. Racial differences in heat tolerance. *Am. J. Physical Anthropol.* 16: 287–305, 1958.
6. Bass, D. E. Thermoregulatory and circulatory adjustments during acclimatization to heat in man. In: J. D. Hardy (ed.) *Temperature, Its Measurement and Control in Science and Industry, vol. 3, part 3, Biology and Medicine.* New York, NY: Reinhold, 1963, pp. 299–305.
7. Bass, D. E., E. R. Buskirk, P. F. Iampietro and M. Mager. Comparison of blood volume during physical conditioning, heat acclimatization and sedentary living. *J. Appl. Physiol.* 12: 186–188, 1958.
8. Bass, D. E., and A. Henschel. Responses of body fluid compartments to heat and cold. *Physiol. Rev.* 36: 128–144, 1956.
9. Bass, D. E., C. R. Kleeman, M. Quinn, A. Henschel and A. H. Hegnauer. Mechanisms of acclimatization to heat in man. *Medicine* 34: 323–380, 1955.
10. Bazett, H. C., F. W. Sunderman, J. Doupe, and J. C. Scott. Climatic effects on the volume and composition of blood in man. *Am. J. Physiol.* 129: 69–83, 1940.
11. Bean, W. B. and L. W. Eichna. Performance in relation to environmental temperature. Reactions of normal young men to simulated desert environment. *Fed. Proc.* 2: 144–158, 1943.
12. Bedrak, E., G. Beer and K. I. Furman. Fibrinolytic activity and muscular exercise in heat. *J. Appl. Physiol.* 19: 469–471, 1964.
13. Bonner, R. M., M. H. Harrison, C. J. Hall and R. J. Edwards. Effect of heat acclimatization on intravascular responses to acute heat stress in man. *J. Appl. Physiol.* 41: 708–713, 1976.
14. Braun, W. E., J. T. Maher and R. F. Byrom. Effect of endogenous *d*-aldosterone on heat acclimatization in man. *J. Appl. Physiol.* 23: 341–346, 1967.
15. Brebner, D. F., and D. McK. Kerslake. The effect of soaking the skin in water on the acclimatization produced by subsequent heat exposure. *J. Physiol., London* 166: 13P–14P, 1963, Abstract.
16. Bridger, C. A. and L. A. Helfand. Mortality from heat during July 1966 in Illinois. *Int. J. Biometeorol.* 12: 51–70, 1968.
17. Burch, G. E. and A. Ansari. Artificial acclimatization to heat in control subjects and patients with chronic congestive heart failure at bed rest. *Am. J. Med. Sci.* 256: 180–194, 1968.
18. Buskirk, E. R., P. F. Iampietro and D. E. Bass. Work performance after dehydration: Effects of physical conditioning and heat acclimatization. *J. Appl. Physiol.* 12: 189–194, 1958.
19. Cadarette, B. S., M. N. Sawka, M. M. Toner and K. B. Pandolf. Aerobic fitness and the hypohydration response to exercise-heat stress. *Aviat. Space Environ. Med.* 55: 507–512, 1984.
20. Chen, W. Y. and R. S. Elizondo. Peripheral modification of thermoregulatory function during heat acclimation. *J. Appl. Physiol.* 37: 367-373, 1974.
21. Cleland, T. S., S. M. Horvath and M. Phillips. Acclimatization of women to heat after training. *Int. Z. Angew. Physiol.* 27: 15–24, 1969.
22. Collins, K. J. Endocrine control of salt and water in hot conditions. *Fed. Proc.* 22: 716–720, 1963.
23. Collins, K. J. The action of exogenous aldosterone on the secretion and composition of drug-induced sweat. *Clin. Sci.* 30: 207–221, 1966.
24. Collins, K. J., G. W. Crockford and J. S. Weiner. Sweat-gland training by drugs and thermal stress. *Arch. Environ. Health* 11: 407–422, 1965.
25. Collins, K. J., G. W. Crockford and J. S. Weiner. The local training effect of secretory activity on the response of eccrine sweat glands. *J. Physiol., London* 184: 203–214, 1966.

26. Collins, K.J., K. Hellman, R. M. Jones and J. B. Lunnon. Aldosterone activity in the urine of men exposed to hot environments. *J. Endocrinol.* 13: viii, 1955, Abstract.
27. Collins, K. J., and J. S. Weiner. Endocrinological aspects of exposure to high environmental temperatures. *Physiol. Rev.* 48: 785–839, 1968.
28. Conn, J. W. The mechanism of acclimatization to heat. *Adv. Int. Med.* 3: 373–393, 1949.
29. Conn, J. W. Aldosteronism in man. Some clinical and climatological aspects. Part I. *J. Am. Med. Assoc.* 183: 775–781, 1963.
30. Conn, J. W., and M. W. Johnston. The function of the sweat glands in the economy of NaCl under conditions of hard work in a tropical climate. *J. Clin. Invest.* 23: 933, 1944, Abstract.
31. Dasler, A. R., S. Karas, J. S. Bowman, and E. Hardenbergh. Adverse effects of supplementary sodium chloride on heat adaptation. *Fed. Proc.* 32: 336, 1973, Abstract.
32. Davies, J. A., M. H. Harrison, L. A. Cochrane, R. J. Edwards, and T. M. Gibson. Effect of saline loading during heat acclimatization on adrenocortical hormone levels. *J. Appl. Physiol.* 50: 605–612, 1981.
33. Davis, T. R. A. Effect of heat acclimatization on artificial and natural cold acclimatization in man. *J. Appl. Physiol.* 17: 751–753, 1962.
34. Edholm, O. G., R. H. Fox, R. Goldsmith, I. F. G. Hampton, and K. V. Pillai. A comparison of heat acclimatization in Indians and Europeans. *J. Physiol., London* 177: 15P–16P, 1965, Abstract.
35. Eichna, L. W., W. F. Ashe, W. B. Bean, and W. B. Shelley. The upper limits of environmental heat and humidity tolerated by acclimatized men working in hot environments. *J. Indust. Hyg. Toxicol.* 27: 59–84, 1945.
36. Eichna, L. W., W. B. Bean, W. F. Ashe, and N. Nelson. Performance in relation to environmental temperature. Reactions of normal young men to hot, humid (simulated jungle) environment. *Bull. Johns Hopkins Hosp.* 76: 25–58, 1945.
37. Eichna, L. W., C. R. Park, N. Nelson, S. M. Horvath, and E. D. Palmes. Thermal regulation during acclimatization in a hot, dry (desert type) environment. *Am. J. Physiol.* 163: 585–597, 1950.
38. Eijkman, C. Some questions concerning the influence of tropical climate on man. *Lancet* 1: 887–893, 1924.
39. Finberg, J. P. M., and G. M. Berlyne. Modification of renin and aldosterone response to heat by acclimatization in man. *J. Appl. Physiol.* 42: 554–558, 1977.
40. Finberg, J. P. M., M. Katz, H. Gazit, and G. M. Berlyne. Plasma renin activity after acute heat exposure in nonacclimatized and naturally acclimatized man. *J. Appl. Physiol.* 36: 519–523, 1974.
41. Forbes, W. H., D. B. Dill, and F. G. Hall. The effect of climate upon the volumes of blood and of tissue fluid in man. *Am. J. Physiol.* 130: 739–746, 1940.
42. Fortney, S. M., and L. C. Senay, Jr. Effect of training and heat acclimation on exercise responses of sedentary females. *J. Appl. Physiol.* 47: 978–984, 1979.
43. Fox, R. H., R. Goldsmith, I. F. G. Hampton, and T. J. Hunt. Heat acclimatization by controlled hyperthermia in hot-dry and hot-wet climates. *J. Appl. Physiol.* 22: 39–46, 1967.
44. Fox, R. H., R. Goldsmith, I. F. G. Hampton, and H. E. Lewis. The nature of the increase in sweating capacity produced by heat acclimatization. *J. Physiol., London* 171: 368–376, 1964.
45. Fox, R. H., R. Goldsmith, D. J. Kidd, and H. E. Lewis. Acclimatization to heat in man by controlled elevation of body temperature. *J. Physiol., London* 166: 530–547, 1963.
46. Fox, R. H., R. Goldsmith, D. J. Kidd, and H. E. Lewis. Blood flow and other thermoregulatory changes with acclimatization to heat. *J. Physiol.*, London 166: 548–562, 1963.
47. Francesconi, R., J. Maher, G. Bynum, and J. Mason. Recurrent heat exposure: effects on levels of plasma and urinary sodium and potassium in resting and exercising men. *Aviat. Space Environ. Med.* 48: 399–404, 1977.
48. Gisolfi, C. V., and J. S. Cohen. Relationships among training, heat acclimation, and heat tolerance in men and women: the controversy revisited. *Med. Sci. Sports* 11: 56–59, 1979.
49. Gisolfi, C. V., and C. B. Wenger. Temperature regulation during exercise: Old concepts, new ideas. In: R. J. Terjung (ed.) *Exercise and Sport Sciences Reviews.* Lexington, MA: Collamore Press, 1984, pp. 339–372.
50. Glaser, E. M., and R. J. Shephard. Simultaneous experimental acclimatization to heat and cold in man. *J. Physiol., London* 169: 592–602, 1963.
51. Gonzalez, R. R., K. B. Pandolf, and A. P. Gagge. Heat acclimation and decline in sweating during humidity transients. *J. Appl. Physiol.* 36: 419–425, 1974.
52. Harrison, M. H. Effects of thermal stress and exercise on blood volume in humans. *Physiol. Rev.* 65: 149–209, 1985.
53. Harrison, M. H., R. J. Edwards, M. J. Graveney, L. A. Cochrane, and J. A. Davies. Blood volume and plasma protein responses to heat acclimatization in humans. *J. Appl. Physiol.* 50: 597–604, 1981.

54. Hellon, R. F., R. M. Jones, R. K. Macpherson, and J. S. Weiner. Natural and artificial acclimatization to hot environments. *J. Physiol., London* 132: 559–576, 1956.
55. Hellon, R. F., and A. R. Lind. Circulation in the hand and forearm with repeated daily exposures to humid heat. *J. Physiol., London* 128: 57P–58P, 1955, Abstract.
56. Henane, R., and J. Bittel. Changes of thermal balance induced by passive heating in resting man. *J. Appl. Physiol.* 38: 294–299, 1975.
57. Henane, R., R. Flandrois, and J. P. Charbonnier. Increase in sweating sensitivity by endurance conditioning in man. *J. Appl. Physiol.* 43: 822– 828, 1977.
58. Henane, R., and J. L. Valatx. Thermoregulatory changes induced during heat acclimatization by controlled hyperthermia in man. *J. Physiol., London* 230: 255–271, 1973.
59. Höfler, W. Changes in regional distribution of sweating during acclimatization to heat. *J. Appl. Physiol.* 25: 503–506, 1968.
60. Hori, S., A. Inouye, H. Ihzuka, and T. Yamada. Study on seasonal variations of heat tolerance in young Japanese males and effects of physical training thereon. *Jap. J. Physiol.* 24: 463–474, 1974.
61. Horowitz, M., Y. Shimoni, S. Parnes, M. S. Gotsman, and Y. Hasin. Heat acclimation: cardiac performance of isolated rat heart. *J. Appl. Physiol.* 60: 9–13, 1986.
62. Horvath, S. M., and W. B. Shelley. Acclimatization to extreme heat and its effects on the ability to work in less severe environments. *Am. J. Physiol.* 146: 336–343, 1946.
63. King, D. S., D. L. Costill, W. J. Fink, M. Hargreaves, and R. A. Fielding. Muscle metabolism during exercise in the heat in unacclimatized and acclimatized humans. *J. Appl. Physiol.* 59: 1350–1354, 1985.
64. Kirby, C. R., and V. A. Convertino. Plasma aldosterone and sweat sodium concentrations after exercise and heat acclimation. *J. Appl. Physiol.* 61: 967–970, 1986.
65. Kirwan, J. P., D. L. Costill, H. Kuipers, M. J. Burrell, W. J. Fink, J. E. Kovaleski, and R. A. Fielding. Substrate utilization in leg muscle of men after heat acclimation. *J. Appl. Physiol.* 63: 31–35, 1987.
66. Knochel, J. P., L. N. Dotin, and R. J. Hamburger. Pathophysiology of intense physical conditioning in a hot climate. I. Mechanisms of potassium depletion. *J. Clin. Invest.* 51: 242–255, 1972.
67. Kok, R. Heat tolerance of Bantu undergoing acclimatization. *So. Afr. Med. J.* 47: 960, 1973, Abstract.
68. Kosunen, K. J., A. J. Pakarinen, K. Kuoppasalmi, and H. Aldercreutz. Plasma renin activity, angiotensin II, and aldosterone during intense heat stress. *J. Appl. Physiol.* 41: 323–327, 1976.
69. Kraning, K. K., and P. A, Lehman. Heat acclimation alters the cholinergic sensitivity of sweat glands. *Physiologist* 30: 147, 1987, Abstract.
70. Kuhlemeier, K. V., J. M. Miller, F. N. Dukes-Dobos, and R. Jensen. Determinants of the prescriptive zone of industrial workers. *J. Appl. Physiol.* 43: 347–351, 1977.
71. Kuno, Y. The loss of water and salt by sweating, their replenishment and changes in the blood. In: Y. Kuno (ed.), *Human Perspiration.* Springfield, IL: C.C. Thomas, 1956, pp. 251–276.
72. Kuno, Y. The acclimatization of the human sweat apparatus to heat. In: Y. Kuno (ed.), *Human Perspiration.* Springfield, IL: C.C. Thomas, 1956, pp.318-335.
73. Laaser, U. Physiologische Reaktionen während eines fünfwöchigen Daueraufenthaltes in einem künstlichen feuchtheissen Klima. *Int. Z. Angew. Physiol.* 25: 279–302, 1968.
74. Ladell, W. S. S. Thermal sweating. *Brit. Med. Bull.* 3: 175–179, 1945.
75. Ladell, W. S. S. Acquired heat tolerance of temperate climate men living in the tropics. Abstracts, XVIII *Int. Physiol. Congr.*, Copenhagen, 1950, pp. 320–322.
76. Ladell, W. S. S. Inherent acclimatization of indigenous West Africans. *J. Physiol.*, London 112: 15P–16P, 1951, Abstract.
77. Ladell, W. S. S. Disorders due to heat. *Trans. Roy. Soc. Trop. Med. Hyg.* 51: 189–216, 1957.
78. Leithead, W. S. S. Water and electrolyte metabolism in the heat. *Fed. Proc.* 22: 901–908, 1963.
79. Lichtlen, P. R., N. Solomon, L. Bernstein, G. C. Friesinger, and R. S. Ross. Inotropic effect of aldosterone on the myocardium of normal dogs. *Fed. Proc.* 23: 358, 1964, Abstract.
80. Lind, A. R., and D. E. Bass. Optimal exposure time for development of acclimatization to heat. *Fed. Proc.* 22: 704–708, 1963.
81. McCance, R. A. The effect of salt deficiency in man on the volume of the extracellular fluids, and on the composition of sweat, saliva, gastric juice and cerebrospinal fluid. *J. Physiol.*, London 92: 208–218, 1938.
82. Macfarlane, W. V., and K. W. Robinson. Seasonal changes in plasma antidiuretic activity produced by a standard heat stimulus. *J. Physiol., London* 135: 1–11, 1957.
83. MacGregor, R. G. S., and G. L. Loh. The influence of a tropical environment upon the basal metabolism, pulse rate, and blood pressure in Europeans. *J. Physiol.*, London 99: 496–509, 1941.

84. Machle, W., and T. F. Hatch. Heat: man's exchanges and physiological responses. *Physiol. Rev.* 27: 200–227, 1947.
85. Mack, G. W., X. Shi, H. Nose, A. Tripathi, and E. R. Nadel. Diminished baroreflex control of forearm vascular resistance in physically fit humans. *J. Appl. Physiol.* 63: 105–110, 1987.
86. Macpherson, R. K. Acclimatization status of temperate-zone man. *Nature* 182: 1240–1241, 1958.
87. Martineaud, J. P., J. Raynaud, P. Duhaze, and H. Vieillefond. Modifications du débit sanguin de la main et de la réactivité vasculaire avec l'acclimation à la chaleur. *J. Physiol., Paris* 72: 233–247, 1976.
88. Mitchell, D., L. C. Senay, C. H. Wyndham, A. J. van Rensburg, G. G. Rogers, and N. B. Strydom. Acclimatization in a hot, humid environment: energy exchange, body temperature, and sweating. *J. Appl. Physiol.* 40: 768–778, 1976.
89. Nadel, E. R., K. B. Pandolf, M. F. Roberts, and J. A. J. Stolwijk. Mechanisms of thermal acclimation to exercise and heat. *J. Appl. Physiol.* 37: 515–520, 1974.
90. Pandolf, K. B., R. L. Burse, and R. F. Goldman. Role of physical fitness in heat acclimatization, decay and reinduction. *Ergonomics* 20: 399–408, 1977.
91. Piwonka, R. W., and S. Robinson. Acclimatization of highly trained men to work in severe heat. *J. Appl. Physiol.* 22: 9–12, 1967.
92. Raynaud, J., J. P. Martineaud, O. P. Bhatnagar, H. Vieillefond, and J. Durand. Body temperatures during rest and exercise in residents and sojourners in hot climate. *Int. J. Biometeorol.* 20: 309–317, 1976.
93. Roberts, M. F., C. B. Wenger, J. A. J. Stolwijk, and E. R. Nadel. Skin blood flow and sweating changes following exercise training and heat acclimation. *J. Appl. Physiol.* 43: 133–137, 1977.
94. Robinson, S., H. S. Belding, F. C. Consolazio, S. M. Horvath, and E. S. Turrell. Acclimatization of older men to work in heat. *J. Appl. Physiol.* 20: 583–586, 1965.
95. Robinson, S., D. B. Dill, J. W. Wilson, and M. Nielsen. Adaptations of white men and Negroes to prolonged work in humid heat. *Am. J. Trop. Med.* 21: 261–287, 1941.
96. Robinson, S., R. K. Kincaid, and R. K. Rhamy. Effects of desoxycorticosterone acetate on acclimatization of men to heat. *J. Appl. Physiol.* 2: 399–406, 1950.
97. Robinson, S., and A. H. Robinson. Chemical composition of sweat. *Physiol. Rev.* 34: 202–220, 1954.
98. Robinson, S., E. S. Turrell, H. S. Belding, and S. M. Horvath. Rapid acclimatization to work in hot climates. *Am. J. Physiol.* 140: 168–176, 1943.
99. Rogers, G. G. Loss of acclimatization to heat in man during periods of no heat exposure. *So. Afr. Med. J.* 52: 412, 1977, Abstract.
100. Rowell, L. B. Human cardiovascular adjustments to exercise and thermal stress. *Physiol. Rev.* 54: 75–159, 1974.
101. Rowell, L. B. Cardiovascular adjustments to thermal stress. In: J. T. Shepherd and F. M. Abboud (eds.) *Handbook of Physiology. The Cardiovascular System. Peripheral Circulation and Organ Blood Flow.* Bethesda, MD: American Physiological Society, Sect. 2, Vol. 3, 1983, pp. 967–1023.
102. Rowell, L. B., K. K. Kraning II, J. W. Kennedy, and T. O. Evans. Central circulatory responses to work in dry heat before and after acclimatization. *J. Appl. Physiol.* 22: 509–518, 1967.
103. Sawka, M. N., K. B. Pandolf, B. A. Avellini, and Y. Shapiro. Does heat acclimation lower the rate of metabolism elicited by muscular exercise? *Aviat. Space Environ. Med.* 54: 27–31, 1983.
104. Sawka, M. N., M. M. Toner, R. P. Francesconi, and K. B. Pandolf. Hypohydration and exercise: effects of heat acclimation, gender, and environment. *J. Appl. Physiol.* 55: 1147–1153, 1983.
105. Sawka, M. N., A. J. Young, B. S. Cadarette, L. Levine, and K. B. Pandolf. Influence of heat stress and acclimation on maximal aerobic power. *Eur. J. Appl. Physiol.* 53: 294–298, 1985.
106. Schaible, T. F., and J. Scheuer. Effects of physical training by running or swimming on ventricular performance of rat hearts. *J. Appl. Physiol.* 46: 854–860, 1979.
107. Schneider, E. G., J. O. Davis, J. S. Baumber, and J. A. Johnson. The hepatic metabolism of renin and aldosterone. A review with new observations on the hepatic clearance of renin. *Circulation Res.* 26/27, Suppl. I: I-175-I-183, 1970.
108. Sciaraffa, D., E. Shvartz, L. C. Keil, P. J. Brock, and J. E. Greenleaf. Heat acclimation and resting blood pressure of normotensive men. *Med. Sci. Sports* 9: 51, 1977, Abstract.
109. Scott, J. C., H. C. Bazett, and G. C. Mackie. Climatic effects on cardiac output and the circulation in man. *Am. J. Physiol.* 129: 102–122, 1940.
110. Senay, L. C., Jr. Changes in plasma volume and protein content during exposures of working men to various temperatures before and after acclimatization to heat: separation of the roles of cutaneous and muscle circulation. *J. Physiol., London* 224: 61–81, 1972.

111. Senay, L. C., Jr. Plasma volumes and constituents of heat-exposed men before and after acclimatization. *J. Appl. Physiol.* 38: 570–575, 1975.
112. Senay, L. C., Jr. Effects of exercise in the heat on body fluid distribution. *Med. Sci. Sports* 11: 42–48, 1979.
113. Senay, L. C., and R. Kok. Body fluid responses of heat-tolerant and intolerant men to work in a hot wet environment. *J. Appl. Physiol.* 40: 55–59, 1976.
114. Senay, L. C., and R. Kok. Effects of training and heat acclimatization on blood plasma contents of exercising men. *J. Appl. Physiol.* 43: 591–599, 1977.
115. Senay, L. C., D. Mitchell, and C. H. Wyndham. Acclimatization in a hot, humid environment: body fluid adjustments. *J. Appl. Physiol.* 40: 786–796, 1976.
116. Shibolet, S., S. Fisher, T. Gilat, H. Bank, and H. Heller. Fibrinolysis and hemorrhages in fatal heatstroke. *New Engl. J. Med.* 266: 169–173, 1962.
117. Shvartz, E., A. Bhattacharya, S. J. Sperinde, P. J. Brock, D. Sciaraffa, and W. Van Beaumont. Sweating responses during heat acclimation and moderate conditioning. *J. Appl. Physiol.* 46: 675–680, 1979.
118. Shvartz, E., and N. Meyerstein. Effect of heat and natural acclimatization to heat on tilt tolerance of men and women. *J. Appl. Physiol.* 28: 428–432, 1970.
119. Shvartz, E., Y. Shapiro, A. Magazanik, A. Meroz, H. Birnfeld, A. Mechtinger, and S. Shibolet. Heat acclimation, physical fitness, and responses to exercise in temperate and hot environments. *J. Appl. Physiol.* 43: 678–683, 1977.
120. Shvartz, E., N. B. Strydom, and H. Kotze. Orthostatism and heat acclimation. *J. Appl. Physiol.* 39: 590–595, 1975.
121. Smiles, K. A., and S. Robinson. Sodium ion conservation during acclimatization of men to work in the heat. *J. Appl. Physiol.* 31: 63–69, 1971.
122. Stein, H. J., J. W. Eliot, and R. A. Bader. Physiological reactions to cold and their effects on the retention of acclimatization to heat. *J. Appl. Physiol.* 1: 575–585, 1949.
123. Streeten, D. H. P., J. W. Conn, L. H. Louis, S. S. Fajans, H. S. Seltzer, R. D. Johnson, R. D. Gittler, and A. H. Dube. Secondary aldosteronism: metabolic and adrenocortical responses of normal men to high environmental temperatures. *Metabolism* 9: 1071–1092, 1960.
124. Strydom, N. B. Some physiological aspects of adaptation to heat. *So. Afr. Med. J.* 28: 112–113, 1954.
125. Strydom, N. B., and C. G. Williams. Effect of physical conditioning on state of heat acclimatization of Bantu laborers. *J. Appl. Physiol.* 27: 262–265, 1969.
126. Strydom, N. B., C. H. Wyndham, C. G. Williams, J. F. Morrison, G. A. G. Bredell, A. J. S. Benade, and M. von Rahden. Acclimatization to humid heat and the role of physical conditioning. *J. Appl. Physiol.* 21: 636– 642, 1966.
127. Tanz, R. D. Studies on the inotropic action of aldosterone on isolated cardiac tissue preparations; including the effects of pH, ouabain and SC-8109. *J. Pharm. Exptl. Therap.* 135: 71–78, 1962.
128. Ullrick, W. C., and R. L. Hazelwood. Influence of aldosterone on isometric tension development in the rat heart. *Am. J. Physiol.* 204: 1001–1004, 1963.
129. Weiner, J. S., and R. E. van Heyningen. Salt losses of men working in hot environments. *Brit. J. Industr. Med.* 9: 56–64, 1952.
130. Whitney, R. J. Circulatory changes in the forearm and hand of man with repeated exposure to heat. *J. Physiol., London* 125: 1–24, 1954.
131. Williams, C. G., C. H. Wyndham, and J. F. Morrison. Rate of loss of acclimatization in summer and winter. *J. Appl. Physiol.* 22: 21–26, 1967.
132. Wood, J. E., and D. E. Bass. Responses of the veins and arterioles of the forearm to walking during acclimatization to heat in man. *J. Clin. Invest.* 39: 825–833, 1960.
133. Wyndham, C. H. Effect of acclimatization on circulatory responses to high environmental temperatures. *J. Appl. Physiol.* 4: 383–395, 1951.
134. Wyndham, C. H. Effect of acclimatization on the sweat rate/rectal temperature relationship. *J. Appl. Physiol.* 22: 27–30, 1967.
135. Wyndham, C. H. The physiology of exercise under heat stress. *Ann. Rev. Physiol.* 35: 193–220, 1973.
136. Wyndham, C. H., A. J. A. Benade, C. G. Williams, N. B. Strydom, A. Goldin, and A. J. A. Heyns. Changes in central circulation and body fluid spaces during acclimatization to heat. *J. Appl. Physiol.* 25: 586–593, 1968.
137. Wyndham, C. H., W. v. d. M. Bouwer, M. G. Devine, and H. E. Patterson. Physiological responses of African laborers at various saturated air temperatures, wind velocities and rates of energy expenditure. *J. Appl. Physiol.* 5: 290–298, 1952.
138. Wyndham, C. H., G. G. Rogers, L. C. Senay, and D. Mitchell. Acclimatization in a hot, humid environment: cardiovascular adjustments. *J. Appl. Physiol.* 40: 779–785, 1976.
139. Wyndham, C. H., N. B. Strydom, A. J. S. Benade, and A. J. Van Resnburg. Limiting rates

of work for acclimatization at high wet bulb temperatures. *J. Appl. Physiol.* 35: 454–458, 1973.

140. Wyndham, C. H., N. B. Strydom, J. F. Morrison, C. G. Williams, G. A. G. Bredell, M. J. E. Von Rahden, L. D. Holdsworth, C. H. Van Graan, A. J. Van Resnburg, and A. Munro. Heat reactions of Caucasians and Bantu in South Africa. *J. Appl. Physiol.* 19: 598–606, 1964.
141. Yoshimura, H. Seasonal changes in human body fluids. *Jap. J. Physiol.* 8: 165–179, 1958.
142. Young, A. J., M. N. Sawka, L. Levine, B. S. Cadarette, and K. B. Pandolf. Skeletal muscle metabolism during exercise is influenced by heat acclimation. *J. Appl. Physiol.* 59: 1929–1935, 1985.

5

Physical Training, Cardiorespiratory Physical Fitness and Exercise-Heat Tolerance

LAWRENCE E. ARMSTRONG, Ph.D.

KENT B. PANDOLF, Ph.D.

OUTLINE

INTRODUCTION

Between 1943 and 1945 researchers at 2 laboratories independently proposed a theoretical connection between physical fitness and the apparent rapid onset of acclimation to heat. Robinson *et al* (41) suggested that men who were in good physical condition could be expected to exercise effectively within a few days after working in a hot climate. Bean and Eichna (9) supported this statement and further suggested that physically fit individuals are more capable of a greater physical work output in the heat than are those in poor physical condition. Unfortunately, the statements of both research teams were based upon general impressions and were not supported by statistically significant differences measured during controlled laboratory trials. This theoretical connection between physical fitness and heat acclimation became widely accepted during the next 14 years; the classic review of heat acclimation mechanisms by Bass *et al* (7) in 1955 reiterated this concept, citing these studies (9,41) as sources. In fact, the first controlled study evaluating this theory was not published until a decade after the Bass *et al* paper.

In 1965, Piwonka *et al* (38) utilized a cross-sectional design to compare 5 highly trained distance runners to 7 untrained males. Both groups were exposed to a standard heat-tolerance test of 85 minutes (min) of moderate exercise in dry heat (40° C db, 23° C wb). The authors concluded that physiological adjustments to heat were substantially better in the trained runners, supporting the research described above (9,17,41). However, they also created the impression that heat acclimation may be achieved without actual exposure to heat stress, by stating that competitive distance runners "behaved as though they were acclimatized to the heat" (38). It was this conclusion that sparked a scientific controversy of major proportions. This controversy spanned more than 2 decades, involved investigators from several highly-respected laboratories, and resulted in heated debates. In fact, the pupil who tracks this topic through the scientific literature cannot help but recognize that it reads like a detective novel.

This controversy first surfaced when the longitudinal study of Strydom *et al* was published in 1966 (50). These investigators stated that it was important to know if heat acclimation may be achieved without actual exposure to heat stress, because of potentially serious consequences in industrial or military settings. Five mine laborers were acclimated to heat (36.1° C db, 33.9° C wb) during 5-hour trials conducted for 21 days, utilizing a moderate exercise intensity (block stepping, 30 cm height, 12 steps $\cdot$ min^{-1}). A second group of 5 subjects performed this same exercise for 13 days under mild environmental conditions (mean: 23.7° C db, 15.5° C wb), to assess the effects of physical training on exercise-heat tolerance. Exposure to a standard heat-tolerance test (5 hours of block stepping in the heat) before and after physical training demonstrated that training had only a limited effect on state of heat acclimation. Because of this, the authors stated that the conclusions of Piwonka and his colleagues were not valid. In addition, Strydom *et al* (50) discounted the work of Lofstedt (29) and Allan (1), who had supported the findings of Piwonka *et al*, by reporting that physical training influenced exercise-heat tolerance considerably.

Strydom's criticisms focused on the methodological difficulties of brief heat exposures and unspecified exercise intensities. In fact, methodological flaws seemed to abound in these early investigations, and even the work of Strydom *et al* (50) has not escaped retrospective criticism. Therefore, an examination of these methodological difficulties is pertinent to a complete understanding of the issues at hand. Such a design analysis (a) suggests that methodological flaws and narrow interpretation of data may have *created* this controversy and (b) explains why authors have labelled this dispute as "difficult to evaluate" (35) and "unnecessary" (19).

Methodological Considerations

There are 5 important methodological considerations which affect interpretation of data in this chapter. First, the phrase "heat tolerance" has been used in a wide variety of contexts, ranging from measurements of physiological responses in the face of mild heat stress, to the description of responses to severe heat stress where subjects could not continue due to impending or actual heat syncope. Table 5-1 illustrates these methodological considerations by presenting selected heat-tolerance tests which have been utilized since 1943.

Considering the range of environmental temperatures, clothing ensembles, and exercise modes/durations/intensities used by researchers, it is necessary to define carefully the conditions to which the phrase "heat tolerance" is intended to apply. In most investigations (Table 5-1), heat-tolerance tests were designed to mimic the demands of specific exercise requirements (e.g. mining, military, athletic endeavors) or environmental conditions (e.g. temperate, harsh). The advantage of such specificity is that tests which model real-world conditions are more likely to be useful in predicting heat tolerance under actual exercise-heat stress than tests which do not. The disadvantage of such specificity is that a heat-tolerance test designed for one situation probably will not be useful in other situations. This is one of the most obvious differences between the work of Piwonka *et al* (38) and Strydom *et al* (50), as described above. Although a strenuous 85 min heat-tolerance test exposed differences between Piwonka's groups (untrained men vs. competitive distance runners), this test would have been ineffective in evaluating the heat tolerance of laborers who were required to work 8-hour shifts in hot-humid conditions (50). Indeed, as evidence mounted during the 1960's and as scientists "took sides" in this controversy, this simple error of interpretation was committed time and again (see below).

Second, studies prior to 1970 rarely distinguished between physical training and cardiorespiratory physical fitness. These studies used the general term "fitness" or a similar variation. Usually, the use of this general term meant that no specific measurement of

TABLE 5-1. *Heat Tolerance Tests Utilized in Pertinent Early Investigations.*

		Ambient Conditions[1]				Exercise Variables		
Authors (year)	n	db (°C)	wb (°C)	rh (%)	Clothing Ensemble	Mode	Duration (min)	Intensity
Bean and Eichna (1943)	56	49	22–27	15–22	shorts, socks, shoes, military fatigues	TM walk	50w/10r	$1.34 m \cdot s^{-1}$
Robinson *et al* (1943)	5	40	23	23	cotton trousers, shirt, tie, socks, shoes	TM walk	exhaustion	$1.56 m \cdot s^{-1}$
Allan (1965)	108	40	32	57	NA	bench 30cm	30w/30r	$12 \cdot min^{-1}$
Piwonka *et al* (1965)	6	40	24	25	khaki pants, shirt, service shoes	TM walk	85	$1.56 m \cdot s^{-1}$
Strydom *et al* (1966)	5	36	34	86	NA	bench 30cm	300	$12 \cdot min^{-1}$
Piwonka and Robinson (1967)	4	50	27	16	khaki pants, shirt, shoes, socks	TM walk	85	$1.56 m \cdot s^{-1}$
Gisolfi and Robinson (1969)	5	50	27	16	shorts, socks, tennis shoes	TM walk	90	$1.56 m \cdot s^{-1}$
Strydom and Williams (1969)	23	34	32	89	NA	bench 30cm	300	$12 \cdot min^{-1}$
Greenleaf *et al* (1972)	7	47–48	32–33	34–41	NA	ergometer	120	28% $\dot{V}O_2max$

[1]—some values have been calculated. db, dry bulb; wb, wet bulb; rh, relative humidity; NA, not available; TM, treadmill; w, work; r, rest.

fitness was made during the experiment and that no clear definition of this term was offered. Recent investigations (6,34) have clearly demonstrated the distinct role which these two factors play in exercise-heat tolerance. Cardiorespiratory physical fitness is best determined by measuring an individual's maximal aerobic power ($\dot{V}O_2max$), and is defined as the maximum amount of oxygen that can be transported to and used by exercising muscles during exhaustive exercise (expressed either as $L \cdot min^{-1}$ or as $ml \cdot kg^{-1} \cdot min^{-1}$).

Maximal aerobic power is limited by pulmonary diffusion, oxygen and carbon dioxide transported by the blood, cardiac function, vascular adaptation (e.g. vasodilation of active tissues and vasoconstriction of inactive tissues), and the capillarization and oxidative capacity of the involved muscles (10). In contrast, physical training involves a program of exercise designed to improve oxygen delivery and to increase substrate utilization for a particular event or mode of exercise. Studies which neglected the distinction between these two factors, involved two covariables which influenced exercise-heat tolerance simultaneously.

Third, studies related to cardiorespiratory fitness have employed both cross-sectional (instantaneous) and longitudinal research designs (Table 5-2). The cross-sectional designs in this chapter are based on a comparison of individuals of differing levels of $\dot{V}O_2max$ and have been criticized for the absence of proper controls. The logitudinal approach utilizes a comparison of subjects serving as their own controls, before and after a physical training program (30) or some other experimental treatment. The cross-sectional de-

TABLE 5-2. *Design Characteristics of Pertinent Investigations Through 1972.*

		Training Program (T_{amb} = 18–22° C)		
Authors (year)	Study Design (Duration)	Mode	Duration (min)	Intensity Category
Bean and Eichna (1943)	longitudinal (1–3 wk)	TM walk, I	235–250	moderate
Robinson *et al* (1943)	longitudinal (10–23 days)		NA	
Allan (1965)	longitudinal (14 days)	gym, I	180	strenuous
Piwonka *et al* (1965)	cross-sectional (2 days)		NA	
Strydom *et al* (1966)	longitudinal (12 days)	bench 30cm	300	mild
Piwonka and Robinson (1967)	cross-sectional (5 days)		NA	
Gisolfi and Robinson (1969)	longitudinal (6 wk)	track, game	60	intense
Strydom and Williams (1969)	longitudinal (3 wk)	bench 30cm	240	mild
Greenleaf *et al* (1972)	cross-sectional (3 days)		NA	

TM, treadmill; I, intermittent; T_{amb}, ambient temperature; gym, gymnastics activities; track, track interval training; bench, bench stepping; NA, physical training in a cool environment not conducted during this study.

sign is more convenient but is restricted to examination of a trend, if one exists. The latter design is able to establish (a) whether a given level of exercise-heat tolerance can be increased by improving cardiorespiratory fitness, and (b) the magnitude of that increase. The studies of Piwonka *et al* (37,38) utilized a cross-sectional design while Strydom *et al* (50), and all other studies prior to 1970, employed a longitudinal design (Table 2).

Fourth, small homogeneous samples (five or less subjects, see Table 5-1) reduced the validity of conclusions in early studies. In addition, not one study presented in Table 5-2 reported the statistical significance of their results.

Fifth, the nature of training programs differed greatly. Table 2 illustrates that a variety of exercise modes (e.g. running, bench stepping, calisthenics, games) were utilized. The duration of training ranged from 30–300 min per day and the intensity of exercise ranged from mild to intense. In this respect, the intense interval training of Piwonka *et al* (38) is difficult to compare to the mild exercise performed by the subjects of Strydom *et al* (50); because the exercise and environmental conditions, which subjects can tolerate, will vary with the duration and intensity of the training program.

1966–1972: THE ISSUES UNFOLD

Following their initial studies (38,50), both Piwonka and Strydom published follow-up investigations which supported their respective positions and further fueled the fires of controversy, yet clarified original findings. Previously published reviews by Wyndham (54), and by Gisolfi and Cohen (19) have examined key studies through 1972 quite thoroughly. Piwonka and Robinson (37), for example, continued observations on four of the highly trained distance runners described above. An attempt was made to acclimatize them further, by daily performance of the same exercise (Table 5-1) in a more intense environment (50° C db, 28° C wb). Four days of this exercise improved their sweat responsiveness (sweat rate per degree rise of rectal temperature) by 11%, and greatly increased cutaneous blood flow; the runners also displayed day-to-day decreases of heart rate and rectal temperature. It was concluded that the intensive training program completely conditioned them for exercise in moderate heat (40° C db, 23.5° C wb) and improved their capacities for acclimatization to a severe heat stress (50° C). In addition, these authors suggested that the low exercise intensity employed by Strydom *et al* (50) may have negated any significant improvement in exercise-heat tolerance.

Accordingly, Strydom and Williams (49) trained 23 young Bantu males four times each day for 12 days, utilizing an increased, yet moderate, exercise intensity ($\dot{V}O_2 \approx 1.5–1.6\ L \cdot min^{-1}$). Their performance was evaluated via a standard 4-hour heat-tolerance test (Table 5-1), administered before and after training. Following physical training, these authors conceded that subjects "could be classified as being partially acclimatized to heat." However, heart rate, rectal temperature, and sweat rate were significantly improved during the first and second hours only. They questioned the adequacy of heat-tolerance tests of less than two-hour duration. Therefore, both parties in this debate had modified their initial positions. Piwonka's subjects, who formerly had responded as though they were fully acclimatized to heat, now exhibited improved capacities for heat acclimatization. Strydom's subjects, who formerly displayed no evidence of heat acclimation after physical training, now exhibited partial heat acclimation.

In 1969, Gisolfi and Robinson (20) pursued the effects of 6 weeks (wk) of intense interval training (1 hour per day, 5 days per week) on exercise-heat tolerance in five untrained males. Following training, heart rate and rectal temperature responses improved (168 vs. 141 beats $\cdot$ min^{-1}, 39.6 vs. 38.7° C) and sweat rate increased approximately 50%. When compared to 3 distance runners who had been training and competing for 6–10 years, these five men responded less favorably and were deemed to be only partially acclimated for exercise in the heat. The thermoregulatory improvements exhibited by the five trained men were attributed to increased evaporative cooling and greater stability of the cardiorespiratory system. However, Gisolfi and Robinson responded to the comments of Strydom (49,50) by noting that the 6-week interval training program, used in this study, provided a more powerful thermal stimulation than the prolonged, moderate exercise performed by Strydom's subjects. They concluded that their training program was more effective for preconditioning physically fit young men for exercise in hot environments. A methodological flaw existed in this study and in the project of Piwonka *et al* (38), however. The distance runners used in both studies had lower body weights and higher $\dot{V}O_2max$ values than their untrained counterparts; distance runners were probably exercising at less than 50 %$\dot{V}O_2max$ while the untrained subjects exercised at ≥ 66 %$\dot{V}O_2max$ (anaerobic exercise), which would place the untrained subjects at a disadvantage. Thus, the debate continued into the 1970's.

Next, Greenleaf *et al* (22) examined the relationship between $\dot{V}O_2$ max and exercise-heat tolerance. Seven young males performed cycle ergometer exercise at a low intensity during three, 2-hour heat acclimation exposures (47.7° C db, 32.7° C wb), separated by one day of rest between each. The relationship between $\dot{V}O_2max$ and tolerance time in the heat was weak and not statistically significant. Thus, Greenleaf and colleagues joined Strydom *et al* (49,50) in criticizing the methods of other researchers. However, these findings must be interpreted in light of the fact that subjects exercised at the same *relative* exercise intensity (28 ± 1 %$\dot{V}O_2max$) which reduced between-subject differences in final rectal temperature (5,42), and a small, homogenous sample was used.

1973–1979: REFINEMENT OF ISSUES BY IMPROVED STUDY DESIGN

Underlying each of the studies published during the period of 1966 to 1972 were the similarities in physiological adaptations resulting from physical training, as opposed to those resulting from heat acclimation. Yet, none of the studies prior to 1972 attempted to discriminate between the adaptations to training and acclimation. The physiological adaptations to physical training in a cool environment include: a lowered threshold temperature for the onset of sweating, increased plasma volume, increased stroke volume, decreased heart rate, decreased skin temperature, decreased rectal temperature, increased $\dot{V}O_2max$, and reduced muscle blood flow in response to a given level of submaximal exercise (2,8,12,13,19,49,51). Except for the fact that heat acclimation lowers metabolic rate, while physical training does not, the physiological adaptations of these processes are similar (23,41). Therefore, longitudinal cross-over studies most effectively partition the adaptations due to training, versus those due to heat acclimation. Such cross-over studies involve one group of subjects who undergo physical training and later undergo heat acclimation, to compare the relative contribution of each factor to exercise-heat tolerance.

Several well-designed longitudinal cross-over studies appeared during the period of 1972 to 1979. Gisolfi (18), for example, reported observations of six young men who performed a standard heat-tolerance test (treadmill walking, 1.56 $m \cdot s^{-1}$; 48.9° C db, 29.7° C wb) at the following five points: before training, after 4, 8 and 11 weeks of interval training, and after 8 days of heat acclimation. The interval training program consisted of intense intermittent treadmill running (30 $min \cdot day^{-1}$, 5 $days \cdot week^{-1}$, at 21° C), with training intensity being progressively increased throughout the training period. Heat acclimation was accomplished via eight consecutive daily exposures to the standard heat-tolerance test. Results indicated that $\dot{V}O_2max$ increased 11.5% after 4 wk of training, 15% after 8 wk, and 15% after 11 wk. However, no significant relationship was found between terminal rectal temperature and $\dot{V}O_2max$. Exercise-heat tolerance improved progressively during the first 8 wk of interval training, with little change thereafter. Heat acclimation (8 days) further improved mean exercise-tolerance time, decreased mean rectal temperature and decreased mean heart rate, when compared to physical training (11 wk). Utilizing the terminal rectal temperature/performance time ratio, the percentage of the total adjustment in heat tolerance was 30, 57, and 65 % after 4, 8 and 11 wk of training, respectively.

In 1974, Nadel *et al* (31) utilized a similar longitudinal design, but focused on the mechanisms by which physical training and heat acclimation affected sweating. Six relatively unfit men were trained for 10 days (1 $hr \cdot day^{-1}$, 70–80 %$\dot{V}O_2max$) in a 22° C db environment. This training ($\dot{V}O_2max$ increased 6.6 $ml \cdot kg^{-1} \cdot min^{-1}$) was followed by 10 days of heat acclimation, which resulted in significant reductions in heart rate and esophageal temperature. Physical training alone increased the sensitivity of the sweat glands 67% (a peripheral effect) without significantly changing the point of zero central sweating drive (a central effect), resulting in a steeper slope of the sweat rate-to-core temperature relationship. Heat acclimation lowered the point of zero sweating drive (a central effect) without altering the sensitivity of the sweat glands (a peripheral effect). The responses were explained as either peripheral effects (e.g. training of the sweating mechanism at the glandular level) or central effects (e.g. changes in the central nervous system point of zero sweating drive).

Henane *et al* (25) employed a cross-sectional research design (six cross-country skiers, four swimmers, three sedentary men) to confirm these findings. Intense interval training by the sedentary men (18–20° C db, 13° C wb) for three months resulted in enhanced $\dot{V}O_2max$ (+18 %). The authors concluded that physical training of this type resulted in increased exercise-heat tolerance, increased sweat sensitivity, and shortened sweating onset. A comparison of the cross-country skiers and swimmers in this investigation appears at a later point in this chapter.

Utilizing a longitudinal cross-over design very similar to that of Nadel *et al* (31), Roberts and co-workers (39) evaluated the sweat sensitivity (chest sweat rate-to-esophageal temperature relationship), as well as the forearm blood flow-to-esophageal temperature relationship. Physical training increased $\dot{V}O_2max$ by 13% in men (n = 4) and 11% in women (n = 4). As a result of this training, forearm blood flow increased because the core temperature threshold for vasodilatation decreased; heat acclimation further reduced this threshold, but neither physical training nor heat acclimation altered the slope of the blood flow-esophageal temperature relationship. Figure 5-1 summarizes the thermoregulatory changes measured during this investigation.

Working in the same laboratory utilized by Strydom and colleagues (49,50), Senay and Kok expanded our understanding of the longitudinal relationship between $\dot{V}O_2max$, physical training, and heat acclimation by observing changes in blood constituents and fluid-electrolyte balance (43). They trained 5 men 4 $h \cdot day^{-1}$ for 43 days on a cycle ergometer. Following training, these subjects underwent an 8-day heat acclimation regimen. A standard exercise tolerance test was performed in both a cool (18–22° C db) and a hot (33.8° C db, 32.4° C wb) environment by bench stepping at 35 W for 4 hours. These hot and cool tests were conducted before training began, at intervals during training, before heat acclimation, and after heat acclimation. Physical training increased mean $\dot{V}O_2max$ by 12%, had little effect on thermoregulation during the cool tests, but did modify heart rate early (< 2 h) in the heat-tolerance tests as previously reported by Strydom *et al* (49,50).

Physical training also clearly resulted in improved maintenance of plasma volume during exercise-heat exposure. Heat acclimation increased the sensitivity of the sweating mechanism (in the heat), further decreased submaximal oxygen consumption (in hot and cool conditions), and further expanded plasma volume. To interpret this study properly, the reader should recognize that low exercise intensities (26–35 %$\dot{V}O_2max$) were utilized during the cool and hot standardized tests, even through daily cycle ergometer training sessions increased to 60–90 %$\dot{V}O_2max$ during the final days of training. These low exercise intensities were probably utilized in an effort to simulate tasks in the mining industry. In addition, Senay and Kok concluded their discussion with a caution similar to that presented above (see Methodological Considerations). "There is little doubt that various regimens of endurance training confer on subjects a certain degree of heat tolerance. The differences which exist in such studies depend on the experimental design of the various authors and the semantics resulting from such choices." (43).

A 1977 report published by Gisolfi and colleagues (21) presented results from 2 carefully designed cross-sectional studies; these studies clarified several aspects of the controversy which had begun eleven years earlier. Utilizing 2 groups matched for physical characteristics, 4 distance runners were compared to 6 college students. Unlike the original work of Piwonka *et al* (38), both groups had similar mean $\dot{V}O_2max$ values (60 vs. 62 $ml \cdot kg^{-1} \cdot min^{-1}$) and mean body mass-to-surface area ratios (38.3 vs. 38.5 $kg \cdot m^{-2}$). After the college students completed 8 wk of interval training, both groups under-

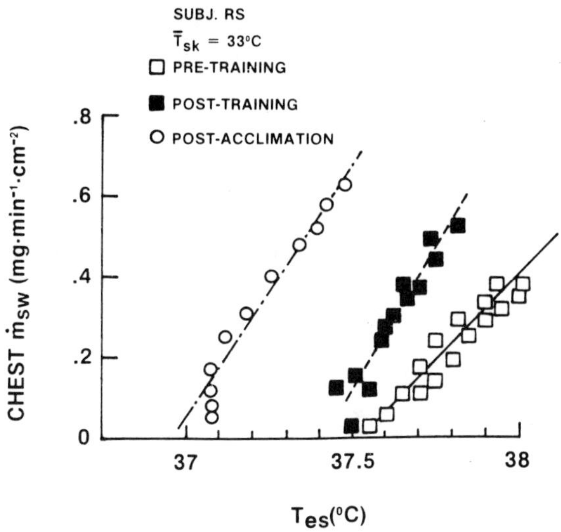

Figure 5-1a. *Chest sweat rate-to-esophageal temperature relationship of subject RS during pre-training, post-training, and post-acclimation trials, redrawn from Roberts* et al *(39). $\dot{m}_{sw}$—sweat rate.*

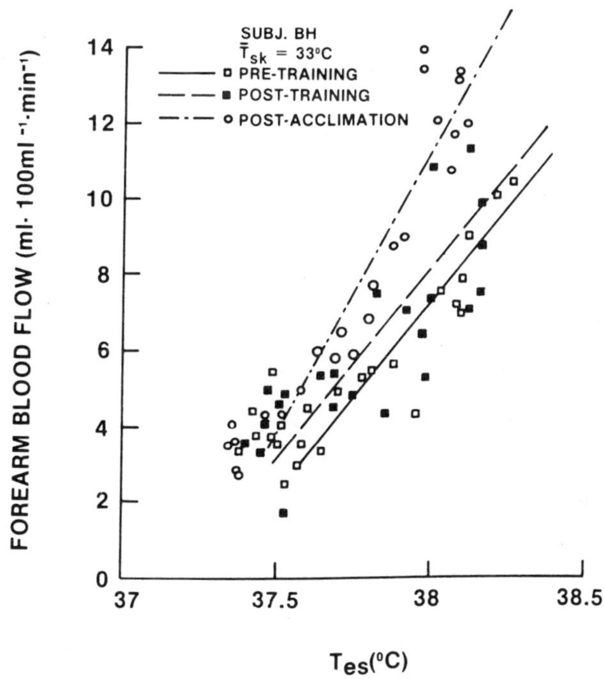

Figure 5-1b. *Forearm blood flow-to-esophageal temperature relationships of subject BH during pre-training, post-training, and post-acclimation trials, redrawn from Roberts* et al *(39).*

took a 100 min standard heat-tolerance test (50° C db, 27° C wb; 1.56 m·s^{-1}, 0% grade).The distance runners maintained significantly lower heart rates and rectal temperatures throughout the chamber exposure, but group sweat rates were very similar (Fig. 5-2).

The lower heart rates of the runners indicated a more stable

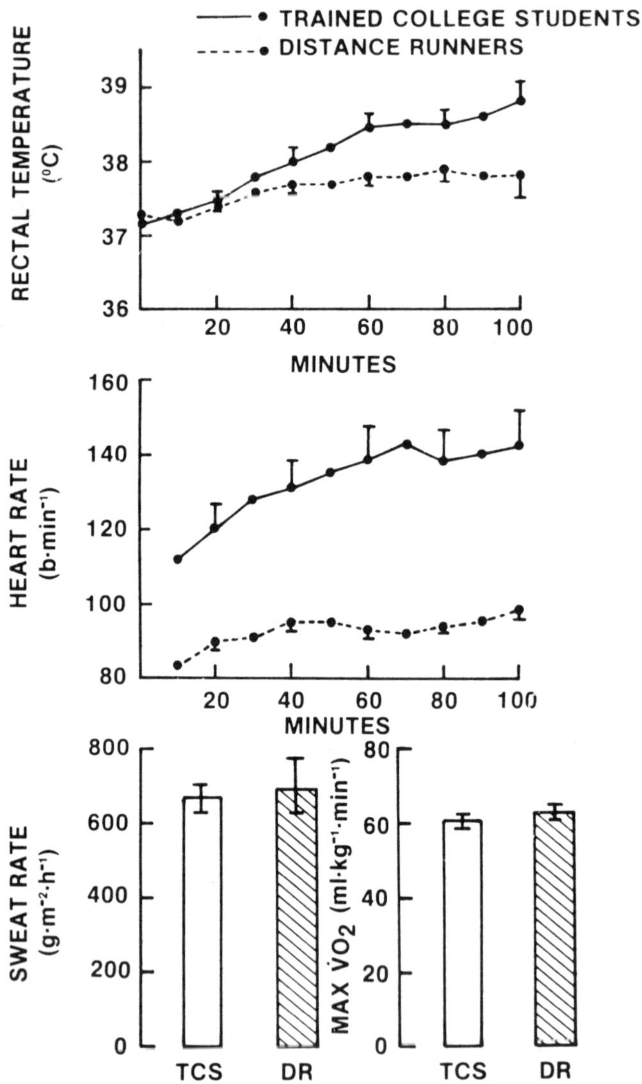

Figure 5-2. *Mean rectal temperature, heart rate, sweat rate and* $\dot{V}O_2$*max of six trained college students and four distance runners during treadmill exercise in dry heat (50° C db, 27° C wb), redrawn from Gisolfi* et al *(21).*

cardiovascular system, suggesting a more adequate blood flow through visceral organs and a more efficient flow of heat from the core to the skin. Evidently, the intensive training of the runners at high exercise intensities (70–80 $\%\dot{V}O_2max$) for 1–2 hours per day for several years produced a more efficient cardiovascular system than the college students who had trained for 8 wk.

The second study reported in this publication (21) focused on a prolonged (5–7 hour) standard heat-tolerance test (walking 1.56 $m \cdot s^{-1}$, 0% grade) in dry heat (49.6° C db, 27.5° C wb), wet heat (36.7° C db, 33.1° C wb), and a cool environment (27.0° C db, 12.5° C wb), each separated by 4–5 wk. Four distance runners had been running 64–80 $km \cdot wk^{-1}$ throughout the experimental period (winter months) and had not been exposed to heat since the preceding summer. The results (Fig. 5-3) clearly indicated that distance runners can tolerate prolonged mild exercise in cool air and dry heat without severe physiological strain, and without prior heat exposure. In the hot-wet environment, the mean rectal temperature and heart rate were higher (Fig. 5-3), probably due to an increased metabolic rate and reduced sweat production (hidromeiosis); but all subjects walked continuously for at least 4 hours with minimal or no signs of severe physiological strain. Therefore, these 2 studies focused on the major criticisms levied by Strydom *et al* (49,50) against the work of Piwonka *et al* (38) (e.g. between-group differences in $\dot{V}O_2max$, short duration of exercise), and upheld the interpretations which Piwonka *et al* had originally published.

Maximal Aerobic Power and Exercise-Heat Tolerance

Shvartz and colleagues published 3 pertinent studies during the period 1973–1979 (45,46,47). Unfortunately, these studies involved combinations of either low intensity exercise, a training program of short duration, or a cross-sectional research design (e.g. large initial between-group differences in $\dot{V}O_2max$) which are the same research design flaws that prevailed during the period 1966 to 1972. The 2 earlier studies (46,47) involved mild physical training ($\dot{V}O_2$ = 1.0 $L \cdot min^{-1}$) in cool and hot environments for 6 days and 8 days only. Therefore, the observation that physical training of short duration at 21–23° C resulted in little or no improvements in exercise-heat tolerance was not surprising. The thorough review of this topic by Pandolf (32) revealed that these 2 studies (46,47) opposed the findings of 7 investigations from other laboratories (15,18,21,25,31,33,39) as well as a 1977 publication by Shvartz *et al* (45).

This 1977 work focused on $\dot{V}O_2max$ by evaluating 26 young men who possessed $\dot{V}O_2max$ values ranging from 29–65 $ml \cdot kg^{-1} \cdot min^{-1}$; these men were classified as either trained, untrained, unfit, or con-

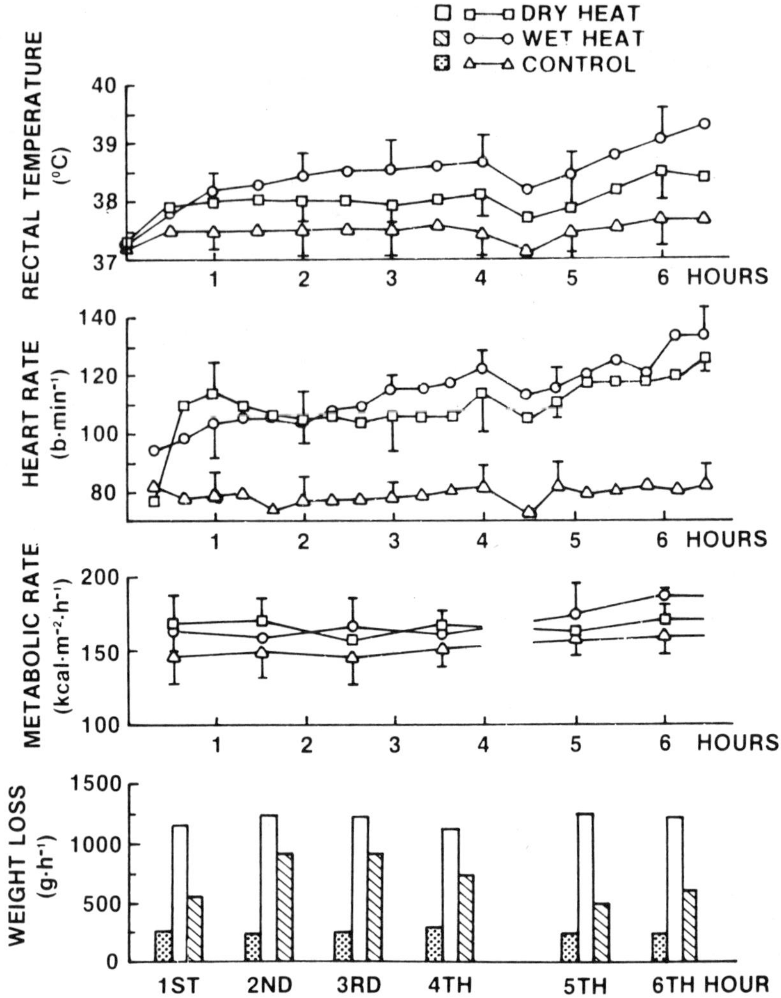

Figure 5-3. *Mean rectal temperature, heart rate, metabolic rate, and sweating weight loss of four distance runners during mild treadmill exercise in dry heat (49.6° C db, 27.5° C wb), wet heat (36.7° C db, 33.1° C wb), and in a cool control environment (27.0° C db, 12.5° C wb), redrawn from Gisolfi* et al *(21).*

trol. Trained subjects (mean $\dot{V}O_2max = 60.1 \pm 3.7\ ml \cdot kg^{-1} \cdot min^{-1}$) exhibited superior heart rate and rectal temperature responses; while unfit subjects (mean $\dot{V}O_2max = 35.7 \pm 3.3\ ml \cdot kg^{-1} \cdot min^{-1}$) exhibited the least favorable responses, during standard exercise tolerance tests. These tests were conducted at 23° C db (bench stepping, 30 cm height, $\dot{V}O_2 = 1.14$ to $1.25\ L \cdot min^{-1}$ or 1.94 to $1.99\ L \cdot min^{-1}$) and at 39.4° C db/30.3° C wb (bench stepping for 3 h, 30 cm height,

$\dot{V}O_2$ = 1.13 to 1.27 $L \cdot min^{-1}$), before and after 8 days of heat acclimation. The correlation coefficient between $\dot{V}O_2max$ and the three-hour rectal temperature indicated a stronger inverse relationship (r = −0.65) than previous investigations, which indicated correlation coefficients of r = −0.35 (55), r = −0.50 (pre-acclimation) and r = −0.41 (post-acclimation) (56) between $\dot{V}O_2max$ and rectal temperature.

Utilizing a heterogeneous sample ($\dot{V}O_2max$ range: 35.5–63.8 $ml \cdot kg^{-1} \cdot min^{-1}$), Pandolf *et al* (33) reported the relationship between $\dot{V}O_2max$ and the number of days to reach the "plateau day" for rectal temperature during heat acclimation (e.g. the point at which 95% of improvements had occurred). The correlation coefficient (r = −0.68) indicated a significant ($p < 0.01$) relationship between the plateau day and $\dot{V}O_2max$. This study was similar to that of Shvartz *et al*, described above (45), in that both research groups calculated that $\dot{V}O_2max$ accounted for essentially the same amount of the variation (42% and 46%, respectively) in their exercise-heat tolerance variables. Figure 5-4 illustrates these relationships from both studies. The reader should recognize, however, that correlation does not imply causation. The $\dot{V}O_2max$, per se, may not be as important in exercise-heat tolerance as the underlying physiological adjustments and functional states of organ systems associated with various fitness levels (32).

THE 1980'S: NEW RESEARCH QUESTIONS DEFINE MECHANISMS FURTHER

As with any research topic, answers to earlier questions generated even more questions. The studies of the 1980's built upon former work by focusing on specific cardiovascular, thermoregulatory, and fluid adaptations induced by physical training and/or heat acclimation. Although it is tempting to explore peripheral issues which arose during this period (e.g. the effects of detraining, diet, or gender on heat tolerance), this chapter will leave that task to other authors (3,24,26,28,33,36).

Shortly after the turn of the decade, Avellini and colleagues published an interesting study (6) which examined physical training on land (22° C), or while immersed to the neck in either warm water (32° C) or cold water (20° C). The effects of these three training media on exercise-heat tolerance were measured via a standard heat-tolerance test (50/10 min exercise/rest cycles, 3 h walk, ≈30 %$\dot{V}O_2max$). Training in all groups (n = five per group, matched for $\dot{V}O_2max$) consisted of pedaling a cycle ergometer for 1 $h \cdot day^{-1}$, 5 $days \cdot wk^{-1}$, for 20 days, at 75 %$\dot{V}O_2max$; weekly readjustments of absolute exercise intensity were made to maintain the same relative

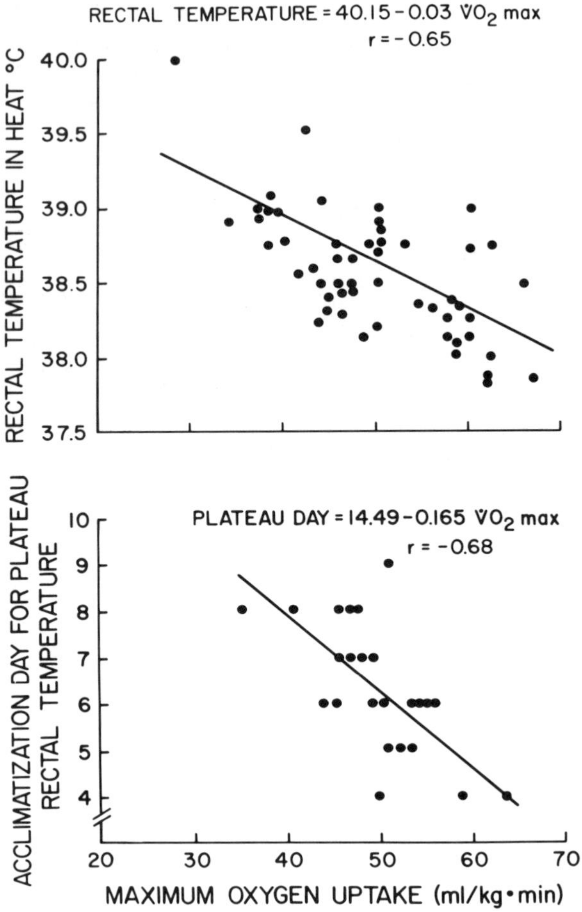

Figure 5-4. *The relationship between $\dot{V}O_2max$ and rectal temperature in hot environments (n = 26) as reported by Shvartz* et al *(45), or between $\dot{V}O_2max$ and the acclimatization day for a plateau in rectal temperature (n = 24) as reported by Pandolf* et al *(33). Figure originally published by Pandolf (32).*

exercise intensity. Following physical training which resulted in similar increases in $\dot{V}O_2max$ in all groups (range: +13% to +16%), subjects underwent a second heat-tolerance test, a 10-day heat acclimation program (2 h · day^{-1}, 49° C db, 28° C wb, 30 %$\dot{V}O_2max$), and a post-acclimation heat-tolerance test. Exercise-heat tolerance times before training were statistically similar between groups, but following physical training were: (a) land, 180 min; (b) 32° C water, 168 min; (c) 20° C water, 122 min. Following heat acclimation, all groups exhibited tolerance times of 180 min. Figure 5-5 illustrates the final mean rectal temperature, mean skin temperature, and heart rate values for all groups for pre-training (HEAT STRESS 1), post-training (HEAT STRESS 2), and post-acclimation (HEAT STRESS 3). Daily training in these three media resulted in mean rectal temperature changes of +1.1° C (land), +0.6° C (warm water), and a steady decline (cold water). During HEAT STRESS 2, in spite of a 15% increase in $\dot{V}O_2max$, the cold water group exhibited virtually no training-induced improvement in acute exercise-heat tolerance other than an 18 beats · min^{-1} decrease in final heart rate; this group also exhibited higher mean rectal temperature and mean skin temperature (Fig. 5-5a). These findings suggest that physical training and improved $\dot{V}O_2max$, per se, are not always associated with improved

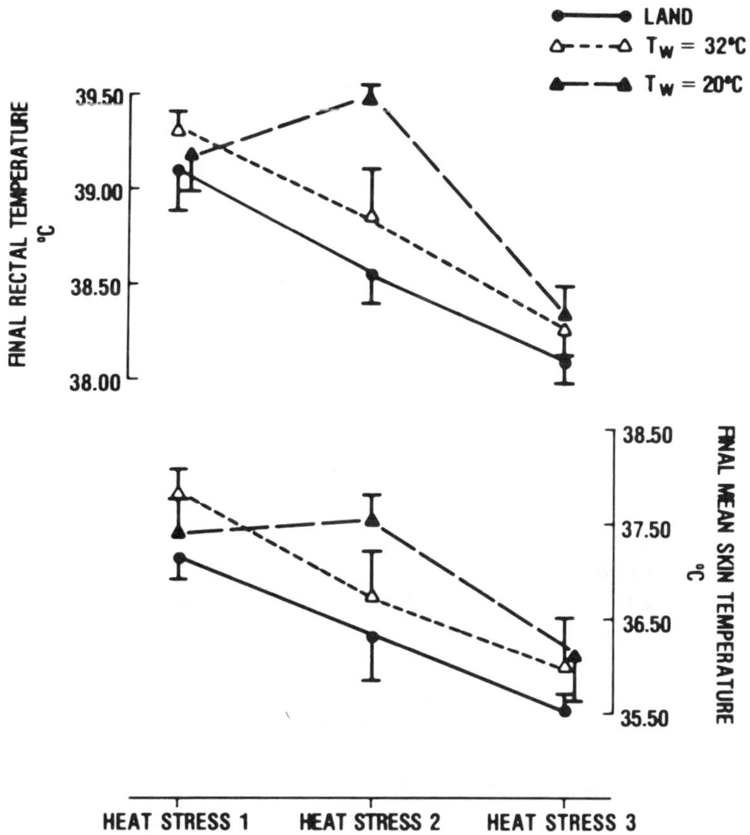

Figure 5-5a. *Mean final rectal temperature and skin temperature of land, warm water (32° C), and cold water (20° C) groups (n = 5 in all groups), as reported by Avellini* et al *(6). Measurements were taken before physical training (heat stress 1), after physical training (heat stress 2), and after heat acclimation (heat stress 3).*

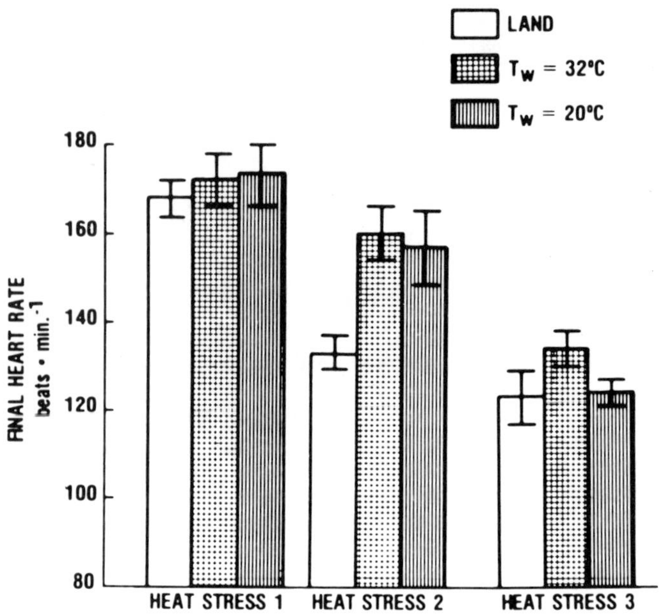

Figure 5-5b. *Comparison of mean final heart rate of land, warm water, and cold water groups, as reported by Avellini* et al *(6). Legends are the same as in Figure 5a.*

exercise-heat tolerance, particularly from a thermoregulatory stand point. Improvements in $\dot{V}O_2$max apparently must be associated with significant elevations in core temperature to improve the thermoregulatory component of exercise-heat tolerance. This conclusion verified a hypothesis stated by Pandolf in an earlier review article (32).

In addition to their discussion on exercise-heat tolerance, Avellini *et al* (6) commented on the thermoregulatory abilities of swimmers, noting that the warm water group increased total body sweat rate after training, while the cold water group did not. The authors attributed this to the absence of a peripheral heat effect in the cold water (20° C) immersion group. Other investigations have also observed differences between swimmers and athletes who trained on land. Nadel *et al* (31) reported that swim training negated sweat gland training. Henane *et al* (25) evaluated the exercise-heat tolerance of swimmers and cross-country skiers, who had been matched for maximal aerobic power (65.8 vs. 66.5 $ml \cdot kg^{-1} \cdot min^{-1}$, respectively). The swimmers exhibited a lesser degree of exercise-heat tolerance, hypothetically explained by the fact that they did not experience marked daily increases in rectal or skin temperatures. This was clearly reflected in weight loss due to sweating (Fig. 5-6). Piwonka *et al* (38) reported findings from one highly trained swimmer who displayed an elevated heart rate and rectal temperature, when compared to land-trained subjects on a standard heat-tolerance test. These findings also support the concept that significant elevations in core temperature must be accompanied by increased skin temperature, to fully induce exercise-heat tolerance improvements.

Recognizing that the maximal heat-tolerance in moderate climates is quite different from that required for hot climates, Armstrong *et al* (4) examined heat acclimatization of highly trained distance runners (n = 5, mean $\dot{V}O_2max$ = 69 $ml \cdot kg^{-1} \cdot min^{-1}$) in the northern United States, by comparing their spring and summer heat-tolerance. These runners undertook a standard heat-tolerance test before (spring) and after (summer) 14.5 wk of outdoor training and competition. This 90 min heat-tolerance test was designed to simulate the hottest days of summer (30.3° C db, 35 %rh) in the northeastern United States, and typical training velocities (200 $m \cdot min^{-1}$, 60–65 %$\dot{V}O_2max$) for these marathoners/ultramarathoners. Thus, the

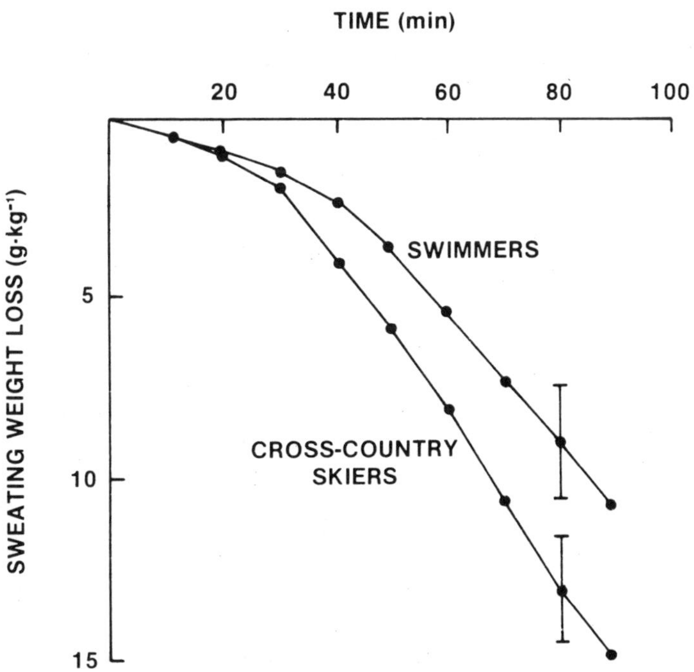

Figure 5-6. *Mean sweating weight loss of six cross-country skiers and four swimmers during passive heating trials (55° C db, 40° C wb), redrawn from Henane* et al *(25). Cross-country skiers and swimmers were matched for $\dot{V}O_2max$ (66.5 vs. 65.8 $ml \cdot kg^{-1} \cdot min^{-1}$).*

reader should recognize that these results may only be applied to a limited population. No longitudinal (spring vs. summer) differences were observed in training distance, mean heart rate, rectal temperature, sweat Na^+ or K^+ (whole-body washdown), plasma Na^+ or K^+, change in plasma volume during exercise, mean weighted skin temperature (except at 50 min of exercise), and sweat rate (except during the initial 30 min segment), but submaximal oxygen consumption decreased ($p < 0.05$) 8.3%. The authors concluded that these distance runners did not require 14.5 wk of summer training in the heat to maintain safe rectal temperatures (≤ 38.4° C) during the spring heat-tolerance test. They recommended, however, that appropriate adjustments be made in heat acclimation procedures if athletes train or race at significantly higher exercise intensities (> 60–65 %$\dot{V}O_2$max) or in hotter climates(> 30.3° C).

The research of Smolander *et al* (48) involved a cross-sectional design, unacclimated men, work clothing, and exercise in conditions similar to those found in hot Finnish industrial sites (30.4° C db, 27.3° C wb and 39.9° C db, 22.3° C wb). Two groups of men, 8 trained and 8 untrained ($\dot{V}O_2$max = 59.1 and 40.6 $ml \cdot kg^{-1} \cdot min^{-1}$, respectively), walked for seven 30-min bouts (30 %$\dot{V}O_2$max), interrupted by 5 min pauses for weighing. On the basis of heart rate and rectal temperature, no statistical differences were found between the groups. The results of this cross-sectional investigation must be interpreted in light of these facts: (a) no distinction was made between $\dot{V}O_2$max and physical training, (b) because all subjects walked at 30 %$\dot{V}O_2$max, it is not surprising that no differences were observed in heart rate and rectal temperature, since both of these factors are strongly correlated with relative exercise intensity (% $\dot{V}O_2$max) (5,42), and (c) the authors made no comment about clothing, but did note that impaired exercise performance was related to circulatory instability caused by increased skin blood flow.

Factors Which Alter the Relationship Between $\dot{V}O_2$max and Exercise-Heat Tolerance

Blood volume and cardiorespiratory fitness were the foci of the 1984 work of Desai and Senay (14), who trained 6 males for 21 days on a cycle ergometer (100 $min \cdot day^{-1}$, 60 %$\dot{V}O_2$max) at 18–20° C db. One week following this training period, the subjects underwent heat acclimation (cycling 60 $min \cdot day^{-1}$, 40 %$\dot{V}O_2$max) at 45° C db, 29° C wb. Mean blood volume was not significantly altered by physical training, but increased 14.4% following heat acclimation. In contrast, post-training $\dot{V}O_2$max increased 18.4%, but remained at that level following heat acclimation. Heat acclimation resulted in a significant hemodilution (−5.8% hemoglobin and −5.3% hematocrit concentration) which, in the opinion of the authors, prevented enhancement of $\dot{V}O_2max$ during heat acclimation.

A recent publication concentrated on the interaction of $\dot{V}O_2max$, hypohydration, and heat acclimation. Cadarette *et al* (11) examined the responses of 8 male and 6 female subjects to a standard heat-tolerance test, before and after heat acclimation, when euhydrated and hypohydrated (−5% body weight), in both a hot-dry (49° C, 20 %rh) and a comfortable (20° C, 40 %rh) environment. The standard heat-tolerance test lasted for 140 min (four, 25/10 min exercise/rest cycles) at 28–30 %$\dot{V}O_2max$. Males and females were selected for similar ages and $\dot{V}O_2max$ values. In the comfortable environment, no significant correlations were calculated between $\dot{V}O_2max$ and final heart rate or rectal temperature. Under hot-dry conditions, however, $\dot{V}O_2max$ correlated inversely ($p < 0.05$) with heart rate during all four treatments (range: $r = -0.57$ to -0.67), and with final rectal temperature in the euhuydrated pre-acclimation state. The following conclusion was drawn: individuals with high levels of cardiorespiratory fitness were under less cardiovascular and thermal strain (mild exercise) than those with lower $\dot{V}O_2max$ values but, following heat acclimation or when hypohydrated, the individuals with high levels of fitness maintained their cardiovascular advantage, yet did not exhibit a thermoregulatory advantage.

In an effort to formulate heat stress limits and to assess the risk of heat injury to laborers in the work force in the South African mining industry, Kielblock (27) evaluated heat stroke mortality records from 1956 to 1985. Kielblock calculated the predictive power of several physical characteristics and test measurements as predictors of exercise-heat tolerance. An unusual feature of these linear regression equations (Table 5-3) is the large number of subjects (n = 266–398), which far outnumbered any other study to date. Final four-hour rectal temperature (TR_4) during a standard heat-tolerance test

TABLE 5-3. *Statistical Correlations and Significance Pertaining to the Relationship Between Four-Hour Rectal Temperature and $\dot{V}O_2max$, %$\dot{V}O_2max$, Age (27).*

Predictor	n	Correlation Coefficient (r^2)	Confidence Level (p)	Percentage of Variance Explained	Prediction Equation
$\dot{V}O_2max$	273	−0.19	< 0.005	3.57	$TR_4 = 40.000 - 0.312\ \dot{V}O_2max$
%$\dot{V}O_2max$	266	+0.35	< 0.001	11.90	$TR_4 = 37.300 - 0.041\ \%\dot{V}O_2max$
Age < 40 yr	273	−0.04	> 0.4 (NS)	0.18	$TR_4 = 39.100 - 0.063$ Age
Age > 40 yr	398	−0.02	> 0.4 (NS)	0.18	$TR_4 = 38.930 - 0.002$ Age

$\dot{V}O_2max$ determined by intermittent, graded treadmill exercise. TR_4, 4-hour rectal temperature after bench stepping at 33.2° C db, 31.7° C wb; %$\dot{V}O_2max$, measured at end of bench stepping exercise (4 hr); NS, not statistically significant.

(bench-stepping at 54 W, 4 h-duration, 33.2° C db, 31.7° C wb) was used as the dependent variable in these calculations (Table 3). The independent variables used to predict TR_4 were: $\dot{V}O_2max$, $\%\dot{V}O_2max$, and age during the four-hour trial. The correlation coefficient between $\dot{V}O_2max$ and TR_4 was established at $r = -0.19$ ($p < 0.005$).

In contrast, Wyndham (56) had earlier reported that $\dot{V}O_2max$ ($L \cdot min^{-1}$) correlated significantly ($r^2 = -0.36$, $p < 0.05$) with 2-hour rectal temperature on a similar heat-tolerance test performed in the same laboratory by 59 men. However, by means of step-wise linear regression analysis, Kielblock determined that the $\dot{V}O_2max$ value accounted for only 3.57% of the variability in TR_4 (compare to Fig. 5-4). This suggested that the physiological mechanism(s) through which exercise-heat tolerance is conferred is more complex than $\dot{V}O_2max$ alone. The relative exercise intensity ($\%\dot{V}O_2max$), measured at the end of the standard heat-tolerance test, exhibited a stronger relationship to TR_4 than $\dot{V}O_2max$, a fact established many years earlier by Saltin *et al* (42); this variable accounted for 11.90% of the variance in TR_4. Advancing age, per se, had no significant effect on heat-tolerance and accounted for only 0.18% of the variability in TR_4.

The question of the interactions between age, $\dot{V}O_2max$, heat acclimation, and physical training also interested Pandolf and colleagues (34). In fact, their experimental design further refined the earliest findings in this controversy, as described at the outset of this chapter. The subjects were nine young men and nine middle-aged men who were matched for body weight, surface area, surface area-to-weight ratio, percent body fat, and $\dot{V}O_2max$ (52.9 vs. 51.3 $ml \cdot kg^{-1} \cdot min^{-1}$). These groups differed in age by 25 years (21.2 ± 2.4 vs. 46.4 ± 4.6 yr), and the young men ran only 4.8 ± 8.0 $mi \cdot wk^{-1}$, while the middle-aged men averaged 24.1 ± 19.7 $mi \cdot wk^{-1}$ ($p < 0.05$). Both groups underwent 10 days of heat acclimation via treadmill walking (1.56 $m \cdot s^{-1}$, 5% grade, ≈44 $\%\dot{V}O_2max$, 50/10/50 exercise/rest/exercise bouts) in ambient conditions of 49° C, 20 %rh. The results of the testing on day 1 indicated that the middle-aged men had longer mean performance times, lower final rectal temperatures, skin temperatures and heart rates, as well as higher whole-body sweat rates. These physiological advantages (depicted in Fig. 5-7) persisted for the first few days of exercise-heat acclimation. By day 10, however, no physiological or performance time differences were observed between groups ($p > 0.05$). The $\dot{V}O_2max$ alone was not hypothesized as the factor which offered the middle-aged men physiological advantages, but rather it was their level of regular weekly physical training. This finding is at odds with previous studies (16,40,52),which did not match groups of middle-aged and young individuals for pertinent morphological factors and/or $\dot{V}O_2max$.

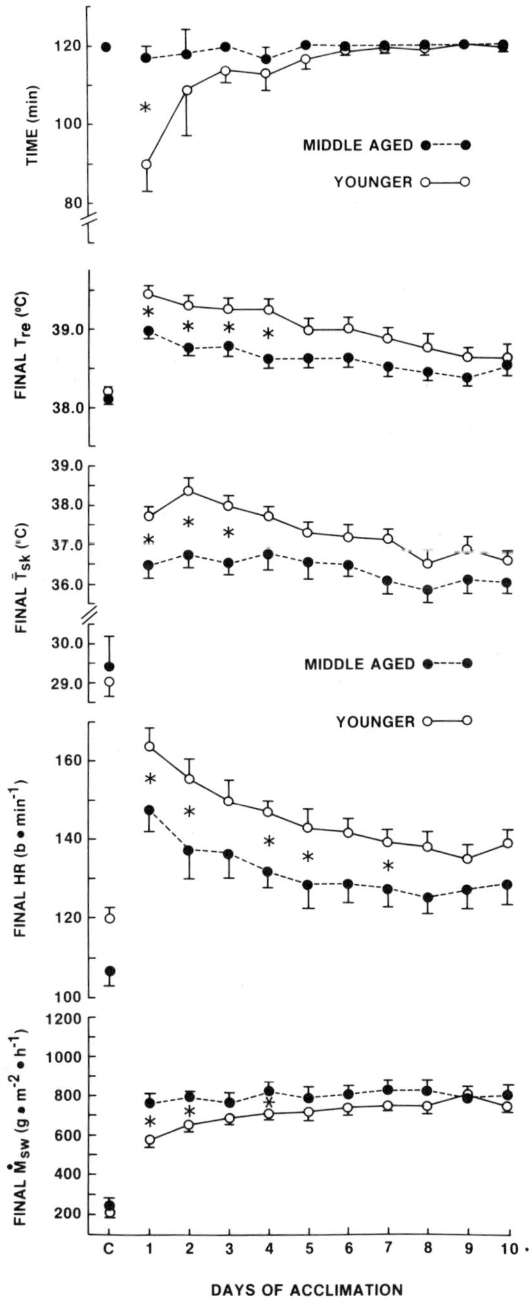

Figure 5-7. *Comparison of exercise-tolerance time, final rectal temperature, skin temperature, heart rate, and sweat rate of nine middle-aged and nine young subjects who were matched for* $\dot{V}O_2max$*, as reported by Pandolf* et al *(34).*

SUMMARY AND RECOMMENDATIONS FOR FUTURE RESEARCH

The importance of physical training and/or cardiorespiratory physical fitness ($\dot{V}O_2$max) in exercise-heat tolerance is a controversial subject, but most authors agree that physical training in a cool environment, or a high $\dot{V}O_2$max, improves physiological responses during exercise at high ambient temperatures and speeds the process of heat acclimation. Table 5-4 summarizes all of the investigations reviewed in this chapter and categorizes them, based upon their interpretation of findings. The studies published prior to 1973 contained investigational design flaws and narrow interpretation of data (9,38,41,50). Probably the most revealing studies were those between 1973 to 1979, which longitudinally tracked groups of subjects undertaking strenuous physical training and then performing heat acclimation trials for 8–10 days (18,31,39,43). Investigations during the 1980's (Table 5-4) rarely disputed the original arguments of Piwonka *et al* (38), and those which did involved only one organ system, methodological flaws, or mild to moderate exercise intensities (14,48).

The relationship between $\dot{V}O_2$max and exercise-heat tolerance is still controversial because the significance of statistical relationships depends heavily on characteristics of the population tested

TABLE 5-4. *Investigations Discussing the Interaction of Endurance Training and/or $\dot{V}O_2$max and Exercise-Heat Tolerance (Modified from Ref. 32).*

Importance of Training or $\dot{V}O_2$max	
Supports	**Disputes**
Bean and Eichna (1943)	Strydom *et al* (1966)
Robinson *et al* (1943)	Strydom and Williams (1969)
Allan (1965)	Greenleaf *et al* (1972)
Piwonka *et al* (1965)	Shvartz *et al* (1973)
Piwonka and Robinson (1967)	Shvartz *et al* (1975)
Gisolfi and Robinson (1969)	Desai and Senay (1984)
Gisolfi (1973)	Smolander *et al* (1987)
Nadel *et al* (1974)	
Gisolfi *et al* (1977)	
Pandolf *et al* (1977)	
Roberts *et al* (1977)	
Shvartz *et al* (1977)	
Henane *et al* (1977)	
Senay and Kok (1977)	
Avellini *et al* (1982)	
Cadarette *et al* (1984)	
Kielblock (1986)	
Armstrong *et al* (1987)	
Pandolf *et al* (1988)	

(e.g. number of subjects, training status) and on the measurements employed to evaluate exercise-heat tolerance. Two early studies (Fig. 5-4) indicated that an individual's $\dot{V}O_2$max accounted for 42–46% of the variability that determined the three-hour rectal temperature in the heat (45), or the acclimation day for a rectal temperature plateau (33). In contrast, other authors reported non-significant relationships between $\dot{V}O_2$max and exercise-tolerance time or final rectal temperature (15,22,27,56,). The most noteworthy of these was the analysis of Kielblock (27), which utilized very large subject samples (Table 5-3) and calculated that $\dot{V}O_2$max accounted for only 3.57% of the variability in four-hour rectal temperature.

Authors on both sides of this issue agree, however, that $\dot{V}O_2$max, per se, may not be as important in determining exercise-heat tolerance as the underlying physiological adaptations (e.g. changes in blood volume, cardiac function, vasodilation/vasoconstriction, intramuscular metabolism) which result in $\dot{V}O_2$max differences between subjects (27,32,33). A clarification of this issue was recently published by Pandolf *et al* (34), who observed that $\dot{V}O_2$max was not the factor which offered middle-aged men physiological advantages over men who were 25 years younger, but rather it was their level of regular weekly physical training. Furthermore, several factors have been identified which may alter the relationship between $\dot{V}O_2$max and exercise-heat tolerance; these include blood volume (14) and hydration status (11).

To achieve optimal results from physical training, researchers suggest the use of strenuous interval training or continuous exercise at an intensity greater than 50 %$\dot{V}O_2$max (18,25,32,39,41). Such training accounts for approximately 50% of the improvements gained by usual heat acclimation procedures (19). Improvements in exercise-heat tolerance produced by mild or moderate exercise (< 50% $\dot{V}O_2$max) are questionable (46). The duration of training must exceed 1 week (31,46,47) while Gisolfi (18) and Henane *et al* (25) reported that the best improvements occurred after 8–12 weeks. It seems that training which increases $\dot{V}O_2$max by 15% or more, maximizes improvements in exercise-heat tolerance (18,25,31,39,46,47). The important aspect of training in a cool environment appears to be a challenge to the body's thermoregulatory apparatus. Maintenance of an elevated core temperature is a prerequisite for optimal results (6,31,32,38). In fact, typical outdoor training/heat acclimation guidelines apparently do not apply to all geographical areas within countries that contain a spectrum of climates (4). This fact was obvious in the work of Avellini *et al* (6), which demonstrated that a training-induced increase in $\dot{V}O_2$max of 15% (training in 20° C water) may coexist with *reduced* exercise-heat tolerance. Similarly, it has been demonstrated that swimmers may increase $\dot{V}O_2max$ greatly, without altering exercise-heat tolerance significantly (31,38).

Future research must be designed to take into account the methodological flaws delineated throughout this chapter. Careful pair-matching of subjects for morphological factors and $\dot{V}O_2max$, and the use of longitudinal study designs are highly recommended. It will also be vital for investigators to understand the sematics used in previous reports, and to define the limits within which their findings may be applied (19,35,43). In terms of specific research areas, we see a need for future investigations as follows:

1. Because $\dot{V}O_2max$, per se, apparently does not account for a large portion of the variation in exercise-heat tolerance, research is needed to define which specific physiological adaptations are most important in optimizing exercise-heat tolerance. In addition, the interactions between body systems are important (e.g. high $\dot{V}O_2max$ offers cardiovascular advantages, but not thermoregulatory advantages, during exercise in the heat) (11).
2. Further studies involving physical training duration, intensity, frequency, and mode effects on exercise-heat tolerance are warranted. The athletic community will continue to seek ways to optimize performance in hot environments.
3. Certain individuals have been identified as being either heat-intolerant (43,53) or hyper-heat tolerant (44). Creative research designs can offer insights into the impact which physical training, $\dot{V}O_2max$, and heat acclimation have on exercise-heat tolerance in these unique populations.
4. Because (a) $\dot{V}O_2max$ is strongly related to blood volume (31), (b) the physiological adaptations during the first 3 days of heat acclimation primarily involve cardiovascular changes (23), and (c) heat intolerant individuals expand and defend vascular volume less effectively than "normals" during exercise-heat exposure (43); body fluid dynamics and vascular components (e.g. protein, Na^+) deserve further attention.
5. Exploration should continue on the effects of hereditary, morphological, and developmental factors (e.g. age, gender, body build) on the relationships between physical training, $\dot{V}O_2max$, heat acclimation, and exercise-heat tolerance. The impact of diet, detraining, and disease also offer fruitful avenues for research.

REFERENCES

1. Allan, J.R. A practical technique for artifical acclimatization. *Ergonomics* 7: 357–358, 1964.
2. Anderson, K.L. The cardiovascular system in exercise. In: *Exercise Physiology*, edited by B. Falls. New York: Academic Press, 1968, p. 79–128.

3. Appenzeller, O. Influences of physical training, heat acclimation, and diet on temperature regulation in man. In: *Heat Stroke and Temperature Regulation,* edited by M. Khogali and J.R.S. Hales. New York: Academic Press, 1983, p. 283–292.
4. Armstrong, L.E., R.W. Hubbard, J.P. DeLuca, and E.L. Christensen. Heat acclimatization during summer running in the northeastern United States. *Med. Sci. Sports Exerc.* 19: 131–136, 1987.
5. Astrand, P.O., and I. Rhyming. A nomogram for calculation of aerobic capacity (ph sical fitness) from pulse rate during submaximal work. *J. Appl. Physiol.* 7: 218–220, 1954.
6. Avellini, B.A., Y. Shapiro, S.M. Fortney, C.B. Wenger, and K.B. Pandolf. Effects on heat tolerance of physical training in water and on land. *J. Appl. Physiol.* 53: 1291–1298, 1982.
7. Bass, D.E., C.R. Kleeman, M. Quinn, A. Henschel, and A.H. Hegnauer. Mechanisms of acclimatization to heat in man. *Medicine* 34: 323–380, 1955.
8. Baum, E., K. Bruck, and H.P. Schwennicke. Adaptive modifications in the thermoregulatory system of long-distance runners. *J. Appl. Physiol.* 40: 404–410, 1976.
9. Bean, W.B., and L.W. Eichna. Performance in relation to environmental temperature. Reactions of normal young men to simulated desert environment. *Fed. Proc.* 2: 144–158, 1943.
10. Bouchard, C., and G. Lortie. Heredity and endurance performance. *Sports Med.* 1: 38–64, 1984.
11. Cadarette, B.S., M.N. Sawka, M.M. Toner, and K.B. Pandolf. Aerobic fitness and the hypohydration response to exercise-heat stress. *Aviat. Space Environ. Med.* 55: 507–512, 1984.
12. Clausen, J.P., O.A. Larsen, and J.T. Jensen. Physical training in the management of coronary artery disease. *Circulation* 40: 143–154, 1969.
13. Convertino, V.A., J.E. Greenleaf, and E.M. Bernauer. Role of thermal and exercise factors in the mechanism of hypervolemia. *J. Appl. Physiol.* 48: 657–664, 1980.
14. Desai, J.B., and L.C. Senay. Influence of endurance training and heat acclimatization on blood volume and maximum aerobic capacity. *Fed. Proc.* 43: 627, 1984.
15. Drinkwater, B.L., J.E. Denton, I.C. Kupprat, T.S. Talag, and S.M. Horvath. Aerobic power as a factor in women's response to work in hot environments. *J. Appl. Physiol.* 41: 815–821, 1976.
16. Drinkwater, B.L., and S.M. Horvath. Heat tolerance and aging. *Med. Sci. Sports.* 11: 49–55, 1979.
17. Eichna, L.W., W.B. Bean, W.F. Ashe, and N. Nelson. Performance in relation to environmental temperature. Reactions of normal young men to hot, humid (simulated jungle) environment. *Bull. Johns Hopkins Hosp.* 76: 25–58, 1945.
18. Gisolfi, C.V. Work-heat tolerance derived from internal training. *J. Appl. Physiol.* 35: 349–354, 1973.
19. Gisolfi, C.V., and J.S. Cohen. Relationships among training, heat acclimation, and heat tolerance in men and women: the controversy revisited. *Med. Sci. Sports* 11: 56–59, 1979.
20. Gisolfi, C.V., and S. Robinson. Relations between physical training, acclimatization, and heat tolerance. *J. Appl. Physiol.* 26: 530–534, 1969.
21. Gisolfi, C.V., N.C. Wilson, and B. Claxton. Work-heat tolerance of distance runners. In: *The Marathon: Physiological, Medical, Epidemiological, and Psychological Studies,* edited by P. Milvy. New York: The New York Academy of Sciences, 1977, p. 139–150.
22. Greenleaf, J.E., B.L. Castle, and W.R. Ruff. Maximal oxygen uptake, sweating and tolerance to exercise in the heat. *Int. J. Biometeor.* 16: 375–387, 1972.
23. Greenleaf, J.E., and C.J. Greenleaf. Human acclimation and acclimatization to heat: a compendium of research. Moffett Field, CA: Ames Research Center, NASA Technical Report No. TMX-62008, 1970, p. 1–188.
24. Haymes, E.M. Physiological responses of female athletes to heat stress: a review. *Physician Sportsmed.* 12: 45–55, 1984.
25. Henane, R., R. Flandrois, and J.P. Charbonnier. Increase in sweating sensitivity by endurance conditioning in man. *J. Appl. Physiol.* 43: 822–828, 1977.
26. Inbar, O., O. Bar-Or, R. Dotan, and B. Gutin. Conditioning versus exercise in heat as methods for acclimatizing 8- to 10-year-old boys to dry heat. *J. Appl. Physiol.* 50: 406–411, 1981.
27. Kielblock, A.J. Heat acclimatization with special reference to heat tolerance. Japan: International Symposium on the Physiology of Stressful Environments, 1986, p. 1–13.
28. Kobayashi, Y., Y. Ando, N. Okuda, S. Takaba, and K. Ohara. Effects of endurance training on thermoregulation in females. *Med. Sci. Sports Exerc.* 12: 361–364, 1980.
29. Lofstedt, B. (Ed.). *Human Heat Tolerance.* Lund: Berlingska Bocktrycheriet, 1966, p. 44.
30. Myles, A.S., and A.K. Chin. Physical fitness and tolerance to environmental stresses: a review of human research on tolerance to and work capacity in hot, cold, and high altitude environments. Downsview, Ontario: Defense and Civil Institute of Environmental Medicine, Report no. 74-R-1008, 1974, p. 1–17.
31. Nadel, E.R., K.B. Pandolf, M.F. Roberts, and J.A.J. Stolwijk. Mechanisms of thermal acclimatization to exercise and heat. *J. Appl. Physiol.* 37: ' 15–520, 1974.

32. Pandolf, K.B. Effects of physical training and cardiorespiratory physical fitness on exercise-heat tolerance: recent observations. *Med. Sci. Sports* 11: 60–65, 1979.
33. Pandolf, K.B., R.L. Burse, and R.F. Goldman. Role of physical fitness in heat acclimatization, decay and reinduction. *Ergonomics* 20: 399–408, 1977.
34. Pandolf, K.B., B.S. Cadarette, M.N. Sawka, A.J. Young, R.P. Francesconi, and R.R. Gonzalez. Thermoregulatory responses of matched middle-aged and young men during dry-heat acclimation. *J. Appl. Physiol.*, 65: in press, 1988.
35. Pandolf, K.B., M.N. Sawka, and Y. Shapiro. Factors which alter human physiological responses during exercise-heat acclimation. In: *Adaptive Physiology to Stressful Environments,* edited by S. Samueloff and M. Yousef. Boca Raton: CRC Press, 1987, p. 91–102.
36. Pivarnik, J.M., and L.C. Senay. Effects of exercise detraining and deacclimation to the heat on plasma volume dynamics. *Eur. J. Appl. Physiol.* 55: 222–228, 1986.
37. Piwonka, R.W., and S. Robinson. Acclimatization of highly trained men to work in severe heat. *J. Appl. Physiol.* 22: 9–12, 1967.
38. Piwonka, R.W., S. Robinson, V.L. Gay, and R.S. Manalis. Acclimatization of men to heat by training. *J. Appl. Physiol.* 20: 379–384, 1965.
39. Roberts, M.F., C.B. Wenger, J.A.J. Stolwijk, and E.R. Nadel. Skin blood flow and sweating changes following exercise, training and heat acclimation. *J. Appl. Physiol.* 43: 133–137, 1977.
40. Robinson, S., H.S. Belding, F.C. Consolazio, S.M. Horvath, and E.S. Turrell. Acclimatization of older men to work in heat. *J. Appl. Physiol.* 20: 583–586, 1965.
41. Robinson, S., E.S. Turrell, H.S. Belding, and S.M. Horvath. Rapid acclimatization to work in hot climates. *Am. J. Physiol.* 140: 168–176, 1943.
42. Saltin, B., and L. Hermansen. Esophageal, rectal, and muscle temperature during exercise. *J. Appl. Physiol.* 21: 1757–1762, 1966.
43. Senay, L.C., and R. Kok. Effects of training and heat acclimatization on blood plasma contents of exercising men. *J. Appl. Physiol.* 43: 591–599, 1977.
44. Shutte, P.C., W.H. Van der Walt, G. Stassen, A.J. Kielblock, and N.B. Strydom. Maximum heat exposure levels for hyper-heat tolerants. Johannesburg, South Africa: Chamber of Mines of South Africa, Research Report No. 16-81, p. 1–19, 1981.
45. Shvartz, E., Y. Shapiro, A. Magazanik, A. Meroz, H. Birnfeld, A. Mechtinger, and S. Shibolet. Heat acclimation, physical fitness, and responses to exercise in temperate and hot environment. *J. Appl. Physiol.* 43: 678–683, 1977.
46. Shvartz, E., E. Sohar, N. Meyerstein and D. Benor. A comparision of three methods of acclimatization to dry heat. *J. Appl. Physiol.* 34: 214–219, 1973.
47. Shvartz, E., N.B. Strydom, and H. Kotze. Orthostatism and heat acclimation. *J. Appl. Physiol.* 39: 590–595, 1975.
48. Smolander, J., R. Ilmarinen, O. Korhonen, and I. Pyykko. Circulatory and thermal responses of men with different training status to prolonged physical work in dry and humid heat. *Scand. J. Work Environ. Health* 13; 37–46, 1987.
49. Strydom, N.B., and C.G. Williams. Effect of physical conditioning on state of heat acclimatization of Bantu laborers. *J. Appl.Physiol.* 27: 262–265, 1969.
50. Strydom, N.B., C.H. Wyndham, C.G. Williams, J.F. Morrison, G.A.G. Bredell, A.J.S. Benade, and M. Von Rahden. Acclimatization to humid heat and the role of physical conditioning. *J. Appl. Physiol.* 21: 636–642, 1966.
51. Varnauskas, E., P. Bjorntorp, M. Fahlen, I. Prerovsky, and J. Stenburg. Effects of physical training on exercise, blood flow and enzymatic activity in skeletal muscle. *Cardiovasc. Res.* 4: 418–422, 1970.
52. Wagner, J.A., S. Robinson, S.P. Tzankoff, and R.P. Marino. Heat tolerance and acclimatization to work in the heat in relation to age. *J. Appl. Physiol.* 33: 616–622, 1972.
53. Wyndham, C.H. The problem of heat intolerance in man. In: *Physiological and Behavioral Temperature Regulation,* edited by J.D. Hardy, A.P. Gagge, and J.A.J. Stolwijk. Springfield, IL: Thomas, 1970, p. 324–341.
54. Wyndham, C.H., The physiology of exercise under heat stress. *Ann. Rev. Physiol.* 35: 193–220, 1973.
55. Wyndham, C.H., N.B. Strydom, A.J.S. Benade, A.J. VanRensberg, A.J.A. Benade, and G. Rogers. Heat stroke risks in unacclimatized and acclimatized men of different maximum oxygen intakes working under hot and humid conditions. Johannesburg, South Africa: Chamber of Mines Research Report No. 12/72, 1972, p. 1–18.
56. Wyndham, C.H., N.B. Strydom, C.G. Williams, and A. Heyns. An examination of certain individual factors affecting the heat tolerance of mine workers. *J. South African Inst. Mining Metallurgy* 68: 79–91, 1967.

6

Body Fluid Responses and Hypohydration During Exercise-Heat Stress

MICHAEL N. SAWKA, Ph.D.

OUTLINE

INTRODUCTION

This chapter describes normal body water distribution and its redistribution as the result of exercise-heat stress. Emphasis is placed on transcapillary pressure gradients in different tissue vascular beds that result in net filtration and net absorption to alter plasma volume. A detailed discussion is provided on how alterations in plasma volume, and its constituents, influence exercise performance in the heat. Attention is directed toward the cardiovascular and thermoregulatory systems as providing the physiological mechanisms responsible for altered exercise performance when body water and its constituents are manipulated.

The distribution of body water and the ability to maintain blood volume are critical factors influencing man's ability to perform exercise in the heat. Water is necessary to support metabolism as it provides the medium for biochemical reactions and body solutes. Likewise, water is the primary component of plasma, which is essential to maintain blood volume. Muscular exercise will routinely increase metabolic rate by 5 to 15 times above resting levels and require an adequate blood volume to, simultaneously, perfuse the metabolically active skeletal muscles, as well as cutaneous vasculature, which enables sensible heat loss. Likewise, plasma provides the precursor fluid for the secreted sweat, which enables insensible heat loss. Therefore, sweat secretion can reduce the blood volume and make it more difficult for the cardiovascular system to support the combined stress of exercise and heat. Fortunately, body water exchanges freely between fluid compartments, but the exchange is dependent upon the gradients of osmotic, colloid, tissue and hydrostatic pressures. During exercise-heat stress, these pressure gradients enable a redistribution of body water to defend blood volume and support the cardiovascular system.

BODY WATER DISTRIBUTION

Water is the largest component of the human body and represents 45 to 70 percent of body weight. Table 6-1 provides the water content of various body tissues for a 75 kg man (102). As indicated in Table 6-1, the composition of an average-sized man includes about 45 ℓ of water, which corresponds to about 60 percent of his body weight. It is important to note that adipose tissue is approximately 10 percent water, while muscle tissue is approximately 75 percent

TABLE 6-1. *Water Content of Various Body Tissues for an Average 75 kg Man*

Tissue	% Water	% of Body Weight	Liters of Water per 75 kg	% of Total Body Water
Skin	72	18	9.72	22
Organs	76	7	3.99	9
Skeleton	22	15	2.47	5
Blood	83	5	3.11	7
Adipose	10	12	0.90	2
Muscle	76	43	24.51	55

water. Therefore, individual differences in total body water are primarily the result of differences in skeletal muscle or lean body mass. Athletes have a greater percentage of total body water than their sedentary counterparts by virtue of a greater skeletal muscle mass and lower percent body fat; likewise, gender and age differences in total body water are related to differences in lean body mass (10,44,91). This relationship enables an accurate estimation of total body water at approximately 72 percent of an individual's lean body mass (54).

The water contained in these body tissues is distributed between the intracellular and extracellular fluid compartments. The intracellular fluid compartment is the larger and contains about 30 ℓ of water, whereas the extracellular fluid compartment contains about 15 ℓ of water for an average 75 kg male (Table 6-2). These compartments are not representative of static volumes, but rather represent a dynamic fluid exchange with different turnover rates. The perturbations of exercise and heat will greatly modify the net volume and turnover rate of water between the fluid compartments.

PRINCIPLES OF WATER EXCHANGE

Extravascular Exchange

Figure 6-1 presents a schematic of the approximate volume of water in each compartment, and the mechanisms for the exchange of fluid between the compartments. At rest, and probably more so during exercise, there is continuous water exchange between fluid compartments, and the volumes depicted represent only net values. Osmosis is the principal mechanism for water exchange between the intracellular and extracellular fluid compartments (39). Osmosis is defined as the passage of water through a membrane from a region of lower to higher solute concentration in an attempt to equalize the 2 solute concentrations. Cell membranes are permeable to water, but only selectively permeable to various solutes. Osmosis is dependent upon the transcellular membrane differences in the num-

TABLE 6-2. *Body Water Distribution Between the Body Fluid Compartments in an Adult Male*

	% of Body Weight	% of Lean Body Mass	% of Body Water
Total Body Water	60	72	100
Extracellular Water	20	24	33
Plasma	5	6	8
Interstitium	15	18	25
Intracellular Water	40	48	67

ber of osmotically active particles regardless of their size or valence. The solute concentrations are expressed in milliosmoles per kilogram of water (osmolality) or milliosmoles per liter (osmolarity). Osmolality is the preferred unit of measurement since it is not influenced by the solution temperature nor the volume that the solutes occupy within the solution. Osmolality is calculated by multiplying the solute concentration by the number of discrete particles into which the solute dissociates (15). Therefore, a mole of sodium and albumin will exert a similar osmotic pressure despite molecular weights of 23 and 69,000 daltons, respectively. As a result, the predominant solutes which exert an osmotic pressure in the body are the electrolytes.

Table 6-3 presents the electrolyte concentration in the intracellular fluid, interstitial fluid and plasma. It should be noted that, as described by the Gibbs-Donnan relationship, the equilibrium concentration of cations and anions is the same within each fluid compartment. The intracellular and extracellular (interstitium) compartments have the greatest electrolyte concentration difference, because they are separated by cell membrane. The transmembrane electrolyte differences are the result of active transport and selective permeability of cell membranes to different solutes (109). Potassium and

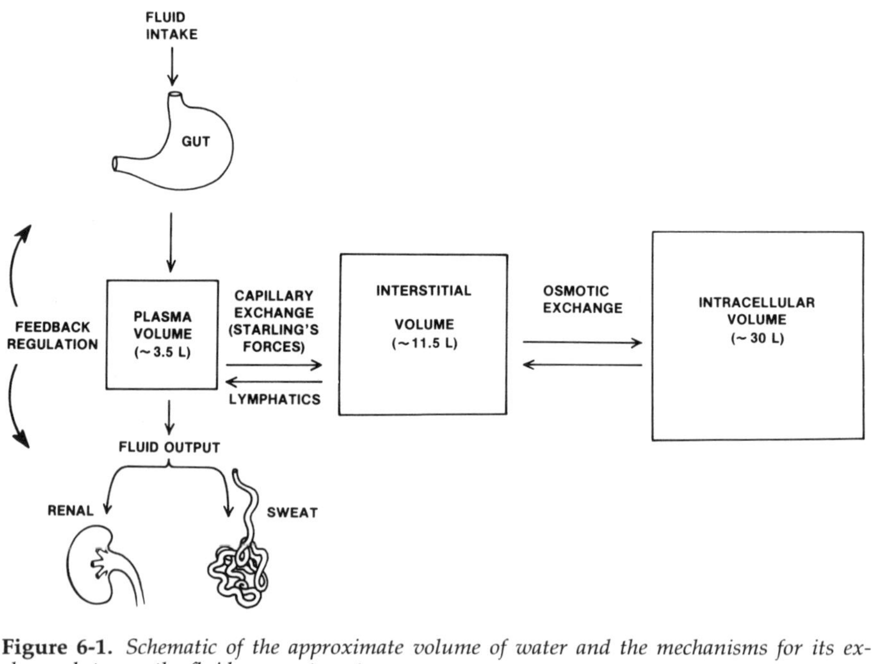

Figure 6-1. *Schematic of the approximate volume of water and the mechanisms for its exchange between the fluid compartments.*

sodium are the predominant intracellular and extracellular cations, respectively. This is the result of the $Na^+ - K^+$ ATPase pump which transports sodium out of the cell and potassium into the cell. The predominant intracellular anions are proteins and organic phosphates, which do not readily permeate the cell membrane. Chloride, the predominant interstitial anion, is small and easily diffuses across the membrane to maintain the Gibbs-Donnan equilibrium.

It is important to remember that Table 6-3 provides only an approximation of the intracellular fluid composition and is based upon skeletal muscle tissue (71). The intracellular fluid composition is probably not homogeneous and will vary between tissues, such as hepatic, neural, cardiac and skeletal muscle. Factors such as tissue differences in subcellular organelles (e.g., mitochondria, ribosomes, endoplasmic reticulum) and functional proteins as well as a poor understanding of the dissociation state of compounds and cation binding within the cell contribute to uncertainty regarding exact intracellular compositions. The metabolic activity of skeletal muscle during exercise will alter the solute composition and water content. Muscular contraction will result in the accumulation of metabolites causing an osmotic gradient favoring net fluid movement into the cell. In fact, several investigators report that a fluid influx accounts for an increased skeletal muscle volume during exercise (49,55,56, 92,101); however, the amount of this fluid accumulating in the interstitium, instead of intracellular space, is unknown. However, Mohsenin and Gonzalez (61) have shown that interstitial fluid pressure will increase in the active muscle and decrease in the inactive muscle during exercise.

The interstitium is composed of solid (sol) collagen fibers and

TABLE 6-3. *Approximate Concentration (mEq per liter) of Electrolytes in the Intracellular Fluid, Interstitial Fluid and Plasma*

	Intracellular (Skeletal Muscle)	Interstitial	Plasma Water
Cations			
Sodium	±10	143	154
Potassium	160	4	4
Calcium	1	4	4
Magnesium	35	2	2
Total	206	153	164
Anions			
Chloride	±2	115	110
Bicarbonate	±8	29	28
Proteins	60	1	16
Organic Phosphates	120	2	2
Other	16	6	8
Total	206	153	164

interstitial fluid in a gelatin-like (gel) state. The collagen fibers, and sometimes elastin fibers, provide the structural latticework and tensile strength for the interstitial spaces. The gel is made from mucopolysaccharide (such as chondroitin sulfate and hyaluronic acid) polymers that are intermittently folded and enmeshed with each other. These mucopolysaccaride polymers provide great resistance (several million fold) to water movement and, therefore, cause interstitial water and solutes to move by diffusion rather than mass movement. Normally, less than 1 percent of the fluid is not of the gel form and can flow through the interstitial spaces (38,39). Excess fluid and solutes that accumulate within the interstitium will enter the lymphatic system to be returned to the intravascular compartment.

The lymphatic system is made of small vessels that reside within the tissue spaces and merge to form the large thoracic ducts that join the vascular system at the juncture of the subclavian and jugular veins. The small lymphatic vessels are very permeable to solutes, since their endothelium is not covered by a basement membrane. As a result, excess interstitial fluid and solutes form lymph, which is unidirectionally transported, because of lymph valves, through the lymph system and returned via the thoracic duct to the intravascular space. Lymph is propelled through the system by a combination of intrinsic smooth muscle activity, the respiratory pump, arterial pulsations and the mechanical compression from skeletal muscle contraction. Initiation of physical exercise increases the ability of the respiratory and skeletal muscle pumps to increase lymph turnover. Also, physical exercise increases interstitial fluid volume and thus lymph formation. The net effect is that physical exercise can increase lymph flow by 4 to 20 fold (32,109).

The thoracic duct is conventionally thought of as the avenue for lymph and protein return from the interstitial to the intravascular space. However, there is also evidence of a direct communication between the lymph nodes and venules (73). These parallel communications may not always be activated, but will be utilized when the main lymph ducts are obstructed (73) or during conditions when there is insufficient lymph production for transport via the thoracic duct system (106). Hemorrhage and dehydration are 2 conditions which may cause lymphatic flow to be transiently decreased. Under such conditions,these parallel communications could be of importance (70).

Capillary Exchange

Fluid exchange between the intravascular and interstitial space occurs at the capillary. Among the diverse tissues in the body, capillaries are unique in the variability of their anatomical structure as well as in their permeability to water and solutes (75). The capillary permeability is affected by both the endothelial and basement membranes; the endothelium may contain clefts (small pores) and fenestrations (large pores), as well as vesicles (75). The clefts readily allow the bulk flow of water and electrolyte diffusion, but they are too small to allow passage of larger molecules. The fenestrations and vesicles, however, will allow the passage of macromolecules such as proteins. These vesicles may be activated by both receptor-mediated and nonspecific, receptor-independent processes for the transport of proteins and other macromolecules (100). Muscle and cutaneous capillaries have no fenestrations and a continuous basement membrane, thus low permeability. Some renal and intestinal capillaries have many fenestrations and a discontinuous basement membrane allowing high permeability. Likewise, the liver and spleen have sinusoidal capillaries which lack a continuous basement membrane and are permeable to macromolecules. As a result, the selective perfusion of tissues, with varying capillary permeabilities, can affect the total body net fluid exchange between the intravascular and interstitial spaces.

The skeletal muscle capillaries will be highly perfused during physical exercise relative to resting conditions. Figure 6-2 is a drawing of typical skeletal muscle microcirculation (112). The microcirculation is a complex of capillaries in both series and parallel arrangements that form a "fishnet" of interconnecting vessels. The precapillary area is composed of arterioles, metarterioles and precapillary sphincters, all of which contain smooth muscle that regulates precapillary resistance. The arterioles primarily influence the total tissue blood flow, whereas the precapillary sphincters control distribution of blood flow within the capillaries. The caliber of the arterioles and metarterioles is primarily controlled by sympathetic vasoconstrictors, but the precapillary sphincters are modified primarily by local control. Post-capillary resistance is regulated by venules controlled primarily by sympathetic constrictor tone.

The precapillary to postcapillary resistance ratio will modify the capillary hydrostatic pressure (59). Normally, the precapillary to postcapillary resistance ratio is about 4, and both resistances can change independently to alter this ratio. The capillary hydrostatic pressure can be described by the following equation (69):

$$P_c = \frac{P_a(R_2/R_1) + P_V}{1 + (R_2/R_1)}$$

where:

P_c is capillary hydrostatic pressure

P_a is arterial pressure

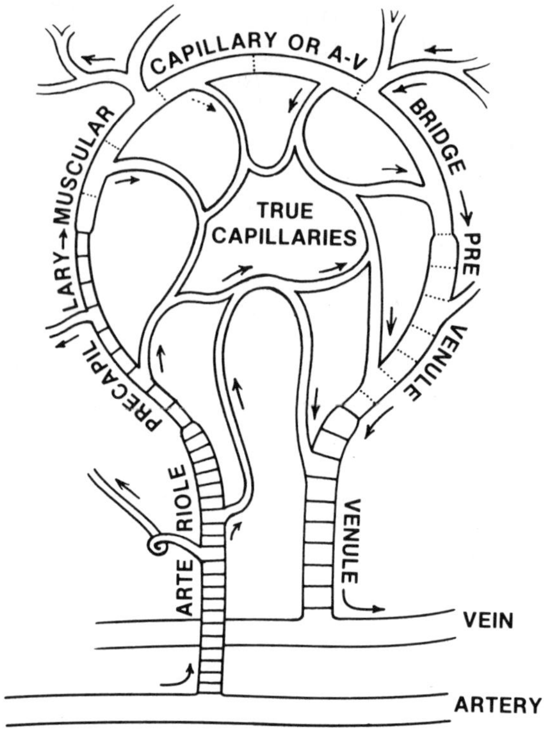

Figure 6-2. *Drawing of a typical skeletal muscle capillary (112).*

P_v is venous pressure
R_1 is precapillary resistance
R_2 is postcapillary resistance

A relative increase in precapillary resistance will lower capillary hydrostatic pressure and a relative increase in postcapillary resistance will increase capillary hydrostatic pressure. Also, since 80 percent of the resistance is normally precapillary resistance, a given change in venous pressure will cause a greater increase in capillary hydrostatic pressure, than a given change in arterial pressure. The capillary hydrostatic pressure is often the most important determinant of whether net filtration or net absorption occurs across a given capillary (53).

Table 6-4 provides the transcapillary pressures which determine net fluid exchange across an average capillary. The approximate matching of filtration and absorption across the capillary is often termed "Starlings Equilibrium." The primary outward force is capillary hydrostatic pressure, which normally decreases from the ar-

TABLE 6-4. *Equilibrium Forces for Fluid Exchange in an Average Capillary*

Pressures (mmHg)	Arterial End	Venous End
Inward Forces		
Plasma Oncotic (O_c)	28	28
	Net 28	28
Outward Forces		
Capillary Hydrostatic (P_c)	26	9
Interstitial Oncotic (O_I)	4	4
Negative Interstitial Hydrostatic (P_I)	7	7
	Net 37	20
Net Filtration (F_n)	+9	−8

$F_n = K_f (P_c - P_I) - (O_c - O_I)$; K_f is the Capillary Filtration Coefficient

terial to venous end of the capillary. Mean capillary pressure is about 17 mm Hg, but will vary between and within tissue vascular beds (38). Interstitial hydrostatic pressure is believed to be negative, providing an outward force for capillary fluid exchanges. However, the existence of a negative pressure for interstitial free fluid is controversial. The final outward force is the interstitial oncotic pressure. Some proteins, such as albumin, leave the intravascular space and collect in the interstitium where they exert an osmotic (oncotic) pressure.

The term "oncotic pressure" is used to denote the osmotic pressure across a capillary membrane, and is the result of protein concentration differences. Plasma proteins are large molecules (e.g., albumin 69,000 daltons and globulin 160,000 daltons) that do not readily pass through the capillary membrane and therefore have a high reflection coefficient. The proteins probably pass through the capillary membrane via fenestrations or the vesicles. The high reflection coefficient results in the plasma protein concentration being greater than the interstitial protein concentration (about $7.0\ g \cdot dl^{-1}$ vs $2.0\ g \cdot dl^{-1}$).

The plasma proteins are primarily composed of albumin ($4.0\ g \cdot dl^{-1}$), globulin ($2.0\ g \cdot dl^{-1}$) and fibrinogen ($0.3\ g \cdot dl^{-1}$) (53). Although albumin constitutes about 60 percent of the plasma proteins' total weight, its osmotic influence is much greater for 2 reasons (77). First, since albumin is a smaller molecule than globulin, a greater number of molecules are needed to provide a gram of protein. Remember, earlier we discussed that osmotic pressure is dependent on the number of solute molecules. Second, at ph 7.4 albumin has 18 negative charges, thus producing a greater osmotic effect than a neutral molecule (110). As described by the Gibbs-Donnan Effect, cations are attracted to the electronegative charges and thus increase the oncotic effect of albumin by 36 percent (77). More importantly, as the protein concentration increases, the Gibbs-Donnan Effect becomes progressively greater. Therefore, each additional gram of plasma protein exerts a disproportionately greater oncotic effect than the previously added gram (53). All of the factors which cause this multiplicative effect of albumin concentration on oncotic pressure are not known, but they will have important implications on fluid exchanges in man during dehydration.

Generally, filtration occurs at the arterial end, while absorption occurs at the venous end of the capillary. Compared to the arterial end, the venous end of the capillary has a greater surface area and permeability (38). Therefore, most capillary networks have a greater absorption than filtration capability. Net filtration (Table 6-4) is the product of the filtration coefficient and the difference between the transcapillary pressures (53). The filtration coefficient is defined as the volume of fluid filtered by 100 g of tissue for a 1 mmHg change in capillary pressure. The filtration coefficient is proportional to the capillary permeability and the total capillary surface area where filtration occurs. As discussed earlier, the permeability is dictated by the structural anatomy of each tissue bed. On the other hand, the capillary surface area available for filtration will be influenced by the metabolic activity level for skeletal muscle (48), the amount of heat stress for cutaneous tissues (50) and the level of sympathetic output for splanchnic tissues (78).

Figure 6-3 provides some examples of how net filtration and net absorption might be changed across an average capillary. Factors that favor net fluid absorption are vasoconstriction, venodilation and hyperoncotic plasma; whereas, vasodilation and venoconstriction favor net fluid filtration from the capillary. Throughout this chapter the term "hemodilution" will indicate a net gain of plasma, and "hemoconcentration" will indicate a net loss of plasma from the intravascular space. Also, the terms "hypervolemia" and "hypovolemia" will define steady-state blood volumes that are greater and less than normal, respectively. Euhydration will refer to "normal" body water content; whereas, hypohydration and hyperhydration refer to body water deficit and excess, respectively. The more common term "dehydration" is used to denote the dynamic loss of body water or the transition from euhydration to hypohydration.

DETERMINANTS OF INTRAVASCULAR FLUID RESPONSES

Premeasurement

As observed by Krebs and Meyer in 1902 (52) and later emphasized by Senay (94), intravascular fluid responses are extremely variable during exercise and heat stress. Various investigators describ-

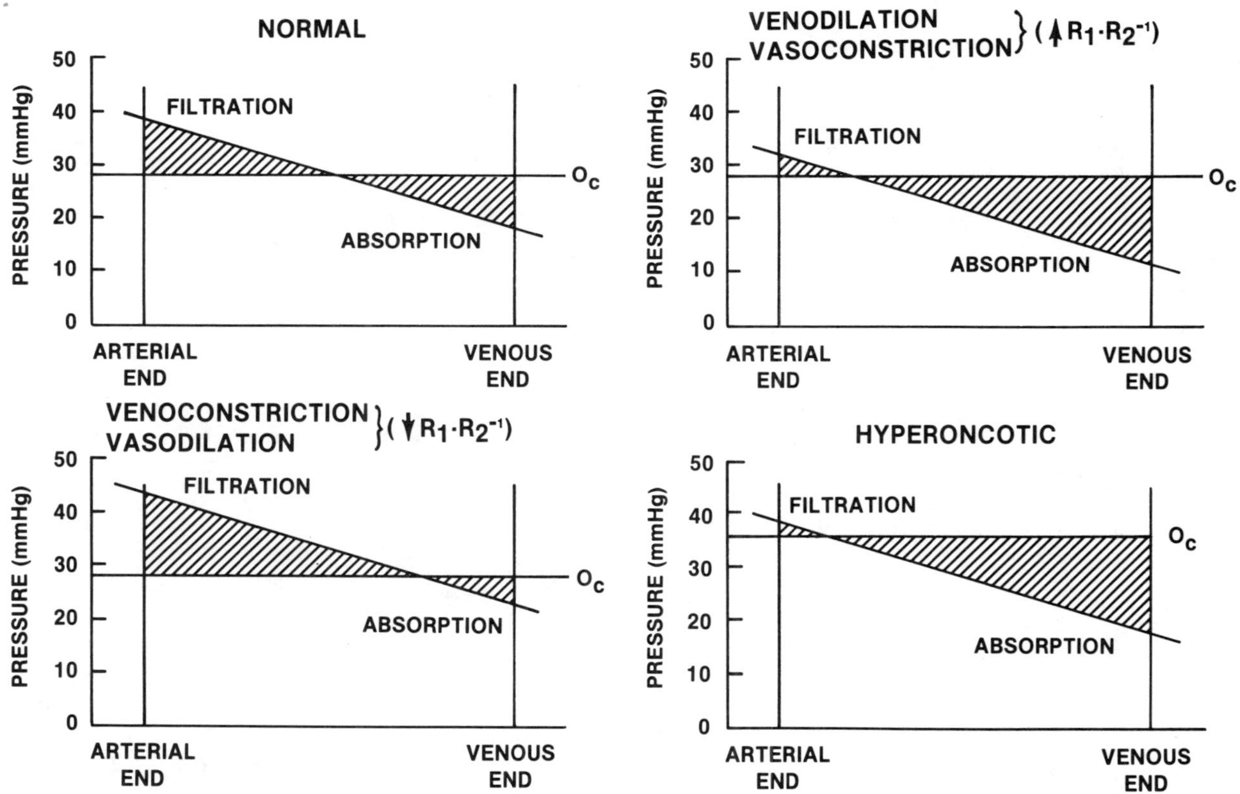

Figure 6-3. *Examples of how net filtration and net absorption might be changed across an average capillary.*

ing apparently similar experiments, concerning intravascular fluid responses, may report a hemodilution, hemoconcentration, or no change in plasma volume (41,94,96). Harrison (41) has provided a systematic analysis of measurement, stressor and subject factors, that when controlled for, will account for many of the apparent discrepancies between experiments. Figure 6-4 provides an outline of some factors which modify the intravascular fluid responses to exercise and heat stress.

The conditions, under which the control blood samples are obtained, provide the greatest source of bias regarding interpretation of subsequent intravascular fluid responses. Generally, investigators calculate the change in plasma volume from a control (usually rest) situation to an exercise and/or heat stress situation. During the control situation, plasma volume may or may not be directly measured, but the percent change in plasma volume is calculated from the control to the subsequent experimental situation based upon the appropriate hemoglobin and hematocrit measurements (16,104). As a result, measurement biases, affecting the control blood sample, will confound interpretation of intravascular fluid responses to the exercise and/or heat stress situation.

The hematocrit of a blood sample may be biased by tourniquet use and arm position. The application of a tourniquet causes venous stasis. This mechanical increase in postcapillary resistance raises mean capillary pressure and causes a net filtration (Figure 6-3). As a result, when venous blood flow is occluded, the hematocrit will increase, and the magnitude of increase is related to the duration of occlusion (62). Tourniquet application can increase the hematocrit by ~0.5 divisions at 1 minute and ~4.0 divisions at 3 minutes of stasis (62). Mollison (62) maintains that tourniquet application for less than 1 minute is acceptable if it is removed soon after the vein is punctured, and before the blood is sampled. Finally, arm position

DETERMINANTS OF INTRAVASCULAR RESPONSES TO EXERCISE-HEAT STRESS

PREMEASUREMENT	×	STRESSOR	×	SUBJECT	=	RESPONSE
POSTURE		POSTURE		FITNESS		DILUTION
SAMPLE		EXERCISE		ACCLIMATION STATE		CONCENTRATION
ARM POSITION		TYPE		HYDRATION LEVEL		
STASIS		INTENSITY				
AMBIENT TEMPERATURE		DURATION				
		THERMAL				
		STRESS				
		STRAIN				

Figure 6-4. *Factors which modify the intravascular fluid responses to exercise and heat stress.*

relative to the trunk will result in an altered hematocrit (20) of the blood sample, and the influence of arm position will be further modified by body posture (113).

Body posture and the duration of that posture greatly alters plasma volume and the concentration of large solutes (such as erythrocytes) that do not easily pass through the capillary membrane. Since about 70 percent of the blood is below heart level and 75 percent of that volume is in the veins, when the subject assumes a more upright posture several cardiovascular responses occur to maintain cerebral perfusion and to avoid syncope (78). Precapillary vasoconstriction is the primary response, which should result in net absorption across the capillary (see Figure 6-3); however, no venoconstriction occurs with these compensatory responses (41). As a result, the columns of venous blood exert a hydrostatic "backpressure" across the capillary, which result in net filtration (41,78). When changing from the supine to standing posture or from the seated to standing posture, the plasma volume will decrease by about 15 and 10 percent, respectively (14,40,108). Likewise, a similar magnitude of hemodilution occurs when changing from standing to seated or supine postures. Immediately upon assuming a new posture the plasma begins to shift and continues to shift for 20 to 40 minutes until a new equilibrium volume is obtained (14,40,108).

Harrison and colleagues (41,42) have shown that small differences (±5° C) in skin temperature will cause large alterations (8–14%) in plasma volume in resting subjects. As a result, the initial intravascular fluid shift during heat/exercise stress can be modified by the skin temperature prior to the experimental stress. These investigators demonstrated that, dependent upon the skin temperature during the control period, either hemodilution or hemoconcentration can be the normal response to heat stress (41,42). These varied responses occur because cutaneous veins are very sensitive to heat and cold (relative to cutaneous arteries and particularly at low core temperatures) and will venoconstrict in response to cooling and venodilate in response to heating (41,42). As described in Figure 6-3, venoconstriction causes net filtration and venodilation causes net absorption across the capillary. If subjects rest in a comfortable (~20° C) environment, they will have cool skin (~30° C) and a somewhat lowered plasma volume during the control blood sample. Therefore, upon being exposed to heat/exercise stress, these subjects are likely to hemodilute relative to the control plasma volume. On the other hand, if the control blood sample is obtained when the skin is warm, no change in plasma volume or a hemoconcentration will be more likely to occur upon heat/exercise stress.

Stressor

During cycle exercise, subjects will immediately lose fluid from the intravascular to extravascular spaces and hemoconcentrate (8,9,41,60). The magnitude of hemoconcentration is proportional to the relative exercise intensity (8,9,60) with maximal plasma loss approximating 15–20 percent of the initial volume (41). However, the posture from which a subject performs cycle exercise will modify the magnitude of the hemoconcentration (14).

Diaz and colleagues (14) studied subjects at rest (supine posture) for 30 minutes in a comfortable environment and then exposed them to a hot environment for 90 minutes. While in the heat, the subjects rested in either a supine, low-sit or upright posture for the initial 45 minutes, and then performed cycle exercise in the same posture during the final 45 minutes of heat exposure. There was a modest change in plasma volume when the subjects maintained the supine posture (−2%), but a marked reduction in plasma volume when they changed to the low-sit (−9%) and upright (−17%) posture for the 45 minutes of resting heat exposure. Subsequently, when they performed light–moderate intensity cycle exercise in the supine posture, a large (−11%) reduction in plasma volume occurred, with a moderate (-7%) and modest (−3%) reduction observed when they exercised in the low-sit and upright postures, respectively. These data indicate that after a maximal posture induced plasma volume reduction, exercise will elicit a minimal additional reduction and vice versa.

During cycle exercise, the initial hemoconcentration for the first 5–10 minutes is probably hydrostatically and not oncotically mediated. Miles *et al.* (60) found that, in a comfortable environment, the amount of plasma volume reduction was related to the increase in mean arterial pressure during arm-crank and cycle exercise. They reasoned that mean arterial pressure changes probably reflect capillary hydrostatic pressure changes (60) and that higher values would favor net filtration from the capillary to interstitium (Figure 6-3). This initial hemoconcentration is not associated with plasma protein loss and, in fact, is sometimes associated with plasma protein gain, especially during light to moderate intensity cycle exercise (8,60). However, plasma protein loss is associated with near maximal intensity cycle exercise in a comfortable environment (8,60).

When cycle exercise is performed for an extended duration or during heat stress, a secondary hemoconcentration is also observed (43). This secondary hemoconcentration occurs more slowly than the initial hemoconcentration and is associated with a loss of plasma protein. The physiological mechanism responsible for this oncotically mediated secondary hemoconcentration is unclear, but it seems to be related to an increase in cutaneous blood flow and/or volume, which occurs to dissipate heat (43).

In comparison to cycle exercise, treadmill exercise elicits modest changes in the intravascular fluid volume. Hemoconcentration is the most often reported intravascular fluid response to treadmill exercise; however, Harrison (43) argues that this hemoconcentration results from inadequate postural controls and not from the treadmill exercise. The pre-exercise blood samples are often obtained while the subject is in a supine or seated posture prior to performing the treadmill exercise which requires an erect posture. The subsequent hemoconcentration from rest to treadmill exercise is similar in magnitude to that expected for the posture change alone. If the subject assumes a standing posture for 15–30 minutes prior to obtaining the initial blood sample, the subsequent treadmill exercise ordinarily will not elicit a hemoconcentration (43).

Regardless of whether or not an individual hemoconcentrates during treadmill exercise, will the exercise intensity modify the magnitude of the intravascular fluid shifts? Most investigators agree that for treadmill exercise (unlike cycle exercise) the exercise intensity will not modify the magnitude of the intravascular fluid shifts (18,72,97); however, one study reports the opposite (114). Additional research is merited examining the influence of treadmill exercise intensity on the magnitude of intravascular fluid shifts.

For euhydrated subjects, the usual response from rest (standing) to treadmill exercise in the heat is hemodilution (18,81,82,89,99). Figure 6-5 presents data illustrating the percent change in plasma volume from rest (while subjects were standing in a comfortable environment) to treadmill exercise in the heat over 10 successive days (98). Note that when the heat acclimated subjects subsequently performed exercise in a comfortable environment, a similar magnitude of hemodilution was observed as during exercise in the heat. Therefore, heat exposure probably will not influence the intravascular fluid shifts which occur in response to treadmill exercise for euhydrated-heat acclimated subjects (18,98).

The hemodilution that is associated with treadmill exercise is probably oncotically mediated (18,81,82,89). With the initiation of exercise, lymph turnover increases, resulting in the addition of protein to the intravascular space. This additional plasma protein exerts an oncotic pressure and causes a proportionate increase in fluid movement from the interstitium to the plasma. Figure 6-6 presents individual data for the relationship between the change in total circulating protein and the change in plasma volume from rest to treadmill exercise for euhydrated, heat-acclimated subjects. The bro-

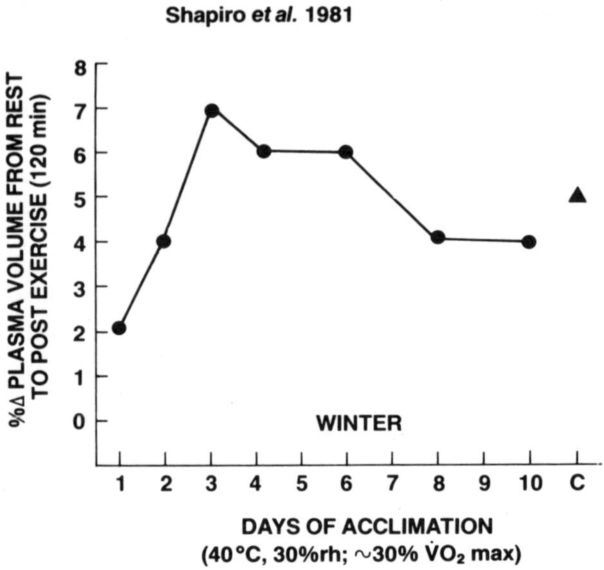

Figure 6-5. *The percent change in plasma volume from rest to treadmill exercise in the heat over 10 successive days. The final experiment (C) was conducted in a comfortable environment (98).*

ken line represents the amount of fluid change that is theoretically associated with a given amount of protein change (90). The usual response to treadmill exercise-heat stress is protein addition and hemodilution; however, some subjects may lose protein and hemoconcentrate. These data emphasize the importance of changes in total circulating protein on plasma volume changes during treadmill exercise.

It is inappropriate to discuss the relationship between plasma protein changes and plasma volume changes during exercise without noting the contributions of Senay and colleagues (94,95,96). These investigators used bench-step exercise, which requires an upright posture, and found that with the initiation of exercise, hemodilution and hemoconcentration were associated with plasma protein addition and loss, respectively. When bench-step exercise is performed in a comfortable environment, hemodilution occurs. In a hot environment, bench-step exercise elicits hemoconcentration for unacclimated subjects and hemodilution for heat acclimated subjects.

Subject Status

Exercise performance in a hot environment is primarily affected by a subject's state of aerobic fitness (68), heat acclimation status (66) and hydration level (83). These subject status factors are dis-

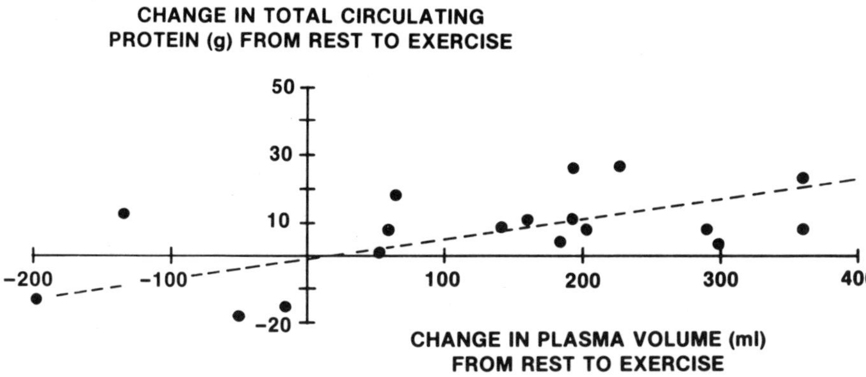

Figure 6-6. *Relationship between the change in total circulating protein and the change in plasma volume from rest to treadmill exercise (82,88). The broken line represents the theoretical relationship between those variables if only oncotic forces resulted in plasma volume changes (90).*

cussed in detail relative to exercise performance within other chapters of this book. Both exercise training and heat acclimation will increase a subject's resting plasma volume (2,9,95,98). The mechanism(s) responsible for this hypervolemia are unclear, as some investigators argue for a crystalloid mediated increase, whereas, others argue for an oncotically mediated increase in plasma volume (41). Regardless, little is known about how these subject status factors modify intravascular fluid shifts during exercise-heat stress.

Convertino *et al.* (9) examined the effects of exercise training on intravascular fluid responses to cycle exercise in a comfortable environment. Their subjects exercise trained for 8 days, at which time their resting plasma volume had increased by 12 percent and their maximal aerobic power had increased by 11 percent. During cycle exercise following this eight-day training, there was a smaller reduction in plasma volume (from rest to exercise) at a given power output, than there was pre-training. However, at a given relative intensity a similar magnitude of hemoconcentration was observed. These observations reinforce the concept that, during cycle exercise, the hemoconcentration is primarily hydrostatically mediated and dependent upon the relationship between mean arterial pressure and relative exercise intensity (60).

Subjects who are heat acclimated exhibit a more stable plasma volume and more consistent intravascular fluid responses to exercise-heat stress than do subjects who are not heat acclimated (81,82,90,95,98). Heat acclimation, however, may modify the intravascular fluid shifts elicited by various types of exercise. In our laboratory, when we employ treadmill exercise in the heat, we find great variability in the intravascular fluid shift for unacclimated subjects (82). However, heat acclimated subjects consistently hemodilute during treadmill exercise in the heat (81,82,88). Consistent with that observation, when bench-step exercise, (which employs an upright posture) is performed in a hot environment, unacclimated subjects hemoconcentrate while heat acclimated subjects hemodilute (94). However, heat acclimation is reported to accentuate the hemoconcentration elicited during cycle exercise (41). More research is necessary to answer questions on the effects of exercise training and acclimation state on intravascular fluid responses to exercise-heat stress.

As noted earlier (2,9,95,98) both exercise training and heat acclimation usually increase plasma volume for a subject at rest. This raises questions concerning the influence of hypervolemia, *per se*, on intravascular fluid responses to exercise-heat stress. Table 6-5 provides a summary of investigations which addressed these questions (23,24,43,84). Most of these investigators found that hypervolemia did not modify the intravascular fluid responses to exercise in the heat. However, Fortney and colleagues (24) reported that after an albumin infusion mediated hypervolemia, cycle exercise elicited an exacerbated (additional 140 ml) hemoconcentration. In that study, the subjects experienced a substantial reduction in total circulating protein (~11 gram) during the hypervolemia, but not during control experiments. The ~11g reduction in total circulating protein will exert sufficient oncotic pressure to account for the ~140 ml difference in plasma volume between the 2 experiments (90). Fortney and colleagues (24) may have observed the combined effects of exercise-induced hemoconcentration, and post-infusion efflux of protein and plasma from the intravascular space after albumin infusion. Hubbard *et al.* (47) assessed the dosage effects of albumin infusion on the extent and duration of plasma volume expansion for resting subjects in both a comfortable and hot environment. In the comfortable environment, they found a large loss of the infused albumin from the intravascular space during the initial 3 hours. In the hot environment, they found a slower and smaller loss of the infused albumin from the intravascular space. Therefore, in the hot environment the albumin infusion was able to expand plasma volume for a longer duration, than in the comfortable environment. Fortney *et al.* (24) infused albumin in a comfortable environment and their subjects then initiated exercise in the heat about 45–75 minutes post-infusion. Therefore, it seems reasonable that they observed the confounding effect of the post-infusion albumin efflux causing an oncotically mediated additional loss of intravascular fluid during the hypervolemia exercise experiments.

TABLE 6-5. *Influence of Hypervolemia on Plasma Volume Responses to Exercise-Heat Stress*

Study	Method	Δ PV, Rest	Exercise Type/Intensity	Environment	Findings
Harrison *et al* 1978	Saline Ingestion	+8%	Cycle/1.2 $1 \cdot min^{-1}$	45° C	−10% Control −12% Hypervolemia } ND*
Fortney *et al* 1981	Blood Infusion	+10%	Cycle/2.1 $1 \cdot min^{-1}$	35° C	−12% (400 ml) Control −11% (390 ml) Hypervolemia } ND
Fortney et al 1981	Albumin Solution	+14%	Cycle/2.7 $1 \cdot min^{-1}$	30° C	−13% (420 ml) Control −15% (560 ml) Hypervolemia } $p < 0.01$**
Sawka *et al* 1984	Albumin Solution	+13%	Treadmill/1.9 $1 \cdot min^{-1}$	45° C	0% Control 0% Hypervolemia } ND***

*We have recalculated the results as they reported a small ΔPV with hypervolemia.

**The hypervolemia experiments had an 11g reduction in total circulating protein.

***The subjects initiated the treadmill exercise after a prolonged period in the heat, and therefore had warm skin. The warm skin helped account for the lack of a hemodilution during the treadmill exercise.

Sawka *et al.* (84) infused albumin to mediate a hypervolemia, and studied the effects of this hypervolemia on the elicited intravascular fluid shifts during subsequent exercise-heat stress. An albumin-saline solution was infused while the subjects were in a hot environment. Treadmill exercise was performed several hours after initiation of the albumin-saline infusion. As a result, subjects exercised while at a fairly stable portion of the albumin kinetics curve, described by Hubbard *et al.* (47). They found that hypervolemia did not affect the intravascular fluid response to exercise-heat stress.

The influence of dehydration, or hypohydration, on intravascular fluid responses, during exercise in the heat, is fairly well documented. However, care is needed in interpreting the results of many of these investigations. For example, some investigators elicit hypohydration with diuretics, which cause an iso-osmotic hypovolemia (7,23,24), which might affect the subsequent intravascular fluid shifts. In addition, diuretics appear to induce a relatively large hypovolemia for a given reduction in total body water, as compared to thermal-exercise induced hypohydration. Finally, some investigators use either a marginal (0.6–1.2% decrease in body weight) hypohydration level (31) or do not report the hypohydration level (30).

Several studies, examining the effects of hypohydration (3–7% decrease in body weight) on the intravascular fluid responses to treadmill exercise in the heat, have been performed in our laboratory (82,89). For heat-acclimated subjects who are euhydrated, we find an increase in total circulating protein and hemodilution during treadmill exercise in the heat. During the hypohydration experiments however, these increments are attenuated or, in some cases, there occurs a loss of total circulating protein with the attendant loss of plasma from rest to treadmill exercise in the heat (82,89). The mechanism responsible for the smaller increase or the decrease in total circulating protein may be related to altered lymph flow (82). Remember, lymph is formed from interstitial fluid. Hypohydration will markedly reduce the interstitial fluid volume (11,17) and may result in a smaller volume of more viscous lymph. The smaller volume of viscous lymph could result in reduced flow during exercise and the return of less protein to the circulatory system. Also, this additional extravascular protein would increase interstitial oncotic pressure and facilitate net filtration (Table 6-4).

Harrison *et al.* (41,43) made similar observations in hypohydrated subjects performing cycle exercise in the heat. When subjects were hypohydrated, they experienced a greater hemoconcentration than they did during euhydration (41); the greater fluid loss was associated with a greater efflux of plasma protein (43). In conclusion, it seems that hypohydration makes it more difficult to retain the initial pre-exercise total circulating protein and plasma volume during exercise-heat stress.

BODY WATER LOSS

In hot environments, body water is lost primarily through eccrine sweat gland secretion, which results in evaporative cooling of the body. Urinary water loss is decreased during heat stress relative to moderate environmental conditions (1). In addition, both exercise and hypohydration have been reported to decrease urine output below control levels (103). For a given person, the sweating rate is dependent on environmental conditions, clothing and physical activity level (1). Body fluid loss by sweat can vary greatly, and sweating rates as high as 1 liter $\cdot$ $m^{-2} \cdot h^{-1}$ are frequently reported.

Thirst may not provide a good index of body water requirements. Numerous investigators report that *ad libitum* water intake results in incomplete water replacement or "voluntary" dehydration during exercise in the heat (1,19,21). Although heat acclimation improves the relationship between thirst and body water needs (1,19,21), voluntary dehydration may still occur.

Distribution

Sweat loss will result in a reduction of total body water if an adequate amount of fluid is not consumed. Remember, total body water constitutes approximately 60 percent of an average adult's body weight (Table 6-2), so a 75-kg person will have a total body water mass of about 45 kg (Table 6-1). Therefore, a fluid loss equal to 5 percent of body weight could constitute 8 percent of total body water for this person. The question arises as to how water loss is partitioned between the body fluid compartments. As a consequence of free fluid exchange, hypohydration should affect each fluid compartment.

Figure 6-7 illustrates the findings of 2 studies on the partitioning of water deficit between fluid compartments during resting conditions (11,17). Costill *et al.* (11) dehydrated subjects by using a combination of cycle exercise and heat exposure. Shortly after completing cycle exercise, their subjects assumed a supine posture, where blood and skeletal muscle samples were obtained. Their data may be biased by the cessation of cycle exercise immediately prior to estimation of water distribution between fluid compartments. Durkot *et al.* (17) dehydrated rats by using a passive heat exposure for several (5 to 11) hours and indicated that their data was slightly biased by the animals' appetite loss. The intent of Figure 6-7 is to present

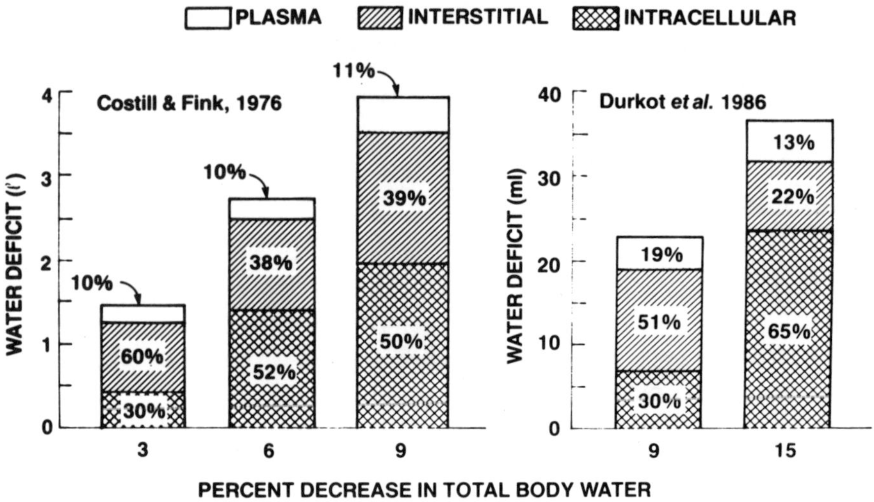

Figure 6-7. *The findings of two studies concerning the partitioning of water deficit between fluid compartments during resting conditions (11,17).*

data trends, and not to imply that a given percent decrease of total body water is similar between man and rats.

At low levels of body water loss, the water deficit primarily comes from the extracellular space. As the level of body water loss increases, a proportionately greater percentage of the water deficit comes from the intracellular space. In fact, the intracellular space can provide the largest percentage of water deficit at the highest levels of body water loss. These studies disagree on the influence of body water reduction on plasma volume. One study shows a linear decrease in plasma volume (11), whereas the other study demonstrates a plasma volume defense at the highest level of body water loss (17).

Figure 6-8 (top) illustrates individual data on the plasma volume reduction associated by a given level of body water loss (89). Hypohydration was achieved by voluntary food and fluid denial combined with exercise in a hot environment. A period of ~15 h was spent resting in a comfortable environment (while still dehydrated) to provide time for the fluid compartments to equilibrate at the achieved hydration level. Plasma volume was estimated after the heat-acclimated subjects stood quietly for about 20 minutes in a comfortable environment. An identical program of exercise-heat stress, but with full fluid replacement, was completed on the day prior to the euhydration measurements (dotted line). The line of identity (solid line) represents a proportional loss of plasma and total body water.

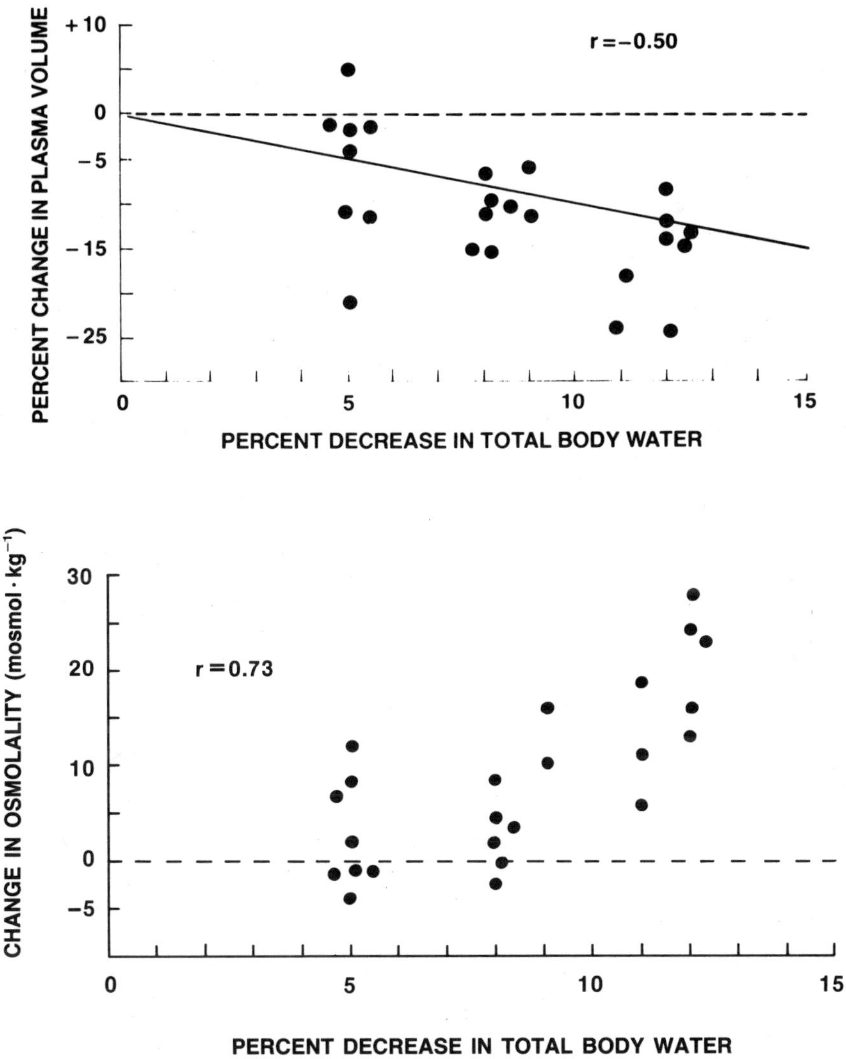

Figure 6-8. *Individual data on the plasma volume reduction and increase in osmolality associated with a given level of body water loss (89).*

The magnitude of plasma volume reduction is somewhat related to the level of total body water loss. Simply, a larger hypovolemia is associated with greater levels of hypohydration. At low levels of body water loss, considerable variability exists for the magnitude of hypovolemia; in fact, in one subject, plasma volume was elevated and in several other subjects plasma volume was not reduced below euhydration levels. The data from the moderate and high level of body water loss could be interpreted as showing either a continued reduction or a defense of plasma volume (similar to Figure 6-7). The mean values for plasma volume reduction between these two levels are similar, but at the high level 2 subjects had a hypovolemia of greater than 20 percent.

A subject's state of heat-acclimation may modify the magnitude of plasma volume reduction at a given level of body water loss. Figure 6-9 presents individual data for the magnitude of hypovolemia associated with a given hypohydration level,[1] both before and after a 10-day heat acclimation program (88). These data represent resting values; hypohydration was achieved by the dehydration procedures (food and fluid denial and exercise in the heat) previously described. Post-acclimation the subjects had a smaller plasma volume reduction at a given body weight loss. The physiological mechanisms responsible for this apparent plasma volume defense are unknown; however, several can be theorized. First, it is known that heat-acclimated subjects secrete a more hypotonic sweat than their unacclimated counterparts (51). Since dehydration was induced by exercise-heat stress, a more hypotonic sweat secretion would result in a greater concentration of solutes, remaining in the plasma and exerting an osmotic pressure counteracting further fluid loss. Secondly, it is known that heat acclimation results in an increased total circulating protein mass (95), which would exert a greater oncotic pressure at a given plasma volume. Therefore, in heat-acclimated individuals, a greater intravascular osmotic and oncotic pressure may permit a redistribution of total body water, enabling better defense of plasma volume at a given level of body water loss. On the other hand, heat acclimation may increase total body water so that a given reduction in body weight would represent a smaller percent reduction in total body water (2,41). In the aforementioned experiments, total body water was not measured; however, the estimated resting plasma volume was not altered by the heat acclimation program.

Adaptation

A recently suggested human adaptation to exercise-heat induced hypohydration is an increase in total circulating protein (89). Figure 6-10 presents individual data on the change (from euhydration levels) in total circulating protein, for subjects at rest, at ~15 h post-dehydration (82,89). Dehydration procedures were identical to those previously described. Note that after ~15 hours of dehydra-

1. The probability level of this comparison was incorrectly presented in that manuscript as being non-significant.

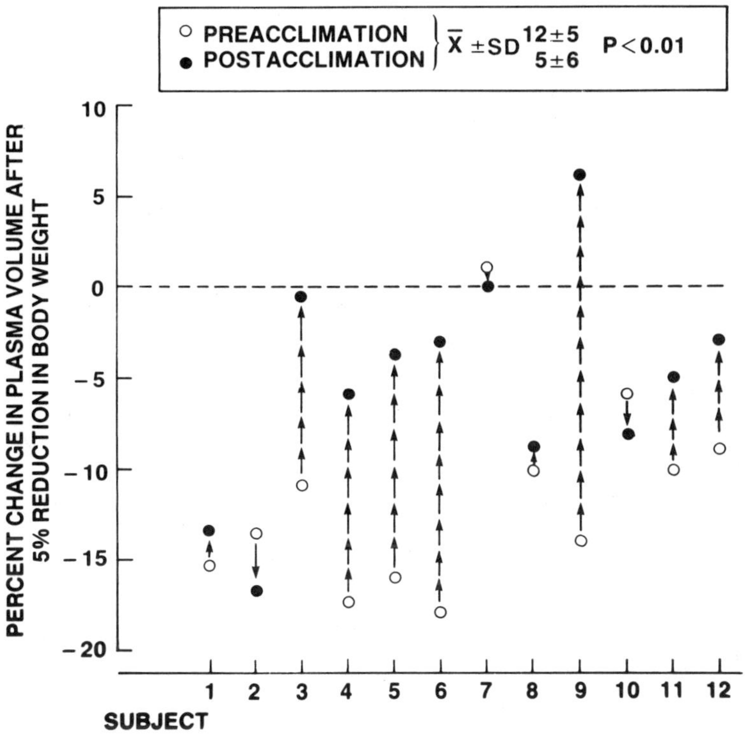

Figure 6-9. *The magnitude of plasma volume reduction, while at rest, associated with a given dehydration level both before and after completing a heat acclimation program (88).*

tion, the total circulating protein is usually increased above euhydration levels, and there is no clear relationship between the magnitude of protein increase and level of body water loss. The physiological stimulus for the increased resting total circulating protein is unclear. Perhaps the protein increments are stimulated by the magnitude of plasma volume reduction (hypovolemia) during the dehydration process. Unfortunately, we did not quantitate the hypovolemia until ~15 hours after achievement of the dehydration level, when the adaptive process had already occurred. All of these subjects were previously heat acclimated, so it is unknown if this adaptation occurs in unacclimated humans. These observations raise the possibility that, in dehydrated humans, acute increases in total circulating protein may occur as an adaptation to aid redistribution of body water into the intravascular space.

Several investigators have recently reported that dehydration elevates total circulating protein in both rats (46) and baboons (115). Hemorrhage, which causes an acute hypovolemic stress similar to

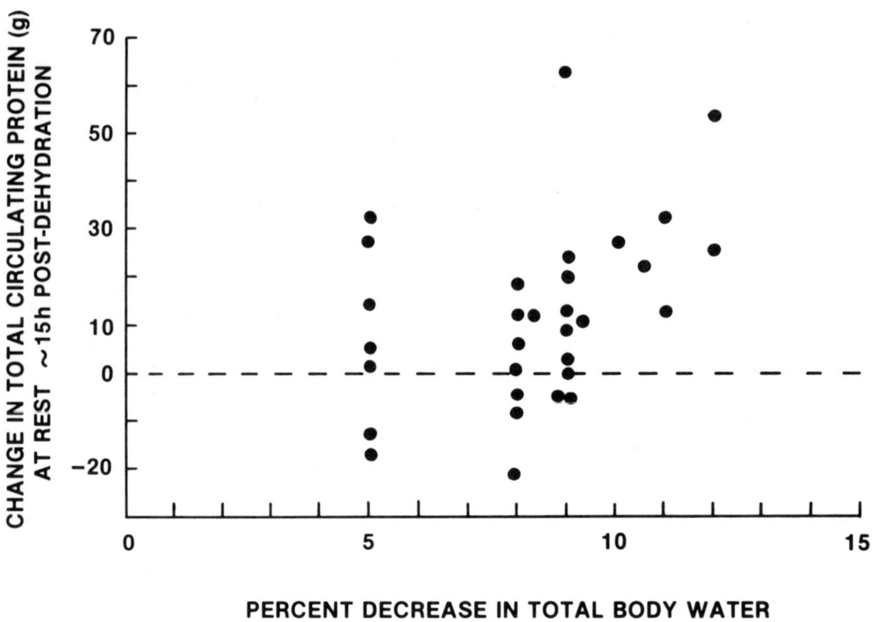

Figure 6-10. *Individual data for the change in total circulating protein from euhydration levels when dehydrated for ~15 hours.*

that of dehydration, also elevates total circulating protein in dogs (57). A common observation in all of these animal studies was an acutely increased hepatic albumin synthesis (46,57,115) and a reduced vascular permeability to albumin (46,115). Whether or not total circulating proteins are increased in dehydrated humans (at rest), by the same mechanisms, needs to be investigated. Another possibility is an improved translocation of protein from the lymph to plasma, while at rest, during severe dehydration, perhaps employing parallel communications between the lymph node and venules (73,106).

It is known that hypohydration will increase the osmotic pressure in the plasma. Eccrine sweat is ordinarily hypotonic relative to plasma (51); therefore, the plasma will become hyperosmotic when hypohydration is induced primarily by sweat output (83,93). For resting subjects, plasma osmolality will increase from about 283 $mosmol \cdot kg^{-1}$ when euhydrated to levels approaching 300 $mosmol \cdot kg^{-1}$ when hypohydrated (Figure 6-8). During exercise-heat stress, plasma osmolality will increase further so that severely dehydrated subjects will routinely have values in excess of 300 $mosmol \cdot kg^{-1}$ (89). Sodium, potassium, and their anions (chloride) are, primarily, responsible for the elevated plasma osmolality during hypohydration (93). Several investigators have reported that plasma hyperosmolality, as well as decreased plasma volume contribute to the less efficient thermoregulatory responses when hypohydration occurs during exercise in the heat (25,83,93). Such fluctuations in plasma osmolality and plasma volume are the triggering mechanism for hormonal adaptations to conserve body water.

Ordinarily, a simultaneous elevation in plasma osmolality and decrease in plasma volume will elicit vasopressin secretion, which reduces urinary water loss and stimulates thirst. Elevations in plasma osmolality will stimulate hypothalamic osmoreceptors to initiate neural responses that increase pituitary vasopressin release. Likewise, a decreased plasma volume during the dehydration process will cause falling central venous and atrial pressures and unload the low-pressure baroreceptors, which stimulate vasopressin release. Robertson *et al.* (76) report that hypovolemia reduces the osmolality threshold for vasopressin release. A dehydrated subject will have reduced mean renal pressure, thus stimulating aldosterone release, causing both water and sodium reabsorption. Smith *et al.* (103) report that both glomerular filtration rate and urine flow levels are linearly decreased with the severity of dehydration (up to an 8-percent reduction in body weight). As expected, Gauer *et al.* (33) report a gradation of increasing plasma aldosterone concentration following graded hemorrhage. Finally, a third factor which may conserve body water in dehydrated subjects is a reduced concentration of natriuretic hormone; however, neither the exact biochemical structure nor driving mechanisms for this hormone(s) have been clearly defined (3).

Numerous investigators indicate that circulating concentrations of the fluid-electrolyte regulatory hormones are elevated in man during heat exposure (26,107). For example, elevations in renin activity, circulating levels of aldosterone, and vasopressin during sedentary heat exposure have been observed. Likewise, in exercising subjects without heat stress, several investigators document increments in the same group of hormones (26,107).

Little research has focused on the fluid-electrolyte regulatory hormone responses in hypohydrated subjects (28,29). Francesconi and colleagues (28) examined the effects of hypohydration (5% reduction in body weight) and heat acclimation on circulating aldosterone and renin activity in several environments. When at rest, in a comfortable environment, hypohydration elevated circulating aldosterone and plasma-renin activity. During exercise, hypohydration increased aldosterone levels even more, but heat acclimation did not modify the aldosterone levels in either euhydrated or hypohydrated subjects. The plasma-renin activity was also elevated during exercise-heat stress and these increments were accentuated by hypohydration, but attenuated by heat acclimation. For euhydrated subjects, some investigators report that heat acclimation will attenuate plasma-renin activity (22), but not plasma aldosterone levels (13,22). However, other investigators report that heat acclimation will attenuate plasma aldosterone levels (51) during exercise-heat stress.

Francesconi and colleagues (29) subsequently examined the effects of graded hypohydration levels (3%, 5% and 7% reduction in body weight) on plasma-renin activity and aldosterone levels during exercise-heat stress, and reported that plasma-renin activity ($r = 0.94$) and aldosterone levels ($r = 0.77$) were related to hypohydration through the 5-percent level. The elevation in these hormones appeared to be related to the reduced plasma volume during exercise. Between 5 percent and 7 percent hypohydration, plasma volume did not decrease further, and likewise, hormone levels did not increase further during exercise-heat stress. In earlier experiments (27) those investigators induced hypervolemia by the infusion of hyperoncotic albumin and observed reduced aldosterone and angiotensin I levels during exercise in the heat. These data all demonstrate the importance of plasma volume on stimulation of the renin-angiotensin-aldosterone system during exercise-heat stress.

RESPONSES TO HYDRATION LEVELS

Symptoms and Performance

Adolph and associates (1) described the symptoms associated with hypohydration in the desert. Thirst was reported to occur after a 2-percent water deficit and did not increase in intensity with greater hypohydration levels. The concept of thirst as a threshold or "all or none" response has been accepted for many years, but recent data have questioned this hypothesis (21). Engell and colleagues (21) report that thirst and many related symptoms increase in intensity with the level of hypohydration. Adolph and associates (1) also report that a 4-percent to 6-percent water deficit is associated with anorexia, impatience and headache; whereas, a 6-percent to 10-percent water deficit is associated with vertigo, dyspnea, cyanosis and spasticity. A person who incurs more than a 12-percent water deficit will be unable to swallow and will need assistance with rehydration. An estimate of the lethal hypohydration range is 15 percent–25 percent (1).

Table 6-6 presents a summary of investigations on the effects of hypohydration on maximal aerobic power and physical work capacity. In the absence of heat stress, a relatively large water deficit

TABLE 6-6. *Effects of Hypohydration on Maximal Exercise Performance*

Study	Year	Test Environ.	%Δ WT	%Δ $\dot{V}O_2$ max	%Δ PWC
Craig & Cummings	1966	Hot	−2%	↓10%	↓22%
			−4%	↓27%	↓48%
Buskirk, *et al.*	1958	Comfort.	−6%	↓?	
Saltin	1964	Comfort.	−7%	=	↓20%

(6%–7%) has a minimal effect on maximal aerobic power (4,79), but reduces physical work capacity (79). In a hot environment, Craig and Cummings (12) demonstrated that small (2%) to moderate (4%) water deficits reduce maximal aerobic power and physical work capacity. In addition, these decrements increase dramatically with the magnitude of water deficit. Consistent with these findings, hypohydration combined with hyperthermia in a moderate environment reduces maximal aerobic power by 6 percent and exercise time by 12 percent from euhydration levels (87). These investigations demonstrate that maximal exercise performance is reduced when hypohydration is combined with thermal strain. Likewise, submaximal endurance exercise is also reduced by dehydration acting through both the thermoregulatory and cardiovascular systems (85,86).

Thermoregulatory and Cardiovascular Effects

In comparison to euhydration, hypohydration results in an increased core temperature during exercise in the heat (1,34,37,89,105). It is generally thought that as the severity of hypohydration increases, there is a concomitant gradation in the elevation of core temperature during exercise. A comparison of results from investigators that examined a single level of body water deficit during exercise-heat stress does not support this concept of graded thermoregulatory responses with hypohydration (83). Conclusions based on inter-investigation comparisons, however, can be tenuous because of differences in subject populations, environmental conditions, exercise intensities and test methodologies (83).

Two studies examined thermoregulatory responses to exercise while hypohydration levels were varied during independent tests in the same subjects (89,105). Strydom and Holdsworth (105) studied 2 miners at 2 hypohydration levels (low, 3%–5%, and high, 5%–8% weight loss) and found higher core temperatures during the high hypohydration level. Sawka *et al.* (89) reported that hypohydration linearly increased the core temperature response during exercise in the heat (range 0.12–0.18° C) for each percent decrease in bodyweight.

Other investigators, reporting a gradation of elevated core temperature with increased water deficits, have interpolated from a single hypohydration level (37) and/or employed prolonged exercise-heat exposure to elicit a progressive dehydration (1,34). Reanalyzing the data of Adolph and associates (1), we find that their subjects had an elevated core temperature of ~0.20° C for each percent decrease in body weight. Their data represents a progressive dehydration under a variety of field and laboratory conditions. Greenleaf and Castle (37) reported that core temperature was elevated 0.10° C for each percent decrease in body weight during exercise (49% $\dot{V}O_2$ max) in a moderate environment. This relationship was based on interpolation from a single hypohydration (5% body weight loss) level. Gisolfi and Copping (34) report that core temperature is elevated by 0.40° C for each percent decrease in body weight after a weight loss of greater than 2% during heavy exercise (74% $\dot{V}O_2$max) in a hot environment. Clearly, exercise intensity and environmental conditions will modify the magnitude of core temperature elevation, associated with hypohydration during exercise-heat stress.

The hypohydration-mediated increase in heat storage is the result of either an increase in metabolic heat production or a decrease in heat loss. Hypohydration does not influence the rate of aerobic or anaerobic metabolism during exercise (37,80,84) and as a result does not cause greater metabolic heat production. Therefore, a decreased heat dissipation must be responsible for the hypohydration-mediated heat storage during exercise. The relative contribution of insensible and sensible heat exchange during exercise depends on the specific environmental conditions, but both avenues of heat loss are adversely affected by hypohydration (83).

Hypohydration is associated with reduced or unchanged sweating rates at a given metabolic rate during exercise in the heat (83). Those investigators, who report no change in sweating rate, will usually still observe an elevated core temperature. Therefore, during hypohydration the sweating rate will be lower for a given core temperature, and the potential for heat dissipation through sweat evaporation will be reduced. Figure 6-11 presents data showing the parallel reduction in sweating rate at a given core temperature during graded hypohydration levels (89). Since core temperature provides 90 percent of the drive for thermoregulatory sweating, this figure indicates that the reduced sweating rate is systematically reduced for a given thermal drive.

The physiological mechanisms mediating the reduced sweating rate response during hypohydration are not clearly defined. Both the singular and combined effects of plasma hyperosmolality (6,25,43,93) and plasma hypovolemia (24,45) have been suggested as mediating this reduced sweating response. Figure 6-12 presents individual data for the change (from euhydration levels) in sweating

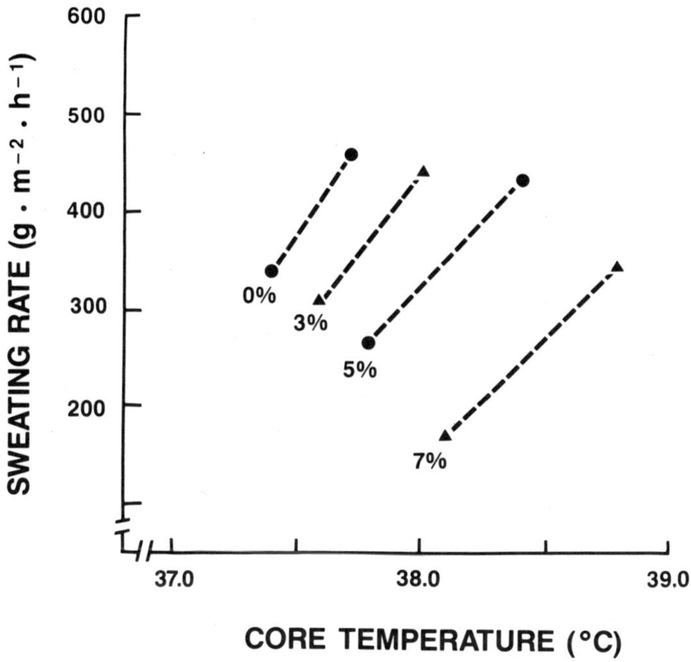

Figure 6-11. *Plot of mean sweating rate and final exercise core temperatures when euhydrated (0%) and hypohydrated by 3%, 5% and 7% of body weight (89).*

rate plotted against the change in osmolality and plasma volume during exercise-heat stress at a variety of hypohydration levels (89). Plasma hyperosmolality maintains a strong ($r = -0.76$) and consistent relationship with the reduced sweating rates during hypohydration (89). Consistent with this, Senay (93) reports an inverse relationship between plasma osmolality and sweating rate ($r = -0.62$) when hypohydration occurs. Also, Harrison *et al.* (43) report that plasma hyperosmolality will elevate core temperature responses during exercise-heat stress, despite the maintenance of euhydration. Hyperosmolality may decrease sweating by a direct central nervous system effect on the hypothalamic thermoregulatory centers (93,99) or by a peripheral effect at the eccrine sweat gland (67).

Fortney *et al.* (24) have provided strong evidence that an iso-osmotic hypovolemia causes a reduced sweating rate and elevated core temperature response during exercise. They used diuretics which induced a 15-percent hypovolemia at an approximate 5-percent reduction in total body water. Note (Figure 6-8) that this is a greater hypovolemia than might be expected for an exercise-heat induced 5% reduction in total body water. Fortney *et al.* (24) theorize that

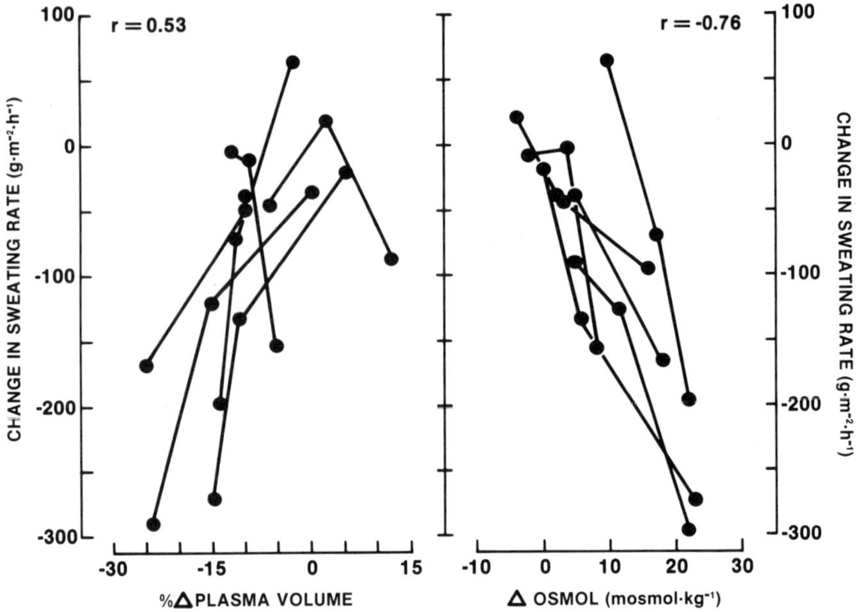

Figure 6-12. *Individual relationship for the change in exercise sweating rate from euhydration levels with changes in plasma volume and changes in osmolality from euhydration levels (89).*

hypovolemia may alter the activity of atrial baroreceptors that have afferent input to the hypothalamus. Therefore, a reduced atrial filling pressure might modify neural information to the hypothalamic thermoregulatory center, controlling sweating rate. Recently, several investigators reported that hypovolemia may be less important than hyperosmolality in eliciting the reduced thermoregulatory effector responses, associated with hypohydration during exercise (6,25,89).

The effects of hypohydration on cardiovascular responses to exercise have been investigated (64,80,85). During submaximal exercise with little thermal strain, hypohydration elicits an increase in heart rate and decrease in stroke volume with no change in cardiac output relative to euhydration levels (80). Apparently, during hypohydration, a decreased blood volume reduces end-diastolic ventricular volume and stroke volume, requiring a compensatory increase in heart rate to maintain cardiac output. During submaximal exercise with moderate (64) or severe (85) thermal strain, hypohydration (3%–4%) elicits an increase in heart rate, decrease in stroke volume, and a decrease in cardiac output relative to euhydration levels. Figure 6-13 presents the cardiovascular responses of subjects

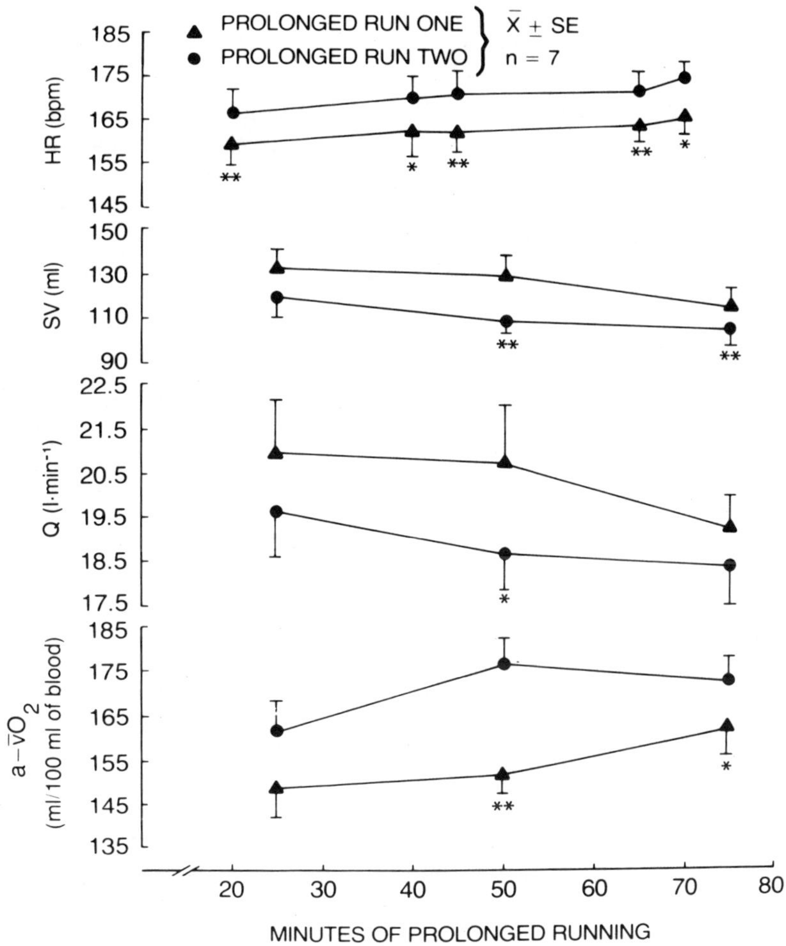

Figure 6-13. *Heart rate, stroke volume, cardiac output, and arteriovenous oxygen difference responses of subjects during two prolonged runs spaced by a rest period (85). During these exercise bouts the subjects experienced progressive dehydration and hyperthermia.* *$p < 0.05$, **$p < 0.01$.

performing 2 prolonged exercise bouts (70% $\dot{V}O_2$ max) that were spaced by a brief rest period. The subjects were progressively dehydrated (to a 4-percent body weight loss) and incurred core temperatures in excess of 40° C. It should be noted that the increasing heart rate does not compensate for the decreasing stroke volume and, eventually, results in a decreased cardiac output.

The combination of exercise and heat strain results in competition between the central and peripheral circulation for a limited blood volume (78). As body temperature increases during exercise, cutaneous vasodilation occurs, thus decreasing venous resistance and pressure. As a result of decreased blood volume and increased blood displacement to cutaneous vascular beds, venous return and, thus, cardiac output will be decreased below euhydration levels (64,85). Nadel *et al.* (64) report that these conditions also reduce cutaneous blood flow for a given core temperature and therefore the potential for sensible heat exchange. Likewise, hyperosmolality, in the absence of hypovolemia, can also reduce the cutaneous blood flow response during exercise-heat stress (25).

Modifying Factors

During exercise in the heat, thermoregulatory responses are primarily influenced by the subject's acclimation state, aerobic fitness and hydration level. Heat acclimated persons, who are aerobically fit and fully hydrated,will have less body heat storage and optimal performance. Little research has focused on the interactive effects of hydration levels with acclimation state and aerobic fitness on subject performance during exercise-heat stress.

Hydration level is particularly important during exercise in the heat because a body water deficit will neutralize the thermoregulatory advantages of heat acclimation (4,88). In addition, some data suggest that hypohydration will negate the thermoregulatory advantage afforded by high aerobic fitness (1,4,5). These aerobic fitness data, however, represent a small number of subjects, who performed only light-intensity (1,5) or moderate-intensity (5) exercise. In addition, the data of Buskirk *et al.* (5) suggest a slight thermoregulatory advantage when subjects are hypohydrated after having completing a physical training program; but methodological problems within that study did not allow those investigators to provide a definitive answer. It is possible that even if aerobically fit subjects lose their thermoregulatory advantage when hypohydrated, they may still maintain an exercise tolerance advantage. Aerobically fit athletes, such as marathon runners, often tolerate very high core temperatures (greater than 40° C) for extended durations with no adverse physical effects, during exercise-heat stress (58,74,85,86). Such high core temperatures would not be expected to be tolerated by less fit or unfit individuals during exercise. Additional research concerning the interaction of aerobic fitness and hypohydration is warranted.

Hyperhydration

If hypohydration reduces performance during exercise-heat stress, can excess body fluids improve performance beyond the levels achieved when euhydrated? Moroff and Bass (63) examined the influence of excessive fluid ingestion on thermoregulatory responses to exercise in the heat. They report that hyperhydration decreases core temperature, while increasing sweating rates above control levels. During the control experiments, however, their subjects were slightly (greater than 1%) hypohydrated. Therefore, these results may demonstrate the effects of hypohydration rather than hyperhydration. More recently, Greenleaf and Castle (37) report that excessive fluid ingestion does not alter core temperature or sweating rate from control levels during exercise in the heat.

If hyperhydration did improve performance during exercise-heat stress, these improvements would most likely be mediated by hypervolemia. In fact, some of the thermoregulatory advantages gained through heat acclimation have been attributed to plasma volume expansion (95). Several recent studies on the effects of artificially expanded blood volume have indicated no differences in core temperature (23,65,84), sweating rate (84) or cutaneous blood flow (23,65) in comparison with normovolemic control levels during exercise-heat stress. Generally, these studies found that plasma volume expansion lowered heart rate responses during exercise in the heat (23,84). In contrast, Fortney *et al.* (24) report that an artificially expanded plasma volume lowered core temperature below control levels during exercise, despite no difference in the sweating rate response. These results indicate that the physiological advantages of hyperhydration are small and may be affected by the protocol used to induce hypervolemia.

Another approach to attempt experimental hyperhydration is to infuse both erythrocytes and saline into subjects prior to exercise-heat stress. Recently it was reported that acute polycythemia provides a small thermoregulatory advantage during exercise in the heat (35,81). Sawka and colleagues (81) infused autologous erythrocytes and found a compensatory reduction in plasma volume, to maintain the same blood volume as during the preinfusion experiments. The euhydrated and nonacclimated subjects tended to reduce heat storage (81) and improved sensible and insensible heat loss after infusion (35). Therefore, an increased erythrocyte volume, independent of blood volume, may confer a small thermoregulatory advantage during exercise in the heat.

CONCLUSION

The distribution of body water and the subsequent ability to maintain blood volume is a critical factor affecting man's ability to perform exercise, as well as survive in hot environments. If sweat output exceeds water intake, hypohydration (body fluid deficit) will occur. This fluid deficit is comprised of water loss from both the intracellular and extracellular fluid compartments, and will result in a decreased plasma volume and increased plasma osmolality at rest. With subsequent exercise, the dehydrated subject may have difficulty maintaining even this reduced plasma volume. Hypohydration during exercise causes a greater heat storage and reduces exercise performance. These decrements are attributed to a reduced ability to maintain cardiac output as well as a decreased ability to dissipate heat. Hyperhydration, or body fluid excess, does not provide a clear advantage during exercise-heat stress, but will delay the development of dehydration.

REFERENCES

1. Adolph, E.F. and Associates. *Physiology of Man in the Desert*. New York: Interscience, 1947.
2. Bass, D.E. and A. Henschel. Responses of body fluid compartments to heat and cold. *Physiol. Rev.* 36:128–143, 1956.
3. Buckalew, V.M. and K.A. Gruber. Natriuretic hormone. *Annu. Rev. Physiol.* 46:343–358, 1984.
4. Buskirk, E.R., P.F. Iampietro and D.E. Bass. Work performance after dehydration: effects of physical conditioning and heat acclimation. *J. Appl. Physiol.* 12:189–194, 1958.
5. Cadarette, B.S., M.N. Sawka, M.M. Toner and K.B. Pandolf. Aerobic fitness and the hypohydration response to exercise-heat stress. *Aviat. Space Environ. Med.* 55:507–512, 1984.
6. Candas, V., J.P. Libert, G. Brandenberger, J.C. Sagot, C. Amoros and J.M. Kahn. Hydration during exercise: effects on thermal and cardiovascular adjustments. *Eur. J. Appl. Physiol.* 55:113–122, 1986.
7. Claremont, A.D., D.L. Costill, W. Fink and P. Van Handel. Heat tolerance following diuretic induced dehydration. *Med. Sci. Sports* 8:239–243, 1976.
8. Convertino, V.A., L.C. Keil, E.M. Bernauer and J.E. Greenleaf. Plasma volume, osmolality, vasopressin, and renin activity during graded exercise in man. *J. Appl. Physiol.* 50:123–128, 1981.
9. Convertino, V.A., L.C. Keil and J.E. Greenleaf. Plasma volume, renin and vasopressin responses to graded exercise after training. *J. Appl. Physiol.* 54:508–514, 1983.
10. Cook, D.R., W.S. Gualtiere and S.J. Galla. Body fluid volumes of college athletes and non-athletes. *Med. Sci. Sports* 1: 217–220, 1969.
11. Costill, D.L., R. Coté and W. Fink. Muscle water and electrolytes following varied levels of dehydration in man. *J. Appl. Physiol.* 40:6–11, 1976.
12. Craig, F.N. and E.G. Cummings. Dehydration and muscular work. *J. Appl. Physiol.* 21:670–674, 1966.
13. Davies, J.A., M.H. Harrison, L.A. Cochrane, R.J. Edwards and T.M. Gibson. Effects of saline loading during heat acclimatization on adrenocortical hormone levels. *J. Appl. Physiol.* 50:605–612, 1981.
14. Diaz, F.J., D.R. Bransford, K. Kobayashi, S.M. Horvath and R.G. McMurray. Plasma volume changes during rest and exercise in different postures in a hot humid environment. *J. Appl. Physiol.* 47: 798–803, 1979.
15. Dick, D.A.T. *Cell Water*. Washington, D.C.: Butterworth, Inc., 1965.
16. Dill, D.B. and D.L. Costill. Calculation of percentage changes in volumes of blood, plasma and red cells in dehydration. *J. Appl. Physiol.* 37:247–248, 1974.
17. Durkot, M.J., O. Martinez, D. McQuade and R. Francesconi. Simultaneous determination of fluid shifts during thermal stress in a small-animal model. *J. Appl. Physiol.* 61:1031–1034, 1986.
18. Edwards, R.J. and M.H. Harrison. Intravascular volume and protein responses to running exercise. *Med. Sci. Sports Exerc.* 16:247–255, 1984.
19. Eichna, L.W., W.B. Bean, W.F. Ashe and N. Nelson. Performance in relation to environmental temperature. *Bull. John Hopkins Hosp.* 76:25–58, 1945.
20. Eisenberg, S. Effect of posture and position of the venous sampling site on the hematocrit and serum protein concentrations. *J. Lab. Clin. Med.* 61:755–760, 1963.

21. Engell, D.B., O. Maller, M.N. Sawka, R.P. Francesconi, L. Drolet and A.J. Young. Thirst and fluid intake following graded hypohydration levels in humans. *Physiol. and Behavior.* 40:226–236, 1987.
22. Finberg, J.P.M. and G.M. Berlyne. Modification of renin and aldosterone response to heat by acclimatization in man. *J. Appl. Physiol.* 42:554–558, 1977.
23. Fortney, S.M., E.R. Nadel, C.B. Wenger and J.R. Bove. Effect of acute alterations of blood volume on circulatory performance in humans. *J. Appl. Physiol.* 50:292–298, 1981.
24. Fortney, S.M., E.R. Nadel, C.B. Wenger and J.R. Bove. Effect of blood volume on sweating rate and body fluids in exercising humans. *J. Appl. Physiol.* 51:1594–1600, 1981.
25. Fortney, S.M., C.B. Wenger, J.R. Bove and E.R. Nadel. Effect of hyperosmolality on control of blood flow and sweating. *J. Appl. Physiol.* 57:1688–1695, 1984.
26. Francesconi, R.P. Endocrinological responses to exercise in stressful environments. *Exercise and Sport Sciences Reviews.* 16:255–284, 1988.
27. Francesconi, R.P., M.N. Sawka, R.W. Hubbard and M. Mager. Acute albumin-induced plasma volume expansion and exercise in the heat: effect on hormonal responses in men. *Eur. J. Appl. Physiol.* 51:121–128, 1983.
28. Francesconi, R.P., M.N. Sawka and K.B. Pandolf. Hypohydration and heat acclimation: plasma renin and aldosterone during exercise. *J. Appl. Physiol.* 55:1790–1794, 1983.
29. Francesconi, R.P., M.N. Sawka, K.B. Pandolf, R.W. Hubbard, A.J. Young and S. Muza. Plasma hormonal responses at graded hypohydration levels during exercise-heat stress. *J. Appl. Physiol.* 59:1855–1860, 1985.
30. Gaebelein, C.J. and L.C. Senay. Influence of exercise type, hydration and heat on plasma volume shifts in men. *J. Appl. Physiol.* 49:119–123, 1980.
31. Gaebelein, C.J. and L.C. Senay. Vascular volume changes during cycling and stepping in women at two hydration levels. *Eur. J. Appl. Physiol.* 48:1–10, 1982.
32. Garlick, D.G. and E.M. Renkin. Transport of large molecules from plasma to interstitial fluid and lymph in dogs. *Am. J. Physiol.* 219:1595–1605, 1970.
33. Gauer, O.H., J.P. Henry and C. Behn. The regulation of extracellular fluid volume. *Annu. Rev. Physiol.* 32:547–595, 1970.
34. Gisolfi, C.V. and J.R. Copping. Thermal effects of prolonged treadmill exercise in the heat. *Med. Sci. Sports* 6:108–113, 1974.
35. Gonzalez, R.R., M.N. Sawka, A.J. Young, S.R. Muza, R.C. Dennis, J.W. Martin, C.R. Valeri and K.B. Pandolf. Effect of erythrocyte reinfusion on local sweating and sensible heat loss during exercise transients. *Fed. Proc.* 45:528, 1986.
36. Greenleaf, J.E., P.J. Brock, L.C. Keil and J.T. Morse. Drinking and water balance during exercise and heat acclimation. *J. Appl. Physiol.* 54:414–419, 1983.
37. Greenleaf, J.E. and B.L. Castle. Exercise temperature regulation in man during hypohydration and hyperhydration. *J. Appl. Physiol.* 30:847–853, 1971.
38. Guyton, A.C. *Textbook of Medical Physiology.* Philadelphia, PA: W.B. Saunders Co., 1981.
39. Guyton, A.C., A.E. Taylor and H.J. Granger. *Circulatory Physiology II: Dynamics and Control of the Body Fluids.* Philadelphia, PA.: W.B. Saunders Co., 1975.
40. Hagan, R.D., F.J. Diaz and S.M. Horvath. Plasma volume changes with movement to supine and standing positions. *J. Appl. Physiol.* 45:414–418, 1978.
41. Harrison, M.H. Effects of thermal stress and exercise on blood volume in humans. *Physiol. Rev.* 65:149–209, 1985.
42. Harrison, M.H., R.J. Edwards, L.A. Cochrane and M.J. Graveney. Blood volume and protein responses to skin heating and cooling in resting subjects. *J. Appl. Physiol.* 54:515–523, 1983.
43. Harrison, M.H., R.J. Edwards and P.A. Fennessy. Intravascular volume and tonicity as factors in the regulation of body temperature. *J. Appl. Physiol.* 44:69–75, 1978.
44. Hays, R.M. Dynamics of body water and electrolytes. In: *Clinical Disorders of Fluid and Electrolyte Metabolism,* edited by M.H. Maxwell and C.R. Kleeman. New York: McGraw-Hill, 1980, p. 1–6.
45. Hertzman, A.B. and I.D. Ferguson. Failure in temperature regulation during progressive dehydration. *U.S. Armed Forces Med. J.* 11:542–560, 1960.
46. Horowitz, M. and J.H. Adler. Plasma volume regulation during heat stress: albumin synthesis vs. capillary permeability. A comparison between desert and non-desert species. *Comp. Biochem. Physiol.* 75A:105–110, 1983.
47. Hubbard, R.W., W.T. Matthew, D. Horstman, R. Francesconi, M. Mager and M.N. Sawka. Albumin-induced plasma volume expansion: diurnal and temperature effects. *J. Appl. Physiol.* 56:1361–1368, 1984.
48. Hudlicka, O., B.W. Zweifach and K.R. Tyler. Capillary recruitment and flow velocity in skeletal muscle after contraction. *Microvasc. Res.* 23:201–213, 1982.
49. Jacobsson, S. and I. Kjellmer. Accumulation of fluid in exercising skeletal muscle. *Acta. Physiol. Scand.* 60:286–292, 1964.

50. Johnson, J.M., G.L. Brengelmann, J.H.S. Hales, P. Vanhoutte and C.B. Wenger. Regulation of the cutaneous circulation. *Fed. Proc.* 45:2841–2850, 1986.
51. Kirby, C.R. and V.A. Convertino. Plasma aldosterone and sweat sodium concentrations after exercise and heat acclimation. *J. Appl. Physiol.* 61:967–970, 1986.
52. Krebs, W. and M. Meyer. Blood findings in sweating procedures. *Z. Phys. Diaet. Ther.* 6:371–384, 1902–1903.
53. Landis, E.M. and J.R. Pappenheimer. Exchange of substances through the capillary walls. In: *Handbook of Physiology. Circulation.* Washington, D.C.: Am. Physiol. Soc., 1963, Sect.2, Vol. 2, Chapt. 29, p. 961–1034.
54. Loeppky, J.A., L.G. Myhre, M.D. Venters and U.C. Luft. Total body water and lean body mass estimated by ethanol dilution. *J. Appl. Physiol.* 42:803–808, 1977.
55. Lundvall, J., S. Mellander, H. Westling and T. White. Dynamics of fluid transfer between the intra- and extravascular compartments during exercise. *Acta. Physiol. Scand.* 80:31A–32A, 1970.
56. Lundvall, J., S. Mellander, H. Westling and T. White. Fluid transfer between blood and tissues during exercise. *Acta. Physiol. Scand.* 85:258–259, 1972.
57. Malt, R.A., C. Wange, Z. Yamazaki and T. Miyakuni. Stimulation of albumin synthesis by hemorrhage. *Surgery.* 66:65–70, 1969.
58. Maron, M.B., J.A. Wagner and S.M. Horvath. Thermoregulatory responses during competitive marathon running. *J. Appl. Physiol.* 42:909–914, 1977.
59. Mellander, S. On the control of capillary fluid transfer by precapillary and postcapillary vascular adjustments. *Microvascular Research.* 15:319–330, 1978.
60. Miles, D.S., M.N. Sawka, R.M. Glaser and J.S. Petrofsky. Plasma volume shifts during progressive arm and leg exercise. *J. Appl. Physiol.* 54:491–495, 1983.
61. Mohsenin, V. and R.R. Gonzalez. Tissue pressure and plasma oncotic pressure during exercise. *J. Appl. Physiol.* 56:102–108, 1984.
62. Mollison, P.L. *Blood Transfusion in Clinical Medicine.* Boston, MA: Blackwell Scientific Publications, 1983.
63. Moroff, S.V. and D.E. Bass. Effects of overhydration on man's physiological responses to work in the heat. *J. Appl. Physiol.* 20:267–270, 1965.
64. Nadel, E.R., E. Cafarelli, M.F. Roberts and C.B. Wenger. Circulatory regulation during exercise in different ambient temperatures. *J. Appl. Physiol.* 45:430–437, 1979.
65. Nadel, E.R., S.M. Fortney and C.B. Wenger. Effect of hydration on circulatory and thermal regulation. *J. Appl. Physiol.* 49:715–721, 1981.
66. Nadel, E.R., K.B. Pandolf, M.F. Roberts and J.A.J. Stolwijk. Mechanism of thermal acclimation to exercise and heat. *J. Appl. Physiol.* 37:515–520, 1974.
67. Nielsen, B., G. Hansen, S.O. Jorgensen and E. Nielsen. Thermoregulation in exercising man during dehydration and hyperhydration with water and saline. *Int. J. Biometerol.* 15:195–200, 1971.
68. Pandolf, K.B. Effects of physical training and cardiorespiratory physical fitness on exercise-heat tolerance: recent observations. *Med. Sci. Sports* 11:60–65, 1979.
69. Pappenheimer, J.R. and A. Soto-Rivera. Effective osmotic pressure of the plasma proteins and other quantities associated with the capillary circulation in the hindlimbs of cats and dogs. *Amer. J. Physiol.* 152:471–491, 1948.
70. Pirkle, J.C. and D.S. Gann. Restitution of blood volume after hemorrhage. *Am. J. Physiol.* 230:1683–1687, 1976.
71. Pitts, R.F. *Physiology of the Kidney and Body Fluids.* Chicago, IL: Year Book Medical Publishers, Inc., 1968.
72. Pivarnik, J.M., E.M. Leeds and J.E. Wilkerson. Effects of endurance exercise on metabolic water production and plasma volume. *J. Appl. Physiol.* 56:613–618, 1984.
73. Pressman, J.J., M.B. Simon, K. Hand and J. Miller. Passage of fluids, cells, and bacteria via direct communications between lymph nodes and veins. *Surg. Gynecol. Obstet.* 115:207–214, 1962.
74. Pugh, L.G.C.E., J.L. Corbett and R.H. Johnson. Rectal temperatures, weight losses and sweat rates in marathon running. *J. Appl. Physiol.* 23:347–352, 1967.
75. Renkin, E.M. Relation of capillary morphology to transport of fluid and large molecules: a review. *Acta. Physiol. Scand. (Suppl.).* 463:81–91, 1979.
76. Robertson, G.L., R.L. Shelton and S. Athar. The osmoregulation of vasopressin. *Kidney Int.* 10:25–37, 1976.
77. Rosenoer, V.M., M. Oratz and M.A. Rothschild. *Albumin Structure, Function and Uses.* New York: Pergamon Press, 1977.
78. Rowell, L.B. Cardiovascular aspects of human thermoregulation. *Circ. Res.* 52:367–379, 1983.
79. Saltin, B. Aerobic and anaerobic work capacity after dehydration. *J. Appl Physiol.* 19:1114–1118, 1964.

80. Saltin, B. Circulatory response to submaximal and maximal exercise after thermal dehydration. *J. Appl. Physiol.* 19:1125–1132, 1964.
81. Sawka, M.N., R.C. Dennis, R.R. Gonzalez, A.J. Young, S.R. Muza, J.W. Martin, C.B. Wenger, R.P. Francesconi, K.B. Pandolf and C.R. Valeri. Influence of polycythemia on blood volume and thermoregulation during exercise-heat stress. *J. Appl. Physiol.* 62:912–918, 1987.
82. Sawka, M.N., R.P. Francesconi, N.A. Pimental and K.B. Pandolf. Hydration and vascular fluid shifts during exercise in the heat. *J. Appl. Physiol.* 56:91–96, 1984.
83. Sawka, M.N., R.P. Francesconi, A.J. Young and K.B. Pandolf. Influence of hydration level and body fluids on exercise performance in the heat. *J. Am. Med. Assoc.* 252:1165–1169, 1984.
84. Sawka, M.N., R.W. Hubbard, R.P. Francesconi and D.H. Horstman. Effects of acute plasma volume expansion on altering exercise-heat performance. *Eur. J. Appl. Physiol.* 51:303–312, 1983.
85. Sawka, M.N., R.G. Knowlton and J.B. Critz. Thermal and circulatory responses to repeated bouts of prolonged running. *Med. Sci. Sports.* 11:177–180, 1979.
86. Sawka, M.N., R.G. Knowlton and R.G. Glaser. Body temperature, respiration and acid-base equilibrium during prolonged running. *Med. Sci. Sport. Exerc.* 12:370–374, 1980.
87. Sawka, M.N., R.G. Knowlton, R.M. Glaser, S.W. Wilde and D.S. Miles. Effect of prolonged running on physiological responses to subsequent exercise. *J. Human Ergol.* 8:83–90, 1979.
88. Sawka, M.N., M.M. Toner, R.P. Francesconi and K.B. Pandolf. Hypohydration and exercise: effects of heat acclimation, gender and environment. *J. Appl. Physiol.* 55:1147–1153, 1983.
89. Sawka, M.N., A.J. Young, R.P. Francesconi, S.R. Muza and K.B. Pandolf. Thermoregulatory and blood responses during exercise at graded hypohydration levels. *J. Appl. Physiol.* 59:1394–1401, 1985.
90. Scatchard, G., A.C. Batchelder and A. Brown. Chemical, clinical and immunological studies on the products of human plasma fractionation. VI. The osmotic pressure of plasma and of serum albumin. *J. Clin. Invest.* 23: 458–464, 1944.
91. Schloerb, P.R., B.J. Friis-Hansen, I.S. Edelman, A.K. Solomon and F.D. Moore. The measurement of total body water in the human subject by deuterium oxide dilution with a consideration of the dynamics of deuterium distribution. *J. Clin. Invest.* 29:1296–1310, 1950.
92. Sejersted, O.M., N.K. Vollestad and J.I. Medbo. Muscle, fluid and electrolyte balance during and following exercise. *Acta. Physiol. Scand.* 128 (Suppl 556):119–127, 1986.
93. Senay, L.C. Temperature regulation and hypohydration: a singular view. *J. Appl Physiol.* 47:1–7, 1979.
94. Senay, L.C. Effects of exercise in the heat on body fluid distribution. *Med. Sci. Sports* 11:42–48. 1979.
95. Senay, L.C., D. Mitchell and C.H. Wyndham. Acclimatization in a hot, humid environment: body fluid adjustments. *J. Appl. Physiol.* 40:786–796, 1976.
96. Senay, L.C. and J.M. Pivarnik. Fluid shifts during exercise. *Exerc. Sports Sci. Rev.* 13:335–387, 1985.
97. Senay, L.C., G. Rogers and P. Jooste. Changes in blood plasma during progressive treadmill and cycle exercise. *J. Appl. Physiol.* 49:59–65, 1980.
98. Shapiro, Y., R. Hubbard, C.M. Kimbrough and K.B. Pandolf. Physiological and hematological responses to summer and winter dry-heat acclimation. *J. Appl. Physiol.* 50:792–798, 1981.
99. Silva, N.L. and J.A. Boulant. Effects of osmotic pressure, glucose and temperature on neurons in preoptic tissue slices. *Am. J. Physiol.* 247 (Regul. Int. Comp. Physiol. 16):R335-R345, 1984.
100. Simionescu, M., L. Ghitescu, A. Fixman and N. Simionescu. How plasma macromolecules cross the endothelium. *News in Physiological Sciences.* 2:97–100, 1987.
101. Sjogaard, G. and B. Saltin. Extra and intracellular water spaces in muscles of man at rest and with dynamic exercise. *Am. J. Physiol.* 243 (Regul. Int. Comp. Physiol. 12):R271–R280, 1982.
102. Skelton, H. The storage of water by various tissues of the body. *Arch. Int. Med.* 40:140–152, 1927.
103. Smith, J.H., S. Robinson and M. Pearcy. Renal responses to exercise, heat and dehydration. *J. Appl Physiol.* 4:659-665, 1952.
104. Strauss, M.B., R.K. Davis, J.D. Rosenbaum and E.C. Rossmeisl. "Water diuresis" produced during recumbency by the intravenous infusion of isotonic saline solution. *J. Clin. Invest.* 30:862–868, 1951.
105. Strydom, N.B. and L.D. Holdsworth. The effects at different levels of water deficit on physiological responses during heat stress. *Int. Z. Angew. Physiol.* 26:95–102, 1968.

106. Takeda, Y. Hormonal effects on lymphatic transport of interstitial albumin in the dog. *Am. J. Physiol.* 207:1021–1029, 1964.
107. Wade, C.E. Response, regulation and actions of vasopressin during exercise: a review. *Med. Sci. Sports Exerc.* 16:506–511, 1984.
108. Waterfield, R.L. The effects of posture on the circulating blood volume. *J. Physiol.* (Lond.) 72:110–120, 1931.
109. Weissmann, G. and R. Claiborne., eds. *Cell Membranes: Biochemistry, Cell Biology and Pathology.* New York: HP Publishing Co., Inc. 1975.
110. White, A., P. Handler and E.L. Smith. *Principles of Biochemistry.* New York: McGraw-Hill Book Co., 1973.
111. White, J.C., M.E. Field and C.K. Drinker. On the protein content and normal flow of lymph from the foot of the dog. *Am. J. Physiol.* 103:34–44, 1933.
112. Wiggers, C.J. The circulation and circulation research in perspective. *In: Handbook of Physiology. Circulation.* Washington, DC: Am. Physiol. Soc., 1962, Sect.2, Vol. I, Chapt. 1, p. 1–10.
113. Wilkerson, J.E. and M.L. Gaddis. Apparent alterations in plasma volume with arm-position changes. *Fed. Proc.* 43:1016, 1984.
114. Wilkerson, J.E., B. Gutin and S.M. Horvath. Exercise-induced changes in blood, red cell, and plasma volumes in man. *Med. Sci. Sports* 9:155–158, 1977.
115. Zurovsky, Y., A. Shkolnik and M. Ovadia. Conservation of blood plasma fluids in hamadryas baboons after thermal dehydration. *J. Appl. Physiol.*57:768–771, 1984.

7

Effect of Gender, Circadian Period and Sleep Loss on Thermal Responses During Exercise

LOU A. STEPHENSON, Ph.D.

MARGARET A. KOLKA, Ph.D.

OUTLINE

INTRODUCTION

Many factors modulate thermoregulatory responses to exercise and/or passive heat stress. The effects of skin temperature, exercise posture, hydration state, heat acclimation and exercise training on thermoregulation are widely known (20, 55, 84, 127). Other factors affecting thermoregulation include the circadian cycle, hypobaria, menstrual cycle, sleep loss, blood doping and drug therapy. In this chapter, we will show how gender, the circadian period and sleep loss affect thermoregulatory responses. Although differences between males and females will be discussed, attention will be paid, particularly, to how the menstrual cycle affects temperature regulation in human females. Before the effects of gender, menstrual cycle, circadian cycle and sleep loss on thermoregulation are presented, a brief review of thermoregulatory control theory will be given.

In some thermoregulatory models, body temperature is thought to be regulated as a proportional control system in which the temperature of the preoptic area and anterior hypothalamus is the regulated temperature (18, 64, 67). Proportional control is defined as a graded response to a signal which increases (or decreases) in proportion to the intensity of the stimulus. Increasing body temperature leads to two efferent heat loss responses. Sweat secretion from the eccrine sweat gland is evaporated on the skin's surface and cools the body. Also, increasing body temperature causes vasodilation of cutaneous vascular beds, transferring heat from the core of the body to its surface. Heat is then transferred from the skin to the environment through radiation and convection, if skin temperature is greater than environmental temperature. In response to decreasing body temperature, eccrine sweating decreases and cutaneous blood vessels constrict. If body temperature continues to decrease, further vasoconstriction occurs which increases the effective thermal insulation. In addition, the metabolic rate increases and shivering produces heat.

The thermoregulatory effectors are not solely responsive to core temperature (T_c), but also respond to local or mean skin temperature ($\bar{T}_{sk}$; 20, 23, 55, 64, 66, 127). Temperature regulation has been described by both multiplicative (23, 69) and linear (20, 55, 66) models. In this chapter we will consider temperature regulation in terms of a linear additive model, using the general equation:

$$R - R_o = a(T_c - T_{co}) + b(\bar{T}_{sk} - \bar{T}_{sko}) \quad \text{(EQ. 1)}$$

where R is the thermoregulatory effector response, R_o is the initial effector response, T_{co} is the initial core temperature, $\bar{T}_{sko}$ is the initial skin temperature and a and b are proportional control constants.

In this chapter we will also discuss thermoregulatory control in terms of threshold and central thermosensitivity (slope or gain) changes, therefore it is important to understand what is meant by each concept or term. The threshold is that temperature above which the effector response is different from the baseline response at rest, in a thermal neutral environment.[1] The slope (gain) or thermosensitivity is considered here as the difference in the effector response per unit change in core temperature. In Figure 7-1, the solid line in the left panel shows that the effector response (R) increases as core temperature (T_c) increases. The dotted lines indicate that a given condition or factor has shifted the core temperature threshold for the onset of the thermoregulatory effector response to a different core temperature. A threshold shift is also called a set point shift. A change in threshold has classically (67) been interpreted as a central nervous system alteration in thermoregulatory control. In the right panel of Figure 7-1, the dotted lines indicate that a condition or a factor has changed the slope of the effector response to increasing core temperature from the solid line. A slope change may indicate that there has been a peripheral alteration, meaning that the effector has been modified locally, or that there has been a central modification which altered the effector response.

[1] The thermal neutral zone is that combination of environmental conditions in which humans regulate with little vasomotor or sudomotor activity and they are in a quasi-state of "thermal comfort" (36, 52).

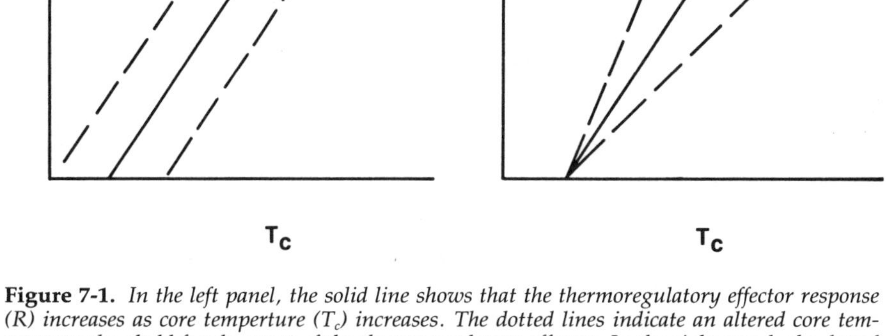

Figure 7-1. *In the left panel, the solid line shows that the thermoregulatory effector response (R) increases as core temperture (T_c) increases. The dotted lines indicate an altered core temperature threshold for the onset of the thermoregulatory effector. In the right panel, the dotted lines indicate altered central thermosensitivity (slope or gain).*

GENDER

Review of the Menstrual Cycle

The components of the female reproductive system—the ovaries, uterus and the hypophysial-hypothalamic tract are controlled and coordinated through the action(s) of various hormones. Figure 7-2 shows the response of estradiol, progesterone, follicle stimulat-

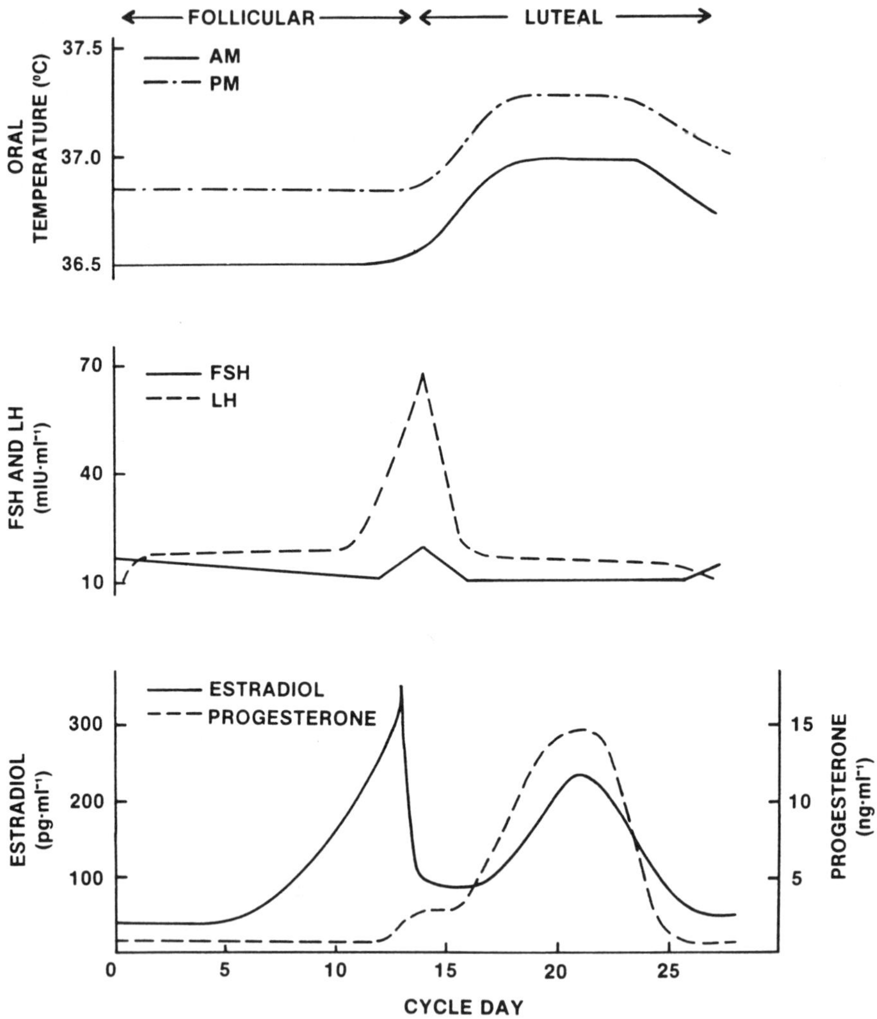

Figure 7-2. *Oral temperature, follicle-stimulating hormone (FSH), luteinizing hormone (LH), progesterone, estradiol and core temperature responses during the follicular and luteal phases of the menstrual cycle. In this schematic, menses occurs between day 0 and 5.*

ing hormone (FSH) and luteinizing hormone (LH) as well as oral temperature in both the morning and afternoon during a complete menstrual cycle. In this diagram, menses occurs on day 1 through day 5. By using the onset of menses (day 1) as a starting point, we can trace the coordinated events of a single menstrual cycle in a mature human female. Several follicles (ova) begin development under the influence of FSH produced in the anterior pituitary (79). During its development, a follicle produces and secretes increasing amounts of estrogen which stimulates the proliferation of the uterine endometrium (58, 79). Additionally, small quantities of progesterone are secreted. This portion of the menstrual cycle, dominated by the developing follicle, is referred to as the follicular phase. Almost all variability in the length of an individual menstrual cycle is related to alteration in the follicular phase length, which can range from 10–21 days (153).

A surge in LH and FSH (days 13–15 in Figure 7-2) from the anterior pituitary gland occurs in response to the release of luteinizing hormone releasing factor (LHRF) from the hypothalamus. LHRF secretion is in response to the increasing estrogen secretion from the developing follicle. FSH induces the production of LH receptors on the dominant follicle. The LH surge causes the dominant follicle to mature. The dominant follicle breaks through the ovarian wall (ovulation) some 16–24 hours after the LH peak. Following ovulation, the supporting structure of the follicle involutes to become the corpus luteum which produces large concentrations of progesterone. This increased concentration of progesterone induces secretory changes in the uterus during this part of the menstrual cycle which is called the luteal phase (79). LH and FSH decline during the luteal phase. If fertilization does not occur, the corpus luteum degenerates, resulting in a sharp decrease in progesterone secretion. FSH begins to increase about this time (days 26–27) and is followed by gradual increases in LH over the next few days (79). The increasing FSH, after progesterone declines, induces other follicles to begin development and the cycle repeats. Without the progesterone support, the uterine endometrium degenerates and menses begins.

Temperature Regulation During the Menstrual Cycle

Human females have a menstrual cycle rhythm in deep body temperature in which the core temperature is ~0.4° C higher during the luteal phase (90). This elevation in resting core temperature occurs during ovulatory cycles when the plasma concentration of progesterone is elevated. Increased progesterone release during the luteal phase has been associated with this increase in core temperature (90, 126). The general pattern of resting core temperature in a eumenorrheic (normal ovulatory menstrual cycle) female, as it relates to the cyclical rhythm in the reproductive hormones during the menstrual cycle, is presented in Figure 7-2 during both the morning and afternoon. The cause of the elevated core temperature during the luteal phase is unknown, but may be related to action(s) of the reproductive hormones (126) or perhaps to the elevated interleukin-1 activity in the luteal phase (24). This elevation in core temperature appears to be defended by a higher core temperature threshold for the onset of heat loss mechanisms (sweating and vasodilation) and lower core temperature threshold for the onset of heat conserving mechanisms (76, 77, 92, 143).

Although there are several ways to investigate whether the menstrual cycle affects thermoregulation, we will examine how the menstrual cycle affects the relationship between the thermoregulatory effector responses and core temperature during the initial stage of exercise or heat stress, when core temperature is changing rapidly (Figure 7-1). Specifically, the data will be examined for any alteration in threshold temperature or central thermosensitivity. It is also important to determine whether the menstrual cycle has a similar effect on sweating, cutaneous vasodilation and shivering (heat production) in order to determine whether thermoregulatory control has been modified by the menstrual cycle.

Convincing evidence has accumulated which indicates that the elevation in core temperature during the luteal phase occurs through a modification of the thermoregulatory set point. In the mid-1960's, Haslag and Hertzman (72) passively heated three women in the early follicular phase, at mid-cycle and again during the luteal phase elevation in core (oral) temperature. The onset of sweating was at a higher oral temperature (approximately 0.4° C) during the luteal phase. No alteration in thermosensitivity was reported. These authors reported no differences in the threshold for onset of forearm blood flow as measured by cutaneous opacity pulses between cycle phases. In another study of passive body heating (16), the threshold for onset of sweating occurred at an elevated core temperature during luteal phase experiments. As in the findings of the earlier study (72), no alteration in the thermosensitivity of sweating was observed between cycle phases (16). Recently (76), a study of passive cooling and passive heating clearly demonstrated that the elevated core temperature during the luteal phase was defended by both an earlier onset of shivering, hence heat production, and a higher core temperature threshold for initiation of heat loss mechanisms (sweating and cutaneous vasodilation).

An evaluation of behavioral thermoregulatory consequences of the luteal phase elevation in core temperature has been conducted

(26, 88). After adapting the skin of subjects to a temperature of 36° C, Kenshalo (88) observed a greater perception of skin cooling in females tested in the luteal phase, compared to the follicular phase. He proposed that this perceptual difference in decreasing skin temperature resulted from the higher regulated core temperature during the luteal phase. He further hypothesized that this difference in perception between menstrual cycle phases was the result of progesterone-induced vasodilation. Cunningham and Cabanac (26) evaluated the perception of a cutaneous challenge (i.e. water bath exposure to 20° or 40° C) during the follicular and luteal phases of the menstrual cycle. They reported that females perceived thermal comfort at a higher level of core temperature (37.6° C vs 37.2° C) during the luteal phase of the menstrual cycle. These authors cited their results as evidence for a behavioral defense of the "elevated set point temperature" during the luteal phase. The differences in thermal perception between the follicular and luteal phases may be partially due to the slightly higher skin temperature during the luteal phase (92, 146).

The threshold or thermosensitivity of heat loss responses associated with the stress of exercise during the menstrual cycle has recently been studied (76, 77, 78, 92, 143). Stephenson *et al.* (143) evaluated the change in forearm cutaneous blood flow and forearm sweating to increasing esophageal temperature, during cycle exercise at 60% $\dot{V}O_2$ peak in a 35° C environment in four women, during both the follicular and luteal phase of their menstrual cycle. The core temperature threshold for sweating and cutaneous vasodilation was shifted to a higher core temperature during the luteal phase. A typical rightward shift in core temperature threshold for sweating onset is shown for an individual subject in Figure 7-3. Forearm sweating is graphed as a function of esophageal temperature during the early stages of exercise in a female subject. The data show that there is a parallel shift (rightward) in the sweating to core temperature relationship during exercise in the luteal phase experiment as compared to the follicular phase. This rightward shift indicates that there is an increased core temperature threshold for the onset of sweating in the luteal phase. The increased threshold temperatures for sweating and vasodilation, observed in this study (143), were affected by the normal circadian rhythm in resting core temperature, which will be discussed in a later section. There was no change in thermosensitivity (slope) for either forearm sweating or forearm blood flow between the menstrual cycle phases. However, an increased thermosensitivity of both chest sweating and forearm and hand blood flow was reported in the studies of Hessemer and Bruck (77). They studied ten female subjects during 70% $\dot{V}O_2$ peak cycle exercise at

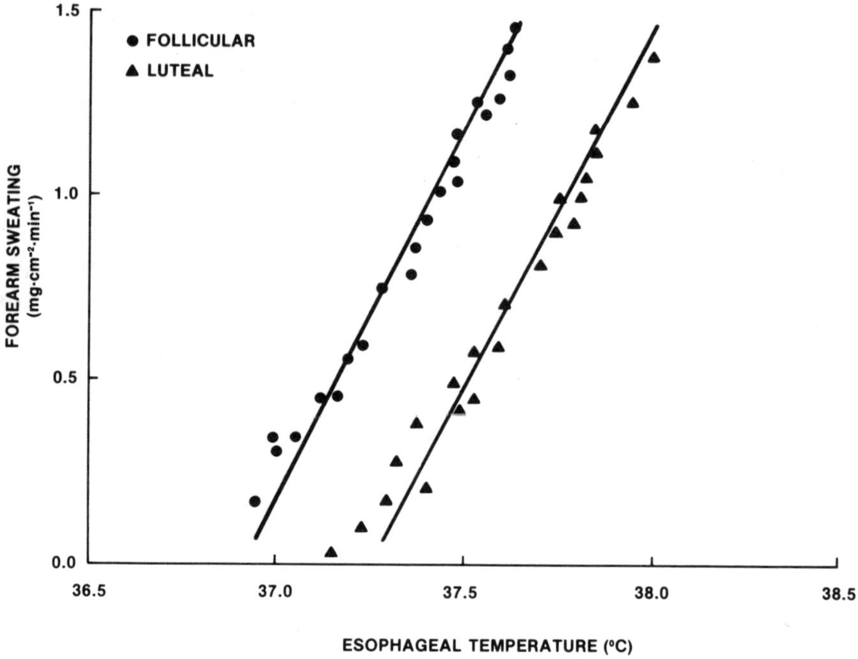

Figure 7-3. *Forearm sweating rate plotted against esophageal temperature (T_{es}) during the exercise transient during both follicular and luteal phase experiments.*

18° C during the nighttime hours (0300 h). These authors reported an average 0.47° C increase in the core temperature threshold for cutaneous vasodilation and chest sweating during luteal phase experiments compared to follicular phase experiments. These investigators also verified that core temperature (77, 143) and plasma progesterone concentration (77) were elevated in the luteal phase.

One recent study has reported the effects of menstrual cycle phase on the control of local sweating to changing core temperature in eight females during both moderate and heavy exercise in hot and extremely hot ambient conditions (92). The control of forearm sweating to increasing core temperature was systematically evaluated during moderate (35° C) and severe (50° C) heat stress during passive heating (+0.8° C core temperature) and moderate (60% $\dot{V}O_2$ peak) and severe (80% $\dot{V}O_2$ peak) exercise in eight women (92). In all conditions studied, the onset of local sweating was shifted to a higher core temperature threshold in the luteal phase. The increased core temperature threshold averaged 0.53° C for all temperatures, exercise intensities and times of day evaluated.

Another study, evaluating peripheral cutaneous blood flow,

demonstrated a higher core temperature threshold for finger vasodilatory onset associated with the luteal phase of the menstrual cycle (78). The control of finger blood flow during low and high (40% and 70% $\dot{V}O_2$ peak) intensity cycle exercise at 20° C was studied in follicular and luteal phase experiments (78). The core temperature threshold for finger vasodilation was increased in the luteal phase, compared to the follicular phase. Thermosensitivity was not different between the follicular and luteal phases. However, thermosensitivity of finger blood flow was reduced during exercise at 70% $\dot{V}O_2$ peak compared to exercise at 40% $\dot{V}O_2$ peak. This reduction in thermosensitivity may have been the result of the higher level of sympathetic activity, associated with the higher exercise intensity (84). Finger blood flow does not respond to increasing core temperature the same as other skin surfaces, such as the forearm, in that the hand does not appear to possess "active vasodilation" (20, 127). That is, any increase in blood flow to the hand in response to body heating is solely the result of decreased vasoconstrictor activity.

To summarize, the data presented thus far have indicated that the menstrual cycle does affect thermoregulation (16, 72, 76, 77, 78, 92, 143).The data indicate that there is an increased core temperature threshold for onset of thermoregulatory effector responses during the luteal phase (16, 72, 76, 77, 92, 143), with one report (77) of increased central thermosensitivity. Furthermore, there was an earlier onset of shivering during the luteal phase (76). The parallel changes in sudomotor and vasomotor responses (76, 77, 143) indicate that a prominent feature of the human menstrual cycle is a central alteration in thermoregulatory control, such that the thermoregulatory set point is increased in the luteal phase.

Menstrual Cycle, Performance

There have been many other studies which have taken a different approach in determining whether the menstrual cycle affects thermoregulation. The analysis of the data from these studies included determining whether there was a difference in sweating rate, core temperature, or exposure time between phases of the menstrual cycle during a given exercise task or heat stress. These studies are presented in the following section.

The menstrual cycle has been reported to exert little (13, 135) or no effect (38, 40, 44, 46, 82, 154, 155) on the "work-heat tolerance" of women (34). The term "work-heat tolerance" should not be confused with the description of thermoregulatory control presented in the previous section. In this context, Drinkwater's conclusion (34) in regard to the "work-heat tolerance" cannot be disputed. Some of the studies previously reviewed will be re-examined to determine why the investigators reported that the menstrual cycle did not affect temperature regulation.

Avellini *et al.* (13) studied four women during low intensity exercise both before and after heat acclimation in a hot, humid environment. The female subjects were tested in both pre-ovulatory and post-ovulatory experiments; surprisingly, resting core temperature was actually lower in the post-ovulatory experiments after heat acclimation. The authors reported a slightly higher heat storage after ovulation before heat acclimation which resulted from a delayed onset time for sweating. There was no difference in the core or skin temperatures at the onset of sweating between pre- and post-ovulatory experiments. After heat acclimation, increased heat storage in the post-ovulatory experiments was no longer apparent. There have been other studies which failed to demonstrate an elevated resting core temperature in the luteal phase (44, 46, 82). One (82) reported elevated serum levels of estradiol and progesterone; however, the elevated core temperature which is the critical defended variable (76, 77, 92, 143) during the luteal phase was overlooked (13, 44, 46, 82, 154, 155). Scrutinization of two other studies (44, 46) which compared pre- and post-ovulatory phases during low intensity exercise in the heat, revealed that data from only one study are actually presented. Their analysis of the end-exercise (last point measured) data suggested that "no significant thermoregulatory differences" were associated with menstrual cycle phase (44, 46). These authors, in effect, were evaluating "work-heat tolerance" during environmental heat stress and there was no difference between pre-ovulatory and post-ovulatory experiments. The unchanging pre- to post-exercise temperature data between menstrual cycle phases support their conclusion. The absence of the elevated core temperature at rest in the luteal phase combined with the failure to analyze definitive effector function during changing core temperature further support this conclusion.

Gender, Control System

Before comparing thermoregulatory differences between genders, it is very important to eliminate or standardize any factor which independently affects thermoregulation, because each could obscure or exacerbate gender differences. For example, Gonzalez *et al.* (56) have reported that maximal aerobic power is the single major determinant of differences in sweat secretion produced by increasing mean body temperature in a large group of individuals (of both sexes and varying ages). Clearly, before thermoregulatory comparisons can be made between genders, maximal aerobic power must be a controlled variable within the study. This section will describe those studies which have investigated differences in thermoregulation between males and females by analyzing core temperature threshold changes for the onset of thermoregulatory effectors or changes in central thermosensitivity.

Cunningham and colleagues (27) evaluated the thermoregulatory responses to air temperature transients in three men and three women. The general responses of the men and women to the thermal stress were similar. However, the onset of sweating was at a higher core temperature threshold for the female subjects. The authors noted that the females were tested in varying phases of their individual menstrual cycle, but menstrual cycle phase was not a controlled variable. The female subjects were also of a lower fitness level than the male subjects (C.B. Wenger, personal communication).

Gonzalez *et al.* (56) later reported that there was no apparent difference in sweat gland responses between males and females of 15–50 years of age during exposure to timed increases and decreases in ambient temperature and dew point temperature. However, their report (56) indicated that females have a greater change in heat production for a given change in mean body temperature than males. The eumenorrheic females were studied in the follicular phase of their menstrual cycles and post-menopausal women were studied at their convenience.

Roberts *et al.* (124) studied males and females during cycle exercise (60–70% $\dot{V}O_2$ peak) before and after a training program and heat acclimation. The females exhibited higher thresholds for sweating and forearm blood flow at all phases of the study. The authors did not control for menstrual cycle phase or aerobic fitness.

In contrast, no gender differences in either the core temperature thresholds for onset of vasodilation and sweating or thermosensitivity of forearm blood flow and sweating were reported in four male and four female subjects during 60% $\dot{V}O_2$ peak cycle exercise at 30° C (94). The females were all studied during the early follicular phase (days 4–7) of the menstrual cycle.

There are also data which indicate that thermoregulatory skin blood flow is compromised in both males and females, presumably to maintain blood pressure regulation (21, 114, 146). During intense exercise in a hot environment a break point occurs in the forearm blood flow to core temperature relationship; such that the increase in forearm blood flow is attenuated after core temperature increases approximately 1° C (21). This relative vasoconstriction of the skin, or break point in the forearm blood flow to core temperature relationship, has been observed in both males (21, 114) and females (146). In males (21, 114), this break point occurred at a core temperature of approximately 38° C and forearm blood flow of 12–15

$ml \cdot 100\ ml^{-1} \cdot min^{-1}$. In females (146), the break point occurred at a core temperature of 37.5–38.0° C and forearm blood flow of 12–15.5 $ml \cdot 100\ ml^{-1} \cdot$ min-1. These data indicate that skin blood flow is altered when core temperature is elevated during exercise in a warm environment in a similar fashion in men and women. This implies that blood presure regulation takes precedence over thermoregulation in both genders. It should be noted that the relative vasoconstriction in women occurred at a significantly higher core temperature in the luteal phase than in the follicular phase (146).

In summary, there are few thermoregulatory differences between men and women, especially women in the follicular phase of the menstrual cycle, if the studies are appropriately controlled for factors which independently affect thermoregulation. There may be differences in thermoregulatory control between men and women in the luteal phase of the menstrual cycle, but there have been no systematic studies investigating this possibility. One notable difference between genders is the greater thermosensitivity in heat production during changing mean body temperature in women, reported by Gonzalez *et al.* (56).

Gender, Performance Differences

Other investigators have studied the effect of gender on thermoregulation using a different method than analyzing whether the set point or the central thermosensitivity was altered. These investigators determined whether the sweating rate or elevation in core temperature induced by a given exercise task or heat exposure was similar in men and women.

Several early studies (42, 68, 164) concluded that females respond to heat stress in a manner qualitatively similar to males; however, females appear to be less heat tolerant than male subjects during both passive heating or exercise in the heat. Several review articles are available summarizing these early efforts along with some more recent studies (33, 34, 37, 73, 75, 87, 115, 130).

The comparison of male and female subjects' responses to environmental heat stress during low intensity treadmill exercise, both before and after heat acclimation, has been reported (14, 136, 137, 138). In one study (14), male and female subjects of unequal fitness levels were studied before and after acclimation to dry heat. Menstrual cycle phase was not considered a critical variable, and was not controlled. Briefly, it was reported that women can acclimate to heat and the responses of female subjects to heat stress are dependent on fitness level. In the data reported by Avellini *et al.* (14), the change in core temperature (ΔT_{re}) was not different during exercise in the heat between the male and female subjects; although, the females exercised at 36% $\dot{V}O_2max$ and the males walked at 29% $\dot{V}O_2max$. Furthermore, there was no difference in whole body sweating between the men and women during heat exposure, post-heat acclimation. However, a breakdown of this same data, in which the less fit male subjects were matched with the more fit female subjects (136), indicated no gender difference in sweating rates. But a similar sweating rate, coupled with higher core temperatures in the female subjects, suggested less efficient evaporation in the female subjects. This was postulated to be due to the larger surface area to mass ratio (136). A comparison of six different environmental conditions evaluated in this study (137) suggested that during humid heat, females had lower core temperatures than the males and conversely, during extreme dry heat, the female subjects had higher core temperatures during steady-state exercise. Shapiro (138) concluded from similar data that women have an elevated set point compared to men for the onset of sweating during low intensity exercise, even though core temperature was lower in women than men during exposure to humid heat. This conclusion was made without an evaluation of the core temperature threshold for onset of sweating. It was suggested that the hormonal changes associated with the female reproductive cycle do not impact on temperature regulation in female subjects (136). Unfortunately, the authors failed to control for menstrual cycle phase, elevation in core temperature or fitness level, and such conclusions are wholly based on an end point temperature. In addition, these authors did not evaluate the effector response to changing core temperature, therefore any conclusions regarding temperature regulation are not appropriate. Another evaluation of the responses of male and female subjects to dry heat stress, during low intensity cycle exercise, reported that females had a greater sweating output than that of male subjects based on pre- to post-exercise core temperature and whole body sweating (81). Again, no attempt to control for cycle phase or to evaluate effector response was made; therefore, conclusions are based solely on descriptive data analysis.

Another group has evaluated temperature regulation in men and women (44, 45, 46) during exposure to both humid and dry heat. Subjects matched for fitness level were studied during low intensity treadmill exercise, before and after a heat acclimation program. Generally, the responses of the subjects were the same to the heat stress, with the exception of greater heat storage in the females before the heat acclimation program. The authors suggested (44, 46) that there were no gender differences in the responses to heat stress. A further evaluation from the same group reported that female subjects had a greater sweating efficiency than male subjects during low intensity treadmill exercise in the humid heat (45). Extrapolation from data presented by these authors would indicate that the sensitivity of sweating to changing core temperature was suppressed in the female subjects, compared to the male subjects during exercise in these humid conditions. This is indicated by a faster increase in core temperature coupled with a lower sweating rate. Thus, what has been described to be an enhanced sweating efficiency of the female subjects during exposure to humid conditions, may actually be a decreased response to increasing core temperature, perhaps resulting from a hidromeiotic (wet skin suppressed further sweat secretion) effect (57). This conclusion is based on our attempt to evaluate thermoregulatory control from presented data, although our conclusion is not in agreement with the conclusions drawn by the authors of the study (45).

The performance of low intensity exercise during dry or wet environmental heat stress elicits similar responses in male and female subjects based on steady-state data analysis. When male and female subjects are matched for fitness level, few differences are apparent in steady-state or post-exercise core temperature or sweating rate. A critical concern at this point is the absence of attention paid to the evaluation of homeostatic control of the core temperature. As pointed out above, different conclusions can be made from the same data depending on how the data are analyzed. Neglecting to control for menstrual cycle phase adds variability to any analysis of those data, especially if responses are to be compared between the genders. It has been presented in an earlier section that heat exchange varies with menstrual cycle phase. How is it that numerous investigations claim no gender differences even when the individual response of the female subjects varied depending on menstrual cycle phase? We suggest that the type of data analysis, perhaps even the perturbation of the system itself, combined with the inherent variability in biological data, obscures these facts.

Fluid Volume Status Affecting Temperature Regulation

Harrison (70) concluded in a recent review that "currently there is no convincing evidence that intravascular volume responses to thermal stress and exercise differ between males and females". Chapters 3, 4 and 6 of this text describe in greater detail the interactions of intravascular volume and thermal stress of exogenous or endogenous origin.

Conflicting descriptions of plasma volume changes during exercise and heat stress in females have been reported by many investigators (35, 40, 51, 132, 135, 154, 155). Hemodilution has been reported during treadmill walking at 50% $\dot{V}O_2$ peak (155) and 30%

$\dot{V}O_2$ peak (132) in women exposed to hot environments. However, hemoconcentration has also been observed during treadmill walking at 30% $\dot{V}O_2$ peak (35) in a hot environment. During cycling exercise, hemoconcentration has been observed in female subjects exercising in warm and hot environments (40, 51, 144, 145). Some of these differences in plasma volume responses have been attributed to body positional changes (70).

Senay has reported no effect of the menstrual cycle on exercise hemoconcentration (40) and that hemoconcentration occurred less rapidly in luteal phase compared to the follicular phase (51). It has also been reported (144) that the contracted plasma volume in the luteal phase (145, 151, 162) was associated with decreased fluid shifted out of the vascular compartment in the luteal phase during an exercise heat stress. The studies listed above estimated plasma volume changes only after the heat or exercise stress. A recent study (144) estimated plasma volume changes at frequent intervals during passive heat stress or exercise heat stress. It was reported that the hemoconcentration during passive heat stress is less in the follicular than luteal phase experiments (144). It was suggested that there may be a lower limit to hemoconcentration during passive heating or exercise expressed as a lower critical plasma volume (70, 144), or reduced central venous pressure (127). In support of this observation, four of five women experienced dizziness, headache or nausea as the plasma volume contraction approached 2.5 ℓ (144). Recent evidence reporting changes in hormones related to fluid volume homeostasis during the menstrual cycle (15, 41, 60, 86, 125, 141) suggests that the understanding of fluid volume maintenance, during heat stress and/or exercise at different phases of the menstrual cycle, is incomplete and necessitates further study.

Summary

Experimental evidence, describing how the menstrual cycle affects heat loss or heat conservation responses to increasing core temperature, is straightforward and clear. These observations (16, 72, 76, 77, 78, 92, 143) support the concept of a shift in the thermoregulatory set point during luteal phase of the menstrual cycle. The evaluation of pre- to post-exercise responses or "steady-state" comparisons indicate that the menstrual cycle does not affect sweating rate, core temperature increases or work-heat tolerance times (13, 38, 40, 44, 46, 82, 135, 154, 155); although, these types of studies may not adequately present the quantitative effects of homeostatic control of body temperature. The changing phases of the menstrual cycle must be a controlled variable whenever females are to be evaluated during thermal stress of endogenous or exogenous origin.

There is very little data to support the theory that gender affects thermoregulatory capability if there is adequate control of independent factors which affect thermoregulation, such as maximal aerobic power, heat acclimation, and even phase of the menstrual cycle. However, one report has indicated that central thermosensitivity of heat production to changing mean body temperature is greater in women than in men (56).

CIRCADIAN PERIOD

Control of Circadian Rhythms

Nearly all physiologic systems in eukaryotic organisms (128, 134) exhibit a circadian periodicity approximating 24 h, which is cued by external rhythmic influences, such as the light/dark cycle (28, 29, 32, 47, 148, 160), sleep/wakefulness cycle, social activity (8, 159) or the feeding/fasting cycle (148). These external rhythmic influences provide time cues which are called Zeitgeber ("time-giver"; 2, 3, 5) and synchronize many circadian rhythms within the organism. Circadian and sleep patterns for core temperature, plasma growth hormone, plasma cortisol and potassium excretion are shown in Figure 7-4. The time of day presented on the horizontal axis represents 0600 h as 6. The circadian rhythms of many physiologic systems[2] are coordinated within the organism (122) by pacemakers, which apparently prepare the organism for a predicted environmental change (107, 128). Normally, Zeitgeber entrain the oscillating systems to a 24 h rhythm by some method of resetting the pacemakers.

In the absence of Zeitgeber, many of the circadian rhythms become free-running (22), independent of the 24 h rhythm (2, 5, 61, 122). In humans, the free-running core temperature rhythm has an average periodicity of 25 h (28, 158), which is an indication that the circadian pacemaker controlling core temperature has an inherent rhythm of 25 h. For the most part, the sleep/wakefulness cycle also follows the same 25 h periodicity in the absence of Zeitgeber. Wever (159) has reported that in 30% of the human experiments analyzed (6, 157, 158), the sleep/wakefulness cycle becomes free-running and is controlled to an average 33 h. However, if Zeitgeber are withheld for at least two months, eventually the core temperature and the sleep/wakefulness rhythms will free-run in different periods (106). Wever (157, 158) originally interpreted this internal desynchronization of the core temperature and sleep/wakefulness rhythms in humans as evidence that there are, at least, two pacemakers, one

---

[2] A comprehensive review of circadian rhythms in many physiologic systems (61, 62, 63, 101, 113, 163) will not be presented in this chapter.

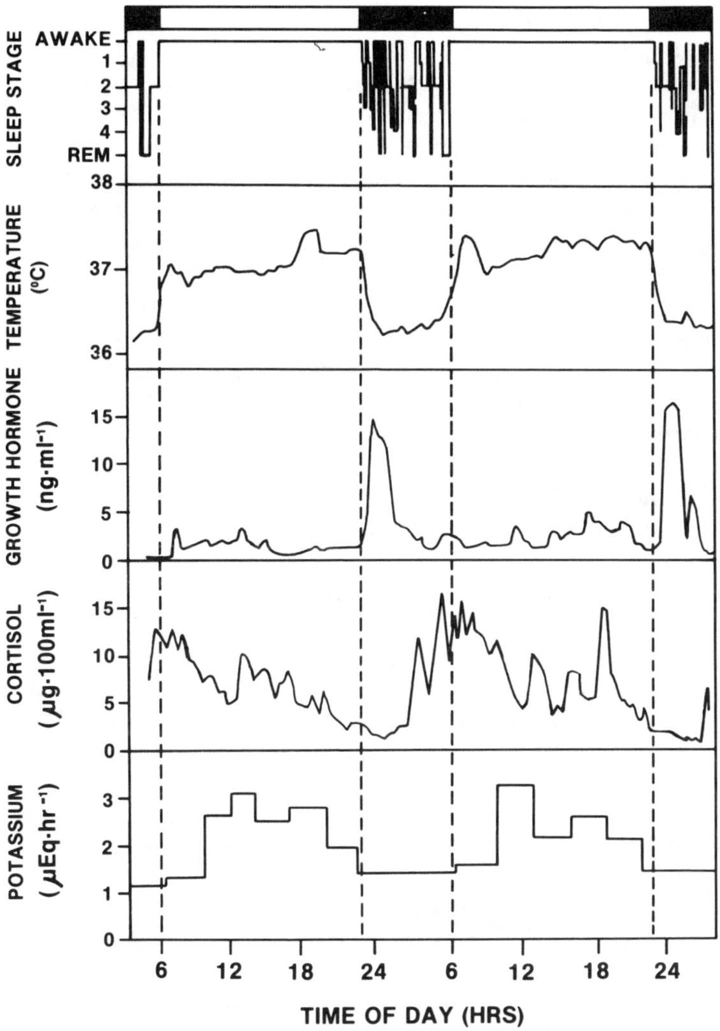

Figure 7-4. *Circadian and sleep patterns of core temperature, plasma growth hormone, plasma cortisol and potassium excretion. Reprinted, by permission of the* New England J. Med. *309:469–476, 1983.*

controlling core temperature (I) and another controlling activity (II). Further evidence to support the two pacemaker theory came from reports that different Zeitgeber entrain individual oscillators (the oscillators themselves are likely controlled by one of the pacemakers). For example, urinary excretion of electrolytes is entrained by the feeding cycle (112), while temperature is entrained by the light/dark cycle (29, 148).

Animal studies have identified the location of a circadian pacemaker (104, 142) for many functions in the suprachiasmatic nucleus (SCN) of the hypothalamus. Further support for the SCN, as being a circadian pacemaker, has come from 1) experiments which have identified a direct neural connection from the retina to the SCN (105), 2) in vitro studies of SCN slices which reported self-sustaining rhythmicity (83), and 3) the observed circadian rhythm in 2-deoxyglucose metabolism of the SCN (149). Yet, the circadian rhythm in core temperature persisted in squirrel monkeys which had an SCN lesion (48), even though the circadian rhythmicity of drinking and activity was disrupted. The retention of the circadian rhythm in core temperature, in the squirrel monkey after the SCN had been destroyed, was interpreted as further evidence that there is more than one pacemaker (48, 106). The SCN, apparently, is the location of the activity rhythm, but it may not contain the pacemaker for core temperature.

Extensive data from human circadian studies has led to further development (53, 96, 106, 113) of the two pacemaker theory describing the operation of circadian cycles originally purposed by Wever (157). Briefly, the two pacemakers have been termed X and Y and each pacemaker coordinates many secondary oscillators. The interrelationship of the two pacemakers and the secondary oscillators controlled by each pacemaker is shown in Figure 7-5. The anatomical location of pacemaker X is unknown, but it controls the circadian rhythm in core temperature, plasma cortisol concentration, urine potassium excretion and rapid eye movement (REM) sleep. The Y pacemaker is located in the SCN and controls skin temperature, growth hormone concentration, urine Ca^{+2} excretion and slow wave sleep. The two pacemakers are mutually coupled (106) as the length of the free-running rhythms of both change after internal desynchronization. The X pacemaker is the stronger of the two pacemakers as its inherent rhythm is closer to the 24 h cycle than the Y pacemaker (106). The coupling strength ratio of the X and Y pacemakers has been estimated as 4:1 (96, 106). Furthermore, interaction between the two pacemakers continued even after internal desynchronization. Minors and Waterhouse (102) presented very interesting data from a single subject that suggested the 'temperature' oscillator was stronger than, and could entrain the 'activity' oscillator. The female subject studied was placed on a 21 h cycle for four days. Her meals, sleep/wakefulness cycle and urinary constituents followed a 21 h rhythm, but her body temperature free-ran in a 25 h rhythm. After four days she was asked to estimate time without Zeitgeber. During the next few days her meals, sleep/wakefulness and urinary rhythms continued on a 21 h rhythm, but body tem-

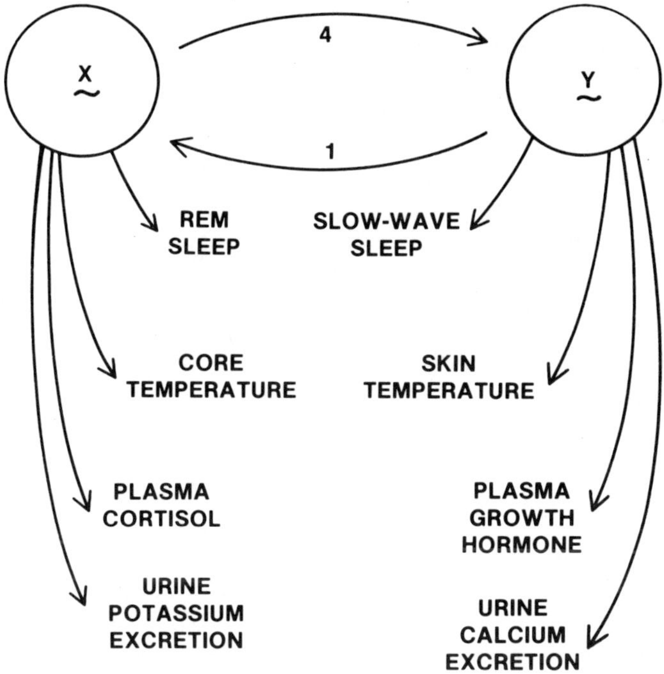

Figure 7-5. *Schematic of a two pacemaker theory for circadian rhythm. The X pacemaker is inherently four times more powerful than the Y pacemaker. Reprinted, by permission of the Federation of American Societies for Experimental Biology from* Federation Proc. *42:2802–2808, 1983.*

perature followed a 25 h rhythm. After four days without a time cue, the 21 h rhythms came into phase with the 25 h rhythm. After that, all rhythms had a 25 h period. Minors and Waterhouse (103) suggested that the 'temperature' oscillator had entrained the 'activity' oscillator.

Although the model of circadian rhythm control discussed above adequately describes much of the experimental data, there is new evidence which suggests that this model is not entirely accurate. In a recent study, Czeisler *et al.* (29) have reported that bright light directly entrains the X pacemaker. They have suggested that this observation may make it necessary to revise current circadian control theory (53, 96), which describes bright light as Zeitgeber for the Y pacemaker, which then is coupled to the X pacemaker.

Temperature Regulation and the Circadian Cycle

In addition to extensively studying the circadian cycle, Jurgen Aschoff has also interpreted how thermoregulation must be affected by the circadian rhythm in core temperature (6, 11). Although he credited Charles Richet with first deducing (1898, as referenced in 6) that the daily rhythm in body temperature was the result of a variation within the central nervous system, Aschoff (4) postulated that the diurnal oscillation in core temperature was regulated by a circadian shift in the set point of the thermoregulatory system.

The regulated circadian rhythm in core temperature must be the result of circadian changes in heat production, heat loss or both. Figure 7-4 shows the relationship between the sleep cycle and the circadian rhythm in core temperature for one individual. Originally, it was assumed that the circadian rhythm in core temperature was simply a function of state of activity. This assumption was mainly the result of the confounding effect of both the activity and core temperature rhythm being entrained to a 24 h period. However, Aschoff and his colleagues reported that the core temperature rhythm changed approximately two hours before the daily change in activity in humans (10). Subsequently, a desynchronization of the two rhythms was observed (9), which clearly indicated that the core temperature rhythm was not contingent upon the activity rhythm. In a normal female subject, however, heat production as measured by oxygen uptake, did indeed have a circadian rhythm which was very similar to that of core temperature (as reported in reference 12). Krieder *et al.* (95) had reported that oxygen uptake was 15% greater in the afternoon than in the morning and Little and Rummel (98) reported that the acrophase of metabolic rate occurred at 2045 h during rest. A more recent study (one male subject) verified the circadian rhythm in oxygen uptake with the acrophase occurring at 1800 h and a trough occurring at 0600 h (123). Again, core temperature and heat production had very similar circadian rhythms. In addition to the data already presented, other observations of the desynchronization of the activity and core temperature cycles (6, 102, 157,159) provided very strong evidence that heat production itself did not explain the circadian rhythm in core temperature.

In 1964, Bloch (17) reported that there were diurnal changes in the skin temperatures of the limbs in patients, and confirmed that this rhythm was also apparent in two normal individuals. Bloch noted that the diurnal change in limb temperature represented changes in skin blood flow. Sasaki and Carlson (131) observed a higher mean skin temperature in four subjects in the afternoon (1330–1630 h) than in the morning (0630–0930 h). Toe blood flow was also increased to a higher level in the afternoon than in the morning (131). Kaneko *et al.* (85) confirmed that limb (forearm) blood flow did vary throughout the day. In resting men, forearm blood flow reached a peak between 1600 and 2000 h with the lowest forearm blood flow measured at 0800 h. No experiments were conducted between 2400 and 0800 h in this study. The circadian rhythm in forearm blood flow persisted during recovery from a 2-min period of sustained submaximal contraction of the forearm muscles. The circadian forearm blood flow, resting heart rate and oral temperature rhythms were similar in magnitude and phase. Kaneko *et al.* (85) also reported that the circadian rhythm in forearm blood flow persisted during vasoconstriction resulting from lower body negative pressure. They postulated that the circadian rhythm in forearm blood flow was due to a change in the set point of peripheral resistance. Other investigators have confirmed that resting dry heat loss changes during the circadian cycle through variation in core to skin conductance (6, 11, 140).

The earliest reports of a circadian cycle in sweating came from the observation that sleep was associated with sweating (31, 97). Day (31) suggested that sleep led to a reduction of the thermostatic set point, while Kuno (97) postulated that sleep decreased the "tonus of a cerebral sweat-inhibitory center thereby releasing the activity of the sweat-excitatory center". Geschickter *et al.* (54) realized that the, so called, sleep-sweating was actually the result of the regulation of body temperature which was linked to the usual sleeping hours. They credited Hammel *et al.* (65, 66) with hypothesizing that differences in core temperature between sleep and wakefulness were consistent with a reduction of the thermoregulatory set point during sleep. Geschickter *et al.* (54), however, made the distinction that the theoretical reduction in set point at night was only associated with sleep and was not dependent upon sleep. A very interesting suggestion made by Geschickter *et al.* came from the observation that sweating occurred during napping in the afternoon, even though the sweating was not enough to decrease core temperature. They suggested that the sleep-sweat linkage may be a "modulating influence that tends to realign the temperature cycle to new sleeping habits". Although it is now known that the core temperature and sleep cycles do desynchronize in the absence of Zeitgeber (10, 30, 157), the suggestion that sleep-sweating realigns the core temperature rhythm may be explained by a coupling link between the X and Y pacemaker.

Further support for the hypothesis that the thermoregulatory set point was lower late at night than in the afternoon (54) came from the observations that sweating began at a lower internal temperature in the early morning than in the late afternoon or evening (25, 43, 150).

One method to critically assess whether there is a circadian shift in the regulated temperature is to measure the thermoregulatory effector responses to a perturbation in the thermal drive. Wenger *et al.* (156) studied the heat dissipation responses to an endogenous heat load (exercise) during the early morning (0400–0530 h) and afternoon (1200–1630 h). They reported a circadian rhythm in the threshold for initiation of both sweating and cutaneous vasodilation, although there was no significant change in the sensitivity of either heat loss effector to changing core temperature. They concluded that the lower esophageal temperature threshold for sweating and vasodilation, which occurred in the early morning compared to the afternoon, accounted for much of the circadian rhythm in core temperature.

A second study from the same laboratory (147) measured the thermoregulatory effector responses to an endogenous heat load (exercise at 65% $\dot{V}O_2$ peak) at six different times of the day and night. Each experiment on a subject was done on a separate day and there were, at least, three days between experiments, if a subject was studied during the hours in which he normally slept. The esophageal temperature thresholds for onset of cutaneous vasodilation and sweating during exercise had a circadian rhythm which was very similar to the resting esophageal temperature (Fig 7-6). The slope of the forearm blood flow to esophageal temperature was significantly lower at 2400 h than at 0400 h, but there was no significant variation in the slope of sweating to esophageal temperature. Since sweating is primarily controlled by the thermoregulatory center and cutaneous vasodilation is affected by thermoregulation and blood pressure regulation, it was postulated that this slope variation could be the result of some confounding variable. It was concluded that the regulation of core temperature, as determined by threshold changes for onset of sudomotor and vasomotor responses, followed a circadian rhythm which was very similar to the circadian rhythm in core temperature.

In a third study of the circadian effect on thermoregulation (143), women were studied in the early morning and late afternoon. There was an increased threshold for initiation of sudomotor and vasomotor responses at 1600 h than at 0400 h in both the follicular and luteal phases of the menstrual cycle. These increases averaged 0.36° C and 0.15° C for forearm blood flow at 0400 h and 1600 h, respectively and 0.74° C and 0.35° C for forearm sweating at 0400 h and 1600 h, respectively. This is consistent with the hypothesis that the circadian modulation in resting core temperature and the core temperature thresholds for initiation of sweating and cutaneous vasodilation in normal humans is the result of a circadian alteration in the thermoregulatory reference signal.

Current circadian control theory (96, 106) defines the circadian core temperature rhythm as a secondary oscillator controlled by the

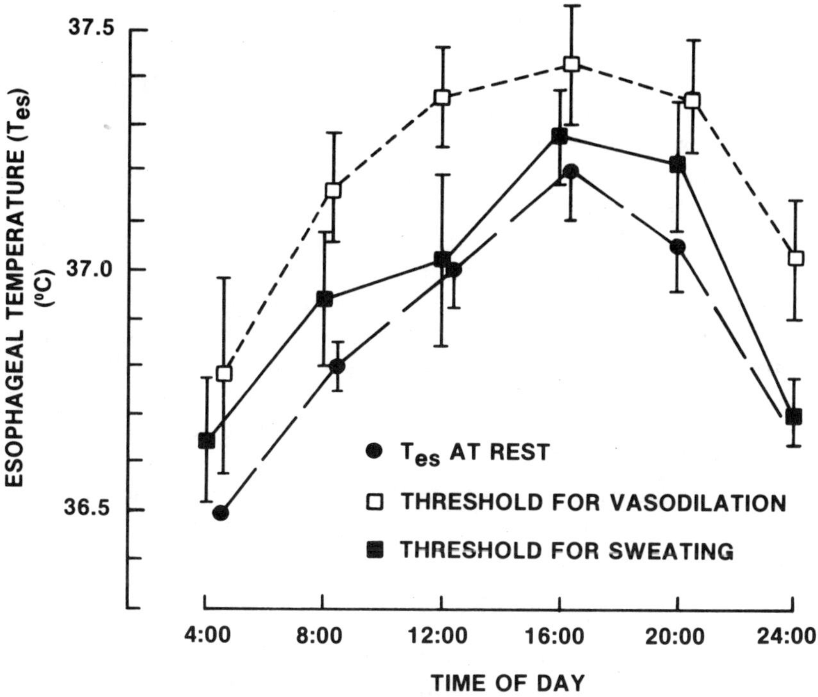

Figure 7-6. *Changing esophageal temperature (T_{es}) at rest as well as the T_{es} threshold for sweating and vasodilation with the circadian period. Redrawn, by permission of the American Physiological Society from* Am. J. Physiol. *246:R321–R324, 1984.*

X pacemaker (Fig. 7-5). The experimental evidence presented above is consistent with the hypothesis that the rhythm in core temperature is due to the circadian modulation of the thermoregulatory reference signal or set point. However, mean skin temperature has been considered as a secondary oscillator under the control of the Y pacemaker (Fig. 7-5). This theory of circadian control features two different oscillators controlling components of the thermoregulatory system. Consequently, the thermoregulatory system could be disrupted via internal desynchronization. Yet, the experimental evidence upon which the circadian control theory is derived appears to be sound. The evidence that the secondary oscillation in skin temperature is controlled by the Y pacemaker is based on experiments on the squirrel monkey (49, 50). Squirrel monkeys were entrained to a LD 12:12[3] cycle (600 lux) in a 28° C isolation chamber

[3] LD 12:12 is commonly used as an abbreviation to note that there was 12 h of light followed by 12 h of darkness. Any change in the ratio of light to darkness can be expressed in this form.

and then exposed to a cold stress (T_a = 20° C) for 6 h at all phases of the 24 h cycle (49). During the entrainment period, colonic temperature had a normal circadian rhythm. Cold exposure at any time of the circadian cycle was well tolerated by the animals as core temperature decreased by only −0.1° C during the 6 h. The monkeys were then maintained in constant light (LL; 600 lux) without any other Zeitgeber available. Core temperature became free-running with an average period of 25.2 h, and the rhythm had a smaller amplitude than that of animals in the LD 12:12 experiment. Cold exposure resulted in an average decrease in core temperature of −1.0° C below the control, a significant difference. The removal of Zeitgeber from the isolation chamber clearly was associated with impairment of the thermoregulatory system. Fuller *et al.* (49) postulated that thermoregulatory failure was probably due to internal desynchronization of the circadian system, resulting in disruption of the timing of the oscillators involved with thermoregulation. A second study by Fuller *et al.* (50) indicated that exposure of the monkeys to LL caused core temperature and tail skin temperature to become less tightly coupled together than in LD 12:12. They reported that core temperature and tail temperature occasionally free-ran with different circadian rhythms, and in some animals core temperature and tail temperature "split", meaning that there were two core temperature rhythms during the time in which tail temperature cycled only once. Fuller *et al.* (50) suggested that thermoregulation is controlled by, at least, two potentially independent oscillators. The observations of the splitting of the core temperature and tail temperature rhythms in some animals and the weaker coupling of the two rhythms in other animals might explain the thermoregulatory impairment during cold exposure (49).

Fuller *et al.* (50) noted that their observations (49, 50) were not consistent with the interpretation that circadian rhythm in core temperature is due to the circadian modulation of a single set point (6, 54, 65, 66, 143, 147, 156). In human studies of circadian effects on thermoregulation, the subjects were always internally synchronized, so the circadian rhythm in sudomotor and vasomotor responses to a heat stress are very similar to, if not the same as, the circadian core temperature rhythm (Fig. 7-6). Hence, the interpretation that the reference signal was modulated in a circadian cycle (6, 54, 65, 66, 143, 147, 156). However, those data are also consistent with the hypothesis that circadian heat loss mechanisms oscillate in a circadian periodicity, which is tightly coupled to the core temperature oscillator. Furthermore, data from human studies do not provide information whether both circadian sudomotor and vasomotor responses are controlled by the same oscillator or by tightly coupled oscillators. Fuller *et al.* (50) clearly demonstrated that the heat loss, due to vasomotor responses, was controlled by a different circadian oscillator than the core temperature rhythm in internally desynchronized monkeys. Presently, there is no specific evidence which indicates that sudomotor responses are controlled independently from, but closely linked to, the circadian oscillator controlling vasomotor responses or to the core temperature oscillator.

One interesting, new area of investigation of the circadian rhythm in core temperature concerns endogenous pyrogen (interleukin-1; IL-1). Ucar *et al.* (152) have postulated that there is a circadian release of an endogenous pyrogen in rats, which is associated with the increased nocturnal body temperature in these animals.[4] They have reported that plasma taken from rats at midnight and injected into other rats at noon, raises the core temperature of the host rats. However, plasma taken from rats at noon and then injected into other rats at noon did not change core temperature. The possibility that the circadian rhythm in thermoregulation is due to a modulation of the thermoregulatory set point by circulating IL-1 is very interesting, but such a mechanism has not been confirmed in humans or other animals.

There has been limited investigation into the possibility that the sleep-promoting factor, which was unequivocally demonstrated by Pappenheimer *et al.* (119), may be involved in the circadian timing system. Wexler and Moore-Ede (161) were unable to detect a phase resetting of the circadian timing system after injection of the active component of sleep-promoting factor (muramyl dipeptide) into squirrel monkeys. Muramyl dipeptide injection during subjective day and night was associated with both significantly decreased wakefulness and elevated core temperature. If sleep-promoting factor is involved in the circadian timing system, the active analogue, muramyl dipeptide does not mimic all effects of sleep-promoting factor.

Summary

In summary, the core temperature rhythm is regulated by circadian changes in the core temperature thresholds for onset of sudomotor and vasomotor responses. These observations are consistent with the hypothesis that the thermoregulatory reference signal(s) oscillate(s) in a circadian rhythm. As long as the heat loss effectors are tightly coupled to the core temperature rhythm (internal synchronization), there is no evidence that the circadian cycle impairs

---

[4] The circadian rhythm of core temperature in rats is different from humans. The acrophase of core temperature is at night, which corresponds to the activity cycle of the rat.

the homeostatic mechanism of thermoregulation. However, there is a potential for disruption of the thermoregulatory system if there is internal desynchronization of the circadian rhythms of the pacemakers, as occurred in the squirrel monkey (49). Additional research is needed to investigate the thermoregulatory problems which may result as a consequence of prolonged jet or space travel (108, 109, 110), shift work (111) or sleeplessness, all of which may induce internal desynchronization. A study of thermoregulatory responses to a heat or cold stress in humans with free-running, desynchronized circadian rhythms would provide new information about thermoregulatory function and perhaps circadian control theory.

SLEEP LOSS

Sleep and Temperature Regulation

Before the effects of sleep deprivation on thermoregulation can be understood, one must have a basic understanding of how thermoregulation varies during the circadian cycle and how sleep itself affects thermoregulation. We will begin this section with a description of how sleep affects thermoregulation, since the circadian cycle has already been discussed.

The hypothesis that sleep affects the thermoregulatory system came from the frequent observation that sleep was associated with bursts of sweating (31, 66, 97). Eventually, the circadian modulation of thermoregulation was distinguished from the effect of sleep per se on sweating (54). Ogawa (116) noted that slow wave sleep (SWS) was associated with the highest sweating rates during sleep in a warm environment. In 1977, Henane *et al.* (74) investigated the effect of sleep on evaporation and body temperatures in man. They reported that core temperature and mean skin temperature fluctuated markedly during sleep and these fluctuations were synchronous with rapid eye movement (REM) sleep. Specifically, evaporation decreased abruptly, immediately before the detection of REM sleep. Mean skin temperature increased during the period of REM sleep and core temperature increased after a phase delay. (The phase delay in core temperature was probably due to the inherent lag in the measurement of core temperature when rectal temperature is used.) At the end of REM sleep, evaporation increased, mean skin temperature decreased and rectal temperature slowly decreased. These cyclical changes in thermoregulatory function associated with REM sleep were superimposed on the normal circadian fluctuation in core temperature. Henane *et al.* (74) suggested that the neural integrative function (oscillator) conditioning the rhythmic patterns of sleep also was involved with phasic cyclic changes in thermoregulation.

Sagot *et al.* (129) claimed that the esophageal temperature thresholds for initiation of sweating during REM and stage 1–2 sleep were significantly greater than during slow-wave sleep. The slope of sweating to esophageal temperature relationship was significantly lower during REM than SWS and stage 1–2 (non-REM, NREM) sleep. The esophageal temperature thresholds for sweating and the slope of the sweating to esophageal temperature relationship were derived from the compilation of sweating and esophageal temperature measurements over the course of eight nights in three different environments. However, these conclusions were based on data from three subjects, as the data from another two subjects had insufficient number of episodes of either REM or SWS sleep at a given time of the circadian cycle to be included in the statistical analysis.

The data presented above (74, 129) are consistent with the hypothesis that REM sleep inhibited the sudomotor response to a given thermal drive. Apparently, REM sleep does not inhibit vasomotor responses since mean skin temperature increased after sudomotor inhibition occurred (74, 129). Esophageal temperature also increased after sudomotor inhibition (129), but it immediately decreased upon cessation of REM sleep when sweating increased. During NREM sleep, heat loss was no different than when the subject was awake (129), but varied as a function of the circadian rhythm in the reference signal controlling the thermoregulatory effectors (54, 66). However, during REM sleep (circadian cycle controlled) the sudomotor response is inhibited (74, 129).This may be the first evidence in humans that the sudomotor effector is controlled by a different oscillator than the vasomotor effector.

Since REM sleep apparently does not affect thermoregulatory vasomotor responses, there is no reason to suspect that REM sleep would impair thermoregulation in a cold environment. In fact, REM sleep is associated with an increased $\dot{V}O_2$ (19). Haskell *et al.* (71) reported that REM sleep in humans was not associated with any marked changes in thermoregulatory or metabolic responses at five different ambient temperatures between 21–37° C. Oxygen consumption, however, increased during REM sleep at ambient temperatures below thermoneutrality. Palca *et al.* (118) reported that REM sleep was associated with an increased forehead and tympanic temperatures and $\dot{V}O_2$ at 21 and 29° C. However, at 21° C, limb temperature decreases during REM sleep. The authors interpreted the increased $\dot{V}O_2$ as probably reflecting increased brain metabolism during REM sleep. Palca *et al.* (118) concluded that REM sleep in the cold did not impair thermoregulation.

Sleep Loss

Loss of sleep is a common occurrence in our society. Recently, the effect of an extended period of sleep loss on the ability of individuals to thermoregulate during subsequent environmental heat or cold exposure has received attention (91, 93, 133). The general response, following sleep loss, is a reduced resting core temperature with time of sleep deprivation (39) with limited effect(s) on any gross, whole body measure of performance (i.e. maximal aerobic power, work time or heart rate) (99). These generalized physiological responses to periods of sleep loss have been recently reviewed (80, 99, 139), and the reader is directed to these extensive review articles.

Sleep loss has pronounced effects on mood state, i.e. irritability and motivation (99). Activity has been shown to decrease sleepiness, which may confound the responses seen during exercise evaluations after periods of sleep loss. However, it is not difficult to understand the varied experimental findings on gross measures of performance, as the motivation state of the individual subject is crucial to performance (99).

Sleep Loss and Temperature Regulation

Generally, when core temperature varies throughout the circadian period or during the menstrual cycle, alterations in heat gain or loss accompany this changing core temperature which appear to "defend" this temperature. Extended periods without sleep (>30 h) are associated with a decreased resting core temperature (39, 89, 91, 93, 121). It would seem that alterations in heat loss or production mechanisms may accompany this change in core temperature, in a manner similar to that which has been reported during the changing core temperature associated with the circadian period and the menstrual cycle. Two recent studies (93, 133) attempted to characterize heat loss mechanisms associated with the sleep loss period, while a third study (91) indirectly evaluated thermoregulatory alterations resulting from sleep loss. The decrease in resting core temperature associated with periods of sleep loss may be related to a phase shift in the circadian rhythm for core temperature (80). Therefore, we might expect to see the appropriate shift in heat loss or conservation mechanisms to defend this temperature (143, 147, 156).

Based on direct analysis of thermogulatory effector responses to increasing core temperature from two recent studies (93, 133), the predominant effect of sleep loss is a decrease in the thermosensitivity of both sweating and peripheral blood flow with no evidence supporting a change in the core temperature threshold. These responses are shown in Figure 7-7 for both evaporative and non-evap-

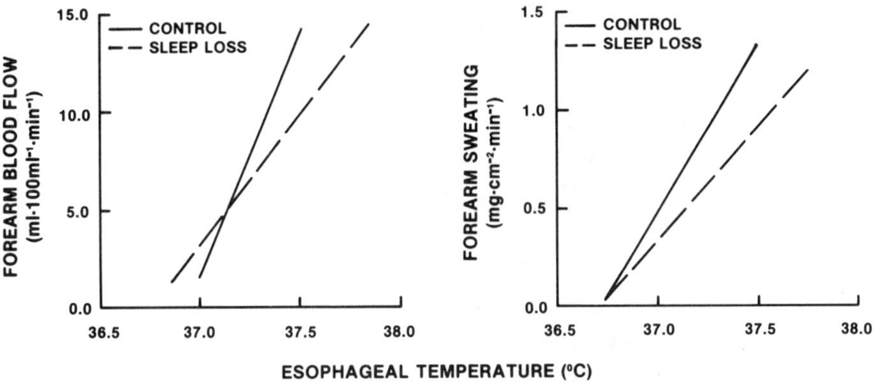

Figure 7-7. *Forearm blood flow (left panel) and forearm sweating rate (right panel) of one representative subject during the control experiment and after 33 hours without sleep.*

orative heat loss in control experiments and after 33 h without sleep. These experiments were conducted in the afternoon, when, during a normal circadian cycle, core temperature would be near its acrophase and initiation of heat loss responses would occur at a higher core temperature compared to early morning (143, 147, 156).

Sawka *et al.* (133) commented that the peripheral alteration in heat loss responses may have occurred at the individual effector, i.e. vascular smooth muscle or sweat gland. It was recently confirmed (93) that the cutaneous vasodilatory response to increasing core temperature (thermosensitivity) was decreased during exercise after sleep loss; this alteration appeared to be a peripheral response (93). Furthermore, it has been implied that central alterations (i.e. hypothalamic) may have occurred as sudomotor outflow loses its synchrony following sleep loss (133). However, even this asyncrony may be a peripherally mediated event, as the response of the individual sweat gland to a sudomotor signal is processed differently after sleep loss. Possible peripheral changes may include a decrease in the concentration of neurotransmitter per sudomotor impulse, a decrease in the cholinergic sensitivity of the sweat gland, or a decrease in blood flow to the gland. The attenuated vasodilatory response to increasing core temperature following 33 h without sleep (93) also indicates a peripheral modulation of the effector similar to that observed for sweating. It is not clear from these data whether these two events are linked in the periphery or if a central alteration in the effector signal has occurred. However, measurements of blood flow to the local sweat gland or evaluation of the quality or quantity of sudomotor or vasomotor impulses to the appropriate effector have not been done.

Indirect evidence from a third study (91) pointed to increased heat storage during the exercise transient phase, as core temperature increased an average 1.5° C in the first 15 min of exercise after sleep loss versus 1.1° C in the control experiment. Metabolic heat production (calculated from oxygen uptake) was unchanged between control and sleep loss experiments at all points during exercise, thus implying that heat dissipation was attenuated after sleep loss.

The possible modulation of the normal circadian variability in heat loss responses by an extended period of sleep loss (>30 hours) is an issue that has received limited experimental attention. One question which remains speculative at this point is whether sleep loss is associated with a phase shift in the circadian rhythm of the thermoregulatory reference signal(s). Akerstedt (1) reported that the trough of the oral temperature circadian rhythm is approximately 0.35° C lower after 72 h of sleep loss. The trough in the circadian cycle of adrenaline excretion was also decreased after 72 h of sleep loss. There were no control data presented however, so it is unknown whether there was a phase shift associated with the change in amplitude of the two circadian rhythms. During the sleep deprivation period it appeared that the phase of both core temperature and adrenaline excretion was unchanged (1).

Data from another investigation, which examined whether the circadian cycle in thyroid stimulating hormone (TSH) was affected by sleep loss, indicated that a phase shift may have occurred. The normal circadian rhythm in TSH has an inverted waveform compared to core temperature, with the core temperature (rectal) nadir and TSH peak in early morning (117). In sleep deprived (64 h) subjects, TSH release continued during the night so that there was no distinct peak at 2400 h as occurs normally (120). Instead, there was a prolonged elevation in TSH lasting throughout the night. TSH decreased at approximately 0800 h and increased again in the evening. Unfortunately, core temperature during these experiments was not reported. Parker *et al.* (120) interpreted the change in waveform as an indication that sleep deprivation had masked the circadian TSH rhythm. However, the data also appear consistent with approximately a 6 h phase delay in the nadir of the waveform, when it was compared to normal individuals.

A decreased resting core temperature is a consistent observation in sleep deprived individuals (80, 89, 91, 93). In one recent study, resting esophageal temperature was 0.3° C lower at 1500 h after 33 h of sleep loss than when the subjects had slept the night before (93). Although the circadian rhythm of esophageal temperature was not monitored in this study, the decreased resting esophageal temperature may indicate a change in amplitude or a phase shift in the esophageal temperature circadian rhythm. If there were a phase shift, the 0.3° C decrease would correspond to approximately a 6 h phase delay in resting esophageal temperature, as reported from another study (147). The possibility that both the TSH and the core temperature rhythms are affected by sleep deprivation, which results in a 6 hr phase delay, is speculative at this point. However, Aschoff (7) has cited research which indicated that performance, urinary electrolyte excretion and rectal temperature were phase delayed after 48 h of sleep deprivation. Although the data are somewhat variable, the calculated acrophase of the rectal temperature data indicated that it was delayed to 2400 h after sleep deprivation, which was 6 h later than occurred with sleep. The effect of sleep deprivation on the normal synchronization of circadian rhythms clearly requires further study.

Fluid Volume Control

The regulation of blood volume has not been extensively studied following periods of sleep loss. In general, the plasma volume is expanded after sleep loss (39, 59, 139), perhaps resulting from acidosis or changes in dietary pattern (139). Plasma renin activity is depressed as a result of the long term postural and blood volume changes. The stress hormone response associated with periods of sleep loss is quite variable (80, 100). Some of the variability in cortisol or catecholamine concentrations reported could be due to circadian variability in these hormones. The reports of expanded plasma volume conflict with the report of an attenuated cutaneous blood flow (93), as attenuated blood flow would not be expected to occur when plasma volume was expanded. The responses of hormones associated with fluid volume homeostasis following periods of sleep loss are not in agreement with changes seen in related thermoregulatory responses either. For example, a significant decreased plasma renin activity should result in less vasoconstrictor activity, which is not consistent with the attenuated forearm blood flow which has been reported (93). Definitive studies in the associated areas of thermoregulatory control and fluid volume homeostasis have not been conducted at this time.

Summary

In the context of this chapter, the extended loss of sleep (>30 hours) is associated with an attenuation in the effector responses for heat dissipation during exercise. In contrast to whole body performance measures which are dependent on motivation status of the test subject (99), when exercise is used only to increase the individual's core temperature, the appropriate heat loss pathways will be initiated, independent of the psychological state of the individual. There are still major areas of investigation needed to complete our understanding of how sleep loss affects humans. It is not known whether sleep loss causes a circadian phase shift in many systems, which normally have a fairly precise circadian rhythm, or whether sleep loss actually masks the circadian rhythms. A definitive study of the effects of sleep loss on fluid volume homeostasis must also be done, perhaps in association with an investigation of thermoregulatory responses during the course of several circadian cycles.

CONCLUSIONS

Problems associated with the evaluation of thermoregulatory responses to changing core temperature are greatly complicated when heat acclimation state, fluid volume state or state of physical training are variable. In this chapter, we have demonstrated how the time of day, the sleep status of the individual and the menstrual cycle can further affect these thermoregulatory responses. The study of human temperature regulation is difficult, especially when time constraints due to academic calendars, acclimation state, trained state, sleep status and changing reproductive function are superimposed. However, the critical evaluation of thermoregulatory responses to an altered thermal drive, whether it is endogenous or exogenous in origin, must include controls of the menstrual and circadian cycles as well as sleep status.

ACKNOWLEDGMENTS

We acknowledge the bibliographic assistance of B.L. Stephenson.

REFERENCES

1. Akerstedt, T. Altered sleep/wake patterns and circadian rhythms. *Acta Physiol. Scand. Suppl.* 469:1–48, 1979.
2. Aschoff, J. Die 24-Stunde-Periodik der Maus unter konstanten Umgebungsbedingungen. Naturwiss. 38:506–507, 1951.
3. Aschoff, J. Zeitgeber der tierischen Tagesperiodik. *Naturwiss.* 41:49–56, 1954.
4. Aschoff, J. Der Tagesgang der Korpertemperatur beim Menschen. *Klin. Wochenschr.* 33:545–551, 1955.
5. Aschoff, J. Exogenous and endogenous components in circadian rhythms. *Cold Spring Harbor Symp. Quant. Biol.* 25:11–27, 1960.
6. Aschoff, J. Circadian rhythm of activity and of body temperature. In:J. D. Hardy, A. P. Gagge and J. A. J. Stolwijk (Eds.) *Physiological and Behavioral Temperature Regulation.* Springfield, IL: Thomas, 1970, pp. 905–919.
7. Aschoff, J. Circadian rhythms: interference with and dependence on work-rest schedules. In: L. C. Johnson, D. I. Tepas, W. P. Colquhoun and M. J. Colligan (Eds.) *Biological Rhythms, Sleep and Shift Work.* New York: Spectrum Publications, 1981, pp. 11–34.
8. Aschoff, J., M. Fatranska and H. Giedke. Human circadian rhythms in continuous darkness: entrainment by social cues. *Sci.* 171:213–215, 1971.
9. Aschoff, J., U. Gerecke and R. Wever. Desynchronization of human circadian rhythms. *Jpn. J. Physiol.* 17:450–457, 1967.

10. Aschoff, J., U. Gerecke and R. Wever. Phasenbeziehungen zwischen den circadianen Perioden der Aktivitat und der Kerntemperatur beim Menschen. *Pflugers Archiv* 295:173–183, 1967.
11. Aschoff, J. and A. Heise. Thermal conductance in man: its dependence on time of day and on ambient temperature. In: S. Itoh, K. Ogata and H. Yoshimura (Eds.) *Advances in Climatic Physiology*. Tokyo: Igaku Shoin Ltd., Ch 20, 1972, pp.335–348.
12. Aschoff, J. and H. Pohl. Rhythmic variation in energy metabolism. *Fed. Proc.* 29:1541–1552, 1970.
13. Avellini, B. A., E. Kamon, and J. T. Krajewski. Physiological responses of physically fit men and women to acclimation to humid heat. *J. Appl. Physiol.* 49:254–261, 1980.
14. Avellini, B. A., Y. Shapiro, K. B. Pandolf, N. A. Pimental, and R. F. Goldman. Physiological responses of men and women to prolonged dry heat exposure. *Aviat. Space Environ. Med.* 51:1081–1085, 1980.
15. Baylis, P. H., B. A. Spruce, and J. Burd. Osmoregulation of vasopressin secretion during the menstrual cycle. In: R. W. Schrier (Ed.) *Vasopressin*. New York: Raven Press, 1985, pp. 241–247.
16. Bittel, J. and R. Henane. Comparison of thermal exchanges in men and women under neutral and hot conditions. *J. Physiol. (London)* 250:475–489, 1975.
17. Bloch, M. Rhythmic diurnal variation in limb blood flow in man. *Nature (London)* 202:398–399, 1964.
18. Boulant, J. A. Hypothalamic control of thermoregulation. In: P. J. Morgane and J. Panksepp (Eds.) *Handbook of the Hypothalamus Vol. 3, Part A, Behavioral Studies of the Hypothalamus*. New York: Marcel Dekker, 1980, pp. 1–82.
19. Brebbia, D. R. and K. Z. Altshuler. Oxygen consumption rate and electroencephalographic stage of sleep. *Sci.* 150:1621–1623, 1965.
20. Brengelmann, G. L. Circulatory adjustments to exercise and heat stress. *Annu. Rev. Physiol.* 45:191–212, 1983.
21. Brengelmann, G. L., J. M. Johnson, L. Hermansen and L. B. Rowell. Altered control of skin blood flow during exercise at high internal temperatures. *J. Appl. Physiol.* 43:790–794, 1977.
22. Bruce, V. G. and C. S. Pittendrigh. Endogenous rhythms in insects and microorganisms. *Am. Nat.* 91:179–195, 1957.
23. Bullard, R. W., M. R. Banerjee, F. Chien, R. Elizondo and B. A. MacIntyre. Skin temperature and thermoregulatory sweating: a control systems approach. In: J. D. Hardy, A. P. Gagge and J. A. J. Stolwijk (Eds.) *Physiological and Behavioral Temperature Regulation*. Springfield, IL: Thomas, 1970, pp. 597–610.
24. Cannon, J. G. and C. A. Dinarello. Increased plasma interleukin activity in women after ovulation. *Sci.* 227: 1247–1249, 1985.
25. Crockford, G. W., C. T. M. Davies and J. S. Weiner. Circadian changes in sweating threshold. *J. Physiol. (London)* 207:26P–27P, 1970.
26. Cunningham, D. J. and M. Cabanac. Evidence from behavioral thermoregulatory responses of a shift in setpoint temperature related to the menstrual cycle. *J. Physiol. (Paris)* 63:236–238, 1971.
27. Cunningham, D. J., J. A. J. Stolwijk, and C. B. Wenger. Comparative thermoregulatory responses of resting men and women. *J. Appl. Physiol.* 45:908–915, 1978.
28. Czeisler, C. A. Internal organization of temperature, sleep-wake and neuroendocrine rhythms monitored in an environment free of time cues. PhD. Thesis, Stanford, CA: Stanford University, 1978.
29. Czeisler, C. A., J. S. Allan, S. H. Strogatz, J. M. Ronda, R. Sanchez, C. D. Rios, W. O. Freitag, G. S. Richardson and R. E. Kronauer. Bright light resets the human pacemaker independent of the timing of the sleep-wake cycle. *Sci.* 233:667–671, 1986.
30. Czeisler, C. A., E. D. Weitzman, M. C. Moore-Ede, J. C. Zimmerman and R. S. Knauer. Human sleep: its duration and organization depend on its circadian phase. *Sci.* 210:1264–1267, 1980.
31. Day, R. Regulation of body temperature during sleep. *Am. J. Dis. Child.* 61:734–746, 1941.
32. DeCoursey, P. J. Phase control of activity in a rodent. *Cold Spring Harbor Symp. Quant. Biol.* 25:49–54, 1960.
33. Drinkwater, B. L. Physiological responses of women to exercise. In: J. H. Wilmore (Ed.) *Exercise Sport Sci. Rev.* Vol. 2. New York: Academic Press, 1973, pp. 125–153.
34. Drinkwater, B. L. Women and Exercise: Physiological Aspects. In: R. L. Terjung (Ed.) *Exercise Sport Sci. Rev.* Vol. 12. Lexington, MA: Collamore Press, 1984, pp. 21–51.
35. Drinkwater, B. L., J. E. Denton, I. C. Kupprat, T. S. Talag, and S. M. Horvath. Aerobic power as a factor in women's response to work in hot environments. J. Appl. Physiol. 41:815–821, 1976.
36. DuBois, E. F., F. G. Ebaugh, Jr. and J. D. Hardy. Basal heat production and elimination

of thirteen normal women at temperatures from 22° C to 35° C. *J. Nutr.* 48:257–293, 1952.
37. Eston, R. G. The regular menstrual cycle and athletic performance. *Sports Med.* 1:431–445, 1984.
38. Fein, J. T., E. M. Haymes, and E. R. Buskirk. Effects of daily and intermittent exposures on heat acclimation of women. *Int. J. Biometeorol.* 19:41–52, 1975.
39. Fiorica, V., E. A. Higgins, P. F. Iampietro, M. T. Lategola, and A. W. Davis. Physiological responses of men during sleep deprivation. *J. Appl. Physiol.* 24:167–176, 1968.
40. Fortney, S. M. and L. C. Senay, Jr. Effects of training and heat acclimation on exercise responses of sedentary females. *J. Appl. Physiol.* 47:978–984, 1979.
41. Forsling, M. L., P. Stromberg and M. Akerlund. Effect of ovarian steroids on vasopressin secretion. *J. Endocrinol.* 95:147–151, 1982.
42. Fox, R. H., B. E. Lofstedt, P. M. Woodward, E. Erickson and B. Werkstrom. Comparison of thermoregulatory function in men and women. *J. Appl. Physiol.* 26:444–453, 1969.
43. Fredericq, L. La courbe diurne de la temperature des centres nerveux sudoripares fonctionnant sous l'infleunce de la chaleur. *Arch. Biol.* 17:577–580, 1901.
44. Frye, A. J. and E. Kamon. Responses to dry heat of men and women with similar aerobic capacities. *J. Appl. Physiol.* 50:65–70, 1981.
45. Frye, A. J. and E. Kamon. Sweating efficiency in acclimated men and women exercising in humid and dry heat. *J. Appl. Physiol.* 54:972–977, 1983.
46. Frye, A. J., E. Kamon, and M. Webb. Responses of menstrual women, amenorrheal women, and men to exercise in a hot, dry environment. *Eur. J. Appl. Physiol.* 48:279–288, 1982.
47. Fuller, C. A. and D. M. Edgar. Effects of light intensity on the circadian temperature and feeding rhythms in the squirrel monkey. *Physiol. Behav.* 36:687–691, 1986.
48. Fuller, C. A., R. Lydic, F. M. Sulzman, E. Albers, B. Tepper and M. C. Moore-Ede. Circadian rhythm of body temperature persists after suprachiasmatic lesions in the squirrel monkey. *Am. J. Physiol.* 241:R385–R391, 1981.
49. Fuller, C. A., F. M. Sulzman and M. C. Moore-Ede. Thermoregulation is impaired in an environment without circadian time cues. *Sci.* 199:794–796, 1978.
50. Fuller, C. A., F. M. Sulzman and M. C. Moore-Ede. Circadian control of thermoregulation in the squirrel monkey, Saimiri sciureus. *Am. J. Physiol.* 236:R153–R161, 1979.
51. Gaebelein, C. J. and L. C. Senay, Jr. Vascular volume dynamics during ergometer exercise at different menstrual phases. *Eur. J. Appl. Physiol.* 50:1–11, 1982.
52. Gagge, A. P., L. P. Herrington, and C-E. A. Winslow. Thermal interchanges between the human body and its atmospheric environment. *Am. J. Hyg.* 26:84–102, 1937.
53. Gander, P. J., R. E. Kronauer and R. C. Graeber. Phase shifting two coupled circadian pacemakers: implications for jet lag. *Am. J. Physiol.* 249:R704–R719, 1985.
54. Geschickter, E. H., P. A. Andrews and R. W. Bullard. Nocturnal body temperature regulation in man: a rationale for sweating in sleep. *J. Appl. Physiol.* 21:623–630, 1966.
55. Gisolfi, C. V. and C. B. Wenger. Temperature regulation during exercise: old concepts, new ideas. In: R. L. Terjung (Ed.) *Exercise Sport Sci. Rev.* Vol. 12. Lexington, MA: Collamore Press, 1984, pp. 339–372.
56. Gonzalez, R. R., L. G. Berglund and J. A. J. Stolwijk. Thermoregulation in humans of different ages during thermal transients. *Sat. 28 Int. Cong. Physiol. Sci.* 32:357–361, 1980.
57. Gonzalez, R. R., K. B. Pandolf and A. P. Gagge. Heat acclimation and decline in sweating during humidity transients. *J. Appl. Physiol.* 36:419–425, 1974.
58. Goodman, H. M. Reproduction. In: V. B. Moncastle (Ed.) *Medical Physiology* Vol. 2. Philadelphia: W. B. Saunders, 1974, pp. 1741–1775.
59. Goodman, J. M. Cardiovascular adjustments to prolonged sleep deprivation at rest and during maximal exercise. M.S. Thesis. Toronto: University of Toronto, 1983.
60. Gray, M. L., K. S. Strausfeld, M. Watanabe, E. A. H. Sims and S. Solomon. Aldosterone secretory rates in the normal menstrual cycle. *J. Clin. Endocrinol.* 28:1269–1275, 1968.
61. Halberg, F. Temporal coordination of physiologic function. *Cold Spring Harbor Symp. Quant. Biolo.* 25:289–308, 1960.
62. Halberg, F. Chronobiology. *Annu. Rev. Physiol.* 31:675–725, 1969.
63. Halberg, F. Biological rhythms. In: L. W. Hedlund, J. M. Franz, and A. D. Kenny (Eds.) *Biological Rhythms and Endocrine Function.* New York: Plenum Press, 1975, pp. 1–41.
64. Hammel, H. T. Regulation of internal body temperature. *Annu. Rev. Physiol.* 31:641–710, 1968.
65. Hammel, H. T., D. C. Jackson, J. A. J. Stolwijk and J. D. Hardy. Hypothalamic temperatures in dog and monkey and thermoregulatory responses to environmental factors. Technical Documentary Report No. AMRL-TDR-63-5, Jan., 1963.
66. Hammel, H. T., D. C. Jackson, J. A. J. Stolwijk, J. D. Hardy and S. B. Stromme. Temperature regulation by hypothalamic proportional control with an adjustable set point. *J. Appl. Physiol.* 18:1146–1154, 1963.

67. Hardy, J. D. Physiology of temperature regulation. *Physiol. Rev.* 41:521–606, 1961.
68. Hardy, J. D. and E. F. DuBois. Differences between men and women in their response to heat and cold. *Proc. Natl. Acad. Sci.* 26:389–398, 1940.
69. Hardy, J. D. and J. A. J. Stolwijk. Partitional calorimetric studies of man during exposure to thermal transients. *J. Appl. Physiol.* 21:1799–1806, 1966.
70. Harrison, M. F. Effects of thermal stress and exercise on blood volume in humans. *Physiol. Rev.* 65:149–209, 1985.
71. Haskell, E. H., J. W. Palca, J. M. Walker, R. J. Berger and H. C. Heller. Metabolism and thermoregulation during stages of sleep in humans exposed to heat and cold. *J. Appl. Physiol.* 51:948–954, 1981.
72. Haslag, Sister W. M. and A. B. Hertzman. Temperature regulation in young women. *J. Appl. Physiol.* 20:1283–1288, 1965.
73. Haymes, E. M. Physiological responses of female athletes to heat stress: a review. *Phys. Sportsmed.* 12(3):45–59, 1984.
74. Henane, R., A. Buguet, B. Roussed and J. Bittel. Variations in evaporation and body temperatures during sleep in man. *J. Appl. Physiol.* 42:50–55, 1977.
75. Hertig, B. A. Human physiological responses to heat stress; males and females compared. *J. Physiol. (London)* 63:270–273, 1971.
76. Hessemer, V. and K. Bruck. Influence of menstrual cycle on shivering, skin blood flow, and sweating responses measured at night. *J. Appl. Physiol.* 59:1902–1910, 1985.
77. Hessemer, V. and K. Bruck. Influence of menstrual cycle on thermoregulatory, metabolic, and heart rate responses to exercise at night. *J. Appl. Physiol.* 59:1911–1917, 1985.
78. Hirata, K., T. Nagasaka, A. Hirai, M. Hirashita, T. Takahata and T. Nunomura. Effects of human menstrual cycle on thermoregulatory vasodilation during exercise. *Eur. J. Appl. Physiol.* 54:559–565, 1986.
79. Hogarth, P. J. *Biology of Reproduction.* New York: John Wiley and Sons, 1978, pp. 54–76.
80. Horne, J. A. A review of the biological effects of total sleep deprivation in man. *Biol. Psychol.* 7:55–102, 1978.
81. Horstman, D. H. and E. Christensen. Acclimation to dry heat: active men vs. active women. *J. Appl. Physiol.* 52:825–831, 1982.
82. Horvath, S. M. and B. L. Drinkwater. Thermoregulation and the menstrual cycle. *Aviat. Space Environ. Med.* 53:790–794, 1982.
83. Inouye, S. T. and H. Kawamura. Persistence of circadian rhythmicity in mammalian hypothalamic "island" containing the suprachiasmatic nucleus. *Proc. Natl. Acad. Sci.* 76:5961–5966, 1979.
84. Johnson, J. M., G. L. Brengelmann, J. R. S. Hales, P. M. Vanhoutte and C. B. Wenger. Regulation of the cutaneous circulation. *Fed. Proc.* 45:2841–2850, 1986.
85. Kaneko, M., F. W. Zechman and R. E. Smith. Circadian variation in human peripheral blood flow levels and exercise responses. *J. Appl. Physiol.* 25:109–114, 1968.
86. Kaulhausen, H., G. Leyendecker, G. Benker and H. Breuer. The relationship of the renin-angiotensin-aldosterone system to plasma gonadotropin, prolactin and ovarian steroid patterns during the menstrual cycle. *Arch. Gynakol.* 225:179–200, 1978.
87. Kenney, W. L. A review of comparative responses of men and women to heat stress. *Environ. Res.* 37:1–11, 1985.
88. Kenshalo, D. R. Changes in the cool threshold associated with phases of the menstrual cycle. *J. Appl. Physiol.* 21:1031–1039, 1966.
89. Kleitman, N. Studies on the physiology of sleep. I. The effects of prolonged sleeplessness on man. *Am. J. Physiol.* 66:67–92, 1923.
90. Kleitman, N. and A. Ramsaroop. Periodicity in body temperature and heart rate. *Endocrinology* 43:1–20, 1948.
91. Kolka, M. A., B. J. Martin and R. S. Elizondo. Exercise in a cold environment after sleep deprivation. *Eur. J. Appl. Physiol.* 53:282–285, 1984.
92. Kolka, M. A. and L. A. Stephenson. Thermoregulation during active and passive heating during the menstrual cycle (Abstract). *Physiologist* 28:368, 1985.
93. Kolka, M. A. and L. A. Stephenson. Exercise thermoregulation after prolonged wakefulness. *J. Appl. Physiol.* (In Press).
94. Kolka, M. A., L. A. Stephenson, P. B. Rock and R. R. Gonzalez. Local sweating and cutaneous blood flow during exercise in hypoxic environments. *J. Appl. Physiol.* 62:2224–2229, 1987.
95. Kreider, M. B., E. R. Buskirk and D. E. Bass. Oxygen consumption and body temperatures during the night. *J. Appl. Physiol.* 12:361–366, 1958.
96. Kronauer, R. E., C. A. Czeisler, S. F. Pilato, M. C. Moore-Ede and E. D. Weitzman. Mathematical model of the human circadian system with two interacting oscillators. *Am. J. Physiol.* 242:R3–R17, 1982.
97. Kuno, Y. *Human Perspiration.* Springfield, IL: Thomas 1956, pp. 180–181.

98. Little, M. A. and J. A. Rummel. Circadian variations in thermal and metabolic responses to heat exposure. *J. Appl. Physiol.* 31:556–561, 1971.
99. Martin, B. J. Sleep deprivation and exercise. In: K. B. Pandolf (Ed.) *Exercise Sports Sci. Rev.* Vol. 14. Lexington, MA: D.C. Heath Co., 1986, pp. 213–229.
100. Martin, B. J. and H.-I. Chen. Sleep loss and the sympathoadrenal response to exercise. *Med. Sci. Sports Exercise* 16:56–59, 1984.
101. Mills, J. N. Human circadian rhythms. *Physiol. Rev.* 46:128–171, 1966.
102. Minors, D. S. and J. M. Waterhouse. Endogenous and exogenous components of circadian rhythms when living on a 21-hour day. *Int. J. Chronobiol.* 8:31–48, 1981.
103. Minors, D. S. and J. M. Waterhouse. Circadian rhythms and their mechanisms. *Experientia* 42:1–13, 1986.
104. Moore, R. Y. and V. B. Eichler. Loss of a circadian adrenal corticosterone rhythm following suprachiasmatic lesions in the rat. *Brain Res.* 42:201–206, 1972.
105. Moore, R. Y. and N. J. Lenn. A retinohypothalamic projection in the rat. *J. Comp. Neurol.* 142:1–14, 1972.
106. Moore-Ede, M. C. The circadian timing system in mammals: two pacemakers preside over many secondary oscillators. *Fed. Proc.* 42:2802–2808, 1983.
107. Moore-Ede, M. C. Physiology of the circadian timing system: predictive versus reactive homeostasis. *Am. J. Physiol.* 250:R735–R752, 1986.
108. Moore-Ede, M. C. Jet lag, shift work and maladaptation. *News in Physiol. Sci.* 1:156–160, 1986.
109. Moore-Ede, M. C., C. A. Czeisler and G. S. Richardson. Circadian timekeeping in health and disease. Part I. Basic properties of circadian pacemakers. *New Engl. J. Med.* 309:469–476, 1983.
110. Moore-Ede, M. C., C. A. Czeisler and G. S. Richardson. Circadian timekeeping in health and disease. Part II. Clinical implications of circadian rhythmicity. *New Engl. J. Med.* 309:530–536, 1983.
111. Moore-Ede, M. C. and G. S. Richardson. Medical implications of shift-work. *Annu. Rev. Med.* 36:607–617, 1985.
112. Moore-Ede, M. C. and F. M. Sulzman. Internal temporal order. In: Aschoff, J. (Ed.) *Handbook of Behavioral Neurobiology: Biological Rhythms.* New York: Plenum Press, 1981, pp. 215–241.
113. Moore-Ede, M. C., F. M. Sulzman and C. A. Fuller. *The Clocks That Time Us.* Cambridge, MA: Harvard University Press, 1982.
114. Nadel, E. R., E. Cafarelli, M. F. Roberts and C. B. Wenger. Circulatory regulation during exercise in different ambient temperatures. *J. Appl. Physiol.* 46:430–437, 1979.
115. Nunneley, S. H. Physiological responses of women to thermal stress: a review. *Med. Sci. Sports* 10:250–255, 1978.
116. Ogawa, T., T. Satoh and K. Takagi. Sweating during night sleep. *Jpn. J. Physiol.* 17:135–148, 1967.
117. O'Malley, B. P., A. Richardson, N. Cook, S. Swart and F. D. Rosenthal. Circadian rhythms of serum thyrotrophin and body temperature in euthyroid individuals and their responses to warming. *Clin. Sci.* 67:433–437, 1984.
118. Palca, J. W., J. M. Walker and R. J. Berger. Thermoregulation, metabolism, and stages of sleep in cold-exposed men. *J. Appl. Physiol.* 61:940–947, 1986.
119. Pappenheimer, J. R., T. B. Miller and C. A. Goodrich. Sleep-promoting effects of cerebrospinal fluid from sleep-deprived goats. *Proc. Natl. Acad. Sci.* 58:513–517, 1967.
120. Parker, D. C., L. G. Rossman, A. E. Pekary and J. M. Hershman. Effect of 64-hour sleep deprivation on the circadian waveform of thyrotropin (TSH): further evidence of sleep-related inhibition of TSH release. *J. Clin. Endocrinol. Metab.* 64:157–161, 1987.
121. Patrick, G. T. W. and J. A. Gilbert. On the effects of loss of sleep. *Psychol. Rev.* 3:468–483, 1896.
122. Pittendrigh, C. S. Circadian rhythms and the circadian organization of living systems. *Cold Spring Harbor Symp. Quant. Biol.* 25:159–182, 1960.
123. Reilly, T. and G. A. Brooks. Investigation of circadian rhythms in metabolic responses to exercise. *Ergonomics* 25:1093–1107, 1982.
124. Roberts, M. F., C. B. Wenger, J. A. J. Stolwijk and E. R. Nadel. Skin blood flow and sweating changes following exercise training and heat acclimation. *J. Appl. Physiol.* 43:133–137, 1977.
125. Robertson, G. L. Osmoregulation of thirst and vasopressin secretion: functional properties and their relationship to water balance. In: R. W. Schrier (Ed.) *Vasopressin.* New York: Raven Press, 1985, pp. 203–212.
126. Rothchild, I. and A. C. Barnes. The effects of dosage and of estrogen, androgen or salicylate administration on the degree of body temperature elevation induced by progesterone. *Endocrinology* 50:485–496, 1952.

127. Rowell, L. B. Cardiovascular adjustments to thermal stress. In: J. T. Shepherd and F. M. Abboud (Eds.) *Handbook of Physiology—The Cardiovascular System III*, Baltimore: Waverly Press, 1983, pp 967–1023.
128. Rusak, B. and I. Zucker. Neural regulation of circadian rhythms. *Physiol. Rev.* 59:449–526, 1979.
129. Sagot, J. C., C. Amoros, V. Candas and J. P. Libert. Sweating responses and body temperatures during nocturnal sleep in humans. *Am. J. Physiol.* 252:R462–R470, 1987.
130. Sargent, F. and K. P. Weinman. Comparative study of the responses and adjustments of the human female and male to hot atmospheres. *Environmental Physiology and Psychology in Arid Conditions. Proceedings of the Lucknow Symposium 1962.* Paris: UNESCO, pp. 157–161, 1964.
131. Sasaki, T. and L. D. Carlson. Effect of a step change in temperature on skin temperature and blood flow. *Proc. Soc. Exp. Biol. Med.* 117:334–338, 1964.
132. Sawka, M. N., R. P. Francesconi, N. A. Pimental and K. B. Pandolf. Hydration and vascular fluids shifts during exercise in the heat. *J. Appl. Physiol.* 56:91–96, 1984.
133. Sawka, M. N., R. R. Gonzalez and K. B. Pandolf. Effects of sleep deprivation on thermoregulation during exercise. *Am. J. Physiol.* 246:R72–R77, 1984.
134. Schweiger, H-G., R. Hartwig and M. Schweiger. Cellular aspects of circadian rhythms. *J. Cell. Sci. Suppl.* 4:181–200, 1986.
135. Senay, L.C., Jr. Body fluids and temperature responses of heat-exposed women before and after ovulation with and without dehydration. *J. Physiol. (London)* 232:209–219, 1973.
136. Shapiro, Y., K. B. Pandolf, B. A. Avellini, N. A. Pimental and R. F. Goldman. Physiological responses of men and women to humid and dry heat. *J. Appl. Physiol.* 49:1–8, 1980.
137. Shapiro, Y., K. B. Pandolf, B. A. Avellini, N. A. Pimental and R. F. Goldman. Heat balance and transfer in men and women exercising in hot-dry and hot-wet conditions. *Ergonomics* 24:375–386, 1981.
138. Shapiro, Y., K. B. Pandolf and R. F. Goldman. Sex differences in acclimation to a hot-dry environment. *Ergonomics* 23:635–642, 1980.
139. Shephard, R. J. Sleep, biorhythms and human performance. *Sports Med.* 1:11–37, 1984.
140. Smith, R. E. Circadian variations in human thermoregulatory responses. *J. Appl. Physiol.* 26:554–560, 1969.
141. Spruce, B. A., P. H. Baylis, J. Burd and M. J. Watson. Variation in osmoregulation of arginine vasopressin during the human menstrual cycle. *J. Clin. Endocrinol.* 22:37–42, 1985.
142. Stephan, F. K. and I. Zucker. Circadian rhythms in drinking behavior and locomotor activity of rats are eliminated by hypothalamic lesions. *Proc. Natl. Acad. Sci.* 69:1583–1586, 1972.
143. Stephenson, L. A. and M. A. Kolka. Menstrual cycle phase and time of day alter reference signal controlling arm blood flow and sweating. *Am. J. Physiol.* 249:R186–R191, 1985.
144. Stephenson, L. A. and M. A. Kolka. Plasma volume during heat stress and exercise in women. *Eur. J. Appl. Physiol.* (In press).
145. Stephenson, L. A., M. A. Kolka and R. R. Gonzalez. Circadian and menstrual cycle variation in blood parameters (Abstract). *The Physiologist* 27:230, 1984.
146. Stephenson, L. A., M. A. Kolka and R. R. Gonzalez. Effect of the menstrual cycle on the control of skin blood flow during exercise (Abstract). *Fed. Proc.* 45:407, 1986.
147. Stephenson, L. A., C. B. Wenger, B. H. O'Donovan and E. R. Nadel. Circadian rhythm in sweating and cutaneous blood flow. *Am. J. Physiol.* 246:R321–R324, 1984.
148. Sulzman, F. M., C. A. Fuller and M. C. Moore-Ede. Environmental synchronizers of squirrel monkey circadian rhythms. *J. Appl. Physiol.* 43:795–800, 1977.
149. Swartz, W. J. and H. Gainer. Suprachiasmatic nucleus: use of 14C-labeled deoxyglucose uptake as a functional marker. *Sci.* 197:1089–1091, 1977.
150. Timbal, J., J. Colin and C. Boutelier. Circadian variations in the sweating mechanism. *J. Appl. Physiol.* 39:226–230, 1975.
151. Turner, C. and S. Fortney. Plasma volume changes during the menstrual cycle (Abstract). *Fed. Proc.* 43:718, 1984.
152. Ucar, D. A., R. J. Tocco and M. J. Kluger. Circadian variation in circulating pyrogen: possible role in resistance to infection. *Proc. Soc. Exp. Biol. Med.* 173:319–323, 1983.
153. Vollman, R. F. *The Menstrual Cycle. Major Problems in Obstetrics and Gynaecology* Vol. 7. Philadephia: W. B. Saunders, 1977.
154. Wells, C. L. and S. M. Horvath. Heat stress responses related the menstrual cycle. *J. Appl. Physiol.* 35:1–5, 1973.
155. Wells, C. L. and S. M. Horvath. Responses to exercise in a hot environment as related to the menstrual cycle. *J. Appl. Physiol.* 36:299–302, 1974.
156. Wenger, C. B., M. F. Roberts, J. A. J. Stolwijk and E. R. Nadel. Nocturnal lowering of thresholds for sweating and vasodilation. *J. Appl. Physiol.* 41:15–19, 1976.
157. Wever, R. The circadian multi-oscillator system of man. *Int. J. Chronobiol.* 3:19–55, 1975.

158. Wever. R. A. *The Circadian System of Man: Results of Experiments Under Temporal Isolation.* New York: Springer Verlag, 1979.
159. Wever, R. A. Fractional desynchronization of human circadian rhythms. A method for evaluating entrainment limits and functional interdependencies. *Pflugers Archiv* 396:129–137, 1983.
160. Wever, R. A., J. Polasek and C. M. Wildgruber. Bright light affects human circadian rhythms. *Pflugers Archiv* 396:85–87, 1983.
161. Wexler, D. B. and M. C. Moore-Ede. Effects of a muramyl dipeptide on the temperature and sleep-wake cycles of the squirrel monkey. *Am. J. Physiol.* 247:R672–R680, 1984.
162. Wilkerson, J. E., E. M. Leeds and G. D. Brown. Hematological differences in regularly cycling females with normal and low progesterone (Abstract). *Fed. Proc.* 44:846, 1985.
163. Wisser, H. and H. Breuer. Circadian changes of clinical chemical and endocrinological parameters. *J. Clin. Chem. Clin. Biochem.* 19:323–337, 1981.
164. Wyndham, C. H., J. F. Morrison and C. G. Williams. Heat reactions of male and female caucasians. *J. Appl. Physiol.* 20:357–364, 1965.

8

The Heat Illnesses: Biochemical, Ultrastructural, and Fluid-Electrolyte Considerations

ROGER W. HUBBARD, Ph.D.

LAWRENCE E. ARMSTRONG, Ph.D.

Outline

INTRODUCTION

Although an excessive rise in core temperature is considered the ultimate cause of heat illness, it (hyperthermia) commonly occurs with "host factors." A brief survey of these host factors indicates that many affect thermoregulation through heat loss mechanisms (e.g. lack of acclimatization, fatigue, lack of sleep, dehydration, skin disorders), while others contribute to heat production (e.g. obesity, lack of physical fitness, dehydration, febrile illness, sustained exercise). Some do both or have unknown mechanisms.

The International Statistical Classification of Diseases (108) lists ten separate categories of heat illness, with heat cramps, heat exhaustion and heat stroke being the most common. This classification scheme facilitates diagnosis and treatment by identifying fairly distinctive signs and symptoms. It also serves our purposes by providing classifications broad enough to include the major disorders without trivializing the list of important host factors: hypohydration, electrolyte imbalance, exhaustive exercise, acidosis, lack of acclimatization and drug effects.

The purpose of this chapter is to move toward a concept which integrates the major host factors and this classification scheme into a new theory of heat pathophysiology at the cellular level. The point of emphasis is the integrity or leakiness of the cell membrane and its resultant effects on energy consumption, heat production and exhaustion. By focusing on the cell, we are attempting to provide a new basis for examining the impact of systemic disorders (anhidrosis, regional ischemia, shock) on the organs and tissues involved. We hope this will lead to new experimental paradigms and to improvements in diagnosis and treatment. Our further intent is to supplement the clinical knowledge of these disorders with selected experimental findings from the fields of exercise physiology, cell biology and neurology.

BODY TEMPERATURE IN SPORTS AND MEDICINE

Body temperature increases with heat stress and exercise and during athletic events rectal temperatures of 40–42° C (104–107.6° F) are not uncommon (63,145,210). Once the limits of cardiac output are reached, combined with the maximum skin blood flow, sweat rate and evaporative cooling, further excess heat production is stored as accumulated heat and body temperature rises. Cellular metabolism increases 13% for each 1° C rise in temperature and is 50% above normal at 40.6° C (43). The use of hyperthermia to treat cancer patients suggests, however, that if blood pressure and pH are carefully controlled, then temperature in the 40–41° C range will often produce no ill effects (40). Some patients have fully recovered from temperatures in the 44–45° C range (26). We have previously reported that a temperature of 40.4° C (100,104) represented a threshold hyperthermia above which heatstroke mortalities occur in exercised heat-stressed rats. The mean (±SE) core temperature at exhaustion which produced a 50% mortality rate within 24h (LD_{50}) was 41.5 ± 0.1° C. These results suggested a "continuum" (104) of increasing risk as the temperature rose above this threshold value. Thus, the probability of mortality appeared directly related to the core temperature at collapse.

This temperature, which is reliably recorded in a research setting, is only rarely measured under "field circumstances." Even in the military, which has trained aid-men accompanying troops during training, the casualty reporting card often does not contain an accurate rectal temperature taken and recorded close to the time of collapse. This is often due to the fact that the cabins and cargo spaces of prime movers will reach 54–60° C under desert conditions and the standard clinical thermometer (mercury in glass) will thermally burst at these temperatures. Metal thermocouples or thermistor probes are a must.

A source of potential confusion is that a temperature of 40.6° C is usually reached twice—once while the temperature is ascending and once while it is descending. Thus, it is important to inquire and record whether the temperature was measured prior to or after collapse. Approximately, one-half of the total heat exposure of untreated, heat-exhausted rats occurs after the point of collapse (104). Thus, any attempt to aggressively limit the intensity or duration of the hyperthermia will reduce morbidity and mortality. The early application of cooling to stricken miners while still underground has reduced the incidence of fatal heatstroke (209). This concept is referred to as Quick-Recognition, Speedy-Treatment (QRST).

A source of potential underdiagnosis was noted by Shibolet (177), in that striking changes in body temperature occur between measurements in the field ($\bar{x}$ = 41.1° C) and upon admission to a hospital ($\bar{x}$ = 37.8° C) where temperatures appeared normal. Another complication involves a delayed and secondary rise in body temperature. In the cases reported by Malamud *et al* (136), the core temperature on admission to the hospital ranged from 36 to 44° C, and nine of those with temperatures below 41° C subsequently rose to hyperthermic levels. In view of the fact that Malamud *et al* reported multiple hemorrhages in 122 of 125 fatal cases of heatstroke, Shibolet (176) has attached special significance to a clinical course of coma, a lucid interval characterized by apparent improvement in vital signs within the first twelve to twenty-four hours, followed by subsequent rapid and relentless deterioration and death. For example, Shibolet reasoned that when death occurs rapidly (70% of the Malamud series within 24hr), . . . "an acute and general derangement of biochemical reactions on a cellular level seems to be implicated" (176). In contrast, when the clinical picture shows the lucid period followed by degeneration, severe fibrinolysis should be suspected. In three fatal cases, the clinical and laboratory findings were consistent with fibrinolysis. Afibrinogenemia and hypoprothrombinemia were found, and post-mortem examination revealed multiple hemorrhages and the absence of arterial thrombi. It was assumed in these cases, free of severe or progressive liver or kidney damage, that multiple hemorrhages especially to vital centers (brain and heart) were the cause of death.

Lack of Pain in Hyperthermia

Finally, the victims of hyperthermia, especially young, highly motivated, healthy individuals are largely unaware of the seriousness of their situation in contrast to the more familiar sensations of fatigue or exhaustion (15). For example, Gilat *et al* (78) presented the rectal temperatures of eight volunteers marching eight hours with a load of 35 kg. Six of the volunteers had rectal temperatures in excess of 40.6° C ($\bar{x}$ = 41.1° C, n = 8) and all were sweating profusely. Three had premonitory signs of restlessness and euphoria (78). So subtle are the symptoms of evolving hyperthermia, that athletes, who are driven by pride and discipline, may voluntarily increase exercise intensity during the latter stages of competition (87). In addition, during many years of research with an animal model for human hyperthermia, exercising animals have never exhibited overt signs which indicated that they were any more painfully stressed at a T_{re} of 42° C than at 40° C, or that passively-heated animals (up to 42.8° C) experienced pain (99,100,101,102).

HEAT ILLNESS SPECTRUM

The heat illnesses represent a spectrum of disorders which range in intensity and severity from mild cardiovascular and CNS disturbances (hypotension, fainting) to evidence of profound cellular damage, involving characteristically the brain, kidneys, liver and blood clotting mechanisms (heatstroke). Costrini (54) hypothesized, based on comparative data, that the pathophysiology of heat exhaustion and heatstroke are so similar that they may represent a continuum of diseases rather than separate, distinct pathophysiologic entities. Callaham (43), in his monograph on emergency management of heat illness, defined heat illness, as a "spectrum of disease" and the sequence of events leading to heat illness have often been referred to as a continuum (43,149,54,100,178). It is recognized, however, that heat exhaustion may occur without symptoms of heatstroke, and *vice versa*.

Heat exhaustion may develop over several days and is a manifestation of the strain on the cardiovascular system (43). It is diagnosed following collapse with either a normal or an elevated temperature (severe cases around 40° C) usually with profuse sweating and a normal mental status. The significant alteration of mental status (coma, disorientation, confusion, psychoses) is the key diagnostic landmark in suspected heatstroke cases. Sweating may or may not be present. As Callaham (43) has pointed out, heat exhaustion is excluded from heatstroke by generally lower temperatures and less severe changes in neurological status (ie. mental function remains intact).

Generally, heat illness denotes some adverse health effect, if even minor and temporary, that indicates a given heat load has exceeded that person's heat tolerance in a given situation (65). Animal data (100,102) has clearly indicated that even within highly inbred species individuals display large variations in heat tolerance. For example, mortality curves (100) demonstrated a continuum of increasing risk with increasing severity of hyperthermia. However, this data translated to at least a 24-fold difference in heat tolerance between a highly inbred rat that succumbed to a five-degree-minute exposure and another of the same strain that survived over 120 degree-minutes of hyperthermia (See Section IX, Temperature and the Cell, for the definition of degree-minutes). We infer from the foregoing that individual differences in tolerance represent different thresholds for heat injury. This demonstrable variability in susceptibility to morbidity and mortality, *changes with the circumstances of the exposure.* (102,100,104).

Indices of Injury

The individual variability in tolerance, especially with changing circumstances, represents aspects of both good news and bad. It suggests that an increase in tolerance correlates with an elevation in the threshold for cellular injury. For example, 40.7° C is the core temperature at which 25% of a run-exhausted rat population will have elevations in serum glutamic oxalacetic transaminase (SGOT) over 1000 U/L at 24h (102) which confirms the occurrence of severe heatstroke (97). By the same token, it takes a core temperature of 42.0° C to give 75% of that population similar elevations in serum transaminase. Clearly, individuals within the population differ in their inherent capacity to withstand heat and exercise. Furthermore, if the situation changes such as with passive heating without exercise, it takes a core temperature of 42.2° C to produce a similar elevation in enzymes within *only 25%* of the heated population.

The increased heat tolerance of the sedentary population is probably related to less severe splanchnic vasoconstriction and therefore relatively less splanchnic ischemia and acidosis (210). This is consistent with the apparent lack of ill effects in cancer patients treated with hyperthermia (40–41° C range) and whose blood pressure and pH are carefully controlled (40). On the other hand, tolerance appears to decline both with intercurrent illnesses such as viral gastroenteritis and prior heatstroke (113). This represents bad news for those who want to predict or prevent heatstroke on other than a population basis and in other than well-defined circumstances.

Reinjury

These data reinforce the assertion that if an individual's cardiovascular capacity is low, there is an increased probability that an early collapse may ensue and thus, a moderate or even severe heat exhaustion might occur and preclude triggering a full-blown case of heatstroke (65). This leaves unanswered, however, whether the individuals who succumb to a bout of heat syncope in the first three days of heat exposure (30) are the same individuals at apparently greater risk as heat acclimation improves within the population, generally, with several days of exposure. Population studies which would follow individuals who suffer febrile convulsions as infants throughout adolescence and adulthood could be helpful. It could be argued that highly efficient and heat resistant mechanisms to shift visceral blood flow away from vital organs (to sustain working muscles) could contribute to ultimately more serious injuries. Further, this type of argument would predict that as a workforce is acclimatized and once the incidence of severe heat exhaustion and heatstroke declines, those incidents of heatstroke which do occur are more likely to be fatal. There is always the danger that an increase in environmental heat stress, due to either natural causes or air conditioning failures, will require additional acclimatization.

Chronic Heat Illnesses

It is widely held that an individual who has suffered a serious heat illness will display reduced tolerance to heat for sometime thereafter (178,129). This is consistent with a prior, but undetected, reduction in tolerance before the episode. It is clear from experimental studies on human beings that heat rash reduces sweating (155) and certainly heatstroke can produce irreversible cellular damage—especially within the central nervous system (178). Dukes-Dobos (65) has classified three types of chronic heat illness according to their etiology: Type I—after effects of acute heat illness. Type II—cummulative effects of long-term exposure and Type III—effects of residence in hot climates (Table 8-1). This review of occupational heat illnesses concluded that primary targets of chronic heat appear to be the cardiovascular and gastrointestinal systems (65).

GENERAL FLUID-ELECTROLYTE CONSIDERATIONS

Water, in normal, healthy humans, comprises 50%–70% of total body weight (mean ~ 61%), depending on one's percentage of body fat. In a temperate climate and with minimal activity, approximately 90% of daily water intake is gained via food moisture and beverages, while approximately 90% of daily water is lost as urine or skin and lung insensible water (82). Daily water turnover in a resting 70kg human totals approximately 2.5kg. Exercise increases sweating and water intake, but the thirst mechanism and hormonal responses regulate total body water in cool environments to within ±0.22% (±150g) of the body weight each day; plasma volume is regulated daily within ±0.7% (±25g) by vascular homeostatic mechanisms (82). A behavioral component of fluid intake (105) apparently works in consort with the Na^+-osmotic-vasopressin pathway and the renin-angiotensin-aldosterone system (83) to stimulate, maintain and terminate human water intake.

TABLE 8-1. *Chronic Heat Illnesses (Type I, II, And III)*

Chronic heat illnesses—Type I, aftereffects of acute heat illnesses.	
Acute illnesses	**Chronic aftereffects**
Prickly heat, heat rash	Reduced heat tolerance; disfunction of sweat glands; reduced sweating capacity
Heat cramp	Reduced heat tolerance; muscle soreness, stiffness; reduced mobility
Heat exhaustion	Reduced heat tolerance; chronic heat exhaustion
Heat stroke	Reduced heat tolerance; cellular damage in different organs, particularly in the central nervous system, heart, kidneys, and liver

Chronic heat illnesses—Type II, cumulative effects of long-term exposure.	
After several months in a hot job	**After many years in a hot job**
Chronic heat exhaustion	Hypertension
Symptoms	Reduced libido
Headache	Sexual impotency
Gastric pain	Myocardial damage
Sleep disturbance	Nonmalignant diseases of the digestive organs
Tachycardia	
Irritability	Hypochromemia
Vertigo	
Nausea	

Chronic heat illnesses—Type III, effects of residence in climatically hot regions.	
Tropics	**Desert**
Frequent skin diseases	Kidney stones
Sleep disturbance	Anhidrotic heat exhaustion
Susceptibility to minor injuries and sickness	
Psychoneurosis (tropical lethargy)	
Anhidrotic heat exhaustion	

During exercise in hot environments, voluntary water intake does not keep pace with water losses; this phenomenon has been named voluntary dehydration, even though it is not the result of conscious behavior (184). Voluntary dehydration has a behavioral component (e.g. palatability, water temperature, availability of water), but inadequate drinking behavior may be overridden by physiological inputs which increase drinking (e.g. thirst, sweating, increased rectal temperature). If exercise in the heat is prolonged (16), water deficits of 4–5% body weight may be incurred (184). Obviously, the turnover of water and electrolytes increases because sweating increases. Most individuals lose 1.0–1.5L of sweat during one hour of moderate or heavy work in the heat (13). Daily water turnover rate in the desert may exceed 15L per person, although much of this is consumed with meals and during the evening (2).

Water lost by the body, except through the lungs, is always accompanied by electrolytes—mainly Na^+ and Cl^-, since Na^+ and Cl^- comprise approximately 80% of the osmotically active particles in extracellular fluid. The primary intracellular electrolytes (K^+, Mg^{++}, HPO_4^{--}) are found in small quantities in plasma and sweat but in higher quantities in urine. Table 8-2 (4,11,18,76) describes selected ion concentrations of different body fluids, and supports the fact that fluid shifts between the intracellular, interstitial and vascular compartments result primarily from Na^+ and K^+ movements. The usual adult daily intake of these electrolytes in the United States is 100–300 mEq Na^+, 75–225 mEq Cl^-, 50–150 mEq K^+, 120 mEq Mg^{++} (75). One investigation (16) demonstrated that whole body sweat electrolyte losses, exclusive of urine losses, were 47–196 mEq Na^+, 9–19 mEq K^+, 37–194 mEq Cl^-, and 0.4–1.2 mEq Mg^{++}, during a 6-hour simulated desert march (1.34 $m.s^{-1}$, 30/30 min work/rest cycles) covering 14.5 km in conditions of 40.6° C dry bulb, 30% rh. These losses of Na^+, Cl^- and K^+ are noteworthy, when compared to the daily dietary intakes described above.

TABLE 8-2. *Selected Ion Concentrations of Body Fluids (mEq/L)*

Ion Species	Intracellular Fluid	Plasma	Sweat	Urine	Mixed Saliva
Na^+	10	135–145	15–120	39–218*	10–75*
Cl^-	4	95–105	5–110	32–224*	11–45*
K^+	140	3.5–5.0	3–10	43–60	11–23
Mg^{++}	58	1.8–2.4	2–5	8–10	0.1–0.4

*—dependent heavily on dietary intake

Is Dietary Sodium Excessive?

The dogma regarding the body's need for elevated NaCl intake in hot environments has been questioned; however, Hubbard *et al* (98) reviewed the literature regarding long-term salt and water deficits. They concluded that previous NaCl intake recommendations of 15–48 g NaCl/day during continuous desert living (47,128,132,197) are unnecessarily high in light of hormonal adaptations (aldosterone) during heat acclimatization, that result in Na^+ and Cl^- conservation in urine and sweat (46). They revealed that current dogma regarding salt supplements in hot environments originated primarily from one study performed by Taylor *et al* in 1944 (197). However, Hubbard *et al.* (98) noted that there were nine major methodological flaws in Taylor's study. In addition to this revelation, nutritional studies conducted throughout the world have identified several healthy populations which consume low levels of salt. Several of these populations inhabit tropical climates, including: the Sahara Bedouins (156), the proud and vigorous Masai warriors of Africa

(151) who eat less that 5 g NaCl/day (85 mEq Na^+/day), the inhabitants of tropical Nigeria (127) who consume less than 7 g NaCl/day (120 mEq Na+/day), and the Galilean Naturalists (183) who ingest only 1.9 g NaCl/day (32 mEq Na^+/day). The factors which allow these groups to survive on low levels of salt include (a) extreme hormonal sparing of salt in sweat and urine, (b) controlled water intake, and (c) careful dietary practices (156).

In contrast to the low salt intakes described above, the average adult in the United States ingests 5.8–17.4 g NaCl/day (100–300 mEq Na^+/day) (75). This relatively large NaCl intake is propagated by food additives, convenience foods and canned goods; it is also well in excess of physiological need. The actual metabolic requirements for salt are remarkably small, generally below 1g/day. Dahl (56) reported that active, ambulatory patients remained healthy on 0.1–0.75 g NaCl/day (2–13 mEq Na^+/day). Except for a few cases of renal impairment, maintenance of Na^+ balance on these intakes was never a problem. Therefore, this question arises, "What are the salt requirements of individuals who leave Western lifestyles in mild climates and move to hot environments?"

Dietary Salt and Successful Heat Acclimation

Few studies have investigated the impact of dietary salt intake and simultaneous entry into a hot climate. The 1949 work of Conn (46) is a classic in this arena. Conn acclimated a group of healthy males for 15 days (*ad lib* food intake), at which point he instituted a controlled diet of 20 g NaCl/day (343 mEq Na^+/day) for 10 days. This was followed by two 10-day periods of 11 g NaCl/day (189 mEq Na^+/day) and 6 g NaCl/day (103 mEq Na^+/day), respectively. At each level of decreasing dietary salt intake, Conn's subjects demonstrated successful NaCl homeostasis, when given 3–10 days to adapt. Two recent papers by Armstrong and colleagues (10,11) reported the effects of high (23.2 g NaCl/day; 399 mEq Na^+/day) and low (5.7 g NaCl/day; 98 mEq Na^+/day) Na^+ diets on physiological adaptations during eight consecutive days of heat acclimation. Subjects ate both diets (separated by 24 days) and exercised for 90 min per day (1.56 $m \cdot s^{-1}$ treadmill walking) at 40.1° C, 23.5 % rh. The low Na^+ diet resulted in significantly higher ($p < .05$, mean heart rate (days 3–5) and rectal temperature (days 3–6), but lower ($p < .05$) plasma osmolality (day 4), plasma volume (day 4), urine Na^+ (days 2–8), and sweat Na^+ (days 4 and 8); blood and sweat samples were measured on days 1, 4 and 8 only. Interestingly, these between-diet differences were not detected by the end of 8 days. The urinary Na^+ levels increased (high Na^+) and decreased (low Na^+) on days 1–3, but converged toward pre-acclimation levels on days

5–8. The authors concluded that the differences between these diets reflected temporary adjustments in fluids and electrolytes mediated by aldosterone, as well as renal escape from the influence of aldosterone (low Na^+ diet). Because days 3–6 contained most between-diet differences, it was concluded that the low-Na^+ trials placed subjects at greater risk of heat injury on those days, especially in light of the plasma volume deficit, increased rectal temperature and increased heart rate.

Considering the fact that low Na^+ diets may predispose humans to heat illness, there is appeal in the advice to ingest high levels of NaCl (15–48 g NaCl per day, see ref. 47,128,132,197) in hot climates. Although a NaCl deficiency may be harmful, the following scientific evidence refutes the need for high dietary levels of NaCl. 1. As noted above, many populations living in tropical areas of the earth survive on very low dietary NaCl levels (e.g. 1.9–7 g NaCl/day). 2. A high NaCl diet results in low aldosterone levels; this hormonal status is exactly the opposite of that desired during the initial 3–5 days of work in heat (98). 3. If two individuals enter a hot environment equilibrated to high (20g/day) and low (3g/day) NaCl diets, respectively, and then eat consistently low NaCl diets (1–3 g NaCl/day) because of loss of appetite, the time required for the former individual to reach Na^+ balance will be 6–8 days, but will be only 2–4 days for the latter individual (98). Thus, humans on high NaCl diets require longer to regulate fluid-electrolyte balance in a hot environment. 4. During the course of 10 days of heat acclimation trials, two high NaCl diets (22.5g NaCl/day and 30g NaCl/day) produced cardiovascular impairment, decreased optimal exercise capacity, decreased metabolic efficiency, impaired heat acclimation adaptations, and increased losses of K^+, Cl^-, Ca^{++} and PO_4^{--}, when compared to a "normal" (15g/day) diet (61,211). Salt loading has been implicated as a hypothetical factor in the production of K^+ depletion, rhabdomyolysis, and heat injury (121). 5. To maintain normal body fluid tonicity, elevated salt consumption increases the daily requirement for water. This need has been determined to be approximately 1 L of additional water for each 5g of supplemental NaCl added to the diet (27).

Heat acclimatization impacts on fluid-electrolyte losses dramatically. In addition to the landmark physiological adaptations of decreased heart rate, rectal temperature, and skin temperature (84), repeated days of heat exposure result in expanded plasma volume, increased sweat rate, and decreased Na^+ and Cl^- concentrations in sweat and urine. Clearly, unacclimatized humans will exhibit a different fluid-electrolyte profile following 8–14 days of exercise-heat exposure. Both Conn's (46) and Armstrong's (10,11) investigations (above) suggest that the body's fluid-electrolyte homeostatic mechanisms can adapt to a wide range of dietary Na^+ intake in a hot environment, when 3–10 days are allowed for equilibration. These investigations emphasize the importance of the coincidental introduction of multiple stressors (e.g. profuse sweating, environmental insult, dietary deficiency, hyperthermia, fatigue) in the etiology of heat injury.

THE ETIOLOGY OF HEAT CRAMPS

Although several reviews have commented on the etiology, diagnosis, treatment and prevention of heat cramps (64,120,43), few systematic scientific observations (126,130,195) have been made on the fluid-electrolyte status of heat cramp patients. The exact mechanism of heat cramps (and other similar maladies such as exercise-induced muscle cramps, nocturnal cramps, writer's cramps) is unknown. Heat cramps are usually unhearalded and occur in the voluntary muscles of the legs, arms, and abdomen of approximately 1% of all laborers (129,130). Steel workers, coal miners, sugar cane cutters, and boiler operators are among the most commonly reported victims (195). Heat cramps usually occur after several hours of hard exercise in individuals who have lost a large volume of sweat, have drunk a large volume of unsalted water, and have excreted a small volume of urine (130,190). Interestingly, the incidence of heat cramps among laborers in steel mills, boiler rooms, or mines is seasonal (129); more than 90% of the hospitalized patients in the Youngstown, Ohio steel district (1929–1934) were admitted between April and October (195). This suggests that laborers adapt to high temperatures in the workplace, but that the additional stress of environmental heat and sweat loss during off-duty hours is enough to induce heat cramps in some individuals.

The Differential Diagnosis of Heat Cramps

Heat cramps may be confused with other forms of muscle cramps. Muscular cramps similar to heat cramps are observed in other hyponatremic states (e.g. cholera) (132) and in dialysis patients from whom large quantities of fluid are removed quickly, even though their serum electrolytes remain normal (43). Another debilitating state which is often confused with heat cramps is hyperventilation-induced tetany. This syndrome, sometimes observed in connection with heat exhaustion, involves metabolic alkalosis and carpopedal spasms (35). Prolonged exercise, such as a 42.1 km foot race, may induce muscle cramps even in cool environments (112). Furthermore, the authors have observed that gastro-intestinal upset may be confused with heat cramps (Armstrong, unpublished observations, 1985).

During a summer field investigation at a military installation in Georgia, three young soldiers reported to a medical aid station with complaints of debilitating abdominal cramps. Diagnosed initially as abdominal heat cramps, it was later determined that these three soldiers had slept in the same tent and that they all had incurred an intestinal virus. Talbott (195) similarly described the difficulty of distinguishing gastro-intestinal upsets from abdominal heat cramps. Therefore, it is important, though not difficult, to diagnose heat cramps properly. The beginning of heat cramps may be ushered-in by feeble twitchings (fasiculations) of the muscle about to be affected. The cramp consists of a localized contraction affecting a few muscle bundles only, which can be accurately located by the patient or by an observer who remembers that heat cramps occur in voluntary skeletal muscle. The entire muscle mass is never affected all at one time. As one bundle relaxes, an adjacent bundle contracts for 1–3 minutes, giving the impression of the cramp wandering over the muscle. The pain of severe heat cramps is excruciating.

Heat Cramps and Heat Acclimatization

Recent reviews of heat cramp etiology do not agree with field/industrial observations regarding the effects of heat acclimatization on heat cramps. The heat cramp reviews of Knochel and Reed (120) and Callaham (43) state that heat acclimatized individuals are the most common victims of this heat illness, but these authors do not cite verifying evidence. The fact that Shibolet (178) observed no cases of heat cramps among Israeli soldiers during many years of heat illness triage, suggests that heat acclimatized individuals are less likely to experience heat cramps. The incidence of heat cramps within the Indian Armed Forces is likewise very low (137). Only two cases with urinary chlorides below 4 grams per liter were observed from 131 cases of confirmed heat illness. Ladell's analysis (126) also supports improved heat cramp resistance with heat acclimation. Furthermore, Talbot observed that the incidence of heat cramps was greatest during the first few days of a heat wave at a Boulder Dam construction site (196), and that steel workers were maximally susceptible to heat cramps during the first few days of a high temperature period in Youngstown, Ohio (195). These findings prompted Talbott to recommend restricted activity and an increased salt intake during the *first few days* of heat exposure (195). The half-life of aldosterone-mediated Na^+ conservation at the kidney is 18 ± 1 hr in mild environments, and is less than 18 hr if Na^+ loss also occurs by sweating (200); aldosterone-mediated sweat gland conservation of Na^+ requires 5–10 days. Therefore, Talbott's advice to increase salt intake during the first few days of heat exposure seems logical, until one realizes that circulating aldosterone levels are thereby likely to be *lowered*, and that Na^+ conservation will be reduced as long as dietary NaCl levels are high. It is clear that an unacclimatized individual will lose more Na^+ than an acclimatized individual per liter of sweat (75 vs 35 mEq Na^+/L, respectively), but the goal of heat acclimatization efforts should be to develop Na^+ conservation, not to lower circulating aldosterone levels. Entering a hot environment to perform intense exercise and concurrently consuming a high salt diet may involve two simultaneous stressors which the body may not handle effectively. A better suggestion is to reduce salt loss via sweating by providing adequate physical work-rest cycles early in the acclimatization process.

Heat Cramps and Electrolyte Depletion

To our knowledge, only one author (178) since 1935 has doubted the connection between heat cramps and salt depletion. This consistency in opinion can be explained by three facts. First, heat cramp victims exhibit hyponatremia, hypochloremia and reduced levels of Na^+ and Cl^- in urine. Table 8-3, redrawn from data published by Leithead and Gunn in 1964 (130), illustrates urinary sodium conservation among heat cramp patients which is indicative of a negative salt balance. Similarly, Talbott (195) observed serum Na^+ levels ranging from 121–140 mEq/L (normal = 135–145 mEq/L) in 32 heat cramp patients. Second, treatment of severe heat cramps is rapid and effective, when IV saline solutions are used (e.g. 0.5–1.0 L normal saline; a bolus of 10–20 ml of 23.5 % hypertonic saline; a bolus of 162 ml of 15 % saline solution administered over 3.5 min) (43,129) or when 0.1 % NaCl solution (two crushed salt tablets in 1 liter of water) is taken orally to treat mild heat cramps. Third, at least three steel mills successfully reduced their incidence of heat cramps to nearly zero by encouraging salt consumption (195). Anectdotal reports from Germany and England indicate that laborers salt their beer to successfully prevent heat cramps; alcohol is a factor of only minor or occasional significance in the etiology of heat cramps (129).

TABLE 8-3. *Urinary Excretion of NaCl by Laborers in British Guiana*

SUBJECTS	No. Random Samples	Urinary g NaCl/L
General Laborers (e.g. housecleaners)	178	12
Sugar Cane Cutters (healthy, working)	40	10
Sugar Cane Cutters with heat cramps	30	4

In addition to reduced Na^+ and Cl^- in plasma and urine, several other clinical signs and symptoms have been reported in heat cramp patients (195,120). For example, increased urinary excretion of nitrogen, inorganic phosphate, CPK (5 times normal), and creatine (75mg/hr) signal muscle damage. However, normal levels of plasma lactate, Ca^{++}, Mg^{++}, and K^+ are also reported. The decreased plasma volume and decreased body weight (−2 kg in 32 heat cramp patients) (195) are consistent with descriptions of fluid-electrolyte shifts in 6-day Na^+ restricted patients described by Romero *et al.* (162). Only a small percentage of all heat cramp patients exhibit other symptoms. These include abr rmal deep reflexes, mild headache, vertigo, giddiness, and nausea (130,195); these same symptoms are also observed in salt depletion heat exhaustion patients.

Heat Cramps: Relationship to Transmembrane Fluid-Electrolyte Flux is Unknown

Published reports indicate that heat cramps occur in 50%, 60% (131,132), and 70 % (126) of salt-depletion heat exhaustion patients. In this type of heat exhaustion, salt depletion receives the emphasis as the primary cause. In pure heat cramps, extracellular salt dilution (drinking pure water) receives the emphasis. In either case, some water will move into cells. It is this intracellular water expansion which most authors believe to be essential in the development of heat cramps (126,129). Indeed, intramuscular water expansion has been observed during heat acclimation on a low Na^+ diet, via analysis of muscle biopsies (11,10). The obvious question here is, "Why do some salt-depletion heat exhaustion patients experience heat cramps while others do not?" One explanation could be that physical training and high levels of strength and endurance may offer protection from cramp-producing conditions. Another explanation involves the possibility of an inherent metabolic deficit (64), especially in those who repeatedly are afflicted with heat cramps (130).

Similarly, one might ask, "Why do heat cramps occur during some salt depletion trials, but not during other periods of equivalent (or greater) salt depletion, in the same subject?" For example, W.S.S. Ladell realized that he was susceptible to heat cramps and subjected himself to 2 hr of exercise-heat exposure during negative salt balance on 26 occassions (126). Cramps occurred during 20 trials, but did not develop unless Ladell became salt deficient to at least 11 g NaCl. However, during two trials (deficits of −18 g and −29 g NaCl) he was free from cramp and felt in excellent condition. Ladell himself offered an explanation. When heat cramps did not develop, he usually became puffy and edematous towards the end of the trial.

This suggested that water had been retained extracellularly and that the intracellular fluid had not been diluted. If fluid-electrolyte movements across the muscle membrane were different during some trials, then some unknown host factor intervened on those days.

Heat Cramps and the Search for a Useful Paradigm

The physiological mechanism of heat cramps is not completely understood, but future research studies are suggested by the results of seven previous publications. 1. The reproduction of heat cramps has been accomplished artificially by using (a) counter-pressure against a flexed muscle, (b) cold air or water, or (c) a mild alternating current (195). 2. Ladell (126) reported an experiment in which he induced heat cramps in both of his own calves and thighs (deficit of 20.84 g NaCl) and then inflated a pneumatic tourniquet (250 mm Hg) above his right knee. Prior to this, he had determined that tourniquet placement for 30 min did not induce muscle cramps. When the tourniquet had been inflated for 15 min, 162 ml of 15 % NaCl solution was injected over a period of 3.5 min. Two min after injection, cramp could not be detected in the left leg (unoccluded), but was present in the occluded leg. Blood samples indicated that plasma Cl^- levels increased from 90.0 to 111.2 mEq/L. Ladell's experiment suggested that heat cramps result from peripheral (not central) disturbances in fluid-electrolyte balance. 3. The pain of heat cramps also apparently originates peripherally, according to Weisenburg (203). He studied a childhood polio victim who experienced paralysis, atrophy, and loss of reflexes in the left leg. As an adult, this man had heat cramps which were as painful and violent in the left leg as in the right leg. Weisenburg concluded that the origin of the pain was in the muscle. 4. Low blood sugar, as the cause of heat cramps, has been ruled-out because no relief from cramps was observed when 6 patients received IV dextrose solutions in varying quantities. 5. Electrolyte disturbances other than Na^+ and Cl^- have not been pursued and offer fertile ground for future research. Potassium fluctuations during prolonged exercise clearly have been demonstrated (95), and the effects of these transient K^+ disturbances may be realized at a variety of tissue elements (e.g. presynaptic motor nerve terminals, contractile apparatus) or in muscle metabolism. 6. It has been suggested that vasoaction and blood flow changes play a role in the development of heat cramps (64). This theory focuses on heat cramps induced after a work shift, during cold air (or cold water) exposure. In this theory, cold water induces a reflex vasoconstriction in the skin, resulting in a redirection of blood flow away from the extremities and the abdominal wall. Using current technology, such a theoretical mechanism could easily be investigated. 7. Known causes of other types of muscle cramps might be investigated to clarify the series of events which lead to heat cramps. These other causes of painful muscle cramps include: insufficient ATP for relaxation, nerve disorders caused by irritation of the nerve roots, neuropathies involving the axon (e.g. diabetes, alcohol abuse, uremia), metabolic glycolytic defects of skeletal muscle (e.g. McArdle's syndrome or phosphorylase deficiency, phosphofructokinase deficiency, abnormalities associated with lipid metabolism (e.g. carnitine palmityl transferase deficiency), and passive shortening of a muscle without resistance or tension (e.g. swimmer's cramps) (191).

FLUID-ELECTROLYTE CONSIDERATIONS IN HEAT EXHAUSTION AND HEAT SYNCOPE

At the onset of upright exercise of moderate intensity, plasma volume decreases as the result of an initial shift of water into muscle; this increases the concentration of all ionic species in plasma. Further losses of sweat decrease the interstitial and intracellular spaces, while plasma volume remains relatively stable. As for intracellular electrolyte content, K^+ shows a small decrease which parallels a small rise intravascularly. The threat of dehydration, however, is potentially more serious than the acute loss of any electrolyte, because the muscle itself is not depleted of Na^+, Cl^-, Mg^{++}, or K^+ and the vascular compartment has a salt excess. Clearly, the acute need of an exercising adult is water, not salt; water depletion may result in heat exhaustion in a matter of hours. The hyponatremia and hypochloremia of heat cramp patients or salt-depletion heat exhaustion patients, in contrast, requires at least 3–5 days to develop (129,140), and usually occurs in the unacclimatized individual who has not developed salt conserving mechanisms fully (183).

Dehydration due to a negative salt balance (either low dietary NaCl, high losses of NaCl, or both) results in a loss of extracellular fluid and a reduction of plasma volume, cardiac output, and blood pressure; it is neither characterized by thirst nor relieved by administration of salt-free fluids. Dehydration due to water deprivation and increased fluid losses is characterized by thirst and oliguria, and is completely relieved by the administration of pure water (71,144). These two forms of dehydration comprise the two widely-recognized categories of heat exhaustion (43,120,128,129,200): salt depletion heat exhaustion (SD) and water depletion heat exhaustion (WD). Severe SD produces a form of peripheral vascular collapse closely resembling traumatic shock; WD, with a comparable decline in extracellular fluid volume, does not produce peripheral vascular collapse, and little or no protein leaves the vascular compartment (71). WD is more likely to lead to increased core temperature and heatstroke than SD (7). Adolph (2) and Vaamonde (200) have both published the signs and symptoms of SD and WD. Table 8-4 includes a compilation of their observations. The most striking aspect of this table is the similarity in the symptoms of SD and WD. Urine and plasma levels of Na^+ and Cl^- seem to be the only consistant diagnostic indices, but Vaamonde (200) cautions that a normal or even increased serum Na^+ may be observed temporarily in SD, depending on the magnitude of water lost and type of fluid replacement. The above facts prompted Dinman and Horvath (64) to state that differentiation between SD and WD was not clinically feasible in the Aluminum Industry. Similarly, Hubbard *et al.* (98) emphasize that pure forms of SD or WD are rare, because usual water and salt losses (e.g. voluntary dehydration, sweating, diarrhea, vomiting, urine) involve mixed SD and WD.

TABLE 8-4. *Signs and Symptoms of Salt Depletion and Water Depletion Heat Exhaustion*

Signs & Symptoms	Salt Depletion Heat Exhaustion	Water Depletion Heat Exhaustion
Thirst	not prominent	yes
Muscle cramps	in most cases	no
Nausea	yes	yes
Vomiting	in most cases	no
Anorexia	yes	yes
Fatigue, weakness	yes	yes
Loss of skin turgor	yes	yes
Mental dullness, apathy	yes	yes
Orthostatic rise in pulse rate	yes	yes
Tachycardia	yes	yes
Dry mucuous membranes	yes	yes
Increased rectal temperature	yes	in most cases
Urine Na^+/Cl^-	negligible	normal
Plasma Na^+/Cl^-	below average	above average

Replacing Deficits During Competition

Heat exhaustion is primarily a fluid volume depletion problem, and rapid recovery occurs when intravenous fluids are administered. Normal, young heat exhaustion victims often require up to 4 L of intravenous fluids (70,43). For example, the 1985 Boston Marathon was run on a warm day in April (24° C) at a time when few runners had acclimatized to heat (average daily maximum temperature in Boston during April, 1986 was 12.5° C). The medical tent at the finish line administered intravenous fluids to 158 cases of heat exhaustion. It was later reported that an estimated 90 % of all of these patients recovered and walked-out of the medical tent under their own power within 15–20 min (146). In events of this duration, WD is much more likely to occur than SD.

One of the largest sweat rates ever recorded in a human being was reported by our research group (14). A world-class distance runner produced sweat at 2.79 $L \cdot hr^{-1}$ in our laboratory (before heat acclimation) and at 3.71 $L \cdot hr^{-1}$ (after 19 days of heat acclimation training in Florida) during the hot, humid 1984 Olympic marathon. This runner's fluid losses (−8.1% of total body weight) were projected prior to the Olympics (Figure 8-1). With his highly trained cardiovascular system and well-developed heat dissipation mechanisms, the major threat to his health was not hyperthermia, but dehydration. Because endurance performance declines when dehydration reaches −3 to −5 % of total body weight (90,9), these −3 % and −5% fluid levels were included in our projections. Actual sweat losses were depicted by the dashed line in Figure 8-1; this athlete was unable to ingest water rapidly enough to maintain body weight above the −5 % level. In contrast, Dancaster and Whereat (59) observed relatively low sweat rates (0.5–1.5 L/hr) in ultra-marathon runners, during an 87 km race conducted in cool weather. Although

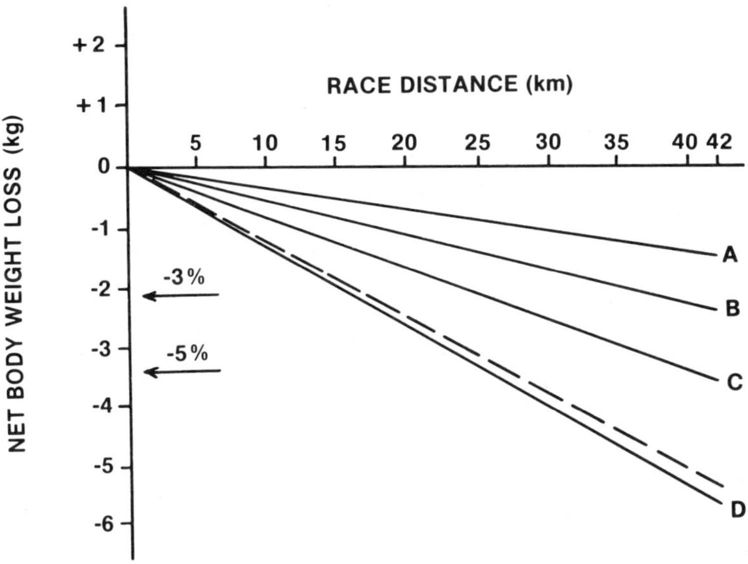

Figure 8-1. *Projected net body weight loss, based on 2.79 $L \cdot h^{-1}$ sweat rate and "worst case" Olympic marathon (OM) conditions. Lines A, B, C, and D represent water consumption rates of 0.5, 0.375, 0.25, and 0.01/5 km, respectively. Dashed lines illustrate this athlete's actual body weight loss (3.71 $L \cdot h^{-1}$ sweat rate), in spite of water intake of 0.223 $L \cdot 5\ km^{-1}$.*

these sweat rates were not large, the total fluid deficits incurred during this 6–11 hr event were noteworthy (range: 4.3–12.8 L). Again, the primary requirement was water, not salt.

At the Medical Aid Station

In mild heat exhaustion cases, oral electrolyte solutions (e.g. 0.1% saline solution) may suffice when patients are conscious. In more severe heat exhaustion cases, one of the following intravenous solutions may be used: 5% dextrose (D5W), 0.9% NaCl (NS), 0.45% NaCl (1/2 NS), or 5% dextrose in 0.45% NaCl (D5 1/2 NS). All of these solutions have been recommended by various authorities for both heat exhaustion and heatstroke; no evidence exists for the clear superiority of any of them (43). The appropriate intravenous solution is determined by measurements of pulse, blood pressure, orthostatic changes, serum Na^+, protein, blood urea nitrogen, and hematocrit. The deficit in total body water (TBW) can be approximated clinically by measuring the change in serum Na^+ (7,139). A sample calculation is presented in Table 8-5 for a 70kg WD patient. This calculation assumes that TBW is equivalent to 60% of body weight and that normal serum Na^+ is 140 mEq/L; this calculation underestimates the TBW deficit to the extent that serum NaCl has been lost along with water in body fluids (primarily sweat and urine). In this example, the patient's hypernatremia due to WD (serum Na^+ = 165 on admission) indicated a 6.4 L deficit which had to be replaced by intravenous and oral fluids.

Acute water losses during exercise in the heat are not distributed evenly between all fluid compartments. A large body water deficit incurred prior to exercise in a hot environment affects performance to a greater extent if it is confined primarily to the vascular space (174). Table 8-6 presents a compilation of data from several previous studies, which induced TBW losses via a variety of investigatory techniques. It is evident that certain dehydration techniques stimulate a larger plasma volume loss (when compared to total body water loss) than others.

TABLE 8-5. *Sample Calculation of Total Body Water Deficit using Serum Na^+ Values*

1. Serum Na^+ upon examination = 165 mEq/L
2. Assumption: Total Body Water (TBW) = 60% of body weight
3. $$\frac{\text{TBW (normal)} * \text{normal serum Na}^+}{\text{observed serum Na}^+} = \text{prevailing volume of TBW}$$
   $$\frac{42 * 140}{165} = 35.6 \text{ L of TBW}$$
4. TBW (normal) − prevailing volume of TBW = water deficit
   42 L − 35.6 L = 6.4 L deficit

TABLE 8-6. *The Effects of Investigatory Protocol on the Relative Contributions of Plasma Volume and Total Body Water to Fluid Losses*

Treatment	Investigation (reference no.)	Ratio of % plasma volume change to % body weight change
Witholding fluids and food from wrestlers for 48 hours and performing moderate exercise.	Allen, Smith, Miller (5)	1.1 to 1
Work dehydration in a hot environment.	Costill, Coté, Fink, VanHandel (52)	2.8 to 1
	Costill, Coté, Fink (51)	2.0 to 1
		1.8 to 1
		2.4 to 1
Work dehydration in a cool environment.	Costill, Fink (53)	3.1 to 1
	Nielsen, Kubica, *et al.* (147)	3.4 to 1
	Dill, Costill (62)	4.0 to 1
Passive thermal dehydration.	Costill, Branam, Eddy, Fink (50)	4.1 to 1
		5.1 to 1
	Dill, Costill (62)	4.5 to 1
Diuretic induced dehydration no work.	Armstrong, Costill, *et al.* (9)	5.6 to 1
	Armstrong, unpublished observations, 1981	6.0 to 1

Salt Depletion Heat Exhaustion—A Case Report

Our research group had a rare opportunity to measure the cardiovascular (and other physiological) responses to progressive SD in a normal, healthy 32-year-old male (S.H.) who was undergoing 8 days of heat acclimation in an environmental chamber maintained at 41.2° C, 39 % rh (17). A large man (180 cm height, 110.47 kg weight, 2.34 m^2 surface area), reported that he routinely lost 4–6 kg of body fluid during training sessions as a youthful college football lineman. The protocol was strenuous and has been fully described elsewhere (12); it involved intermittent running (57–66 % $\dot{V}O_2max$) on a treadmill for 56 min out of each daily 100 min heat exposure. S.H. exhibited *Δ* heart rate, sweat rate, *Δ* rectal temperature, and *Δ* skin temperature responses on days 1 through 4 which were indicative of successful early heat acclimation. Between days 4 and 5, however, the intervention of an unknown host factor was indicated. (To simulate an actual training program, two days of rest were given between heat acclimation days 4 and 5. Upon his return to daily exercise-heat exposure on day 5, S.H. had gained 2.39kg over the weekend.) Between days 5–8 of heat acclimation, his body weight decreased 5.44kg in 72 hours. The *Δ* heart rate, *Δ* rectal temperature, and *Δ* skin temperature increased on days 5–8, at a time when low, stable values were expected. His loss of plasma volume (Δ PV%) during the trial on day 8 was 5 % greater than on day 1 (62). Most significantly, S.H. experienced extreme difficulty in completing the 100 min trial on day 8. S.H. was removed from the environmental chamber after 98 min by the medical monitor because rectal temperature exceeded 39.5° C and because he displayed classical symptoms of SD (Table 8-4): vomiting, muscular weakness, fatigue, and abdominal cramps (17). Cortisol, a marker of general body stress, and beta-endorphin, a sensitive index of heat stress (8), both were greatly elevated on day 8 (post-exercise). However, the Na^+ deficit which S.H. incurred was a mild one (< 0.1g NaCl/kg body weight), when compared to Marriott's "early" salt deficit (0.5g NaCl/kg body weight) (139). This raised questions regarding Marriott's widely-accepted categories of salt depletion.

The Gut—Vulnerable to Ischemia and Pathogens

Two hypothetical constructs may explain S.H.'s heat exhaustion and the rapid turn of events on days 5 through 8. These hypotheses involve (a) the intestines as a fluid-electrolyte reservoir and (b) distribution of cardiac output during strenuous exercise in the heat. First, there is little appreciation for the fact that animal research (212) has shown that the intestine assumes the role of a temporary water and salt depot, facilitating the preservation of water. When animals are repeatedly exposed to the influence of high environmental temperature (33°–40° C), an adaptation occurs, which results in smaller losses of water through sweat and urine. Because the human small intestine absorbs 25–35 g of Na^+ per day, which amounts to approximately one-seventh of all bodily Na^+, one can easily understand that the body can be depleted to a lethal level within hours, when fluid and electrolyte absorption at the intestine is disrupted by bacterial or viral infections (85). In fact, Israeli researchers have published a case report involving a soldier who was predisposed to two heatstroke episodes by gastroenteritis (113). Therefore, if absorption of fluid and electrolytes in the gastrointestinal tract of S.H. had been disturbed by a pathogen between days 4 through 8, his fluid-electrolyte and plasma volume profiles would be logical; such a pathogen also could explain the vomiting and abdominal cramps, which he experienced on day 8 as well as the acute body weight gain (2.39 kg, days 4–5) presumed to be fluid in the gut.

Second, many years of work by Rowell (165) have formulated a clear picture of cardiac output distribution during exercise in a hot environment. During such exercise, cardiac output is divided between internal organs, exercising muscle, and the skin. As heart rate increases, blood flow to the liver, kidney and gastrointestinal tract decreases; as heart rate approaches maximal values, liver blood flow is reduced by as much as 70%. Ischemia is imminent at this time, and may explain the well-known diarrheal distress suffered by long distance runners who compete on warm days (143). The data of Sullivan *et al.* (192) indicate that circulating levels of regulatory peptides also may explain the abdominal cramps and vomiting which S.H. experienced on day 8 of heat acclimation. Furthermore, there is some evidence that relative ischemia of the digestive organs, due to perfusion of skin, results in increased nonmalignant digestive diseases after many years of industrial exposure to high temperatures (65).

Heat Exhaustion—A Diagnosis of Exclusion

Callaham (43) stated that clinical determination of heat exhaustion is a diagnosis of exclusion, because heat exhaustion signs and symptoms are often vague and may be markedly different in different work environments. Clinical descriptions of heat exhaustion include various combinations of headache, dizziness, fatigue, hyperirritability, anxiety, tachycardia, hyperventilation, diarrhea, piloerection, chills, hypotension, nausea, vomiting, syncope, heat cramps, as well as "heat sensations" in the head and upper torso (15).

In cases of severe heat exhaustion, the levels of serum creatine phosphokinase (CPK), serum glutamic pyruvic transaminase (SGPT), and lactate dehydrogenase (LDH) may be elevated above normal levels. In fact, the degree of serum enzyme elevation may be used to distinguish between heat exhaustion and heatstroke. Other distinguishing landmarks include: spontaneous body cooling (not prominent in severe heatstroke), disrupted mental state (unconsciousness, coma present in severe heatstroke), and elevated rectal temperature (the boundary between heat exhaustion and heatstroke is usually defined as 39.4–40.0° C). However, there is a small subgroup of heat injury patients who have high temperatures, hypotension, confusion, and who fall in the middle ground between heat exhaustion and heatstroke. When the diagnosis is in doubt, these patients should be treated for heatstroke, since this injury may be fatal. The unexpected heat exhaustion episode of S.H. was a minor one—probably because rectal temperature elevation was limited to 39.5° C and because he was rapidly removed from the environmental chamber. The circulating levels of CPK, SGPT, and SGOT were normal before exercise on day 8. Only CPK (at 2–3 min post-exercise) rose above normal levels (153 U/L), but fell back to normal limits by 5.5 hr (119 U/L) and 21 hr (109 U/L) post-trial. These CPK levels may have been elevated solely by exercise.

The Incidence of Symptoms

We recently reported (15) the signs and symptoms of heat exhaustion experienced by unacclimatized males who were undergoing heat acclimation concurrently with subject S.H. These trials were valuable because controlled studies involving any form of heat injury did not exist prior to this (43), and because clinicians and scientists rarely have the opportunity to observe the preliminary course of heat exhaustion. Table 8-7 presents the signs and symptoms of heat exhaustion during 8 days of heat acclimation via intermittent, high-intensity exercise; these signs and symptoms occurred in 17.8% of all trials and were not significantly correlated with physical characteristics (e.g. age, height, mass, surface area). The most common signs and symptoms were "heat sensations" (including flushed head and upper torso) and chills. Interestingly, these two symptoms have been associated with heatstroke in previous reports, but not with heat exhaustion. By day 8, subjects were deemed to be heat acclimated, based upon improvements in heart rate, rectal temperature, skin temperature, and plasma volume. A comparison of days 1–4 vs days 5–8 (Table 8-7) indicated that heat acclimation reduced the number of signs and symptoms from 23 to 10. The authors concluded that cardiovascular adaptations played a major role in this decrease of symptoms. Roberts and Wenger (160) similarly postulated that the symptoms of heat strain may result from peripheral pooling of blood and decreased central venous pressure.

TABLE 8-7. *Signs and Symptoms of Heat Exhaustion (n = 14) During Intermittent, Intense Exercise*

Signs &	EXPERIMENTAL DAYS								
Symptoms	1	2	3	4 **	5	6	7	8	TOTAL
"heat sensations"[+]	1	1	2	2		1			7
chills	1	2	1	1	1	1			7
abdominal cramps		2						3*	5
piloerection	2	2							4
resting tachycardia[++]	1	1	1	1					4
extreme muscular weakness				1				1*	2
vomiting & nausea								2*	2
dizziness				1					1
hyperirritability							1		1
DAILY TOTALS	5	8	4	6	1	2	1	6	33

[+]—included flushed skin on head and torso
[++]—> 160 beats/min for 5 min of resting quietly
*—represents symptoms experienced by subject S.H.
**—two days of rest were given between days 4 and 5 (weekend)

Heat Syncope

Heat syncope occurs most often on the first 5 days of heat exposure (129). Figure 8-2 has been redrawn from the data of Bean and Eichna (30), in which 45 subjects lived in a hot environment for 24 hr each day and underwent exercise trials. The cardiovascular adaptations during heat acclimation clearly resulted in a rapid decline in syncopal events after day 1. Similarly, the sudden onset of severe humidity during an otherwise dry summer in Kuwait was associated with several cases of heat syncope among Arab construction workers (129). Most heat injury reviews categorize heat syncope as a syndrome distinct from heat exhaustion (7,43,64, 117,120,129), because water and salt depletion do not always contribute to this form of heat illness (43,64,129). Heat syncope is also related to the shunting of a large volume of blood through dilated cutaneous vessels, postural pooling of blood, diminished venous return to the heart, reduction of cardiac output, and cerebral ischemia. Heat syncope usually occurs when an individual performs stren-

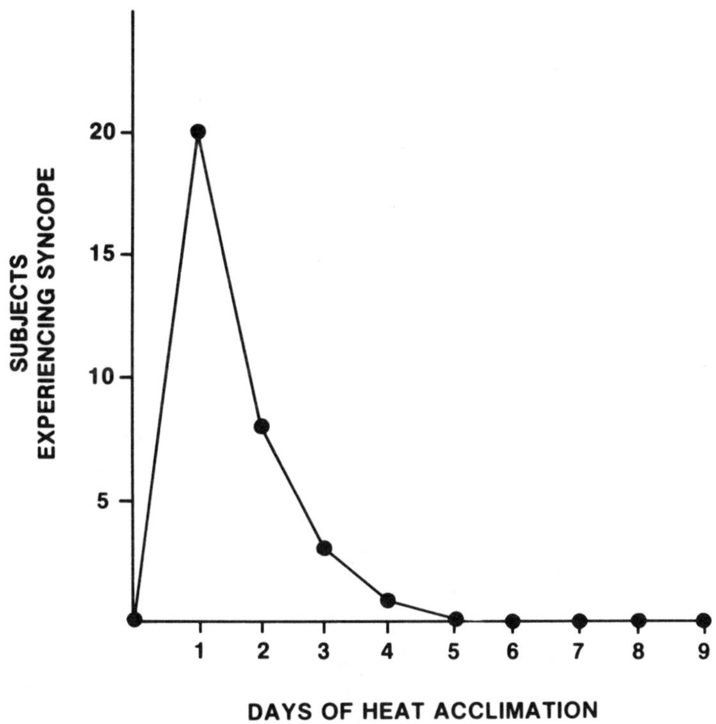

Figure 8-2. *Incidence of syncope among 45 subjects living in a hot environment for 24h each day and undergoing exercise trials. Data redrawn from Bean and Eichna (30).*

uous and unaccustomed work or exercise, or when there has been a sudden rise in ambient temperature or humidity. Recumbency, rest, avoidance of sudden or prolonged standing are adequate treatment (43,64,129).

THE CARDIOVASCULAR DEMANDS OF SEVERE HEAT STRESS AND EXERCISE

Vasodilation and increased blood flow to skin act together to increase the amount of blood pooled in peripheral vessels, which thereby reduces the central blood volume. Blood flow to muscles may increase ten fold and can be as high as 14 liters per minute (118). Cardiac output may double or increase up to 4 fold (80). Cutaneous vasodilation may increase the peripheral blood flow from 5% to 20% of total cardiac output (20). The intravascular volume is contracted as a result of fluid losses from the intravascular volume due to sweating (0.5 to 3.0 L/hr), reduced fluid intake due to voluntary dehydration, and further losses of fluid into cells due to osmotic forces (73). Thus, at a time when greater cardiac output is required, peripheral vasodilation is increasing the apparent vascular capacity and other factors are acting to reduce the available blood supply through dehydration and fluid shifts. Clearly, compensation is required (116). The decrease in the effective blood volume results in a decrease in venous return limiting cardiac filling and produces a fall in stroke volume (167,7). In order to maintain cardiac output, compensatory increases in pulse rate are required (207). Thus, pulse rate continues to rise with little or no change apparent in cardiac output.

Heat and Shock

Knochel (118) has reemphasized the concept that hard physical exercise in a hot environment may lead to a serious deficit of effective arterial volume and profound shock would occur, were it not for intense splanchnic vasoconstriction. Shock is clinically characterized by, among other things, a systolic blood pressure less than 90 mmHg, a urine output less than 30 ml per hour, and in the absence of exercise, an arterial blood lactate concentration exceeding 1.4 mM per liter (58). Barger *et al* (28) have shown that cutaneous blood flow decreases drastically as exhaustion appears during exercise in the heat. We also have reported that running rats display a drastic drop in tail temperature that coincides with an explosive rise in rectal temperature just prior to exhaustion (99). Under circumstances where exercise and, therefore, heat production are continuing, the decline in skin temperature is interpreted as a fall in peripheral blood flow. This significant event signals that exhaustion and collapse are imminent and that the underlying cardiovascular compensations supporting physical and thermoregulatory performance have failed (116,99). A spiralling increase in rectal temperature during prolonged physical effort has been observed in humans with both high (1) and low (123) physical performance characteristics.

Kielblock and associates, in a very significant paper, have explored the sequence and nature of the circulatory events during the prodromal period of heatstroke in rats (116). The laboratory rat is widely accepted as a valid model for heatstroke (102) and the regulation of cardiac output is qualitatively and quantitatively identical to that in man (86). Kielblock (116) characterized the response pattern as three distinct stages—compensation, crisis, and failure (Table 8-8). As described above, the first stage was comprised of peripheral vasodilation and the resultant fall in total peripheral vascular resistance (TPVR). Functional hypovolemia is avoided by compensatory vasoconstriction of the splanchnic circulation (57,166), nonexercising muscle and the kidney (158). During this phase, mean arterial pressure was relatively stable and the cardiovascular responses suggested well-adjusted compensation (116). As has been

TABLE 8-8. *Hemodynamic Status During Exercise Hyperthermia*

	Central Circulation		Peripheral	Tre	Central Nervous System Status
Compensation	↑ Cardiac Output to 20L/min ↑ SV, ↑ HR Tachycardia Resp Alkalosis	↓ Low SPBF (Splanchnic Blood Flow) ↓ Pl Volume (1.5–2.5L sweat losses/hr)	↓ Low TPVR ↑ Skin BF ↑ Muscle BF (4L/min)	37.0° C- to 39.5	Premonitory Signs Dizziness Headache Euphoria, Psychoses
Crises	↑ ↓ C.O., ↑ CI (16L) ↑ MABP, ↓ SV ↑ ↑ HR Tachycardia (180 bpm) Metabolic acidosis	↑ ↓ SPBF ↓ Pl Volume Moderate CVP	↓ TPVR ↑ ↓ Skin BF	39.5° C- 41.5° C	Cerebral congestion ↓ Cerebral Edema ↓ Intracranial Hypertension
Failure	↓ ↓ CO ↓ ↓ MABP ↑ HR Tachycardia Metabolic Acidosis (14.7 m M/L)	↑ ↑ SPBF (autoregulatory escape) High CVP but low if hypovolemic	↓ TPVR ↓ Low Skin BF	41.5° C	Coma, Decreased Cerebral Perfusion ↓ Cerebral Ischemia ↓ Neurologic damage, Seizures

noted in man wearing garments that inhibit evaporation (154), the convergence of core and skin temperatures signal an impending crisis.

The second stage (116) was characterized by the development of a hyperkinetic circulation (tachycardia). Gold (80) demonstrated a hyperkinetic circulatory stage after acute exposure to dry heat (54.4° C) in human volunteers. In normal men exercising with thermal stress (54.4° C), Rowell *et al* (167) reported that as heart rate increased, stroke volume, cardiac output, and central venous pressure declined. After maximal stroke volume is reached, the only way to increase or maintain cardiac output is to increase heart rate and, therefore patients with heatstroke usually display tachycardia (to 180 beats per minute) (43). The hyperdynamic circulatory pattern in heatstroke, described by O'Donnell and Clowes (149) with well defined low peripheral resistance and high circulatory demand, resembled a similar state in post-traumatic, post shock, and septic states (44). Under the prevailing circumstances, Kielblock *et al* regard the hyperkinetic circulation as a "supranormal response" (116).

The third stage was characterized by cardiac failure heralded by a fall in mean arterial pressure (116). To quote the authors, "Although circulatory failure could ultimately be ascribed to cardiac failure, the trigger in all probability was the excessive reduction in total vascular resistance following the abolishment of compensatory splanchnic vasoconstriction" (116). It is interesting to note that a failure in myocardial energy transformation reactions was also postulated.

Stine has also discussed three different hemodynamic states in heatstroke: hyperdynamic, hypodynamic, and hypovolemic (190). In each case, the total peripheral resistance is low due to peripheral vasodilation (Table 8-8). In the hyperdynamic state, cardiac output remains high (up to 20 liters per minute) but because of the low total peripheral vascular resistance, the blood pressure is normal or low. In contrast, the hypodynamic state occurs when cardiac output drops and there is a state of cardiogenic shock (149). Hypotension is common and depending upon the degree of hypoperfusion of vital organs, there is shock. Because peripheral vasodilation and the low peripheral vascular resistance are major factors in the shock state, peripheral cooling with resultant vasoconstriction should improve the cardiovascular compensation (reduce hypotension). By improving blood pressure, cooling should lower the risk of overenthusiastic fluid administration (pulmonary edema). These events are consistent with the hypothesis of Adolph and Fulton (3), who believed heatstroke to be the result of circulatory failure leading to shock. Vigorous early therapy appears to prevent the myocardial complications of heatstroke (44,149).

HEAT AND THE CNS

Although heat affects every organ system and exercise in the heat requires profound cardiovascular compensation (Table 8-8), the central nervous system is particularly vulnerable. CNS dysfunction is manifested (Table 8-8) by a range of symptoms from irritability to coma but the diagnosis of heatstroke requires a major alteration. The range of symptoms reported includes, potentially: oculogyric crises, trauma, dystonia, muscle rigidity, decerebrate posturing, and hemiplegias (178). Cabral *et al* (41) have reported reversible profound depression of cerebral electrical activity with fixed and dilated pupils and a flat EEG.

After maximal stroke volume is reached, the only way to increase cardiac output is to increase heart rate. Autoregulatory escape of the intense splanchnic vasoconstriction causes the early compensation to fail, producing a sudden decline in effective arterial volume. This increase in blood flow to the viscera triggers a rapid reduction in cardiac output, a fall in mean arterial blood pressure, a reduction in peripheral blood flow, and hypotension. This is a state of high output failure and many patients have elevated central venous pressures (149).

Intracranial Hypertension and Reduced Perfusion Pressure

Since the brain stores little energy, it is dependent upon a constant supply of blood to provide both oxygen and glucose (150). Recently, Shih *et al* (180) have studied the early pathogenesis of heatstroke in conscious rabbits exposed to an ambient temperature of 40° C. Heatstroke was identified by the presence of coma. The data demonstrate that hyperthermia occurs concurrently with cerebral edema and cerebral vascular congestion. At the onset of heatstroke (coma), intracranial pressure increased from 14 to 49 Torr. The intracranial hypertension (Table 8-8) is a significant event which in combination with the reduction in MABP conspires to produce a dramatic fall in cerebral perfusion pressure (CPP). In the rabbit experiments, the CPP declined from 80 to 19 Torr during this period. Analysis confirmed that cerebral water content increased from 78 to 83% and histological examination revealed brain congestion but a lack of cerebral or ventricular hemorrhages. This is consistent with the description of Osler (152) (i.e. "the venous engorgement is extreme, particularly in the cerebrum").

Cerebral Ischemia

Decreased cerebral perfusion pressure results in cerebral ischemia (180). Seizures are common in heatstroke and occur in approximately 70% of cases (178). During seizures there is increased electrical activity in the brain which requires increased transmembrane ion flux and energy expenditure (150). Since brain blood vessels dilate during seizures, cerebral blood flow is dependent upon the mean arterial pressure which is falling. Clearly, key system failures (the loss of splanchnic vasoconstriction and a reduction in MABP) in combination with the direct cellular effects of heat on brain tissue, set the stage for serious injury to occur.

The damaging impact of heatstroke on the brain, like ischemia, depends upon the severity and duration of the insult. The outcome in terms of tissue injury depends in large part on the metabolic consequences of events at the cellular and subcellular level (179). In heat injury, the *accumulation of heat* as well as *the rate of heat storage* produces metabolic change. At this point, other reactions or states are achieved that produce morbidity and mortality.

In brain ischemia, cell infarction can be prevented if a critical minimum blood flow can be exceeded. A concept of "ischemic flow thresholds" (25) has defined critical levels of blood flow at which cellular energy depletion results in a failure of synaptic transmission and the Na^+-K^+ pump. This concept impacts treatment by postulating that affected areas, although non-functional, are still viable with appropriate intervention. Hypothermia, by reducing the demands for glucose and oxygen consumption, provides protection from ischemia (21) and perhaps should be explored in experimental heatstroke.

Sodium Flux and Brain Metabolism

The metabolic demands of the brain are in large part due to "leakage" of Na^+ and K^+ ions (204). Membrane stabilization with reduced ion flux occurs with *hypo*thermia and large doses of lidocaine (160mg per kg), and this delays the rate of ATP depletion (25). Ouabain inhibition of the Na^+-K^+ pump in brain slices reduced basal metabolic rate by about 50% (204). Similar results were obtained with intact brain by inhibiting Na^+-K^+ ATPase with ouabain and by inhibiting the Na^+-K^+ leakage with lidocaine (23). A central conclusion from these results is that the cellular compartment of the electrically inactive (barbituate coma), but intact brain is a "leaky system" (25). Barbituates produce a flat EEG by inhibiting synaptic transmission but have no effect on Na^+-K^+ leakage or active transport (25). Lidocaine by inhibiting the sodium channels will reduce the metabolism of barbiturate-inhibited brain by another 15–20% (23,25).

Potassium and Choline Release and Brain ATP Levels

In ischemic brain, K^+ efflux is caused by a failure of the Na^+-K^+ transport due to ATP depletion (22). This may provide a clue to the cause of K^+ efflux in hyperthermic liver (34). Zeisel (213) has shown that choline release from rat brain phosphatidylcholine occurs when ATP concentrations within the brain are low and presumably is due to increased phospholipase A activity. The nerve agents, soman and sarin, will significantly elevate brain choline levels (74). This effect is unrelated to the inhibition of acetylcholinesterase by these nerve agents, but can be blocked by the anticonvulsant diazepam. Diazepam is effective in controlling heatstroke convulsions (178) and in combination with atropine has a positive ergogenic effect in rats (103).

Sweadner and Goldin (193) have drawn special attention to the critical role of the sodium-potassium pump in the function of the brain. Although agents which reduce intracellular ATP levels (42), such as cyanide or DNP block acid extrusion from nerve cells (32), Na^+ influx down its electrochemical gradient could provide the energy necessary for coupled acid expulsion (163). This is consistent with the more general concept (193) that secondary transport systems, "symports" use the sodium gradient to transport nutrients across the gut (55). In the brain, the reuptake of neurotransmitter is linked to a secondary transport system (110).

Seizures and Hyperkalemia

Glial cells play a fundamental role in maintaining the ionic homeostasis of brain by removing excess K^+ released from neurones during peak activity (91). Failure to prevent elevated potassium levels from depolarizing neural membranes results in epileptic seizures (91). Hyperkalemia (101), seizures (43), convulsions, and brain edema with swelling of nerve cells are common in heatstroke (178) and the degeneration of neurones with replacement by glial cells has been described (178).

Brain Swelling and Hypoglycemia

An increase of brain intracellular sodium has been reported in passively heated dogs (185) and the swelling of brain cells with hypoxia and hypoglycemia has been described (178). Costrini *et al* (54) found a significant difference in serum glucose in pretreatment specimens between exertional heatstroke and severe heat exhaustion patients; five of 12 patients with heatstroke, but only one of 13 with heat exhaustion, had serum glucose levels of less than 65 mg %. This observation suggests that depletion of muscle glycogen and subsequent hypoglycemia may play a role in the pathophysiology of heatstroke, especially in the CNS. Severe progressive hypoglycemia leads to EEG flattening, ATP depletion (133), membrane failure and K^+ efflux (24).

Intracellular Calcium and Neurotransmitter Release

Sweadner and Goldin (193) have also drawn attention to the role of elevated intraterminal calcium concentration in facilitating the release of neurotransmitter (179). This elevated Ca^{++} concentration may result from a slowing of outward transport of calcium due to a decrease in the transmembrane Na^+ gradient (92). A calcium increase in the cytosol will activate phospholipase A_2. A Ca^{++} increase will induce vasoconstriction in vascular smooth muscle cells and may worsen oxygen availability (31). Additionally, neuronal activity, membrane depolarization, dissipation of the Na^+ gradient and ATP depletion all facilitate Ca^{++} accumulation within the cell (179). The resultant increase in calcium activity induces calcium sequestration and binding by cellular components (mitochondira, synaptic vesicles, phospholipids and proteins). The triggering of abnormal proteolytic and lipolytic reactions would represent potential new markers for cytotoxic events. The increased K^+ efflux prior to any change in intracellular Ca^{++} indicates ion homeostasis is differentially sensitive to energy depletion (194).

TEMPERATURE AND THE CELL

The mechanisms by which cellular injury and death result from heat are not well understood or characterized, but new hypotheses are forthcoming (103). Threshold temperatures for defining the lower limits of heatstroke hyperthermia have been estimated for animals (104,175) and man (129). Dose-response curves of percent mortality and indices of cellular morbidity (102,104) have been compared as a function of maximum core temperature. Insight as to how injury is produced has been enhanced by the hyperthermic area concept (the integrated time-intensity relationship above a threshold core temperature of 40.4° C). This technique (175) serves to expand the useful range of heat stress units from about four units of temperature (ie. from 40 to 44° C) to about 200 units of hyperthermic area (deg-min). In constrast to either the duration or intensity of body heating, the hyperthermic area technique improves the resolution for quantitating subtle differences in morbidity and mortality (100).

Heat Content and Ion Flux

An increase in the heat content of the body produces more profound effects than is generally indicated by a few units of change on a thermometer. Heat increases the kinetic energy of ions in solution, increases diffusion, and increases the permeability of the cell (103). Cellular metabolism increases 13% for each 1° C rise in temperature and is 50% above normal at 40.6° C (43). The energy cost of force development in muscle increases approximately threefold (161) for each 10° C rise in temperature and the direct effects of heat are manifested by an increased metabolic rate (38,69). As the temperature of an *in vitro* muscle preparation increases, *an increase in stimulation frequency is also necessary to develop the same force* (173). Mitochondrial respiratory control (38), neural transmitter release and resynthesis, metabolic rate, force development and fatigue are all impacted by heat.

Heating Rate and Lactate Accumulation

More recently, MacDougal *et al* (134) altered the core temperatures of human subjects with a water perfused unit during prolonged treadmill running. The most rapid rise in core temperature was associated with the hyperthermal condition which increased the $\dot{V}O_2$, but *depressed the performance time* from 91 to 48 min. Stroke volumes declined with time and this was accelerated in the hyperthermal condition. Most striking were the similarities in positive slopes of minute ventilation and venous lactate accumulation (Fig. 8-3). A reduction in the efficiency of energy metabolism was proposed as a possible explanation of the results.

Heating Rate and Exhaustion

We have recently summarized a decade of research relating the rat's endurance capacity (the amount of physical work done) to the average rate of heat storage prior to exhaustion (103). The results (Fig. 8-4) appear to describe a "continuum" characterized by physical exhaustion (high total work output, low rate of temperature rise) at one extreme (upper left of curve) and heat collapse (low total work output, high rate of temperature rise) at the other (lower right). By varying both the body weight (237 to 500g) and the environmental temperature (5 to 26° C), it was possible to demonstrate that the rate of rise of rectal temperature, rather than the metabolic rate, was a major determinant of physical work capacity. Furthermore, as the rate of temperature rise was increased (even independently of metabolic rate), work capacity dropped off markedly (Fig. 8-4). In support of this concept, recent data on exhaustive treadmill ex-

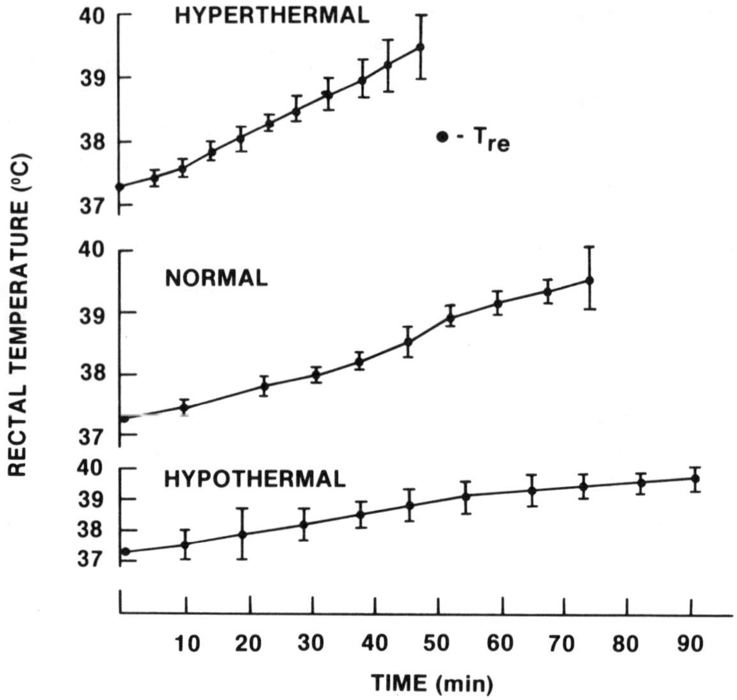

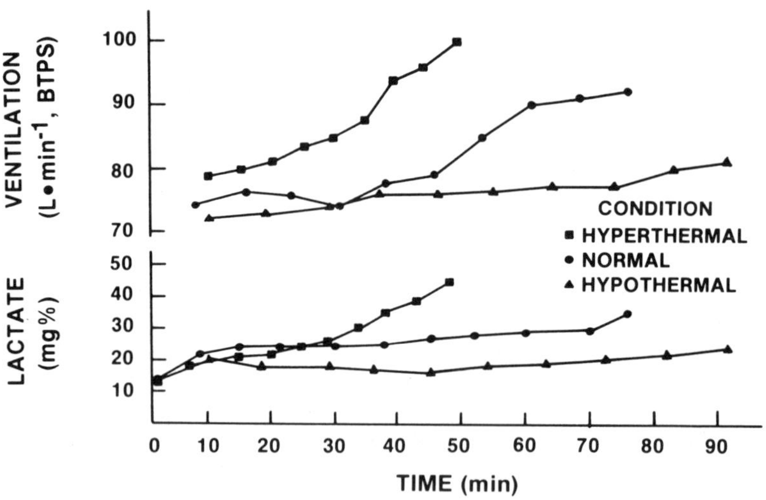

Figure 8-3. *Rectal temperature (n = 4), mean venous lactate concentrations and minute ventilatory volumes (n = 6) of six subjects during prolonged exhaustive treadmill running under three thermal conditions: normal, hyperthermal, and hypothermal. Treadmill speed was identical under each condition and was set at approximately 70% of each subject's maximum aerobic power.*

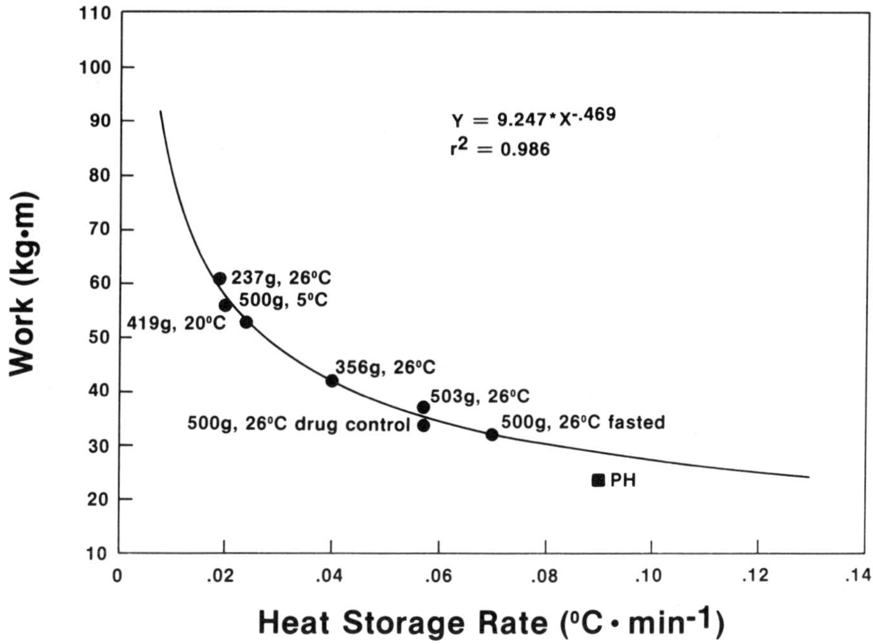

Figure 8-4. *Work done (kg · m) vs. heat stored (°C · min^{-1}) for groups of rats of different weights exercised at different temperatures. One group of rats was administered physostigmine (PH) prior to running. Redrawn from reference 103.*

ercise in dogs (125) in which external cooling was applied (ice packs) demonstrated increased exercise duration. The authors found that cooling the dogs during exercise resulted in lower blood lactic acid levels, higher muscle ATP concentrations, and lower muscle AMP concentrations (122). Clearly, evidence is mounting that the heat the cell sees (i.e. stored heat) has a major impact on the efficiency of energy metabolism.

Heating Rate, Symptoms, and Injury

Shibolet has suggested that "the rate of temperature change is also of importance in causing clinical signs" (178). For example, heat induced hyperventilation with dyspnea and tingling sensations of the hands and feet has been described (3). Few studies have dealt with this subject, but Iampietro and colleagues (106,107) measured several parameters thought related to the symptoms ranging from slight tingling of the fingers and toes to carpopedal spasm. The percentage of subjects exhibiting symptoms (20–90%) was inversely related to the mean exposure time (173–27 minutes). A direct relationship between the *rate of change* of increase in rectal temperature, increase in pH, decline in CO_2 content and fall in CO_2 tension, and incidence of symptoms was described. No correlation between the absolute change in these parameters and the incidence of symptoms was found. The authors concluded that when the rate of change was high, there was an "imbalance between intra- and extracellular compartments," which may have induced the symptoms. The initial loss of CO_2 is through the lungs (hyperventilation) and secondarily through the skin (sweat) which increases blood pH. Hyperventilation in heatstroke is a universal finding with rates up to 60 breaths per minute (43).

Paradigms which contrast absolute change from rate of change are continuing to provide new insights into the effects of heat on the cell. For example, in a cleverly designed experiment, two groups of rats were subjected to similar sub-lethal elevations in mean body temperature, but disimilar rates of body heat gain (138). Membrane permeability as assessed by elevations in serum CPK and LDH increased dramatically with the higher rate of body temperature change.

THE HYPOTHETICAL SITE OF HEATSTROKE PATHOPHYSIOLOGY

These prior findings will hopefully serve to refocus attention on the insult to the cell from increased rates of heat storage without diminishing the clinical significance of the well known system failures common to heatstroke pathophysiology. In keeping with this approach, we have hypothesized (103) that the cellular site of action for initiating the pathophysiology of heatstroke should accomodate the hypothetical characteristics set forth in Table 8-9.

TABLE 8-9. *Site of Cellular Heatstroke Injury: Hypothetical Characteristics*

* Common feature of all cells—esp nerves and muscles
* Temperature sensitive
* Related to cell volume changes
* Functionally related to endurance training
* Functionally related to the acclimatization response
* Functionally related to tolerance and fatigue
* Ability to generate heat
* Potential for inducing irreversible change

Cholinergic Stimulation During Work in the Heat and Ion Flux

Cholinergic nerves control both muscarinic heat loss effectors (sweat glands), as well as heat-producing reactions at the neuromuscular junctions (voluntary contractions, shivering). During heat acclimation, centrally mediated adaptations are transmitted by sympathetic cholinergic pathways. Exercise is thought to be essential in achieving a fully-acclimated, heat-tolerant condition. Endurance training increases the capacity to withstand heat while sustaining exercise, possibly by altering muscle energy metabolism (77). Antimuscarinic drugs such as atropine, which can traverse the blood-brain barrier, produce a "central anticholinergic syndrome" which bears a striking resemblance to heat illness (6,103). Edwards (67,68) has reviewed the many steps leading to a voluntary contraction of skeletal muscle. This "chain of commands" includes many potential sites of fatigue including motivation, spinal transmission, motor neuron recruitment, peripheral nerve function, and neuromuscular junction transmission (67,68). Sodium and potassium gradients necessary for the propagation of these neural impulses are maintained by a specific enzyme in the cell membrane, the sodium and potassium-dependent adenosine triphosphatase (sodium-potassium-ATPase) (193). Temperature affects these processes (68,69) by altering the tendency to fatigue by altering ATP turnover, lactic acid accumulation, and neuromuscular transmission.

The Sodium Pump, Basal Metabolism and Cell Volume Regulation

Sodium-potassium-ATPase activity is regarded as the enzymatic equivalent of the sodium pump (182). Active sodium transport is present in all cells and accounts for a high proportion of the total energy use (66) in resting cells (20 to 45%). Thus, the sodium pump not only regulates intracellular ion composition but also acts as a major source of heat for mammals (66). The regulation of sodium transport is considered crucial to the evolution of homeothermy (189).

The rapid sodium influx into the cell via either acetylcholine-gated (ACh) ion channels of the muscle membrane or voltage-gated sodium channels of the nerve membrane stimulates the sodium-potassium electrogenic pump. The active transport of either ion requires the presence of the other on the opposite side of the membrane (79) and takes place against both concentration and electrical gradients. The active transport is tightly coupled and requires that one molecule of ATP be hydrolyzed in order to return three sodium ions to the exterior of the cell in exchange for two potassium ions. Each cycle of the pump removes a net positive charge from inside the cell, increases ionic concentration outside the cell, and draws water out. Thus, the sodium-potassium pump is important in maintaining cell volume.

The velocity of the transport system depends upon the concentration of the cations at their respective binding sites (193) and is remarkably sensitive to an excess of sodium ions inside the cell membrane. Since pumping activity increases approximately in proportion to the third power of the sodium concentration, a doubling in internal sodium concentration would result in an eightfold increases in ATP hydrolysis (85). Thus, any factors or conditions which would stimulate the flux of sodium into the cell would increase ATP utilization, would increase heat production and would constitute an energy drain upon the organism.

Dehydration, Sodium Flux, and Heat Production

Dehydration, particularly if accompanied by hypernatremia results in an increase in body temperature and a combination of voluntary dehydration with excessive losses of hypotonic sweat predisposes to heat injury (105). In primary water depletion heat exhaustion, the deficit in total body water is assumed to be inversely proportional to the elevation of serum sodium (7). Since the membrane potential that would serve to oppose the movement of sodium into the cell (Nernst potential) is approximately +61 mV, and since the actual membrane potential is −65 mV, sodium tends to diffuse into the cell. The rate of net diffusion is proportional to the concentration outside minus the concentration inside (85) and hypernatremia may thereby stimulate the sodium pump. This effect has been studied *in vivo* (135) and can be prevented by ouabain in tissue preparations suggesting that hypernatremia increases intracellular sodium (148). Heat production could be stimulated by either an increased rate of neural stimulation (eg. hyperthermia, carbamates) or an increased leakage of sodium into the cell (hyperthermia, hypernatremia).

Heat, per se, increases the kinetic energy of ions in solution, increases diffusion and thereby, increases the permeability of the cell to sodium ions. Neuro-and cardiovascular surgeons recognize that *hypo*thermia provides clinical protection from circulatory arrest by lowering the basal metabolic rate (25). Cooling restricts Na^+ channels, delays energy depletion and K^+ efflux, and stabilizes the cell membrane (25). Lidocaine, by blocking the Na^+ channels, provides an additive effect. Other sodium dependent, ouabain sensitive, calorigenic states include thyroid thermogenesis, chronic cold exposure, chronic ethanol administration, and the thermogenic effect of norepinephrine on brown fat (66). Physostigmine, a reversible cholinesterase inhibitor, causes acetylcholine to accumulate at cholinergic receptor sites. This produces an exaggeration of the normal response to ACh equivalent to excessive stimulation of the cholinergic receptors throughout the body. This has the effect of stimulating increased sodium leakage into the cell and sodium pump activity, especially at nicotinic type effectors, and superimposes an inefficient energy drain at the cellular level. Rats, treated with physostigmine and run to exhaustion, demonstrated a significant increase in heating rate (Fig. 8-4) coupled with a significant reduction in endurance capacity (103). Intracellular acidosis stimulates an energy dependent Na^+-H^+ exchange across the cell membrane that, while regulating intracellular pH (163), increases Na^+ permeability. A summary of these drivers of the cell membrane Na-K^+ pump is shown in Table 8-10.

TABLE 8-10. *Drivers of the Cell Membrane Na^+-K^+ Pump*

Kinetic Energy	—Heat increases diffusion and permeability
Dehydration	—Increase in extracellular Na^+
	—Decrease of extracellular water
Acidosis	—Increase in transmembrane Na^+-H^+ exchange
Sweating/Salivation	—Stimulation of muscarinic receptors
Exercise/Shivering	—Stimulation of nicotinic receptors
Drugs	—Cholinomimetic
	—Anticholinesterases
Altered Membrane Permeability	—Toxic agents: alcohol, metabolites, enzymes, etc.
	—Ionic imbalance, Li^+

Fatigue, Lactic Acidosis, and Heatstroke

Muscle fatigue is defined as a failure to sustain force or power output (157) and is used synonomously with a decline in tension (171). Slowing of relaxation is a feature of fatigued muscle (171). Relaxation rate bears close relationship to calcium uptake by sarcoplasmic reticulum (36) which is also consistent with a reduction of the extracellular—intracellular sodium gradient. The latter favors accumulation of calcium ions in muscle cytoplasm (193). The reuptake of calcium with a decline in cytosolic calcium concentration and the release of cross-bridges is an energy dependent process (169). In extremes of energy deficiency, the muscle remains contracted and cannot relax (rigor). Rigor does not normally occur (169) except when glycolysis is inhibited such as in muscles poisoned with iodoacetic acid (IAA) or in McArdles Syndrome (phosphorylase deficiency). Although feelings of fatigue and muscle weakness as symptoms of heat illness are defined more subjectively, *muscle rigidity and rapid rigor mortis are classic features of heatstroke death* (208). Furthermore, the extensibility of muscle during the development of rigor mortis appears linearly related to the muscle ATP content (29). The stimulation of IAA poisoned muscle resulted in pronounced decrements in both ATP *and the total adenine nucleotide content* (about 50%), which correlated with early fatigue and rigor (171). Does this suggest that the energy from glycolysis is vital in maintaining the impermeability of the membrane to nucleotide depletion? The stimulation of unpoisoned muscle resulted in decreased tension (−50%), increased relaxation time (+250%), and a 15-fold increase in muscle lactate. Hydrogen ions were formed in an equivalent amount to lactate and reduced the pH from 7.10 to 6.76. In contrast, muscles poisoned with IAA had glycolysis almost completely inhibited and as a result there was no accumulation of protons. The relaxation time of poisoned muscles was unchanged even when tension had declined by 50% (171). Does this suggest that a Na^+-H^+ exchange occurred in unpoisoned muscle which altered the extracellular-intracellular sodium gradient sufficiently to raise the cytosolic calcium concentration?

Patients with exertional heatstroke usually have metabolic acidosis with whole blood lactate concentrations in the range from 6 to 13 mM and in severe cases exceeding 20 mm (120). Blood lactate levels in exertional heatstroke do not appear predictive of patient outcome (118). In contrast, in classical, non-exertional heatstroke (89) modest elevations in blood lactate correlated with permanent neurological deficits (5.4 ± 0.83 mM/liter) and 4 deaths (9.650 ± 2.54 mM/liter).

In 1970, Weil and Afifi (202) reported on the significance of arterial lactate as a predictor of mortality and survivability in shock. Of 142 patients who presented with clinical manifestations of circulatory shock, 62 survived and 80 died. In this series, as lactate increased from 2.1 to 8.0 mM, the estimated probability of survival decreased from 90 to 10%. Discriminant function analysis confirmed the percentage probability of misclassification based on lactate was 12%. The combination of excess lactate and the lactate/pyruvate ratio with lactate failed to improve discrimination. In contrast to the presence of shock, single or multiple bouts of maximal exercise can elevate blood lactate above 20 mM/liter (81).

Lactic Acidosis, Inhibition of Glycolysis and the Na^+-H^+ Exchange

An increase in lactate concentration is accompanied by an equimolar increase in hydrogen ion and a decrease in intracellular pH (pH_i). A decrease in muscle pH can also be accomplished by elevation of CO_2 tension. CO_2 induced acidosis resulted in decreased tetanic tension, prolonged relaxation time, a decrease of phosphocreatine content but no change in lactate content (172). Because lactic acid (pk = 3.9) is nearly totally dissociated at physiological pH, its hydrogen ions must be immediately buffered or an inhibition of glycolysis and fatigue will result. The main regulatory step in inhibiting glycolysis appears to be phosphofructokinase activity (199) and lowering the pH_i causes marked inhibition. The activation of phosphorylase "b" to "a" (glycogenolysis) is also pH sensitive and a reduction of pH_i depresses phosphorylase conversion (60). The inhibition of glycolysis by acidosis also helps contribute to the regulation of pH_i by reducing the production of pyruvate (199).

Acidosis reduces myocardial (187) contractility (negative inotropic effect), wherease alkalosis has a positive inotropic effect. Acidosis also reduces the myocardial responsiveness to catecholamines (205) and depresses the activity of myofibrillar ATPase (72). The short term regulation of intracellular pH involves several relatively rapid mechanisms that consume acid including physico-chemical buffering, consumption of non-volatile acids, and the transfer of acid and base between cytosol and organelles (163). For example, phosphocreatine (PCr) acts as an energy source and as a temporary buffer against lactic acidosis. During high rates of glycolysis, there is a utilization of PCr via the creatine kinase reaction(s) (168):

$$H^+ + PCr + ADP \rightarrow ATP + Cr$$

Likewise, the conversion of lactic acid to glucose or to CO_2 and water results in the loss of intracellular hydrogen ions. However, as Roos and Boron (163) explain these are only short term measures of limited capacity.

In order to restore pH_i and the intracellular buffering capacity, the long term solution requires the transmembrane extrusion of H^+ by an *energy consuming* reaction. The Na^+-H^+ exchange requires external HCO_3^- and internal Cl^-, and experiments with ion specific electrodes and isotopes confirm that acid extrusion is accompanied by the influx of Na^+ and the efflux of Cl^- (163). Recovery from cellular acidosis represents an added energy drain upon the cell. It is interesting to note that acidosis increases and alkalosis decreases serum potassium independently of total body potassium (39) and for every 0.1 unit change in pH, there is an inverse 0.6 mEq/liter change in potassium. Acidosis therefore creates an ionic imbalance across the muscle membrane with a tendency toward increased intracellular sodium and extracellular potassium. A change in the ionic composition of fatigued muscle in this direction has been reported (170,181).

Temperature, pH, and Membrane Permeability

An extensive literature indicates that there is an inverse relation between temperature and pH (pH falls as temperature is raised (159)). It is therefore consistent with the foregoing that a rise in body temperature to about 39° C increases serum potassium (45). This may occur despite hyperventilation, a fall in PCO_2, and acute respiratory alkalosis (120). It is interesting to note that highly trained marathoners (164), and heat acclimatized individuals (119) may show low-normal serum potassiums at rest.

Isolated rat livers perfused at normal and elevated temperatures demonstrated that K^+ leakage was the earliest sign of tissue injury (34). Bile flow, an indicator of liver function, was linear at 37° C, but decreased as perfusion temperature increased to 43° C. Bile flow and K^+ leakage had a strong inverse relationship. Livers perfused at 41° C had reduced glycogen but normal structural appearance. The leakage of liver transaminase enzymes did not occur until perfusion temperature reached 42° C and this correlated with signs of structural injury including loss of endothelium, reduced numbers of microvilli, and marked hepatocellular vacuolization (34). These signs are consistent with the hepatocellular damage observed in rat and human heatstroke (33). The direct efflux of potassium and muscle enzymes into the blood stream has been correlated with exhaustive work (198) and depletion of intracellular ATP (206). A decline in muscle endurance with an increase in muscle temperature is correlated with an increased ATP turnover rate for a given force maintenance (68). The maintenance of cell function and membrane integrity is therefore related to the energy content of the cell. If potassium is leaking out of the cell down its concentration gradient, is it not logical to assume that sodium is leaking into the cell also?

Lactate Metabolism

Over 64 years ago, A.V. Hill (93) advanced the hypothesis that lactate accumulation during exercise represented an anaerobic supplementation of energy supplies due to an oxygen deficit. The lack of oxygen within the tissue was thought to cause a slowing of electron transport with an increase in the intramitochondrial ADP concentration. The intramitochondrial ADP coupled with an increased reducing potential (reduced coenzymes) within the cell, increased glycolysis and lactate production within the cytoplasm. This classic view appears to relegate lactate production to a back-up position and implies a more fundamental role for aerobic mechanisms in serving cell function.

Lactate accumulation is considered an inefficent use of glycogen, because it produces only 3 ATP per glucosyl unit as opposed to 37 with complete oxidation (169). However, the continued production and catabolism of lactate during rest (96,124) as well as exercise (96,109) attacks the fundamental premise that lactate reflects an oxygen deficit or an anaerobic component. Whereas lactate production could reflect an anaerobic component *in vitro*, an *in vivo* accumulation could, in part, represent a redirection of blood flow away from the liver and kidneys during splanchnic vasoconstriction. A recent review (37) has emphasized that lactate production is not necessarily associated with muscle anaerobiosis. Furthermore, muscle (37) is seen to play a profound role in the carbohydrate metabolism of resting individuals by converting much of the glucose undergoing glycolysis to lactate, which serves as a substrate for glycogen synthesis by the liver (new "Glucose to Liver Glycogen Pathway").

Sahlin (169) has pointed out that lactic acid formation has two distinct advantages over aerobic ATP production: a) The maximal rate of ATP formation by anaerobic glycolysis is about 2-fold higher than from oxidative phosphorylation (142) and b) ATP production from glycolysis can accelerate from a resting to a maximal rate in a shorter time (<5s) than maximal power can be generated by aerobic ATP production (2–3 min). Sahlin concluded (169) that anaerobic ATP production would occur either when the energy demand was high or rapidly increasing. This appears analogous to either a high heat content or rise in heating rate.

Does Glycolysis Primarily Serve the Sodium-Potassium Pump?

Active sodium transport accounts for a significant portion (up to 45%) of the basal metabolic rate (66) and approximately 40–50% of the lactate produced at rest is oxidized (37). During sustained, steady-state exercise, the proportion of lactate oxidized increases further (75%+), especially in the active muscle (37). These data suggest that lactate production and oxidation are direct contributors to the immediate energy needs of the cell both at rest and during voluntary muscular work. This view stands in direct opposition to the classical position of supplemental energy source.

Furthermore, the rapidity (169) with which ATP production can be increased via glycolysis appears uniquely suited to an ion pumping role during membrane repolarization. Since the Na^+-K^+ ATPase is located within the cell membrane, a soluble enzyme system to produce ATP could theoretically provide better coverage of the inner membrane surface than a number of point sources (mitochondria) within the cell. This spatial geometry should probably not be ignored, especially since the ADP and P_i liberated by pump activity will stimulate glycolysis directly (169,186). This model does not require that oxygen availability be decreased or electron transport be slowed to account for active glycolysis. Lactate production would occur as reducing equivalents (reduced coenzymes) accumulated within the cytoplasm, perhaps reflecting that the mitochondrial oxidative capacity had been exceeded by the rate of glycolysis (114).

Askew (19) has recently reviewed nutrition and performance under adverse environmental conditions and has cited the research of Consolazio *et al* (48), describing an increased rate of energy metabolism at higher environmental temperatures. There was approximately a 10% increase in metabolic rate for work at 37.8° C compared to work at 21.1° C (49). Although no specific requirements for high carbohydrate diets in the heat appear to have been identified (19), our hypothesis suggests that part of the increased energy demand would be met by production of lactate via glycolysis. This hypothesis also suggests that any attempt to reduce the carbohydrate content of hot weather rations be approached cautiously.

Lactate and Heating Rate Revisited

Although lactate concentration is elevated both at rest and during submaximal exercise by environmental heat (37), *heating rate* receives little or no comment in discussions of lactate metabolism (37,81,186,201). Yet, we are struck by how closely lactate tracks heating rate (Fig. 8-3) in humans during constant work (134). A similar finding (Fig. 8-5) was evident in the recent data of Owen *et al* (153), but received no comment by the authors. Jacobs (111) has pointed out that the lactate variable is more highly correlated with running performance than other variables including the maximal aerobic power. This is consistent with the highly significant inverse relationship between endurance capacity and heating rate in humans and rats and the high correlation between heating rate and lactate production. This suggests that part of the observed variability in individual anaerobic threshold lactate concentrations (188) could be related to different rates of heat storage. Furthermore, since the lactate response to exercise is used as a longitudinal marker of adaptation to endurance training (111), it should be correlated with heat acclimation as well.

Heating Rate, Lactic Acidosis, the Sodium Pump, and Fatigue

If one accepts the validity of the inverse relationship between physical work output to exhaustion and heating rate, then a potential mechanistic relationship between heating rate, lactate accumulation, and fatigue makes sense. The key factors in this relationship are all "coupled drivers" of the Na^+-K^+ pump and include: a) heat increases acidity and acidity stimulates the Na^+-H^+ exchange, b) heat increases diffusion and diffusion increases sodium permeability, c) heat increases dehydration and dehydration increases sodium permeability by increasing the extracellular sodium concentration, d)

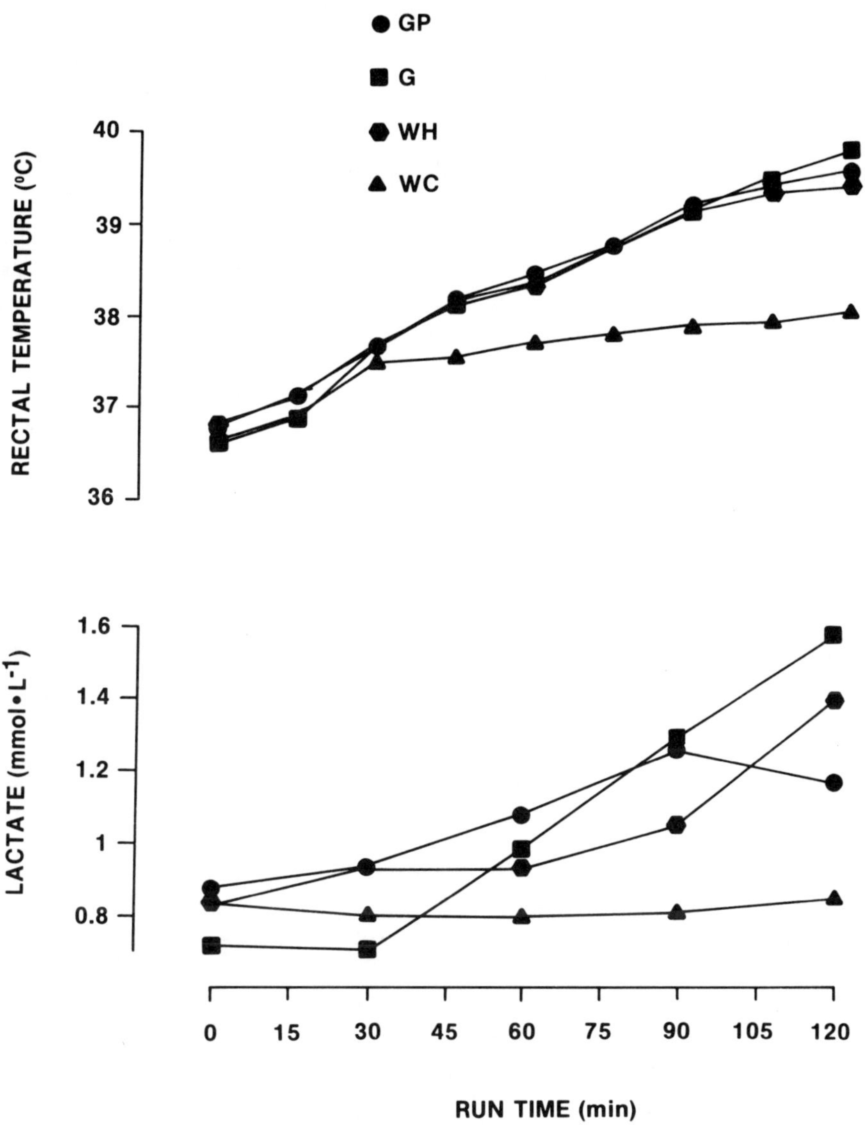

Figure 8-5. *Effects of four different 2-hr treadmill runs at 65% $\dot{V}O_2$ max on concentration of plasma lactate. Three of the runs were performed in the heat (Tdb = 35° C) and included receiving 200ml every 20 min of either a glucose polymer (GP), glucose (G), or water (WH) drink. The fourth run was performed in a cool environment (Tdb = 25° C) and included receiving 200 ml every 20 min of a water drink (WC).*

heat increases the stimulation frequency necessary to maintain muscle force and this increases sodium-potassium flux, e) heat and work produce regional ischemia (early splanchnic; late cerebral) which increases acidosis and Na^+ flux, and finally, f) heat and work increase cholinergic stimulation which increases Na^+ flux. All of these factors and probably more yet to be determined are contributing to fatigue.

THE ENERGY DEPLETION MODEL

It is likely that until we see heat illness as something more than a problem in thermal balance, we will never fully understand it. A good beginning might be substituting "energy" for "thermal." By focusing on the cell, we are attempting to take our concept of heatstroke and the other heat illnesses beyond the simple clinical criteria we use to define them. The concept that heatstroke develops whenever "elevated body temperature in itself becomes a noxious agent," (78) is *a message without a mechanism.* Our purpose then is to work toward achieving a new theory of heat pathophysiology at the cellular level (108) and to produce new experimental paradigms for future research. The intent is to supplement the vast clinical data base that already exists with experimental research from the fields of exercise physiology, cell biology, and neurology.

We have compiled the hypothetical characteristics of a cellular site (Table 8-9) which would mediate the physical effects of heat stress into the physiological manifestations of heat strain. The cell membrane and the sodium pump (Fig. 8-6) appear to fulfill many of the hypothetical properties sought. Moreover, they are in direct contact with the hot blood, lymph, and extracellular fluid (milieu interiéur) and must function to balance its destabilizing effects on the intracellular milieu. Astrup (21) has reviewed the evidence that ischemia or severe hypoglycemia can result in cellular energy depletion, membrane failure and the consequences shown in Table 8-11.

We have reported that the incidence of morbidity and mortality in both rats and humans (97) is higher with exercise—induced hyperthermia than with *equivalent heat loads* in the absence of physical

TABLE 8-11. *Characteristics of the Energy-Depleted Cell*

* Depletion of ATP
* Failure of Na^+-K^+ transport
* Efflux of K^+ and influx of Na^+
* Cell swelling
* Shrinkage of extracellular space
* Membrane failure

effort. We have hypothesized (103) that a thermally driven energy drain is superimposed on that caused by exhaustive physical work. This combination exceeds the hypothetical threshold for cellular injury and explains why an equivalent degree of hyperthermia in the absence of work produces less injury. The focal point for this hypothesis is that energy depletion is worsened by a collaboration between heat and work. Since a significant portion of the energy drain due to heat, per se, could be anaerobically derived, this would not necessarily require an immediate increase in oxygen consumption. We have recently stressed the central role that cholinergic mechanisms play in endurance training, the heat acclimation response, and improvement in heat tolerance (103). We have diagrammed this increased rate of neurotransmitter and neuromuscular activity as part of a dynamic relationship between rate of heat gain, cellular energy depletion, membrane permeability, and heat/exercise tolerance (103).

This model takes the form of a vicious circle (Fig. 8-6) whose ultimate outcome at the cellular level would depend on the duration, intensity and rate of heating as well as variations in regional and local circulation within the affected tissue. For example, brain cells (neurones vs. glial cells) differ in their susceptibility to injury and neurones display a regional susceptibility to damage (179). The liver is a prime target for heatstroke damage (115), perhaps through

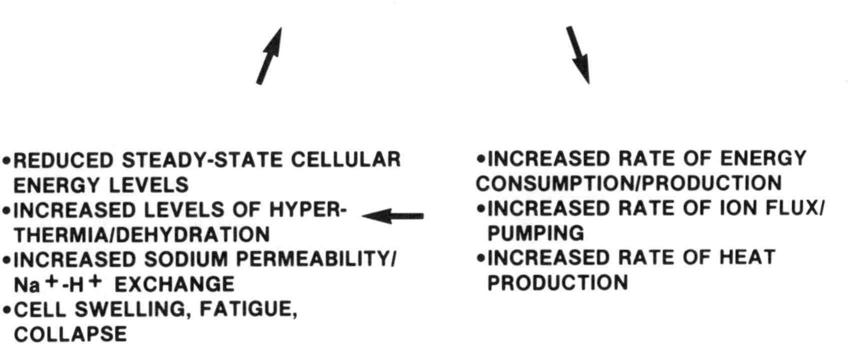

Figure 8-6. *Energy depletion model characterized as a viscious circle and taken from Hubbard* et al. *(103).*

a combination of higher temperature, energy depletion, tissue acidosis, functional ischemia, and reperfusion with oxygen (141). Hepatocytes along the vascular pole and endothelial lining cells are particularly susceptible to damage (115).

Lactic acidosis in conjuction with muscular work, an increased heating rate (134,153), hypohydration (105), and the Na^+-H^+ exchange (163) all act to increase Na^+ permeability and K^+ efflux. Cell swelling (21), stimulation of the sodium pump (16), further energy utilization, and heat production (21), in combination with *depressed mitochondrial respiratory function* (94), reduce cellular energy levels. ATP depletion and failure of the Na^+-K^+ transport system leads to intracellular Ca^{++} accumulation (179), which stimulates neurotransmitter release and, along with voluntary contractions, closes the circle.

Siesjo and Weilock (179) have described how the deterioration of the cellular energy state causes the initiation of dissipative ion fluxes, metabolic cascades, and reactions leading to irreversible cell damage (Fig. 8-6). Depletion of adenine nucleotides by deamination and dephosphorylation pathways (179) can lead to substrates for free radical formation thought to be destructive to membranes and endothelial cells (141). Xanthine oxidase, a major source of free radicals, is found in high concentrations in liver, intestine, and lung (141).

This Energy Depletion Model suggests that cellular and metabolic processes are initiated during hyperthermia that are still operating for some time after a patient's temperature has returned to the normal range with cooling. Since a significant portion of the energy dissipated by these processes could be anaerobically derived via lactate production, oxygen uptake measurements would not provide a complete picture of the energy demand or the efficiency with which it was met (103). Newer indices of cell injury and the metabolic pathways involved must be borrowed from other fields (21,179,213) and applied to the heat illnesses before this model can be verified or discounted.

REFERENCES

1. Adams, W.C., R.H. Fox, A.J. Fry, and J.C. MacDonald. Thermoregulation during marathon running in cool, moderate, and hot environments. *J. Appl. Physiol.* 38:1030–1037, 1975.
2. Adolph, E.F. *Physiology of Man in the Desert*. New York: Interscience, p.1–357, 1947.
3. Adolph, E.F., and W.B. Fulton. The effects of exposure to high temperature upon the circulation in man. *Am. J. Physiol.* 67:573–588, 1924.
4. ALBA's Medical Technology. Anaheim, CA: Berkley Scientific Publications, p.55–60, 1980.
5. Allen, T.E., D.P. Smith, and D.K. Miller. Hemodynamic response to submaximal exercise after dehydration and rehydration in wrestlers. *Med. Sci. Sports* 9:159–163, 1977.
6. Aquilonius S.M. Physostigmine in the treatment of drug overdose. In: *Cholinergic Mechanisms in the Treatment of Drug Overdoses*, D.J. Jenden, (ed.). NY: Plenum Press, p.817–825, 1977.

7. Anderson, R.J., G.R. Hart, W.G. Reed, and J. Knochel. In: *Early Assessment and Management—Heat injuries. Emergency Medicine Annual,* v. 1, B. Wolcott and D.A. Rund (eds.). Norwalk, CT: Appleton-Century-Crofts, p.117–140, 1982.
8. Appenzeller, O., M. Khogahli, D.B. Carr, K. Gumaa, M.K.Y. Mustafa, A. Jamajoom, and B. Skipper. Makkah Hajj: Heatstroke and endocrine responses. *Ann. Sports Med.* 3:30–32, 1986.
9. Armstrong, L.E., D.L. Costill, and W.J. Fink. Influence of diuretic-induced dehydration on competitive running performance. *Med. Sci. Sports Exerc.* 17:456–461, 1985.
10. Armstrong, L.E., D.L. Costill, and W.J. Fink. Changes in body water and electrolytes during heat acclimation: Effects of dietary sodium. *Aviat. Space Environ. Med.* 58:143–148, 1987.
11. Armstrong, L.E., D.L. Costill, W.J. Fink, D. Bassett, M. Hargreaves, I. Nishibata, and D.S. King. Effects of dietary sodium on body and muscle potassium content during heat acclimation. *Eur. J. Appl. Physiol.* 54:391–397, 1985.
12. Armstrong, L.E., R.W. Hubbard, J.P. DeLuca, and E.L. Christensen. Self-paced heat acclimation procedures. Natick, MA: USARIEM Technical Report No. T8-86, p.1–28, 1986.
13. Armstrong, L.E., R.W. Hubbard, J.P. DeLuca, and E.L. Christensen. Heat acclimatization during summer running in the northeastern United States. *Med. Sci. Sports Exerc.* 19:131–136, 1987.
14. Armstrong, L.E., R.W. Hubbard, B.H. Jones, and J.T. Daniels. Preparing Alberto Salazar for the heat of the 1984 Olympic marathon. *Physician Sportsmed.* 14:73–81, 1986.
15. Armstrong, L.E., R.W. Hubbard, W.J. Kraemer, J.P. DeLuca, and E.L. Christensen. Signs and symptoms of heat exhaustion during strenuous exercise. *Ann. Sports Med.,* 3:182–189, 1988.
16. Armstrong, L.E., R.W. Hubbard, P.C. Szlyk, W.T. Matthew, and I.V. Sils. Voluntary dehydration and electrolyte losses during prolonged exercise in the heat. *Aviat. Space Environ. Med.* 56:765–770, 1985.
17. Armstrong, L.E., R.W. Hubbard, P.C. Szlyk, I.V. Sils, and W.J. Kraemer. Heat intolerance, heat exhaustion monitored: a case study. *Aviat. Space Environ. Med.,* in press, 1987.
18. Armstrong, L.E., W.T. Matthew, P.C. Szlyk, R.W. Hubbard, I.V. Sils, J.P. DeLuca, H.J. Hodenpel, and P. Evans. Evaluation of a field expedient technique for sweat sample collection. Natick, MA: USARIEM Technical Report, T2-85, p.1–45, 1984.
19. Askew, E.W. Nutrition and performance under adverse environmental conditions. In *Nutrition in Exercise and Sport.* J. Hickson, and I. Wolinsky (eds.). Cleveland: CRC Press, Inc., in press, 1987.
20. Astrand P.O., and K. Rodahl. *Textbook of Work Physiology: Physiological Basis of Exercise,* ed 2. NY: McGraw-Hill Co, p.55–128, 1977.
21. Astrup, J. Energy-requiring cell functions in the ischemic brain: Their critical supply and possible inhibition in protective therapy. *J. Neurosurg.* 56:482–497, 1982.
22. Astrup, J., G. Blennow, and B. Nilsson. Effects of reduced cerebral blood flow upon EEG pattern, cerebral extracellular potassium and energy metabolism in the rat cortex during bicuculline-induced seizures. *Brain Res.* 177:115–126, 1979.
23. Astrup, J., P. Moller Sorensen, and H. Rahbek Sorensen. Inhibition of cerebral oxygen and glucose consumption in the dog by hypothermia, pentobarbital and lidocaine. *Anesthesiology* 55:263–268, 1981.
24. Astrup, J., S. Rehncrona, and B.K. Seisjo. The increase in extracellular potassium concentration in the ischemic brain in relation to the preischemic functional activity and cerebral metabolic rate. *Brain Res.* 199:161–174, 1980.
25. Astrup, J., B.K. Siesjo, and L. Symon. Thresholds in cerebral ischemia—the ischemic penumbra. *Stroke* 12:723–725, 1981.
26. Atkins, E. Fever. In: *Signs and Symptoms.* C. MacBryde, and R. Blacklaw (eds.). New York: JB Lippincott, pp.451–474, 1970.
27. Baker, E.M., I.C. Plough, and T.H. Allen. Water requirements of men as related to salt intake. *Am. J. Clin. Nutr.* 12:394–398, 1963.
28. Barger, A.C., W.F. Greenwood, J.R. DiPalma, J. Stokes, and L.H. Smith. Venous pressure and cutaneous reactive hyperemia in exhausting exercise and certain other circulatory stresses. *J. Appl. Physiol.* 2:81–96, 1949.
29. Bate-Smith, E.C., and J.R. Bendall. Changes in muscle after death. *Br. Med. Bull.* 12:230–235, 1956.
30. Bean, W.B., and L. W. Eichna. Performance in relation to environmental temperature. *Fed. Proc.* 2:144–158, 1943.
31. Betz, E., and J. Csornia. Ionic actions on cerebral vessels. *Int. J. Neurol.* 11:243–258, 1977.
32. Boron, W.F., and P. DeWeer. Active proton transport stimulation by CO_2/HCO_3 blocked by cyanide. *Nature (Lond)* 259:240–241, 1976.
33. Bowers, W.D. Jr., R.W. Hubbard, I. Leav, P. Daum, M. Conlon, M.P. Hamlet, M. Mager,

and P. Brandt. Alterations of rat liver subsequent to heat overload. *Arch. Path. Lab. Med.* 102:154–157, 1978.

34. Bowers, W. Jr., I. Leav, P. Daum, M. Murphy, P. Williams, R. Hubbard, and M. Hamlet. Insulin and cortisol improve heat tolerance in isolated perfused rat liver. *Lab. Invest.* 51:675–681, 1984.
35. Boyd, A.E., and G.A. Beller. Acid-base changes in heat exhaustion during basic training. *Proc. Army Sci. Conf.* 1:114–125, 1972.
36. Briggs, F.N., J.L. Poland, and R.J. Solaro. Relative capabilities of sarcoplasmic reticulum in fast and slow mammalian skeletal muscles. *J. Physiol. (Lond)* 266:587–594, 1977.
37. Brooks, G.A. The lactate shuttle during exercise and recovery. *Med. Sci. Sports Exerc.* 18:360–368, 1986.
38. Brooks, G.A., K.J. Hittelman, J.A. Faulkner, and R.E. Beyer. Temperature, skeletal muscle mitochondrial functions, and oxygen debt. *Am. J. Physiol.* 220:1053–1059, 1971.
39. Burnell, J.M., M.F. Villamil, B.T. Uyino, and B.H. Scribner. The effect in humans of extracellular pH change on the relationship between serum potassium concentration and extracellular potassium. *J. Clin. Invest.* 35:935–939, 1956.
40. Bynum, G.P., K.B. Pandolf, W.H. Schuette, R.F. Goldman, D.E. Lees, J. Whang Peng, E.R. Atkinson, and J.M. Bull. Induced hyperthermia in sedated human and the concept of critical thermal maximum. *Am. J. Physiol.* 235:R228–R236, 1978.
41. Cabral, R., P.F. Prior, D.F. Scott, and J.B. Brierley. Reversible profound depression of cerebral activity in hyperthermia. *Electroencephal. Clin. Neurophysiol.* 42:697–701, 1977.
42. Caldwell, P.C. The phosphorous metabolism of squid axons and its relationship to the active transport of sodium. *J. Physiol. (Lond)* 152:545–560, 1960.
43. Callaham, M.L. Emergency management of heat illness. *Emergency Physician Series.* North Chicago, IL: Abbott Laboratories, p.1–23, 1979.
44. Clowes, G.H.A., Jr, M. Vucinic, and M.G. Weidner. Circulatory and metabolic alterations associated with survival or death in peritonitis: Clinical analysis of 25 cases. *Ann. Surg.* 163:866–885, 1966.
45. Coburn, J.W., R.C. Reba, and F.N. Craig. Effect of potassium depletion on response to acute heat exposure in unacclimatized man. *Am. J. Physiol.* 211:117–124, 1966.
46. Conn, J.W. Acclimatization to heat. *Adv. Intern. Med.* 3:373–393, 1949.
47. Consolazio, C.F. Nutrient requirements of troops in extreme environments. *Army Res. Develop. Mag.* 11:24–27, 1966.
48. Consolazio, C.F. Energy metabolism and extreme environments (heat, cold, high altitude). Proc. Eighth Internat. Cong. Nutr., *Excerpta Medica International Congress Series* No. 213:324–326, 1969.
49. Consolazio, C.F., L.O. Matoush, R.A. Nelson, J.B. Tones, and G.J. Isaac. Environmental temperature and energy expenditures *J. Appl. Physiol.* 18:65–68, 1963.
50. Costill, D.L., G. Branam, D. Eddy, and W.J. Fink. Alterations in red cell volume following exercise and dehydration. *J. Appl. Physiol.* 37:912–916, 1974.
51. Costill, D.L., R. Cote, and W.J. Fink. Muscle water and electrolytes following varied levels of dehydration in man. *J. Appl. Physiol.* 40:6–11, 1976.
52. Costill, D.L., R. Cote, W.J. Fink, and P. Van Handel. Muscle water and electrolyte distribution during prolonged exercise. *Int. J. Sports Med.* 2:130–134, 1981.
53. Costill, D.L., and W.J. Fink. Plasma volume changes following exercise and thermal dehydration. *J. Appl. Physiol.* 37:521–525, 1974.
54. Costrini, A.M., H.A. Pitt, A.B. Gustafson, and D.E. Uddin. Cardiovascular and metabolic manifestations of heatstroke and severe heat exhaustion. *Am. J. Med.* 66:296–302, 1979.
55. Crane, R.K. The gradient hypothesis and other models of carrier-mediated active transport. *Rev. Physiol. Biochem. Pharmacol.* 78:99–159, 1977.
56. Dahl, L.K. Salt and hypertension. *Am. J. Clin. Nutr.* 25:234–244, 1972.
57. Daily, W.M., and T.R. Harrison. A study of the mechanism and treatment of experimental heat pyhexia. *Am. J. Med. Sci.* 215:42–55, 1948.
58. daLuz, P.L., M.H. Weil, V.Y. Liu, and H. Shubin. Plasma volume prior to and following volume loading during shock complicating acute myocardial infarction. *Circulation* 49:98–105, 1974.
59. Dancaster, C.P., and S.J. Whereat. Fluid and electrolyte balance during the comrades marathon. *South African Med. J.* 45:147–150, 1971.
60. Danforth, W.H. Activation of glycolytic pathway in muscle. In: *Control of Energy Metabolism,* B. Chance., R.W. Estabrook, and J.B. Williamson (eds.). New York: Academic Press, 1965, p.287–298.
61. Dassler, A.R., S. Karas, J.S. Bowman, and E. Hardinbergh. Adverse effects of supplementary sodium chloride on heat adaptation. *Fed. Proc.* 32:336A, 1973.
62. Dill, D.B., and D.L. Costill. Calculation of percentage changes in volumes of blood, plasma, and red cells in dehydration. *J. Appl. Physiol.* 37:247–248, 1974.

63. Dill, D.B., L. Scholt, and D. Maclean. Capacity of young males and females for running in desert heat. *Med. Sci. Sports* 9:137–142, 1977.
64. Dinman, B.D., and S.M. Horvath. Heat disorders in industry: A reevaluation of diagnostic criteria *J. Occupat. Med.* 26:489–495, 1984.
65. Dukes-Dobos, F.N. Hazards of heat exposure: A review. *Scand. J. Work Environ. Health* 7:73–83, 1981.
66. Edelman, I.S. Thyroid thermogenesis. *New Eng. J. Med.* 290:1303–1308, 1974.
67. Edwards, R.H.T. Human muscle function and fatigue. In: R. Porter and J. Whelan (eds.). *Human Muscle Fatigue: Physiological Mechanisms,* Ciba Foundation Symposium No 82, pp1–18 (Pitman Medical, London, 1981).
68. Edwards, R.H.T. Interaction of chemical with electromechanical factors in human skeletal muscle fatigue. *Acta Physiol. Scand.* 128 (Suppl 556):149–155, 1986.
69. Edwards, R.H.T., R.C. Harris, E. Hultman, L. Kaijses, D. Koh, and L-O Nordesjo. Effect of temperature on muscle energy metabolism and endurance during successive isometric contractions, sustained to fatigue, of the quadriceps muscle in men. *J. Physiol.* 220:335–352, 1972.
70. Eichler, A., A. McFee, and H. Root. Heatstroke. *Am. J. Surg* 118:855–863, 1969.
71. Elkinton, J.R., T.S. Danowski, and A.W. Winkler. Hemodynamic changes in salt depletion and in dehydration. *J. Clin. Invest.* 25:120–129, 1946.
72. Fabiato A., and F. Fabiato. Effects of pH on the myofilaments and the sarcoplasmic reticulum of skinned cells from cardiac and skeletal muscle. *J. Physiol. (Lond)* 276:233–255, 1978.
73. Felig, P., C. Johnson, and M. Levitt. Hypernatremia induced by maximal exercise. *JAMA* 248:1209–1211, 1982.
74. Flynn, C.J., and L. Wecker. Elevated choline levels in brain. A non-cholinergic component of organophosphate toxicity. *Biochem. Pharmacol.* 35:3115–3121, 1986.
75. Food and Nutrition Board. *Recommended Dietary Allowances,* 9th ed. Washington, D.C.: National Academy of Sciences, p.1–185, 1985.
76. Geigy Scientific Tables. C. Lentner (ed.). Basle, Switzerland: CIBA-GEIGY, 1981.
77. Gibson, H., and R.H.T. Edwards. Muscular exercise and fatigue. *Sports Med.* 2:120–132, 1985.
78. Gilat, T., S. Shibolet, and E. Sohar. The mechanism of heatstroke. *J. Trop. Med. Hyg.* 66:204–212, 1963.
79. Glynn, M. Membrane adenosine trephosphatase and cation transport. *Br. Med. Bull.* 24:165–169, 1968.
80. Gold, J. Development of heat pyrexia. *JAMA* 173:1175–1182, 1960.
81. Gollnick, P.D., W.M. Bayly, and D.R. Hodgson. Exercise intensity, training diet, and lactate concentration in muscle and blood. *Med. Sci. Sports Exerc.* 18:334–340, 1986.
82. Greenleaf, J.E. The Body's Need for Fluids. In: *Nutrition and Athletic Performance,* Haskell, W., J. Scala, and J. Whittam (eds.). Palo Alto, CA: Bull Publishing Co, p. 34–50, 1981.
83. Greenleaf, J.E., P.J. Brock, L.C. Keil, and J.T. Morse. Drinking and water balance during exercise and heat acclimation. *J. Appl. Physiol.* 54:414–419, 1983.
84. Greenleaf, J.E., and C.J. Greenleaf. Human acclimation and acclimatization to heat—A compendium of research. Moffett Field, CA: NASA Technical Report TMX-62008, p.1–188, 1970.
85. Guyton, A.C. *Textbook of Medical Physiology,* 7th ed. Philadelphia: W.B. Saunders, pp.88–119, 1986.
86. Guyton, A.C. *Textbook of Medical Physiology,* 6th ed. Philadelphia: W.B. Saunders, pp.801–835, 1981.
87. Hanson, P.G., and S.W. Zimmerman. Exertional heatstroke in novice runners. *JAMA* 242:154–157, 1979.
88. Harrison, M.H. Heat and exercise effects on blood volume. *Sports Med.* 3:214–223, 1986.
89. Hart, G.R., R.J. Anderson, C.P. Crumpler, A. Shulkin, G. Reed, and J.P. Knochel. Epidemic classical heatstroke: Clinical characteristics and course of 28 patients. *Med.* 61:189–197, 1982.
90. Herbert, W.B. Water and electrolytes. In: *Ergogenic Aids in Sport.* Champaign, IL: Human Kinetics Publishers, 1983, p. 56–98.
91. Hertz, L. Biochemistry of glial cells. In: *Cell Tissue and Organ Culture in Neurobiology.* Federoff S. and L. Hertz, (eds.). New York: Academic Press, p.39–71, 1977.
92. Heuser, D., and H. Guggenberger. Ionic changes in brain ischaemia and alterations produced by drugs. *Br. J. Anaesth.* 57:23–33, 1985.
93. Hill, A.V., and H. Lupton. Muscular exercise, lactic acid and the supply and utilization of oxygen. *Q.J. Med.* 16:135–171, 1923.
94. Hillered, L., L. Ernster, and B.K. Siesjo. Influence of in vitro lactic acidosis and hypercapnia on respiratory activity of isolated rat brain mitochondria. *J. Cereb. Blood Flow. Metab.* 4:430–437, 1984.

95. Hnik, P., F. Vyskocil, E. Ujec, R. Vejsada, and H. Rehfeldt. Work-induced potassium loss from skeletal muscles and its physiological implications. *Biochem. Exerc.* 16:345–364, 1986.
96. Hubbard, J. The effect of exercise on lactate metabolism. *J. Physiol.* 231:1–18, 1973.
97. Hubbard, R.W. Effects of exercise in the heat on predisposition to heatstroke. *Med. Sci. Sports* 11:66–71, 1979.
98. Hubbard, R.W., L.E. Armstrong, P.K. Evans, and J.P. DeLuca. Long-term water and salt deficits: A military perspective. In: *Predicting Decrements In Military Performance Due to Inadequate Nutrition.* Washington, D.C.: National Academy Press, p.29–48, 1986.
99. Hubbard, R.W., W.D. Bowers, and M. Mager. A study of physiological, pathological and biochemical changes in rats with heat- and/or work-induced disorders. *Israel J. Med. Sci.* 12:884–886, 1976.
100. Hubbard, R.W., W.D. Bowers, W.T. Matthew, F.C. Curtis, R.E.L. Criss, G.M. Sheldon, and J.W. Ratteree. Rat model of acute heatstroke mortality. *J. Appl. Physiol.* 42:809–816, 1977.
101. Hubbard, R.W., M. Mager, W.D. Bowers, I. Leav, G. Angoff, W.T. Matthew, and I.V. Sils. Effect of low-potassium diet on rat exercise hyperthermia and heatstroke mortality. *J. Appl. Physiol.* 51:8–13, 1981.
102. Hubbard, R.W., W.T. Matthew, R.E.L. Criss, C. Kelly, I. Sils, M. Mager, W.D. Bowers, and D. Wolfe. Role of physical effort in the etiology of rat heatstroke injury and mortality. *J. Appl. Physiol.* 45:463–468, 1978.
103. Hubbard, R.W., C.B. Matthew, M.J. Durkot, and R.P. Francesconi. Novel approaches to the pathophysiology of heatstroke: The energy depletion model. *Ann. Emerg. Med.* 16:1066–1075, 1987.
104. Hubbard, R.W., W.T. Matthew, J.D. Linduska, F.C. Curtis, W.D. Bowers, I. Leav, and M. Mager. The laboratory rat as a model for hyperthermic syndromes in humans. *Am. J. Physiol.* 231:1119–1123, 1976.
105. Hubbard, R.W., B.L. Sandick, W.T. Matthew, R.P. Francesconi, J.B. Sampson, M.J. Durkot, O. Maller, and D. Engell. Voluntary dehydration and alliesthesia for water. *J. Appl. Physiol.* 57:868–875, 1984.
106. Iampietro, P.F. Heat-induced tetany. *Fed. Proc.* 22:884–886, 1963.
107. Iampietro, P.F., M. Mager, and E.B. Green. Some physiological changes accompanying tetany induced by exposure to hot, wet conditions. *J. Appl. Physiol.* 16:409–412, 1961.
108. International Classification of Diseases, revision 8. Public Health Service publication 1693. Washington DC: US Government Printing Office, p.443, 1983.
109. Issekutz, B. Jr., W.A.S. Shaw, and A.C. Issekutz. Lactate metabolism in resting and exercising dogs. *J. Appl. Physiol.* 40:312–319, 1976.
110. Iversen, L.L., and J.S. Kelly. Uptake and metabolism of gamma-aminobutyric acid by neurones and glial cells. *Biochem. Pharmacol.* 24:933–938, 1975.
111. Jacobs, I. Blood lactate: Implications for training and sports performance. *Sports Med.* 3:10–25, 1986.
112. Jones, B.H., P.B. Rock, L.S. Smith, M.A. Teves, J.K. Casey, K. Eddings, L.H. Malkin, and W.T. Matthew. Medical complaints after a marathon run in cool weather. *Physician Sportsmed.* 13:103–110, 1985.
113. Keren, G., Y. Epstein, and A. Magazanik. Temporary heat intolerance in a heatstroke patient. *Aviat. Space Environ. Med.* 52:116–117, 1981.
114. Keul, J., E. Doll, and D. Keppler. The substrate supply of human skeletal muscle at rest, during and after work. *Experientia* 23:974–979, 1967.
115. Kew, M., I. Bersohn, H. Seftel, and G. Kent. Liver damage in heatstroke. *Am. J. Med.* 49:192–202, 1970.
116. Kielblock, A.J., N.B. Strydom, F.J. Burger, P.J. Pretorius, and M. Manjoo. Cardiovascular origins of heatstroke pathophysiology: An anesthetized rat model. *Aviat. Space Environ. Med.* 53:171–178, 1982.
117. King, B.A., and M.E. Barry. The physiological adaptations to heat-stress with a classification of heat illness and a description of the features of heat exhaustion. *South African Med. J.* 36:451–455, 1962.
118. Knochel, J.P. Environmental heat illness. An eclectic review. *Arch. Intern. Med.* 133:841–864, 1974.
119. Knochel, J.P., L.N. Dotin, and R.J. Hamburger. Pathophysiology of intense physical conditioning in a hot climate: 1. Mechanisms of potassium depletion. *J. Clin. Invest.* 51:242–255, 1972.
120. Knochel, J.P., and G. Reed. Disorders of Heat Regulation. In: *Clinical Disorders, Fluid & Electrolyte Metabolism.* by C.R. Kleeman, M.H. Maxwell, and R.G. Narin, (eds.). New York: McGraw Hill Co., p.1197–1232, 1987.
121. Knochel, J.P., and R.M. Vertel. Salt loading as a possible factor in the production of potassium depletion, rhabdomyolysis, and heat injury. *Lancet* 1:659–661, 1967.
122. Kozlowski, S., Z. Brzezinska, B. Kruk, H. Kaciuba-Uscilko, J.E. Greenleaf, and K. Nazar.

Exercise hyperthermia as a factor limiting physical performance: Temperature effect on muscle metabolism. *J. Appl. Physiol.* 59:766–773, 1985.
123. Kozlowski, S., and J. Domaniecki. Thermoregulation during physical effort in humans of different physical performance capacity. *Acta Physiol. Polon.* 28:816–825, 1972.
124. Kreisberg, R.A., L.F. Pennington, and B.R. Boshell. Lactate turnover and gluconeogenesis in normal and obese humans. *Diabetes* 19:53–63, 1970.
125. Kruk, B., H. Kaciuba-Uscilki, K. Nazar, J.E. Greenleaf, and S. Kozlowski. Hypothalamic, rectal and muscle temperatures in exercising dogs: Effect of cooling. *J. Appl. Physiol.* 58:1444–1448, 1985.
126. Ladell, W.S.S. Heat cramps. *Lancet* 2:836–839, 1949.
127. Ladell, W.S.S. Disorders due to heat. *Trans. Royal Soc. Tropical Med. Hygiene* 51:189–216, 1957.
128. Ladell, W.S.S., J.C. Waterlow, M.F. Hudson. Desert climate physiological and clinical observations. *Lancet* 2:491–497, 1944.
129. Leithead, C.S., and A.R. Lind. *Heat Stress and Heat Disorders,* Philadelphia: Davies, 1964.
130. Leithead, C.S., and E.R. Gunn. The aetiology of cane cutter's cramps in British Guiana. In: *Environmental Physiology and Psychology in Arid Conditions.* Leige, Belgium: UNESCO, 1964, p.13–17.
131. Leithead, C.S., J. Guthrie, S. DeLaPlace, and B. Maegraith. Incidence, aetiology, and prevention of heat illness on ships in the Persian Gulf. *Lancet* 2:109–112, 1958.
132. Leithead, C.S. L.A. Leithead, and F.D. Lee. Salt-deficiency heat exhaustion. *Ann. Tropical Med. Parasitol.* 52:456–467, 1958.
133. Lewis, L.D., B. Ljunggren, R.A. Ratcheson, and B.K. Siesjo. Cerebral energy state in insulin-induced hypoglycemia, related to blood glucose and to EEG. *J. Neurochem.* 23:673–679, 1974.
134. MacDougal, J.D., W.G. Reddan, C.R. Layton, and J.A. Dempsey. Effects of metabolic hyperthermia on performance during heavy prolonged exercise. *J. Appl. Physiol.* 36:538–544, 1974.
135. Mager, C., and S. Nissan. Alterations in the basal oxygen consumption of rats attendant upon three types of dehydration. *Ann. Surg.* 154 (Suppl):51–64, 1961.
136. Malamud, N., W. Haymaker, and R.P. Custer. Heatstroke: A clinico-pathologic study of 125 fatal cases. *Mil. Surg.* 99:397–449, 1946.
137. Malhotra, M.S., and Y. Venkataswamy. Heat casualties in the Indian Armed Forces. *Ind. J. Med. Res.* 62:1293–1302, 1974.
138. Manjoo, M., F.J. Burger, and A.J. Kielblock. A relationship between heat load and plasma enzyme concentration. *J. Thermal. Biol.* 4:221–225, 1985.
139. Marriott, H.L. *Water and Salt Depletion.* Springfield, IL: Charles C. Thomas, p.1–80, 1950.
140. McCance, R.A. Experimental sodium chloride deficiency in man. *Proc. Royal Soc. London.* 119:245–268, 1936.
141. McCord, J.M. The role of superoxide in post ischemic tissue injury. In: *Superoxide Dismutase,* Oberley, L.W. (ed.). Vol III, Boca Raton, FL: CRC Press, 1985, p.143–150.
142. McGilvery, R.W. The use of fuels for muscular work. In: *Metabolic adaptation to prolonged physical exercise* H. Howald and J.R. Poortsman (eds.). Basel: Brauheuser Verlag, pp.12–30, 1973.
143. Milvy, P. (ed.). *The Marathon: Physiological, Medical, Epidemiological, and Psychological Studies.* New York: New York Academy of Sciences, 1977.
144. Nadal, J.W., S. Pedersen, and W.G. Maddock. A comparison between dehydration from salt loss and from water deprivation. *Am. J. Physiol.* 134:691–703, 1941.
145. Nadel, E.R., C.B. Wenger, M.F. Roberts, J.A.J. Stolwijk, and E. Cafarelli. Physiological defenses against hyperthermia of exercise. *Ann. N.Y. Acad. Sci.* 301:98–109, 1977.
146. Nash, H.L. Treating thermal injury: disagreement heats up. *Physician Sportsmed.* 13:134–144, 1985.
147. Nielsen, B., R. Kubica, A. Bonnesen, I.B. Rasmussen, J. Stoklosa, and B. Weiik. Physical work capacity after dehydration and hyperthermia. *Scand. J. Sports Sci.* 3:2–10, 1981.
148. Nissan, S., A. Aviram, J.W. Czaczkes, L. Ullman, and T.D. Ullman. Increased O_2 consumption of the rat diaphragm by elevated NaCl concentrations. *Am. J. Physiol.* 210:1222–1224, 1966.
149. O'Donnell, T.F. Jr., and G.H.A. Clowes, Jr. The circulatory abnormalities of heatstroke. *New Engl. J. Med.* 287:734–737, 1972.
150. Olson, K.R., and N.L. Benowitz. Experimental and drug-induced hyperthermia. *Emerg. Med. Clin. North Am.* 2:459–474, 1984.
151. Orr, J.B., and J.L. Gilks. Studies of Nutrition: The Physique and Health of two African Tribes. London: Medical Research Council Report, 1931, p.1–82.
152. Osler, W. *The Principles and Practice of Medicine. Designed for the Use of Practioners and Students of Medicine.* 7th ed. New York: D. Appleton and Co, p.1017–1018, 1893.
153. Owen, M.D., K.C. Kregel, P.T. Wall, and C.V. Gisolfi. Effects of ingesting carbohydrate

beverages during exercise in the heat. *Med. Sci. Sports Exerc.* 18:568–575, 1986.
154. Pandolf, K.B., and R.F. Goldman. Convergence of skin and rectal temperature as a criterion for heat tolerance. *Aviat. Space Environ. Med.* 49:1095–1101, 1978.
155. Pandolf, K.B., T.B. Griffin, E.H. Munro, and R.F. Goldman. Persistence of impaired heat tolerance from artificially induced miliaria rubra. *Am. J. Physiol.* 239:R226–R232, 1980.
156. Paque, C. Saharan Bedouins and the Salt water of the Sahara: A model for salt intake. In: *Biological and Behavioral Aspects of Salt Intake.* New York: Academic Press, p.31–47, 1980.
157. Porter, R., and J. Whelan. (eds.). *Human Muscle Fatigue Physiological Mechanisms.* Ciba Foundation Symposium No. 82, London: Pitman, 1981.
158. Radigan, L.R., and S. Robinson. Effects of environmental heat stress and exercise on renal blood flow and filtration rate. *J. Appl. Physiol.* 2:185–191, 1949.
159. Reeves, R.B. The interaction of body temperature and acid-base balance in ectothermic vertebrates. *Ann. Rev. Physiol.* 39:559–586, 1977.
160. Roberts, M.F., and C.B. Wenger. Control of skin circulation during exercise and heat stress. *Med. Sci. Sports* 11:36–41, 1979.
161. Rome, L.C., and M.J. Kushmeric. Energetics of isometric contractions as a function of muscle temperature. *Am. J. Physiol.* 224:C100–C109, 1983.
162. Romero, J.C., R.J. Stameloni, M.L. Dufau, R. Dohmen, A. Binia, B. Kluman, and J.C. Fasciolo. Changes in fluid compartments, renal hemodynamics, plasma renin and aldosterone secretion induced by low sodium intake. *Metabolism* 17:10–19, 1968.
163. Roos, A., and W.F. Boron. Intracellular pH. *Physiol. Rev.* 61:296–434, 1981.
164. Rose, K.D. Warning for millions: Intense exercise can deplete potassium. *Physician Sportsmed.* 3:67–70, 1975.
165. Rowell, L.B. Hepatic metabolism during hyperthermia and hypoxemia. In: *Biochemistry of Exercise VI.* Champaign IL: Human Kinetics Publishers, 1986, p203–305.
166. Rowell, L.B., J.R. Blackmon, R.H. Martin, J.A. Mazzarella, and R.A. Bruce. Hepatic clearance of indocyanine green in man under thermal and exercise stresses. *J. Appl. Physiol.* 20:384–394, 1965.
167. Rowell, L.B., J. Marx, R.A. Bruce, R.D. Conn, and F. Kusumi. Reductions in cardiac output, central blood volume and stroke volume with thermal stress in normal men during exercise. *J. Clin. Invest.* 45:1801–1816, 1966.
168. Sahlin, K. Intracellular pH and energy metabolism in skeletal muscle of man. With special reference to exercise. *Acta Physiol. Scand.* 103 (Suppl 455):3–56, 1978.
169. Sahlin, K. Muscle fatigue and lactic acid accumulation. *Acta Physiol. Scand.* 128 (Suppl 556):83–91, 1986.
170. Sahlin, K., A. Alvestrand, R. Brandt, and E. Hultman. Intracellular pH and bicarbonate concentration in human muscle during recovery from exercise. *J. Appl. Physiol.* 45:474–480, 1978.
171. Sahlin, K., L. Edstrom, H. Sjoholm, and E. Hultman. Effects of lactic acid accumulation and ATP decrease on muscle tension and relaxation. *Am. J. Physiol.* 240:C121–C126, 1981.
172. Sahlin, K., L. Edstrom, and H. Sjoholm. Fatigue and phosphocreatine depletion during carbon dioxide-induced acidosis in rat muscle. *Am. J. Physiol.* 245:C15–C20, 1983.
173. Segal, S.S., J.A. Faulkner, and T.P. White. Skeletal muscle fatigue in vitro is temperature dependent. *J. Appl. Physiol.* 61:660–665, 1986.
174. Senay, L.C., G. Rogers, and P. Jooste. Changes in blood plasma during progressive treadmill and cycle exercise. *J. Appl. Physiol.* 49:59–65, 1980.
175. Shapiro, Y., T. Rosenthal, and E. Sohar. Experimental heatstroke: A model in dogs. *Arch. Intern. Med.* 131:688–692, 1973.
176. Shibolet, S., S. Fisher, T. Gilat, H. Bank, and H. Heller. Fibrinolysis and hemorrhages in fatal heatstroke. *New Eng. J. Med.* 266:169–173, 1962.
177. Shibolet, S., and T. Gilat. The clinical picture of heatstroke. *Proc. Tel-Hashomer Hospital.* 1:80–83, 1962.
178. Shibolet, S., M.C. Lancaster, and Y. Danon. Heatstroke: A review. *Aviat. Space Environ. Med.* 47:280–301, 1976.
179. Siesjo, B.K., and T. Wielock. Cerebral metabolism in ischaemia: Neurochemical basis for therapy. *Br. J. Anaesth.* 57:47–62, 1985.
180. Shih, C.J., M.T. Lin, and S.H. Tsai. Experimental study on the pathogenesis of heatstroke. *J. Neurosurg.* 60:1246–1252, 1984.
181. Sjogaard, G., R.P. Adams, and B. Saltin. Water and ion shifts in skeletal muscles of humans with intense dynamic knee extension. *Am. J. Physiol.* 248:R190–R196, 1985.
182. Skou, J.C. Enzmatic basis for active transport of Na^+ and K^+ across cell membrane. *Physiol. Rev.* 45:596–617, 1965.
183. Sohar, E., and R. Adar. Sodium requirements in Israel under conditions of work in hot climate. In: *USESCO/India Symposium on Environmental Physiology and Psychology.* Lucknow: UNESCO, 1962, p.129–135.

184. Sohar, E., J. Kaly, and R. Adar. The prevention of voluntary dehydration. In: *UNESCO/India Symposium on Environmental Physiology and Psychology*. Lucknow: UNESCO, 1962, p.129–135.
185. Spurr, G.B., and G. Barlow. Tissue electrolytes in hyperthermic dogs. *J. Appl. Physiol.* 28:13–17, 1970.
186. Stainsby, W.N. Biochemical and physiological bases for lactate production *Med. Sci. Sports Exerc.* 18:341–342, 1986.
187. Steenbergen, C., G. Deleeuw, T. Rich, and J.R. Williamson. Effects of acidosis and ischemia on contractility and intracellular pH of rat heart. *Circ. Res.* 41:849–858, 1977.
188. Stegmann, H., W. Kinderman, and A. Schnabel. Lactate kinetics and individual anaerobic threshold. *Int. J. Sports Med.* 2:160–165, 1981.
189. Stevens, ED. The evolution of endothermy. *J. Theor. Biol.* 38:597–611, 1973.
190. Stine, R.J. Heat Illness. *JACEP* 8:154–160, 1979.
191. Strauss, R.H. Skeletal muscle abnormalities associated with sports activities. In: *Sports Medicine*. Philadelphia: W.B. Saunders., p.159–164, 1984.
192. Sullivan, S.N., M.C. Champion, N.D. Christofides, T.E. Adrian, and S.R. Bloom. Gastrointestinal regulatory peptide responses in long-distance runners. *Physician Sportsmed.* 12:79–82, 1984.
193. Sweadner, K.J., and S.M. Goldin. Active transport of sodium and potassium ions: mechanism, function and regulation. *New Eng. J. Med.* 302:777–783, 1980.
194. Symon, L. Flow thresholds in brain ischaemia and the effects of drugs. *Br. J. Anaesth.* 57:34–43, 1985.
195. Talbot, J.H. Heat Cramps. In: *Medicine*. Baltimore: Williams and Wilkins, p.323–376, 1935.
196. Talbot, J.H., and J. Michelsen. Heat cramps. *J. Clin. Invest.* 12:533–535, 1933.
197. Taylor, H.L., A. Henschel, O. Michelson, and A. Keys. The effect of sodium chloride intake on the work performance of man during exposure to dry heat and experimental heat exhaustion. *Am. J. Physiol.* 140:439–451, 1944.
198. Thomson, W.H.S., J.C. Sweetin, and J.D. Hamilton. ATP and muscle enzyme efflux after physical exertion. *Clin. Chimica Acta* 59:241–245, 1975.
199. Trivedi, B., and W.H. Danforth. Effect of pH on the kinetics of frog muscle phosphofructokinase. *J. Biol. Chem.* 241:4110–4114, 1966.
200. Vaamonde, C.A. Sodium depletion. In: *Sodium: Its Biological Significance*. Boca Raton, FL: CRC Press, 1982, p.208–234.
201. Wasserman, K, W.L. Beaver, and B.J. Whipp. Mechanisms and patterns of blood lactate increase during exercise in man. *Med. Sci. Sports Exerc.* 18:344–352, 1986.
202. Weil, M.H., and A.A. Affi. Experimental and clinical studies on lactate and pyruvate as indicators of the severity of acute circulatory failure (Shock). *Circulation* 41:989–1001, 1970.
203. Weisenberg, T.H. Nervous symptoms following sunstroke. *JAMA* 58:2015, 1912.
204. Whittam, R. The dependence of the respiration of brain cortex on active cation transport. *Biochem. J.* 82:205–212, 1962.
205. Wildenthal, K.S., D.S. Mierzwiak, R.W. Myers, and J.H. Mitchell. Effects of acute lactic acidosis on left ventricular performance. *Am. J. Physiol.* 214:1352–1359, 1968.
206. Wilkinson, J.H., and J.M. Robinson. Effect of ATP on release of intracellular enzymes from damaged cells. *Nature* 249:662–663, 1974.
207. Williams, C.O., G.A.G. Bredell, C.H. Wyndham, N.B. Stydom, J.R. Morrison, J. Peter, P.W. Flemming, and J.S. Ward. Circulatory and metabolic reactions to work in the heat. *J. Appl. Physiol.* 17:625–638 1962.
208. Wilson, G. The cardiopathology of heatstroke. *JAMA* 114:557–558, 1940.
209. Wyndham, C.H. A survey of causal factors in heatstroke and their prevention in the gold mining industry *J. South African Inst. Mining Metallergy* 1:245–258, 1966.
210. Wyndham, C.H. Heatstroke and hyperthermia in marathon runners. *Ann. N.Y. Acad. Sci.* 301:128–138, 1977.
211. Wyndham, C.H., N.B. Strydom, A.J. Benade, and W.H. van der Walt. The effect on acclimatization of various water and salt replacement regimens. *South African Med. J.* 47:1773–1779, 1973.
212. Yunusou, A.Y. Water-salt metabolism in hot climates. In: *Environmental Physiology and Psychology in Arid Conditions*. Liege, Belgium: UNESCO, 1964, p.357–361.
213. Zeisel, S.H. Formation of unesterified choline by rat brain. *Biochemica Biophysica Acta* 835:331–343, 1985.

ACKNOWLEDGEMENTS

The authors gratefully acknowledge the technical expertise and administrative assistance of Susan Henry and Diane Danielski in the preparation and editing of the manuscript.

9

Physiological Adjustments of Man to the Cold

MICHAEL M. TONER, Ph.D.

WILLIAM D. MCARDLE, Ph.D.

OUTLINE

INTRODUCTION TO ENVIRONMENTAL EXTREMES OF COLD

Body temperature, or more specifically, the temperature of the deeper core is in dynamic equilibrium as a result of a balance between factors that add and subtract body heat. This balance is maintained by the integration of mechanisms that alter heat transfer to the periphery or shell, regulate evaporative cooling, and vary the body's rate of heat production. Normally, the gradient for heat transfer is from the body to the environment, and core temperature is maintained without excessive physiologic strain. As such, the mechanisms for thermoregulation are geared primarily to protect against overheating. In the inadequately clothed human, however, excessive body heat can be lost to the environment. In this situation, heat production is increased and heat loss is retarded, as physiologic adjustments are made to prevent a fall in internal temperature.

As a physiologic challenge, cold ranks high among the differing terrestrial environments of potentially lethal human consequences. Because heat conduction in water is about 25 times greater than in air, immersion in water of only 28 to 30° C can provide a considerable thermal stress which brings about thermoregulatory adjustments in a relatively short period of time (98,126). Despite these adjustments, immersion in ice water may cause total body cooling at rates exceeding $6° C \cdot h^{-1}$ with death predicted at between 45 min to three h (67). Although the thermal strain during exposures to cold air is less than similar water temperatures (135), the consequences of continuous cold exposure in air are as severe. However, the cooling rates are lower, yet localized tissue damage is more pronounced.

This chapter reviews the physiologic responses of humans exposed to both the natural extremes of cold and the artificial environments imposed by a laboratory setting. It is our intent to focus discussions primarily on the thermal adjustments of humans to cold water. Also included is a discussion of the thermoregulatory strain induced by cold-air exposure. When appropriate, the interaction of the thermoregulatory system with other physiologic responses in maintaining thermal balance is presented; in addition, some discussion of heat transfer between the body and the cold environment is presented. Other delimits include bodily responses to core strain no greater than 35° C, and skin temperatures above freezing. Conditions that exceed these limits and lead to injury are primarily covered in other chapters.

Recent expeditions to the North Pole exemplify the ability of humans to tolerate extremes of environmental conditions (44,136).

Ambient temperatures encountered by these explorers were in the low range of about −56° C. Interacting with, and magnifying the thermal challenge of most cold environments in the natural setting are the hardships of physical exertion and wear on clothing (136). Despite the expertise of the explorers, the technologic advancements in equipment, and existing knowledge of cold stress, such expeditions are still prone to both mild (44) and potentially serious (136) incidents of cold injury. Whereas extremes of cold air challenge the equipment and stamina of the individual, immersion in cold water exacerbates the physiologic challenge to a much greater extent than air. Concern for the potentially lethal hazards from this form of thermal stress are vividly illustrated in the recent swim, by a highly trained female, between North America and the Soviet Union, in the frigid waters of the Bering Strait.

Although these illustrations do not necessarily reflect the limits of individual tolerance to cold environments, they do highlight the extremes to which individuals are subjected, often voluntarily. For the most part, however, concern for human cold exposure is greater under temperature conditions that are less extreme, yet potentially quite dangerous as in the case of accidental cold exposure. Examples of such cold exposure routinely reported by the media include cases of an individual or group becoming lost in the snow, or a child breaking through ice into frigid waters. Such accidents constitute the main source of cold injury encountered by the civilian population. The potential for danger, on the other hand, is quite different for the military population (60). As reported by Hanson and Goldman (60), military personnel during tactical operations may spend more than 30 percent of a day exposed to extreme temperatures, compared with civilians who spend up to five to 10 percent. As will be discussed, the length of cold exposure also becomes a critical factor in heat loss from the body as a whole as well as from regional body parts.

EXPOSURE TO COLD WATER

Metabolic and Thermal Adjustments

Water is an excellent medium in which to study physiologic adjustment to cold because body heat is lost about two to four times as fast in cool water as in air at the same temperature (134). Even with moderate exercise in cold water, the metabolic heat generated is often insufficient to counter the large, thermal drain. This is especially true during swimming, because heat transfer by convection markedly increases with the movement of water past the skin surface (107).

Factors Contributing to Heat Gain

All cells of the body require energy liberation to maintain a constancy in normal processes. This energy for homeostasis is ultimately derived from the combustion of carbohydrates, fats, and proteins with the resultant release of significant metabolic heat. Though considered in some mechanical systems as a detriment to effective function, cellular metabolism provides the necessary energy source to maintain an optimal homeothermia for bodily function.

In thermoneutral water, body temperature is maintained through a balance between heat lost to the environment and heat gained through metabolic heat production. This energy metabolism is principally determined by the size (25) and composition (26) of the body's tissues, as well as the state of its endocrine system (13), and the age (26), and race (62,126) of the individual. To explain the balance in heat exchange, physiologists have closely linked the metabolic rate with both surface area (130) and body weight (120,130). Regardless of the factors influencing energy metabolism, the basal heat production for adult Americans generally varies between 40 and 49 $W \cdot m^{-2}$ (96). Based on reported values of resting subjects, immersion in thermoneutral water, per se, does not appear to alter metabolic rate (16,126,135) compared with air.

Critical Water Temperature. Rennie and colleagues (126), by use of a water bath for immersion, determined the critical water temperature of Korean and Caucasian diving women. Critical water temperature (126) was used analogously to the lower critical temperature in air described by Scholander (131), and established the environmental temperature below which there is an increase in energy metabolism above the resting level. Smith and Hanna (135) evaluated the critical temperature of subjects in both air and water, and related these values to skinfold thickness. Their findings, shown in Figure 9-1, indicate a high negative correlation between subcutaneous fat and critical temperature in water ($r=-0.95$) and only a moderate relationship ($r=-0.49$) in air. The relatively lower correlation in air is not surprising and is most likely due to less uniform skin temperatures of the subjects for a given environmental temperature due to variations in air movement, posture, and radiant heat loss.

Figure 9-1 also illustrates the difference in critical temperature in air compared with water (135). It is clear that the critical temperature for metabolism is substantially higher in water than in air at any skinfold thickness. The highest recorded critical temperature in water was approximately 35° C observed for a lean subject (135). For the majority of subjects, however, values for critical water tem-

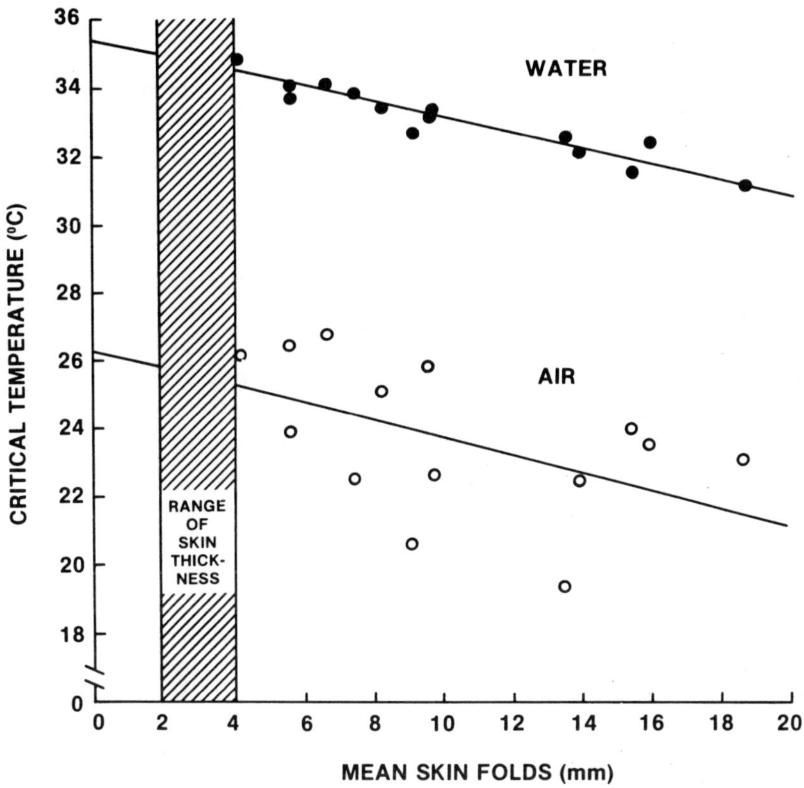

Figure 9-1. *Lower critical temperatures in air and water in relation to subcutaneous body fat. (From Smith and Hanna (135)).*

perature range between 30° C and 34° C (59,126,135). The exceptions are for those individuals with large subcutaneous and total body fat; the critical water temperature may be lower than 30° C (126). As the environmental temperature falls below the critical temperature, there is an increase in energy metabolism proportional to the decrease in water temperature (135). However, this increased metabolic rate contributes minimally to the maintenance of core temperature in cold water (85).

The use of critical water temperature based on metabolism to assess the limits of thermal neutrality has been the subject of some debate (14,35). Similar to Rennie (126), Craig and Dvorak (35) reported that metabolism became elevated if water temperature was 30° C or lower; however, there was a reduction in core temperature if the water temperature was below 35.6° C. This suggests that the neural drive for temperature regulation in water may inadequately stimulate metabolism to balance heat loss in the highly conductive water environment.

The resultant mismatch between neural stimulation of heat conservation mechanisms (vasoconstriction and shivering metabolism) and heat loss to the water may require higher water temperatures to maintain thermal balance. Therefore, it is possible that indices other than the critical temperature for increased metabolism might be required to assess thermoneutral temperatures in water. Boutelier (16 as reported by 14), for example, uses the following criteria to determine the thermoneutral zone in water: (a) a constant metabolic rate, equal to that observed in thermoneutral air; (b) a core/shell conductance close to that found in thermoneutral air; (c) a slight fall in rectal temperature during the first 90 min of immersion with a stable temperature at 36.7 to 36.8° C; (d) regional skin temperatures representing circulatory constriction in that the extremities are coolest and the trunk and abdomen warmest, with the average skin temperature about 34° C; and (e) a sensation of thermal comfort for the subjects with the absence of shivering.

Although the data of Craig and Dvorak (35) and Boutelier (14) suggest that the thermoneutral zone in water may be overestimated when defined solely on the basis of the critical temperature for metabolism, the direct calorimetry determinations of Burton and Bazett (22) indicate that thermal balance can be maintained to about 33° C. The use of core temperature data as an indication of thermal balance in the cold also has been questioned (102). Minard (102) argued that changes in rectal temperature fail to reflect the true body heat loss in transient cooling. This is somewhat illustrated by Greenleaf and colleagues (53), who showed that rectal temperature declined slightly during the first two h of immersion in 34.5° C water; however, after eight h, the rectal temperature was statistically higher than the preimmersion values.

Thermogenesis Below Critical Temperature. During the body's initial contact with water below thermoneutral, the metabolic rate increases and then returns within 10 min to a lower steady state value still above the air control value (141). This initial peaking of metabolism and its subsequent return to a lower, yet elevated, steady level appears to parallel the firing rate pattern of the skin's thermoreceptors in response to a rapid change in skin temperature (72).

The alteration in energy metabolism in response to environmental cold stress is physiologically a function of the efferent information leaving the thermoregulatory center. This, in turn, is determined by the integration of afferent input from the skin and deep body receptors, as well as the local temperature of the hypothalamic thermoregulatory center and the general hormonal state of the body

(13). Therefore, both neural drive and hormonal state must be influenced by the physical components of the water temperature and water movement (107), the individual's subcutaneous and deep body fat (97,135), the level of core temperature (42,103), and the gender (97), race (26), and metabolic characteristics (77) of the person. Also, the insulation provided by clothing (14) has partial influence.

By use of a neuronal animal modelling approach, Hayward and colleagues (69) developed equations to predict human energy metabolism during cooling in water from skin and either rectal or tympanic temperatures as follows:

$$M = 0.0314(\bar{T}_{sk} - 42.2)(T_{re} - 41.4)$$

or

$$M = 0.0356(\bar{T}_{sk} - 41.8)(T_{ty} - 41.0)$$

where, M is the metabolic rate in $W \cdot kg^{-1}$, $\bar{T}_{sk}$ is mean skin temperature, T_{re} is rectal temperature, and T_{ty} is tympanic temperature in °C. The equation is delimited by metabolic rates between 1.1 $W \cdot kg^{-1}$ and 6.4 $W \cdot kg^{-1}$. Cross validation of equations on more heterogeneous samples will verify the general applicability of such equations for thermoregulatory predictions.

Maximal metabolic response during non-exercise immersion conditions has been reported in ice water to average 419 W (67) to about 700 W, reported for a competitive long distance swimmer in water at 12° C (50). Most reports have operationally defined maximal shivering thermogenesis as the highest metabolic rate achieved by all subjects during immersion at a single cold temperature (49,51,67). While these approaches may elicit a peak rate of metabolism during shivering, they fail to systematically provide for a clarification of the factors limiting shivering capacity. Consequently, a true maximum for shivering thermogenesis in humans has yet to be defined. Within these limitations, however, Golden and colleagues (49) reported that both submaximal and maximal shivering oxygen uptake is directly and significantly correlated to an individual's maximal aerobic power ($\dot{V}O_2$ max). The highest oxygen uptake during shivering represented about 50% of $\dot{V}O_2$ max in humans (49) and, as reported by Lucas (95), about 60% in dogs during immersion. While $\dot{V}O_2$ max may be one important factor influencing the magnitude of thermogenesis during cold stress, its role may be more a reflection of the muscle mass available for shivering (95) as well as the specific metabolic level attained before the onset of anaerobic metabolism.

An understanding of the physiologic factors contributing to heat production during cold-water stress must be considered within the framework of the absolute core temperature.

While this review does not consider hypothermia, it must be noted that when core temperature falls below a certain level, the relationship between metabolic rate and the fall in core temperature is reversed. As described by Molnar (104), the metabolism of a partially anesthetized human declines with core temperatures below 35° C, whereas Dill and Forbes (42) observed this phenomenon at about 30° C. This response also has been observed in studies with dogs (12,55). This makes it difficult to predict the actual metabolic rate below a core temperature of 35° C. At core temperatures above 35° C, however, the increase in neural driven shivering metabolism competes with and overrides the direct slowing effect on metabolism of decreasing tissue temperature (1). Below a particular core temperature, however, both shivering and tissue metabolic rates decline with decreasing body temperature which leads to an overall drop in the total metabolic rate, and an acceleration of core cooling (24,48).

Factors Contributing to Heat Loss

Systematic research applying thermodynamic principles to human temperature regulation was reported in the mid-1930s (19,20,22,61). Since then, investigators have applied physical principles to the study of human heat transfer in both cold air and cold water. Heat transfer models and equations based on physical laws are now numerous for water immersion. As pointed out by Burton (20,23), the transfer of heat from the body's core to the environment is a function of two components: the external transfer of heat from the skin to the environment, and the internal heat transfer from the core to the skin. Physiologists have attempted to describe the responses of the immersed body within the framework of these two components.

External Heat Loss

Water Temperature and Duration of Exposure. Primary factors contributing to the loss of body heat are the water temperature and the duration of the exposure (48). As pointed out in the discussion of critical water temperature, immersion in cold water rapidly and profoundly affects thermal response compared to exposure at a similar air temperature. While duration of exposure also augments body cooling, the precise relationship appears to be complex.

The preponderance of studies of cold-water immersion have evaluated thermal responses during immersion between 20 and 90 min (35,68,83,87,143). In estimating the effects of longer immersion durations, extrapolations have been extended based on the patterns observed during the immersion period (68). Clearly, this procedure assumes that the fall in core temperature during cold-water exposure follows a linear pattern as duration progresses. This assumption of linearity, though supported in very cold water (3), may be questionable at higher temperatures (104). In reporting the unpublished data of Spealman, for example, Molnar (104) described the linear reduction in rectal temperature for up to 60 min of immersion. It was noted, however, that in water above 20° C, rectal temperature leveled off as the immersion duration progressed. Cannon and Keatinge (29) also observed that individuals were able to stabilize rectal temperature at values below 36° C during prolonged immersion. Subsequent work (70) indicated that a person could eventually stabilize core temperature in water as cold as 5° C. Though the importance of exposure duration on heat loss from the body cannot be denied, more data are required to precisely define the magnitude and shape of this relationship.

Morphology and Mass. The shape, size, and contour of the body are physical factors that influence external heat transfer. Since the principal avenue for heat transfer in water is through convective heat loss (excluding heat loss from the head and respiratory passages), then the contribution of surface area to the loss of body heat can be described by Newton's law of cooling as it pertains to convection as:

$$H_c = h_c(\bar{T}_{sk} - T_a)A_D$$

where, H_c is rate of convective heat loss (W), h_c is the convective heat transfer coefficient ($W \cdot m^{-2} \cdot {}^\circ C^{-1}$), $\bar{T}_{sk}$ is the average temperature of the skin's surface (° C), T_a is the ambient temperature, and A_D is the body's surface area (m^2). Consequently, with similar heat transfer coefficients and temperature gradients between individuals, the absolute heat loss is greater in a large person with a corresponding large surface compared to that of a smaller person.

The consequence for body temperature of this greater absolute convective heat loss for the larger person, however, must be considered within the framework of alterations in total body heat content. The change in the mean body temperature (ΔT_b) can be determined through partitional calorimetry by:

$$\Delta T_b = \frac{M - H_c}{m \cdot c}$$

where, m is the body mass (kg), c is the specific heat of the body ($W \cdot h \cdot kg^{-1} \cdot {}^\circ C^{-1}$), and H_c is the rate of convective heat loss (W).

It should be noted that the equation is simplified by the exclusion of both convective and evaporative respiratory heat loss and the heat loss from the head which generally is not immersed. From a size perspective, this equation relates the change in body temperature to surface area (within H_c term) and mass. Consequently, body mass relates inversely to the change in body temperature during cold-water stress. This is consistent within the temperature regulation literature where the surface area-to-mass ratio is considered as a physical factor contributing to heat loss. In the thermal model for heat transfer during immersion proposed by Strong and colleagues (138), for example, the surface area-to-mass ratio of the various body compartments is incorporated in predicting skin and core temperatures during cold stress.

Though surface area is important in the transfer of heat from the body's surface to the environment, its relevance as a factor in explaining thermal differences between individuals immersed at the same water temperature is unclear. Using correlational analysis, Sloan and Keatinge (133) and Kollias and colleagues (87) established statistical relationships with heat conservation between surface area and mass. Those individuals with the greatest surface area-to-mass ratio generally showed the largest drop in core temperature during cold-water immersion. Others observing differences in thermal responses between dissimilar populations also postulated that the surface area-to-mass ratio was operating as a factor in heat transfer (43,99). However, one study did attempt to experimentally examine this question by maximizing differences in both mass and surface area-to-mass ratio between two groups within a similar population (144). Although delimited by the 26° C water experiments, the two groups differing significantly on the average in mass, surface area, and surface area-to-mass ratio, had similar skin and rectal temperature responses throughout the immersion period (144). In colder water, however, the surface area-to-mass may become an important factor in human thermoregulation (135).

The influence of surface area-to-mass ratio in explaining individual differences in body cooling during cold-water immersion may also become apparent when considering extreme differences in body size, as was observed with comparisons between younger and older children (133). However, within the normal range of variation in surface area-to-mass ratio for a particular group, the observed core temperature differences are not convincingly attributed to body size. The precise role of this physical parameter in body cooling during water immersion may not be easily established. This is because a crucial factor in determining body cooling is not simply external surface per se, but rather the effective surface of the body in contact with the water, and the distribution of the body heat to that surface.

Surface Insulation. Table 9-1 provides a contrast of convective heat transfer coefficients under a variety of conditions. This coefficient represents the amount of heat transferred to the water across each square meter of skin surface for a thermal gradient of 1° C between skin and water. As such, the heat transfer coefficient is the reciprocal of insulation. Although it may be difficult at first to consider that water provides insulation to the body, it must be realized that a layer of water essentially adheres to the skin's surface in such a way as to be still. In actuality, body heat is first conducted through this thin layer of water and then is convected away with increasing distance from the surface. The relatively low conduction of heat through this warmer boundary layer provides the basis for the insulation.

TABLE 9-1. *Heat transfer coefficients (h_c, $W \cdot m^{-2} \cdot °C^{-1}$) and convective heat losses (H_c, $W \cdot m^{-2}$) at various water temperatures (T_w,° C) and water velocities (V_w,$m \cdot s^{-1}$) during immersion.*

T_w	V_w	$(T_{sk} - T_w)$	Activity	h_c	H_c	Reference
20	0.005		(manikin)	137		Witherspoon(155)
20	0.15		(manikin)	206		Witherspoon(155)
20	0.5		(manikin)	588		Witherspoon(155)
20	0.8		(manikin)	1434		Witherspoon(155)
33–44	0	0.94	still	42	40	Boutelier(15)
33–44	0.05	0.70	still	63	44	Boutelier(15)
33–44	0.1	0.54	still	81	44	Boutelier(15)
33-44	0.25	0.43	still	147	63	Boutelier(15)
35	0	0.14	still	38	5	Gee(47)
32	0	0.34	still	96	33	Gee(47)
28	0	0.75	still	208	156	Gee(47)
24	0	0.97	still	358	347	Gee(47)
20	0	1.49	still	537	800	Gee(47)
18–33	0		still	230		Nadel(107)
18–33	0.5–0.75		still	460		Nadel(107)
18–33	0.5–0.75		exercise	580		Nadel(107)

As is indicated in Table 9-1, the insulation provided by the boundary layer at the skin-water interface varies with water conditions. Absolute values for the heat transfer coefficient also vary between authors when measurements are made under similar conditions. From data obtained on a nude copper manikin immersed in water, Witherspoon (155) determined that the convective heat transfer coefficient was 137 $W \cdot m^{-2} \cdot ° C^{-1}$, which represented a surface insulation of 0.0073° $C \cdot m^2 \cdot W^{-1}$. This coefficient is dependent of course upon the thickness of the boundary layer. Clearly, with increasing body or water movement, the still boundary layer is disrupted and its insulatory benefit reduced. As predicted, the convective heat transfer coefficient increases up to a value of 1434 $W \cdot m^{-2} \cdot ° C^{-1}$ in water flowing at 0.8 $m \cdot s^{-1}$ (155). In both thermoneutral (15) and cool (154) water, despite the decreased insulation at the skin surface with increasing water movement, the overall convective heat loss does not appear to change (15,107). This is partially explained by the observed lowering of skin temperature with a subsequent reduction in the thermal gradient between skin and water (Table 9- 1,(152)).

The surface insulation also is reduced in cold water (155) as the heated water in the boundary layer begins to rise, thus creating a form of natural or free convection (as indicated by an elevated coefficient). In contrast to the effect of increasing water velocity on the heat transfer coefficient, decreasing water temperature elevates both the heat transfer coefficient and the overall convective heat loss (47,107). Exercise also appears to modify the surface insulation. Nadel and colleagues (107) reported higher convective heat losses and heat transfer coefficients during swimming, compared to rest. Also, the exercise heat transfer coefficients were similar at any swimming speed, water velocity, or water temperature (107); this suggests that the body movements, per se, during swimming maximize the removal of the boundary layer.

Internal Heat Loss. The transfer of body heat from core to the water environment represents a complex interaction of several physical and physiologic processes. From the cellular site of metabolic heat production, thermal energy moves toward the surface in a variety of ways. The components of internal heat transfer include the mechanisms of both conductive and convective heat transfer.

One basic approach in formulating research strategies during cold-water immersion is to begin with a theoretical framework based on presently known physiological information. The conceptual model presented in Figure 9-2, which is slightly modified from that originally described by Bullard and Rapp (18), divides the body into areas that have large vasomotor regulation, including the limbs and extremities (9,65,70), and areas that have limited or no vasoconstrictive ability, such as the head (65,147) and trunk (65,147). Within the core of these compartments, the heat input is a function of metabolism (M_R, resting metabolic rate; M_{EX}, exercise metabolism; M_{TG}, thermogenesis). Transfer of heat within each area becomes a function of a circulatory output (adjustable circulatory insulation) from core to skin, and a conductive output (fixed insulation) from the site of metabolism and peripheral vasculature through the overlying tissue to the surface. Alterations in the temperatures of the trunk and extremity cores are based on heat transfer and the specific heat (C_1 and C_2) of each area. Information on the state of heat transfer and distribution can roughly be determined by the core-to-skin (T_c-T_{sk}) and skin-to-water (T_{sk}-T_w) temperature gradients.

Regional Heat Flow. The first line of defense against heat loss to

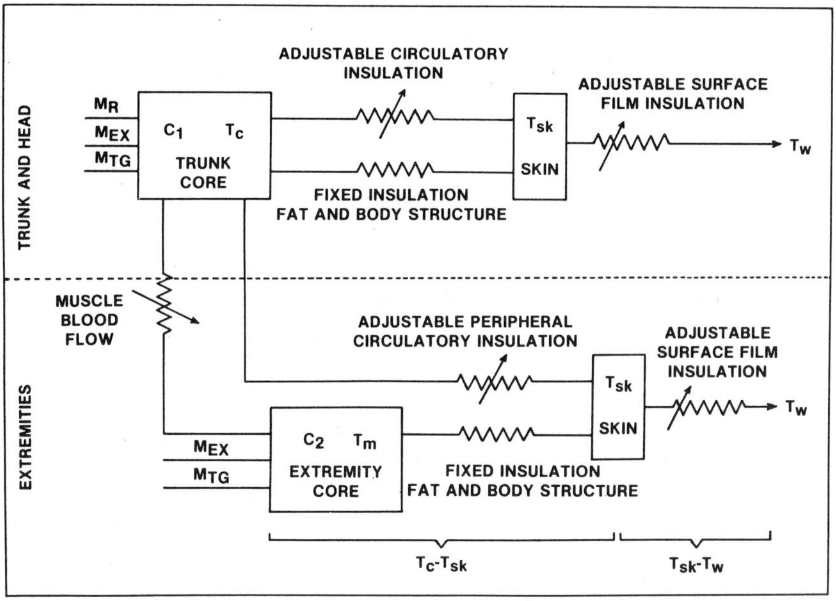

Figure 9-2. *Physiologic and physical components of internal heat transfer. (Modified from Bullard and Rapp (18)).*

the environment is through vasoconstriction of peripheral and extremity vessels. Figure 9-2 illustrates the adjustable circulatory resistance to the extremities. In effect, this represents a redistribution of blood away from the extremities and skin, thus reducing the internal effective surface for heat transfer. This vasoconstriction takes place when the individual is in water either above the critical temperature (35) or at a water temperature within 4° C of rectal temperature (22).

Craig and Dvorak (35) in their evaluation of 10 men during immersion in water between 24° C and 37° C, determined that the finger temperature was reduced to near water temperature when the bath temperature was below 34° C. In contrast, graded lowering of the water bath temperature elicited progressively increasing skin-to-water temperature gradients on the abdomen. This suggested to Craig that the periphery of the body was not at a uniform temperature (35). Several studies, including thermographic evaluations, have demonstrated that there are considerable local temperature differences throughout the body during cold exposure (65,151). These differences create a patchwork of different temperature profiles at the body's surface (148).

Non-uniform skin temperatures have been consistently noted

during cold- air exposure. This provides the basis for the large number of skin temperature sites required to determine a mean weighted skin temperature in the cold. This non-uniformity in surface skin temperature suggests that the underlying peripheral tissue temperature parallels the response of skin temperature. Since vasoconstriction at the skin retracts the circulation below the level of peripheral tissues, then heat transfer from the core depends upon conduction through the overlying muscle and fat tissues. As some have noted, the distribution of subcutaneous fat is quite variable both within and between individuals (4). This variation in fatness as well as in muscle tissue within an individual makes the determination of an average peripheral temperature extremely difficult.

Based on considerations of peripheral insulation, one might suspect that heat loss across the skin surface is also not uniform. Therefore, at any surface location (assuming a uniform convective heat transfer coefficient at the skin surface and temperature gradient), heat loss would depend upon the interaction of local vasomotor tone, the quantity and distribution of fat and muscle tissue, and the underlying core temperature.

Although there may be variations in temperature and heat flow within any particular area of the body, evidence strongly suggests that heat loss varies from region to region. Hong and colleagues (73), for example, found that for resting subjects in cool water (30° C-33° C) the heat flow was greatest in the abdomen, compared to the extremities. Heat loss through the limbs, however, was also considerable. Similar to these observations, Hayward and Keatinge (70) reported that the insulation values for men resting in water at a temperature allowing stability of the core temperature was highest in the feet and hands and lowest in the trunk. The two to three times greater insulation in the extremities could not be explained by differences in subcutaneous fat in the two body regions. Obviously, skinfold thickness contributes little to retard heat transfer from the limbs to the water at rest (73,145). More than likely, the greater heat conservation in the limbs is the result of either reduced blood flow or insulation provided by lean tissue. In addition to the trunk and extremities, specific areas of high heat loss have been described using infrared thermography. The neck, lateral thorax, upper chest, and groin appear to be areas of high heat loss (65), although these findings have been questioned by Reed (124).

Another potential avenue of heat transfer to the water was observed by Lewis (94) during immersion of the fingers in ice water. As described, the fingers initially vasoconstricted, then dilated (94). This cold-induced vasodilation has been substantiated by the classic work of Yoshimura and Iida (157,158) and observed by most investigators who examined the extremities during extreme cooling. Simply, the finger, when adequately chilled, begins a cycling of vasodilation and vasoconstriction. The pattern and frequency of this cycling has been the interest of several investigators. In one report, Keatinge (81) examined finger temperature, blood flow, and heat loss as a function of local cooling of the fingers in conjunction with whole body cooling and heating.

The results indicate that finger cold-induced vasodilation occurs in individuals who are chilled by cold water, though the time to onset of vasodilation is longer, and the temperature of the finger is lower than when the subject is heated. There is general agreement that as the core temperature becomes depressed, cold vasodilation is eliminated. However, there have been reports that have demonstrated vasodilation under extreme water temperatures (29). For example, Cannon and Keatinge (29) have demonstrated increases in finger temperature in cold water during mild hypothermia. Such dilation in the fingers could further contribute to the overall body heat loss.

Body Fatness. The experimental data presented in Table 9-2 show that fat provides a significantly greater thermal resistance than either skin or wet muscle tissue. With respect to the combined effects of skin and subcutaneous fat, Veicsteinas and colleagues (147) determined the relationship between the superficial shell thickness and the afforded peripheral insulation as,

$$I_{ss} \max = 0.0048\ d - 0.0052\ (r = 0.95)$$

where, I_{ss} max is the maximal superficial shell insulation ($°\,C \cdot m^2 \cdot W^{-1}$), and d is the thickness of the shell (mm).

Subcutaneous and deep body fat are major components determining the rate of internal heat transfer during water immersion. This insulatory benefit is represented by the fixed resistances in the trunk and extremities as presented in the model in Figure 9-2. It was apparent to investigators that a subject's skinfold thickness and

TABLE 9-2. *Thermal resistivity ($°C \cdot m^2 \cdot W^{-1} \cdot cm^{-1}$) of fat, muscle, and skin.*

Insulator	Resistivity	Condition	Reference
skin	0.026	in vivo	Burton(21)
skin	0.028	in vivo (T_w)	Lefevre(93)
skin	0.032	in vivo (T_a, 30° C)	Reader(123)
skin	0.022	in vivo (CWT)	Veicsteinas(147)
fat	0.114	in vivo	Lefevre(93)
fat	0.048	in vivo (CWT)	Veicsteinas(147)
fat	0.048	in vitro	Hatfield(64)
muscle (wet)	0.026	in vitro	Hatfield(64)

Data from Veicsteinas *et al.* (147), and Hatfield and Pugh (64).

total body fat were essential measurements for explaining heat conservation during cold-water experiments. Carlson and colleagues (32) provided some of this information on the relationships between body fatness and tissue insulation, and between maximal tissue insulation and water temperature. From these data it was clear that body fatness and insulation were linearly related and positively sloped. Two years later, Keatinge (83) reported a strong inverse relationship between skinfold thickness and the fall in rectal temperature during 30 min of immersion in 15° C water. Since then, the benefits of body fat in the maintenance of core temperature during immersion have been amply and consistently verified (29,70,97,135). A relatively fat person with large subcutaneous fat and minimal perfusion of the skin and extremities can withstand the challenge of cold water more effectively then a leaner counterpart. Similarly, the larger the quantity of body fat, the less reduction in core temperature at any specific cold-water temperature (107).

Exercise and Thermal Balance

Although significant metabolic heat is generated through shivering, the greatest potential contribution of muscle to defend against cold occurs during physical activity. In air as low as −30° C, exercise energy metabolism can sustain a constant core temperature without the need for heavy restrictive clothing. However, exercise with little clothing in extremes of cold water may accelerate the drop in rectal temperature when compared to still conditions because the heat loss stimulated by both the movement of water and exercise exceeds the heat production of the exercise (66,84).

As mentioned earlier, the drive for heat production for the resting state is mediated by neural output from the thermoregulatory center, and therefore is not directly determined by the heat production in the body, per se. Thus, shivering is observed even during exercise if core and skin temperatures are low. As a result, in light and moderate exercise, oxygen uptake is proportionally higher in cold water than it is during the same exercise in a warmer environment (36,98,99). This additional oxygen uptake is directly related to the added energy cost of shivering as the body attempts to combat heat loss to the cold water.

Based on the model in Figure 9-2, a facilitated heat loss during exercise compared with rest is a function of: (a) the increased transfer of heat from the trunk and head core to the extremity core via increased muscle blood flow, (b) the increased heat production in the extremities versus the trunk when compared to the non-exercise condition, (c) the increased movement of the arms and legs which reduces the insulatory benefit of the boundary layer at the skin-water interface, and (d) the increased effective surface area for heat transfer provided by the redistribution of blood from the trunk core to the extremity core. These alterations in heat gain/heat loss dynamics in the extremities, where surface area-to-mass ratio is high, decrease heat storage during exercise and could account for the reported lower core temperatures during exercise compared with still conditions in cold water (66,84).

Various studies have attempted to evaluate the contribution of exercise to thermal balance during cold-water immersion. The specific nature of the research including, where available, subject characteristics, exercise mode and intensity, environmental conditions, and general conclusions are summarized in Table 9-3. Crittenden and colleagues (39), for example, observed a lowering of rectal temperature during exercise in water at both 20° C and 26° C. However, their experimental protocol included a 10 min pre-cooling period prior to exercise. This procedure could lower the temperature in the limbs, thereby accelerating the rate of body cooling by increasing blood flow and subsequent heat loss in the limbs as exercise is initiated.

Keatinge (84) contrasted thermoregulatory responses between rest and arm-leg exercise of 12 subjects in water ranging between 5° C and 35° C. In 35° C water, rectal temperature increased during 40 min of exercise and was 0.4° C higher than the resting value during immersion for the same duration. Core temperatures were similar between rest and exercise only in water at 25° C. Subjects resting in 5° C and 15° C water, however, maintained higher body temperatures than if they had exercised. In fact, the difference in rectal temperature was 0.75° C in 5° C water. Under all these conditions, the fatter individuals maintained higher core temperatures, again verifying the insulatory benefits of body fat. Additionally, in contrasting rest and exercise to evaluate the stabilization of core temperature at the lowest possible water temperature, Hayward and Keatinge (70) reported that when water was too cold to allow for the maintenance of deep body temperature at rest, exercise usually intensified cooling by increasing conduction to the poorly insulated, highly perfused active peripheral areas without greatly affecting heat loss from the trunk. Furthermore, Hayward and colleagues (66) showed that slow swimming movements that elevated heat production by 2.5 times rest increased body cooling in 10.5° C water to a 35 % greater rate than when subjects remained still.

Despite the reports of lower core temperatures during exercise in cold water compared with still conditions, several studies demonstrate a beneficial thermoregulatory effect of moderate exercise. Observations on Channel swimmers in the 1950s showed that rectal temperatures during ocean swimming ranged between 34.0° C and

TABLE 9-3. *Role of physical activity in thermoregulation during cold water immersion.*

Reference	Subjects	Fat	Exercise	Approx. Exercise Intensity	Duration (min)	T_w (°C)	Thermoregulatory Effect
Cannon & Keatinge (29)	8M	$\bar{X}$-4 sites 6.5–27 mm	Arm/leg row	low?	30–70	lowest to maintain T_{re}(5–20)	Physical exertion always accelerated the fall in T_{core} compared to rest.
Craig & Dvorak (36)	10M	14%	Arm/leg cycle	0.65–0.98*	60	24,26,28, 30,32,34	Exercise contributed to maintenance of T_{re} in cold water.
Craig & Dvorak (37)	2M	$\bar{X}$-6 sites 15.5mm	Arm/leg cycle	0.6–2.2*	14	22–34	Exercise prevented a fall in T_{ear} in cold water.
Crittenden et al. (39)	5M	13.7%	Submerged scuba	1.57–1.68*	17	20,26	T_{re} dropped faster during exercise than during immersion without exercise.
Hayward & Keatinge (70)	7M 7F	$\bar{X}$-48 sites 4.7 mm 5.8 mm	Arm/leg cycle	1.5*	45	lowest to maintain $T_{re}(T_{stab})$ (12–32)	Exercise is of benefit during immersion for those individuals with low shivering response at T_{stab}.
Keatinge (84)	12M	$\bar{X}$-4 sites 7.95 mm	Arm/leg row	0.34–1.37*	20–40	5,15,25, 35,37.8	Physical exertion accelerated body cooling in cold water compared to rest. This occurred whether the men were clothed or not, or whether the work was hard or moderate.
McArdle *et al.* (99)	10M 8F	15.9% 21.9%	Arm/leg cycle	1.70*	60	20,24,28	Exercise in all water temperatures either prevented

							or retarded the fall in T_{re} observed during rest at the same T_w.
McMurry & Horvath (110)	11M	8.8%	Swim	2.5*	30	20,25,30	Exercise in all water temperatures prevented a drop in T_{re}.
Moore *et al.* (106)	8M	18%	Leg	graded	25	16,22,30	No significant effect of exercise on T_{re} in cold water.
Nadel *et al.* (107)	3M	9%	Swim flume	1.65–4.13*	20	18,26,33	Exercise of greater intensity appeared to prevent the fall in T_{es} compared to similar exercise of lower intensity during rest.
Pugh *et al.* (117)	11M	19.6%	Swim	2.0–3.3*	180–1020	16–18	Exercise metabolism provides thermorequlatory benefits among fatter compared to leaner Channel swimmers.
Sloan & Keatinge (133)	12M 16F	1 site 8.7 mm 9.9 mm	Swim	1.0*	18–40	20.3	Exercise failed to prevent a fall in T_{oral} in young swimmers.
Toner *et al.* (146)	8M	13.4%	Arm cycle Leg cycle Arm/leg	1.5–2.0*	60	20,26,33	The greatest benefit of exercise in thermoregulation was observed in leg only exercise compared with arm only and arm/leg exercise.
Toner *et al.* (145)	9M	16.8%	Leg cycle	1.6*	60	18,20,30	Exercise more effective than rest in maintaining T_{core} in both cool and cold water.

*M = male; F = female; $*\ell\ O_2 \cdot min^{-1}$

38.3° C for exposures of nearly 18 h in water as cold as 16° C (117). The ability of these swimmers to maintain such high core temperatures despite prolonged cold exposure was related to their relatively large amounts of subcutaneous fat and the high metabolic rates generated during swimming (117). Furthermore, in contrasting the thermoregulatory responses of a thin and fat subject, Pugh and Edholm (116) showed during immersion in 16° C water that, for the thin subject, exercise accelerated the drop in rectal temperature when compared with still conditions, whereas swimming maintained rectal temperature in the fat subject.

The comprehensive work of Craig and Dvorak (36) systematically examined thermal responses of 10 men during arm-leg exercise ($\approx$0.75 ℓ $O_2 \cdot min^{-1}$) with immersion for 60 min in water between 24° C and 34° C on six occasions. During low-intensity exercise, rectal temperature remained unchanged; thermal balance was achieved in water as low as 26° C. As exercise intensity increased, rectal temperature also increased; thermal balance was achieved in water as low as 24° C. In a more recent study, McArdle and colleagues (99) documented the beneficial effects of the identical mode of submaximal exercise ($\approx$1.7 ℓ $O_2 \cdot min^{-1}$) on temperature regulation during cold-water immersion down to 20° C for both men and women. Nadel and colleagues (107) observed only a slight drop (0.25° C) in esophageal temperature when swimming in water at 18° C. Others utilizing moderate to heavy exercise also have shown similar results (116,133,146).

The type of exercise recently has been suggested as a noteworthy factor in human heat transfer in cold water. Toner and colleagues (146) contrasted strict leg exercise with both arm-leg and strict arm exercise at similar levels of heat production. Leg exercise maintained core temperature in cold water at a higher value compared with the other exercise forms. As shown in Figure 9-3, the thermoregulatory benefits of leg-only exercise were reflected in measures of both rectal temperature and heat flow. The reduction in rectal temperature is greatest when the arms are engaged in exercise. The less effective heat conservation with arm compared to leg exercise may be due to either less insulation or a greater surface area-to-mass ratio in the arms, or both. An increased circulatory output to the extremities, thereby enlarging the shell, may make this exercise far less effective in maintaining thermal balance than leg only exercise. In this regard, Toner and colleagues (145) also showed, in contrast to rest, that leg exercise maintained esophageal temperature in water down to 18° C. They speculated that leg exercise transferred a portion of the exercise heat production to the core.

The available data indicate that water temperature, body fat,

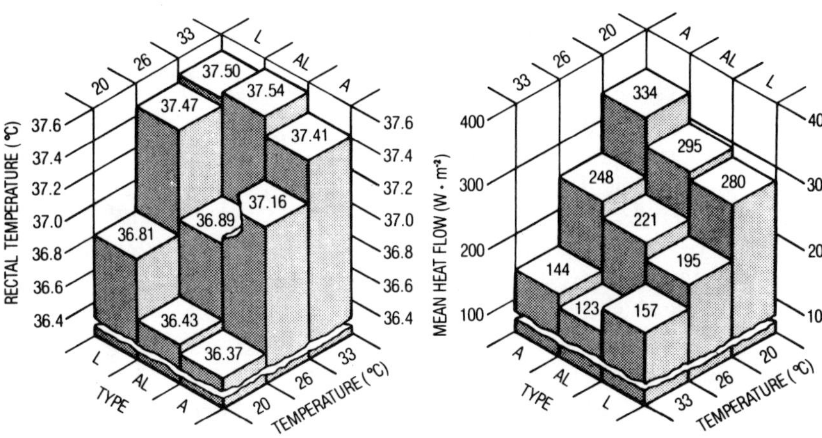

Figure 9-3. *Final rectal temperatures and mean weighted heat flows during arm (A), leg (L), and arm-leg (AL) exercise (≈1.5ℓ $O_2 \cdot min^{-1}$) in water at 20, 26, and 33° C. (Modified from Toner* et al. *(146)).*

and exercise intensity and type, as well as an individual's ability to maintain high levels of exercise are important considerations in evaluating the potential benefits of exercise in offsetting a drop in core temperature with cold-water immersion. Certainly, a complex interaction exists between these variables. It appears that for each individual a water temperature exists at which the heat conserved via insulation and circulatory adjustments and the heat generated by shivering or exercise do not offset the heat flow induced by an elevated conductive and convective heat transfer in cold water.

EXPOSURE TO COLD AIR

The dynamics of human heat exchange have been systematically observed over the past 50 years within varied and extreme environments including air (hot-wet, hot-dry, cold-wet, cold-dry), water, and hypobaric and hyperbaric conditions. Under most environmental conditions the application of the heat balance equation has been used to account for heat exchange between the organism and the environment. Though there are similarities with the dynamics of heat balance in water and air, thermoregulation during cold stress in air is further complicated by evaporative and radiative heat transfers. The following discussion focuses on the physiologic adjustments of humans during cold-air exposure with consideration for the impact of both exercise and clothing as they pertain to temperature regulation and thermal balance.

Sources of Heat Gain and Loss in Cold Air

Non-exercise Metabolism. As with the case of water immersion, metabolic rate is used by some investigators as a means to determine thermoneutrality in air (16). As was shown in Figure 9-1, the lower critical temperature in air is much lower than that observed in water, and ranges between 20° C and 27° C. This appears to be on the low side of previously reported values of 27° C to 29° C (5). The increase in metabolism observed as the air temperature decreases below the critical temperature has been attributed to: (a) the increase in metabolism without shivering, and (b) shivering thermogenesis.

Nonshivering thermogenesis. Over the years, there has been speculation that a portion of the increased metabolism observed in humans during cold stress was due to nonshivering thermogenesis similar to that clearly observed in a number of small mammals. Specifically, nonshivering thermogenesis is defined as an increased metabolic heat production from sources other than muscular contraction. These sources in animals include both elevated rate of aerobic metabolism of brown adipose tissue as well as the stimulating effects of the metabolic endocrines. For humans, the heat production in infants during cold stress is partly attributed to the futile metabolism of brown fat (41). However, controversy exists with respect to either the prevalence of this tissue or its contribution as a significant source for heat production in adults. Dawkins and Hull (41) describe the conversion of brown adipose tissue of the infant into white adiposity by the time adulthood is reached. However, through the injection of sympathomimetic drugs which stimulate the activity of brown fat (80,127), as well as the direct determination of mitochondrial activity in the adipose tissues of outdoor workers (75), a number of reports have argued that brown fat does exist in the adult. Additional observations may confirm this, and clarify the precise contribution of brown fat to nonshivering heat production in humans and its role in thermoregulation during cold stress. However, it currently appears that its role is either minimal or non-existent.

Other potential heat sources from nonshivering thermogenesis during cold exposure include the metabolic effects of the circulating endocrines, such as the adrenal catecholamines, the glucocorticoids, and thyroxine. A number of reports have shown increased levels of catecholamines with cold exposure (8,79,92,118,159). Recently, Young and colleagues (159) exposed subjects clothed in swim trunks to air at 5° C for 90 min. During this brief exposure, plasma norepinephrine increased by a factor of two ($\bar{X}_{pre} = 485\ ng \cdot \ell^{-1}$, $\bar{X}_{post} = 969$

ng $\cdot \ell^{-1}$), whereas epinephrine levels remained unchanged at about 62 ng $\cdot \ell^{-1}$. Elevated norepinephrine levels are most likely the result of enhanced sympathetic stimulation in the peripheral tissues where pronounced vasoconstriction is evident during cold stress.

The catecholamines have long been shown to stimulate metabolism in most tissues, especially when thyroxine is also elevated (54). Despite the vast evidence documenting the calorigenic effect of endocrines in many animal species, this does not appear to account for a nonshivering elevated metabolism in cold exposed, unacclimated humans (71), though some have reported a modest relationship between norepinephrine and metabolic rate (89). Considering the elevated catecholamine production during cold stress in the unacclimated individual, it appears that elevated adrenocorticoids may suppress the production of thyroxine (118) and thus limit the calorigenic effects of the catecholamines (45).

Shivering thermogenesis. An early report of one carefully controlled case suggested that muscular tensing without noticeable shivering contributed 36% to the increase in heat liberation upon exposure to cold (139). This has been further substantiated (7) and suggests that it may be difficult for a subject to eliminate all muscular tension upon cold exposure. The term "preshivering" has been used in the literature (149) to denote metabolism of this nature.

To determine the maximal shivering capacity in air, Iampietro and colleagues (76) exposed 16 nude men to air temperatures between −18° C and 32° C, with wind speeds between $< 0.7\ \mathrm{m \cdot s^{-1}}$ and $4.5\ \mathrm{m \cdot s^{-1}}$. As expected, the lowest temperature and the highest wind condition elicited the highest metabolic rate which averaged 493 W. This is somewhat lower than that reported for peak shivering in cold water. Values for metabolic rate during exposures in air between 0° and 10° C ranged between 1.3 to 4.2 times the resting metabolism (2,31,74,76).

Exercise Metabolism. In direct contrast to cold water, exercise in air consistently provides an optimal means for maintaining thermal balance. A classic demonstration of the minimal effect of cold air on core temperature during exercise was performed by Nielsen in the late 1930s (109), and later confirmed by Nielsen in the early 1960s (108). In the latter, reported air temperatures varied between 5° C and 30° C, while subjects exercised at 150 W for one h. The results showed that esophageal temperature after one h averaged only 0.1° C lower in air 5° C compared with tests at 30° C. Subsequent research has confirmed these findings in principle; although, as contrasting environments establish a greater range for air temperatures, some thermoregulatory differences are evident. Claremont and colleagues (33), for example, showed an average rectal temperature difference of 0.5° C between exercise at 35° C (T_{re}, 38.75° C) and 0° C (T_{re}, 38.25° C) air conditions. The benefit of exercise in maintaining core temperature and comfort in cold air also appears to be sensed by the cold-stressed subjects. Subjects permitted to choose an exercise intensity to maintain thermal comfort in 10° C air regulated exercise intensity in a manner that maintained skin and core temperature as well as the feeling of comfort (30).

The value of exercise in maintaining regional temperatures during cold-air exposure is also an important consideration. Stromme and colleagues (137) had subjects perform exercise after precooling in air at about 5° C. At rest, the finger temperature of the unacclimated subjects cooled to an average of 12° C. During subsequent bouts of no-load and 50 W exercise, performed on a separate day under identical conditions, finger temperature remained unchanged compared to values in thermoneutral air; exercise at 100 W caused a precipitous rise in temperature to between 20° C and 34° C within 20 min. The abrupt rise in finger temperature clearly illustrated vasodilation in the finger pad. Unfortunately, this study provided no other temperature data (137). Similar results have been reported using an identical protocol with exercise at 100 W being the threshold for dilation of the hand. In this research the inter-individual time to peripheral dilation was greater, ranging from 10 to 35 min (6). This effect of exercise on regional blood flow during cold-air exposure can be of benefit to peripheral areas prone to cold injury.

The potential for heat loss via increased exercise ventilation has concerned a few researchers who realized that respiratory evaporative heat loss represents about 9% of the heat production at rest with the magnitude of evaporation being in proportion to the ventilatory volume (17). Hartung and colleagues (63), however, reported no significant rectal temperature differences between air breathing at 24° C compared to −35° C, when subjects rested or performed exercise at 60% of $\dot{V}O_2$ max. At 70% of maximal aerobic power, however, subjects breathing -35° C had a 0.3° C lower rectal temperature. Based on available data, it appears that the increase in ventilation during exercise in the cold has a minimal effect on core cooling. In considering the effects of cold air on airway and esophageal temperatures, Jaeger and colleagues (78) showed that breathing air at −40° C resulted in a substantial reduction of upper esophageal temperature during moderate exercise. Some have concluded that there is a quantitative association between heat loss from the airways during exercise and the subsequent constriction of the bronchioles in the patient who experiences exercise-induced asthma (28,100).

Body Fat and Cold Air

As was the case with water immersion, body fat provides significant protection during cold-air exposure. Daniels and Baker (40) examined the responses of 21 men, varying in body fat between 1.7% to 18.2%, by skinfold assessment. For individual skin temperatures during exposure to 15° C, they reported significant negative correlations between calf, thigh, back, chest, and upper arm skinfolds and percent body fat (r ranging between −0.42 to −0.60). The highest correlation (r = −0.80) with body fatness was established with mean weighted skin temperature (11), whereas no significant relationship was observed for measures of the extremities including the head (40). Metabolic rate after two h exposures also was significantly and inversely correlated (r = −0.60) with body fat; this relationship was stronger than that observed between body fat and either skin or rectal temperature (40).

LeBlanc (90) established positive relationships between skinfold thickness and insulation index (resistance of heat transfer from the core to the skin). These relationships became stronger with decreasing air temperatures. This suggests that the insulation afforded by body fat is more beneficial during severe cold stress than when the air temperature is relatively warmer. This is supported by data during heat stress where the insulation provided by body fat,per se, does not contribute to heat gain (24). More than likely, the insulatory effect of body fat is negated by the circulatory output to the skin where the fat is bypassed.

In further studies of the thermoregulatory benefits of body fat, metabolic and body insulative responses of 12 men and three women were determined during two to four h exposures in air at 10° C. Percent body fat was linearly and inversely related to metabolism ($M, W \cdot m^{-2} = 75.2 - .60\%$ BF). Insulations of the chest, upper arm, and lateral thigh were twice as large for individuals possessing more than 30% body fat as compared to those under 30% (27). These observations also have been verified in other research (87). Wyndham and colleagues (156), in observing Bantu and Caucasians (N=11 each), suggested that the metabolic response to a fall in skin temperature of Caucasians upon cold-air exposure is adjusted to the insulation of the subcutaneous fat in a way that rectal temperature will actually rise.

Further studies by Buskirk and colleagues (26) identified a linear, positive relationship between minimal rectal temperature and % body fat in 10° C air in which T_{re}, min = 36.33 + 0.011 % BF. The earlier work of Baker and Daniels (10) supports these findings. In considering the data of Allen *et al.* (4), Buskirk and colleagues (26) also noted that the greater the individual's total fat, the greater the percentage of that fat distributed to subcutaneous tissue. Consequently, it may be more appropriate to consider the subcutaneous fatness rather than total body fat during cooling in cold air, especially when group variations in body fatness are considerable.

It is clear that subcutaneous fat reduces body cooling in air by providing resistance to heat transfer from the body's core to the environment. The insulatory benefits of subcutaneous fat are brought into play when the circulation is retracted below the level of this fat, thus minimizing conductive heat transfer to the skin. This reduces the need for increased heat production in the core.

Interaction of Clothing and Cold Air

Unlike exposure to heat, behavioral regulation appears to be much more effective in preventing cold-induced injuries. In this regard, humans have the ability to insulate themselves from the environment by wearing clothing appropriate for particular environmental conditions and the physical activities performed. This clothing can be worn with little increase in energy cost or reduction in the scope of activity. This is in contrast to heat exposure where man's tolerance is primarily based on physiologic capacity for heat dissipation in relation to the thermal stress of the environment.

Though it provides a means to maintain body temperature during cold exposure, clothing often complicates thermoregulation in terms of both its insulation qualities and permeability characteristics which allow for effective movement of moisture through the clothing. Clothing selection is under control of behavioral factors, as opposed to thermoregulatory control. What becomes most important in relation to clothing during cold exposure is the micro-environment established at the skin surface, and between the skin and the shell of the clothing. This was illustrated by the recent Antarctic expedition where measurements of temperature and humidity were obtained within the clothing ensembles (110).

The individual remains protected from the environment when heat gain and heat loss are balanced by the selection of clothing with an appropriate configuration for insulation and water permeability. Unfortunately, even under relatively constant environmental conditions, this balance is rarely achieved because of the wide range of physical activity levels. As previously noted, metabolism during exercise in the cold has the potential to increase body heat storage. Wearing similar clothing insulations as worn at rest and at thermal balance, exercise causes an individual to store body heat which elevates core temperature. This, in turn, stimulates sweat production.

Despite low environmental temperatures, exercising individuals often become heat stressed, as the microclimate system of the clothing becomes saturated with water vapor. Depending on the extent of the heat gain and the exposure duration, the liberation of sweat into the clothing can be substantial. Minimal wetting of the clothes is of no major consequence to cooling rate, whereas a modest to severe wetting of the clothing can reduce the overall insulation of the garment. This occurs by the replacement of the insulative air layer with water, or the compression of the wetted material. The effect of clothing wetness is most noticeable when the individual returns to a resting state and the metabolic heat from exercise is no longer contributing to thermal balance. Concern for the thermal state of the individual is greatest at this time, when cooling and perceptual chilling are pronounced. Careful consideration for the interaction between clothing and both environmental extremes and activity patterns should be given by those interested in human temperature regulation.

Blood Flow and Distribution

During cold-air exposure the transfer of body heat to the environment is largely a function of the convective heat transfer from core to skin via the circulation. Of significance to the heat transfer process are two alterations representing: (a) a reduction in the diameter of peripheral vessels, and consequently, (b) a redistribution of blood from the cutaneous and peripheral vasculature to more central locations.

Cutaneous vessels are under neural control that is influenced by skin and core temperatures (13), as well as baroreflexes (129). Regulation of these vessels during cold stress operates to increase vasomotor tone. In this regard, the blood flow and volume to the skin is reduced. This action produces several effects depending upon the initial blood flow to the skin at the time of cooling. In the heated individual, cutaneous blood flow can be great as is the case for the volume of blood in the peripheral veins. With mild cooling, peripheral venoconstriction displaces blood from the cutaneous vessels to the deeper veins (129) and, in turn, increases central blood volume. This volume displacement is less pronounced in both the normothermic and slightly cooled individual because they are less peripherally dilated before cold application.

During more severe cooling, the blood flow to the skin is reduced by pronounced vasoconstriction (128). Redistribution of the blood volume to the thorax could provide, through stretch receptors, the reflex depression of ADH secretion and cold-diuresis (118). Despite increases in urine flow, both the plasma and blood volume appear to go unaltered with acute-cold stress (111). Depending upon the degree of the redistribution of blood and the magnitude (cool air or ice-cold showers) and location (cold-pressor test of the hand or face in ice water) of the cold stress, cardiac output, stroke volume, and blood pressure can be elevated (121) or remain the same (122). The heart rate responses to cold exposure are also mixed with reported increases (2,121), decreases (8), and no changes (112). The total peripheral resistance is believed to increase during cold exposure due to the pronounced increase in cutaneous and peripheral resistance; however, some researchers have reported a reduction in resistance and attributed this to decreased resistance in the total splanchnic region (121).

As noted, for cold exposure in both water and air, a decrease in peripheral blood flow allows a person to maximize the benefits of the insulation provided by subcutaneous tissue. Furthermore, the reduction in peripheral blood flow is highly specific, being most pronounced in the extremities including the arms, legs, hands, and feet. The control of blood flow in the extremities can be considered from three major sources: central neural drive (13), local reflexes (including spinal) (13,150), and local physical factors including direct effects of temperature (86) and endocrines on the vasculature (13,113).

Factors Contributing to Lowering of Body Temperature

Metabolic and other regulatory adjustments determine an individual's ability to resist core cooling. As discussed in Chapter 10 on human adaptation to cold, physiologists have noted that some races can thermoregulate at lower core temperatures compared to Caucasians (58). Other factors contributing to the hypothermia in cold air were identified for walkers (113,114), climbers and campers (114), and cave explorers (88). The detailed report by Pugh (114), comprised observations of 23 cold-related incidents that resulted in 25 deaths, five cases of unconsciousness with recovery, and 58 milder cases. The environments were generally near freezing temperatures with wind and wet-cold conditions. With respect to fatalities, the two principal factors were wet clothing and walking to the point of collapse (114). At the onset of the walk, it appeared that the people were clothed appropriately for the prevailing environmental conditions. However, unanticipated wettness of clothing, climatic temperatures, or extended length of exposure precipitated hypothermia even among the experienced people (113). The incidents involving cave explorers suggest that there is an increased risk of hypothermia when coordination and skill are required in maintaining high metabolic rates especially when individuals are fatigued (88).

Pugh also has identified fatigue as a factor precipitating the onset of hypothermia in air (113). Although individuals became wet, core temperature appeared to be relatively high if energy metabolism was high. As the individual fatigued, however, exercise intensity decreased; this lowered the rate of heat production and contributed to hypothermia (115). Such data suggest that the ability to maintain a high metabolic rate during cold stress in air combats the onset of debilitating hypothermia. Factors determining capacity for sustained muscular exercise such as the lactate threshold (56), or the highest submaximal exercise intensity at which the energy is generated principally by aerobic metabolism, may be important in reducing the risk of hypothermia in the wet individual when sustained muscular exercise is possible. Care must be taken in the exclusive usage of a single physiologic assessment as it relates to the cold. Progressive hypothermia and its associate alteration in coordination and state of consciousness might well contribute to fatigue.

PREDICTION OF THE PHYSIOLOGIC STRAIN

Hypothermia In Water And Survival Prediction

It is clear that the unprotected individual accidentally immersed in sufficiently cold water will progress to hypothermia in a relatively short period of time. The challenge to the physiologist and physicist is to further identify and quantify factors that contribute to heat loss in water and predict the thermal responses and ultimate survival for a particular individual immersed in cold water. Numerous reviews and reports considering cold-water survival have been published throughout the years, among them are the works of Keatinge (85), Boutelier (14), and Molnar (104).

Several approaches have been applied to the problem of survival prediction during cold-water immersion (68,104,134,142). One of the original systematic examinations of the problem was performed by Molnar in the mid-1940s (104). Of course, the limitation of this approach, as noted by Molnar, is that the information on survival has numerous uncontrollable factors including the estimated sea temperatures and immersion durations, the depth of immersion, the clothing variations of those immersed, level of physical exertion, and the actual cause of death (hypothermia, drowning, injury). Such lack of objective control makes it difficult to interpret the report of Critchley (38 as reported by 104) that only one of 10 survived one-half hour in sea water at 27° C.

Figure 9-4 shows a modification of the illustration presented by Hayward (68), and displays estimates of cold-water survival from a number of sources. As described by Hayward (68), the dotted lines delineate three zones established on the basis of probability for sur-

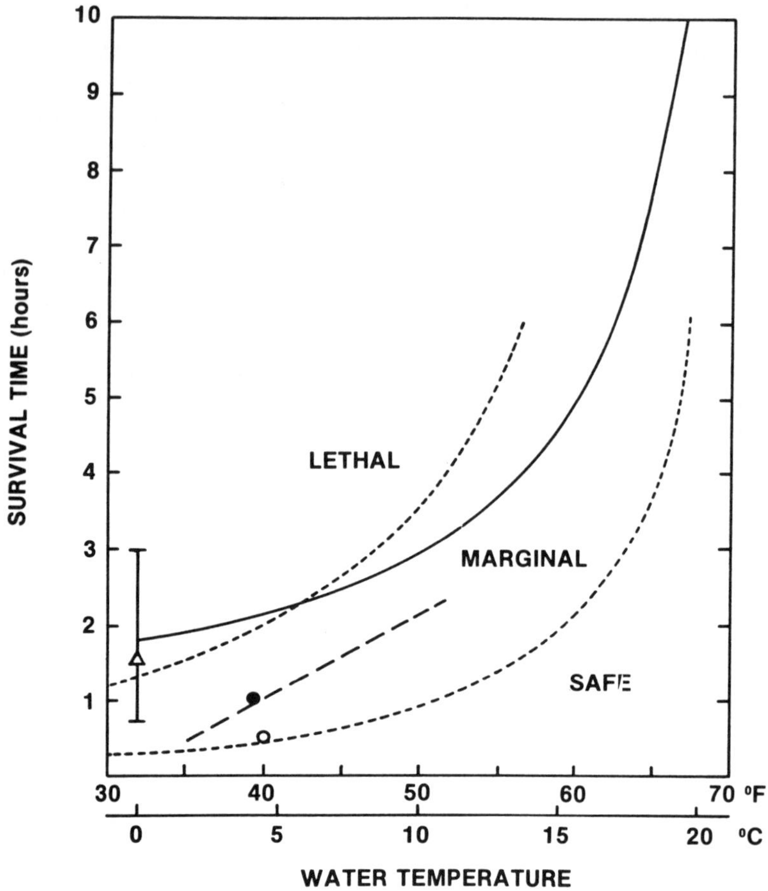

Figure 9-4. *Estimated survival time during immersion in water of varying temperatures. (Modified from Hayward (68)).*

vival and used by the U. S. Navy for general guidance. The "safe zone," based on the original data of Molnar (104), represents a combination of water temperature and immersion duration for which there is no likelihood of death. The "marginal zone" indicates those water temperatures and immersion durations where some fatalities (50% expectancy of unconsciousness that will probably result in drowning) are expected, whereas the "lethal zone" predicts no survivors from the exposures. The dashed line is Molnar's estimate of survival based on the Dachau data. The solid line represents the prediction of 50% survival based on hypothermia alone (68), and complements the guidelines used by the Navy. The open circle is based on the prediction of useful activity (mean body temperature

= 32.8° C) by Hall (57) for a clothed person. Closed circle is the prediction of Smith and Hames (134) for 50% survival at a mean body temperature of 31° C.

The open triangle and vertical bars reflect recent immersion data for lightly clothed individuals (67). Based on these observations for 0° C water, all subjects would survive for at least a 45 min exposure, 50% would survive 95 min, and no subject would survive beyond three h of immersion. Considering strictly death from hypothermia, one must conclude that the safety limits based on Molnar are conservative estimates of survival time for the clothed person. Additional information on extremes of core temperatures during both cooling and resuscitation from immersion, as well as a clearer picture of the factors contributing to core cooling for extended durations are needed to provide more accurate estimates of cold-water survival.

Others have taken a thermal modelling approach to predict body cooling in cold water (52,105,138,140,152). These models are based on thermodynamic principles applied to a representation of the human body in shape, size, and composition. In most heat transfer models, the shape of the body is considered a combination of cylinders and spheres. These shapes have determinable heat transfer coefficients based on their dimensions, and the direction and magnitude of the water flow (119). A recently developed thermal model for humans configured the body into 25 compartments including five cylindrical (trunk, arm, hand, leg, and foot) and one spherical (head) segments, each containing four compartments (core, muscle, fat, and skin) plus a central blood compartment. This configuration is based on the Montgomery model (105) and is similar to most other models (52,152). The basic difference in the models lies in variations with respect to the number of segments and compartments, the types of mathematical solutions, and the use of passive or controlling systems. Metabolic and thermal responses are computed based on these considerations.

The complexity of human physiologic responses and the characteristics of heat transfer within the body and between the body and the environment make it difficult to achieve consistently accurate predictions with any of the available models (153). Prediction for cold water conditions is further confounded because of the variety of exposure and clothing conditions, activity levels, and the physiologic diversity of individuals.

The modelling approach will ultimately achieve the best results in prediction of thermal responses of humans immersed in water. As described by Wissler (154), a mathematical model of the human thermal system can provide the means to: (a) evaluate physiologic phenomena (counter-current heat exchange) that are not possible to obtain experimentally, (b) predict thermal behavior without the need of human experimentation, and (c) predict thermal responses that are beyond the range of human experimentation. It is necessary that the physiologist and the physical modeller collaborate on experimental studies that provide data contributing to the refinement of these models.

Surface Cooling in Air: The Wind Chill Index

One dilemma in evaluating the thermal stress of an environment is that ambient temperature alone is not always a valid indication of potential cold strain. Another important factor is the wind. On a windy day, air currents magnify heat loss as the warmer insulating air layer surrounding the body is replaced by colder ambient air.

Siple and Passel (132) developed the wind chill formula for estimating the cooling power of the environment as a function of the ambient temperature and the wind velocity. Their approach was to determine the time of cooling of a given volume of water to the point of freezing, over a variety of wind velocities and air temperatures. These cooling rates were then related to exposure conditions that had evinced frostbite of unprotected human flesh. The wind chill formula given by Siple is,

$$K_c = 1.16(10v^{0.5} + 10.45\text{-}v)(33\text{-}T_a)$$

where, K_c is the wind chill ($W \cdot m^{-2}$), v is the wind velocity ($m \cdot s^{-1}$), and T_a is the air temperature (° C). The point of tissue freezing was estimated on a short term basis to be at a rate of heat loss of about 1625 $W \cdot m^{-2}$ (132), although some have argued that rates of heat loss between 580 $W \cdot m^{-2}$ and 1045 $W \cdot m^{-2}$ can be dangerous when exposure is prolonged (60). A more detailed discussion of cold injuries is presented in Chapter 11. Figure 9-5 gives a nomogram for determining the wind chill effect based on the formula of Siple. A number of reports have identified limitations concerning the wind chill effect (34,104). Among these, Molnar (104) argues that the formula: (a) applies only to an unprotected surface, (b) does not consider respiratory heat loss, (c) cannot be used at wind speeds exceeding 20 $m \cdot s^{-1}$, (d) is strictly empirical and not interpretable by thermodynamic laws, and (e) overestimates cooling power based on the effects of wind speed for naked surfaces and underestimates cooling for clothed surfaces. Despite these and other limitations, the wind chill continues to be used as a reference for assessing the convective cooling power of the air. It also provides a convenient relative index for comparison between various experimental protocols and field situations. Some suggestion has been given toward improving the information provided by wind chill. As pointed out by Boutelier (14), the application of the globe temperature to the index would include the important component of radiant heat transfer within the cold environment.

In addition to relating both air temperature and wind speed with wind chill, the wind chill nomogram (Figure 9-5) provides a means to link perceptual information and wind chill indexing. Although there appears to be some problem with perceptual information in the zone described by "bitterly cold" (the index underestimates the

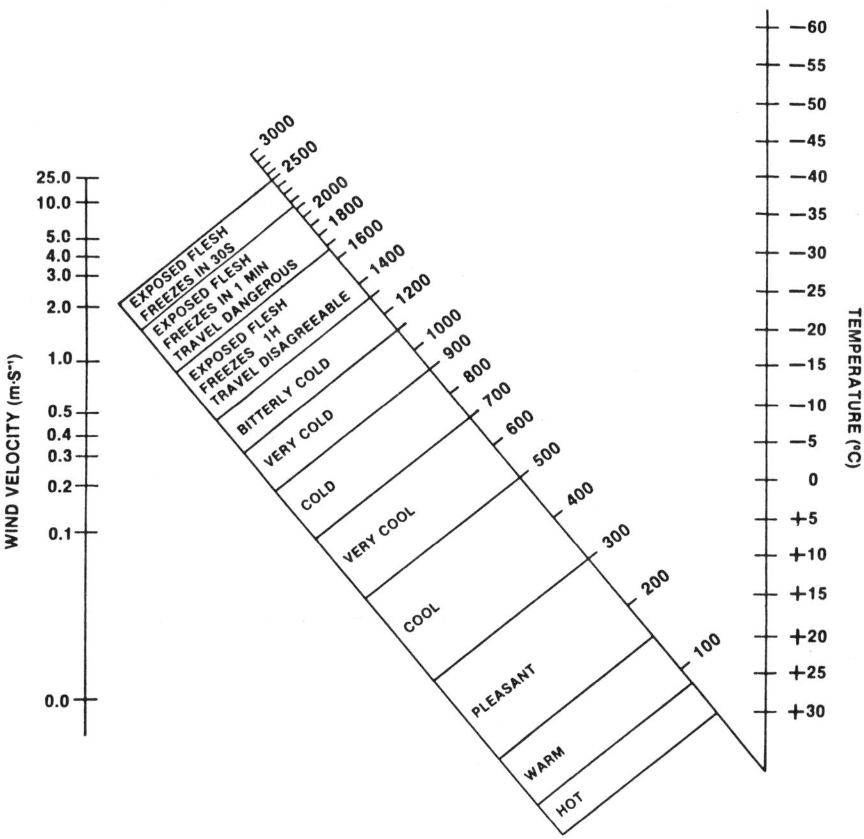

Figure 9-5. *The wind chill nomogram: interrelation between ambient temperature and wind velocity, and perceptual sensations and rate of heat loss (kcal $\cdot$ m^{-2} $\cdot$ h^{-1}) from the skin. Under conditions of bright sunshine, cooling is reduced by about 200 kcal $\cdot$ m^{-2} $\cdot$ h^{-1}. Expressions of relative comfort are based upon a lightly clad, inactive individual. To convert kcal $\cdot$ m^{-2} $\cdot$ h^{-1} to W $\cdot$ m^{-2} multiply by 1.163.*

cooling effect by about 10° C compared to physiologic assessments), the perceptual and physiological data fit well for the "dangerously cold" zone (91).

REFERENCES

1. Abramson, D. I., A. Kahn, S. Tuck, Jr., G. A. Turman, H. Rejal, and C.J. Fleischer. Relationship between a range of tissue temperature and local oxygen uptake in the human forearm. I. Changes observed under resting conditions. *J. Clin. Invest.* 37: 1031–1038, 1958.
2. Adolph, E. F., and G. W. Molnar. Exchanges of heat and tolerance to cold in man exposed to outdoor weather. *Amer. J. Physiol.* 146: 507–537, 1946.
3. Alexander, L. The treatment of shock from prolonged exposure to cold, especially in water. CIOS 24 Med., Combined Intelligence Objectives sub-committee G-2 Div., SHAEF (Rear), APO 413, 1946.
4. Allen, T. H., M. T. Peng, K. P. Chen, T. F. Huang, C. Chang, and H.S. Fang. Prediction of total adiposity from skinfolds and the curvilinear relationship between external and internal adiposity. *Metabolism* 5: 346–352, 1956.
5. Andersen, K. L. Thermogenetic mechanisms involved in man's fitness to resist cold exposure. In: *The Physiology of Cold Weather Survival,* edited by A. Borg and J. H. Veghte. AGARD Report No. 620, Neuilly-Sur-Seine, France: NATO, 1973.
6. Andersen, K. L., and J. S. Hart, H. T. Hammel, and H. B. Sabean. Metabolic and thermal response of Eskimos during muscular exertion in the cold. *J. Appl. Physiol.* 18: 613–618, 1963.
7. Andersen, K. L., V. Loyning, J. D. Nelms, O. Wilson, R. H. Fox, and A. Bolstad. Metabolic and thermal response to a moderate cold exposure in nomadic Lapps. *J. Appl. Physiol.* 15: 649–659, 1960.
8. Arnett, E. L., and D. T. Watts. Catecholamine excretion in men exposed to cold. *J. Appl. Physiol.* 15: 499–500, 1960.
9. Aschoff, J., and R. Wever. Modellversuche zum Gegenstrom-Warmeaustausch in der Extremitat. *Ztschrift. f.d.Ges. Exper. Med..* 130: 385–395, 1958.
10. Baker, P. T., and F. Daniels, Jr. Relationship between skinfold thickness and body cooling for two hours at 15° C. *J. Appl. Physiol.* 8: 409–416, 1956.
11. Baker, P. T., F. Daniels, Jr., R. F. Byrom, and E. H. Munro. Relationship between skinfold thickness and body cooling at 59° F. Technical Report, EP-14, U. S. Army Quartermaster Research & Development Center, Natick, MA, 1955.
12. Bigelow, W. G., W. K. Lindsay, R. C. Harrison, R. A. Gordon, and W. F. Greenwood. Oxygen transport and utilization in dogs at low body temperatures. *Am. J. Physiol.* 160: 125–137, 1950.
13. Boulant, J. A. Hypothalamic control of thermoregulation: Neurophysiological basis. In: *Handbook of the Hypothalamus,* edited by P. Morgane and J. Panksepp. New York: Marcel Dekker, 1981.
14. Boutelier, C. Survival and protection of aircrew in the event of accidental immersion in cold water. AGARD Report No. AG-211, Neuilly Sur Seine, France: NATO, 1979.
15. Boutelier, C., L. Bougues, and J. Timbal. Experimental study of convective heat transfer coefficient for the human body in water. *J. Appl. Physiol.* 42: 93–100, 1977.
16. Boutelier, C., J. Colin, and J. Timbal. Determination de la Zone de Neutralite Thermique dans l'Eau. *Revue de Med. Aero. et Spat.* 10: 25–29, 1971.
17. Brebbia, D. R., R. F. Goldman, and E. R. Buskirk. Water vapor loss from the respiratory tract during outdoor exercise in the cold. *J. Appl. Physiol.* 11: 219–222, 1957.
18. Bullard, R. W., and G. M. Rapp. Problems of body heat loss in water immersion. *Aerosp. Med.* 41: 1269–1277, 1970.
19. Burton, A. C. Human calorimetry. II. The average temperature of the tissues of the body. *J. Nutrition* 9: 261–280, 1935.
20. Burton, A. C. The application of the theory of heat flow to the study of energy metabolism. *J. Nutrition* 7: 497–533, 1934.
21. Burton, A. C. The direct measurement of thermal conductance of the skin as an index of peripheral blood flow. *Am. J. Physiol.* 129: 326–327, 1940.
22. Burton, A. C., and H. C. Bazett. The study of the average temperature of the exchanges of heat and vasomotor responses in man by means of a bath calorimeter. *Am. J. Physiol.* 117: 36–54, 1936.
23. Burton, A. C., and O. G. Edholm. *Man in a Cold Environment.* London: Arnold, 1955.
24. Buskirk, E. R., O. Bar-Or, and J. Kollias. Physiological effects of heat and cold. In: *Obesity,* edited by N. L. Wilson. Philadelphia: F. A. Davis, 1969, pp. 119–139.

25. Buskirk, E. R., and J. Kollias. Total body metabolism in the cold. *Bull, N.J. Acad. Sci.* (Spec. Symp. Issue): 17–25, 1969.
26. Buskirk, E. R., R. H. Thompson, and G. D. Whedon. Metabolic response to cooling in the human: role of body composition and particularly of body fat. In: *Temperature, Its Measurement and Control in Science and Industry,* edited by C.M. Herzfeld. New York: Reinhold, Vol 3, Part 3, 1963, pp. 429–442, 1963.
27. Buskirk, E. R., R. H. Thompson, and G. D. Whedon. Metabolic response to cold air in men and women in relation to total body fat content. *J. Appl. Physiol.* 18: 603–612, 1963.
28. Busse, W. Exercise induced asthma. *Am. J. Med.* 68: 471–476, 1980.
29. Cannon, P., and W. R. Keatinge. The metabolic rate and heat loss of fat and thin men in heat balance in cold and warm water. *J. Physiol. London* 154: 329–344, 1960.
30. Caputa, M., and M. Cabanac. Muscular work as thermal behavior in humans. *J. Appl. Physiol.* 48: 1020–1023, 1980.
31. Carlson, L. D. The adequate stimulus for shivering. *Proc. Sc. Exptl. Biol Med.* 85: 303–305, 1954.
32. Carlson, L. D., A. C. L. Hsieh, F. Fullington, and R. W. Elsner. Immersion in cold water and body tissue insulation. *J. Aviat. Med.* 29: 145–152, 1958.
33. Claremont, A. D., F. Nagle, W. D. Reddan, and G. A. Brooks. Comparison of metabolic, temperature, heart rate and ventilatory responses to exercise at extreme ambient temperatures (0° C and 35° C). *Med. Sci. Sport* 7: 150–154, 1975.
34. Court, A. Wind chill. *Bull. Am. Meteor. Soc.* 29: 487–493, 1948.
35. Craig, A. B., Jr., and M. Dvorak. Thermal regulation during water immersion. *J. Appl. Physiol.* 21: 1577–1585, 1966.
36. Craig, A. B., Jr., and M. Dvorak. Thermal regulation of man exercising during water immersion. *J. Appl. Physiol.* 25: 28–35, 1968.
37. Craig, A. B., Jr., and M. Dvorak. Comparison of exercise in air and in water of different temperatures. *Med. Sci. Sports* 1: 124–130, 1969.
38. Critchley, Mc D. *Shipwreck-Survivors: A Medical Study.* London: J. and A. Churchill, 1943.
39. Crittenden, G., J. F. Morlock, and T. O. Moore. Recovery parameters following underwater exercise. *Aerosp. Med.* 45: 1225–1260, 1974.
40. Daniels, Jr., F., and P. T. Baker. Relationship between body fat and shivering in air at 15° C. *J. Appl. Physiol.* 16: 421–425, 1961.
41. Dawkins, M. J. R., and D. Hull. The production of heat by fat. *Sci. Amer.* 213: 62–67, 1965.
42. Dill, D. B., and W. H. Forbes. Respiratory and metabolic effects of hypothermia. *Am. J. Physiol.* 132: 685–697, 1941.
43. Epstein, Y., Y. Shapiro, and S. Brill. Role of surface area-to mass ratio and work efficiency in heat intolerance. *J. Appl. Physiol.* 54: 831–836, 1983.
44. Etienne, J.L. Skiing alone to the pole. *Nat. Geo.* 170: 319–323, 1986.
45. Fregly, M. J. Hormonal interactions in body temperature regulation. In: *Neural Integration of Physiological Mechanisms and Behavior,* edited by G. J. Mogenson, and F. R. Calaresu. Toronto: University of Toronto Press, 1975, pp. 309–325.
46. Froese, G., and A. C. Burton. Heat losses from the human head. *J. Appl. Physiol.* 10: 235–241, 1957.
47. Gee, G. K., and R. F. Goldman. Heat loss of man in total water immersion. *Physiologist* (abstract) 16: 318, 1973.
48. Golden, F. St C. Immersion hypothermia. In: *The Physiology of Cold Weather Survival,* edited by A. Borg and J. H. Veghte. NATO Report, AGARD-R-620, London: Technical Editing and Reproduction, LTD, 1974.
49. Golden, F. St C., I. F. G. Hampton, G. R. Hervey, and A. V. Knibbs. Shivering intensity in humans during immersion in cold water. *J. Physiol. London* (abstract) 290: 48P, 1979.
50. Golden, F. St. C., I. F. G. Hampton, and D. Smith. Lean long distance swimmers. *J. Roy. Naval Med. Serv.* 66: 26–30, 1980.
51. Golden, F. St C., and G. R. Hervey. A class experiment on immersion hypothermia. *J. Physiol. London* (abstract) 227: 35–36, 1972.
52. Gordon, R. G., R. B. Roemer, and S. M. Horvath. A mathematical model of the human temperature regulatory system-transient cold exposure response. *IEEE Trans. Biomed. Eng. BME* 23: 434–444, 1976.
53. Greenleaf, J. E., E. Shvartz, S. Kravik, and L. C. Keil. Fluid shifts and endocrine responses during chair rest and water immersion in man. *J. Appl. Physiol.* 48: 79–88, 1980.
54. Griffith, F. R., Jr. Fact and theory regarding the calorigenic action of adrenaline. *Physiol. Rev.* 31: 151–187, 1951.
55. Gutierrez, G., A. R. Warley, and D. R. Dantzker. Oxygen delivery and utilization in hypothermic dogs. *J. Appl. Physiol.* 60: 751–757, 1986.
56. Hagberg, J. M., and E. E. Coyle. Physiological determinants of endurance performance as studied in competitive racewalkers. *Med. Sci. Sports Exerc.* 15: 287–289, 1983.

57. Hall, J. F., Jr. Prediction of tolerance in cold water and life raft exposures. *Aerosp. Med.* 43: 281–286, 1972.
58. Hammel, H. T. Terrestrial animals in cold: recent studies of primative man. In: *Handbook of Physiology, Section 4, Adaptation to the Environment,* edited by D. B. Dill, E. F. Adolph, and C. G. Wilber. Washington, DC: The American Physiological Society, 1964, pp. 413–434.
59. Hanna, J. M., and S. K. Hong. Critical water temperature and effective insulation in scuba divers in Hawaii. *J. Appl. Physiol.* 33: 770–773, 1972.
60. Hanson, H. E., and R. F. Goldman. Cold injury in man: a review of its etiology and discussion of its prediction. *Military Med.* 134: 1307–1316, 1969.
61. Hardy, J. D., and G. F. Soderstrom. Heat loss from the nude body and peripheral blood flow at temperatures of 22° C to 35° C. *J. Nutrition* 16: 493–510, 1938.
62. Hart, J. S. Surface cooling versus metabolic response to cold. *Fed. Proc.* 22: 940–942, 1963.
63. Hartung, G. H., L. G. Myhre, and S. A. Nunneley. Physiological effects of cold air inhalation during exercise. *Aviat. Space Environ. Med.* 51: 591–594, 1980.
64. Hatfield, H. S., and L. G. C. Pugh. Thermal conductivity of human fat and muscle. *Nature* 168: 918–919, 1951.
65. Hayward, J. S., M. L. Collis, and J. Eckerson. Thermographic evaluation of relative heat loss areas of man during cold water immersion. *Aerosp. Med.* 44: 708–711, 1973.
66. Hayward, J. S., J. Eckerson, and M. L. Collis. Effect of behavioral variables on cooling rate of man in cold water. *J. Appl. Physiol.* 38: 1073–1077, 1975.
67. Hayward, J. S., and J. D. Eckerson. Physiological responses and survival time prediction for humans in ice-water. *Aviat. Space Environ. Med.* 55: 206–212, 1984.
68. Hayward, J. S., J. D. Eckerson, and M. L. Collis. Thermal balance and survival time prediction of man in cold water. *Can. J. Physiol. Pharmacol.* 53: 21–32, 1975.
69. Hayward, J. S., J. D. Eckerson, and M. L. Collis. Thermoregulatory heat production in man: prediction equation based on skin and core temperatures. *J. Appl. Physiol.* 42: 377–384, 1977.
70. Hayward, M. G., and W. R. Keatinge. Roles of subcutaneous fat and thermoregulatory reflexes in determining ability to stabilize body temperature in water. *J. Physiol. London* 320: 229–251, 1981.
71. Hemingway, A. Shivering. *Physiol. Rev.* 43: 397–422, 1963.
72. Hensel, H. Thermoreceptors. *Ann. Rev. Physiol.* 36: 233–249, 1974.
73. Hong, S. K., C. K. Lee, J. K. Kim, S. H. Song, and D. W. Rennie. Peripheral blood flow and heat flux of Korean women divers. *Fed. Proc.* 28: 1143–1148, 1969.
74. Horvath, S. M., G. B. Spurr, B. K. Hutt, and L. H. Hamilton. Metabolic cost of shivering. *J. Appl. Physiol.* 8: 595–602, 1956.
75. Huttunen, P., J. Hirvonen, and V. Kinnula.The occurrence of brown adipose tissue in outdoor workers. *Eur. J. Appl. Physiol.* 46: 339–345, 1981.
76. Iampietro, P. F., J. A. Vaughan, R. F. Goldman, M. B. Krieder, F. Masucci, and D. E. Bass. Heat production from shivering. *J. Appl. Physiol.* 15: 632–634, 1960.
77. Jacobs, I., T. Romet, J. Frim, and A. Hynes. Effects of endurance fitness on responses to cold water immersion. *Aviat. Space Environ. Med.* 55: 715–720, 1984.
78. Jaeger, J. J., E. C. Deal, Jr., D. E. Roberts, R. H. Ingram, Jr., and E.R. McFadden, Jr. Cold air inhalation and esophageal temperature in exercising humans. *Med. Sci. Sports Exerc.* 12: 365–369, 1980.
79. Johnson, D. G., J. S. Hayward, T. P. Jacobs, M. L. Pollis, J. D. Eckerson, and R. H. Williams. Plasma norepinephrine responses of man in cold water. *J. Appl. Physiol.* 43: 216–220, 1977.
80. Jung, R. T., P. S. Shetty, W. P. T. James, M. A. Barrand, and B. A. Callingham. Reduced thermogenesis in obesity. *Nature* 279: 322–323, 1979.
81. Keatinge, W. R. The effect of general chilling on the vasodilator response to cold. *J. Physiol. London* 139: 497–507, 1957.
82. Keatinge, W. R. The effect of low temperatures on the responses of arteries to constrictor drugs. *J. Physiol. London* 142: 395–405, 1958.
83. Keatinge, W. R. The effects of subcutaneous fat and of previous exposure to cold on the body temperature, peripheral blood flow and metabolic rate of men in cold water. *J. Physiol. London* 153: 166–178, 1960.
84. Keatinge, W. R. The effect of work and clothing on the maintenance of body temperature. *Quart. J. Exp. Physiol.* 46: 69–82, 1961.
85. Keatinge, W. R. *Survival in Cold Water,* Oxford, UK: Blackwell, 1969.
86. Keatinge, W. R., and M. Clare Harman. *Local Mechanisms Controlling Blood Vessels.* Monographs of the Physiological Society. London: Academic Press, 1980.
87. Kollias, J., L. Bartlett, V. Bergsteinova, J. S. Skinner, E. R. Buskirk, and W. C. Nicholas. Metabolic and thermal responses of women during cooling in water. *J. Appl. Physiol.* 36: 577–580, 1974.

88. Krieder, M. B. Physical and physiological factors in fatal exposures to cold. *Natl. Speleol. Soc. Bull.* 29: 1–11, 1967.
89. Lamke, L.O., S. Lennquist, S. O. Liljedahl, and B. Wedin. The influence of cold stress on catecholamine excretion and oxygen uptake of normal persons. *Scand. J. Cli. Lab. Invest.* 30: 57–62, 1972.
90. LeBlanc, J. Subcutaneous fat and skin temperature. *Can. J. Biochem. Physiol.* 32: 354–358, 1954.
91. LeBlanc, J., B. Blais, B. Barabe, and J. Cote. Effects of temperature and wind on facial temperature, heart rate, and sensation. *J. Appl. Physiol.* 40: 127–131, 1976.
92. LeBlanc, J. A., and G. Nadeau. Urinary excretion of adrenaline noradrenaline in normal and cold-adapted animals. *Can. J. Biochem. Physiol.* 39: 215–217, 1961.
93. Lefevre, J. *Chaleur Animale et Bioenergetique.* Paris: Masson & Cie, 1911.
94. Lewis, T. Observations upon the reactions of the vessels of the human skin to cold. *Heart* 15: 177–208, 1930.
95. Lucas, A. , A. Therminarias, and M. Tanche. Maximum oxygen consumption in dogs during muscular exercise and cold exposure. *Pfluegers Arch.* 388: 83–87, 1980.
96. McArdle, W. D., F. I. Katch, and V. L. Katch. *Exercise Physiology* Philadelphia: Lea & Febiger, 1986.
97. McArdle, W. D., J. R. Magel, T. J. Gergley, R. J. Spina, and M. M. Toner. Thermal adjustment to cold-water exposure in resting men and women. *J. Appl. Physiol.* 56: 1565–1571, 1984, p. 132.
98. McArdle, W. D., J. R. Magel, G. R. Lesmes, and G. S. Pechar. Metabolic and cardiovascular adjustment to work in air and water at 18, 25, and 33° C. *J. Appl. Physiol,* 40: 85–90, 1976.
99. McArdle, W. D., J. R. Magel, R. J. Spina, T. J. Gergley, and M. M. Toner. Thermal adjustment to cold-water exposure in exercising men and women. *J. Appl. Physiol.* 56: 1572–1577, 1984.
100. McFadden, J., and R. H. Ingram. Exercise induced asthma: observation on the ventilatory stimulus. *New Engl. J. Med.* 301: 763–768, 1978.
101. McMurray, R. G., and S. M. Horvath. Thermoregulation in swimmers and runners. *J. Appl. Physiol.* 46: 1086–1092, 1979.
102. Minard, D. Body heat content. In: *Physiological and Behavioral Temperature Regulation,* edited by J. D. Hardy, A. P. Gagge, and J. A. J. Stolwijk. Springfield, IL: C. C. Thomas, 1970, pp. 345–357.
103. Molnar, G. W. An evaluation of wind-chill. In: *Cold Injury,* edited by S.M. Horvath. New York: J. Macy Jr. Foundation, 1960, pp. 175–222.
104. Molnar, G. W. Survival of hypothermia by men immersed in the ocean. *J. Am. Med. Assoc.* 131: 1046–1050, 1946.
105. Montgomery, L. D. A model of heat transfer in immersed man. *Ann. Biomed. Eng.* 2: 19–46, 1974.
106. Moore, T. O., E. M. Bernauer, G. Seto, V. S. Park, S. K. Hong, and E.M. Hayashi. Effect of immersion at different water temperatures on graded exercise performance in man. *Aerosp. Med.* 41: 1404–1408, 1970.
107. Nadel, E. R., I. Holmer, U. Bergh, P.-O. Åstrand, and J. A. J. Stolwijk. Energy exchanges of swimming men. *J. Appl. Physiol.* 36: 465–471, 1974.
108. Nielsen, B., and M. Nielsen. Body temperature during work at different environmental temperatures. *Acta Physiol. Scand.* 56: 120–129, 1962.
109. Nielsen, M. Die Regulation der Korpertemperatur bei Muskelarbeit. *Skand. Arch. Physiol.* 79: 193–230, 1938.
110. Oakley, E. H. N. The first Antarctic winter in tents: the joint services expedition to Brabant Island. *J. Roy. Nav. Med. Serv.* 73: 43–50, 1987.
111. Oddershede, I. R., and R. S. Elizondo. Body fluid and hematologic adjustments during resting cold acclimation in rhesus monkey. *J. Appl. Physiol.* 52: 1024–1029, 1982.
112. O'Hanlon, Jr., J. F. Changing physiological relationships in men under acute cold stress. *Can. J. Physiol. Pharmacol.* 48: 1–10, 1970.
113. Pugh, L. G. C. Deaths from exposure on Four Inns Walking Competition, March 14–15. *Lancet* I: 1210–1212, 1964.
114. Pugh, L. G. C. E. Accidental hypothermia in walkers, climbers and campers. *Brit. Med. J.* 1: 123–129, 1966.
115. Pugh, L. G. C. E. Cold stress and muscular exercise with special reference to accidental hypothermia. *Brit. Med. J.* 2: 333–337, 1967.
116. Pugh, L. G. C., and O. G. Edholm. The physiology of channel swimmers. *Lancet* II: 761–768, 1955.
117. Pugh, L. G. C. E., O. G. Edholm, R. H. Fox, H. S. Wolff, G. R. Hervey, W. H. Hammond, J. M. Tanner, and R. H. Whitehouse. A physiological study of channel swimming. *Clin. Sci.* 19: 257–273, 1960.

118. Radomski, M. W., and C. Boutelier. Hormone response of normal and intermittent cold-preadapted humans to continuous cold. *J. Appl. Physiol.* 53: 610–616, 1982.
119. Rapp, G. M. Convective mass transfer and the coefficient of evaporative heat loss from human skin. In: *Physiological and Behavioral Temperature Regulation,* edited by: J. D. Hardy, A. P. Gagge, and J.A.J. Stolwijk. Springfield, IL: 1970, pp. 55–80.
120. Rasch, P. J., and W. R. Pierson. The relation of body surface area, mass and indices to energy expenditure. *Rev. Can. Biol.* 21: 1–6, 1962.
121. Raven, P. B., I. Niki, T. E. Dahms, and S. M. Horvath. Compensatory cardiovascular responses during an environmental cold stress, 5° C. *J. Appl. Physiol.* 29: 417–421, 1970.
122. Raven, P. B., J. E. Wilkerson, S. M. Horvath, and N. W. Bolduan. Thermal, metabolic, and cardiovascular responses to various degrees of cold stress. *Can. J. Appl. Pharmacol.* 53: 293–298, 1975.
123. Reader, S. R. The effective thermal conductivity of normal and rheumatic tissue in response to cooling. *Clin. Sci.* 11: 1–12, 1952.
124. Reed, L. D., S. D. Livingstone, and R. E. Limmer. Patterns of skin temperature and surface heat flow in man during and after cold water immersion. *Aviat. Space Environ. Med.* 55: 19–23, 1984.
125. Rennie, D. W., B. G. Covino, M. R. Blair, and K. Rodahl. Physical regulation of temperature in Eskimos. *J. Appl. Physiol.* 17: 326–332, 1962.
126. Rennie, D. W., B. G. Covino, B. J. Howell, S. H. Song, B. S. Kang, and S. K. Hong. Physical insulation of Korean diving women. *J. Appl. Physiol.* 17: 961–966, 1962.
127. Rothwell, N. J., and M. J. Stock. A role of brown adipose tissue in diet-induced thermogenesis. *Nature* 281: 31–35, 1979.
128. Rowell, L. B. Human cardiovascular adjustments to exercise and thermal stress. *Physiol. Rev.* 54: 75–159, 1974.
129. Rowell, L. B. Cardiovascular aspects of human thermoregulation. *Circ. Res.* 52: 367–379, 1982.
130. Schmidt-Nielsen, K. Energy metabolism, body size, and problems of scaling. *Fed. Proc.* 29: 1524–1532, 1970.
131. Scholander, P. F., R. Hock, V. Walters, F. Johnson, and L. Irving. Heat regulation in some arctic and tropical mammals and birds. *Biol. Bull.* 99: 237–258, 1950.
132. Siple, P. A., and C. F. Passel. Measurements of dry atmospheric cooling in subfreezing temperatures. *Proc. Am. Philosoph. Soc.* 89: 177–199, 1945.
133. Sloan, R. E. G., and W. R. Keatinge. Cooling rates of young people swimming in cold water. *J. Appl. Physiol.* 35: 371–375, 1973.
134. Smith, G. B., and E. G. Hames. Estimation of tolerance times for cold water immersion. *Aerosp. Med.* 33: 834–840, 1962.
135. Smith, R. M., and J. M. Hanna. Skinfolds and resting heat loss in cold air and water: temperature equivalence. *J. Appl. Physiol.* 39: 93–102, 1975.
136. Steger, W. North to the Pole. *Nat. Geo.* 170: 289–317, 1986.
137. Stromme, S., K. L. Andersen, and R. W. Elsner. Metabolic and thermal responses to muscular exertion in the cold. *J. Appl. Physiol.* 18: 756–763, 1963.
138. Strong, L. H., G. K. Gee, and R. F. Goldman. Metabolic and vaso-motor insulative responses occurring on immersion in cold water. *J. Appl. Physiol.* 58: 964–977, 1985.
139. Swift, R. W. The effects of low environmental temperature upon metabolism. II. The influence of shivering, subcutaneous fat, and skin temperature on heat production. *J. Nutrition* 5: 227–249, 1932.
140. Tikuisis, P., R. R. Gonzalez, and K. B. Pandolf. Human thermo-regulatory model for immersion in cold water. *J. Appl. Physiol.* 64:719–727, 1988.
141. Timbal, J., C. Boutelier, M. Loncle, and L. Bougues. Comparison of shivering in man exposed to cold in water and in air. *Pfluegers Arch.* 365: 243–248, 1976.
142. Timbal, J., M. Loncle, and C. Boutelier. Mathematical model of man's tolerance to cold using morphological factors. *Aviat. Space Environ. Med.* 47: 958–964, 1976.
143. Toner, M. M., L. L. Drolet, and K. B. Pandolf. Perceptual and physiological responses during exercise in cool and cold water. *Percept. Mot. Skill* 62: 211–220, 1986.
144. Toner, M. M., M. N. Sawka, M. E. Foley, and K. B. Pandolf. Effects of body mass and morphology on thermal responses in water. *J. Appl. Physiol.* 60: 521–525, 1986.
145. Toner, M. M., M. N. Sawka, W. L. Holden, and K. B. Pandolf. Comparison of thermal responses between rest and leg exercise in water. *J. Appl. Physiol.* 59: 248–253, 1985.
146. Toner, M. M., M. N. Sawka, and K. B. Pandolf. Thermal responses during arm and leg and combined arm-leg exercise in water. *J. Appl. Physiol.* 56: 1355–1360, 1984.
147. Veicsteinas, A., G. Ferretti, and D. W. Rennie. Superficial shell insulation in resting and exercising men in cold water. *J. Appl. Physiol.* 52: 1557–1564, 1982.
148. Wade, C. E., and J. H. Veghte. Thermographic evaluation of the relative heat loss by area in man after swimming. *Aviat. Space Environ. Med.* 48: 16–18, 1977.

149. Wagner, J. A., and S. M. Horvath. Influence of age and gender on human thermoregulatory responses to cold exposures. *J. Appl. Physiol.* 58: 180–186, 1985.
150. Wenger, C. B., R. B. Bailey, M. F. Roberts, and E. R. Nadel. Inter-action of local and reflex thermal effects in control of forearm blood flow. *J. Appl. Physiol.* 58: 251–257, 1985.
151. Werner, J. and T. Reents. A contribution to the topography of temperature regulation in man. *Eur. J. Appl. Physiol.* 45: 87–94, 1980.
152. Wissler, E. H. A mathematical model of the human thermal system. *Bull. Math. Biop.* 26:147–166,1964.
153. Wissler, E. H. An evaluation of human thermal models: Part A. Technical Report AFOSR-82-0214, U. S. Air Force Office of Scientific Research, Bolling AFB, D. C., 1984.
154. Wissler, E. H. Mathematical simulation of diver performance. Technical Report N00014-76-C8953, Office of Naval Research, Arlington, VA, 1982.
155. Witherspoon, J. M., R. F. Goldman, and J. R. Breckenridge. Heat transfer coefficients of humans in cold water. *J. Physiol.* Paris 63: 459–462, 1970.
156. Wyndham, D. H., J. F. Morrison, J. S. Ward, G. A. G. Bredell, M.J. E. Von Rahden, L. D. Holdsworth, H. G. Wenzel, and A. Munro. Physiological reactions to cold of Bushmen, Bantu, and Caucasian males. *J. Appl. Physiol.* 19: 868–876, 1964.
157. Yoshimura, H., and T. Iida. Studies on the reactivity of skin vessels to extreme cold. Part 1. A point test on resistance against frosbite. *Jap. J. Physiol.* 1: 147–159, 1950–1951.
158. Yoshimura, H., and T. Iida. Studies on the reactivity of skin vessels to extreme cold. Jap. J. Physiol. 2: 177–185, 1952.
159. Young, A. J., S. R. Muza, M. N. Sawka, R. R. Gonzalez, and K. B. Pandolf. Human thermoregulatory responses to cold air are altered by repeated by cold water immersion. *J. Appl. Physiol.* 60: 1542–1548, 1986.

10

Human Adaptation To Cold

ANDREW J. YOUNG, Ph.D.

INTRODUCTION

Human adaptations to chronic cold stress have not been as extensively researched as have adaptations to other environmental extremes. Compared to the number of persons showing adaptations to heat or high altitude, relatively few humans exhibit cold adaptations. Until the development of modern refrigeration and air conditioning technology, there was very little that inhabitants of hot climates could do to limit their exposure to heat stress. Likewise, high-altitude residents cannot avoid exposure to hypoxia. The chronic stress of these environments provides the stimulus for the adaptations that lessen the physiological strain. In contrast, there are many behavioral and/or cultural strategies (e.g. migration, clothing, shelter or fire) available to cold-climate inhabitants, even primitive societies, which reduce the degree of environmental stress to which they are exposed. As observed by LeBlanc, man in the cold is not necessarily a cold man.

Human cold adaptations are described as acclimatization, acclimation or habituation. The terms acclimatization and acclimation refer to related, but distinct physiological processes. Acclimatization is the adaptation resulting from changes in the natural environment, while acclimation is an adaptation induced by an unusual (experimental) alteration in environmental conditions. For the purpose of this chapter, the definition of acclimatization will be further limited to refer to adaptations shown by persons experiencing a lifetime of cold stress. Habituation is a desensitization or damping of the normal response to a stressor. Differences in findings among investigations of cold adaptation may be somewhat resolved by determining which of the processes was being studied. Cold adaptations are further categorized according to the physiological processes involved. Thus, metabolic adaptations describe altered body heat production, hypothermic adaptations describe maintenance (tolerance) of a lower body temperature and insulative adaptations describe improvements in the ability to conserve heat. This chapter will consider all of these adaptive processes.

NATURAL ADAPTATIONS TO COLD

Primitive People

A number of investigations have studied the thermoregulatory responses of primitive people. Primitive people can be described as being geographically and culturally isolated from the mainstream modern society and its technological amenities. The primitive people showing cold adaptations have different ethnic backgrounds and come from different climates, but they share the common disadvantage that their clothing and shelter are inadequate to provide complete protection from repeated cold stress. The arctic Eskimos are the most notable primitive people residing in a cold environment. However, even in regions noted for warm climates there are often nighttime air temperatures well below the body temperature of humans.

Some of the earliest investigations of cold adaptations of primitive people were those of C.S. Hicks and his co-workers who studied the central Australian Aborigines between 1930 and 1937 (24). The central Australian desert is a dry region with temperature variations between 0 and 23° C in the winter and 20 and 40° C in the summer. The clear atmosphere enhances radiant cooling at night. At the time that Hicks studied them, the Aborigine people were nomadic tribes who wore no clothing and lived out of doors. They slept on the ground at night and their only protection from the night cold was a small fire at their feet and a windbreak made of light brush. Despite the difficulties in performing field studies of primitive people, Hicks reported several important observations of the Aborigines' cold adaptations (24). The basal metabolic rate of the Aborigine, measured by open circuit spirometry under temperate conditions, was observed to be about the same as that of Europeans having the same body surface area. Typically, unadapted persons exposed to cold air increase metabolism by shivering. In contrast, the Aborigine's resting metabolic rate remained unchanged with falling ambient temperature. Although core temperatures could not be measured, skin temperatures were observed to be lower in the sleeping Aborigine compared to Europeans sleeping in comparable conditions. Radial arterial pulse pressure measurements at different skin temperatures indicated that the aborigine showed a greater cutaneous vasoconstictor response to cold than Europeans. By reducing skin blood flow in response to cold, body heat could be conserved.

More detailed studies of the thermoregulatory responses of the Australian Aborigine were reported by Scholander *et al.* (60) and Hammel *et al.* (20) whose data are shown in Figure 10-1. These investigators both confirmed that the central Australian Aborigine ex-

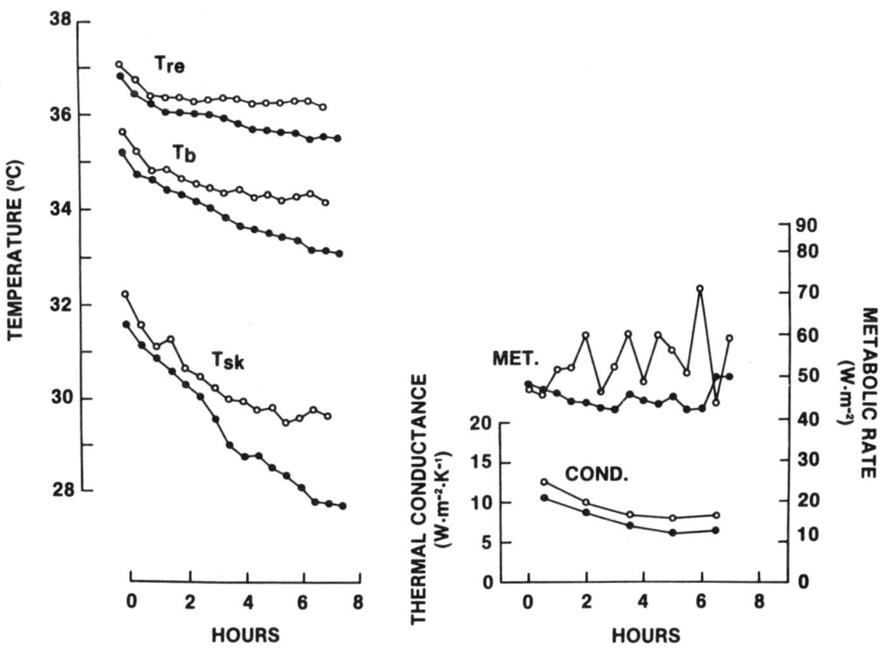

Figure 10-1. *Thermal and metabolic responses of subjects of European descent (open circles) and Australian aborigines (N = 7, closed circles) while sleeping naked at 5° C. Redrawn from Hammel* et al. *(20).*

hibited lower skin temperatures and little or no increase in metabolic heat production while sleeping naked in a cold environment as compared to Europeans exposed to comparable conditions. They also extended the earlier observations by obtaining core, in this case rectal, temperature measurements of their subjects. The Aborigine exposed to cold experiences a greater fall in rectal temperature than the European. Thermal conductance between core and skin was less in the Aborigine than in the European, despite a similar or greater subcutaneous fat thickness in the latter.

The nomadic South African residents of the Kalahari desert are another primitive people whose responses to cold stress have been studied. These people have been referred to as the Bushmen. The climate of the Kalahari desert is similar to that of the central Australian desert, and the Aborigines and Bushmen are very similar in their minimal use of clothing and shelter. Wyndham and Morrison (65) did note, however, that the Bushman slept wrapped in a cloak, thereby achieving some protection from the nighttime cold. On the other hand, the Bushmen are extremely lean and have less subcutaneous fat than the early Aborigines or persons of European descent. Thus, the magnitude of the cold stress and, therefore, the adaptations that Bushmen demonstrate differ in some respects from those of the Aborigine. Whereas the Aborigine showed no visible shivering or increased metabolism during cold exposure (19,20,24,60), the Kalahari Bushmen did increase metabolism and shiver (25,60,64,65).

The increase in metabolic rate observed by Hammel and coworkers (25,60) was not as pronounced as in the control subjects of European descent. Although Ward *et al.* (65) reported that the metabolic response of Bushmen and unadapted control subjects exposed to cold did not differ, the discrepancy with the earlier observations (60) appears related to the choice of the control subjects (bigger, fatter, lower metabolic rate) rather than an absence of adaptation in the Bushmen. Like the Australian Aborigine, the Bushmen experienced a greater fall in rectal temperature during sleep on a cold night compared to control subjects, but in contrast to the Aborigine the Bushmen did not have lower skin temperatures.

Arctic and Subarctic natives such as the Eskimos, the Lapps and the American Indians of northern Canada live in regions having colder climates than any other populated area. The traditional lifestyle of these people is similar to that of the Bushmen and the Aborigine in that considerable time is spent hunting and working outdoors. Unlike the Bushmen and the Aborigine, however, the Circumpolar natives have relied on excellent clothing to protect themselves from the elements. They are also considerably less nomadic and live in heated shelters formerly constructed of animal hides and ice, but today made largely from modern materials. These people have not been subjected to the repeated whole body cold exposure that the primitive people described above have experienced. Therefore, it is not surprising that when exposed to moderate whole-body cold exposure the response of these people differs from that of the Aborigine and the Bushmen.

In general, Circumpolar residents respond to cold stress in a pattern similar to temperate-climate residents. Eskimos (1,23,60) and Native American Indians of the arctic (15,33) have been observed to increase metabolic heat production by shivering when they are exposed resting to moderate whole-body cold-air stress. The Lapps living in the arctic regions of Norway also increase metabolic heat production in response to cold, but their response may be less pronounced than in control subjects (2). Andersen *et al.* (1) reported that when Eskimo subjects were moved from a warm (~25° C) to a cold (~5° C) room, their resting metabolic rate increased by about 27%, as compared to an increase of over 60% in the unadapted control subjects. The smaller percent increase in metabolic rate for the Eskimo indicates that they shiver less than the control subjects. A diminished shivering response to cold stress fits the definition of a habituation type of adaptation. However, the Eskimo (1,55) and possibly the Norwegian Lapp (2) maintain warmer skin temperatures during whole-body cold exposure. Thus, there is the possibility that these people may actually be cold acclimatized.

At least 2 mechanisms might account for differences in shivering response and warmer skin temperatures in the Eskimo. First of all, the resting metabolic rate of Eskimos (1,55,60) and arctic Indians (11,33) in warm temperatures has been observed to be higher than that of control subjects. Therefore, the Eskimo can sustain a greater rate of body heat loss and still maintain constant body temperature without shivering. Despite a smaller increment in the Eskimo's metabolism upon moderate whole-body cold exposure, their steady-state metabolic rates reach about the same value as unadapted control subjects (60). The higher resting metabolic rate of the Eskimo has been attributed to dietary induced thermogenesis resulting from the high protein content of the Eskimo diet, as opposed to cold adaptation (58). In addition to a greater basal metabolic heat production, the Eskimo also appears to transfer heat from core to skin more readily than unadapted persons (55,60).

Rennie *et al.* (55) analyzed the circulatory (convective) and noncirculatory (conductive) components of the Eskimos' overall tissue thermal conductance in the cold. Thermal and metabolic responses during the last 30 minutes of a 2-hour immersion in 33° C water of

5 Eskimos were compared with those of five non-Eskimos. Under these nearly steady-state conditions, finger blood flow (used as an indicator of whole body cutaneous blood flow) was almost unmeasurable in both the Eskimo and non-Eskimo subjects. Furthermore, skeletal muscle blood flow was considered to be minimal because the subjects were not shivering and their oxygen uptake remained unchanged and the same as in 35° C water. Therefore, the thermal conductance measured under these conditions was assumed to be a relatively pure reflection of the physical conduction of heat. Tissue thermal conductance averaged 17 $W \cdot m^{-2} \cdot K^{-1}$ for the Eskimos as compared to 11 for the non-Eskimos. However, the particular Eskimo subjects studied were leaner and had less subcutaneous fat than the non-Eskimos. Figure 10-2 shows the thermal conductance of the subjects plotted against subcutaneous body fat, and indicates that the differences in thermal conductance observed between Eskimos and non-Eskimos can be largely accounted for by differences body composition.

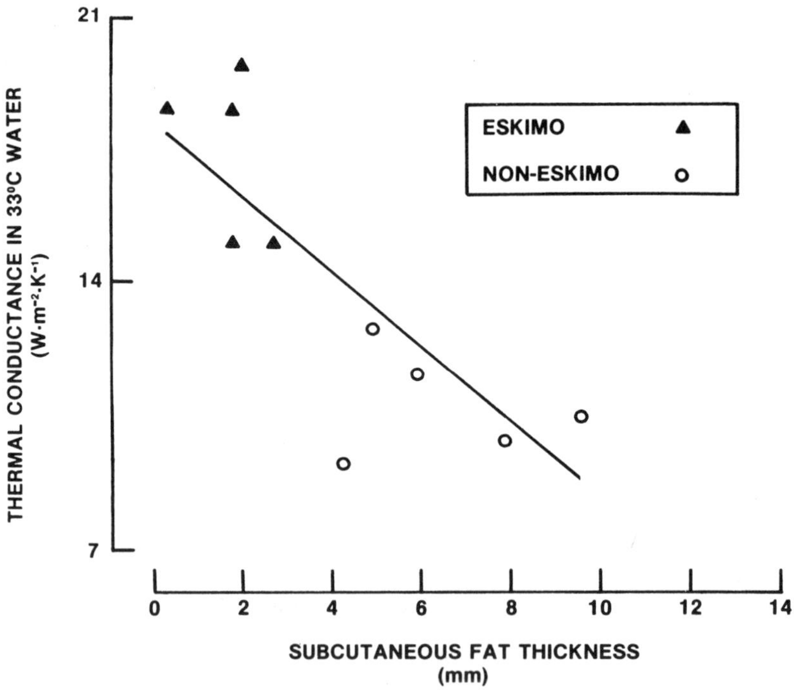

Figure 10-2. *Relationship between subcutaneous fat thickness and thermal conductivity of Eskimo (closed triangles) and non-Eskimo subjects (open circles). Measurements were made during the last 30 minutes of a 1-hour resting immersion in 33° C water. Redrawn from Rennie* et al. *(55).*

Overall, it appears that Eskimo responses to whole-body cold exposure do not reflect any physiological adaptations. This probably reflects the fact that they rarely, if ever, experience this type of stress. Periodic short-term exposure of small body areas such as the hands and face would, however, be fairly commonplace, and a greater thermal conductance and warmer skin would be advantageous for protection from local tissue injury.

Despite the apparent lack of adaptation in the arctic residents' responses to whole-body cold stress, there is evidence that hand blood flow responses to cold are adapted. Brown and Page (6) reported that when Eskimo subjects immersed their hand in cold (5° C) water, hand blood flow was reduced more slowly than in non-Eskimos. This suggests a blunting or habituation of the initial vasoconstrictor response to cold in Eskimos. Also, as shown in Figure 10-3, the steady-state hand blood flow during prolonged cold water immersion was maintained higher in the Eskimos compared to non-Eskimos. Prolonged exposure to extreme cold evokes a vasodilation after the initial vasoconstriction. Cold-induced vasodilation can be sustained or periodic in which case there are cyclic fluctuations in blood flow and finger temperature. These fluctuations were observed to be greatest in the Eskimos (6), and Krog *et al.* (42) showed

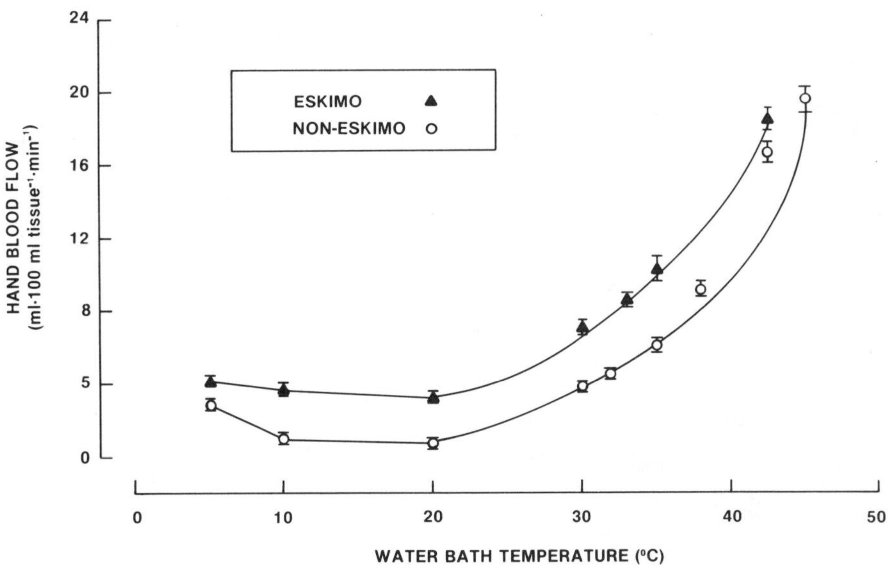

Figure 10-3. *Steady-state blood flow to the hand during immersion in water of various temperatures. The figure shows the mean ± SE of Eskimo (closed triangle) and non-Eskimo (open circle) blood flow measurements made at 5 minute intervals throughout a 2-hour (90 minutes for the 5° C test) hand immersion period. Redrawn from Brown and Page (6).*

that the onset of cold-induced vasodilation was more rapid in arctic Lapps and fishermen from northern Norway. Elsner *et al.* (16) compared the responses to hand immersion in cold (5° C) water of arctic Indians and control subjects and estimated that heat flow to the hands via the circulation was about twice that of control subjects. A greater blood flow to the hands probably accounts for the finding that Eskimos maintained warmer finger temperatures than non-Eskimo subjects during bare-hand exposure to cold (−3 to −7° C) air (49). This localized type of adaptation is not due to racial or genetic differences since, as will be discussed in the next section, it is has also been observed in certain modern day people whose occupations require frequent hand immersion in cold water.

In summary, people from various primitive societies differ from each other and from subjects from modern societies in their thermoregulatory responses to whole-body cold stress. Both the Aborigines and Bushmen show evidence of cold habituation since they maintain a lower metabolic rate at any given skin temperature, and allow a greater fall in rectal temperature during cold exposure. Evidence for cold acclimatization, has been clearly documented only in the Australian Aborigine who showed a greater cutaneous vasoconstrictor response to body cooling than control subjects. The development of an insulative acclimatization in the Aborigine versus a hypothermic habituation in the Bushmen may reflect a lesser degree of chronic cold stress in the latter due to the more effective use of clothing and shelter to protect themselves from the environment (65). For that same reason, Eskimos and other residents of extremely cold climate regions show little adaptation in their responses to whole body cooling, although arctic residents may experience an adaptation in control of blood flow to localized body areas at risk for cold injury.

For the most part, the different groups of people that have been discussed are no longer culturally isolated. They have integrated into modern society and availed themselves of its technological advances. It would be interesting to determine whether any cold adaptations are still demonstrable in these people and their children. This would enable the effects of race and natural selection to be clearly separated from the acclimatization process. This approach has been employed to study cold acclimatization in a group of modern people.

Modern People

Cold adaptations exhibited by modern people, like those of the primitive people, are specific to the type of chronic cold stress that they experience. As previously mentioned, persons whose occupations necessitate frequent immersion of the hands into cold water exhibit alterations in their response to local cold stress. One such group are the Gaspe' fishermen of Quebec, Canada, whose responses to cold have been extensively studied by Leblanc and his co-workers (44,46,47). During the fishing season (April through November), these fishermen spend several hours a day immersing their bare hands in cold (9 to 12° C) sea water. Similar to observations of arctic residents, LeBlanc *et al.* (46) found that the finger temperatures of 14 of these fishermen were higher during a 10-minute hand immersion in 2.5° C water as compared to 14 control subjects. This same adaptation, as well as a more rapid onset of cold-induced vasodilation was documented in fish filleters who work 4 to 8 hours a day with one hand continually immersed in ice water (51). In addition, the Gaspe' fishermen showed a smaller rise in mean arterial blood pressure upon immersing their hand in the cold water (46). This latter observation substantiated the suggestion that, in addition to the local control of blood flow, the overall systemic vascular responses to immersion of the hands in cold water had become adapted.

Further evidence that the Gaspe' fishermen exhibited both a local as well as central adaptation to the specific type of cold stress that they experienced was observed when LeBlanc (44) studied their responses to immersing the foot in cold water. The fishermen's feet are presumably better protected from chronic cold exposure than are their bare hands, and indeed there was little difference in skin temperature on the feet of fishermen and control subjects during a 5-minute immersion of the foot in 2.5° C water. However, the systemic arterial pressure response was again less pronounced in the fishermen as compared to the controls, and in fact, systolic pressures in the fishermen were actually lower after 5-minutes of foot immersion than before cold exposure.

The observation that the Gaspe' fishermen exhibited separate central as well as peripheral adaptations to chronic severe local cold exposure stress, encouraged LeBlanc to study their responses to whole-body exposure to moderate cold stress (47). Seven fishermen and 7 control subjects rested nude for 1 hour at an air temperature of 15° C. Metabolism, determined by open circuit spirometry, increased in all subjects with cold exposure, and there were no differences in metabolic rate between the groups. Likewise, there were no differences between the groups in rectal temperature which remained unchanged during cold exposure. Mean skin temperatures of the fishermen averaged ~1° C higher than those of the control subjects throughout the cold exposure, indicating that cutaneous blood flow was maintained greater in the former. The slightly greater thermal conductance of fishermen without a drop in deep body temperature was surprising since their metabolic rate was the same as that of the control subjects. LeBlanc *et al.* speculated that the adapted fishermen responded to cold with redistribution of body heat stores from the core to the periphery. This, however, seems unlikely since the subjects were all shivering, and the authors contended that the fishermen showed a more pronounced intensity of shivering. Shivering does increase blood flow, thus convective heat flux, to the peripheral skeletal muscle, but concomitantly, shivering increases heat conduction from skin to air (9). On the other hand, shivering was assessed using a subjective rating, and the lack of a difference in metabolic rate between the groups would seem to indicate a similar degree of shivering in fishermen and control subjects. Therefore, it remains uncertain whether or not repeated exposure of small areas of the body to intense cold can produce adaptation of shivering or the regulation of body heat distribution.

Thus, it seems that frequent prolonged immersion of the hands in cold water can produce two types of cold adaptations. The blunting of the overall systemic arterial pressure response to local cold exposure and warmer skin temperatures during whole-body exposure to moderate cold are both indicative of habituation. There is also an adaptation of the specific local body region repeatedly exposed to cold since the fishermen's hands, but not their feet are maintained warmer during immersion in cold water than unadapted subjects' hands. This may represent both a habituation (i.e. less vasoconstriction) as well as acclimatization (i.e. enhanced cold-induced vasodilation). These adaptations are advantageous in terms of comfort, sensitivity and dexterity for persons who must work with exposed hands in the cold. An increased thermal conductance would be a disadvantage in a cold environment were it not for the fact that clothing offsets the loss in insulation over most of the body and enables body heat stores to be conserved. In the next group of naturally cold-adapted people to be considered, a lack of adequate clothing protection is apparently the key factor in the development of an entirely different type of adaptation.

The professional breath-hold divers of Korea are a comparatively modern group of people who, in contrast to similar divers in Japan, work daily throughout the entire year in water temperatures ranging from 10° C in the winter to about 25° C in summer (29). Traditionally, the divers, all of whom are women, wore only a lightweight cotton bathing suit offering little or no insulative protection. Beginning in 1977, divers began wearing wet suits which provide nearly complete protection from cold stress. An extensive series of studies on the thermoregulatory responses to cold exhibited by both the traditional and wet-suit divers have been carried out. The results of these investigations provide fairly clear confirmation of the existence of a human cold acclimatization process, as opposed to habituation.

As would be predicted, the amount of time cotton-suit divers spent in the water varied with the water temperature (37). On summer days, they would complete 3 bouts of diving each lasting about 45 minutes, with the time between bouts spent rewarming by a fire. On winter days, only a single diving session lasting about 15 minutes would be completed. Kang *et al.* (37,38) documented the thermal stress experienced by these divers working throughout the year. In both summer and winter, the divers voluntarily terminated diving sessions when their oral temperature had declined to about 34° C, and in the summer the divers did not re-enter the water until the oral temperature had returned to normal (37). Figure 10-4 shows data from a subsequent study in which rectal temperatures during diving were measured (38). As shown, rectal temperatures fell to about 35° C during 30 minutes of winter and 45 minutes of summer diving. Thus, the divers experienced a somewhat greater drop in rectal temperature than had been observed in the Aborigines sleeping in the cold air (see Figure 10-1). Furthermore, the divers' mean body temperature (calculated using weighting factors of 40% and 60% for skin and rectal temperatures, respectively, and assuming that skin and water temperatures were equal) fell to much lower temperatures than experienced by the Aborigine. It is clear that the traditional cotton-suit Korean diving women experienced a greater degree of chronic cold stress than any other group thus far considered.

One of the first cold adaptations noted (other than the ability of the divers to repeatedly subject themselves to this stress) was an alteration in the divers' basal metabolic rate (37). Basal metabolic rates of 20 divers and 20 non-diving women from the same community were measured in the winter, spring, summer and fall. Basal metabolic rates of the non-divers remained constant throughout the year. Basal metabolic rates of the divers varied with the seasons (water temperatures), being lowest in the summer when they were the same as in non-divers, increasing through fall, and reaching a peak in winter when they were 35% higher than in the non-divers. Urinary nitrogen excretion of both groups was the same and substantially lower than observed in Eskimo populations, apparently ruling out increased dietary protein as a cause of the higher metabolism in the divers. The possibility that increased thyroid gland activity accounted for the increase in the divers' metabolism in winter was discounted by results of later studies showing no difference in the rate of iodine uptake by the thyroid in divers and non-divers, however, the data did suggest that tissue uptake of thyroid hormone

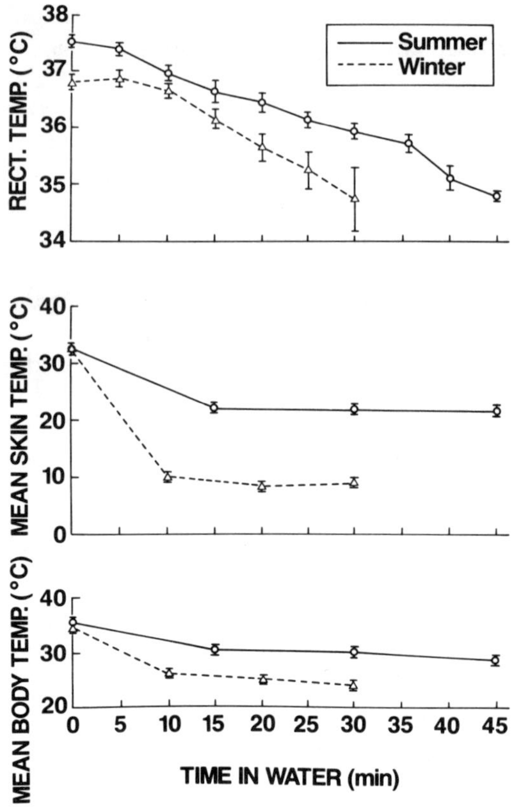

Figure 10-4. *Thermoregulatory responses (mean ± SE) of 5 Korean diving women wearing cotton suits during a diving shift. From Kang* et al. *(38).*

could be greater in divers than non-divers (27). It was also observed that the divers increased urinary norepinephrine excretion during winter while non-divers did not (36). When infused with norepinephrine in the winter, the divers showed a small increase in resting metabolism which was not seen in the non-divers; neither group responded to norepinephrine infusion during the summer (36). This latter finding was taken as evidence that the metabolic processes enabling nonshivering thermogenesis had not developed to any meaningful extent in the divers.

Thus, the Korean diving women did demonstrate a form of metabolic acclimatization to cold in that basal metabolic rates were highest during the time of the year that they were exposed to the lowest water temperatures. The physiological mechanism for this adaptation remains unclear. Furthermore, as noted by the investigators (27,29,37), the practical value of an increased basal metabolic rate for protection against body cooling is negligible, given the tremendous conductive heat transfer capacity of water. In fact, in another investigation of seasonal cold acclimatization in man, it was observed that the metabolic response to acute cold stress (i.e. the increased heat production) was actually diminished in winter as compared to summer (13). The divers, however, exhibit other cold adaptations which may provide an advantage for conservation of body heat.

At this point, it is necessary to review some terminology. Thusfar in this discussion, the term "tissue conductance" has been used to quantify an individual's tendency to lose body heat to the environment. For example, it was observed that the tissue conductance of the Aborigine sleeping in cold air was lower than that of control subjects of equal subcutaneous fat thickness. In the studies of the Korean diving women, the investigators considered that it was more useful to evaluate an individual's ability to resist loss of body heat. This is expressed as "tissue insulation" which is quantified simply as the reciprocal of conductance.

As mentioned earlier in this chapter, shivering has been shown to increase thermal conductance, or rather, decrease insulation (9). Increased muscle blood flow with shivering increases convective heat transfer from body core to shell, and movement of the skin surface disrupts stationary boundary layers above the surface thereby enhancing conductive heat transfer from skin surface to the external environment. Thus, one mechanism of enhancing tissue insulation would be by a decrease in shivering. This possibility was investigated by measuring the threshold for shivering in the Korean women divers, and making comparisons to the shivering thresholds measured in non-divers. Shivering threshold was measured by a standardized test which determined a subject's critical water temperature (T_{cw}), defined as the lowest temperature of water a resting subject could remain immersed in for 3 hours without shivering. The rationale for this protocol was that the prolonged slow cooling period allowed maximal vasoconstriction to develop, and the effects of skin cooling on shivering would be eliminated thereby allowing the "central" setpoint for shivering to be identified (29).

Identification of T_{cw} also allowed the investigators to depict differences by comparing the water temperature at which 50% of a particular subject group shivered. The shivering responses of 10 diving women were compared with those of 21 non-diving Korean women from the same community; responses of 12 non-diving males were also studied to determine if there were any gender-specific differences in shivering (26). As shown in Figure 10-5, the diving women had a higher shivering threshold (i.e. a lower T_{cw}) than both male and female non-divers. The water temperature at which 50% of the divers shivered was 28.2° C, as compared to 29.9° C for the female and 31.1° C for the male non-divers, respectively. When T_{cw} was plotted as a function of subcutaneous fat thickness, an inverse relationship was found in the data from the non-diving population, and the difference between the non-diving males and females could be accounted for by differences in subcutaneous fat. The diving women, however, had lower shivering thresholds and T_{cw} even when compared to non-divers of similar fat thickness (26,27). Hong (27) also reported that rectal temperatures at the shivering threshold (T_{cw}) of the divers was lower in comparison to non-divers. The lowering of the rectal temperature at the T_{cw} has been cited as strong evidence that the central set-point for shivering in the divers has been reduced as a result of cold acclimatization (29). However, a decrease in shivering sensitivity would seem to better fit the definition of a cold habituation.

Additional evidence of cold adaptation in the cotton-suit divers was observed in the measurements of maximal tissue insulation (27). Figure 10-6 shows the regression lines for the relationship between maximum tissue insulation and subcutaneous fat thickness of the

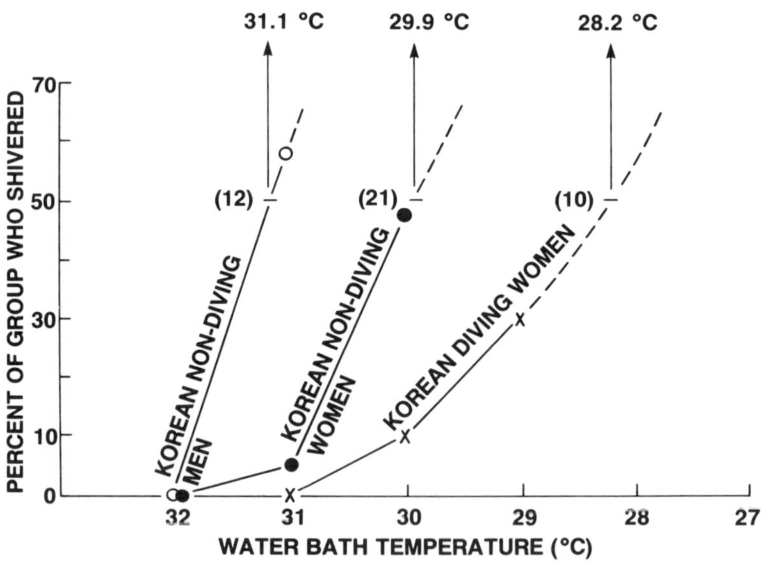

Figure 10-5. *Incidence of shivering in 3 groups of Korean subjects during water immersion at different temperatures. The water temperatures indicated at top are the temperatures at which 50% of the subjects in a group shivered. Redrawn from Hong (26).*

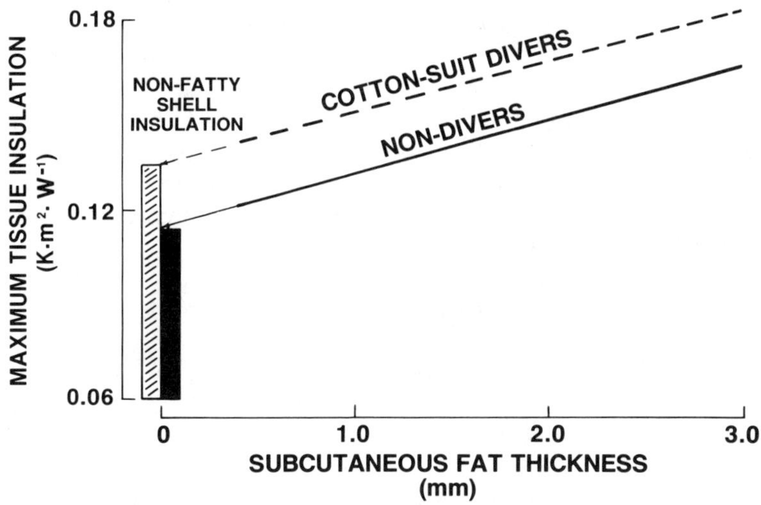

Figure 10-6. *Regression lines for the relationship between maximum tissue insulation and subcutaneous fat thickness in 89 non-diving Koreans (both men and women) and 30 traditional (cotton-suit) Korean diving women. Histogram bars depict the insulative value of the non-fatty shell of the divers (hatched bar) and non-divers (solid bar). Redrawn from Hong (27).*

Korean diving women compared to non-diving Koreans. As shown, the divers have a greater maximum insulation than non-divers of comparable fat thickness. The slopes of the 2 regression lines are not different, simply reflecting the fact that the insulating value of fat is the same for all people. The intercepts of the regression lines, however, were significantly different. By extrapolating the regression lines back to the zero fat point on the axis, the insulation provided by the non-fat components (i.e. skin and muscle) of the body shell can be visualized. Assuming that skin thickness is the same for both groups and that its contribution to total insulation is negligible, it is apparent that the divers derive more insulation from their muscular shell than do the non-divers. The greater non-fatty insulative shell of the divers may simply be due to the elevated shivering threshold. On the other hand, it may reflect a vascular acclimatization involving an enhanced vasoconstriction and/or development of a more improved countercurrent heat exchange mechanism in the limb musculature.

In subsequent experiments, Hong *et al.* (28) measured regional heat flux and limb blood flow of divers and non-divers immersed in cold water.

Average heat flux from 6 separate body regions was the same in the divers as non-divers, indicating that the divers had not developed the ability to preferentially vasoconstrict a particular region to restrict heat loss. The divers did, however, maintain a higher limb blood flow during cold water immersion than did the non-divers, despite the similar heat flux. This finding tends to support the suggestion that the diver has developed a more efficient countercurrent heat exchange mechanism in their limbs. An improved countercurrent heat exchange mechanism enables the maintenance of a warmer and more highly perfused muscle shell without increased conductive heat loss from the shell to the environment. Maintenance of improved muscle blood flow and temperatures would optimize metabolic and contractile functions. Unfortunately, measurements of the muscle temperatures maintained by the divers under conditions eliciting maximum tissue insulation were never made, and as will be discussed next, it is now too late to perform such studies.

As mentioned previously, the Korean breath-hold divers were allowed by their union to work while wearing wet-suits beginning in 1977. Studies of other professional divers working in cold water indicated that the use of wet suits probably had prevented the development of any cold adaptations similar to those experienced by the Korean cotton-suit divers (63). Therefore in 1980, studies of the thermoregulatory responses of the Korean breath-hold divers were reinitiated to determine whether or not the adoption of the use of the wet suit had any measurable effect on the cold adaptations previously observed in these divers.

One of the most obvious effects of wet-suit use was that the duration of daily diving work had been greatly prolonged. Whereas traditional cotton-suit divers had worked one 16-minute session in winter and three 45-minute sessions in summer, the contemporary wet-suit divers performed only a single session a day which lasted about 2 hours in winter and about 3 hours in summer (29). Furthermore, in contrast to the cotton-suit divers whose work sessions appeared to be limited by the time it took for their core (rectal) temperature to fall to about 35° C, the divers wearing wet suits experienced only a 0.2° C drop in rectal temperature during a summer work session, and a 0.6° C drop during a winter session (39). Kang *et al.* (39) evaluated the thermal insulation provided by the wet suits, and the effects of its use on body heat loss during a diving session. Despite a considerable increase in the amount of time spent working in the water, the heat loss during a diving shift while wearing a wet suit was 40 (summer) to 60% (winter) of the heat loss during a shift while wearing a cotton suit. The decrease in heat loss during diving was due to a greater total insulation in the wet-suit diver compared to the traditional cotton-suit divers (39). Total insulation of the wet-suit divers was estimated to be 2 to 3 times that of the cotton-suit divers. These measurements were, however, made on subjects resting on the surface between dives and the insulative value of wet suits is known to be decreased by both exercise and compression with increasing depth (29).

The wet suit does provide some increment in total insulation for the modern diver, particularly during periods of rest or low activity while in the water. However, the greatest contribution to the increased insulation of the modern diver derives from their substantially greater subcutaneous fat thickness (53) as compared to the traditional diving women who had been rather lean (56). Irrespective of the reason for the increased insulation, the modern diver clearly does not experience the same level of chronic cold stress as did the traditional cotton-suit diving women.

The effects of the reduction in the level of chronic cold stress on the thermoregulatory functions of the breath-hold divers were investigated in a longitudinal study carried out from 1980 to 1982 and reported by Park *et al.* (53). In contrast to the earlier divers, the basal metabolic rate of the contemporary divers did not vary with season (water temperature) and was not significantly different from non-diving controls at any time during the year. As mentioned above, the contemporary diving women were considerably fatter (~26% body fat) and had greater subcutaneous fat thickness (~9 mm) than the earlier divers studied (~11% and ~2 mm, respectively). The relationship between T_{cw} and subcutaneous fat thickness in the modern divers was observed to change during the period of the longitudinal study (53). In 1980, the divers still had an elevated shivering threshold (as indicated by the T_{cw}) compared to non-divers of comparable fat thickness, although the difference between divers and non-divers (~2° C) was less than had been observed in the earlier studies (~4° C); measurements made in 1981 show that the difference between the two groups was even smaller, and by 1982 there was less than a 0.2° C difference between the T_{cw} of divers and nondivers. There was also an increase in the divers' central set-point for shivering (i.e. the core temperature at the shivering threshold), so that by 1982 the relationship between steady-state rectal temperature and skin temperature after 180 minutes of immersion in water at T_{cw} was the same in divers and non-divers. Finally, the increased tissue insulation due to the non-fatty shell observed in the cotton-suit divers (see Figure 10-6), was completely absent in the divers studied in 1980. The observation that this adaptation was lost more rapidly than the elevation of the shivering threshold, suggests that some additional mechanism apart from the reduced shivering (e.g. improved countercurrent heat exchange in the limbs) contributed to the increased tissue insulation of the cotton-suit divers.

In summary, the professional breath-hold divers of Korea have exhibited some of the most pronounced natural adaptations in thermoregulatory responses to cold. Traditionally, these divers wore only cotton bathing suits offering no protection from the cold water in which they worked year-round. Several different thermoregulatory adaptations had developed in apparent response to the chronic cold stress to which the divers were subjected. A seasonal metabolic acclimatization was indicated by a increase in basal metabolic rate during winter when water temperatures are coldest. An elevated shivering threshold was evidence that the divers had experienced a cold habituation. Most important were the insulative adaptations the divers exhibited. Although the increased insulation provided by the divers non-fatty shell was partially due to the reduced shivering, there was also evidence that cold acclimatization had resulted in an improved countercurrent heat exchange mechanism in the peripheral vasculature of their limbs.

While Hammel *et al.* (20) suggested that racial or genetic factors did influence thermoregulatory responses to cold of coastal Australian Aborigines having little previous cold exposure, this was a subjective interpretation of their data without strong quantitative basis. There is experimental evidence suggesting that racially inherited factors have little or no role in determining differences in responses to whole-body cold exposure (21,32) although there may be racial differences in the responses to local (i.e. finger) cold exposure (32). Genetic factors related to a natural selection process may indeed play a role in determining differences in responses to cold between groups of people. For example, it has been suggested that the absence of males among the professional breath-hold divers of Korea may have come about due to the generally lower subcutaneous fat thickness of males resulting in an inherently poorer tolerance to cold stress as compared to females (29). In any event, the relatively rapid disappearance of differences between divers and non-divers, concomitant with the adoption of the use of wet suits, strongly indicates that adaptations in response to chronic cold stress had occurred and were independent of genetic factors. The next section of this chapter will consider to what extent cold acclimation can be induced in humans lacking such a lifetime experience of cold stress.

COLD ACCLIMATION

Metabolic Adaptation

As discussed in the preceding chapter, nonadapted humans respond to acute cold stress with marked shivering and elevated metabolic heat production, in addition to cutaneous vasoconstriction. Depending on the intensity of the cold stress and the degree to which increased metabolic heat production offsets body heat loss, core temperature will increase, remain constant or decline. Consensus is lacking regarding the effects of cold acclimation procedures on thermoregulatory responses of humans exposed to acute cold stress. Several different types of cold acclimation have been reported to occur, and a number of different procedures and cold stress programs have been employed to induce cold acclimation.

It is well documented that the increase in metabolic heat production elicited by acute cold exposure is more pronounced in animals that have been chronically exposed to cold than control animals lacking prior cold exposure (22,45). There is, however, only scant evidence that a metabolic type of cold acclimation can be induced in humans. Scholander *et al.* (59) studied a group of 8 Norwegian students who had spent a 6-week period camping in the mountains during the autumn months when ambient temperatures were moderately cold, particularly at night. The men were provided with only light-weight summer style clothing and a single blanket for use at night. Food intake and activity were completely *ad libitum*. At the end of the 6-week period, the metabolic responses, skin and rectal temperatures of these subjects were measured while they slept exposed to ambient temperatures of 20° C (nude) and 3° C (in a sleeping bag). Compared to the responses of eight control subjects, the campers maintained warmer skin temperatures during sleep, especially under the cold conditions. The possibility that warmer skin temperatures were due to thinner subcutaneous fat thickness cannot be ruled out, since body composition of the subjects was not assessed. The principal evidence indicating the occurrence of a metabolic acclimation was that the campers had higher basal metabolic rates and exhibited a greater increment in metabolic heat production upon cold exposure than the controls.

There are 2 possible mechanisms for the increased body heat production associated with metabolic cold acclimation. The shivering response to cold could become more pronounced, but Scholander *et al.* (59) could discern no change in the shivering activity of their cold acclimated subjects, and, in fact suggested that shivering was not apparent during sleep. This suggests that an increase in nonshivering metabolic thermogenesis might be the underlying mechanism for metabolic cold acclimation. Humans, however, are generally considered to lack the capacity for nonshivering thermogenesis (22). The basis of this supposition is that adult humans lack significant amounts of brown adipose tissue which is thought to be the major site of nonshivering thermogenesis. Yet, it has been reported that outdoor workers in northern Finland do retain depots of brown adipose tissue (31).

The presence or absence of brown adipose tissue notwithstanding, there is some experimental evidence for a cold-induced nonshivering thermogenesis in man. Davis (12) reported that a 31-day cold acclimation program (eight hours per day in a chamber at ~12° C) resulted in a large decrease in cold-induced shivering activity without a concomitant decrease in metabolic heat production. This observation was interpreted as indicating substitution of nonshivering thermogenesis for shivering, but shivering (indicated by muscle electrical activity) was assessed in only the thigh and upper arm. Continued shivering of torso muscle groups may have accounted for the sustained metabolic heat production. On the other hand, Joy (35) reported that following a similar cold acclimation program, subjects showed an increased metabolic response to norepinephrine infusion suggesting that increased sympathetic nervous activity mediated a cold-induced nonshivering thermogenesis. There are other reported studies (14,34) purportedly demonstrating the presence of cold-induced nonshivering thermogenesis in humans, but these have employed visual assessment of shivering activity which cannot be considered to be reliable. Nevertheless, the possibility that a nonshivering thermogenesis develops in humans as a form of metabolic cold acclimation cannot be dismissed.

Hypothermic Adaptation

It is unfortunate that the experimental design employed by Scholander *et al.* (59) did not include measurements made on the campers before and again after the 6-week acclimation period. Numerous other investigations of similar (43,48) or much more rigorously controlled (7,12,40) cold exposure regimines that employed a repeated measures experimental design (where each subject serves as his own control) have failed to demonstrate the development of a metabolic type of cold acclimation as described by Scholander *et al.* (59). In fact, as will be discussed next, one of the most commonly reported effects of cold acclimation programs is a reduction in the shivering and metabolic heat production elicited by cold stress.

A completely different pattern of cold acclimation which is often observed is referred to as hypothermic. Although the existence of a hypothermic pattern of cold acclimation was first postulated by Carlson *et al.* (10), very little substantiating data were presented. A more detailed description of this pattern of adaptation was reported by LeBlanc (43). Metabolic rate and body temperatures during a 1-hour standardized cold air (9° C) exposure were measured in 10 soldiers at the beginning and again after 2 and 4 months of exposure to Arctic winter conditions. The men's duties required them to spend about 12 hours a day outdoors. By the end of winter, there was a small, but significant decrease in metabolic heat production and the men experienced a significantly greater (0.4° C) fall in rectal temperature during the cold exposure. These findings were subsequently confirmed by the results of a more controlled laboratory acclimation study reported by Keatinge (40), in which 5 subjects spent 7 hours a day in a cold chamber at 6° C while wearing only shorts and shoes. The exposures were repeated on 19 out of 21 days, and metabolism and body temperatures were measured on the first and last day. A decrease in metabolic heat production and a greater fall in rectal temperature were observed at the end of the acclimation program. Similarly, Kreider *et al.* (41) observed that subjects dressed only in shorts and living continuously in a cold chamber at 16° C experienced a greater fall in rectal temperature while they slept on the 14th night as compared to the first night. More recently, Bruck *et al.* (7) reported that as few as four 60-minute cold (5° C) air exposures resulted in a decrease in the esophageal and mean body temperature at which shivering and metabolic heat production was increased.

In the aforementioned studies, the skin temperatures maintained during cold exposure were unaffected (40,41,43) or possibly even increased (10) by the acclimation program. Carlson *et al.* (10) hypothesized that cold acclimation had produced a circulatory readjustment which resulted in a redistribution of body heat from the core to the peripheral shell. This redistribution of body heat can be visualized by considering that in the calculation of mean body temperature there would be a decrease in the core temperature weighting factor and an increase in the skin temperature weighting factor. Thus, the extremities could be maintained warmer without an increase in metabolic heat production. This interpretation was challenged by LeBlanc (43), who suggested that, as a result of repeated cold stress, the central set point for shivering had been reduced. Maintenance of warm (relatively) skin temperature with decreased metabolic heat production resulted in loss of body heat stores and a fall in core temperature. In this case, the hypothermic pattern of adaptation would probably be better described as a habituation rather than an acclimation. Radomski and Boutelier (54) observed that an acclimation program which produced the characteristic hypothermic pattern of adaptation also produced a reduction in the sympathetic response to cold exposure. This observation indicates that the hypothermic pattern of adaptation is indeed due to a habituation to cold stress.

Insulative Adaptation

In contrast to the studies in which hypothermic habituation was observed, Davis (12) observed a reduction in mean skin as well as rectal temperatures during cold exposure in 6 subjects following a 31-day cold acclimation program. These changes suggest the development of a cold acclimation rather than habituation. The lower skin temperatures during cold exposure could reflect a lower thermal conductivity resulting from the reduction in shivering activity (9). As discussed previously, Davis (12) attributed the reduction in shivering to a metabolic cold acclimation involving the development of nonshivering thermogenesis. Alternatively, the lower skin temperatures could reflect an insulative acclimation due to more pronounced vasoconstriction or development of an improved countercurrent heat exchange mechanism as described by Hong *et al.* (28) in the Korean diving women.

Throughout this chapter, it has been stressed that the stimulus for cold adaptation is not cold exposure *per se* but rather the magnitude of physiological strain produced by the cold exposure. On this basis, it was recently suggested that the type of cold acclimation (i.e.metabolic, hypothermic, insulative) achieved by a acclimation program is determined by the frequency and duration of core temperature reduction experienced by the subjects (66). Subjects repeatedly exposed to cold air generally demonstrate metabolic (59) or hypothermic (7,40,41,43) patterns of adaptation. Rapid and more pronounced core temperature reductions without risk of freezing injury to the tissues are possible during cold water immersion than in cold air, because of water's greater capacity for heat conduction. This, combined with observations regarding thermoregulation in Korean diving women, provided the rationale for an investigation in which the effects of a program of repeated cold water immersion were studied.

The results of an investigation in which an insulative type of cold acclimation was produced in a relatively short-term program were recently reported by Young *et al.* (66,67) and Muza *et al.* (50). Seven male subjects completed an acclimation program consisting of a daily 90-minute immersion in cold (18° C, stirred) water, repeated 5 consecutive days each week for 5 successive weeks. The subjects, who were all Caucasians and permanent residents of the United States, performed a standardized cold air (5° C, 30% relative humidity) exposure 2 days before and 2 days after completion of the acclimation program. During all acclimation sessions and standardized cold air tests the subjects reclined in a semisupine position while wearing only swim trunks. The subjects sustained a reduction in rectal temperature of about 1° C during each of the 90-minute immersion periods. There were no changes in thermoregulatory (66), vascular fluid (67), or cardiorespiratory (50) responses to cold water immersion during the acclimation period. Therefore, while cold water immersion provided what was thought to be a sufficient stimulus to induce cold acclimation, any adaptations in the responses to cold which had occurred were apparently inadequate to offset the severity of the stress imposed by the large conductive heat transfer capacity of the aqueous environment. The possibility that evidence for cold acclimation would be masked during water immersion by the magnitude of the cold stress had been anticipated by the investigators, and was the reason for inclusion of the cold air exposures in the experimental design.

In contrast to the responses to cold water immersion, the thermoregulatory responses to cold air exposure did indicate that significant adaptations had resulted from the acclimation program (66). Some of the adaptations observed were characteristic of the hypothermic pattern of cold habituation. As shown in Figure 10-7A for example, rectal temperatures were significantly lower following acclimation both before and at all times during the cold air exposure. A reduction in basal core temperature had also been observed in the Korean women divers during winter (see Figure 10-4), and might reflect a downward resetting of the normally regulated core temperature as a result of cold acclimation. The validity of this conclusion remains to be demonstrated, since experimental artifact may have contributed to the reduction in basal temperature observed by Young *et al.* (66). Nevertheless, even after correcting for differences in basal temperatures between the tests (Figure 10-7B), it was clear that there was a greater and more rapid fall in rectal temperature during the 90-minute cold air exposure following acclimation as compared to before acclimation. The effect of the acclimation program on the metabolic responses to the cold air exposure were also characteristic of a hypothermic habituation. Metabolic heat production during the initial ten minutes of cold air exposure was lower after acclimation, which was interpreted as a delay in the onset of shivering. Furthermore, despite lower mean body temperatures following acclimation, metabolic rates achieved similar levels before acclimation by the 30th minute of cold exposure.

Other thermoregulatory adaptations observed by Young *et al.* (66) were more characteristic of the development of an insulative type of cold acclimation. The effect of the acclimation program on the change in skin temperature during cold air exposure is shown in Figure 10-8. The mean weighted skin temperature maintained during cold air exposure was about 4° C lower after acclimation. It

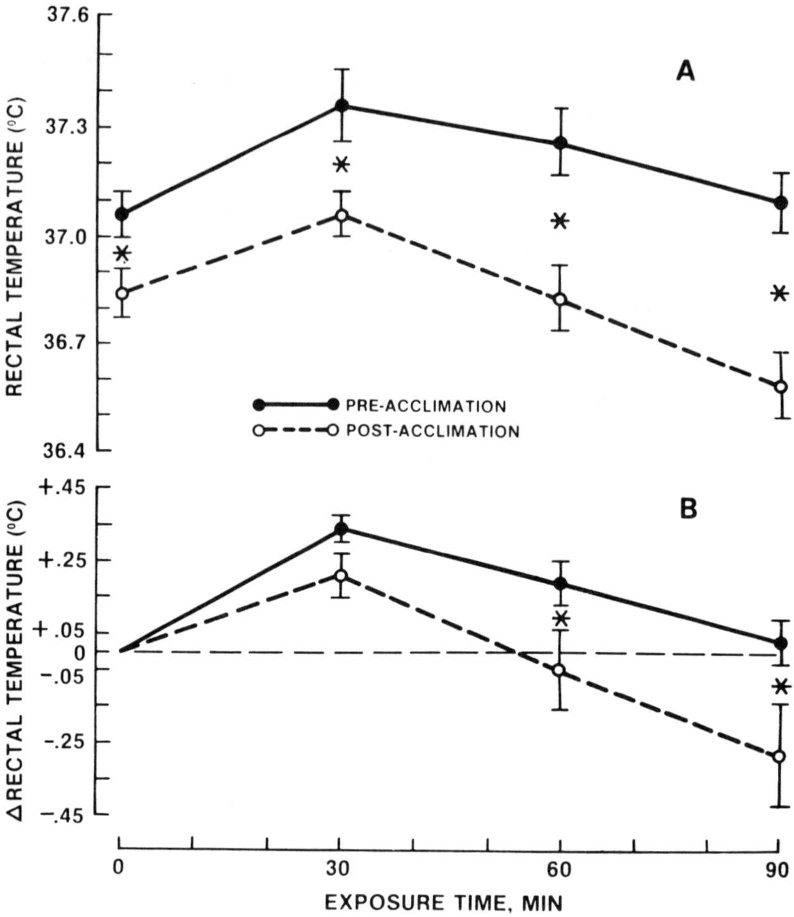

Figure 10-7. *Rectal temperature before (0 min) and during a 90-minute resting exposure to cold air (A), and the change, relative to initial values, in rectal temperature during cold exposure. Values are means ± SE, N = 7. *Significant (P < 0.01) difference pre- versus postacclimation. From Young* et al. *(66).*

was suggested that the lower skin temperatures following acclimation were due to a greater cutaneous vasoconstriction and reduced skin blood flow, although skin blood flow was not actually measured. On the other hand, the increment in plasma norepinephrine concentration during cold air exposure was two-fold greater following the acclimation program, indicating an enhanced sympathetic nervous response to cold stress. An increased sympathetic responsiveness to cold stress is one mechanism whereby a more pronounced vasoconstriction could be elicited, thus accounting for the lower skin temperatures.

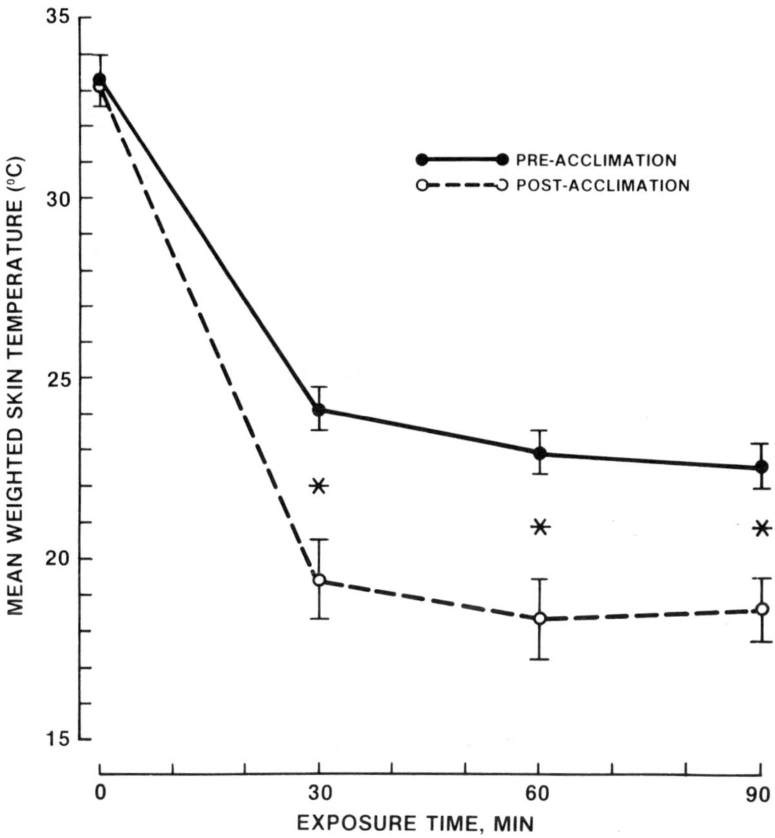

Figure 10-8. *Mean weighted skin temperatures before and during a 90-minute resting cold air exposure. Values are means ± SE, N = 7. * Significant (P < 0.01) difference pre- versus postacclimation. From Young* et al. *(66).*

There was no effect of repeated water immersion on skin temperature response to cold exposure observed in the study reported by Radomski and Boutelier (54). As previously mentioned, however, they observed a decreased sympathetic responsiveness to cold following the adaptation program. There are at least 2 explanations for the difference in the findings of Young *et al.* (66) and Radomski and Boutelier (54). The acclimation program employed by Young *et al.* (66) was considerably more severe than that used by Radomski and Boutelier (54), thus resulting in the development of an insulative acclimation instead of a hypothermic habituation. In addition, Radomski and Boutelier (54) reported that a 2-week period separated the completion of their adaptation program and the experimental cold exposure, during which time the subjects may have deacclimated. Finally, the observation that repeated cold water immersion results in a lower skin temperature during cold air exposure was confirmed in a study recently reported by Bittel (5). Therefore, a reduction in the skin temperature maintained during cold air exposure can be produced by acclimation programs which provide a sufficiently intense stimulus, and the mechanism for this adaptation appears to be an enhanced sympathetic responsiveness to cold.

The body's insulative shell can be subdivided into two regions, the superficial shell (skin + subcutaneous fat) and the subcutaneous muscle layer, both of which contribute to total shell insulation. As discussed elsewhere in this chapter, the insulation provided by the subcutaneous fat is proportional to its thickness (see Figures 10-2 and 10-6). Since the study reported by Young *et al.* (66) was completed over a relatively short time period and total body mass remained constant, it can be assumed that the subcutaneous fat thickness and, therefore, the insulation it provided remained relatively constant. However, the insulation provided by the skin and muscle is primarily determined by the blood flow to these regions, thus there is a mechanism by which shell insulation can be regulated. With acute cold exposure, cutaneous vasoconstriction augmented by reduced perfusion of skeletal muscle provide an immediate increase in shell insulation. The reduction in the skin temperature maintained during cold exposure following acclimation has 2 important implications. First, insulation is improved due to the reduction in the thermal gradient for heat transfer between the skin and the environment. Second, the reduction in skin temperature (Figure 10-8) during cold air exposure resulting from acclimation was considerably larger than the concomitant reduction in rectal temperature response (Figure 10-7). Therefore, as shown in Figure 10-9, the rectal-to-skin temperature gradient is enlarged following acclimation.

An enlarged thermal gradient between core and skin would facilitate the transfer of body heat between core and subcutaneous muscle shell. The enhanced insulation of the superficial shell due to greater cutaneous vasoconstriction would limit heat loss from the muscle shell. Thus, insulative cold acclimation may enable the maintenance of a warmer and better perfused muscle shell, and the reduction in rectal temperature during cold exposure may reflect a redistribution of body heat stores from the core to muscle shell. This is essentially a modification of the hypothesis first proposed by Carlson *et al.* (10). However, Bittel (5) recently reported that core temperatures, while lower initially, declined less during a cold air exposure following a 2-month repeated cold water immersion program as compared to before. In agreement with the results reported by

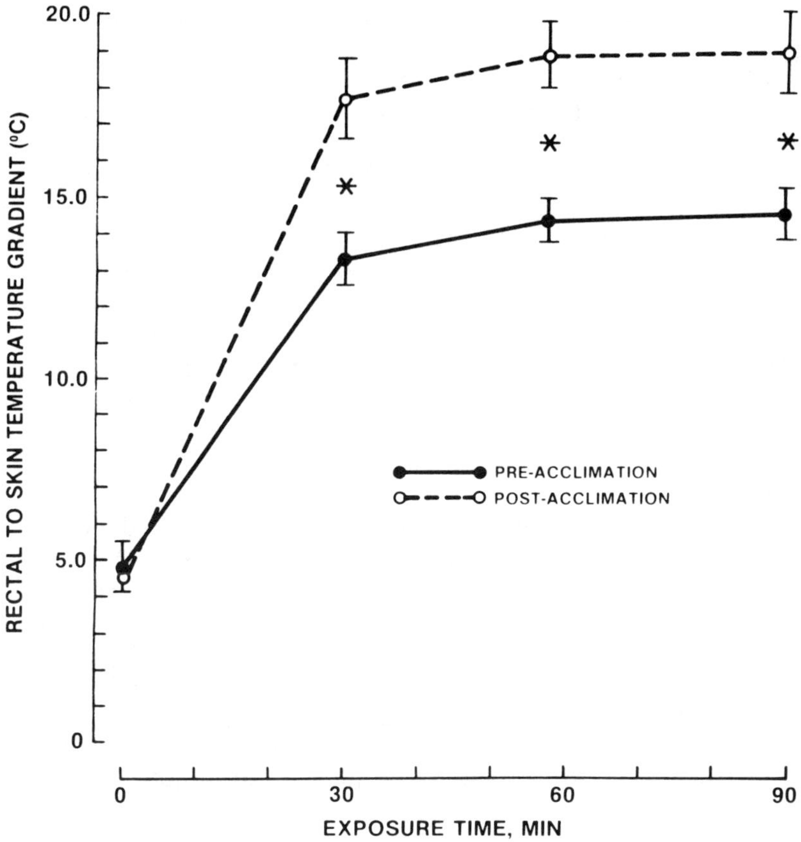

Figure 10-9. *Difference between rectal and mean weighted skin temperature before and during a 90-minute exposure to cold air. Values are means ± SE, N = 7. *Significant (P < 0.01) difference pre- versus postacclimation. From Young* et al. *(66).*

Young *et al.* (66), cold acclimation resulted in lower skin and mean body temperatures during cold air exposure, but metabolic heat production was unchanged (5). Possibly, the longer acclimation period resulted in a more complete development of insulative adaptations. The pattern of cold acclimation observed in these 2 studies (5,66) employing repeated water immersion programs does confer a physiological advantage. Bittel (5) calculated that heat debt at the end of a 120-minute cold exposure was over 30% lower following acclimation.

There is not necessarily any contradiction indicated by the different patterns of cold acclimation reportedly observed. Skreslet and Aarefjord (62) studied the thermoregulatory responses of 3 men exposed to a standardized cold water immersion test at the beginning and at 2-week intervals throughout a 45-day period, during which the men were scuba diving daily in cold (2-4° C) sea water. The results of the study were inconclusive in that only 3 subjects were studied and no statistical analyses were attempted. Nevertheless, the data appeared to be consistent with the hypothesis. The response to the initial cold stress test was clearly an increase in metabolic heat production which, in 2 out of 3 subjects maintained rectal temperature constant. After 2 weeks of cold water diving, the metabolic response to cold exposure was depressed and rectal temperatures declined, indicative of the acquisition of the hypothermic habituation to cold described above. During the final standardized tests, there appeared to be a trend for skin temperatures (measured underneath a wet suit) on the torso and thigh surfaces to be somewhat cooler, and rectal temperatures to be maintained closer to basal levels as compared to during the initial test. These observations were interpreted as indicating the establishment of an insulative cold acclimation. Thus, Skreslet and Aarefjord (62) hypothesized that the different types of cold acclimation did not represent development of mutually exclusive physiological states, but rather, different successive stages in the development of complete cold acclimation.

An alternative explanation for observations of different types of cold acclimation has recently been proposed by Bittel (5). As already discussed, acclimation appeared to result in an enhanced vasconstriction with only a small change in the metabolic response during cold exposure. This interpretation was based on average responses and statistical analyses of pooled data from all 9 subjects studied. However, when each individual's responses were considered separately, only 4 of the 9 subjects exhibited an acclimation pattern reflected by the group averages. One subject showed no evidence of an insulative adaptation, but a greatly enhanced metabolic response to cold following acclimation. Three of the 9 developed both an enhanced insulation and a greatly increased metabolic response to cold. One subject showed a decreased metabolic response to cold but enhanced insulation following acclimation. Because of the small number of subjects studied, factors accounting for differences in individual patterns of cold acclimation could not be identified, but body composition and physical fitness were suggested as important determinants.

Effects of Cold Acclimation on Body Fluid Regulation

Thusfar, the focus of this chapter has been exclusively on adaptations in thermal responses produced by chronic or repeated cold exposure. However, as discussed in the other chapters of this book, heat acclimation, high altitude acclimatization, and exercise training have all been shown to influence vascular fluid regulation and cardiorespiratory responses to environmental extremes. Although the effects of acute cold stress on these responses have been studied, the influence of cold acclimation on body fluid and cardiorespiratory control has received little attention.

It has been suggested that seasonal changes in blood volume are related to chronic cold exposure (4). On the other hand, seasonal changes in blood volume may actually be due to changes in physical activity (18), length of daylight (61), or induction or decay of heat acclimation (3) rather than cold adaptation. The results of animal studies indicate that chronic cold exposure may influence body fluid regulation. Monkeys exposed continuously to cold air for 29 days experience an expansion of blood and plasma volume and a decreased intracellular volume (52). The mechanism for this expansion of cardiovascular volume was not clear, but the authors suggested that an increased sympathoadrenal activity and the increased basal metabolic rate both contributed to the adaptation. In humans exposed to continuous cold for 1 to 7 days, a reduction in plasma and blood volume has been observed (17), however, a negative water balance was maintained during this period. The reduction in plasma and blood volume was, therefore, likely to be an effect of hypohydration. Furthermore, 1 to 7 days of cold exposure may not be a sufficient duration for the development of significant cold acclimation.

Continuous cold exposures of longer (more than 11 days) duration are generally observed to have no effect on body fluid regulation in humans (17). Likewise, Young *et al.* (67) reported that repeated cold water immersion appeared to have no effect on basal blood or plasma volumes. As shown in Figure 10-10, the magnitude of the plasma volume reduction in response to acute cold exposure was also unaffected by cold acclimation. There is one report suggesting that plasma volume responses to cold stress are altered by cold acclimation. Rochelle and Horvath (57) reported that surfers, who chronically immersed themselves in cold water, experienced a smaller plasma volume reduction during cold stress as compared to control subjects. However, when the data presented in that report were re- examined, it was found that differences in body fat between the surfers and control subjects may have accounted for the difference in fluid responses to cold (67). Although it has been suggested that cold habituation may reduce the diuresis associated with acute cold exposure (54), alterations in body fluid regulation do not appear to be a major aspect of cold acclimation.

Effects of Cold Acclimation on Cardiorespiratory Responses

Very few studies have been reported in which the effects of acclimation or acclimatization on cardiovascular and respiratory re-

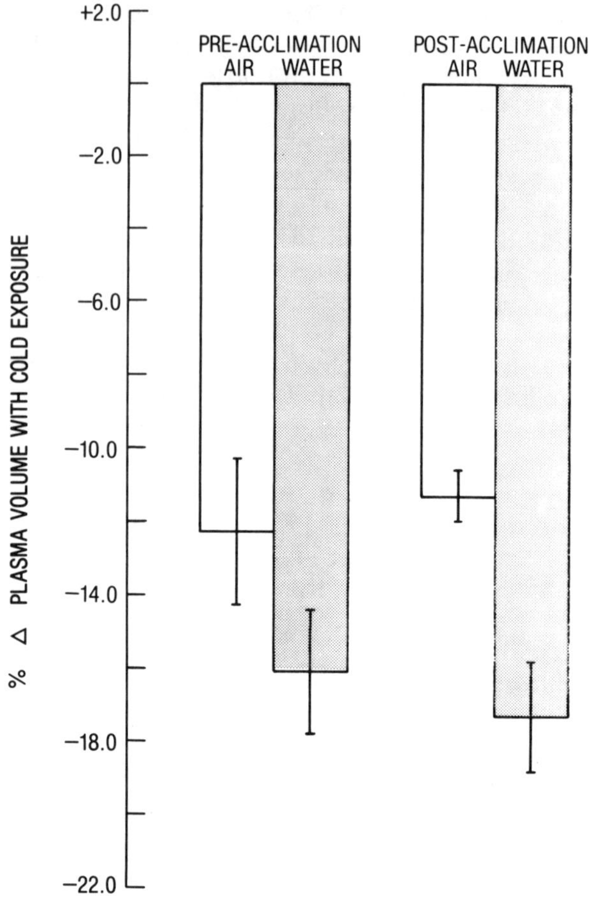

Figure 10-10. *Percent changes in plasma volume during a 90-minute exposure to cold air and cold water. Values are mean ± SE, N = 7. From Young* et al. *(67).*

sponses to cold were evaluated. Hong *et al.* (30) reported a trend, though not statistically significant, for the Korean women divers to develop a greater bradycardia while diving in winter compared to summer. Budd and Warhaft (8) reported that the increase in systolic blood pressure during whole body cold (10° C) air exposure was attenuated by a period of residence in the Antarctic. The most comprehensive evaluation of the effect of acclimation on cardiorespiratory responses to cold was reported by Muza *et al.* (50). Cardiac output, heart rate, blood pressure, minute ventilation, tidal volumes and respiratory rate measurements were obtained during cold air and cold water immersion before and again after the 5 week cold acclimation program described by Young *et al.* (66,67). With the exception of the blood pressure response to cold water immersion, there were no changes in the cardiorespiratory responses to cold observed following acclimation. Blood pressure measurements indicated, in agreement with findings of Budd and Warhaft *et al.* (8), that the pressor response to cold was diminished (50). A diminished pressor response to cold may be a habituation effect or result from an increased perfusion of vascular beds in the skeletal muscle. While the mechanism for the diminished pressor response remains to be determined, it seems safe to conclude that cold acclimation does not produce any major cardiovascular or respiratory adaptations.

SUMMARY

The descriptive studies of naturally cold acclimatized people and the controlled investigations reporting the development of cold acclimation in persons lacking a lifetime experience of cold stress clearly show that humans can adapt physiologically in response to cold. These cold adaptations follow several patterns. Some evidence suggests that a metabolic cold acclimatization/acclimation exists whereby a more pronounced increase in heat production is elicited by cold exposure. The possibility that a shift from shivering to non-shivering thermogenesis accounts for this adaptation has been suggested. By far, the most commonly observed adaptation is a cold habituation which is primarily characterized by a greater drop in core temperature prior to the onset of shivering. An insulative type of adaptation appears to develop only in individuals who experience frequent and severe cold stress conditions which result in a reduction in body heat stores. Improved vasoconstrictor responses to cold and development of enhanced avenues of countercurrent heat exchange in the limbs have been suggested as mechanisms for this type of adaptation. The existence of a cold acclimatization process in humans is an interesting phenomenon for physiologists. However, the advantage provided by these adaptations in terms of conservation of body heat is small, particularly in comparison with the protective effects available from increases in body fat and modern clothing.

REFERENCES

1. Andersen, K.L., J.S. Hart, H.T. Hammel and H.B. Sabean. Metabolic and thermal response of Eskimos during muscular exertion in the cold. *J. Appl. Physiol.* 18:613–618, 1963.
2. Andersen, K.L., Y. Loyning, J.D. Nelms, O. Wilson, R.H. Fox and A. Bolstad. Metabolic and thermal response to a moderate cold exposure in nomadic Lapps. *J. Appl. Physiol.* 15:649–653, 1960.
3. Bass, D.E. and A. Henschel. Responses of body fluid compartments to heat and cold. *Physiol. Rev.* 36:128–144, 1956.

4. Bazett, H.C., F.W. Lunderman, J. Doupe and J.C. Scott. Climatic effects on the volume and composition of blood in man. *Am. J. Physiol.* 129:69–83, 1940.
5. Bittel, J.H. Heat debt as an index for cold adaptation in men. *J. Appl. Physiol.* 62:1627–1634, 1987.
6. Brown, G.M. and J. Page. The effect of chronic exposure to cold on temperature and blood flow of the hand. *J. Appl. Physiol.* 5:221–227, 1952.
7. Bruck, K., E. Baum and H.P. Schwennicki. Cold-adaptive modifications in man induced by repeated short-term cold exposures and during a 10-day and -night cold exposure. *Pflugers Arch.* 363:125–133, 1976.
8. Budd, G.M. and N. Warhaft. Body temperature, shivering, blood pressure and heart rate during a standard cold stress test in Australia and Antarctica. *J. Physiol.* (Lond.) 186:216–232, 1966.
9. Burton, A.C. and H.C. Bazett. A study of the average temperature of the tissues, of the exchanges of heat and vasomotor responses in man by means of a bath calorimeter. *Am. J. Physiol.* 117:36–54, 1936.
10. Carlson, L.D., H.L. Burns, T.H. Holmes and P.P. Webb. Adaptive changes during exposure to cold. *J. Appl. Physiol.* 5:672–676, 1953.
11. Crile, G.W. and D.P. Quiring. Indian and Eskimo metabolisms. *J. Nutr.* 18:361–368, 1939.
12. Davis, T.R.A. Chamber cold acclimatization in man. *J. Appl. Physiol.* 16:1011–1015, 1961.
13. Davis, T.R.A. and D.R. Johnston. Seasonal acclimatization to cold in man. *J. Appl. Physiol.* 16:231–234, 1961.
14. Doi, K., T. Ohno, M. Kurahashi and A. Kuroshima. Thermoregulatory nonshivering thermogenesis in man, with special reference to lipid metabolism. *J. Appl. Physiol.* 29:359–372, 1979.
15. Elsner, R.W., K.L. Andersen and L. Hermansen. Thermal and metabolic responses of Arctic Indians to moderate cold exposure at the end of winter. *J. Appl. Physiol.* 15:659–661, 1960.
16. Elsner, R.W., J.D. Nelms and L. Irving. Circulation of heat to the hands of Arctic Indians. *J. Appl. Physiol.* 15:662–666, 1960.
17. Fregley, M.J. Water and electrolyte balance during exposure to cold. *Pharmac. Ther.* 18:199–231, 1982.
18. Greenleaf, J.E. Physiological responses to prolonged bed rest and fluid immersions in humans. *J. Appl. Physiol.* 57:619–633, 1984.
19. Hammel, H.T. Terrestrial animals in cold: recent studies of primitive man. In: *Handbook of Physiology, Section 4: Adaptation to the Environment.* D.B. Dill, E.F. Adolph, C.G. Wilber (eds.), Washington, D.C.: American Physiology Society, 1964. p.413–437.
20. Hammel, H.T., R.W. Elsner, D.H. LeMessurier, H.T. Anderson and F.A. Milan. Thermal and metabolic responses of the Australian aborigine exposed to moderate cold in summer. *J. Appl. Physiol.* 14:605–615, 1959.
21. Hanna, J.M. and R.M. Smith. Responses of Hawaiian born Japanese and Caucasians to a standandized cold exposure. *Hum. Biol.* 47:427–440, 1975.
22. Hart, J.S. Metabolic alterations during chronic exposure to cold. *Fed. Proc.* 17:1045–1054, 1958.
23. Hart, J.S., H.B. Sabean, J.A. Hildes, F. Depocas, H.T. Hammel, K.L. Andersen, L. Irving and G. Foy. Thermal and metabolic responses of coastal Eskimos during a cold night. *J. Appl. Physiol.* 17:953–960, 1962.
24. Hicks, C.S. Terrestrial animals in cold: exploratory studies of primitive man. In: *Handbook of Physiology, Section 4: Adaptation to the Environment.* D.B. Dill, E.F. Adolph, C.G. Wilber (eds.), Washington, D.C.: American Physiology Society, 1964, p.405–412.
25. Hildes, J.A. Comparison of coastal Eskimos and Kalahari Bushmen. *Fed. Proc.* 22:843–845, 1963.
26. Hong, S.K. Comparison of diving and nondiving women of Korea. *Fed. Proc.* 22:831–833,1963.
27. Hong, S. K. Pattern of cold adaptation in women divers of Korea (Ama). *Fed. Proc.* 32:1614–1622, 1973.
28. Hong, S.K., C.K. Lee, J.K. Kim, S.H. Hong and D.W. Rennie. Peripheral blood flow and heat flux of Korean women divers. *Fed. Proc.* 28:1143–1148, 1969.
29. Hong, S.K., D.W. Rennie and Y.S. Park. Cold acclimatization and deacclimatization of Korean women divers. *Exer. Sport Sci. Rev.* 14:231–268, 1986.
30. Hong, S.K., S.H. Song, P.K. Kim and C.S. Suh. Seasonal observations on the cardiac rhythm during diving in the Korean Ama. *J. Appl. Physiol.* 23:18–22, 1967.
31. Huttunen, P., J. Hirvonen and V. Kinnula. The occurrence of brown adipose tissue in outdoor workers. *Eur. J. Appl. Physiol.* 46:339–345, 1981.
32. Iampietro, P.F., R.F. Goldman, E.R. Buskirk and D.E. Bass. Response of negro and white males to cold. *J. Appl. Physiol.* 14:797–800, 1959.
33. Irving, L.K., K.L. Andersen, A. Bolstad, R. Elsner, J. A. Hildes, Y. Loyning, J.D. Nelms,

L.J. Peyton and R.D. Whaley. Metabolism and temperature of Arctic Indian men during a cold night. *J. Appl. Physiol.* 15:635–644, 1960.
34. Jessen, K. An assessment of human regulatory nonshivering thermogenesis. *Acta Anaesth. Scand.* 24:138–143, 1980.
35. Joy, R.J.T. Responses of cold acclimatized men to infused norepinephrine. *J. Appl. Physiol.* 18:1209–1212, 1963.
36. Kang, B.S., D.S. Han, K.S. Paik, Y.S. Park, J.K. Kim, C.S. Kim, D.H. Rennie and S.K. Hong. Calorigenic action of norepinephrine in the Korean Women divers. *J. Appl. Physiol.* 29:6–9, 1970.
37. Kang, B.S., S.H. Song, C.S. Suh and S.K. Hong. Changes in body temperature and basal metabolic rate of the Ama. *J. Appl. Physiol.* 18:483–488, 1963.
38. Kang, D.H., P.K. Kim, B.S. Kang, S.H. Song and S.K. Hong. Energy metabolism and body temperature of the Ama. *J. Appl. Physiol.* 20:46–50, 1965.
39. Kang, D.H., Y.S. Park, Y.D. Park, I.S. Lee, D.S. Yeon, S.H. Lee, S.Y. Hong, D.W. Rennie and S.K. Hong. Energetics of wet-suit diving in Korean women breath-hold divers. *J. Appl. Physiol.* 54:1702–1707, 1983.
40. Keatinge, W.R. The effect of repeated daily exposure to cold and of improved physical fitness on the metabolic and vascular responses to cold air. *J. Physiol.* (Lond.) 157:209–220, 1961.
41. Kreider, M.B., P.F. Iampietro, E.R. Buskirk and D.E. Bass. Effect of continuous cold exposure on the nocturnal temperature of man. *J. Appl. Physiol.* 14:43–45, 1959.
42. Krog, J., B. Folkow, R.H. Fox and K.L. Andersen. Hand circulation in the cold of Lapps and North Norwegian fishermen. *J. Appl. Physiol.* 15:651–658, 1960.
43. LeBlanc, J. Evidence and meaning of acclimation to cold in man. *J. Appl. Physiol.* 9:395–398, 1956.
44. LeBlanc, J. Local adaptation to cold of Gaspe' fishermen. *J. Appl. Physiol.* 17:950–952, 1962.
45. LeBlanc, J. Man in the cold. Springfield, IL: C. Thomas, 1975.
46. LeBlanc, J., J.A. Hildes and O Heiroux. Tolerance of Gaspe' Fishermen to cold water. *J. Appl. Physiol.* 15:1031–1034, 1960.
47. LeBlanc, J., M. Pouliot and S. Rheaume. Thermal balance and biogenic amine excretion in Gaspe' fishermen exposed to cold. *J. Appl. Physiol.* 19:9–12, 1964.
48. Milan, F.A., R.W. Elsner and K. Rodahl. Thermal and metabolic responses of men in the Antarctic to a standard cold stress. *J. Appl. Physiol.* 16:401–404, 1961.
49. Miller, L.K. and L. Irving. Local reactions to air cooling in an Eskimo population. *J. Appl. Physiol.* 17:440–445, 1962.
50. Muza, S.R., A.J. Young, M.N. Sawka, J.E. Bogart and K.B. Pandolf. Respiratory and cardiovascular responses to cold stress following repeated cold water immersion. *Undersea Biomed. Res.* (In Press) 1988.
51. Nelms, J.D. and D.J.G. Soper. Cold vasodilation and cold acclimatization in the hands of British fish filleters. *J. Appl. Physiol.* 12:444–448, 1962.
52. Oddershede, I.R. and R.S. Elizondo. Body fluid adjustments during resting cold acclimation in the rhesus monkey. *J. Appl. Physiol.* 52:1024–1029, 1982.
53. Park, Y.S., D.W. Rennie, I.S. Lee, Y.D. Park, K.S. Paik, D.H. Kang, D.J. Suh, S.H. Lee, S.Y. Hong and S.K. Hong. Time course of deacclimatization to cold water immersion in Korean women divers. *J. Appl. Physiol.* 54:1708–1716, 1983.
54. Radomski, M.W. and C. Boutelier. Hormone response of normal and preadapted humans to continuous cold. *J. Appl. Physiol.* 53:610–616, 1982.
55. Rennie, D.W., B.G. Covino, M.R. Blair and K. Rodahl. Physical regulation of temperature in Eskimos. *J. Appl. Physiol.* 17:326–332, 1962.
56. Rennie, D.W., B.G. Covino, B.J. Howell, S.H. Hong, B.S. Kang and S.K. Hong. Physical insulation of Korean diving women. *J. Appl. Physiol.* 17:961–966, 1962.
57. Rochelle, R.D. and S.M. Horvath. Thermoregulation in surfers and non-surfers immersed in cold water. *Undersea Biomed. Res.* 5:377–390, 1978.
58. Rodahl, K. Basal metabolism of the Eskimo. *J. Nutr.* 48:359–368, 1952.
59. Scholander, P.F., H.T. Hammel, K.L. Andersen and Y. Loyning. Metabolic acclimation to cold in man. *J. Appl. Physiol.* 12:1–8, 1958.
60. Scholander, P.F., H.T. Hammel, J.S. Hart, D.H. LeMessurier and J. Steen. Cold adaptation in Australian aborigines. *J. Appl. Physiol.* 13:211–218, 1958.
61. Sjorstrand, T. Volume and distribution of blood and their significance in regulating circulation. *Physiol. Rev.* 33:202–228, 1953.
62. Skreslet, S. and F. Aarefjord. Acclimatization to cold in man induced by frequent scuba diving in cold water. *J. Appl. Physiol.* 24:177–181, 1968.
63. Viecsteinas, A. and D.W. Rennie. Thermal insulation of and shivering threshold in Greek sponge divers. *J. Appl. Physiol.* 52:845–50, 1982.
64. Ward, J.S., G.A.C. Bredell and H.G. Wenzel. Responses of Bushmen And Europeans on

exposure to winter night temperatures in the Kalahari. *J. Appl. Physiol.* 15:667–670, 1960.
65. Wyndham, C.H. and J.F. Morrison. Adjustment to cold of Bushman in the Kalahari desert. *J. Appl. Physiol.* 13:219–225, 1958.
66. Young, A.J., S.R. Muza, M.N. Sawka, R.R. Gonzalez and K.B. Pandolf. Human thermoregulatory responses to cold air are altered by repeated cold water immersion. *J. Appl. Physiol.* 60:1542–1548, 1986.
67. Young, A.J., S.R. Muza, M.N. Sawka and K.B. Pandolf. Human vascular fluid responses to cold stress are not altered by cold acclimation. *Undersea Biomed. Res.* 14:215–228, 1987.

11

Human Cold Injuries

MURRAY P. HAMLET, D.V.M.

OUTLINE

I. HISTORY
II. EPIDEMIOLOGY
    A. Environmental Factors
        1. Temperature
        2. Precipitation
        3. Wind
    B. Host Factors
        1. Smoking
        2. Previous Cold Injuries
        3. Race
        4. Rank
        5. Malnutrition
        6. Fatigue
        7. Environmental Protection
        8. Leadership
        9. Other Injuries
        10. Drugs
III. CLASSIFICATION OF COLD INJURIES
    A. Cold/Wet Injuries
        1. Chilblain
        2. Pernio
        3. Trenchfoot (Immersion Foot)
    B. Cold/Dry Injuries
        1. Frostnip
        2. Frostbite
    C. Hypothermia
IV. PATHOPHYSIOLOGY OF LOCALIZED COLD/WET AND COLD/DRY INJURIES
V. INITIAL WHOLE BODY RESPONSE TO COLD
VI. DEFINITIONS OF HYPOTHERMIA
VII. PATHOPHYSIOLOGY OF HYPOTHERMIA
    A. Cardiovascular Effect
    B. Respiratory System
    C. CNS Changes
    D. Blood Sugar
    E. Autopsy Findings

HISTORY

Although cold injury occurs only sporadically in civilian settings, it is of major significance to armies operating in the cold. References to frostbite appear in the writings of Hippocrates, Aristotle, and Gaelen, and cold injury played a major role in the campaigns of the Greek armies of Xenophon operating in Armenia in the early 3rd century B.C. (51). Alexander the Great's army suffered significant casualties both in troops and pack animals. During the Revolutionary War, there were approximately 15,000 frostbite cases resulting in over 1,000 fatalities (30). Larrey's description of cold injury in Napoleon's army, however, is the best description of the impact of cold on a large fighting force (26). His use of the term "congelation" gives us a first clue in the pathophysiology of frostbite. His description of the impact of the sudden change of weather causing large numbers of casualties has been substantiated by subsequent armies. In the winter of 1854 in the Crimean War, there were almost 2,000 cold injuries among 50,000 men; 457 were fatal (29). In the two-year period starting in 1854, 5,215 out of 309,000 French troops were victims of frostbite; 1,178 were fatal. In just two consecutive nights, there were 2,800 cases of frostbite and 900 deaths.

In WWI, the differentiation of trenchfoot from frostbite first became evident (17). British soldiers' combined injuries from these two problems amounted to 115,000. Morbidity rates were between 27% and 33% for 1914 and 1915. The U.S. Medical Department lists 2,061 admissions for trenchfoot with a loss of 97,000 man-days or approximately 47 days per case. The French army had 80,000 trenchfoot casualties, and the Italians 38,000. During WWII, 10% or 90,000 American soldiers suffered cold injury (51). In November and December of 1941, the Germans had 100,000 cold injuries requiring 15,000 amputations, approximately 10% of their total fighting force on the Russian front (51). In Korea, there were approximately 9,000 cold injuries to United States soldiers, again 10% of our force was wounded. More recently, the British experience in the Falklands claimed trenchfoot as the major medical problem of the war: the Argentines recorded 273 trenchfoot injuries from the same conflict (38).

The combined impact of acute loss of manpower, the long hospitalization time (50 days), the loss of fighting strength, and subsequent compensation after the war, makes cold injury a significant problem for all military units fighting in the cold.

EPIDEMIOLOGY

The host and the environmental setting combine to produce cold injuries. Both cold/wet and cold/dry conditions combined with a series of host factors and environmental factors produce these injuries. Environmental conditions, such as temperature, precipitation, and wind, along with the activity of the individual, the duration of exposure, amount of protection, level of fitness and individual susceptibility all contribute to overall susceptibility (5).

Civilians seldom put themselves in high risk settings thus limiting the numbers of cold injuries. Outdoor forms of recreation including boating, are the major contributors in an urban setting. Street people also often end up as cold injury patients. However, the military in combat, be it offensive, defensive or retrograde provides the most significant threat scenario.

Temperature, precipitation, and wind combine to increase the rate of body heat loss. Low temperature and low humidity favor the development of frostbite. Higher temperatures, along with water, favor the development of trenchfoot. Wind accelerates body heat loss and, if wind and water are involved, there is a higher risk of hypothermia. Water immersion produces the greatest surface heat loss, a significant threat of hypothermia and subsequent drowning if floatation is not available.

Environmental Factors

Temperature. Historically, as the temperature drops, there is a higher likelihood of cold injuries in a combat setting. This may not be a direct cause and effect relationship for trenchfoot, however.

The exposure period for trenchfoot is approximately three days, whereas, when the temperature drops below freezing, frostbite cases increase significantly because the exposure period can be as short as a few minutes. If the temperature is near the freezing point, it may take a number of hours to produce frostbite. However, if it is severely cold, it can occur quickly.

Precipitation. Wetness greatly increases the incidence of cold injury. It is usually in the form of precipitation, rain and snow, but can occur from perspiration soaking the insulation layers of clothing. Although Larrey first described the impact of precipitation on increased production of cold injuries (26), it was not until WWI that the British really understood the role that wetness played in the production of trenchfoot casualties (17). The combination of freezing and thawing, which occurs night and day with subsequent wetting of the insulation of boots, produces significant cold injuries. The wetting of skin over a long period of time produces excessive hydration of cells and plays an important role in the production of the trenchfoot injury. When thawing occurs during the day with subsequent freezing temperatures at night, and troops are forced to move without shelter or the ability to take care of their feet, major injury occurs.

Wind. Wind greatly accelerates the loss of body heat. It has proven difficult to apply statistical methods to analyzing the role that wind plays in the historical production of cold injuries; however, it can be addressed in general terms. Wind increases convective heat loss from the surface of clothing because it penetrates loose-fitting clothing or openings in clothing to remove still warm air layers trapped in insulation and it increases evaporative cooling when that insulation material is wet from perspiration. This can account for up to 80% of total body heat dissipation (7). Paul Siple, an Antarctic explorer, developed a term "wind chill" to define the combined cooling effect of moving air and low temperature on the skin (7). His wind chill formula is used extensively to describe the impact of the cooling power of wind chill on the human body.

Host Factors

Age has a fairly significant impact on the production of cold injury; the young are quite resistant, the old seem more susceptible. In the militarily relevant age groups, however, there appears to be little difference in susceptibility. Elderly individuals often are more inactive, poorly dressed, inadequately nourished and often have concommitant diseases that impact on cold susceptibility. Medications often can enhance this susceptibility. Children and young adults seem to be quite resistant to freezing injuries, tend to have fewer post-freeze injury sequellae, and show better overall clinical results. Neonates, with a large surface area and small body mass, tend to cool very fast, for this reason, so-called "dumpster babies" are generally hypothermic but do well with resuscitative efforts.

Smoking. There is some evidence that the vasoconstrictive action of nicotine causes increased cooling of the extremities potentially increasing the likelihood of cold injury. Infrared studies of tobacco use show a marked, rapid peripheral vasoconstriction which reduces cold induced vasodilation over an extended period of time (Ahle, N., Unpublished Data). The supposition from this work is that smokers would incur a higher incidence of cold injury. A Korean study of 9,000 cold injuries, however, did not support this supposition (36). Of the 506 frostbite patients studied, 17.2% were nonsmokers; 13.1 of the 237 men who never suffered a cold injury were nonsmokers. The percent of nonsmokers to smokers was essentially the same in injured and noninjured groups. A small garrison study in Alaska, however, indicated that smokers were at higher risk and that the most severe injuries tended to occur in people who smoke up to one pack of cigarettes a day (44). Smoking more than two packs involved a smaller injury percentile possibly attributed to a decreased sensitivity of the peripheral vasculature to the effects of nicotine. Smoking also may play a part in the level of fitness with fit individuals tending to deal with cold better than unfit individuals. Although clear statistical data are not available, it appears that the role of tobacco in peripheral vasoconstriction and overall physical fitness suggests avoidance of tobacco use during cold weather exposure. German medical officers perceived that smoking had a significant impact on the susceptibility of individuals to cold injury. Killian, in 1952 stated, "I conclude from these pharmacological tests from my experience with cold injury that the use of tobacco should be contraindicated under conditions of cold exposure" (23).

Previous Cold Injuries. Of 932 frostbite casualties during the Korean study, 14.9% had prior cold injuries (36). No doubt, previous cold injury predisposes a soldier to further trauma from the cold (51). Every author on cold injury states unequivocally that even minor prior cold injury predisposes one to a subsequent injury (23,27,44,48,51). Recurrent first degree frostbite clearly decreases peripheral blood flow, increases cold sensitivity to the point observed with Raynaud's disease, and increases the risk to subsequent frostbite. Prior trenchfoot injuries, even though mild, render the individual highly susceptible to injury with a second experience in a cold/wet environment. Relatively minor trenchfoot will make that individual intolerant of intense marching even in warm weather. Data from Italy in WWII indicates that those soldiers with even relatively minor cold injuries should result in the stricken individual being reassigned to a less threatening zone unless the injuries are extremely mild first degree injuries. These individuals often prove to be a burden rather than an asset to their organizations (51). In the German army (winter of 1941–42), pathologists observed that frostbite was present in about 1/3 of all autopsies (23).

Race. Differences in racial susceptibility to cold injury has been described from WWII, Korea, and recent garrison experience in Alaska. An all black unit assigned to the 92nd Infantry Division in Italy in 1944–1945 contributed one out of every three cold casualties in that campaign (36). Even considering that they were a new unit, lacking combat training and experience with cold, the casualty numbers appear slightly high relative to the total casualties produced. An epidemiologic survey team on cold injury headed by LTC Kenneth Orr, MC, between 1950 and 1951 in Korea, found that blacks were at a significantly higher risk (six times more) for frostbite than other soldiers when other environmental factors were equal (36). At the regimental level, blacks had an incidence of 35.9 per thousand compared to 5.8 per thousand for whites. In the European campaign from January to March, 1956, Major General Gorby reported that, although the winter was significantly milder than 1955, most of the cold injuries occurred in blacks for that time period (35). Cold casualties in February, 1956 exceeded the total casualties for 1954 and 1955, and blacks made up the largest portion of those injuries.

The Korean study (36) of the 9th regiment, 2nd Division involved a mixture of blacks and whites. Of the 167 cases of frostbite, 138 or 82% were black and 29 or 17.4% were white. However, the unit structure averaged 33% black with a range of 26 to 41% during this period.

A recent study conducted with a mixed military population of black and white individuals examined the rates of cooling and the rewarming responses. This study failed to clearly define a difference in susceptibility between blacks and whites (21). There is a wide range of response to dipping the hand in cold water, and blacks show a distinct population of good responders. There is, however, a larger population of blacks who are poor responders to a specific cooling challenge. Their hands tend to cool closer to the bath temperature and stay colder longer while lacking a good cold induced vasodilation (CIVD) response.

Anatomic differences in the structure of the extremities also may play a role (3). Anthroprometric studies show that blacks tend to have longer, thinner fingers and feet than whites (25). In Jackson's study, however, the home of origin rather than race seemed to play a more significant role in one's hand rewarming response. Individuals from northern tier states with more cold weather experience and knowledge had a much better rewarming response after a cold challenge. Individuals from southern tier states, both black and white, were the worst responders in their respective ethnic groups. In the Korean study of 464 cold injury casualties, 69% were from states that had an average daily minimum temperature of 20° F or above.

Rank. Rank in a military setting predisposes individuals to having cold injuries. Injuries tend to be centered in lower ranking individuals with greater duration of cold exposure. Frontline riflemen appear to be in the highest threat setting. In the Korean study, 99.8% of the 1,846 frostbitten soldiers were enlisted men (36). The lowest enlisted ranks have the highest exposure potential and the highest percentage of injury. Exposure potential is a major factor in susceptibility. There also might be inadequate training in that lower ranks have less experience in foot hygiene and improper provision of clothing.

Malnutrition. It is difficult to define the impact of malnutrition on the production of cold injuries. In fact, the Korean study drew no conclusions concerning this factor. In a study by Miller (31), men with low caloric intakes were inactive and responded with less serious effort to situations in which their survival was at stake. This projected to the lack of desire to do necessary things, such as change socks and massage the feet. During November and December of 1944, the 121st Infantry Regiment was cut off from proper supplies for a two-week period; the men ate only K-rations. Their injuries went from 25 per week to 86 the first week, and to 89 the second week. When they were relieved, their cases went down to 15 per week (51). "It can be said, therefore, that inadequate nutrition was apparently a factor in bringing about an increase in cold injury in this experienced unit during a period of isolation" (51). In the Falklands War, approximately 2,200 Argentine soldiers were cut off from support for the duration of the war. They had three fatalities from malnutrition and a significant number of trenchfoot casualties in this unit (38).

Fatigue. As with nutrition, it is difficult to document the isolated impact of fatigue on the production of cold injuries. Mental and physical fatigue, however, renders one unlikely to perform simple preventive measures such as changing socks, massaging feet and exercising to restore circulation. This lack of awareness of the threat clearly makes one more susceptible to both trenchfoot and frostbite. In WWII, out of 1,018 hospitalized patients, 70% with trenchfoot had been in combat eight days or more and 44% of those who had contracted trenchfoot had fought for 15 days without rest.

Environmental Protection. Cold injury prevention involves the rational use of clothing and shelter. Cold weather clothing should be worn loosely, and in layers. It should provide moisture transfer from the body to the outer layers, and should be easily ventilated depending on one's work and sweat output. Keeping clothing dry is of utmost importance in the field. A current approach to cold weather clothing is to use a thin, polypropylene or similar synthetic material next to the skin for high moisture transfer (37). Multiple layers of either synthetic pile, wool, or bunting materials should be layered concentrically over the body. This is then followed by an outer windproof water-repellent fabric that breathes. It is advantageous to have a vapor permeable outer clothing layer to allow moisture to travel through the clothing to the outside. The inner layer and outer clothing layer should stay intact and the insulation layers should be adjusted for thickness, depending on one's workload and the outside temperature.

There is a significant amount of training and experience required for proper use of clothing under a wide range of situations. It is particularly difficult to deal with a cold, wet, windy situation. Footwear remains the single greatest challenge to both designers and users for cold weather operations. The proper design and layering of footgear to allow for moisture transfer, flexibility, abrasion protection and insulation for road marching involves a series of compromises and a significant individual effort to provide foot protection in the cold. The approach involves sealing the insulation with vapor barrier envelopes while providing layered insulation that can be stripped apart and dried easily in a field setting. This appears to be the most useful and functional approach for military units.

The proper fitting of footwear cannot be over-emphasized and one's march boots may be significantly different in size than one's garrison footwear. Foot measurements change both from load bearing and from the trauma of marching. The inside volume of the boot can be changed to meet this requirement using different layers of socks and altering the thickness of the insole. Having the appropriate footwear ensemble and intensive training can significantly decrease the incidence of cold injuries to the feet and also reduce the severity of the injuries that occur. Handwear is the second greatest challenge; again, multiple layers of insulation are the best approach. The complexity of the fingers and their dexterity requirements remain an engineering challenge. Single needle-knitted wool handwear, both gloves and mittens, along with an outershell appears to be the best handwear ensemble. In military settings, the availability of changes of socks and the ability to issue new portions of uniform to replace old, worn materials have historically been key parts in the prevention of cold injury. This point is repeatedly forgotten by armies deploying into the cold. A sound, functional, logistical, clothing support system is essential for conducting military operations in a cold environment.

Keeping the clothing dry from the outside rain, snow, and mud along with moisture transfer from the skin to prevent perspiration wetting the insulation layers requires significant training effort. Rubberized overboots contribute significantly to protection from trenchfoot injury. Constant vigilance and intense command influence must be exerted to see to it that tired, combat-weary, cold, wet individuals protect themselves from injury.

Finding shelter to protect oneself from the rigorous outdoor environment is not often easy or possible. One's thoughts should be toward eliminating the duration and severity of cold exposure. Individuals trapped out in the open or having to live out of foxholes are at significantly greater risk to all cold injuries than those in even sparse shelter, such as bombed outbuildings, barns, or tents. The Germans were repeatedly frustrated by the Russians ability to improvise shelter to get out of the weather and their clever use of sparse materials to protect themselves from the winter (48). The Falklands experience again indicates the increased incidence of cold injury associated with troops unable to have tents or shelter other than foxholes and ponchos (15).

Leadership. A high standard of leadership significantly influences an army's ability to protect itself from cold injury. Strong leadership implies discipline, sound training, and attention to detail. Commanders must assure an appropriate supply of food, fuel, replacement clothing, medical facilities, and rotation policies, and the use of field warming facilities to decrease the likelihood of injury on a large scale. Well-trained, competently led troops are more alert to the threat and are more likely to take preventive measures. This effort is more likely to be sustained during periods of high stress, combat and extreme fatigue. Numerous examples exist for units fighting on the same terrain, the same situation, and the same time period, having significantly different levels of cold injury directly relatable to the quality of their leadership (51). As soldiers become tired and the intensity of combat demoralizes them, morale collapses and foot discipline disappears. If leadership is poor in a small unit, it will have a significant number of cold injuries. "Taking care of oneself" is the bottom line and the higher the quality of the training and leadership, the farther the individual's training is likely to carry him in the prevention of cold injuries. There is rapid feedback for both good and bad leadership disciplines in the cold.

Other Injuries. In a military setting, concomitant injuries with rapid blood loss increase susceptibility to hypothermia and to other cold injuries, particularly involving the extremities. During the transport of injured individuals, there is an increased risk of frostbite in the extremities, particularly the feet. Extra care must be taken during evacuation to prevent this occurrence. Auxiliary heat is often necessary if the transportation time is extended. Dehydration, causing hypovolemia which decreases the circulation to the extremities, increases the likelihood of cold injury. New ration systems containing dehydrated foods add significantly to the water requirement in the cold. Dehydration tends to cause pooling of blood in the core, limiting peripheral blood flow while increasing susceptibility to cold injury. Cold induced diuresis and limited availability of water for combat soldiers add to this problem. The impact of dehydration can be significant in the production of cold injuries. Disease, illness, cold urticaria, chryoglobulinemia, and other coagulopathies may increase cold injuries but are of relatively minor significance.

Drugs. Drugs that modify central temperature regulation, the autonomic system, or that modify judgement can have significant impact on cold injury and the individual's ability to protect himself from the cold. Historically, alcohol use has been the single leading factor in production of hypothermia in a civilian community and also can be of significance in a military setting.

CLASSIFICATION OF COLD INJURIES

Cold injury can be differentiated into major categories of cold/wet, cold/dry, and subsequently subdivided into localized and generalized (hypothermia) injury. Cold/wet injuries are in ascending order of severity: chilblain, pernio, and trenchfoot (immersion foot). Cold/dry injuries are frostnip, frostbite, and high altitude frostbite. Hypothermia is whole body cooling, and can terminate in total body freezing.

Cold/Wet Injuries

Cold/wet injuries are a time/temperature phenomenon. The warmer it is, the longer the time required to produce damage. The colder the water, the shorter the duration required to produce injury.

Chilbain. Chilbain can occur both to the hands and to the feet. Most lesions are found on the dorsum of the extremities between the joints. They generally result from cold and dampness experienced over a long period of time. Areas of moderate cold with high humidity are generally thought to produce this injury. Repeated exposure to a cold, damp environment produces superficial injury. First, the area is red, swollen, tender, and is hot to the touch. Vasodilation and subcutaneous edema and swelling are characteristic of chilblain; usually, the only symptom at this time is itching. In chronic lesions, the vasodilation is no longer present but the swelling may increase and the tissue may become more turgid. Color changes ranging from light grey to reddish-purple are common and blister formation with subsequent ulceration may occur. In a chronic form, itching is replaced by pain, often lasting many years, and is not responsive to non-steroidal analgesics and pain medication. Open lesions are slow to heal, indicating the compromise of superficial blood flow. The temperature of exposure is usually described from freezing to as high as 60° F (16° C), and is associated with high humidity. This situation can occur inside rubberized boots and footwear where the humidity is often very high.

Pernio. Pernio is a more severe form of chilblain. The swelling and itch of chilblain is replaced with superficial burning and pain. There are superficial circumscribed patches of necrotic skin present, usually on the dorsum of the hands and feet. These patches escharify and eventually slough, leaving tender, painful tissue underneath. These lesions are split-skin thickness and the damage is slow to heal. As the depth of injury increases with duration of exposure, this injury eventually would be termed trenchfoot.

Trenchfoot. Trenchfoot, or immersion foot is usually described as resulting from a longer exposure, in excess of 12 hours, to water temperatures below 50° (10° C) but above freezing. Immersion in water does not produce frostbite. Ungley described this injury as peripheral vasoneuropathy after chilling (47). This injury can be produced by having wet feet in a relatively mild, even tropical, environment but it tends to take a much longer time to be produced. Shipwrecks and individuals floating in life rafts for long periods of time are generally a major source for this injury. Pathophysiologically, immersion foot is indistinguishable from trenchfoot although other factors have been implicated (1,4). Ungley is of the opinion that the effect of immersion lies in keeping the extremities cold, although wetness also may play a part in the pathophysiology of this injury (46).

There are three stages of immersion foot: first, a prehyperemic phase lasting a few hours or days in which the extremity is cold, numb, swollen, discolored and usually pulseless. Next, there occurs a hyperemic phase which lasts two to six weeks in which tingling pain, vasomotor disturbances and wide temperature gradients appear on the skin. There may be blistering with subsequent ulceration and occasionally gangrene. The third stage involves a post-hyperemic phase lasting weeks or months. The patient may be asymptomatic until exposed to cold again or there may be development of Raynaud's disease. Edema is the most prominent physical finding followed by tingling pain, itching, numbness, and loss of skin sensitivity. There may be progressive changes in skin color from mildly cyanotic to deep blue and eventually to black.

Trenchfoot, although similar to immersion foot, generally occurs on land and at temperatures from freezing to approximately 50° F (10° C). It is often associated with immobilization or dependency of the extremities, but trauma also may play a part. Walking, standing, or sitting for long periods of time with wet footwear, combined with apathy, malnutrition, and depression can lead to large numbers of trenchfoot casualties.

Cold/Dry Injuries

Frostnip. Frostnip is a freezing of superficial layers of the skin with no subsequent damage or tissue loss. Repeated frostnip to the same spot may produce dry, scaly, desquamation of skin that will fissure and crack while being quite sensitive to the cold. This is not considered a serious cold injury. Repeated frostnip predisposes to subsequent frostbite injury (40).

Frostbite. Frostbite is the actual crystallization of tissue fluids in the skin or subcutaneous tissue after exposure to temperatures below freezing. The temperature, wind velocity, and duration of exposure varies from a few minutes to a few hours. Since WWI, frostbite has become the most common cold injury in military operations. The clinical appearance is dry, waxy, skin,whitish in appearance that is insensitive to the touch. The skin is hard and fixed over joints and nondepressable. Rewarming results in intense pain, hyperemia, and swelling. Blister formation also may occur depending on the severity. (More detailed descriptions of degrees of frostbite will appear in a following section.)

Hypothermia

Hypothermia is a condition associated with a lowering of body core temperature below 94° F (34.4° C). This life-threatening situation usually results from long-term, dry-land whole body cold exposure, or immersion in cold water. Freezing temperatures are not necessary to produce hypothermia, but the combination of cool air, rain, and a light wind are sufficient to produce devastating hypothermic situations.

PATHOPHYSIOLOGY OF LOCALIZED COLD/WET AND COLD/DRY INJURIES

The initial response to cold is a peripheral vasoconstriction. This constriction is an attempt to conserve core heat at the expense of the extremities. As cooling time increases, cold induced vasodilation intervenes to cause peripheral dilation. This alternating constriction and dilation, termed CIVD or "Lewis Hunting Wave" phenomenon, functions to conserve core heat and yet intermittently saves function of the hands and feet. Superficial vessels react first causing local constriction followed by a transient general vasoconstriction. The latter is caused by a return of venous blood from the cooled skin and the lowering temperature of the general circulation.

As the skin temperature drops to 50° F (10° C), it becomes numb, and loses its sense of touch and pain. Perception of cold changes to pain at about 59° F (15° C). The impact of cold has its greatest effect on the vascular system. Cold increases cellular permeability, particularly in the endothelial cells, causing release of fluid and localized inflammation. Skin actually freezes between 28° F (−17.8° C) and 32° F (0° C). Supercooling can occur so freezing actually occurs between 25° F (−16° C) and 14° F (−10° C). After tissue freezes and thaws, there is an initial movement of blood back into the skin vessels. They dilate markedly, increasing blood flow or causing hyperemia. There is loss of integrity of endothelial cells, fluid extravasates into the tissues causing edema and vesicle formation. The loss of fluid causes hemoconcentration and sludging of red cells.

During the initial stages of cooling there is itching, tenderness, coldness, and some discoloration of the skin. After thawing, it is followed in more severe cases by general swelling, blistering, and ulceration of tissues. Lewis believed that much of the inflammatory process associated with cold was from the release of histamine-like substances from skin cells injured by ice crystals (27). Blood flow stagnates, more metabolic toxic products accumulate which further impair cellular function and lead to cell death.

Cold injuries to the skin are generally indurative and slow healing. They are often resistant to skin grafting because of the compromise of the underlying vascular bed. Kreyberg's early theory of extravasation of fluid with concentration of red cells producing stasis and without intervascular coagulation has been supported and clarified in subsequent studies (24). Fibrin split products are not found early in frostbite injury which would indicate a lack of participation of the clotting processs in the pathogenesis of the injury (51). However, damage to endothelial cells, later in the course, initiates platelet aggregation, clumping, and vascular coagulation. Kreyberg felt that the necrosis was the primary result of a vascular stasis and anoxia, and not influenced by cold per se. The theory holds that the combination of tissue anoxia, vascular damage, and loss of tubular integrity of the vascular system produces the resultant injury. Complete loss of capillary networks in frozen thawed tissue creates a loss of nutritive pathways (11). Although large vessel flow may be maintained for a short period of time, the nutritive pathways are often nonexistent.

Cell types differ in their susceptibility to cold. Nerve tissue and muscle tissue are probably the most highly susceptible. Skin is quite resistant to freezing, but being the first line of defense, it is subjected to the coldest temperatures during the cold injury process. Lasting damage in nerve and bone can occur with skin showing relatively minor changes. Small bone infarcts with subsequent necrosis and lysis of bone occurs in second and third degree frostbite injury (16,41). Long term cooling of cells without freezing produces significant impact on cellular function. Cellular metabolism is markedly reduced changing membrane permeability. The exact cause of cell death has not been clearly defined. Freezing of cells, however, produces a massive disruption of membranes, and loss of intercellular integrity.

In chronic cooling, the release of peroxides, histamines, serotonin-like substances, and free radicals can significantly affect cellular function. Cold has a major impact on sympathetic nerves controlling blood vessel diameter and the subsequent ischemia associated with frostbite and trenchfoot injuries. The inability of sympathetic pathways to control blood vessel diameter, along with the local release of vasoconstrictive compounds such as histamine or seratonin, are added to the destruction of endothelial cells. This causes a substantial reduction in nutritive flow to tissues. The destruction of small capillary beds, more proximal arterial and venous pathways, and subsequent compromise of endothelial linings causes blood flow to retreat to a line of demarcation. This line of demarcation, dividing viable and nonviable tissue, is variable and may progress proximally if initiated by vasoconstrictive substances in the area.

The amount of functional tissue retained after frostbite injury is determined by the extent of nutritive vessel destruction, the impact of vasoconstriction, and the extent of the neurovascular injury. Tissue will die up to the level where adequate nutritive blood flow occurs. Tissue salvage may be determined by the ability to chemically intervene and extend arterial-venous flow as far distal as possible. A standard approach is to decrease anoxia by relieving vasoconstriction, decrease local sludging, dissolve clots in vessels, and improve overall blood flow. The distal extent of this salvage may be determined by the actual line of ice crystal formation deep in the tissue.

The complete pathogenic process has eluded investigators for many years. The metabolic effect of cold on individual cells is quite easy to understand and define. Although this impact varies from one cell type to another, it is the combination of responses from different tissues that ultimately inhibit function, particularly blood flow. Frostbite and trenchfoot are primarily vascular injuries. As endothelial cells become metabolically depressed by cold, they lose their ability to hold fluid in the vascular space. Edema is seen after rewarming as a result of this loss of fluid. The post-rewarming increase in blood flow appears to be due to dilation of a greater percentage of vessels than would normally be open for flow. This sudden surge of blood flow is compromised, however, by the loss of fluid and sludging of red cells in the smaller vessels. A rapid loss of flow due to peripheral vasoconstriction may be either neurogenic or chemical in origin, or both. The release of vasoactive amines, seratonin or histamine, further exacerbate this constriction, although centrally mediated constriction may also play a part in this process. This vasospastic response extends proximally depending on the severity of injury. As constriction progresses, anoxia further increases the severity and extent of the injury. Some time after rewarming, there is an inability of the vessels to improve nutritive flow. Physical destruction of capillary beds, from ice crystal formation or plugging of the beds, shunts blood to major vessels.

The tremendous variability in animal studies, both intra- and interspecies, probably results from differences in vascular responsiveness to a specific cold insult. If all vessels not physically damaged could be dilated to their maximum extent, then the individual variability in response to a specific freezing insult might be minimized. A wide range of substances have been utilized to counteract various aspects of this pathophysiologic process. Most are without substantial effect. This is probably the result of the dynamic changes occurring in vessel function over the wide range of vascular responsiveness to this insult. The sharp line of demarcation between dry/mummifying tissue and the proximal viable tissue is probably the result of a combination of the farthest extent of vasoconstriction and the most proximal advancement of the freeze front in the extremity. Tissue proximal to this can be lost if the vasospasm is significant enough to produce tissue anoxia and ischemia.

Inroads into management of frostbite will have to alter the interface between the natural line of demarcation in severe frostbite and the more proximal viable tissue. A key to frostbite therapy involves the ability to increase blood flow at this interface in order to decrease tissue metabolites, to increase nutritive flow, and increase oxygenation during the early demarcation process. The lack of an early prognosticator has seriously hampered the ability both clinically and experimentally to judge the efficacy of a treatment modality used for frostbite. Thermography may be beneficial in this area and in identifying cold sensitive individuals.

INITIAL WHOLE BODY RESPONSE TO COLD

The initial response to cold involves a reduction in skin temperature, a shudder, piloerection, and the formation of goose bumps, along with peripheral vasoconstriction. The sympathetic discharge results in increased respiration, heart rate, basal metabolic rate, and shivering as skin cools (39). With longer cooling, core temperature drops and there is a decrease in respiration, heart rate, and blood pressure. Cold diuresis results from an increase in fluid in the core after peripheral vasoconstriction. There is also a significant volume depletion, with increased blood viscosity, hemoconcentration, and a shift to the left in the oxygen hemoglobin disassociation curve. Peripheral vascular resistance increases conduction velocities in the heart resulting in specific arrythmias termed J-wave, or Osbourne waves. As general suppression of organ function occurs, insulin and corticosteroid secretion is increased. Shivering is produced by involuntary muscle contraction which can increase metabolic rate two to five times (20). Oxygen consumption can increase 3.7 times in 15 minutes and will continue to increase until shivering becomes constant, and glycogen depletion or exhaustion renders the muscles incapable of further contraction. Non-shivering thermogenesis from metabolic activity can increase thyroid production, adrenal corticoids, insulin, and catecholamines. Central nervous system release of norepinephrine has been documented in humans.

Man's ability to adapt to cold can involve all five mechanisms of heat loss. Radiation can account for 55 to 65% of total heat loss. Subcutaneous blood flow, insulation material, and huddling can change this radiant loss. Conduction and convection account for about 15%. Conductive heat loss can increase 25 times in cold water. Disruption of still water due to shivering or swimming in cold water increases convective heat loss significantly. Evaporation and respiration account for 20 to 30% of the heat loss and are affected by both temperature and humidity. Wind and rain can account for significant heat loss and have contributed to the production of numerous hypothermia incidents.

DEFINITIONS OF HYPOTHERMIA

Accidental hypothermia is the accidental depression of core temperature to 95° F (35° C) or lower. Immersion hypothermia involves a rapid cooling from prolonged immersion in cold water, usually defined as 21° C or colder. Chronic hypothermia results from an exposure to dry/cold for six hours or more in duration and generally does not involve immersion. The term, "urban hypothermia", describes a setting composed of predisposing factors such as age, drug use, and alcohol use.

Acute immersion hypothermia follows a series of time related events. In the first 20 minutes, the core temperature remains generally normal and the effects are the result of reflex responses to cooling the skin. A second period of longer duration starts when the body begins to lose control of core temperature.

Plunging into cold water induces hyperventilation, or the gasp response. If this occurs underwater, there is likelihood of laryngeal spasm, asphyxiation and drowning. When hyperventilation occurs, an increased respiration rate (60 or 70 breaths per minute) causes a drop in arterial carbon dioxide pressure with a resultant cerebral vasoconstriction and a reduced cerebral blood flow. This, along with hypothermia, can result in confusion, loss of consciousness, and coma. The respiratory alkalosis from hyperventilation can result in tetany and the inability to conduct coordinated swimming responses.

Ventricular arrhythmia, such as premature ventricular contractions, can result from cold-water immersion and can sometimes have fatal consequences. There is a vagally induced bradycardia and the diving reflex associated with immersing the face in cold water. Peripheral cooling that results in the inability of muscles to perform coordinated motion is a result of decreased nerve conduction velocity, and the direct effect of cold on muscle physiology, another reason for the lack of ability to swim after plunging into cold water (18,22).

The amount of clothing, body size and type, alcohol use, and the water temperature all play a role in the rate of cooling. An alcohol induced peripheral vasodilation increases heat loss significantly in cold water. If the individual is unable to stay afloat because of cooling of the shell and loss of coordinated muscle activity, he will become a drowning victim. If a flotation device is worn, the individual will cool slowly, depending on the temperature of the water, to a point that renders him/her unconscious. At this time, in high seas he/she will flip over and become a drowning victim. However, if a life jacket keeps the individual's head above water in a reasonably rough sea, then cooling will continue and the victim eventually will die of hypothermia.

PATHOPHYSIOLOGY OF HYPOTHERMIA

Cardiovascular Effect

The combination of cold and cold diuresis produces an initial decrease in circulating blood volume, an increase in peripheral resistance and blood viscosity. There is an initial tachycardia which is subsequently followed by a slow, progressive decrease in heart rate. The cardiac cycle lengthens, with changes in pH, oxygen delivery, electrolytes (particularly potassium) and nutritive flow impacting on conduction velocity across the myocardium. Prolonged cooling results in a positive deflection in the RT segment, called J-wave. This J-wave may be the result of delayed depolarization or early repolarization of the left ventricle. Atrial rhythm disfunctions commonly occur below 32° C core temperature. Atrial fibrillation is common. Ventricular fibrillation and asystole are the life threatening arrhythmias of hypothermia. Asystole is more common in young patients, while older patients often suffer ventricular fibrillation. Patients with both metabolic and respiratory acidosis along with hyperkalemia often progress to asystole. Mixed metabolic and respiratory alkalosis commonly produces ventricular fibrillation. Asystolic patients may convert to ventricular fibrillation prior to establishment of effective rhythm on rewarming.

Respiratory System

During cold exposure as with cardiac function, there is an initial increase in respiratory rate followed by a slow, progressive decrease. There is a decrease in tidal volume proportional to the drop in metabolic demand. With each 8° C drop in core temperature, carbon dioxide production is halved. Above 32° C, ventilation is adequate to remove CO_2; however, below that value, carbon dioxide retention occurs producing respiratory acidosis. Bronchorea combined with a depressed cough reflex results in airway obstruction. Cardiogenic pulmonary edema may ensue and result in post rewarming pulmonary complications. Metabolic acidosis is a common finding in hypothermia. Alkalosis may also occur although shivering increases lactic acid production. Shivering, increased peripheral resistance, and arterial hypoxemia generally lead to enhanced metabolic acidosis. Both hyper- and hypokalemia have been described, but hyperkalemia appears more commonly in long-term hypothermia.

CNS Changes

As cooling occurs, a progressive blunting of CNS function develops. Below a core temperature of 32° C, patients become apathetic, lethargic, withdrawn, and retreat inward psychologically. They become reclusive, less sociable, and belligerent. Disturbances in gait and coordination follow with lethargy progressing to stupor, and eventually collapse. Bizarre behavior patterns often occur in individuals exposed in a hypothermia setting. Common events include the loss of clothing and forgetting to do simple protective measures, such as putting on a hat or using a sleeping bag which would be life saving. Paradoxical undressing by hypothermia victims is well documented (50). Women often enter the medical/legal system as assault victims but are actually victims of hypothermia. Complete loss of CNS function resulting in coma, absence of corneal and retinal reflexes occurs below a core temperature of 28° C. Fixed and dilated pupils, muscle rigidity and a flat EKG often lead to a misdiagnosis of death. Failure to respond to rewarming is the only criteria for death in hypothermia; this leads to the adage "No one is cold and dead until warm and dead." Resuscitative efforts must be made to rewarm the hypothermic.

Blood Sugar

Both hyper- and hypoglycemia can occur in hypothermia. Hypoglycemia results from glycogen depletion caused by shivering. Long-term shivering burns tremendous amounts of glucose. Hypoglycemia may be more common in acute immersion type hypothermias. Alcoholic patients are often presented as hypoglycemic because of the alcohol blockade of gluconeogenesis and also the cessation of shivering. Hyperglycemia can occur because of a general decreased utilization of glucose during sustained hypothermia and increased circulating catecholamines producing increased circulating gluconeogenic precursors.

Autopsy Findings

Pancreatic lesions occur in 80% of hypothermia victims, and gastric erosion occurs in 45 to 86% of these cases (28,14). Microscopic degeneration of the myocardium and cyanotic red discoloration of skin are the most common findings in hypothermia and patients have a high incidence of pulmonary change, pulmonary edema, interstitial hemorrhage, and focal emphysema.

DIAGNOSIS OF HYPOTHERMIA

The key to survival for hypothermic victims is recognition at every level of medical care that they are treatable. Failure to recognize this often results in a fatal outcome. The clinical history as to time, place and antecedent condition of the victim combined with a simple rectal temperature measurement is often enough to arrive at a diagnosis. A cold, cyanotic, apneic patient with fixed and dilated pupils, and little or no cardiac electrical activity often is perceived to be dead. Resuscitation of this patient must be attempted. Apnea occurs below 27° C core temperature and asystole occurs below 22° C. Successful resuscitations have occurred after three hours of asystole which may, in fact, be many hours after initial apnea occurred (43). The principles of resuscitation involve rewarming the body's core, returning cardiovascular function to normal, controlling acid-base balance, relieving the hyperkalemia, and slowly increasing circulatory volume. Attempts should be made to prevent ventricular fibrillation and improve respiratory function during the rewarming process.

The patient may appear to have extensive muscle rigidity. Associated dermatologic changes may include edema and blotchy cyanotic patches. Severe bradycardia and hypotension are the hallmarks of hypothermia. Heart sounds are muffled, if at all audible. Peripheral pulses may be absent because of peripheral vasoconstriction and cold, tense skin. The abdomen is often sunken, doughy, and quite rigid. Hyporeflexia occurs below 32° C core temperature. In a field setting, mild hypothermia victims should be prevented from further cooling and, if conscious, given warm, sweet drinks to return circulating volume. Also, allow them to rewarm with large muscle activity such as walking. Unconscious hypothermia victims should be considered to be in life-threatening situations. They should be handled carefully, insulated, and transported to definitive medical care for physiologic monitoring and controlled rewarming. Attempts to rewarm deep hypothermia victims in the field should be avoided. In a hospital setting, physicians skilled at electrolyte, metabolic and cardiovascular problems, are necessary for resuscitation. Intensive care is usually necessary in the initial management phase.

TREATMENT OF LOCALIZED AND GENERALIZED COLD INJURIES

Trenchfoot Treatment

Trenchfoot, or immersion foot, is a disease of the limbs associated with being cold and wet for an extended period of time. As described earlier in this chapter, this injury progresses through three stages: first, a hyperemic phase lasting a few hours to a few days when the extremity is cold, discolored, and numb. Secondly, a hyperemic phase which lasts from two to six weeks and defined as a tingling pain with extensive swelling, blister formation with desquamation, ulceration and gangrene. This is the most active time for treatment and the only hope for increase of input on the trenchfoot injury. Third, a post-hyperemic phase occurs lasting many months or even a lifetime in which the limb is warm, but possesses an increased sensitivity to cold or emotional stress.

Management of trenchfoot remains controversial as few treatments have shown any impact on tissue survivability (1). This injury produces a moist liquification gangrene. Patients often have systemic infections with elevated CPK's and require early surgical intervention. Initial management involves careful washing and drying of feet and elevation and cooling with a fan. Although chemical sympathectomy may have some advantage in the treatment, therapeutic value has not been established. Surgical sympathectomy should be avoided. Anti-sludging agents such as dextran, and nonsteroidal anti-inflammatory medications may be effective. Pain medications are often necessary. Fever, and elevated CPK's disseminate intervascular coagulation, and liquification of tissue requires early surgical intervention. Sequential amputation may be necessary before all necrotic tissue has been removed. Keeping the feet cool and decreasing metabolic tissue demands, may be wise in the management of trenchfoot. Post-injury sequelae include cold sensitivity, Raynaud's disease, hyperhydrosis and tender swollen feet. It should be stressed that prevention of trenchfoot is of paramount importance because treatment is relatively ineffective. An outline of the advisable steps in the treatment of trenchfoot is presented in Table 11-1.

TABLE 11-1. *Outline Of Steps In The Treatment Of Trenchfoot Injuries*

1. Dry feet carefully. Warm the body, but keep feet cool.
2. Bed rest with feet elevated. Fan the feet.
3. Medication for pain and infection.
4. Improve blood flow, chemical sympathectomy, dextran and heparin may be useful but are unproven.
5. Improve nutritional status, vitamin supplementation and high protein diet.
6. Avoid early surgical intervention. High CPK, DIC and fever are indications for amputation.

Frostbite Treatment

The foundation of frostbite treatment is initial rapid rewarming and next improving blood flow as far distal as possible (12,19). Frostbite starts distally, and superficially, and then moves proximally, and deeply with the freeze front advancing in the tissue as a bullet-shaped front progressing up the limb. Tissue distal to the freeze front is usually not salvageable. The interface tissue is under jeopardy from vasospasm and ischemia, local release of cellular toxic substances and anoxia. As freezing moves more proximal, tissue is less susceptible to permanent damage; however, vasospasm can produce enough anoxia to cause sluffing some distance from the freeze front.

Rapid Rewarming. The decision to rewarm frostbite is often logistical, i.e., in the field one must analyze the evacuation chain. If there is a chance the tissue will freeze a second time, it is inadvisable to thaw at the site, because the freeze/thaw/refreeze is a devastating injury. If there is no likelihood of refreeze, the tissue should be rapidly rewarmed, insulated and the patient transported to definitive medical care. In the hospital setting, frozen extremities should be immersed in warm water, preferably a whirlpool, between 102° to 110° F to which provodone iodine has been added. The patient should be encouraged to do a full range of motion exercise of the involved limb during the rewarming process and in subsequent daily treatment. Twice daily whirlpooling in 94° F water should follow. This procedure produces great pain, but it is the best way to salvage tissue. If the individual has thawed prior to entering the hospital, this procedure is not advisable. For a relatively minor frostbite injury in the field, the individual can place the hands in the armpits or groin to obtain rewarming. Feet can be put under a companion's coat next to the abdomen. Other methods of rewarming have proven to be too harsh and severe for good tissue preservation (51). The limb is anesthetic while frozen, and the use of excessive heat to rewarm produces a devastating secondary injury. History of having used hot water, an open fire, stove or oven, forced hot air such as from an exhaust pipe of a vehicle, or a kerosene heater should be carefully recorded in the medical record. This produces a burn on top of the cold injury with early tissue destruction and liquification. Other methods of rewarming have been proposed. None appear to be more effective than warm-water immersion.

Vasodialators. A variety of vasodialators have been used to improve distal blood flow in a frozen extremity. Most beneficial has been the norepinephrine depleter Reserpine. This drug, however, is no longer available for injection. Half milligram injections interarterially proximal to the frozen extremity yield rapid vasodilation which lasts many hours or days. Indications to use Reserpine have been if (a) it is perceived to be a serious frostbite injury, (b) it is to relieve pulsatile pain, and (c) if infection ensues late in the treatment process. If the entire injection is made interarterially there are no systemic blood pressure changes, no tranquilization effect, and no other untoward side effects associated with this procedure. Patients get immediate relief of pain, a warming of the extremity from a vascular flush of the skin and a relief of pulsatile pain. Oral use has not proven effective in frostbite management. Research is currently underway to find a replacement for Reserpine. A complete outline of advisable steps in the treatment of frostbite is presented in Table 11-2.

TABLE 11-2. *Outline Of Steps In The Treatment Of Frostbite*

1. Remove constrictive clothing, protect the extremity from further cooling, trauma, and transport.
2. Obtain rectal temperature and exposure history.
3. Measure baseline laboratory studies and implace intravenous lines.
4. Initiate rewarming procedures. Rapid rewarming in warm water, 43.3° C whirlpooling.
5. Use interarterial Reserpine or similar chemical sympathectomy, Ringer's lactate and low molecular dextran.
6. Twice daily, utilize 30-minute whirlpooling at body temperature water. Full range of motion exercise in the water.
7. Use Morphine or Demerol for pain.
8. Loose dressings or towels over the extremities. Tent feet.
9. Do not rupture blebs.
10. Antibiotics only if infected.
11. Avoid early surgical intervention. Allow for self-demarcation.
12. Fever, elevated CPK and DIC are indications for early surgical interventions.

Hypothermia Treatment

Prehospital Care. Core temperature measurements in the field are extremely helpful in diagnosis of hypothermia. Low reading rectal thermometers are commercially available. The goal of the field treatment is to prevent further heat loss. More severely hypothermic patients should be carefully insulated to prevent further heat loss and care should be taken to prevent an irritated myocardium and hypovolemia from producing asystole or ventricular fibrillation. Intravenous fluids consisting of 500 cc's or 5% dextrose should be given if possible. Initiating intravenous injections in these patients is extremely difficult as peripheral vasoconstriction and hypovolemia render the vessels difficult to locate. The removal of wet clothing is important, but care must be taken during this process. Field rewarming procedures should not be attempted unless evacuation is delayed as these processes often hinder the rescue effort and compromise the rescuers. Heated humidified oxygen is appropriate if available but should not be considered as a major heat input. This process prevents respiratory heat loss although such losses are quite small.

Other methods of field rewarming have not proven to be effective, and are often time-consuming, require specialized equipment, and can produce serious cardiovascular changes that often cannot be monitored or treated in the field. The airways must be maintained in these patients if they are unconscious. Determination of peripheral pulse is difficult or impossible. A depressed cardiac output and slow heart rate combined with a cold, stiff chest may be perceived as cardiovascular collapse. Pathogenic ventricular fibrillation can result if chest compression is prematurely started on these patients. If cardiac monitoring is available, asystole or ventricular fibrillation should be verified and attempts to defibrillate should be made if the patients core temperature is above 30° C. If unsuccessful, rewarming and CPR should be continued until the patient reaches the hospital. Fluid replacement can be lifesaving by decreasing blood viscosity. Cardiac output is increased significantly but early overload can occur. Any respiratory support is helpful and positive pressure of respiration in apneic patients often produces gasping respiration. Even if monitoring equipment is not available and cardiac function cannot be determined, this gasping respiration indicates presence of some cardiac function.

Rewarming Procedures. The basic premise of rewarming the hypothermic patient is to get heat to the body core. There have been a variety of techniques and procedures used, but they are generally divided between active rewarming or passive rewarming. Active rewarming may be internal or external. Passive methods of rewarming include wrapping in blankets, sleeping bags or reflective films to prevent further heat loss and to allowing shivering thermogenesis to produce a rise in core temperature. Patients above 90° F (32° C) core temperature generally can be allowed to rewarm passively. External, active rewarming procedures utilize water immersion in a tub, water circulating blankets, hot water bottles or heating pads applied to the groin and axilla. It should be noted that the method of rewarming, approach, and procedure used generally reflects the capability of the medical care available at the time hypothermia is diagnosed. The most effective and efficient method of adding heat to the core should be selected depending on the circumstances. Analysis of data from different rewarming procedures indicate the skill and capability of the treatment team is more important than the choice of rewarming procedures. Active, internal methods include heated humidified air or O_2 inhalation, warm peritoneal dialysis, intravenous solutions, gastric and colonic lavage, extracorporeal rewarming and microwave. Conscious patients below 90° F (32° C) core temperature should have some active, external method of rewarming along with careful monitoring and maintenance of fluid balance. Unconscious patients below 90° F require more care, and an active, internal method of rewarming. The coldest patients, below 70° F (21° C), require extracorporeal rewarming to attain a high success rate.

Water Immersion. During hypothermia, peripheral vasoconstriction decreases blood flow to the extremities and sequesters blood in the core, raising arterial pressure. This rise in arterial pressure is perceived by the baroreceptors which initiate a centrally mediated diuresis, termed cold diuresis. With long-term cold exposure, this diuresis occurs repeatedly decreasing circulating volume. The blood in the periphery, however, is colder than the core and is highly acidotic from shivering and later hyperkalemic. Rewarming the periphery causes surface blood vessel dilation and a central hypervolemia, termed "rewarming shock." There is a rapid return of colder, more acidotic, hyperkalemic blood to the core which often initiates a drop in core temperature resulting in ventricular fibrillation or standstill. To mitigate this potential, it is wise to warm the torso only, keeping blood sequested in the arms, legs, and head during the rewarming process. As torso temperature rises, warm blood from the core will start peripheral vasodilation and the cold peripheral blood will be added to the central circulation slowly. Gastric and colonic lavage provide too small a surface area for a significant heat exchange. Warm intravenous fluids, although wise, provide a small heat input to the large torso mass. CPR procedures are impossible in water-immersed patients.

Heated Humidified Inhalation Rewarming. Air or oxygen can be humidified and heated to 47° C and used without danger to mucous membranes. This procedure offers great promise, but with drawbacks. The majority of heat input is from the precipitation of water in the alveoli and the technique provides a small, total heat input. It can be perceived more as a prevention of further heat loss than adding significantly to the increase in core temperature. Since the total heat input is small, it should be considered as an adjunct to other active rewarming procedures and not a major method of rewarming (6,33).

Peritoneal Dialysis. Warm peritoneal lavage has been used successfully and has significant advantages (42). It can easily be performed in emergency departments, and provides a high surface area for heat transfer along with some control of pH, electrolytes and toxic substances. The dialysis solution should be heated to approximately 45° C, and results in a 1 to 2° C temperature rise per hour in a 70-kilogram man. The technique involves emptying both the bladder and stomach, the introduction of two infraumbilical catheters, one in each colonic gutter and injection of 500 cc of fluid carefully into the abdominal cavity. Warm fluid is instilled in one and drained via the other (46). Peritoneal dialysis has demonstrated a high success rate although peritoneal infection may complicate the procedure.

Microwave Rewarming. Microwave rewarming has great promise, particularly for field management. The use of multihead microwaves to directly rewarm the heart and core organs could be highly advantageous (2). Although still in the experimental stages, technology in microwaves holds potential for providing heat deep under the skin in a controlled noninvasive fashion. This technique awaits controlled studies.

Emergency Department Management. The key to resuscitation is recognition at every level of medical care that these patients are treatable. Prehospital diagnosis depends on a low reading clinical thermometer. Although continuous rectal temperature frequently has been used, esophageal temperature probes are probably the most effective and efficient method of in-hospital management for monitoring core temperature. After temperature measurement is started, cardiac monitoring is applied and catheters for fluid management are placed. Some hypothermics may require cut downs to attain arterial and venous catheterizations. Central venous lines and pulmonary artery catheters may precipitate arrhythmias and should be left out until significant rewarming has been accomplished. Differential diagnoses of injury, cerebral vascular accident, and diabetes should be considered. Screening for thyroid disfunction, alcohol and drug misuse, or combinations of medications should be standard laboratory evaluations. Arterial blood gases, complete blood count, electrolytes, clotting times, platelet counts, and fibrinogen levels should be checked, keeping in mind the differentials.

Fluid Management. Mild hypothermics require only minimal intravenous replacement, while long-term deep hypothermics require more rigorous fluid management because of hemoconcentration, increase in blood viscosity, and decrease in flow rates. The combination of cold diuresis, and an increase in vascular permeability leads to significant volume shifts and hemoconcentration, with a loss of effective perfusion. Intravenous fluid replacement is lifesaving in hypothermia. The choice of fluid should be crystalloid, such as dextrose and saline, 250–500 cc. Lactate solutions should not be used because the liver is incapable of metabolizing lactate below a core temperature of 33° C. Fluid replacement decreases viscosity and increases cardiac output and should be administered prior to field transportation if possible. Fluid can be warmed to 45° C prior to administration. In patients below 33° C core temperature, central venous pressure should be kept low during initial fluid replacement because endothelial cells are incapable of holding fluid in the vascular space. Small incremental increases in CVP are indicated during rewarming as an indication of the fluid remaining in the vascular space. Early, inadvertent, overzealous fluid replacement has resulted in significant edema, compartment syndromes, and fatal cerebral swelling. Use of military antishock trousers should be reserved for stabilization of water immersion patients but not used routinely for hypothermia. Persistent cardiovascular instability may reflect poor intervascular replacement. Pulmonary wedge catheters may induce arrhythmias and ventricular fibrillation during this process. Whole blood products are usually not necessary during initial resuscitation but may be indicated later.

Electrolyte Management. Although recommended for many years, correction of blood gases to temperature is not necessary. After initial respiratory alkalosis from hyperventilation, the process is towards a progressive acidosis caused by respiratory depression. As blood gases cool, cellular and blood CO_2 increases. Overzealous correction of pH is not indicated. Dilution and increased ventilation steadily improve the pH during the rewarming process. Rapid corrections in pH may lead to a decrease in oxygen release to the myocardium with an increased danger of ventricular fibrillation. Small amounts of bicarbonate may be used initially, however. Serum potassium levels may vary significantly during cooling and rewarming.

The lowered core temperature and acidosis often lead to hyperkalemia. With peripheral rewarming and vasodilation there may be a return of large amounts of hyperkalemic blood to the core. Temperature and acidosis severely affect the cellular membrane pumps. Dilution alone improves the hyperkalemia. Rewarming and glucose improve the function of the sodium pump, moving potassium back into the cells. Hypokalemia can occur with rapid rewarming procedures from a rebound production of insulin and a sudden shift of potassium back into muscles. Regular electrolyte screening is indicated particularly with administration of bicarbonate, insulin, and glucose during rewarming. Although hyper- and hypoglycemia have been demonstrated on admission, hypoglycemia is the most threatening. Long-term shivering without oral replacement can lead to hypoglycemia.

Chronic cold exposure often leads to long-term shivering, exhaustion, and glycogen depletion. Hypothermia, however, decreases glucose utilization at the cellular level. Pancreatic release of insulin is reduced below a core temperature of 30° C. The target cells for insulin are insulin resistant and, with this decreased utilization and ineffectiveness of insulin, hyperglycemia may result. Utilization of insulin during resuscitation to improve the sodium cellular pump function, to move potassium back into the cells, and to improve basal metabolic function should be tempered by frequently monitoring glucose. It is advisable to use insulin only if blood sugar measurements do not drop during rewarming. Acute hypothermia causes elevation in blood glucose. Care should be used to prevent hypoglycemia during late rewarming from overuse of insulin earlier in treatment.

Platelet counts and determination of fibrinogen levels are necessary in all but mild hypothermia. Hypercoagulability with disseminated intravascular coagulation (DIC) syndromes are common in severely hypothermic patients. Thrombocytopenias and leukopenia are seen in patients undergoing induced hypothermia. Where these cells go is not clearly understood. Bone marrow suppression and sequestration in both the spleen and liver have been proposed (35). DIC is best managed with frozen blood products although it often reverts on rewarming. Hemodilution during initial rewarming reduces the hyperkalemia, the acid-base balance, increase in viscocity, and DIC present during rewarming.

CPR. Initiation of CPR in those perceived to be pulseless, but unmonitored in a field setting is unwise. There are wide ranges of both time and core temperature in which the individual will be perceived pulseless, but actually has a functional rhythm. Chest compression at that point can initiate ventricular fibrillation. Severe bradycardias are common along with marked peripheral vasoconstriction. Also, a reliable determination of cardiac inactivity is difficult without monitoring devices. Initiation of CPR significantly compromises the rescue effort and the rescuers. When the patient and rescuers are in a vehicle being transported, the decision to initiate CPR may be more clinical than situational. Chest compression on a cold, stiff chest wall; a cold, solid dilated heart, and highly viscous blood portends for minimal cardiac output. American Heart Association CPR standards for hypothermia are as follows: initiate CPR in accidental hypothermia unless: a) a do not resuscitate status is documented and verified; b) obviously, lethal injuries are present; c) chest wall depression is impossible; d) any signs of life are present; and, e) rescuers are endangered by evacuation delays or altered triage conditions. Usual signs of death such as muscle rigidity, mimicking rigor mortis, and fixed, dilated pupils should not be criteria for witholding CPR. An initial screening procedure of small volume positive pressure ventilation will often initiate gasping respiration in the perceived pulseless hypothermic. This procedure can be accomplished throughout evacuation without compromising the rescuers, while providing oxygenation and a change in circulating pH, which are helpful to there suscitation process. Observed cardiac arrest during evacuation necessitates CPR even if it cannot be done continuously.

Emergency room monitored asystole or ventricular fibrillation indicate initiation of CPR. Although the rate and technique is currently unknown, half-rate with a prolonged filling time, because of increased blood viscosity and decreased cardiac compliance, may be indicated. As rewarming proceeds, rate should increase as the chest wall excursion increases and tissue demands increase. If asystole or fibrillation occurs during rewarming, massage at one half the usual compression rate is indicated until the heart is rewarmed to approximately 31° C core temperature at which time post defibrillation can be tried. Attempts to difibrillate below this temperature are often unsuccessful. Early oxygenation, decreasing viscosity, lowering serum potassium and improving the acidosis shield the heart from ventricular fibrillation or standstill during rewarming. Drugs have not been effective in blunting the effect of hypothermia on cardiovascular irritability and collapse.

Pulmonary Support. As previously mentioned, early oxygenation is essential in hypothermia management. Intubation, although once thought to be dangerous in hypothermia, has been proven to be a quite benign procedure. Nasotracheal intubation was accomplished in 40 patients without incident (10,32). Protecting the airways by intubation prevents aspiration and improves ventilatory control. Arterial gases should be used as a guideline for determining oxygen concentration for resuscitation. Continuous, positive airway pressure may be necessary for severely hypothermic and respiratorily depressed patients. Respiration rate and depth should increase as the patient rewarms. Respiratory alkalosis with subsequent ventricular fibrillation is a result of over-ventilation, particularly late in resuscitation.

Pharmacology of Resuscitation. Parenteral medications are poorly absorbed because of peripheral vasoconstriction, and compromised hepatic and renal functions by impairments in organ blood flow. Cold depression of metabolism adds to the suppression of pharmacologic activity of all drugs. A basic dictum should be that medications are not indicated during resuscitation. Attempts to modify cardiovascular tone, blood pressure, and respiratory drive usually are not successful. Low dose dopamine may be used judiciously to control blood pressure. Atrial fibrillation converts spontaneously during rewarming and does not require pharmacological intervention. Correction of fluid, acid-base, and electrolyte balances will improve atrial rhythm.

Minor ventricular dysrhythms should not be thought of as life-threatening and ventricular arrhythmias will disappear with oxygenation and small increments of fluid replacement. Lidocaine has not been effective, either for prophalaxis or treatment of ventricular arrhythmias, while Bertilium tosylate has been extremely effective both in animal and human studies (8,9,13). Limited case reports of beneficial effects of quinidine and magnesium sulphate have been reported. Procainamide, and propanolol should be avoided in these cases. Thyroxine should not be given unless clearly indicated and diagnosed by thryoid function. Corticosteroids have not been shown to increase survivability. Pre-existing adrenal disease may indicate later supplementation.

Infection. Hypothermia appears to induce compromises in the host defenses and subsequent bacterial and/or viral infections. A combination of atalectasis, bronchorea and poor pulmonary function lead to compromising normal pulmonary cleansing. Neutrophil suppression can occur for 12 hours or more with phagocytic activity reduced significantly by low temperature. Although prophylatic antibiotics are generally not necessary, a liberalization from normothermia is indicated. Careful screening with cultures and sensitivities should be continued throughout the first week of management of a hypothermia patient. An outline of the steps in treating the hypothermic victim is illustrated in Table 11-3.

TABLE 11-3. *Outline Of Steps In The Treatment Of Hypothermia*

1. Place EKG and esophageal temperature probe.
2. Intubate after preoxygenation if unconscious and airway problems are present. Start heated humdified oxygen.
3. Implant venous and arterial catheters. Do not touch the heart. Place Foley catheter.
4. Instill 200–400 cc of warmed intravenous fluids. Do not raise CVP quickly in early management.
5. Sodium bicarbanate if pH is below 7.2 or if pH does not rise on rewarming. Insulin if blood sugar is over 400 mg.
6. Initiate rewarming procedures, peritoneal dialysis, torso water immersion, and rewarming blankets.
7. Monitor electrolytes, (potassium), blood gases, glucose, every 15 minutes or every 3° F rise in temperature.
8. Screen for other medical problems, drugs, injuries, and endocrine levels.
9. Perform cultures for potential infections.

SUMMARY

Epidemiological considerations in the development of cold injury involve a number of environmental and host factors. Environmental factors discussed involve temperature, precipitation, and wind. Host factors of importance include smoking, previous cold injury, race, rank, malnutrition, fatigue, environmental protection, leadership, other injuries, and drugs. Human cold injuries may be classified into categories of localized cold/wet injuries, localized cold/dry injuries, and generalized whole body hypothermia. The localized cold/wet injuries are chilblain, pernio, and trenchfoot. Localized cold/dry injuries are frostnip and frostbite. The pathophysiology of localized cold/wet and cold/dry injuries, and hypothermia is presented. This chapter concludes with a section on the treatment of trenchfoot, frostbite, and hypothermia.

REFERENCES

1. Adnot, J. and C.W. Lewis. Immersion foot syndromes. *J. Assoc. Mil. Dermatol.* 11:87–92, 1985.
2. Auerback, P.S., and E.C. Geehr (eds.). *Management of Wilderness and Environmental Emergencies.* New York, NY: Macmillan Publishing Co., 1983, pp. 27–63.
3. Baker, P.T. American negro-white differences in the thermal insulative aspects of body fat. *Human Biol.* 31:316–324, 1959.
4. Blackwood, W. Studies in the pathology of human "immersion foot". *Brit. J. Surg.* 31:329–350, 1944.
5. Boswick, J.A., Jr., J.D. Thompson, and R.A. Jonas. The epidemiology of cold injuries. *Surg. Gynecol. Obstet.* 149:326–332, 1979.
6. Collis, M.S., A.M. Steinman, and R.D. Chaney. Accidental hypothermia: An experimental study of practical rewarming methods. *Aviat. Space Environ. Med.* 48:625–632, 1977.
7. Court, A. Wind chill. *Bull. Amer. Meteor. Soc.* 29:487–493, 1948.
8. Danzl, D.F. Bretylium in hypothermia. *Wilderness Med. Soc.* 4:5–6, 1987.
9. Danzl, D.F., M.B. Sowers, S.J. Vicario, D.M. Thomas and J.W. Miller. Chemical ventricular defibrillation in severe accidental hypothermia. *Ann. Emerg. Med.* 11:698–699, 1982.
10. Danzl, D.F. and D.M. Thomas. Nasal tracheal intubations in the emergency department. *Crit. Care Med.* 8:677–682, 1980.
11. Daum, P.S., W.D. Bowers, Jr., J. Tejada, and M. Hamlet. Vascular casts demonstrate microcirculatory insufficiency in acute frostbite. *Cryobiology* 24:65–73, 1987.
12. Doolittle, W.H. Disturbances due to cold. In: H.F. Conn (eds.). *Current Therapy 1977.* Philadelphia, PA: W.B. Saunders Co., 1981, pp. 917–922.
13. Dronen, S., R.M. Nowak, and M.C. Tomlanovich. Bretylium tosylate and hypothermic ventricular fibrillation. *Ann. Emerg. Med.* 9:335, 1980.
14. Duguid, H., R.G. Simpson, and J.M. Stowers. Accidental hypothermia. *Lancet* 2:1213–1219, 1961.
15. Francis, T.J. Non-freezing cold injury: A historical review. *J. Roy. Nav. Med. Serv.* 70:134–139, 1984.
16. Franz, D. Studies on the response of bone and skin blood flow to acute cold exposure. Baylor College of Medicine, Doctoral Dissertation, 1980.
17. Grattan, H.W. *General History of Medical Services Hygiene of War.* Vol II, London: His Majesty's Stationary Office, 1931.
18. Golden, F. StC., and P.T. Hardcastle. Swimming failure in cold water. *J. Physiol.* 330:60P–61P, 1982.
19. Hedblom, E.E. Disturbance due to cold. In: H.F. Conn (eds.). *Current Therapy.* Philadelphia, PA: W.B. Saunders Co., 1965, pp. 691–696.
20. Iampietro, P.F., J.A. Vaughan, R.F. Goldman, M.B. Kreider, F. Masucci, and D.E. Bass. Heat production from shivering. *J. Appl. Physiol.* 15:632–634, 1960.
21. Jackson, R.L., F.T. Fay, M.W. Sharp, L.J. Fulco and E. Kraus. Cardiovascular (CV) and temperature responses of blacks and whites to hand cooling. *Physiologist* 30:167, 1987. (Abstract)
22. Keatinge, W.R., C. Prys-Roberts, K.E. Cooper, and J. Haight. Sudden failure of swimming in cold water. *Brit. Med. J.* 1:480–483, 1969.
23. Killian, H. Cold injuries with special reference to German experiences during WWI. Unpublished Monograph, Bureau of Medicine and Surgery Navy Dept., Washington, D.C., 1952.
24. Kreyberg, L. Tissue damage due to cold. *Lancet* 1:338–340, 1946.
25. Krogman, W.M. and M.Y. Ischam (eds.). *Human Skeleton in Forensic Medicine.* Springfield, IL: Charles C. Thomas, 1986.
26. Larrey, D.J. Memories of Military Surgery and campaigns of the French Armies on the Rhine, in Corsica, Catalonia, Eygpt and Syria, and Boulogne, Ulm and Austerlitz, in Saxony, Prussia, Poland, Spain and Austria. Translated by Richard Wilmart Hall, 1st Amer. Edition, Baltimore, MD: Joseph Cushing, 1814, pp. 153–164; 205–224.
27. Lewis, T. Observations on some normal and injurious effects of cold upon skin and underlying tissues: Reactions to cold, and injury of normal skin. *Brit. Med. J.* 2:795–797, 1941.
28. Mant, A.K. Autopsy diagnosis of accidental hypothermia. *J. Forensic Med.* 16:126–129, 1969.
29. *Medical and Surgical History of the British Army During War Against Russia in the years 1854-1856.* Vol. II: London. Her Majesty's Stationary Office, 1858.
30. *Medical and Surgical History of the War of the Rebellion, Surgical History.* Vol. II. Washington Government Printing Office, 1883, pp. 670–679.
31. Miller, A.J. Physical fitness for strenuous work in relation to the survival situation in a cold environment. Project 14, 1st Arctic Aeromed. Lab., Ladd Air Force Base, Alaska, 1948.
32. Miller, J.W., D.F. Danzl and D.M. Thomas. Urban accidental hypothermia: 135 cases. *Ann. Emerg. Med.* 9:456–461, 1980.

33. Mills, W.J., Jr. Out in the cold. *Emerg. Med.* 8:134–139, 1976.
34. Myers, R.A., J.S. Britten, and R.A. Cowley. Hypothermia: Quantitative aspects of therapy. *JACEP.* 8:523–527, 1979.
35. O'Brien, H., J.A. Amess, and D.L. Mollin. Recurrent thromobocytopenia erytheroid hypoplasia and sideroblastic anemia associated with hypothermia. *Brit. J. Hematol.* 51:451–456, 1982.
36. Orr, K.D. and D.C. Fainer. Cold injuries in Korea during winter of 1950–51. *Medicine* 31:177–220, 1952.
37. Pandolf, K.B. (ed.). *Exercise and Sport Sciences Reviews.* New York, NY: Macmillan Publishing Co., 1987, pp. 261–295.
38. Personal correspondence and contacts, Hamlet, M.P., 1987.
39. Pozos, R., D. Israel, R. McCutcheon, L.E. Wittmers, Jr., and D. Sessler. Human studies concerning thermal induced shivering, post-operative shivering and cold induced vasodilation. *Ann. Emerg. Med.* 16:1037–1041, 1987.
40. Riddell, D.I. Is frostnip important? *J. Roy. Nav. Med. Serv.* 70:140–142, 1984.
41. Selke, A.C., Jr. Destruction of phalangeal epiphyses by frostbite. *Radiology* 93:859–60, 1969.
42. Soung, L.S., L. Swank, T.S. Ing., R.A. Said, J.W. Goldman, J. Perez and W.P. Geis. Treatment of accidental hypothermia with peritoneal dialysis. *Can. Med. Assoc. J.* 117:1415–1416, 1977.
43. Southwick, F.S., and P.H. Dalglish, Jr. Recovery after prolonged asystolic cardiac arrest and profound hypothermia. *JAMA.* 243:1250–1253, 1980.
44. Sumner, D.S., T.L. Criblez and W.H. Doolittle. Host factors in human frostbite. *Military Med.* 141:454–461, 1974.
45. Toppe, A. Frostbite problems in the German army during WWII, notes compiled 29 August 1950.
46. Ungley, C.C., G.D. Channell and R.C. Richard. Immersion foot syndrome. *Brit. J. Surg.* 33:17–31, 1945.
47. Ungley, C.C. and W. Blackwood. Peripheral vasoneuropathy after chilling "immersion foot and immersion," with note on morbid anatomy. *Lancet* 2:447–451, 1942.
48. Wagstaff, M.A. and R.J. Pethyridge. Cold Injuries Norwegian Winter Deployment 85. Report #20/86. Alverstoke Hants, Great Britain: Institute of Naval Medicine, 1986.
49. Wagstaff, M.A. and R.J. Pethyridge. Cold Injuries Norwegian Winter Deployment 85. Report #9/87, Alverstoke Hants, Great Britain: Institute of Naval Medicine, 1987.
50. Wedin, B., L. Vanggaard and J. Hirvonen. 'Paradoxical undressing' in fatal hypothermia. *J. Forensic Sci.* 24:543–553, 1979.
51. Whayne, T. and M. DeBuakey. Cold injury, ground type. U.S. Army Med. Dept. Superintendent of Documents, U.S. Government Printing Office, Washington, D.C., 1958.

12

Human Performance and Acute Hypoxia

CHARLES S. FULCO, M.A.T.

ALLEN CYMERMAN, Ph.D.

INTRODUCTION

Men have ventured into high mountainous regions from the beginning of recorded history for many reasons, not the least of which was to escape from the ravages of their fellow man. Religious reasons and natural resource exploitation have also had a large influence on early travels to remote high environments. The first recorded description of the possible ill effects due to oxygen deficiency was described in Alexander's troops crossing into India in 326 B.C.

(86). But it was not until the Renaissance that man's increasing curiosity about himself and his surroundings resulted in new knowledge of the properties and chemistry of his natural environment.

It has now been about 300 years since the importance of "oxygen" in mammalian respiration was dramatically shown by John Mayow in mouse-bell jar experiments (45). Significant steps in our knowledge of oxygen and its effects were slowly made over the next 200 years, but a giant step was made in 1878 with the studies of Paul Bert. Bert, the father of modern high-altitude physiology, demonstrated that oxygen deficiency resulted in specific symptoms which were due to the decrease in the partial pressure of oxygen. Bert showed that this decrease could be caused by a reduction of the total barometric pressure from either chamber decompression or mountain ascension (45).

Man is the only animal native to sea level who purposefully subjects himself to the rigors of hypoxic environments, such as those found in the mountainous regions of the world for reasons other than pure survival. In recent times, ascent by sea-level residents using modern conveyances to high altitude regions of the world for trekking, tourism, mountaineering, business, and scientific and military reasons has increased many fold. It is estimated that 4000 tourists ascend daily via cable car from Chamonix, France (1000 m) to Arguille du Midi (3800 m) in 20 minutes, remain for several hours, and then descend (6). Similar numbers are found in the Western Hemisphere but to an even higher altitude. In the 1987 season, approximately 500,000 tourists ascended Pikes Peak, Colorado (1830 m to 4300 m) in about one hour by car, bus, or railway (Personal Communication, National Park Service). The number of skiing and mountain enthusiasts that enjoy their hobbies after rapid translocation from sea-level originations is inestimable. Because of the large numbers of individuals, the elevations to which they are ascending, the speed of ascention, and the relatively high level of associated physical exertion, there is no question that increased numbers of people are being affected by the relative hypoxia.

This chapter considers the acute human physiological responses to hypoxia from both the temporal (<4 hours) and severity domains (inspired oxygen pressure from approximately 110 torr or 2440 m elevation to 70 torr or 5500 m). The following two chapters discuss longer hypoxic exposures and the potential medical problems, respectively. A number of recent reviews, in part or in their entirety, expertly discuss various aspects of the acute responses to reduced partial pressures of oxygen (20,39,69,88). The present chapter will be limited to a discussion of the effects of acute exposure to hypoxia on human performance.

In the present context, "acute" will be defined as 4 hours or less. It should be understood that the designation of 4 hours is quite arbitrary. It is certainly not an objective point in time denoting a significant physiological change. Rather, the 4-hour designation allows this chapter to include much of the information gathered during early periods of hypobaric chamber exposures as well as the information gathered during hypoxic gas breathing. The information presented from field studies will be limited since altitude field studies imply longer exposures, possible acclimatization effects, and uncontrollable factors such as exertion, nutrition, dehydration, etc.

The term "acute" is also often used to mean "of short duration." But does short duration mean seconds, minutes, hours, days, or weeks? Scientific investigations studying hypoxia have been as short as seconds or as long as 4 months. In many of these, "acute" has been useful as a relative term to discribe the earlier period of study in contrast to a later period in the same experiment, usually described as "chronic" or "long-term." While this time-period classification system may be adequate for intra-experimental comparisons, it is not appropriate for comparing "acute" phases of different experiments. For example, in a study lasting several months, "acute" may be defined for descriptive purposes as any time period up to weeks, whereas in a study lasting an hour acute may be defined as minutes.

The problem of defining acute hypoxia is more than an exercise in semantics. Many physiologic responses to hypoxia are initiated almost immediately and are continually readjusting for the entire duration of the exposure. To compare an acute period of exposure lasting minutes to one lasting days would surely lead to inconsistant and contradictory conclusions. Unfortunately, there does not exist an operational definition of "acute" as it relates to hypoxia research, although attempts have been made (22). But, as pointed out by Welch (88), the idea was never widely accepted. It may be appropriate to mention the concepts discussed by Bouverot (6), who used the term "accommodation" to refer to hypoxic exposures resulting from simulated or real exposures lasting from seconds to hours depending on the severity. We will use the term "acute" since, medically speaking, "accomodation" connotes a relatively rapid return toward a normal state as occurs with the visual, olfactory, and subthreshold nerve stimulation. "Acclimation" and "acclimatization" can be accurately associated with simulated and real exposures lasting from days to months, respectively.

The study of the effects of hypoxia requires a reduction in the partial pressure of ambient oxygen by decreasing either the oxygen concentration of the inspired gas or the ambient barometric pressure. While each procedure is thought to elicit the same responses (33,70), it is often desirable to use one method in preference to another. In experiments where immediate responses to hypoxia are being studied, breathing a hypoxic gas mixture may be warranted. If acclimatization to hypoxia is of primary interest, experiments conducted in a high-altitude field laboratory for a number of weeks would be appropriate. Hypobaric chambers, capable of housing a limited number of subjects, are ideal for: (a) experiments lasting from several hours to several days; (b) experiments where the level of hypoxia will be varied; or (c) for use of experimental procedures too risky to perform in a field laboratory.

Basic Concepts

An uninterrupted supply of oxygen must be made available to man in order for him to function for more than a few minutes. Each cell of the body receives oxygen and rids itself of carbon dioxide. However, because of the distance between most cells and the ambient air, the cells cannot exchange these gases directly. They must rely on a delivery and exchange system that includes: movement of gases between the ambient air and the lungs; matching of ventilation with blood flow; diffusion between alveolar air and capillary blood; vascular transportation from the lungs to the tissues; and finally, diffusion between capillary blood and the tissues.

At each stage of gas exchange, oxygen and carbon dioxide readily diffuse "down" their pressure gradients. At sea level, the partial pressure of oxygen (PO_2) in the ambient air is 159 torr (760 torr * 20.95% O_2). As the inspired air passes through the respiratory passages, where it becomes totally saturated with water, the PO_2 will have been reduced by the partial pressure of water vapor (47 torr at a body temperature of 37° C) to a value of 149 torr ((760 − 47 torr) * 20.95%), and is designated as P_IO_2. P_IO_2 is further reduced to approximately 104 torr in the alveolar air because of incomplete replacement of alveolar air with atmospheric air as well as the fact that oxygen is constantly diffusing out of the alveoli into pulmonary capillaries. Since the PO_2 of venous blood entering the pulmonary capillary is 40 torr, the 64 torr pressure gradient (104 − 40 torr) causes oxygen to rapidly diffuse into the capillary. In approximately 0.25 sec, pulmonary arterial hemoglobin will become almost completely saturated with oxygen (89). Even with a reduction in transit time through the capillaries, as would occur with an increase in cardiac output during exercise, the PO_2 on the arterial side of the capillary will be almost equal to the PO_2 of the alveolar air (89). Because of the presence of veno-arterial shunts, unequal matching of ventilation with perfusion, and to a lesser extent, the actual physical diffusion barrier of the pulmonary and circulatory membranes, not all of the blood becomes totally oxygenated. The PO_2 of the blood that is actually pumped by the left heart into the aorta and the peripheral arteries is reduced to approximately 95–97 torr. Oxygen then diffuses rapidly from the blood into the interstitial fluid where the PO_2 is 38–40 torr and continues into the cells where the average intracellular PO_2 is 20–25 torr (1,47,48).

Carbon dioxide diffuses in the opposite direction at each level of gas exchange. But because carbon dioxide diffuses 20 times as rapidly as oxygen, the pressure differences are much less than are required for oxygen to effect adequate transfer. Thus, the intracellular partial pressure of carbon dioxide (PCO_2) of about 46 torr provides a sufficient pressure gradient for diffusion of CO_2 into the interstitial spaces (45 torr) and subsequently into the tissue capillaries where the PCO_2 of the arterial blood is 40 torr. The PCO_2 of the venous blood leaving the tissue capillaries and entering the pulmonary capillaries is about 45 torr. The CO_2 then diffuses into the alveolar air (40 torr) and is subsequently released into the atmosphere (1,47,48). CO_2 is transported in a physically dissolved state (7–10%), by the bicarbonate ion after conversion from carbonic acid (60–70%), and by its combination with hemoglobin and plasma proteins (23–30%) (1,89).

As man ascends from sea level to high altitude, the P_IO_2 is diminished in direct proportion to the reduction in atmospheric barometric pressure. For example, at 4300 meters, the atmospheric pressure is about 448 torr, and the PO_2 of the tracheal air is 84 torr ((448 − 47 torr) * 20.95%) and the arterial PO_2 (P_aO_2) is approximately 40–45 torr. Thus, the oxygen pressure gradient for diffusion into the pulmonary capillary is 39–44 torr, considerably less than the sea-level value. The PO_2 gradient at each level of gas exchange is also reduced. The result is an impaired ability to transport oxygen from the atmosphere to the cell. Table 12-1 lists the partial pressures of oxygen at several stages of the oxygen cascade for five different al-

TABLE 12-1. *Typical acute reduction in arterial PO_2 and saturation with increasing altitude for unacclimatized men.*[a]

Elevation m	Elevation ft	P_B torr	P_IO_2 torr	P_AO_2 torr	P_aO_2 torr	P_ACO_2 torr	S_aO_2 %
0	0	760	149	104	96	40	96
1600	5248	627	122	82	69	36	94
3100	10169	522	100	62	57	33	89
4300	14104	448	84	53	40	30	84
5500	18000	379	70	38	35	27	75

[a]references (46, 48, 64, and 76).

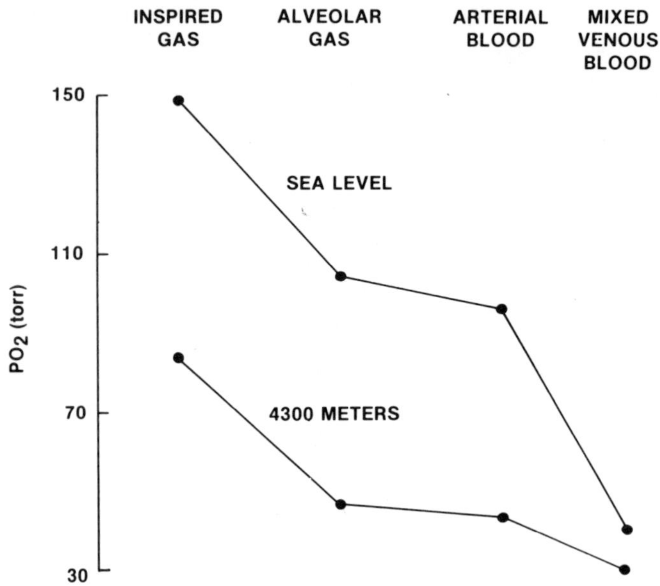

Figure 12-1. *Oxygen transport cascade at sea level and at 4300 m. Modified from West (89).*

titudes. Figure 12-1 graphically illustrates the reduction in the pressure gradient at each step of the oxygen transport chain for an altitude of 4300 m.

RESTING VENTILATION

An increase in resting minute pulmonary ventilation can be detected when the P_IO_2 drops below 110 torr or the P_aO_2 is less than 60 torr (20,51,65). Beyond this hypoxic "threshold", resting ventilation rises as the level of hypoxia increases (20,51,62). The increase in ventilation serves to increase the partial pressure of alveolar O_2 (P_AO_2) and reduce the partial pressure of alveolar CO_2 (P_ACO_2). Consequently, the PO_2 gradient between atmosphere and alveolar air is narrowed but the PO_2 gradient to which the capillary blood is exposed is widened allowing an increased arterial oxygen saturation.

If minute ventilation is recorded within the first few minutes of a rapid exposure to an ambient PO_2 of less than 100 torr ($P_aO_2 < 55$ torr) and the values obtained are compared to values measured after the first few minutes (while at the same level of hypoxia), then the latter values will be lower but still above the ventilatory values obtained in normoxia (51,65). This blunting of the ventilatory response is thought to be caused by hyperventilation-induced hypocapnia which decreases both the central and peripheral hypoxic drives (20,47,51,79,87) and varies widely from person to person (61). Furthermore, a recent study by Easton *et al.* (25) indicated that the blunting of ventilation still occurs even when hypocapnia is prevented. Thus, although the exact mechanisms and interactions responsible for the relative reduction in ventilation are still to be determined, there appears to be a balance between peripheral and central mechanisms increasing and inhibiting ventilation (20). Excellent reviews of the acute modulation of ventilation by hypoxia in humans and animals are given by Lahiri (53), Lahiri and Gelfand (52), and Dempsey and Forester (20).

The initial increase in pulmonary ventilation during hypoxic exposure is associated primarily with an increase in tidal volume. If the exposure is extended or the level of hypoxia more severe, the frequency of breathing will also increase (33,65). Enhancement of the depth and frequency of breathing causes a more complete turnover of alveolar gases with atmospheric air, effectively limiting the drop in the P_AO_2.

Chemoreceptor Control

A myriad of both direct and indirect experimental evidence indicates that the peripheral chemoreceptors within the carotid and aortic bodies mediate the increase in ventilation during acute exposure to hypoxia (20). No other known mechanism is involved in the initial phase of exposure. The most compelling evidence, perhaps, is that total denervation of these structures significantly attenuates or completely abolishes the ventilatory response to acute hypoxia (53). Chemoreceptors are stimulated when the P_aO_2 and not the arterial oxygen content (C_aO_2) is reduced (1). Blood flow through the chemoreceptors is extremely high, allowing any reduction in P_aO_2 to be sensed immediately with a resultant increase in chemoreceptor afferent firing activity that is proportional to the decrease in P_aO_2 (20). The transmission rate to the central respiratory center is greatly increased when the P_aO_2 is reduced below 60–65 torr (20). This is approximately the PO_2 level marking the beginning of the "steep" portion of the oxygen equilibrium curve.

The increase in ventilation due to hypoxic stimulation of the peripheral chemoreceptors causes a reduction in P_aCO_2 with a resultant alkalosis which tends to oppose and limit the rise in ventilation. As will be discussed in the next chapter, the respiratory center gradually (over a period of days) becomes adapted to the reduction in P_aCO_2 resulting in a greater ventilatory response to any submaximal level of oxygen consumption.

The exact mechanism by which a low P_aO_2 stimulates the chemoreceptors is not known. Direct stimulation of the respiratory center with hypoxic arterial blood has no effect (53). Neural output from the chemoreceptors to the medulla for a given reduction in P_aO_2 is not diminished with prolonged periods of hypoxia (53). There is also a large inter-individual range of ventilatory responses to hypoxia suggesting that differing sensitivities of chemoreceptors exist from individual to individual (55).

GAS EXCHANGE AND TRANSPORT

Once oxygen has entered the alveolar sacs, the next step is diffusion across the alveolar-capillary membrane to oxygenate the arterial blood. There are two important factors that can affect how complete the gas exchange will be and, thus, how well the arterial blood will be oxygenated. One is the actual process of diffusion occurring between functional alveoli and their immediately adjacent capillaries. The other is an optimal matching of ventilation with blood flow within various regions of the lung.

Diffusion and Ventilation/Perfusion at Rest

The pressure gradient between P_AO_2 and the P_aO_2 plays an important role in the rate of diffusion of oxygen through the alveolar-capillary membrane. During rest at sea level, the P_AO_2 is approximately 100–104 torr and that of the venous capillary PO_2, 40 torr. Within about 0.25 sec, before the blood exits the capillary, the PO_2 of the arterial side of the capillary will be in equilibrium with the P_AO_2. With acute exposure to altitude, the partial pressure gradient across the alveolar-capillary membrane is lessened. At 448 torr or 4300 m, the P_AO_2 is reduced to 53 torr. The venous capillary PO_2 is correspondingly decreased to about 20–30 torr, reducing the pressure gradient to the capillary to 23–33 torr. The rate that oxygen diffuses across the membrane is directly related to the PO_2 gradient and, therefore, is lessened in hypoxia. Because of the reduced rate of diffusion, the potential for pulmonary arterial blood to become equilibrated, while still in proximity to the functional alveoli, is also lessened. However, the time that it takes to become equilibrated, although longer in hypoxia (approximately 0.5 sec), is still less than the total pulmonary transit time estimated to be 0.75 seconds. Thus, in the normal lung during acute exposure to a moderate level of hypoxia, the ability to reach equilibrium does not seem to be a problem. This makes it unlikely that a diffusion limitation to the transport of oxygen exists at rest (80). At extreme altitudes (>7600 m) however, a diffusion limitation at rest is detectable in some individuals (85). Figure 12-2 illustrates the difference in equilibrium times at rest between sea level (A) and 4300 m (B).

Edema formation would also interfer with the rate of diffusion and increase the time necessary for pulmonary arterial blood to become as fully oxygenated as possible. There is some evidence which suggests that subclinical pulmonary edema may occur after as little as 2 hours of decompression to 447 torr (12).

The efficiency with which gas exchange occurs is also dependent upon the distribution of the ratio of alveolar ventilation to blood perfusion ($\dot{V}_A/\dot{Q}$) throughout the lung (55). At rest, not all alveoli are ventilated equally nor is blood flow through the alveolar capillaries the same for each alveolus. Therefore, alveolar ventilation and alveolar capillary blood flow are usually not distributed uniformly. The upper lobes of the lung are ventilated more but are perfused less than the lower lobes. The $\dot{V}_A/\dot{Q}$ ratios of the various regions within the lung thus range from high values in the upper lobes to

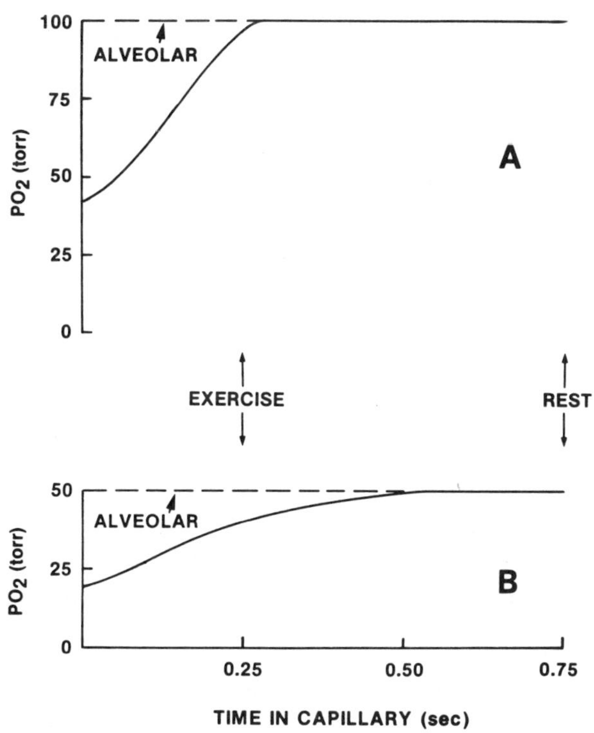

Figure 12-2. *Time course for arterial PO_2 to become equilibrated with alveolar PO_2 in a pulmonary capillary at rest and exercise, at sea level (A) and altitude (B). Modified from West (89).*

lower values in the lower lobes. The measured, overall $\dot{V}_A/\dot{Q}$ ratio, then, actually represents a mean value of many regional and intraregional $\dot{V}_A/\dot{Q}$ relationships throughout the lung.

At altitudes above 2500 m., the increase in alveolar ventilation is matched by an increase in pulmonary perfusion in poorly perfused areas of the lung (e.g., apices), secondary to an increase in pulmonary artery pressure (56,89). The net effect is an enlarged surface area for gas exchange. While an improvement in perfusion would seem to lead to an enhancement of the $\dot{V}_A/\dot{Q}$ distribution, such is not always the case, suggesting that the functional, intraregional $\dot{V}_A/\dot{Q}$ relationships worsen (33,40). In other words, the more uniform, overall topographical distribution of perfusion is offset by a number of individual intraregional $\dot{V}_A/\dot{Q}$ mismatches so that the overall $\dot{V}_A/\dot{Q}$ relationship at rest does not change with altitude exposure (33,84).

Diffusion and Ventilation/Perfusion during Exercise

During heavy exercise at sea level, arterial oxygen saturation (S_aO_2) is slightly, if at all, reduced from resting values (1). In hypoxia, heavy exercise causes a marked desaturation (33,42,78,84). The magnitude of the desaturation during exercise under hypoxic conditions is directly proportional to the level of hypoxia (see Figure 12-3). Desaturation due to an increase in alveolar-arterial PO_2 difference [$(A\text{-}a)DO_2$] during heavy exercise at altitude, can be attributed to either a diffusion limitation, a worsening $\dot{V}_A/\dot{Q}$ relationship, or an increased amount of shunted blood bypassing the oxygenation process. While it is theoretically possible that each of these factors plays a role in the observed increase in $(A\text{-}a)DO_2$, it has been difficult to distinguish their relative contribution.

In recent years, the expanded use of the multiple inert gas elimination technique has allowed such discrimination (33,41,80,84). Gale *et al.* (33) investigated the effects of acute exposure to three levels of hypobaria (632, 523 and 429 torr) on $\dot{V}_A/\dot{Q}$ relationships during exercise. They reported increases in $\dot{V}_A/\dot{Q}$ mismatching at 523 and 429 torr but not at 632 torr. At 429 torr, during exercise, one third of the $(A - a)DO_2$ was attributed to a $\dot{V}_A/\dot{Q}$ mismatch and the remainder to alveolar-end-capillary diffusion disequilibrium. They also reported that the $\dot{V}_A/\dot{Q}$ mismatch was completely reversed by breathing 100% O_2. This finding suggested that the degree of $\dot{V}_A/\dot{Q}$ inequality is related to the degree of pulmonary hypertension, possibly due to nonuniform hypoxic pulmonary vasoconstriction. Torre-Bueno *et al.* (80) showed that when oxygen uptake ($\dot{V}O_2$) exceeded $1.0\ l \cdot min^{-1}$ at 523 torr and $0.5\ l \cdot min^{-1}$ at 429 torr, the $(A\text{-}a)DO_2$ widened. The contribution of a diffusion limitation was found

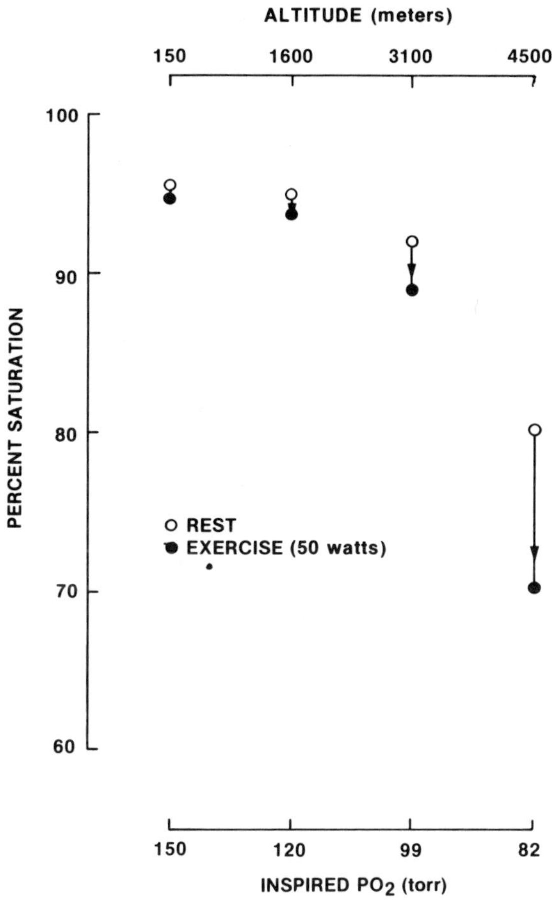

Figure 12-3. *Effect of increasing altitude (or reducing P_IO_2) on arterial saturation at rest (open circles) and during exercise (closed circles). Modified from Banchero (2).*

to be slightly greater than the $\dot{V}_A/\dot{Q}$ mismatch to the increase in (A-a)DO_2. Post-pulmonary shunt was excluded as a contributing factor to the increase in (A-a)DO_2.

The increasing importance of a diffusion limitation was further demonstrated by Wagner *et al.* (84) who showed that the contribution of the diffusion limitation to the total (A-a)DO_2 increases progressively with graded exercise at 523 and 429 torr, while the contribution of the $\dot{V}_A/\dot{Q}$ mismatch to (A-a)DO_2 became proportionally less. They were also able to show that $\dot{V}_A/\dot{Q}$ was related to mean pulmonary arterial pressure and to confirm previous work (85) showing that the contribution of post-pulmonary shunt towards the

$(A\text{-}a)DO_2$ is minimal at altitude. The data from all of these studies suggest that during exercise at altitude, the primary pulmonary factor interfering with gas exchange and increasing the $(A\text{-}a)DO_2$ is an alveolar-end-capillary diffusion limitation.

Transportation of Oxygen in the Blood

Oxygen diffuses into the pulmonary capillary where it is carried by the blood to the tissues. The oxygen is carried both dissolved in the aqueous phase, and in a loose, reversible chemical combination with hemoglobin. Transportation of oxygen to the tissues in physical solution accounts for about 1% of the total oxygen requirement in normoxia and is even less of a factor with increasing levels of hypoxia. The amount of oxygen in the dissolved state is insufficient to supply tissue needs, even at rest. It is, therefore, evident that most (>99%) of the oxygen is transported in combination with hemoglobin.

Hemoglobin combines extremely rapidly with oxygen in the lungs (less than 0.01 sec) and transports the oxygen to the tissues where the oxygen is unloaded. A sigmoidal-shaped oxyhemoglobin equilibrium curve describes the relationship between the P_aO_2 and the completeness with which the hemoglobin is saturated with oxygen. Blood temperature, blood pH, arterial PCO_2 (P_aCO_2), and 2,3 diphosphoglycerate (2,3 DPG) can all affect the position of the curve and thereby alter the relationship between the P_aO_2 and saturation.

Acute exposure to hypoxia causes a shift of the standard (pH, 7.4; PCO_2, 40 torr; temperature, 37° C) oxygen equilibrium curve to the right. The amount of displacement is proportional to the level of hypoxia (54). The increase in ventilation occurring during exposure will concurrently decrease P_ACO_2 and P_aCO_2 with a resultant respiratory alkalosis, all of which cause a leftward shift. However, as a consequence of the hyperventilatory alkalosis, one of the rate controlling enzymes of red cell glycolysis, phosphofructokinase (PFK) is stimulated and causes an increase in 2,3 DPG production which, in turn, results in a rightward shift in the oxygen equilibrium curve. Thus, the right shift in the curve observed in hypoxia is actually the net result of 2 competing influences. The increase in 2,3 DPG is thought to be the main cause of the right shift, because without it the shift does not occur (55). However, it should also be noted that without an increase in pH, an increase in 2,3 DPG will not occur (54).

At first, it would appear that a rightward shift would be advantageous since oxygen release to the tissues would be enhanced. For any given P_aO_2, a rightward shift would present to the tissues a greater PO_2. However, an increase in 2,3 DPG also makes it more difficult for oxygen to combine with the hemoglobin in the lungs. Thus, the advantage at the tissues is lessened by the disadvantage in the lungs. At levels of hypoxia less than that found at altitudes of 3500 m, there is a net advantage of a right-ward shift (55). At very high altitudes, the inability to pick up oxygen at the lungs may outweigh the advantage provided to the tissues.

Blood to Tissue Oxygen Transport

Oxygen travels from the peripheral capillaries, through the interstitial and intracellular fluids, and finally into the mitochondria. Close to the mitochondria, the PO_2 may be 10 torr in normoxia, and about 5 torr at 5500 meters. To adequately support oxidative reactions, a minimum of 1 to 3 torr oxygen pressure gradient is required between the cytoplasm and the mitochondria while a 10-torr PO_2 difference may be needed between the plasma and the cytoplasm (11,69). Apparently, only during heavy exercise in extreme hypobaria (240 torr) are these values approached and to be considered as possibly limiting (67).

EXERCISE

Maximal Cardiorespiratory Responses

Man's ability to perform muscular work is usually evaluated relative to his maximal oxygen uptake ($\dot{V}O_2max$) measured during severe exercise leading to exhaustion in a brief period of time (typically 3 to 6 minutes). During acute exposure to altitudes greater than 1500 m (720 torr) or during breathing of hypoxic gas mixtures with less then 17.5% O_2, the maximal oxygen uptake ($\dot{V}O_2max$) of sea-level residents is reduced (39,73,88). Buskirk (10), after compiling data from a number of studies, established a general relationship between altitude and $\dot{V}O_2max$. As illustrated in Figure 12-4, $\dot{V}O_2max$ is not measurably altered between sea level and about 1500 meters. Above 1500 meters, there is a linear decrease in $\dot{V}O_2max$ at the rate of 10% per 1000 meters. Thus, at an altitude of 4300 meters (or by breathing a gas mixture containing approximately 12.25% oxygen), $\dot{V}O_2max$ will be reduced by an average of 27–28%. It should be strongly emphasized however, that at any given level of hypoxia, there is wide range in $\dot{V}O_2max$ decrements among individuals. For example, Young *et al.* (90) compiled and analyzed the $\dot{V}O_2max$ data collected from several different studies conducted at an altitude of 4300 meters. All subjects (n = 51) were males, young (18 to 31 years), and sea-level residents. The average $\dot{V}O_2max$ value was 48 $ml \cdot kg^{-1} \cdot min^{-1}$ at sea level (range: 36 to 60 $ml \cdot kg^{-1} \cdot min^{-1}$). At 4300 meters, there was an average decrement in $\dot{V}O_2max$ of 27% with a

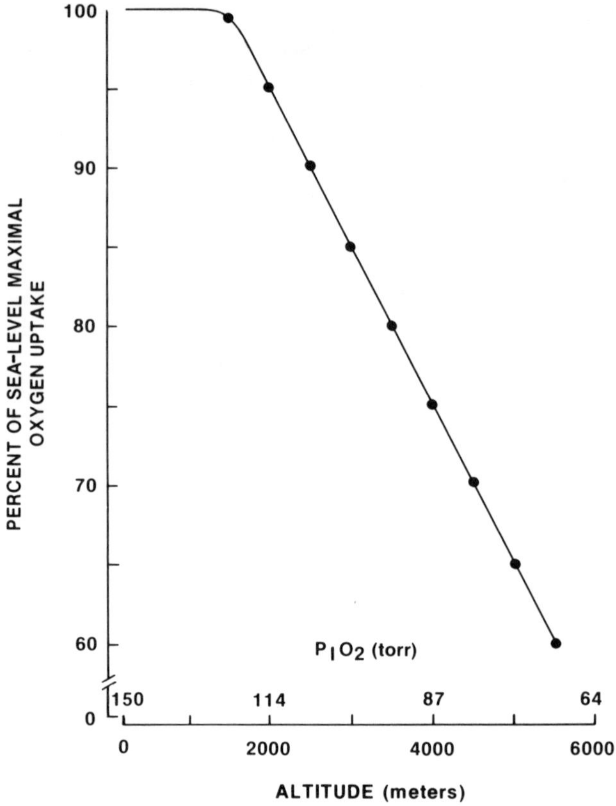

Figure 12-4. *Percent reduction in $\dot{V}O_2max$ as a function of altitude exposure. Note that $\dot{V}O_2max$ is not measurably altered until the altitude exceeds 1500 m (or inspired PO_2 is less than 120 torr). Above 1500 m, there is a linear decrease in $\dot{V}O_2max$ at the rate of 10% per 1000 m. Adapted from Buskirk (10).*

distribution of values ranging from 9% to 54%. Young *et al.* (90) also showed that the decrement in $\dot{V}O_2$max is directly related (albeit weakly) to an individual's fitness level at sea level.

Whether females have the same magnitude of hypoxic-induced decrement in $\dot{V}O_2$max as males is the subject of some debate (24,28,63). Recent data reported by Paterson *et al.* (63) suggests that females may be better able than males to limit the reduction in $\dot{V}O_2$max at altitude. While breathing a gas mixture containing 11.81% oxygen (approximating 4500 meters), the three males in the study (63) had a reduction in $\dot{V}O_2$max of 29.5% and the four females, a 24% reduction. Although there was a significant statistical difference in decrement between genders, caution must be used in making conclusions from these results for three major reasons. First, the number of individuals studied was very small. Second, the males and females were not matched in terms of $\dot{V}O_2$max at sea level. The males had a much higher $\dot{V}O_2$max value than the females in normoxia (62.5 $ml \cdot kg^{-1} \cdot min^{-1}$ vs. 49.8 $ml \cdot kg^{-1} \cdot min^{-1}$). The implication of this observation is that the larger decrement in $\dot{V}O_2$max of the males may be a reflection of a more aerobically fit individual incurring a larger decrement in $\dot{V}O_2$max under hypoxic conditions rather than a gender difference *per se* (90). Third, the mean difference (5.5 percentage points) between the males and females does not seem to be so large that it cannot be accounted for by a normal intersubject variation. For example, in two studies (24,28) which were performed in approximately the same level of hypoxia (12.6–12.8% O_2; 3962–4100 meter equivalents), the mean decrements in $\dot{V}O_2$max for the females were 26.7% in one (24) and 18.0% in the other (28) suggesting that females, like males, have a large intersubject variation in $\dot{V}O_2$max decrement. When these three points are taken into consideration, and the comparative data are looked at closely, it is likely that there are no gender differences in terms of $\dot{V}O_2$max decrement in hypoxia.

Oxygen uptake is the product of the total quantity of blood presented to the tissues (cardiac output) and the difference in the oxygen content of the arterial blood (C_aO_2) and venous blood (C_vO_2). With acute exposure to hypoxia, the maximal values for heart rate, stroke volume, cardiac output and C_vO_2 are not reduced from sea level values (30,35,39,42,74,82). Therefore, the reduction in $\dot{V}O_2$max must be directly related to the amount of oxygen in the arterial blood available to the working muscles. This is indeed the case. Stenberg *et al.* (74) showed that subjects exposed to 4000 meters had a 28% reduction in $\dot{V}O$max and a 26% reduction in C_aO_2. Similar results have been observed repeatedly (30,35,83). C_aO_2 is determined by hemoglobin concentration and the saturation of hemoglobin. Since the hemoglobin concentration is not altered during acute exposure to hypoxia, C_aO_2 is directly related to the saturation of hemoglobin which is a function of the characteristics of the oxygen equilibrium curve and the P_aO_2.

When the P_aO_2 is greater than about 60 torr (on the flat portion of the curve), hemoglobin will be greater than 90% saturated with oxygen. At this saturation level and above, C_aO_2 (and thus, $\dot{V}O_2$max) is affected slightly. Once the P_aO_2 drops below 60 torr (steep portion of the curve), a small decrease in P_aO_2 will cause a large reduction in hemoglobin saturation resulting in a relatively large drop in C_aO_2.

Exercise (1), as well as a hypoxic exposure (54,55), causes a rightward shift of the oxygen equilibrium curve. The shift is pronounced in blood at the site of the working muscle owing to a local increase in temperature, a reduction in pH, and a release of phosphate ions (1). This local shift obviously enhances the release of oxygen from hemoglobin to the working muscles at a lower mean capillary partial pressure of oxygen (55).

Maximal pulmonary ventilation during acute exposure to hypoxia is similar to sea level values (24,28,30,32,63,74) even at extreme altitudes (19). Since $\dot{V}O_2max$ is reduced at altitude, the maximal ventilatory equivalent ($\dot{V}_E/\dot{V}O_2$) will be higher. Although inefficient from a metabolic point of view, such hyperventilation is thought to be essential in defending P_AO_2 and thus facilitating the diffusion of oxygen (19).

Although the maximal values for heart rate, stroke volume, cardiac output, C_vO_2, ventilation as well as lactate (4,74) obtained during an acute exposure to hypoxia do not differ from normoxia, each of these values are reached at a lower maximal exercise intensity. Not surprisingly, the magnitude of the reduction in maximal exercise intensity is similar to the magnitude of reduction in $\dot{V}O_2max$ (32,56).

Submaximal Cardiorespiratory Responses

During rest and submaximal exercise in acute hypoxia, the reduced C_aO_2 is compensated for by an increase in cardiac output such that oxygen uptake at rest or for any fixed (i.e., absolute) exercise intensity does not differ from sea level (42,74,84). The increase in submaximal cardiac output is due primarily to an increase in heart rate since stroke volume during acute exposure is only slightly affected (5,42,51,74,84). The increase in heart rate above sea-level values both at rest and during submaximal exercise is evident within the first hour of hypobaria (31,74). The initial increase in heart rate seems to result from stimulation of the cardiac beta-adrenergic receptors by the cardiac sympathetic nerves (39,68).

The oxygen uptake elicited by the fixed exercise intensity actually represents a greater fraction of the $\dot{V}O_2max$ in hypoxia relative to sea level. Therefore, the relative stress on the body will be greater to a degree that is proportional to the hypoxia-induced reduction in $\dot{V}O_2max$. For example, cycling at an exercise intensity of 150 watts requires a $\dot{V}O_2$ of about 2.1 $l \cdot min^{-1}$ in both normoxic and hypoxic conditions (1). If a person who has a $\dot{V}O_2max$ of 4.0 $l \cdot min^{-1}$ in normoxia and a $\dot{V}O_2max$ of 2.8 $l \cdot min^{-1}$ in hypoxia (measured while breathing 12% O_2 or while at a barometric pressure of 430 torr), pedals an ergometer at an exercise intensity of 150 watts under both conditions, he will be exercising at 53% of his $\dot{V}O_2max$ in normoxia and at 75% of his $\dot{V}O_2max$ during hypoxia. Obviously, because of the greater relative stress, endurance time for performing submaximal exercise at a fixed exercise intensity will be reduced in hypoxia (27).

Although the $\dot{V}O_2$ will be identical with a fixed exercise intensity, the submaximal values for ventilation, heart rate, cardiac output, O_2-deficit, O_2-debt and lactate accumulation will be higher during the hypoxic condition (1,30,39,50,74,88). The difference in response of each of these variables between the two ambient conditions becomes more marked as the exercise intensity or the level of hypoxia is increased (50,74). The response of pulmonary ventilation to an increase in exercise intensity during exposure to altitude is shown in Figure 12-5.

If an exercise intensity is reduced so that the associated $\dot{V}O_2$ elicits the same percentage of $\dot{V}O_2$max in acute hypoxia as it did in normoxia, the values for heart rate, ventilation, O_2-deficit, O_2-debt,

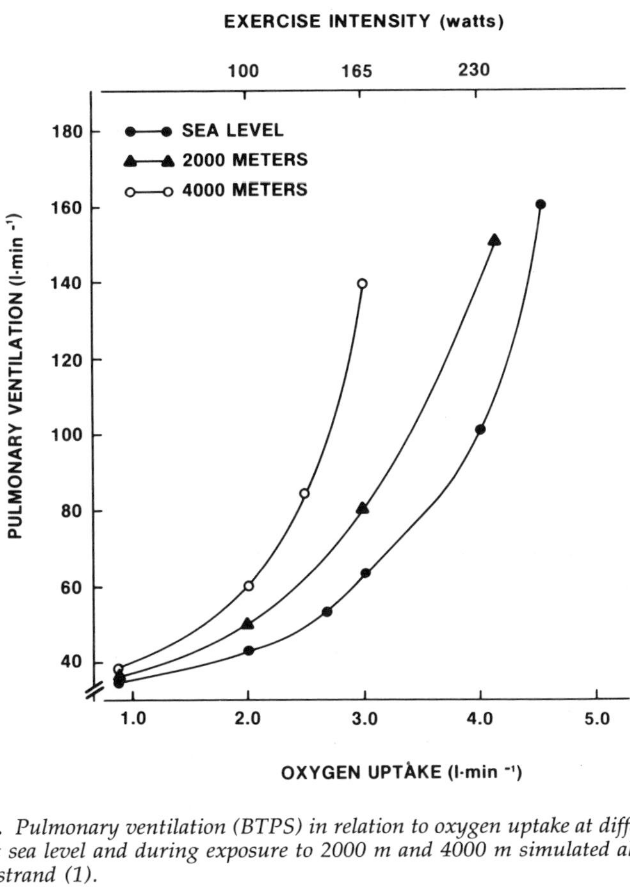

Figure 12-5. *Pulmonary ventilation (BTPS) in relation to oxygen uptake at different exercise intensities at sea level and during exposure to 2000 m and 4000 m simulated altitude. Modified from Åstrand (1).*

respiratory quotient, lactate accumulation, and endurance time to exhaustion will be similar between normoxic and hypoxic conditions (4,29,50,56).

Muscular Strength and Endurance

There is little information available about the effects of acute hypoxic exposure on muscle strength and muscle endurance. Furthermore, much of the information that is available is contradictory (7,9,26,91). Bowie and Cumming (7) found that sustained handgrip strength was unaffected by hypobaric exposure to 480 torr (P_IO_2 = 101 torr) but when two of the subjects breathed 9% O_2 (P_IO_2 = 68 torr), grip time actually increased above the normoxic baseline by 12%. Burse *et al.* (9) found isometric handgrip strength to be increased and handgrip endurance time to fatigue reduced during acute exposure to 4300 meters (448 torr). Eiken and Tesch (26) studied the influence of hypoxia (11% O_2, P_IO_2 = 83 torr) on the quadriceps muscle to repeated dynamic and sustained static contractions (knee extensions). They reported that hypoxia impaired muscle performance in both dynamic and sustained static exercise. Young *et al.* (91) have conducted what has been the most complete strength assessment on humans during acute exposure to hypobaria (429 torr). A series of seven strength assessments, which included measures of both static and dynamic strength of a number of muscle groups, were performed. Measurements were made of isokinetic (180°, 36°, and 0° sec^{-1}) knee extensor strength; isometric strength of the upper torso, knee, and trunk extensor muscles; and strength endurance of knee extensor and elbow flexor muscles during repeated isokinetic contractions. The results indicated that none of the strength measurements were statistically different from normobaria. Thus, the authors concluded that muscle strength was unimpaired by acute hypobaric exposure.

NEUROENDOCRINE AND METABOLIC RESPONSES TO HYPOXIA

Rest

Exposure to acute hypoxia causes a number of complex and interrelated physiological responses affecting not only respiratory, cardiovascular, and metabolic functions, but endocrine functions as well. Such widespread and orchestrated control implies that the adjustments to hypoxia involve the nervous system. This is indeed correct. Primary responsibility seems to lie with the sympathetic portion of the autonomic nervous system. Numerous studies have shown that catecholamine concentration in the plasma and urine are increased within a day of exposure to hypoxia and altitude (18,31,57). Furthermore, the increase in catecholamine levels is almost totally due to an increase in norepinephrine and not epinephrine (18,31). This observation indicates that the rise in norepinephrine is due to increased release from terminal endings of the sympathetic nerves only. A rise in epinephrine as well as norepinephrine would suggest that adrenal medulla release is also involved. Thus, a rise in circulating plasma norepinephrine has been determined to be a useful, though indirect estimation of overall sympathetic hyperactivity (36).

At rest, a significant increase in plasma catecholamines to a hypoxic stress is usually not measurable within the first several hours of exposure. This is not to imply that the activity of the sympathetic nervous system is not enhanced during the onset of hypoxia. On the contrary, an increase in sympathetic activity stimulating the cardiac beta-adrenergic receptors through the cardiac sympathetic nerves is apparent within minutes of exposure to hypoxia (68). The reason for a lack of a measurable increase in plasma norepinephrine during early exposure to hypoxia may be that the rate of removal of norepinephrine exceeds the rate of secretion. During the first several hours of exposure, the small but gradually increasing amount of norepinephrine that is secreted does not exceed the rate of removal due to the combined effects of re-uptake of the released norepinephrine by the nerve endings and the degradation by the enzymes monoamine oxidase (in nerve endings) and catechol-O-methyl transferase (in liver). In fact, it has been shown that splanchnic removal rate of norepinephrine, either during normoxia or hypoxia (10% O_2), rises in proportion to its arterial concentration (3). When the rate of noradrenergic outflow greatly exceeds the rate of re-uptake and degradation, norepinephrine concentration in the blood is significantly raised. This scenario may partially explain why an increase in resting plasma levels of catecholamines is not usually observed until 14–18 hours of exposure to 4300 m (31,57). At plasma concentrations above approximately 1.8 $ng \cdot ml^{-1}$, the transmitter has a more widespread action, similar to a circulating vasoconstrictor hormone (71).

Other metabolic and endocrine responses to acute exposure to altitude or hypoxia are also little affected during rest. Blood concentrations of glucose, lactate, free fatty acids, cortisol, glucagon, 4-androstenedione, testosterone, luteinizing hormone, insulin, growth hormone, arginine vasopressin, prolactin, and thyroxine have been found to be unaltered from normoxic resting values (4,43,44,57, 66,70,75). Resting levels of plasma renin activity, aldosterone, and angiotensin II have been found to be reduced slightly or unchanged (43,44,57).

Exercise

Exercise during normoxic conditions causes marked changes in sympathetic, metabolic, and hormonal activity (1,34). During incremental exercise, plasma levels of norepinephrine, epinephrine, lactate, ammonia, testosterone, cortisol, growth hormone, 4-androstenedione, and luteinizing hormone are increased and insulin is decreased at the highest exercise intensities relative to rest and the lower exercise intensities (1,4,8,29). Plasma levels of glucose, glucogon and free fatty acids are altered slightly from resting values (1,4,34). With prolonged exercise at a given exercise intensity, most of the aforementioned hormones, metabolites, as well as the catecholamines are also altered from resting values (1,34,57,66,75). The magnitude of change of these parameters is directly related to the duration and intensity of exercise (1,34,66,75).

Whether these responses to exercise are altered during acute exposure to altitude or hypoxia is questionable (4,29,57,59,66,70, 75,77). Sutton (75) reported that blood concentrations of glucose, lactate, free fatty acids, cortisol and growth hormone were higher, and the concentration of insulin lower in subjects pedalling for 20 minutes while at a simulated altitude of 4550 m compared to normoxia. However, the subjects exercised at the same absolute exercise intensity (125 watts) during each condition. Because the exercise intensity used at altitude was not reduced in proportion to the reduction in $\dot{V}O_2max$, the subjects at altitude were working at a significantly higher percentage of their sea-level $\dot{V}O_2max$. At sea level, the exercise intensity was between 30% and 50% $\dot{V}O_2max$, but at altitude the exercise intensity was 70% to 90% of their sea level $\dot{V}O_2max$. In other words, the differences between the normoxic and hypoxic condition may have been totally due to the difference in relative intensity and not to the effects of hypoxia *per se*.

Bouissou *et al.* (4) compared metabolic and endocrine responses to graded exercise during normoxic and hypoxic conditions (3000 m) after adjusting for the 17% hypoxia-induced reduction in $\dot{V}O_2max$. Comparisons were made, therefore, at the same relative intensity (40%, 60%, 80% $\dot{V}O_2max$ and at $\dot{V}O_2max$). Each exercise intensity lasted 5 minutes. As the intensity of the exercise increased, blood levels of lactate, cortisol, 4-androstenedione, testosterone, epinephrine, and norepinephrine also increased. However, except for a small difference in norepinephrine at $\dot{V}O_2max$, no environmentally-induced differences were found in these blood values as well as in blood glucose, free fatty acids, glucagon, insulin, and luteinizing hormone. It was concluded that acute, moderate hypoxia does not affect metabolic and hormonal responses to short periods of exercise performed at similar relative exercise intensities.

Similar results are also found at higher levels of hypoxia. Escourrou *et al.* (29) had subjects pedal an ergometer at intensities requiring 40%, 60% and 75% of their $\dot{V}O_2max$ for 15 minutes at each intensity during normoxia and while breathing a hypoxic gas (11–12% O_2 in N_2; 4500–5000 m equivalent). The exercise intensities were adjusted during hypoxia to reflect the 24% reduction in $\dot{V}O_2max$. It was found that the concentrations of epinephrine and norepinephrine were higher for any absolute level of $\dot{V}O_2$ during hypoxia. When the catecholamine values were compared to $\dot{V}O_2$, expressed as a relative percent of $\dot{V}O_2max$, no differences were found between the two environmental conditions.

While the majority of endocrine and sympathetic responses to exercise seem to be similar under normoxia and acute hypoxic conditions when the intensity of exercise is expressed in relative terms, there does exist an example where there are differences between the environments (77). During exercise at sea level or in normoxia, plasma renin activity is increased due to a redistribution of blood to the working muscles and away from the juxtaglomerular cells of the kidney. Renin is secreted when the cells sense a reduction in perfusion pressure or flow. These cells, being directly innervated by sympathetic nerves, also release renin when sympathetic activity is high as it is during exercise. Renin reacts with the plasma protein angiotensinogen to form angiotensin I, which, in turn, reacts with angiotensin converting enzyme (ACE) in the lung. After losing two amino groups, angiotensin I is converted to angiotensin II. Aldosterone is secreted from the adrenal cortex when stimulated by angiotensin II. Both angiotensin II and aldosterone have a direct effect on the kidneys to reabsorb sodium and water thereby increasing extracellular fluid.

There is much evidence to suggest that hypoxia affects the renin-angiotensin-aldosterone-system (RAAS) differently (13,14,57, 59,60,70,72). Slater *et al.* (72), after measuring an increase in plasma renin activity and a reduction in aldosterone at rest, concluded that altitude must affect the control mechanism. Maher *et al.* (57) confirmed and extended this conclusion to include exercise at altitude. Interest in determining how or why the RAAS becomes altered during acute exposure to hypoxia grew when Hogan *et al.* (44) reported that the subjects, who experienced the most severe symptoms of acute mountain sickness, also showed the greatest reduction in aldosterone secretion.

More recent studies have attempted to determine the point at which the RAAS control mechanism(s) differ from sea level. Milledge and Catley (59) had subjects exercise on an ergometer for 2 hours at 50 watts. Ambient, normoxic air was breathed the first hour and a hypoxic gas mixture (12.8% O_2, 4000 m equivalent) for the second hour. Throughout the entire 2 hours, plasma renin activity (estimated by measuring angiotensin I), aldosterone, and ACE were monitored by obtaining blood samples from a forearm vein. During normoxia, plasma renin activity and aldosterone rose together while ACE concentration was not altered. With the onset of hypoxia, plasma renin activity continued to rise (50% above control values), while aldosterone and ACE fell significantly. The authors concluded that the reduction in ACE activity resulted in a reduction in angiotensin II concentration which directly affected aldosterone secretion. These results were later confirmed in a similar study at 3100 m altitude by some of the same authors (60).

Shigeoka *et al.* (70) also reported similar findings but provided a different explanation. By reversing the sequence of the daily presentation of normoxic and hypoxic (17% O_2) conditions to the same test subjects during exercise, from one testing day to another, they found that aldosterone concentration did not relate to changes in plasma renin activity in normoxia when the normoxic condition followed the hypoxic condition. They, therefore, suggested that an inhibitor, possibly of angiotensin I secretion of aldosterone, was responsible. In an effort to determine if an inhibitor does indeed exist at this point of the RAAS, these authors (70) infused angiotensin II into their subjects at different rates while the subjects were normoxic and while moderately hypoxic (90% saturation). It was hypothesized that if the aldosterone response to infusion of angiotensin II was inhibited during hypoxia and not during normoxia, then an inhibitor between angiotensin II and the release of aldosterone existed. They found that hypoxia did not significantly inhibit the aldosterone response to angiotensin II and had to conclude that hypoxemia does not cause release of an inhibitor of angiotensin II-mediated aldosterone release.

In summary, there seems to be little doubt that the RAAS is disrupted during exercise in acute exposure to hypoxia (77). However, exactly where in the RAAS the uncoupling occurs is more questionable. It should be noted that the disruption of the RAAS is observed only during acute exposures to hypoxia and altitude. With a more prolonged exposure, aldosterone secretion once again becomes directly linked to the renin-angiotensin system (57).

NEUROPSYCHOLOGICAL IMPAIRMENT

The neuro-psychological effects of immediate, or gradually progressive hypoxia were observed from the earliest experiences dealing with low pressure chambers, high-altitude balloon flights, and motorized aviation (see ref. 45). There is a progression in the sensitivity of various nervous tissues to oxygen deprivation. The higher cortical centers appear to be the most sensitive followed in order by the cerebellum, medulla, spinal cord, and sympathetic ganglia (23,38). Relatively small reductions in arterial oxygen saturation can significantly impair mental and motor coordination, personality, and judgement (16). This is especially true at hypoxic levels below a P_IO_2 of 110 torr where hemoglobin saturation lies on the beginning of the steep portion of the oxygen equilibrium curve (see Table 12-1 for the progressive reduction in arterial saturations).

Initial hypoxia affects progress from subtle, almost imperceptible, changes in personality, judgement, and short-term memory to gross derangements. In susceptible individuals, measurable decrements can occur with a P_IO_2 as high as 100 torr (equivalent altitudes of 3,000 m). A number of individuals may experience euphoria and be cheerful and complacent, while others appear less tolerant and may become irritable and uncooperative. When these behavioral changes are acknowledged by cohorts, some individuals will attempt to disprove and deny that hypoxia affects them, in much the same manner as an individual who is intoxicated and tries to hide the fact that he/she has imbibed alcohol (17). This is of special concern when a person is in a potentially dangerous situation (e.g. mountain climbing or piloting a non-pressurized plane). Furthermore, the poor mental and motor performances and personality changes usually precede, but can be independent of, any feelings of sickness (17). Symptoms of impending illness usually begin after four or more hours exposure to a P_IO_2 less than 110 torr. Illness caused by hypoxic exposure is presented in detail in Chapter 14.

Human performance on different types of tasks are affected at different levels of hypoxia. Impairment on a specific task relates to factors such as: (a) the level of hypoxia, (b) the type of task or function being performed (sensory or motor), (c) the complexity of the task, (d) the familiarity of the individual with the task, and (e) whether the task has been learned prior to or during the exposure (15,16,21,37). These factors plus the large interindividual variability in personality traits and motivation obviously have the potential to bias the results of experiments where a high degree of consistency, motivation, and attentiveness is required (16).

Through the years, a number of experiments have been per-

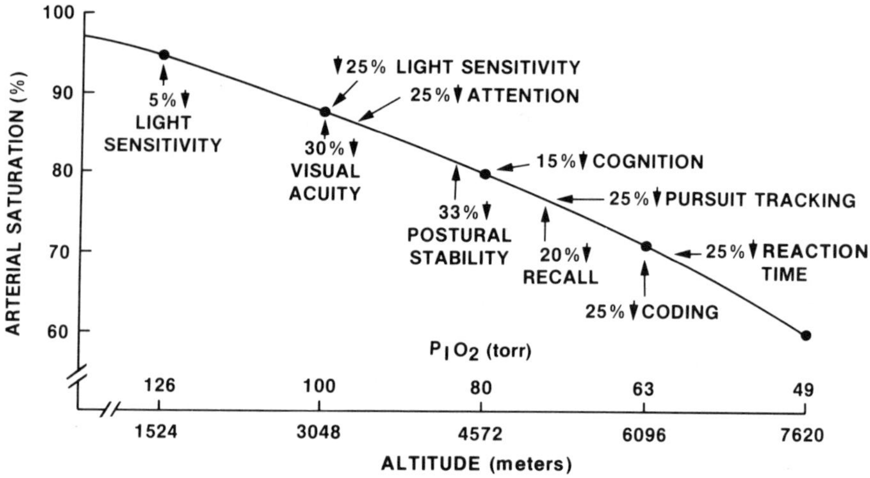

Figure 12-6 *Effect of increasing in altitude (or reduction in P_IO_2) on sensory and mental functions. Redrawn from data presented by McFarland (58) and Cudaback (17).*

formed to determine how various levels of hypoxia affect a number of different psychomotor and cognitive tasks (15,16,21,37,38,49). From these studies and others (81), some generalizations have been made (17,58). These are briefly summarized below and illustrated in Figure 12-6.

Sensory

Decreases in light sensitivity (the amount of light required for perception of a given stimulus) can be detected at a P_IO_2 of 138 torr. At a P_IO_2 of 86 torr, twice as much light is required for perception (14,55). Also at approximately a P_IO_2 of 86 torr, central field extent is reduced 10%, central brightness is reduced 30%, and dark adaptation, 34%. Visual acuity (ability to resolve a given target at a given distance) is affected at a P_IO_2 of about 118 torr and reduces to 50% of the normoxic value at a P_IO_2 of 83 torr, if the illumination is dim. If the illumination is bright, there is little or no impairment until a P_IO_2 of 80 torr is attained (58). In tasks requiring an alignment response to a changing input (e.g., visual pursuit tasks), no decrement from normoxic values are measured until the P_IO_2 drops below 83 torr. Below this level of hypoxia, performance falls off dramatically (21,58). It is generally accepted that auditory sensitivity is little affected by acute exposure to hypoxia (58).

Mental Functioning

The ability to do mathematical calculations, make decisions, and perform coding and conceptional reasoning tasks are affected beginning at a P_IO_2 of 110 torr, with a steady decline in performance down to 86 torr. Below about 86 torr, the quality of performance decreases extremely rapidly (37,38,49). Tasks take longer to perform, and errors are more frequent, in direct proportion to the reduction in P_IO_2 (17). Short-term memory is affected at levels of hypoxia beginning at a P_IO_2 as high as 118 to 127 torr and declines rapidly when the P_IO_2 is less than 101 torr (15,17). At a P_IO_2 of 94 torr, performance during word association, position and pattern as well as immediate recall and delayed recall (15 and 30 min) are reduced by 17%. At 80 torr, there is a 25% reduction in short-term memory (15,58). The ability to detect a reduction in attentiveness occurs at approximately a P_IO_2 of 110 torr with a linear decline thereafter (49).

In summary, percent decrements can be assigned to general types of tasks which demonstrate sensitivity to hypoxia. Figure 12-6 illustrates the progressive decline in vision and psychological tests with reductions in P_IO_2. No decrement in sensory or mental performance can be detected when the P_IO_2 is greater than 133 torr (17,49,58). In general, there is a 25% reduction in sensory performance at 101 torr which may be an acceptable decrement in performance, and a 40% reduction at 94 torr which is usually unacceptable (58). Performance on many mental tasks can be improved significantly at lower levels of hypoxia if the tasks are well learned in normoxia (21,37). Tests of mental function show a 10% reduction at about a P_IO_2 of 94 torr and a 25% reduction at about 83 torr (38).

SUMMARY

The effects of a reduction in ambient oxygen pressure occurring within 4 hours are dependent to a large extent on one principal factor: the degree of hypoxia and the resultant arterial hypoxemia. As the degree of hypoxia increases, a series of compensatory responses occur in various organ systems which are directly proportional to the severity. Below a P_IO_2 of approximately 110–125 torr, demonstrable changes become evident in ventilation, cardiac output, circulation, blood endocrine levels, and sensory and mental function. With a P_IO_2 below 97 torr, the changes can become considerable and can be discerned for almost any physiological and psychological function. The problems are further accentuated when exercise is superimposed on the hypoxia.

With physical performance, reductions of both maximal aerobic power and submaximal endurance exercise capacity are observed almost immediately due to the reduction in oxygen content of arterial blood despite potentially beneficial changes in alveolar ventilation, distribution of ventilation/perfusion ratios, cardiac output distribution, shifts in the oxygen-equilibrium curve, increases in sympathetic nervous system activity, and changes in fluid control hormones. Nevertheless, individuals can function reasonably well due to compensatory responses, which act to minimize the effect of the reduction in oxygen pressure.

With regard to the senses it appears that vision is particularly sensitive to hypoxia. Mental functioning and performance on complex tasks are complicated by many factors, but in general the better a task is learned in normoxia, the less will be the decrement in hypoxia. With moderately severe, acute hypoxic exposures there are demonstrable changes in personality and behavior which may be subtle but recognizable.

REFERENCES

1. Åstrand, P.O. and K. Rodahl. *Textbook of Work Physiology,* New York:McGraw-Hill Book Co., 1986.
2. Banchero, N., F. Sime, D. Penaloza, J. Cruz, R. Gamboa, and E. Marticorena. Pulmonary pressure, cardiac output, and arterial oxygen saturation during exercise at high altitude and at sea level. *Circulation* 33:249–262, 1966.
3. Blackmon, J.R. and L.B. Rowell. Hepatic splanchnic function in acutely hypoxemic humans at rest. *Am. J. Physiol.* 251:R887–R892, 1986.
4. Bouissou, P., F. Peronnet, G. Brisson, R. Helie, and M. Ledoux. Metabolic and endocrine responses to graded exercise under acute hypoxia. *Eur. J. Appl. Physiol.* 55:290–294, 1986.
5. Boutellier, U. and E.A. Koller. Propranolol and the respiratory, circulatory and ECG responses to high altitude. *Eur. J. Appl. Physiol.* 46:105–119, 1981.
6. Bouverot, P. *Adaptation to Altitude-Hypoxia in Vertebrates.* Berlin: Springer-Verlag, 1985, pp. 1–17.
7. Bowie, W., and G.R. Cumming. Sustained handgrip-reproducibility; effects of hypoxia. *Med. Sci. Sports* 3:24–31,1971.
8. Buono, M.J., T. R. Clancy, and J.R. Cook. Blood lactate and ammonium ion accumulation during graded erercise in humans. *J. Appl. Physiol.* 57:135–139, 1984.
9. Burse, R.L., A. Cymerman, and A.J. Young. Respiratory response and muscle function during isometric handgrip exercise at high altitude. *Aviat. Space Environ. Med.* 58:39–46, 1987.
10. Buskirk, E.R. Decrease in physical work capacity at high altitude. In: *Biomedicine of High Terrestrial Elevations,* A.H. Hegnauer (ed.), Natick, MA.: U.S. Army Research Institute of Environmental Medicine, 1969, p. 204–222.
11. Chance, B., P. Cohen, F. Jobsis, and B. Schoener. Intracellular oxidation-reduction states in vivo. *Science* 137:499–508, 1962.
12. Coates, G., G. Gray, A. Mansell, C. Nahmias, A. Powles, J. Sutton, and C. Webber. Changes in lung volume, lung density, and distribution of ventilation during hypobaric decompression. *J. Appl. Physiol.* 46:752–755, 1979.
13. Colice, G.L. and G. Ramirez. Effect of hypoxemia on the renin-angiotensin-aldosterone system in humans. *J. Appl. Physiol.* 58:724–730, 1985.
14. Colice, G.L. and G. Ramirez. Aldosterone response to angiotensin II during hypoxemia. *J. Appl. Physiol.* 61:150–154, 1986.
15. Crow, T.J. and G.R. Kelman. Effect of mild acute hypoxia on human short-term memory. *Brit. J. Anaesth.* 43:548–552, 1971.
16. Crow, T.J. and G.R. Kelman. Psychological effects of mild acute hypoxia. *Brit. J. Anaesth.* 45:335–337, 1973.
17. Cudaback, D.D. Four-KM altitude effects on performance and health. *Pub. Astronom. Soc. Pac.* 96:463–477, 1984.
18. Cunningham, W.L., E.J. Becker, and F. Kreuzer. Catecholamines in plasma and urine at high altitude. *J. Appl. Physiol.* 20:607–610, 1965.
19. Cymerman, A., P.B. Rock, J.R. Sutton, J.T. Reeves, B.M. Groves, M.K. Malconian, P.M.

Young, P.D. Wagner, and C.S. Houston. Operation Everest II: Importance of ventilation in defense of maximal oxygen uptake at extreme altitude. *J. Appl. Physiol.* (In Review).

20. Dempsey, J.A. and H.V. Forester. Mediation of ventilatory adaptations. *Physiol. Rev.* 62:262–346, 1982.
21. Denison, D.M., F. Ledwith, and E.C. Poulton. Complex reaction times at simulated cabin altitudes of 5,000 feet and 8,000 feet. *Aerospace Med.* 37:1010–1013, 1966.
22. Dill, D.B. Physiological adjustments to altitude change. *J. Am. Med. Assoc.* 25:747–753, 1968.
23. Drinker, C.K. *Carbon Monoxide Asphyxia.* New York:Oxford University Press, 1938.
24. Drinkwater, B.L., L.J. Folinsbee, J.F. Bedi, S.A. Plowman, A.B. Loucks, and S.M. Horvath. Response of woman mountaineers maximal exercise during hypoxia. *Aviat. Space Environ. Med.* 50:657–662, 1979.
25. Easton, P.A., L.J. Slykerman, and N.R. Anthonisen. Ventilatory response to sustained hypoxia in normal adults. *J. Appl. Physiol.* 61:906–911, 1986.
26. Eiken, O., and P.A. Tesch. Effects of hyperoxia and hypoxia on dynamic and sustained static performance of the human quadriceps muscle. *Acta Physiol. Scand.* 122:629–633, 1984.
27. Ekblom, B., R. Hout, E.M. Stein, and A.T. Thorstensson. Effects of changes an arterial oxygen content on circulation and physical performance. *J. Appl. Physiol.* 39:71–75, 1975.
28. Elliott, P.R. and H.A. Atterbom. Comparison of exercise responses of males and females during acute exposure to hypobaria. *Aviat. Space Environ. Med.* 49:415–418, 1978.
29. Escourrou, P., D.G. Johnson, and L.B. Rowell. Hypoxemia increases plasma catecholamine concentrations in exercising humans. *J. Appl. Physiol.* 57:1507–1511, 1984.
30. Fagraeus, L., J. Karlsson, D. Linnarsson, and B. Saltin. Oxygen uptake during maximal work at lowered and raised ambient air pressures. *Acta Physiol. Scand.* 87:411–421, 1973.
31. Fulco, C.S., A. Cymerman, P.B. Rock, and G. Farese. Hemodynamic responses to upright tilt at sea level and high altitude. *Aviat. Space Environ. Med.* 56:1172–1176, 1985.
32. Fulco, C.S., P.B. Rock, L.A. Trad, V. Forte, Jr., and A. Cymerman. Maximal cardiorespiratory responses to one- and two-legged cycling during acute and long-term exposure to 4300 m. *Eur. J. Appl. Physiol.* 1988, (In Press).
33. Gale, G.E., J.R. Torre-Bueno, R.E. Moon, H.A. Saltzman, and P.D.Wagner. Ventilation-perfusion inequality in normal humans during exercise at sea level and simulated altitude. *J. Appl. Physiol.* 58:978–988, 1985.
34. Galbo, H. *Hormonal and Metabolic Adaptation to Exercise.* New York: Thieme-Stratton Inc., 1983.
35. Gleser, M.A. Effects of hypoxia and physical training on hemodynamic adjustments to one-legged cycling. *J. Appl. Physiol.* 34:655–659, 1973.
36. Goldstein, D.S., R. McCarty, R. J. Polinski, and I. J. Kopin. Relationship between plasma norepinephrine and sympathetic neural activity. *Hypertension* 5:552-559, 1983.
37. Green, R.G. and D.R. Morgan. The effects of mild hypoxia on a logical reasoning task. *Aviat. Space Environ. Med.* 56:1004–1008, 1985.
38. Greene, R. Mental performance in chronic anoxia. *Brit. Med. J.* 1 (5026):1028–1031, 1957.
39. Grover, R.F., J.V. Weil, and J.T. Reeves. Cardiovascular adaptation to exercise at high altitude. In: *Exercise and Sport Sciences Reviews,* K.B. Pandolf (ed.), New York:Macmillan, 1986, p.269–302.
40. Haab, P., D.R. Held, H. Ernst, and L.E. Farhi. Ventilation-perfusion relationships during high altitude adaptation. *J. Appl. Physiol.* 26:77–81, 1969.
41. Hammond, M.D., G.E. Gale, K.S.Kapitan, A. Ries, and P.D. Wagner. Pulmonary gas exchange in humans during normobaric hypoxic exercise. *J. Appl. Physiol.* 61: 1749–1757, 1986.
42. Hartley, L.H., J.A. Vogel, and M. Landowne. Central, femoral, and brachial circulation during exercise in hypoxia. *J. Appl. Physiol.* 34:87–90, 1973.
43. Heyes, M.P., M.O. Farber, F. Manfredi, D. Robertshaw, M. Weinberger, N. Fineberg, and G. Robertson. Acute effects of hypoxia on renal and endocrine function in normal humans. *Am. J. Physiol.* 243:R265–R270, 1982.
44. Hogan, R.P., T.A. Kotchen, A.E. Boyd, and L.H. Hartley. Effect of altitude on renin-aldosterone system and metabolism of water and electrolytes. *J. Appl. Physiol.* 35:385–390, 1973.
45. Houston, C.S. *Going Higher, The Story of Man and Altitude,* Boston, MA:Little, Brown and Company, 1987.
46. Jaeger, J.J., J.T. Sylvester, A. Cymermen , J.J. Berberich, J.C. Denniston, and J.T. Maher. Evidence for increased intrathoracic fluid volume in man at high altitude. *J. Appl. Physiol.* 47:670–676, 1979.
47. Kellogg, R. Oxygen and carbon dioxide in the regulation of respiration. *Fed. Proc.* 36:1658–1663, 1977.
48. Kellogg, R. Central chemical regulation of respiration. In: *Handbook of Physiology: Respiration,* Section 3, W.O. Fenn and H. Rahn (eds.). Washington, D.C.: Am. Physiol. Soc., 1964, pp. 507–534.

49. Kelman, G.R., T.J. Crow, and A.E. Bursill. Effect of mild hypoxia on mental performance as assessed by a test of selective attention. *Aerospace Med.* 40:301–303, 1969.
50. Knuttgen, H.G. and B. Saltin. Oxygen uptake, muscle high-energy phosphates, and lactate in exercise under acute hypoxic conditions in man. *Acta Physiol. Scand.* 87:368–376, 1973.
51. Laciga, P. and E.A. Koller. Respiratory, circulatory, and ECG changes during acute exposure to high altitude. *J. Appl. Physiol.* 41:159–167, 1976.
52. Lahiri, S. and R Gelfand. Mechanisms of acute ventilatory responses. In: *Lung Biology in Health and Disease. Regulation of Breathing.* T. Hornbein (ed.). New York:Decker, 1981, pp.773–844.
53. Lahiri S. Physiological responses and adaptations to high altitude. In: *International Review of Physiology. Environmental Physiology II,* D. Robertshaw (ed.). Baltimore:University Park Press, 1977, pp. 217–251.
54. Lenfant, C., J.D. Torrance, and C. Reynafarje. Shift of the O_2-Hb dissociation curve at altitude: mechanism and effect. *J. Appl. Physiol.* 30:625–631, 1971.
55. Lenfant, C. and K. Sullivan. Adaptation to high altitude. *New Eng. J. Med.* 284:1298–1309, 1971.
56. Maher, J.T., L.G. Jones, and H. Hartley. Effects of high-altitude exposure on submaximal endurance capacity of men. *J. Appl. Physiol.* 37:895–898, 1974.
57. Maher, J.T., L.G. Jones, L.H. Hartley, G.H. Williams, and L.I. Rose. Aldosterone dynamics during graded exercise at sea level and high altitude. *J. Appl. Physiol.* 39:18–22, 1975.
58. McFarland, R.A. Review of experimental findings in sensory and mental functions. In: *Biomedicine of High Terrestrial Elevations,* Hegnauer, A.H., (ed.). Natick, MA.: U.S. Army Research Institute of Environmental Medicine, 1969, p. 250–265.
59. Milledge, J.S. and D.M. Catley. Renin, aldosterone, and converting enzyme during exercise and acute hypoxia in humans. *J. Appl. Physiol.* 52:320–323, 1982.
60. Milledge, J.S., D.M. Catley, E.S. Williams, W.R. Withey and B.D. Minty. Effect of prolonged exercise at altitude on the renin-aldosterone system. *J. Appl. Physiol.* 55:413–418 ,1983.
61. Moore, L.G., S.Y. Huang, R.E. McCullough, J.B. Sampson, J.T., Maher, J.V. Weil, R.F. Grover, J.K. Alexander, and J.T. Reeves. Variable inhibition by falling CO_2 of hypoxic ventilatory response in humans. *J. Appl. Physiol.* 56:207–210, 1984.
62. Moore, L.G., P. Brodeur, O. Chumbe, J. D'Brot, S. Hofmeister, and C. Monge. Maternal hypoxic ventilatory response, ventilation, and infant birth weight at 4300 m. *J. Appl. Physiol.* 60:1401–1406, 1986.
63. Paterson, D.J., H. Pinnington, A.R. Pearce, and A.R. Morton. Maximal exercise cardiorespiratory responses of men and women during acute exposure to hypoxia. *Aviat. Space Environ. Med.* 58:243–247, 1987.
64. Powles, A.C., J.R. Sutton, G.W. Gray, A.L. Mansell, M. McFadden, and C.S. Houston. Sleep hypoxemia at altitude: Its relationship to acute mountain sickness and ventilatory responsiveness to hypoxia and hypercapnia. In: *Environmental Stress: Individual Human Adaptations,* L.J. Folinsbee, J.A. Wagner, J.F. Borgia, B.L., Drinkwater, J.A. Gilner and J.F. Bedi (eds.). New York:Academic Press, 1978, pp. 373–389.
65. Rahn, H. and A.B. Otis. Alveolar air during simulated flights to high altitude. *Am. J. Physiol.* 150:202–221, 1947.
66. Raynaud, J., L. Drouet, J.P. Martineaud, J. Bordachar, J. Coudert, and J. Durand. Time course of plasma growth hormone during exercise in humans at altitude. *J. Appl. Physiol.* 50:229–233, 1981.
67. Reeves, J.T., J.R. Sutton, P.D. Wagner, A. Cymerman, M.K. Malconian, P.B. Rock, P.M. Young, J.K. Alexander, and C.S. Houston. Oxygen transport during exercise at extreme altitude: Operation Everest II. *Ann. Emerg. Med.* 16:993–998, 1987.
68. Richardson, D.W., H.A. Kontos, A.J. Raper, and J.L. Patterson. Modification by beta-adrenergic blockade of the circulatory responses to acute hypoxia in man. *J. Clin. Invest.* 46:77–85, 1967.
69. Schoene, R.B. and T.F. Hornbein. High altitude adaptation. In: *Textbook of Respiratory Medicine,* J.F. Murray and J.A. Nadel (eds.). Philadelphia, PA, W.B. Saunders, 1988.
70. Shigeoka, J.W., G.L. Colice, and G. Ramirez. Effect of normoxemia and hypoxic exercise on renin and aldosterone. *J. Appl. Physiol.* 59:142–148,1985.
71. Silverberg, A.B., S.D. Shaw, M.W. Hammond, and P.C. Cryer. Norepinephrine: Hormone and neurotransmitter in man. *Am. J. Physiol.* 234:E252–E256,1978.
72. Slater, J.D.H., R.E. Tuffley, E.S. Williams, C.H. Beresford, P.H. Sonksen, R.H.T. Edwards, R.P. Ekins, and M. McLaughlin. Control of aldosterone secretion during acclimatization to hypoxia in man. *Clin. Sci.* 37:327–341, 1969.
73. Squires, R.W. and E.R. Buskirk. Aerobic capacity during acute exposure to simulated altitude, 914 to 2286 meters. *Med. Sci. Sports Exer.* 14:36–40, 1986.
74. Stenberg, J., B. Ekblom, and R. Messin. Hemodynamic response to work at simulated altitude, 4000 m. *J. Appl. Physiol.* 21:1589–1594, 1966.

75. Sutton, J.R. Effect of acute hypoxia on the hormonal response to exercise. *J. Appl. Physiol.* 42: 587–592, 1977.
76. Sutton, J.R., G.W. Gray, M.D. McFadden, A.C. Bryan, E.S. Horton, and C.S. Houston. Nitrogen washout studies in acute mountain sickness. *Aviat. Space Environ. Med.* 48:108–110, 1977.
77. Sutton, J.R. and M.P. Heyes. Endocrine responses to exercise at altitude. In: *Exercise Endocrinology*, K. Fotherby and S.B. Pal (eds.), New York:Walter de Gruyter & Co., 1985, pp. 239–262.
78. Sylvester, J.T., A. Cymerman, G. Gurtner, O. Hottenstein, M. Cote, and D. Wolfe. Components of alveolar-arterial O_2 gradient during rest and exercise at sea level and high altitude. *J. Appl. Physiol.* 50:1129–1139, 1981.
79. Tenny, S.M., P. Scotio,, L.C. Ou, D. Bartlett, Jr., and J.E. Remmers. Suprapontine influences on hypoxia ventilatory control. In: *High Altitude Physiology: Cardiac and Respiratory Effects*, R. Porter and J. Knight (eds.). London: Churchill Livingstone, 1971, pp. 89–102.
80. Torre-Bueno, J. R., P. D. Wagner, H.A. Saltzman, G. E. Gale, and R.E. Moon. Diffusion limitation in normal humans during exercise at sea level and simulated altitude. *J. Appl. Physiol.* 58:989–995, 1985.
81. Tune, G.S. Psychological effects of hypoxia: Review of certain literature from the period 1950 to 1963. *Percept. Mot. Skills* 19:551–562, 1964.
82. Vogel, J.A., L.H. Hartley, J.C. Cruz and R.P. Hogan. Cardiac output during exercise in sea level residents at sea level and high altitude. *J. Appl. Physiol.* 36:169-172, 1974.
83. Vogel, J.A., J.E. Hansen and C.W. Harris. Cardiovascular responses in man during exhaustive work at sea level and high altitude. *J. Appl. Physiol.* 23:531–539, 1967.
84. Wagner, P.D., G.E. Gale, R.E. Moon, J.R. Torre-Bueno, B.W. Stolp, and H.A. Saltzman. Pulmonary gas exchange in humans exercising at sea level and simulated altitude. *J. Appl. Physiol.* 61:260–270, 1986.
85. Wagner, P.D., J.R. Sutton, J.T. Reeves, A. Cymerman, B.M. Groves, and M.K. Malconian. Pulmonary gas exchange throughout a simulated ascent of Mt. Everest: Operation Everest II. *J. Appl. Physiol.* (In review).
86. Ward, M. *Mountain medicine: A clinical study of cold and high altitude.* Great Britain:Crosby Lockward Staples, 1975.
87. Weiskopf, R.B. and R.A. Gabel. Depression of ventilation during hypoxia in man. *J. Appl. Physiol.* 39:911–915, 1975.
88. Welch, H. G. Effects of hypoxia and hyperoxia on human peformance. *Exercise and Sport Sciences Reviews.* K.B. Pandolf (ed.). New York: Macmillan, 1987, p.191–221.
89. West, J.B. *Respiratory Physiology: The Essentials.* Baltimore: Williams and Wilkins Co., 1979.
90. Young, A.J., A. Cymerman and R.L. Burse. The influence of cardiorespiratory fitness on the decrement in maximal aerobic power at high altitude. *Eur. J. Appl. Physiol.* 54: 12–15, 1985.
91. Young, A.J., J. Wright, J. Knapik, and A. Cymerman. Skeletal muscle strength during exposure to hypobaric hypoxia. *Med. Sci. Sports Exer.* 12:330–335, 1980.

13

Human Acclimatization to High Terrestrial Altitude

ANDREW J. YOUNG, Ph.D.

PATRICIA M. YOUNG, Ph.D.

OUTLINE

INTRODUCTION

The lowland resident who travels to high altitude experiences a number of physiological responses that counter the effects of hypobaric hypoxia. The acute effects of hypoxia seen upon initial arrival at high altitude have been discussed previously. This chapter will focus on the process of acclimatization to high altitude, with particular emphasis on adaptations occurring during the first few months that the lowland native resides at altitude. In addition, physiological responses of the high-altitude native will be compared to those of the acclimatizing lowlander to provide a sort of "yardstick" to gauge the degree of acclimatization. Throughout the chapter, the reader should bear in mind that the effects of altitude acclimatization are extremely variable among individuals. Most of the scientific information concerning altitude acclimatization has been obtained from studies of relatively young, highly fit male mountaineers and soldiers. Wherever possible throughout the chapter, the modifying influence on acclimatization of factors such as age, physical fitness and gender will be discussed.

Altitude acclimatization encompasses a wide variety of physiological adaptations. Ventilation is increased. Mechanisms of oxygen transport and delivery, as well as metabolism become more efficient in satisfying energy requirements of rest and exercise. Sensation of effort is lessened. The summated effect of all the adaptations to chronic hypoxia is to dramatically improve exercise tolerance and performance of the well-acclimatized lowlander compared to the new arrival at high altitude. It is from this perspective that the components of the altitude acclimatization process will be discussed.

PULMONARY GAS EXCHANGE

As discussed in the preceding chapter, hypoxia stimulates an increase in ventilation, but this increase is limited during initial exposure to high altitude. One of the key features of altitude acclimatization is that this initial "blunting" of the ventilatory response is overcome. As shown in Figure 13-1, resting ventilation of the acclimatized lowlander is greater than that of the new arrival at high altitude. Furthermore, resting ventilation does not fall back to sea-level values when the acclimatized subject breathes normoxic gas. Rahn and Otis (96) collated data from several investigations in order to compare resting alveolar gas composition of acclimatized persons at various altitudes with that of unacclimatized subjects acutely exposed to comparable reductions in inspired oxygen. Those data are shown in Figure 13-2, along with results of two more recently reported investigations (84,141). Several important points are illustrated concerning this early phase of ventilatory acclimatization. First of all, the increased ventilation lowers alveolar CO_2 pressure (P_ACO_2), but more importantly, raises alveolar O_2 pressure (P_AO_2) compared to the alveolar gas composition initially on arrival at that altitude. Secondly, ventilation increases during acclimatization (i.e. falling P_ACO_2), even at moderate altitudes where P_AO_2 is not below the threshold to stimulate increased ventilation in unacclimatized subjects.

The time dependency of this adaptation probably varies as a function of the exposure altitude. Figure 13-1 shows that at 4300 meters (m), ventilation increased progressively between the first and fourth day of residence, after which it remained stable (61). Caution must be exercised in accepting four days as the requisite for ven-

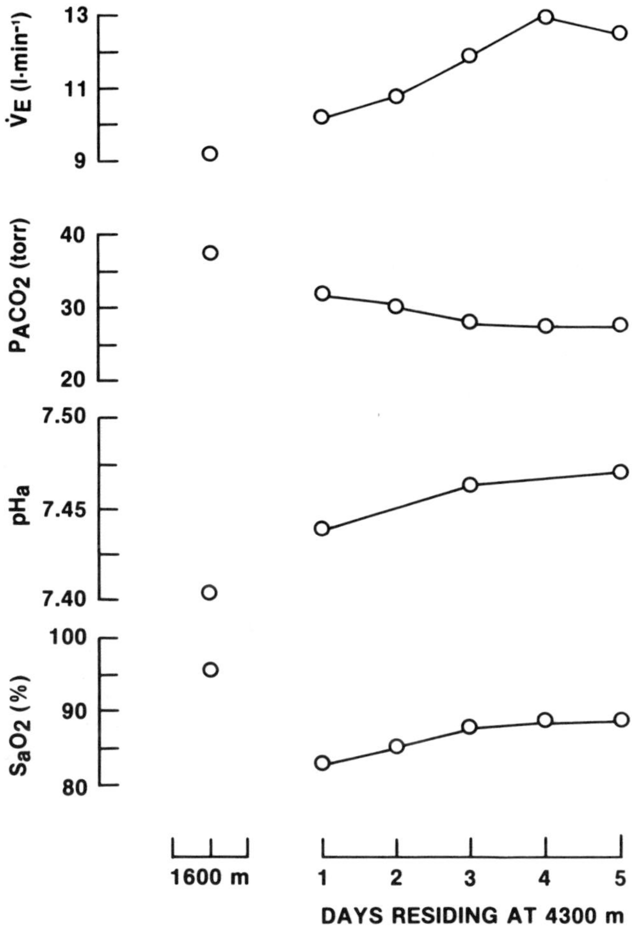

Figure 13-1. *Changes in the resting ventilation (minute ventilation, $\dot{V}_E$), alveolar PCO_2 (P_ACO_2), arterial pH (pH_a), and arterial oxygen saturation of hemoglobin (S_aO_2) measured in 12 lowland residents during the first five days at high altitude (4,300 m). Modified from Huang* et al. *(61).*

tilation to increase at 4300 m, since the subjects of that investigation (61) were permanent residents of Denver, CO (1600 m), and were likely to have been already partially acclimatized. Residents of sea-level regions would probably require a somewhat longer period at that altitude since Rahn and Otis (96) found that it took four days at 3000 m for ventilation to stabilize. Nevertheless, the time frame during which ventilation increases is on the order of several days at moderate altitudes, but this adaptation appears to require considerably longer at extreme altitudes. West *et al.* (141) measured al-

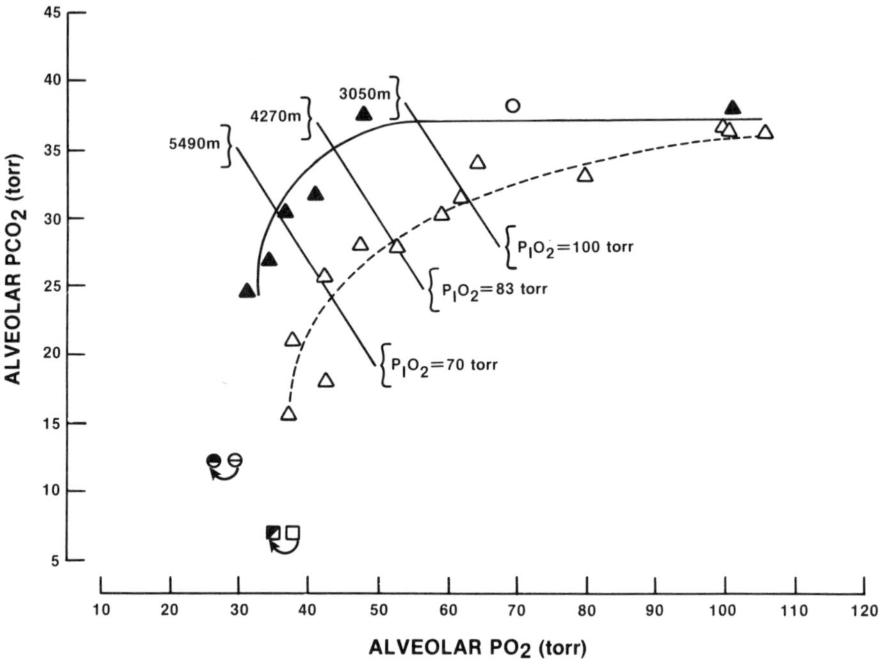

Figure 13-2. *Effect of altitude acclimatization on alveolar gas composition. Closed symbols are unacclimated, open are acclimated. Redrawn from original data of Rahn and Otis (96) shown as triangles, and additional data points added from West* et al. *(141) and Malconian* et al. *(84) shown as squares and circles, respectively. Partially closed circles and squares show the effect of correcting the measured alveolar oxygen pressure assuming a more reasonable (0.85) value of respiratory exchange ratio than actually measured. Isoaltitude lines were calculated using the ideal alveolar gas equation (17) and an assumed respiratory exchange ratio of 0.85.*

veolar gas composition at the summit of Mt. Everest, albeit on only one subject. Figure 13-2 shows that the P_AO_2 (35 torr) and P_ACO_2 (7.5 torr) which they measured agree extremely well with values predicted by extrapolating data on acclimatized subjects at lower altitudes (96). However, alveolar gas composition (N = 6) measured in a hypobaric chamber at an inspired oxygen pressure (P_IO_2) of 42 torr, which is approximately equivalent to the summit of Mt. Everest (84), agrees more closely with predictions extrapolated from the data for unacclimatized subjects. The difference between the observations in these two studies probably reflects differing degrees of ventilatory acclimatization. In the hypobaric chamber experiment (84), the subjects were decompressed over a 35–40 day period with only about 20 days spent above 5500 m (simulated). West *et al.* (141) studied members of an expedition sojourning about five months at high altitude, most of this time at elevations above 5400 m.

The increased ventilation experienced by lowlanders sojourning several weeks at high altitude represents an intermediate phase and not the completion of ventilatory acclimatization. Classically, this has been demonstrated by comparison of the ventilatory responses of the permanent residents with those of lowland sojourners at high altitude. In general, long-term residents appear to have a lower resting ventilation than lowlanders who have spent several weeks at high altitude (15,131). A change in the inspired CO_2 pressure (P_ICO_2) produces equivalent changes in the ventilation of the highlanders and lowlanders (78,121), but resting P_ACO_2 (78) and P_aCO_2 (120) in the highlanders is similar or greater than in the lowlanders. Thus, ventilatory sensitivity to CO_2 of highlanders is similar to acclimated lowlanders, but the highlanders' ventilatory "set-point" for P_ACO_2 regulation is adjusted downward (113). More importantly, ventilatory responsiveness to hypoxia is attenuated in the highlander compared to lowlander (38,77,78,113,121,131,137).

It should be noted that there are dissenting observations regarding the hypoxic sensitivity of highlanders compared to acclimatized lowlanders. Huang *et al.* (62) compared resting ventilatory responses of a group of lowland Chinese men who had lived at 3890 m for at least one year with a group of native Tibetans who were lifelong residents of altitudes above 3500 m. The two groups, who were of similar ages and occupational backgrounds, showed no differences in ventilatory responses. Hackett *et al.* (50) report that if their relatively small body size is taken into account, the Himalayan Sherpas actually have a greater resting ventilation than the acclimatized lowlander. On the other hand, data presented in a report (131) on Andean highlanders allows correction for body size differences, and resting ventilation is still found to be less than in lowlanders. Furthermore, Hackett *et al.* (50) found that hypoxic ventilatory drive in their Sherpa subject group declined as a function of age and residence altitude. Therefore, one factor contributing to the differences in observations of Himalayan and American highlanders may be the age of the population studied. Another factor may be the difference in residence altitudes of the different populations. Hackett *et al.* (49) observed in another report that the Sherpas tend to reside at altitudes between 2500 and 3800 m, although they make short trips to much higher altitudes; many South Americans reside in cities at altitudes above 4000 m. The possibility that ventilatory acclimatization differs between Himalayan and other high-altitude populations cannot be dismissed, but it is likely that hypoxic insensitivity is common to all high-altitude natives.

Another controversial issue pertains to the question of whether all highlanders exhibit depressed hypoxic ventilatory drive. Birth at high altitude (113,120,121), arrival at altitude at a very early age (38) and specific genetic/selection factors (113) have all been suggested as requisite for attenuated hypoxic sensitivity. The influence of specific genetic factors is evidenced by racial differences in hypoxic ventilatory drive. Cruz *et al.* (22) have reported that Mestizo Indian populations are significantly less sensitive to hypoxia than Caucasions. While genetic factors cannot be dismissed, there is ample evidence that the reduced hypoxic sensitivity of highland residents is an acquired adaptation. Lahiri *et al.* (76), however, demonstrated that infants born at altitude were not insensitive to hypoxia but developed this characteristic later in childhood. Others have reported that hypoxic and hypercapnic ventilatory responses of nine- and ten-year old high-altitude natives were the same as lowland children of the same age (14). Forster *et al.* (38) reported that second generation high-altitude natives of European ancestry demonstrated a reduced hypoxic sensitivity. These investigators also observed that hypoxic sensitivity of sojourning lowlanders increased during the first month at high altitude and thereafter tended to decline. Finally, as shown in Figure 13-3, Weil *et al.* (137) found hypoxic ventilatory drive was depressed in non-native residents at high altitude who had arrived during adulthood and remained many years, but not to the same extent as in natives. In summary, ventilatory acclimatization to high

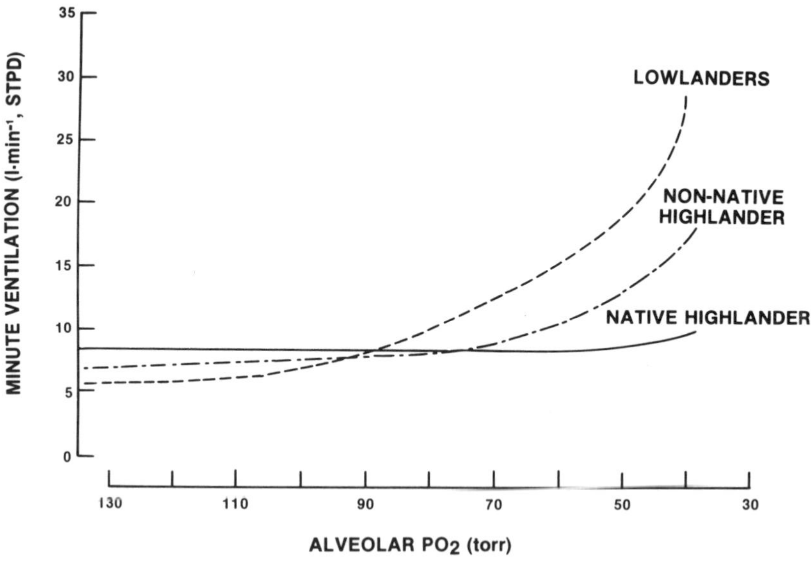

Figure 13-3. *Hypoxic ventilatory response of lowlanders and high altitude residents. Modified from Weil* et al. *(137).*

altitude appears to progress through the intermediate "hyperventilatory" phase to a phase where hypoxic sensitivity is diminished and resting ventilation is reduced. Differences in ventilatory responses to hypoxia among highland populations from various geographic regions may be due to genetic effects, differences in magnitude of hypoxia (elevation), years of altitude exposure and age on arrival.

Up to this point, this discussion has focused on changes in resting ventilation during altitude acclimatization, but acclimatization effects on exercise ventilation have also been investigated. As with the studies on resting ventilation, differences in the elevation studied, birth altitude of subjects, age on arrival at altitude and duration of altitude exposure make interinvestigation comparison and generalization difficult. Differences among studies in exercise mode and/or intensity further complicate interpretation. Finally, because of the nature of this research problem, many of the reported studies of exercise ventilation at altitude, particularly at very high altitudes, have been based on experimental observations of only two or three subjects.

As discussed in the preceding chapter, acute hypoxia increases ventilation during exercise as compared to normoxia. At any submaximal exercise intensity, minute ventilation (BTPS) is greater at high altitude than at sea level (27,45,95,104,122). The increment in ventilation increases as altitude increases (95), and differences in exercise ventilation between sea level and high altitude become more pronounced at the higher exercise intensities (27,45,104). Just as with resting ventilation, there is an additional increase in exercise ventilation with altitude acclimatization. Reeves *et al.* (104) report exercise ventilation of sojourning lowlanders measured between the first and third days of residence at 3100 m was nearly the same as measured after 16 to 18 days of residence. However, during this same time frame at 4300 m, Hansen *et al.* (54) observed that ventilation at each of three exercise intensities increased. Dejours *et al.* (27) measured ventilation during exercise performed once while the subjects inspired hypoxic gas mixture comparable to 3800 m, and again on the third day of sojourn at 3800 m. These investigators observed increased ventilation with acclimatization as compared to acute hypoxia. The discrepancy between the findings of Reeves *et al.* (104) and the other reports (27,54) suggest an altitude dependency on the magnitude of changes in exercise ventilation. Data from the report of Cruz *et al.* (21) suggest that measurements of exercise ventilation made three days after arrival at altitude cannot be considered to represent the unacclimatized response. Figure 13-4 shows changes in resting and exercise ventilation at various intervals during a

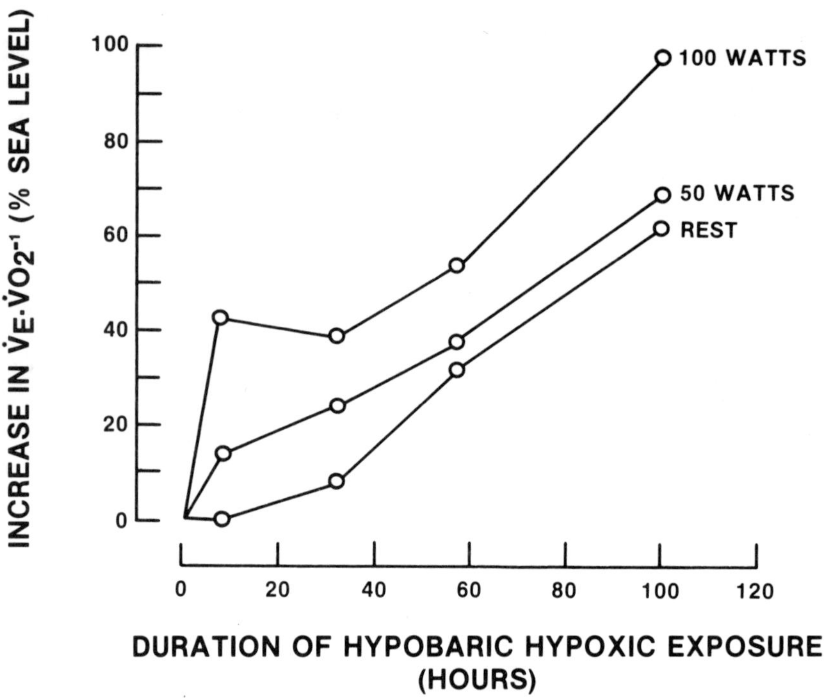

Figure 13-4. *Increase in ventilation measured in four lowlanders during 100 hours at a simulated altitude of ~4100 m. Ventilatory equivalent for oxygen ($\dot{V}_E \cdot \dot{V}O_2^{-1}$) is expressed as percent of sea-level ventilation at rest and during exercise at two different intensities. Drawn from data of Cruz* et al. *(21).*

100-hour hypobaric chamber exposure at a simulated altitude of 4100 m. As shown in the figure, changes in exercise ventilation are apparent much earlier during acclimatization than changes in resting ventilation.

Whether or not sojourning lowlanders at altitude can acclimatize to the point that their exercise ventilation is the same as native highlanders is not completely resolved. Lahiri *et al.* (78) reported that exercise ventilation at 4880 m of Sherpa subjects was lower than acclimatized lowlanders, especially at high exercise intensity, but that conclusion was based on data from only two lowlanders and four (at times three) Sherpa subjects. No statistical evaluation of the data was reported, and the data indicate considerable overlap between the lowlanders and highlanders in the range of ventilation at a given intensity. Grover *et al.* (43) studied ten natives of 4300 m and six acclimatized lowlanders at that altitude and reported that the highlanders had a lower ventilation than acclimatized lowlanders at each of three different intensities of treadmill exercise. Similarly, Pugh *et al.* (95) reported that ventilation at several exercise intensities was observed to be markedly lower in one Sherpa at 5800 m compared to two acclimatized lowlanders. In contrast, studies at lower altitudes suggest little difference in the exercise ventilation of acclimatized lowlanders and highlanders (45,51). Figure 13-5 shows ventilation of highlanders and lowlanders exercising at several different submaximal intensities completed at 308 and 3100 m. These data were obtained from two groups of subjects, each consisting of five men. The groups were closely matched for age, anthropometric measures and physical fitness. As the figure shows, there was little difference between the highlanders and the acclimatized lowlanders in exercise ventilation either at 308 or 3100 m. The failure to observe lower exercise ventilation in the highlanders as compared to lowlanders may reflect the lower altitude considered in this study. Alternatively, the highlanders were relatively young (~17 years) and may not have developed hypoxic insensitivity. Additional research is needed regarding exercise ventilation of persons native to elevations above 4,000 m.

The physiological mechanisms mediating ventilatory acclimatization remain unresolved. Although hypoxia stimulates ventilation, hypocapnia and the associated systemic alkalosis blunt ventilation on arrival at high altitude (21,61). Lowlanders experience an increase in renal bicarbonate excretion at high altitude (15), and it was once thought that this renal compensation for the respiratory alkalosis would return arterial pH to normal (i.e. sea-level values), providing additional stimulus for ventilation (96). Although arterial pH does eventually return to sea-level values, it remains alkalotic compared to sea level during the time frame that ventilation increases (29).

The failure of the time course of renal compensation for respiratory alkalosis to correlate with the increase in ventilation or the increased ventilatory sensitivity to inspired CO_2 suggested to Severinghaus *et al.* (114) that central ventilatory chemoreceptor stimulation was altered with acclimatization. Severinghaus *et al.* (114) reported that lumbar CSF pH, (assumed to reflect the primary stimulus for the medulary chemoreceptor) of four male subjects became alkalotic upon arrival (day 2) at 3800 m, but returned to normal sea-level values after eight days of acclimatization. Those investigators postulated that the CSF alkalosis upon arrival at altitude reduced medulary chemoreceptor stimulation, partially offsetting the increased peripheral chemoreceptor stimulation due to arterial hypoxemia. With the return of CSF pH to normal sea-level values (a compensation they believed due to active transport of bicarbonate

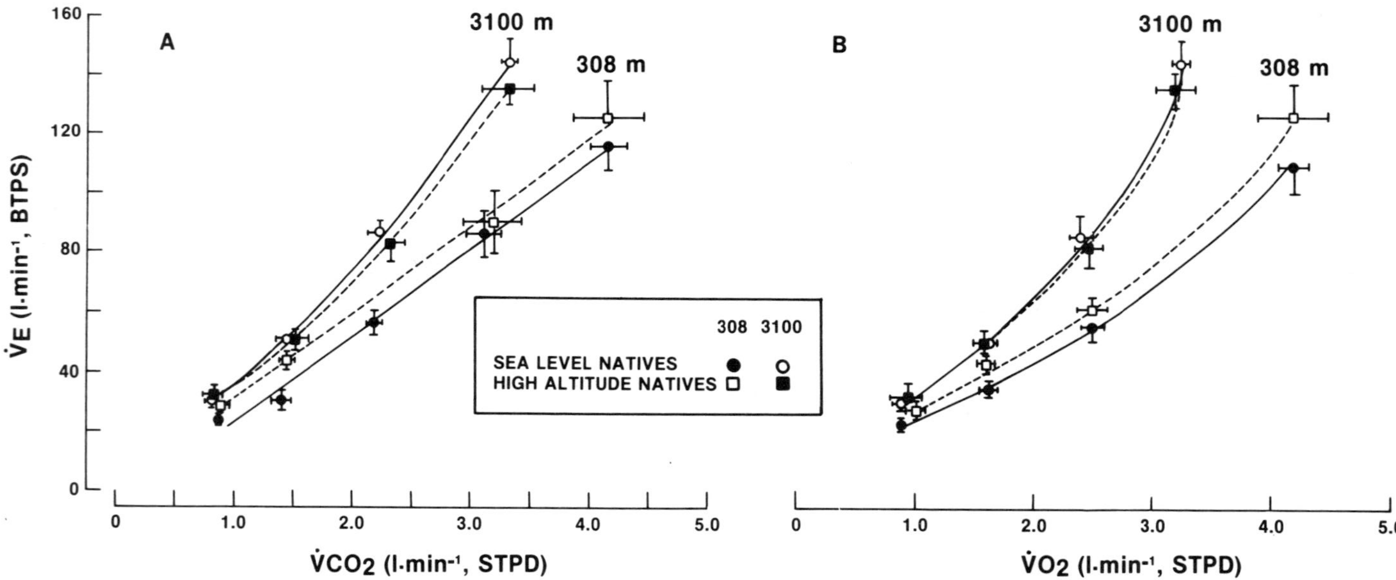

Figure 13-5. *Minute ventilation ($\dot{V}_E$) of acclimatized lowlanders (circles) and highlanders (squares) resting and exercising (treadmill running) at 308 m and 3100 m. Each group consisted of five men of very similar age, size and physical fitness level. The closed symbols depict the response at the residence altitude. Ventilation is shown as a function of carbon dioxide production ($\dot{V}CO_2$) on the left side (A) and oxygen uptake ($\dot{V}O_2$) on the right side (B). Redrawn from data of Reeves* et al. *(104) and Grover* et al. *(45).*

from CSF to blood), the medulary chemoreceptors were again stimulated and ventilation increased. Subsequent investigations failed to confirm that there was a complete compensation of CSF alkalosis concomitant with ventilatory acclimatization. Both Dempsey *et al.* (30) at moderate elevation (3,100 m) and Forster *et al.* (39) at higher elevation (4,300 m) observed that during the period when lowlanders experienced ventilatory acclimatization to altitude, their lumbar CSF remained alkaline despite a reduction in CSF bicarbonate comparable with that in blood. These investigators postulated that ventilatory acclimatization was not mediated by an alteration in CSF pH stimulation of medulary chemoreceptors.

The discrepancy between the findings of Severinghaus *et al.* (114) and those of Dempsey *et al.* (30) and Forster *et al.* (39) stimulated a number of experiments employing chemoreceptor denervation, ventricular-cisternal fluid sampling and other surgical interventions which necessitated the use of animal models rather than human subjects. The details of these animal experiments are beyond the scope of the present discussion and have been reviewed elsewhere (29), but some consideration of these findings is warranted. Results of a number of animal experiments suggest the pH measurements of human lumbar CSF probably do not reflect the true stimulus level at the medulary chemoreceptor. Hypoxia increases brain tissue lactate production (37) so that cerebral interstitial fluid pH becomes acidotic, despite an alkaline pH in the large cavity CSF. However, as discussed by Dempsey and Forster (29), before accepting that cerebral interstitial fluid acidosis results in increased ventilation with chronic hypoxia: 1) it must be established that the time course of the cerebral interstitial pH changes correspond to ventilatory acclimatization; and 2) that medulary chemoreceptor responsiveness remains unchanged during chronic hypoxia.

Different techniques have been used to separate or eliminate effects of either peripheral or central chemoreceptors and then study ventilatory acclimatization. It has been reported that ventilatory acclimatization can proceed to some degree in the absence of either peripheral (118,119) or central chemoreceptor (7,11) stimulation. The data suggest some role for both, but the studies are subject to criticism regarding the appropriateness of the animal model for human comparison (118,119) or the physiological criterion for judging ventilatory acclimatization (7).

Another approach to the study of the role of hypocapnia and alleviation of alkalosis in ventilatory acclimatization has employed human subjects. Those studies have employed voluntary hyperventilation or altering normal P_ICO_2 to separate the effects of hypocapnia and hypoxia. Eger *et al.* (35) reasoned that if ventilatory acclimatization at high altitude represented an adaptation to hypocapnia allowing hypoxic ventilatory drive to be unopposed, then ventilatory acclimatization would be prevented by maintaining P_ACO_2 at sea-level values during an 8-hour hypoxic exposure period. Eger *et al.* (35) used changes in the ventilatory response to altered P_ACO_2 as a measure of ventilatory acclimatization. They found that changes in ventilatory responsiveness to CO_2 were greatest when hypoxia was accompanied by hypocapnia and least when P_ACO_2 was maintained at or above normal sea-level values during hypoxia. Despite the fact that hypoxic exposure without hypocapnia altered ventilatory responses to CO_2, the authors' conclusion was that ventilatory acclimatization to altitude represented an adaptation to hypocapnia which was potentiated by hypoxia (35).

Using the same rationale, Cruz et al. (21) studied ventilatory acclimatization in two groups of four subjects exposed 100 hours to simulated high altitude. The P_ICO_2 for one group was elevated to prevent hypocapnia, while P_ICO_2 was normal for the other group. Ventilation at rest and during exercise was increased after eight hours of altitude exposure in the group with increased P_ICO_2, but not in the other group. The subjects with supplemental CO_2 showed no further changes in ventilation between 8 and 100 hours of hypoxic exposure, but the group with normal P_ICO_2 exhibited a progressive increase in ventilation (both resting and exercise) during this time period. Isocapnic hypoxic sensitivity followed the same pattern of change, that is, an immediate increase with no further change in the CO_2 supplemented group, and slow progressive increase in the non-supplemented group. The authors concluded that the CO_2 supplemented subjects did not experience ventilatory acclimatization, because hypocapnic alkalosis had been prevented. As noted by others (29), however, the control group did not appear to experience any hypocapnia until after 27 hours of hypoxia. No data was reported regarding systemic alkalosis during the early (days 1-3) part of the sojourn, although data reported elsewhere (20) indicate that the control subjects were slightly more alkalotic by the fourth day of exposure. Furthermore, Cruz *et al.* (21) did not measure ventilation after restoration of normoxic normocapnia. Therefore, it is not certain that the CO_2 supplemented group did not experience ventilatory acclimatization. It does appear that mild hypercapnia may have altered the time course and magnitude of ventilatory acclimatization to 100 hours of hypoxia.

Consistent with the idea that hypocapnia and hypoxia exert separate but synergistic effects on ventilatory control are the data of Dempsey *et al.* (31), who observed an increased resting ventilation of eight subjects following 26 hours of voluntary hyperventilation at sea level (normoxic hypocapnia). The increase in ventilation was greater when the 26-hour hyperventilation period was performed under conditions of hypoxia. These studies (21,31,35) indicate that ventilatory acclimatization at high altitude undoubtedly involves some degree of adaptation to both hypoxia and hypocapnia.

The increase in ventilation experienced by the lowlander during the first weeks of altitude acclimatization does raise P_AO_2 (see Figure 13-2), but not back to sea-level values. Thus, other adaptive mechanisms to lessen the effects of hypoxia must explain the dramatic recovery in endurance which is observed during this phase of acclimatization. One such possible adaptation is that the lung's efficiency as a gas exchanger is increased during altitude acclimatization. It has been generally observed that high-altitude natives have a greater pulmonary diffusing capacity than sea-level residents (26,48,132). The greater pulmonary diffusing capacity in highlanders may be a result of a natural selection process at high altitude favoring persons with this trait. In addition, animal studies indicate that pulmonary diffusing capacity increases at high altitude due to structural and functional adaptions in the lung resulting from chronic hypoxia during early childhood while the lung is still growing (64). Acclimatized lowlanders at high altitude show no change compared to sea-level values in their pulmonary diffusing capacity during either rest (26,48,72,132) or exercise (26,140). Changes in pulmonary diffusing capacity and (A-a) PO_2 gradients which have been reported were observed during the first few days (<72 hours) at altitude (106,132,139). These transient changes are exacerbated by exercise and probably reflect impairment of lung function due to development of sub-clinical levels of pulmonary edema (73) or pulmonary hypertension (73,139). These maladaptive responses to altitude have been discussed elsewhere in this book. In summary, the efficiency of the sojourning lowlander's lung as a gas exchanger is not enhanced by altitude acclimatization.

PHYSICAL WORK CAPACITY

For the lowlander sojourning at high altitude, one of the most important consequences of the acclimatization process is the improvement in exercise tolerance. Endurance, that is the length of time a given intensity of exercise can be sustained, is decreased initially upon arrival at high altitude, but within two to three weeks there is a dramatic improvement. For example, Buskirk *et al.* (13) reported that after four to five weeks of residence at 4000 m, the endurance of lowland athletes had recovered to the point that they could compete equally in soccer with athletes native to that altitude.

Subjective observations such as this and other anecdotal reports of performance of lowlanders sojourning at altitude have been verified by more quantifiable experimental observations. Maher *et al.* (81) observed a 45% increase in endurance (cycle ergometry) between the second and twelfth day of acclimatization at 4300 m, and Horstman *et al.* (60) observed a 60% increase (treadmill running) between the second and sixteenth day of acclimatization at the same altitude. Both studies were carefully designed to preclude the occurrence of any training or detraining effects during the altitude sojourn.

Endurance is an inverse function of the relative exercise intensity (41). The relative exercise intensity relates the metabolic rate required to sustain a given absolute intensity to the individual's maximal aerobic capacity. Relative exercise intensity is defined by expressing the steady-state oxygen uptake ($\dot{V}O_2$) as a percent of the maximal oxygen uptake ($\dot{V}O_2$ max). Figure 13-6 depicts the typical change in these parameters for a sea-level resident on arrival at high altitude, in this case 4300 m. As discussed in the preceding chapter, the $\dot{V}O_2$ max decreases with acute altitude exposure, but the steady-state $\dot{V}O_2$ required for a given absolute exercise intensity is the same

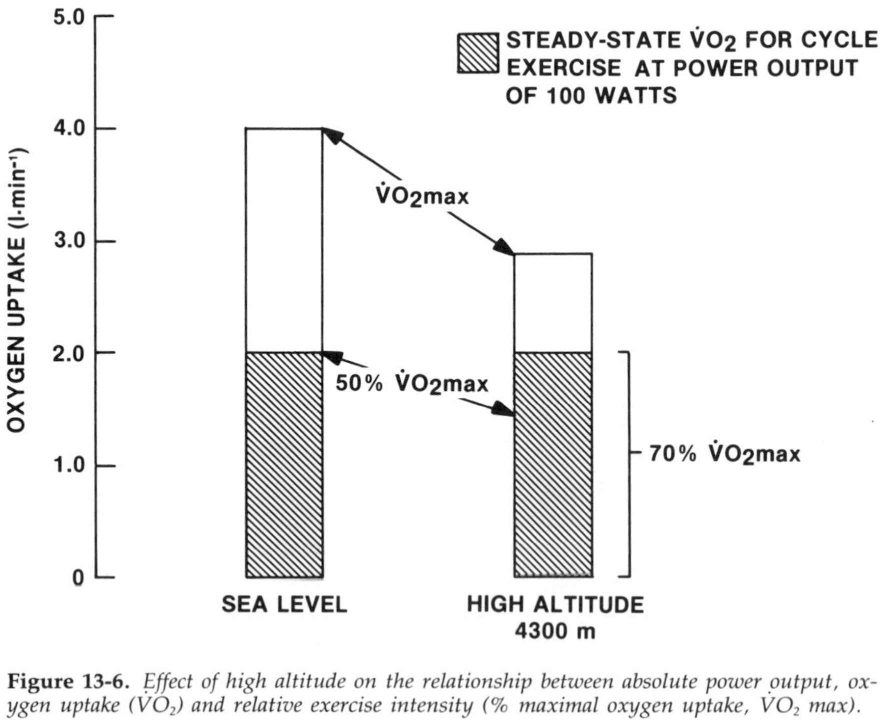

Figure 13-6. *Effect of high altitude on the relationship between absolute power output, oxygen uptake ($\dot{V}O_2$) and relative exercise intensity (% maximal oxygen uptake, $\dot{V}O_2$ max).*

as at sea level (69,108,111). Since the corresponding relative exercise intensity is greater at high altitude, endurance is decreased. Figure 13-6 also shows that to maintain relative exercise intensity the same as at sea level, the power output (absolute $\dot{V}O_2$) must be reduced. It has been demonstrated that endurance is the same with acute high-altitude exposure as at sea level if the relative exercise intensity is the same (60,81).

The foregoing discussion should suggest that one possible explanation for the improved endurance is that relative exercise intensity for a given power output is reduced by acclimatization. Either a reduction in steady-state $\dot{V}O_2$ for a given absolute intensity, an increased $\dot{V}O_2$ max or both during altitude acclimatization would lower relative intensity. There is ample evidence that steady-state $\dot{V}O_2$ during submaximal exercise at high altitude is the same as at sea level, both acutely (104) as well as after acclimatization (69,98,104,108).

There is some disagreement, however, regarding the effects of altitude acclimatization on $\dot{V}O_2$ max. Several investigations studying $\dot{V}O_2$ max during altitude acclimatization are summarized in Table 13-1. Studies reporting changes in $\dot{V}O_2$ max with altitude acclimatization (or in response to any chronic stressor) are often confounded by changes in physical activity resulting in training/detraining effects. It is reasonably clear that at moderate altitude the decrement in $\dot{V}O_2$ max does not worsen with increasing duration of sojourn as long as the subjects remain physically active (60, 68,104,146,148). Sedentary subjects do experience an additional decrement in $\dot{V}O_2$ max during two weeks of high-altitude residence, and it has been suggested that "detraining" may be accelerated at high altitude (148). A progressive worsening in the decrement in $\dot{V}O_2$ max during prolonged sojourn, particularly at extreme altitude, could result from loss of lean body mass (anorexia and accelerated muscle catabolism) or development of chronic mountain sickness, pulmonary or cerebral edema. These pathophysiological conditions, however, do not reflect the normal acclimatization process. In fact, the increase in alveolar P_AO_2 and other physiological adaptations yet to be discussed would suggest that the initial decrement in $\dot{V}O_2$ max at high altitude might be ameliorated with acclimatization.

As shown in Table 13-1, Horstman *et al.* (60) reported that $\dot{V}O_2$ max during treadmill running increased during three weeks of residence at 4300 m. Unfortunately, these investigators expressed the $\dot{V}O_2$ max data relative to total body mass of the subjects and it is unclear to what extent changes in body weight due to diuresis or decreasing body fat may have contributed to the 3.5 $ml \cdot kg^{-1} \cdot min^{-1}$ increase in $\dot{V}O_2$ max observed (60). In contrast, two recent studies also performed at 4300 m and employing a design obviating any

TABLE 13-1. *Effect of altitude acclimatization on maximal oxygen uptake ($\dot{V}O_2$ max).*

Reference	Altitude	Duration	Protocol	Results	Comment
Reeves *et al.* (104)	3100 m	3 wks	Continuous, progressive intensity treadmill walking to exhaustion	No change	Activity constant; $\dot{V}O_2$ max determinations confirmed by constant-grade walks on separate days
Klausen *et al.* (68)	3800 m	5 wks	Continuous, progressive intensity cycling to exhaustion	4% increase	Confounding effect of increased activity at altitude; no physiological criterion for $\dot{V}O_2$ max
Saltin *et al.* (108)	4300 m	2 wks	Cycle exercise at intensity selected to exhaust subjects in 3–5 min	No change	Non-standard test with no physiological criterion of $\dot{V}O_2$ max; reduced activity.
Horstman *et al.* (60)	4300 m	3 wks	Discontinous, progressive intensity treadmill running	10% increase	Carefully controlled activity program precluded training/detraining. Strict physiological criterion for $\dot{V}O_2$max.
Young *et al.* (146)	4300 m	3 wks	Discontinuous, progressive intensity cycle exercise	No change	Carefully controlled activity program precluded training/detraining. Strict criterion for $\dot{V}O_2$ max.
Young *et al.* (148)	4300 m	3 wks	Continuous, progressive intensity cycle exercise	No change/ decrease	Subjects maintaining activity levels at altitude show no change in $\dot{V}O_2$ max, but in sedentary subjects $\dot{V}O_2$ max decreased.

training effects reported no change in $\dot{V}O_2$ max during cycle exercise (expressed absolutely in terms of $\ell \cdot min^{-1}$) after 2–3 weeks of acclimatization (146,148). It is possible that differences between observations of Horstman *et al.* (60) and the latter studies (146,148) relate to the different exercise modes. Thus, $\dot{V}O_2$ max is probably little changed, if at all, by short-term altitude acclimatization, per se. Grover *et al.* (45) observed that highland and lowland natives matched for age, height, weight and $\dot{V}O_2$ max at sea level experienced the same decrement in $\dot{V}O_2$ max at 3100 m as compared to 308 m. However, comparison of residents from higher (4350 m) altitudes (135) with lowland residents (134) suggests a smaller decrement in $\dot{V}O_2$ max between sea level and high altitude for those highlanders. The differences between the observations of Grover *et al.* (45) and Vogel *et al.* (134,135) may relate to differences in subjects (athletes versus non-athletes), residence altitude, or may indicate that $\dot{V}O_2$ max of high-altitude residents is less affected by altitude than that of lowlanders.

A factor contributing to the uncertainty concerning the effect of acclimatization on the decrement in $\dot{V}O_2$ max at high altitude is the considerable interindividual variability in the magnitude of this decrement at any given altitude. For example, Figure 13-7 shows that the magnitude of the decrement in $\dot{V}O_2$ max, experienced by sea-level residents at 4300 m, is normally distributed over a wide range, even though the mean decrement for the group closely fits predictions based on relationships described by Buskirk *et al.* (12). The reason for the large variability is that this decrement in $\dot{V}O_2$ max is not determined by the response of a single physiological parameter to altitude exposure, but the combined effect of the responses of the many components of the systemic oxygen transport system. The maximal rate of oxygen uptake is in direct proportion to the maximal rate of systemic oxygen transport. In the next section, the changes in the components of systemic oxygen transport that occur during altitude acclimatization will be discussed. Because the major components of systemic oxygen transport change in opposing and offsetting directions during acclimatization, the net effect on total systemic oxygen transport can vary between individuals.

SYSTEMIC OXYGEN TRANSPORT AND DELIVERY

The previous chapter discussed how the reduction in arterial oxygen content (C_aO_2) with acute hypoxia necessitated an increased cardiac output at any given level of $\dot{V}O_2$. Maximal cardiac output is unchanged upon arrival at high altitude (122), but is achieved at a lower exercise intensity than at sea level. Acute arterial hypoxemia

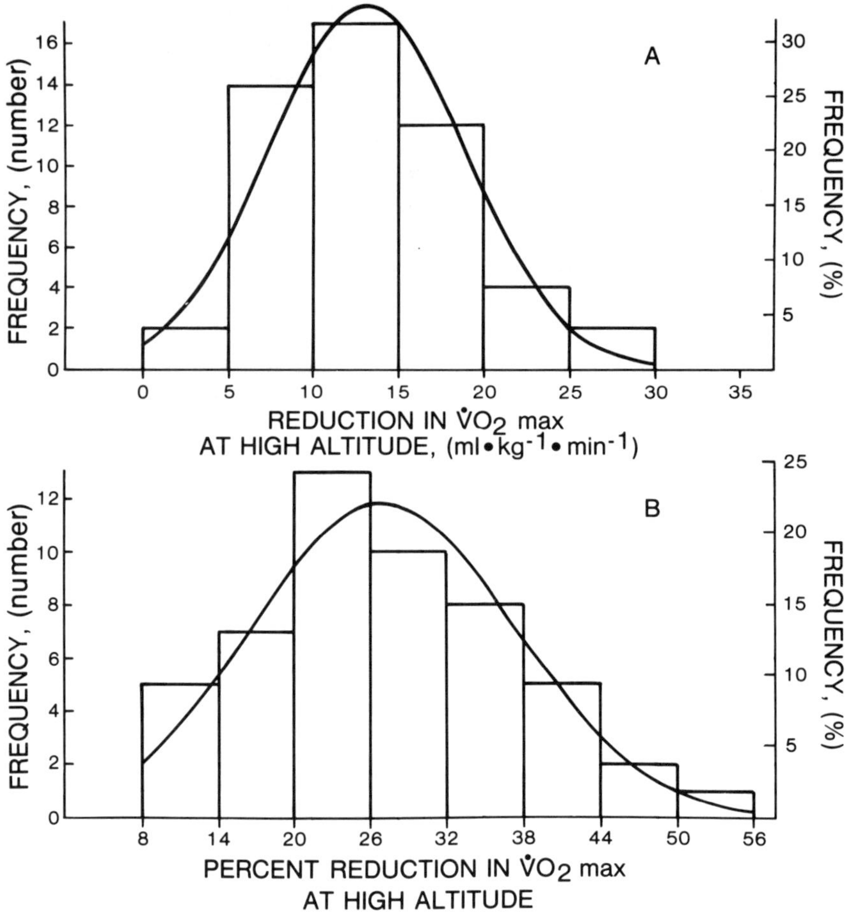

Figure 13-7. *Variability in the decrement in maximal oxygen uptake ($\dot{V}O_2$ max) experienced by 54 individuals at an altitude of 4300 m. From Young* et al. *(144).*

compromises tissue oxygen extraction, and $\dot{V}O_2$ max is reduced in direct proportion to the reduction in C_aO_2 under acute high-altitude conditions. The increased cardiac output during submaximal exercise is an effective means of sustaining $\dot{V}O_2$ requirements under acute hypoxic conditions. However, this response is very stressful for the sojourner in that the work of the heart is greatly increased. With altitude acclimatization, the increased demand on the heart is alleviated.

The effect of altitude acclimatization on sea-level residents' cardiac output is depicted in Figure 13-8. Cardiac output at rest (67,134) and during submaximal (67, 134) and maximal (67,108,134) exercise

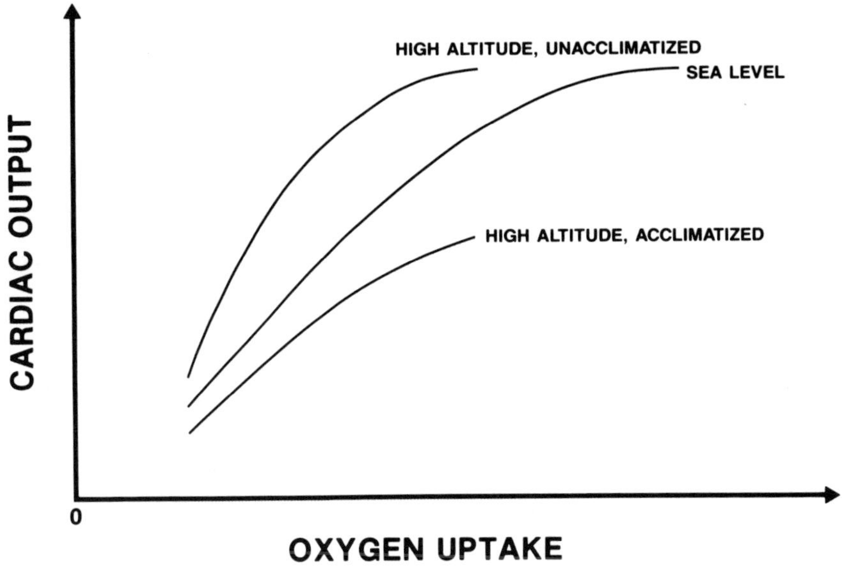

Figure 13-8. *Effect of acclimatization on the relationship between cardiac output ($\dot{Q}$) and oxygen uptake ($\dot{V}O_2$) at high altitude.*

is reduced after acclimatization at moderate altitudes. Most reports indicate that this reduction in cardiac output exceeds the magnitude of the initial increment observed with acute altitude exposure. Thus, with acclimatization, cardiac output for any givel level of $\dot{V}O_2$ is lower at altitude than at sea level (2,67,134). The reduction in cardiac output becomes observable after about two days at high altitude (67,134) and progresses for eight to ten days (2,134). In Figure 13-9, the cardiac output of lowlanders exercising at sea level and at 4350 m after ten days of acclimatization is compared with that of highland natives exercising at the same altitudes. Note that after ten days of acclimatization, the cardiac output of the lowlanders exercising at altitude is similar or even somewhat lower than the highland resident (134,135). Similar conclusions are reached by comparing responses of residents of 3100 m to acclimatized lowlanders at that altitude (2,56). This suggests that, at least at moderate altitudes, ten days is sufficient for full development of this particular adaptation. Figure 13-9 also shows that the difference in maximal cardiac output between sea level and 4350 m is smaller for high-altitude residents than for the lowland sojourner, thus accounting for the smaller decrement in $\dot{V}O_2$ max experienced by the highlanders (134,135).

Changes in heart rate during rest and exercise at altitude are less important than stroke volume changes in determining the re-

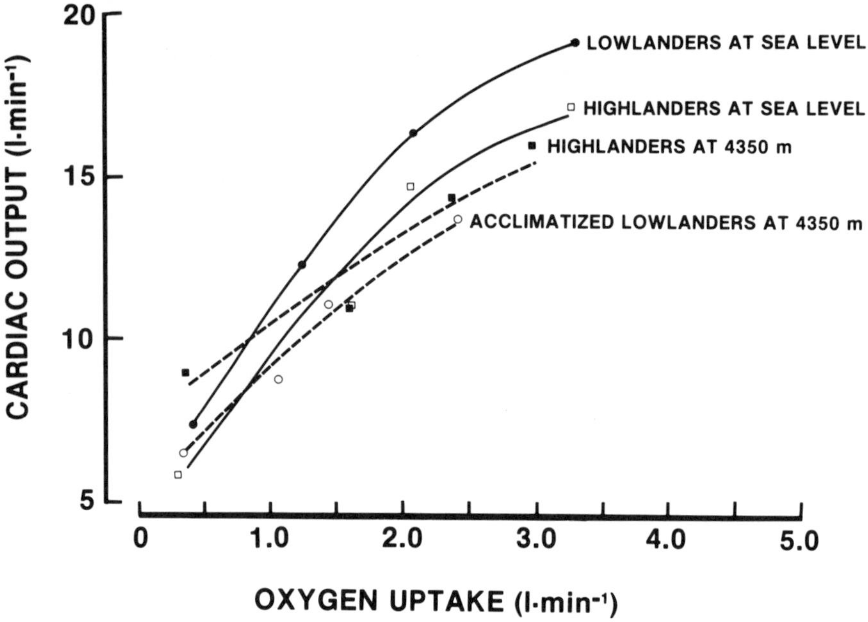

Figure 13-9. *Cardiac output ($\dot{Q}$) as a function of oxygen uptake ($\dot{V}O_2$) during exercise in lowlanders (circles) and highlanders (squares) at sea level and 4350 m. Closed symbols represent data at residence altitude. Drawn from data of Vogel* et al. *(134,135).*

duction in cardiac output with acclimatization. Grover *et al.* (47) have reviewed a number of studies reporting the effects of acclimatization on heart rate responses to exercise at altitude, and have noted considerable intra-individual variability. Whether or not the exercise tachycardia occurring on initial arrival at altitude abates, persists or increases, probably depends on the elevation, exercise mode and exercise intensity being considered. It seems safe to conclude that acclimatization to altitude does not reduce heart rate below sea-level values at least until exercise intensity approaches maximal. For example, Figure 13-10 shows data reported by Moore *et al.* (90) indicating that at 4300 m tachycardia is greatest at low exercise intensity, but above about 90% $\dot{V}O_2$ max a bradycardia is observed.

The results of studies of stroke volume changes with altitude acclimatization are in closer agreement than the studies on heart rate changes. Stroke volume at rest (2,67,90,134) and during submaximal (2,46,134) and maximal (46,60,108,134) exercise at altitude is reduced following acclimatization. The time course for this adaptation matches that of the change in cardiac output; after about two days at altitude, a measurable reduction, compared to sea-level

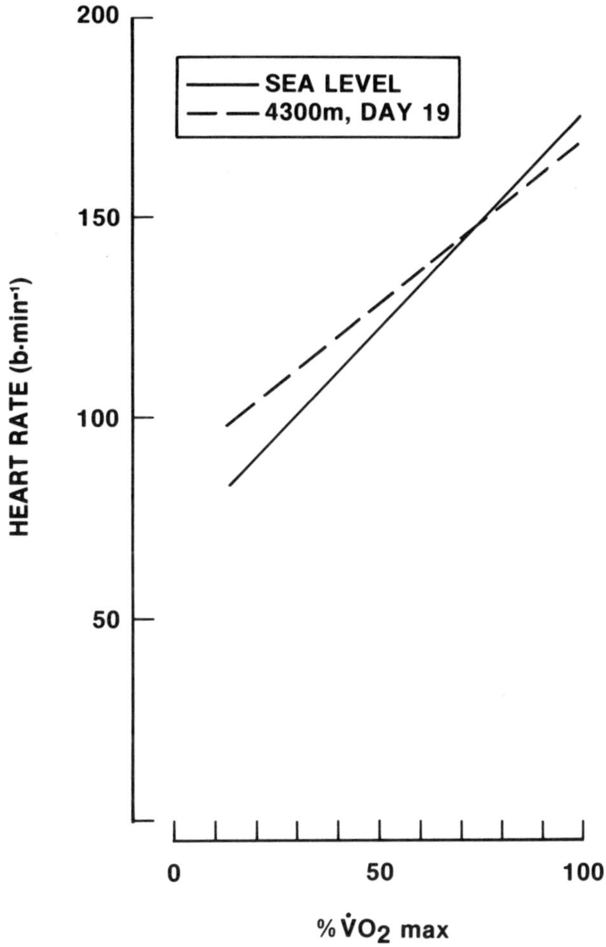

Figure 13-10. *Heart rate of lowlanders at rest and during exercise at sea level and at 4300 m after 19 days. Data are expressed as a function of relative exercise intensity (% maximal oxygen uptake, $\dot{V}O_2$ max). Modified from Moore* et al. *(90).*

values (134), is apparent. Stroke volume changes appear complete by about the tenth day (2,134). Either reduced cardiac filling or depressed myocardial contractility could cause a decrease in stroke volume. Recently, it was reported that stroke volumes for a given right atrial pressure were the same during exercise at various simulated altitudes as at sea level (105). This observation was interpreted as evidence that reduced ventricular filling accounted for reduced stroke volume at altitude, and that chronic severe hypoxia did not impair myocardial contractility. Tachycardia during exercise would contribute to a reduction in cardiac filling. Probably more importantly, a reduction in total blood volume occurs during altitude acclimatization concomitant with the decrease in cardiac output. A reduction in total blood volume would reduce stroke volume by the Frank-Starling relationship. Comparisons of acclimatized lowlanders (2,134) with native highlanders (56,135) indicate that both maintain similar cardiac outputs during exercise at altitude, but the native tends to maintain a higher stroke volume and lower heart rate than the sojourner.

It should be noted that the cardiac output changes described above were observed at relatively moderate (<5000 m) altitudes. Recent evidence from a 40-day simulated ascent of Mt. Everest (Operation Everest II) shows that cardiac output does not fall during sojourn at extreme altitude (>6000 m) (105). The acclimatization process was probably different during the progressive decompression used in Operation Everest II than in prolonged sojourn at a constant elevation. It appears, however, that the fall in stroke volume during exercise at extreme altitudes is offset by an increased heart rate. The subjects in the Operation Everest II study were extremely physically fit as would be any mountaineer attempting to climb Mt. Everest. Thus, in persons sojourning at very high altitudes who have sufficient cardiac reserve capacity, cardiac output for a given $\dot{V}O_2$ can be maintained the same as at sea level.

A decreased cardiac responsiveness to sympathetic nervous stimulation due to altitude acclimatization (133) may contribute to reduced cardiac output at rest and during exercise at moderate altitude but the adaptation is primarily mediated by an increased C_aO_2. According to the Fick equation, a reduction in cardiac output necessitates an increased tissue oxygen extraction (i.e., widened arterial-venous oxygen content difference) to sustain an unchanged $\dot{V}O_2$. Direct measurements have confirmed that the arterial-venous oxygen content difference at rest and during submaximal exercise widens with altitude acclimatization (56). The arterial-venous oxygen content difference during maximal exercise is the same (54,60) or slightly greater (32) following altitude acclimatization than at sea level. During mild to moderate submaximal exercise ($\dot{V}O_2 < 2$ $\ell \cdot min^{-1}$), oxygen content of venous blood draining the exercising muscle is about the same following two to three weeks of acclimatization at 3100 m as at sea level (32). The increased extraction with acclimatization is primarily enabled by an increase in C_aO_2 nearly or completely back to sea-level values (54,60,134), although venous oxygen content during heavy exercise may be somewhat lower after acclimatization than at sea level (32). The increased C_aO_2 reduces demand for cardiac output via local effects on autoregulation of tissue blood flow. Thus, there is a greater total peripheral resistance at rest (2) and during exercise (2,134) at high altitude following acclimatization than at sea level. Natives of both 3100 m (56) and 4350 m (135) maintain total peripheral resistance the same as acclimatized lowlanders at those altitudes, despite a tendency for higher cardiac output (see Figure 13-9). In this way, the resident maintains a lower systemic arterial pressure than the sojourner at high altitude, but at their respective residence altitudes highlanders and lowlanders maintain similar systemic arterial pressures.

The importance of the increased C_aO_2 in the overall acclimatization process experienced by the lowlander sojourning at high altitude has been demonstrated. To understand the mechanism by which C_aO_2 increases, the effect of acclimatization on the sub-components of arterial oxygen content will be considered. Arterial oxygen content is defined by the equation:

$$C_aO_2 = S_aO_2 \times ([Hb]) \times A),$$

where [Hb] is the hemoglobin concentration in the blood, the constant A[1] is the maximum amount of oxygen which can bind to a unit of hemoglobin, and S_aO_2 is the percent saturation of hemoglobin with oxygen. Acclimatization increases the lowlander's C_aO_2 through two effects. As discussed in the first section, ventilatory acclimatization raises the sojourning lowlander's P_AO_2 and, therefore, P_aO_2. An increase in P_aO_2 produces an increase in S_aO_2. Concomitantly, blood hemoglobin concentration increases due to an increase in hematocrit.

The increase in hematocrit experienced by lowlanders during the first few weeks at high altitude represents a hemoconcentration with plasma volume decreasing and red cell volume remaining constant (2,63,65,126). Total blood volume is therefore reduced during the first weeks at altitude. Table 13-2 summarizes the findings of four studies in which the changes in plasma volume reported were based on actual measurements of plasma volume, rather than inferred from changes in hematocrit and/or hemoglobin concentration. In general, acclimatization appears to produce a rather substantial (10–20%) plasma volume reduction. The table suggests that there is a trend toward larger reductions at the higher altitudes, but the relationship needs to be better quantified. At 3800 m, hemoconcentration is observed after as little as one day of acclimatization

---

[1] In reality A is a constant (1.39 ml O_2 per gram of hemoglobin) only when considering a solution containing pure hemoglobin. *In vivo*, the presence of methemoglobin and other forms of hemoglobin having reduced ability to bind oxygen cause A to vary (1.34 to 1.36 ml O_2 per gram of hemoglobin) between individuals.

TABLE 13-2. *Reduction in plasma volume with altitude acclimatization.*

Reference	Altitude/Duration	ΔPlasma Volume* ml	% sea level	Subjects
Alexander, *et al.* (2)	3100 m/10 day	398 ± 55	11%	8 sea-level residents
Jain, *et al.* (63)	3500 m/12 day	415 ± 10	17%	18 sea-level residents
Jung, *et al.* (65)	3800 m/3–4 day	541 ± 135	14%	8 sea-level residents
Surks, *et al.* (126)	4300 m/8 day	674 ± 92	22%	5 Denver (1610 m) residents

*Values are mean ± SE (ml), or mean (% sea level); all values are negative, i.e.,reductions in plasma volume.

and is complete by about the fifth day (33,65). This time course may differ at other elevations. In their review, Grover *et al.* (47) compared data from different studies of men and women at high altitude and concluded that women may hemoconcentrate more slowly than men. This possibility remains to be fully investigated. It has also been suggested that the hemoconcentration may be attenuated or delayed in older subjects (65), but this may really represent an effect of physical fitness rather than age *per se* (33). Finally, the degree of arterial hypocapnea and alkalosis experienced during altitude acclimatization has been shown to affect the magnitude of hemoconcentration (46). In a five-day hypobaric chamber exposure, two groups of men were made equally hypoxic (P_AO_2 = 55 torr); one group received supplemental CO_2 in the inspired air sufficient to prevent alkalosis and hypocapnia, and the other did not. The reduction in plasma volume during the five days was estimated (using changes in hematocrit) to be 9% in the CO_2 supplemented group and 25% in the non-supplemented group.

The reduction in plasma volume elevates C_aO_2, alleviating the demand for blood flow to supply oxygen, and thus is an important adaptation for the lowlander during early periods of acclimatization. Erythropoetin levels in blood of the sojourning lowlander are also elevated during the first week or so at high altitude (142). Stimulation of erythropoesis would require several weeks before an effect on circulating erythrocyte volume would be expected, and during this time erythropoetin returns to normal sea-level values (142). It is not known whether acclimatization allows the lowlander at altitude to maintain a normal C_aO_2 without reduced plasma volume. Ideally, erythrocyte volume would be increased, and the associated problems of increased blood viscosity would be avoided by a concomitant increase in plasma and, therefore, total blood volume. This is possible to some extent, since residents of 3100 m have greater plasma as well as erythrocyte volumes than sea-level residents, with only a modestly elevated hematocrit (138). Vascular anatomy, however, provides the upper limit for any increase in total blood volume. Residents of 4300 m or higher have greater total blood and erythrocyte volume but reduced plasma volume compared to sea-level residents (110).

Changes in the oxygen affinity of hemoglobin are another possible way that systemic oxygen transport could be altered by altitude acclimatization. An increased oxygen affinity favors the binding of oxygen to hemoglobin in the lung under hypoxic conditions, while decreased affinity favors unloading of oxygen to tissue. The shape and position of the oxygen equilibrium curve reflect the affinity of hemoglobin for oxygen, and there are at least two factors exerting significant effects on this curve at high altitude. Systemic alkalosis tends to increase the oxygen affinity of hemoglobin via the Bohr effect. However, hypoxia and associated alkalosis (25,80) stimulate an increase in erythrocyte concentration of 2,3-diphosphoglycerate (2,3-DPG), although the role of alkalosis has been questioned (25). Hemoglobin affinity for oxygen is decreased by 2,3-DPG. Thus, the net effect of altitude acclimatization on hemoglobin affinity for oxygen is dependent upon the degree to which the effects of alkalosis and increased 2,3-DPG offset each other.

Hemoglobin affinity for oxygen has been observed to be decreased in acclimatized lowlanders as well as residents at high altitude when compared to sea-level values (79,131). The decreased oxygen affinity is apparent in as little as 15 hours of altitude exposure (80). There are, however, dissenting reports that hemoglobin oxygen affinity is the same at high altitude as at sea level both in lowland sojourners (109) as well as residents (109,142). Winslow *et al.* (143) have pointed out that the oxygen equilibrium curve is usually measured under standardized *in vitro* conditions (37° C, pH = 7.4, PCO_2 = 40 torr). When the oxygen equilibrium curves were corrected to the subjects' actual arterial pH measured at the time the blood was obtained, there was no difference between high-altitude and sea-level residents in the oxygen affinity of hemoglobin (143). In contrast, it has been reported that even when the *in vivo* arterial pH values are used to correct the oxygen-equilibrium curves, subjects sojourning ten days at 3100 m experience a reduction in hemoglobin affinity for oxygen (44). It seems that there is still some question regarding changes in hemoglobin affinity for oxygen during acclimatization.

All the aforementioned reports considered resting individuals. With exercise at sea level, there is an increase in femoral venous temperature and a decrease in venous pH, both of which contribute to a small decrease in *in vivo* hemoglobin oxygen affinity (130). Acclimatization to 3100 m has no effect on the decrease in hemoglobin-oxygen affinity during exercise (32); however, it has yet to be determined whether or not acclimatization to higher altitudes has any effect on *in vivo* hemoglobin affinity changes during exercise. Theoretical considerations indicate that such a decrease in affinity during exercise at very high altitudes (>5000 m) would be detrimental, especially when oxygen diffusion in the lung is limited (6).

In summary, altitude acclimatization does alter systemic oxygen transport. With adaptation, an increased C_aO_2 enables a given submaximal level of oxygen uptake to be sustained with a widened arterial-venous oxygen content difference and lower cardiac output. Thus, submaximal steady-state exercise at high altitude is performed with less circulatory strain following acclimatization. Despite the increased arterial oxygen content, extraction of oxygen cannot signficantly exceed that achieved during maximal exercise at sea level. Maximal cardiac output falls with acclimatization to moderate altitudes. Whether a given individual experiences some slight recovery, no change or further decrease in maximal oxygen uptake during altitude acclimatization varies, and depends on to what degree the changes in the various components of systemic oxygen transport offset each other. It seems likely that the reduction in circulatory strain during submaximal exercise contributes more to the improvement in exercise tolerance with altitude acclimatization than any change in maximal oxygen uptake.

BODY COMPOSITION AND METABOLISM

As discussed in the previous section, hemoconcentration with a reduction in plasma and blood volumes occurs during the first weeks of altitude acclimatization. Plasma volume reduction can result from either a loss of total body water, or redistribution of plasma water to extra-vascular spaces, or both. Several investigators have attempted to measure total body water during chronic high-altitude exposure and have reported it to increase (52), decrease (19,74) and remain constant (75,126). Differences among the findings of these studies remain unresolved and understandably reflect technical difficulty in the measurement of total body water. For example, some of these studies used deuterium oxide (D_2O) to measure total body water (75,126), and the D_20 technique has been shown to be especially unreliable during the first week of exposure to high altitude (52).

Changes in body fluid compartments observed in nine resting subjects during acclimatization at 4300 m are shown in Figure 13-11 (52). During the first three to seven days at 4300 m, plasma and extracellular volumes decreased while intracellular volume in-

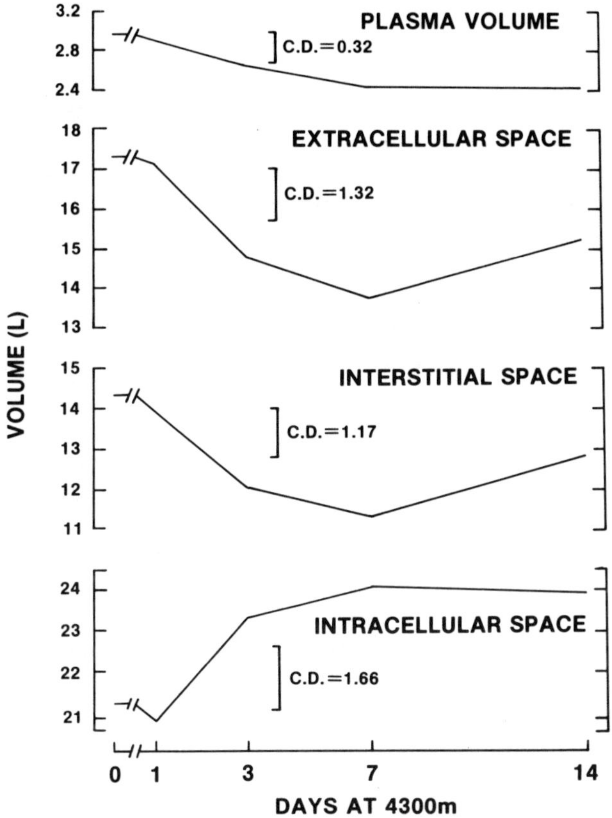

Figure 13-11. *Changes in the volume of various body fluid compartments during seven days of altitude acclimatization. Redrawn from Hannon* et al. *(52).*

creased. Following seven days exposure to 4300 m, 20% (approximately 3.5 liters) of the initial extracellular fluid volume had shifted to intracellular spaces (52). After 14 days at 4300 m, there was a small, but significant recovery of extracellular volume. Concurrently, intracellular volume remained increased, and total body water (measured by the 4-aminoantipyrine technique) was only slightly increased. The observation that total body water of sea-level natives is relatively unchanged during altitude acclimatization is consistent with determinations in highlanders. Residents of 4,500 m have a total body water similar to sea-level residents, whether expressed as percent total body weight (103) or adjusted to height, weight and age (116). Also, as in acclimatized lowlanders at high altitude, the extracellular fluid compartment volume is larger in the high-altitude natives than sea-level residents (103).

The reduction in plasma volume during acclimatization is associated with changes in plasma protein concentrations. During the first 24 to 48 hours of exposure to 4100 m altitude, there is an increase in the concentration of albumin and other blood proteins (124). The increase in protein concentration is primarily due to the rate of intravascular water loss exceeding the rate of equilibration of proteins between the intravascular and extravascular compartments (124). There is, however, also evidence of increased albumin catabolism during the first three days of exposure to 4100 m, and after eight days of acclimatization, the rate of albumin synthesis was decreased (124). This decrease in the rate of albumin synthesis may be important in oncotic regulation of the decreased plasma volume in acclimatized lowlanders at altitude.

Unacclimatized lowlanders sojourning at high altitude often experience a loss of body weight. Body weight loss may be particularly pronounced in mountaineers and trekkers at extreme (>5000 m) altitudes. Certainly, any reduction in total body water would contribute to changes in weight (19,126). However, participants in most high-altitude expeditions and field studies increase their level of physical activity which, combined with inadequate caloric intake (due to anorexia [19,126] or restricted rations) would result in caloric deficits and weight loss. Weight loss has been attributed to a loss of body fat (75,126) and/or lean body mass (19,107). The amount of weight lost during altitude sojourns has been inversely correlated with the initial amount of body fat; leaner subjects may experience the greatest weight loss (126).

When fed *ad libitum* at high altitude, male sea-level natives can lose as much as 3 kg during the first week at altitude (19,75,124,126). During a 40-day hypobaric chamber study that simulated the ascent of Mt. Everest, seven male subjects lost an average 7.4 kg (8.9% of their initial weight). The weight loss was greater than could be accounted for by comparing calculated energy expenditures to actual caloric intake, and fluid balance data showed no evidence of hypohydration. Data from computed tomography scans indicated that most of the weight loss was from muscle mass (107). A decrease in lean body mass at altitude would result in a decrease in total body water since muscle is ~75% water.

Consistent with loss of lean body mass at altitude, a negative nitrogen balance has been reported in humans during exposure to high altitudes (19,124). When compared to sea level, five young male subjects sojourning eight days at 4300 m ingested approximately 30% fewer calories per day and decreased their mean protein intake from 1.32 $g \cdot kg^{-1} \cdot day^{-1}$ to 0.90 $g \cdot kg^{-1} \cdot day^{-1}$ (19). Nitrogen balance was −2.33 $g \cdot day^{-1}$ while body weights were decreased an average 2.24

kg or 3.01% of initial sea-level weight (19). Loss of body fat has also been reported to account for weight loss during altitude sojourn (75,126).

These reports of changes in body composition have prompted investigations of the effects of altitude acclimatization on energy substrate metabolism. Basal $\dot{V}O_2$ increases when low-altitude residents go to high altitude (42,125). During the first two and a half days at 4300 m, increases on the order of 7 (42) to 14% (125) percent above sea-level values have been reported. However, by the fourth day at 4300 m, basal $\dot{V}O_2$ of sojourners appeared to be declining, and after eight days it was observed to be nearly the same as it had been at low altitude (125). The early suggestion that the increased basal $\dot{V}O_2$ at high altitude reflected the increased work of breathing (42), is not supported by the observation that $\dot{V}O_2$ appears to decline during acclimatization, while ventilation is increasing. Basal $\dot{V}O_2$ might decrease with acclimatization due to a progressive increase in the relative contribution of fat oxidation (decreased respiratory quotient) for satisfying resting energy requirements.

There is evidence that resting fat metabolism is altered during prolonged exposure and/or acclimatization to high altitude. Comparison of socioeconomically matched South American high- and low-altitude natives showed that highlanders had lower total blood cholesterol levels, particularly in the low density lipoprotein-cholesterol fraction, as compared to lowlanders (28). It was unclear to what extent these presumably beneficial adaptations in the highlanders were due to altitude effects *per se*, rather than differences in chronic activity levels and physical fitness. Resting fat metabolism also appears altered in lowlanders sojourning at high altitude. For example, sea-level residents experience a progressive increase in resting plasma free fatty acid concentrations during 14–18 days at 4300 m (66,146). Suggestions that altitude exposure is a factor contributing to increased high density lipoprotein (HDL) levels in mountaineers (94) have not been confirmed. In a recent 40-day hypobaric chamber study that simulated ascent of Mt. Everest, test subjects experienced increased (85%) plasma triglyceride levels and decreased total and high density lipoprotein-cholesterol (Young, P.M., unpublished observations). Further study of sojourners and highlanders is needed; however, it is clear that resting fat metabolism is altered by high-altitude sojourn.

Energy substrate metabolism during exercise is also affected by acclimatization to high altitude. As discussed earlier in this chapter, a substantial improvement in endurance for submaximal exercise has been reported to take place during two to three weeks of acclimatization at 4300 m (60,81). For submaximal exercise at 65–85% $\dot{V}O_2$

max, exhaustion at sea level coincides with depletion of glycogen stores in the active muscle (18). By delaying the depletion of muscle glycogen stores, endurance has been shown to increase at sea level. There is evidence that improvements in endurance during altitude acclimatization are also a result of a delaying of the depletion of muscle glycogen stores. Maher *et al.* (81) observed that the plasma lactate levels of eight sea-level residents, following 30 min of steady-state submaximal exercise, (75% $\dot{V}O_2$ max) were reduced after 12 days acclimatization at 4300 m compared to the second day at altitude or at sea level. It was suggested that muscle glycogenolysis following acclimatization resulted in reduced lactate production (81); however, increased removal and oxidation of lactate by the liver and nonexercising muscles could also result in a lower plasma lactate concentration. As shown in Figure 13-12, Young *et al.* (146) observed a reduction in muscle glycogen utilization during 30 min of exercise (85% $\dot{V}O_2$ max) following 18 days at 4300 m, in addition to the reduction in plasma lactate accumulation. Although the effects of altitude acclimatization on muscle glycogen use during prolonged (>30 min) exercise remain to be investigated, it appears that a reduced rate of muscle glycogenolysis during exercise contributes to the improved endurance with altitude acclimatization.

An increase in fat oxidation during exercise apparently allowed the "sparing" of muscle glycogen in the study of Young *et al.* (146). Respiratory exchange ratio during exercise may decrease with acclimatization (146), but this is not a consistent observation (60,148). Changing ventilation during acclimatization lessens the utility of the respiratory exchange ratio as an indicator of substrate metabolism. Changes in free fatty acid and glycerol levels during exercise are affected by acclimatization and provide stronger evidence of greater reliance in fat oxidation. Figure 13-12 shows the effects of altitude and acclimatization on changes in serum free fatty acids and glycerol during exercise. Resting levels of free fatty acids in the blood increased with acute exposure to 4300 m and were further increased by the eighteenth day of acclimatization. Note that there was a large increase in serum glycerol with exercise on the eighteenth day at 4300 m, but no change in glycerol during exercise at sea level or with acute high-altitude exposure. With lipolysis, glycerol and free fatty acids enter the blood in a one to three molar ratio; yet free fatty acids were unchanged, despite the large increase in glycerol during exercise on the eighteenth day of acclimatization. Thus, lipolysis and increased mobilization of free fatty acids had been balanced by removal and oxidation during exercise, so the circulating free fatty acid levels remained unchanged from resting levels. Similar observations were reported for a separate study (148), confirm-

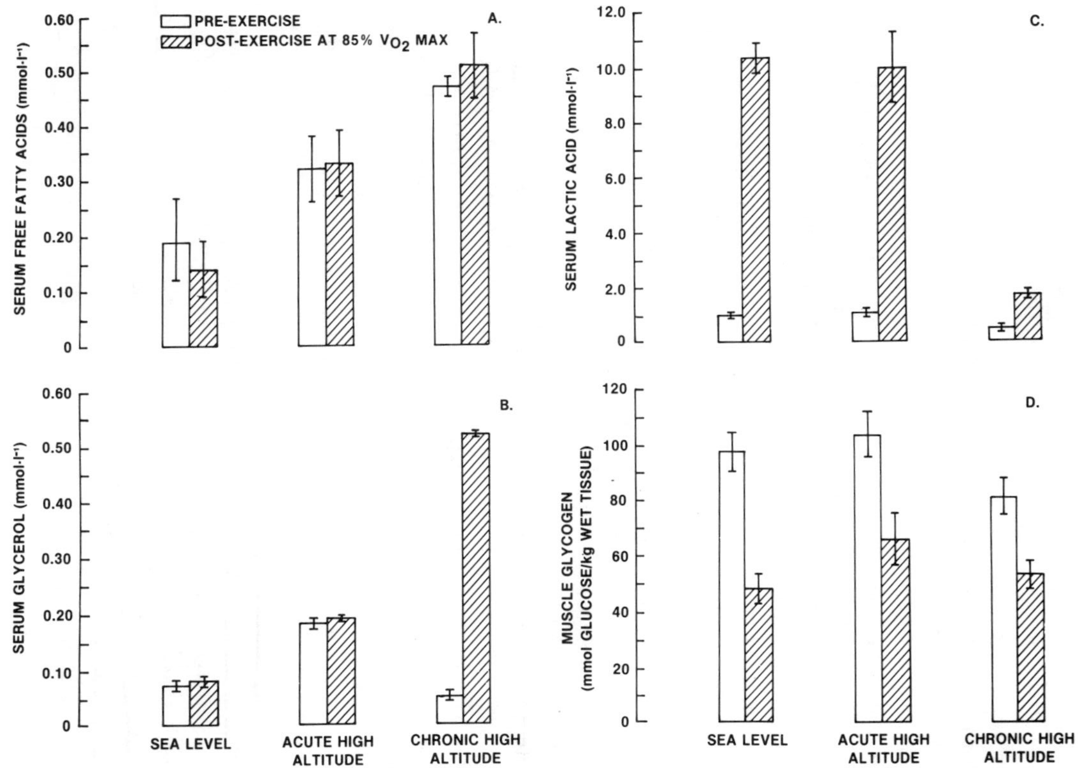

Figure 13-12. *Change (mean ± SE, N = 8 male, sea-level residents) in serum free fatty acid (A), glycerol (B), lactate (C) and muscle glycogen (D) concentration during exercise at sea level and altitude, before and after acclimatization. From Young* et al. *(146).*

ing that the mobilization and utilization of free fatty acids during exercise is increased after chronic altitude exposure.

The mechanisms for sparing of muscle glycogen and a shift to fat metabolism at altitude remain unknown, but there are several possibilities. Unlike physical training programs at sea level, altitude acclimatization does not induce synthesis of skeletal muscle oxidative or glycolytic enzymes, at least, during the first three weeks (147). It has been suggested that a low carbohydrate and/or hypocaloric diet combined with high energy expenditure could reduce the altitude sojourner's glycogen stores forcing a shift to fat metabolism (129). Such an adaptation does take place in response to hypocaloric/low carbohydrate diets at sea level, but the period required for subjects to adapt is on the order of six weeks (18). As will be discussed, there is an increase in sympathetic activity during the first weeks of altitude acclimatization which is associated with an increase in plasma norepinephrine. Increased sympathetic activity stimulates adipose tissue lipolysis (58) releasing free fatty acids and glycerol into the blood. The free fatty acids circulate to the muscle and are taken up via mass action.

Altitude acclimatization has been recently reported to reduce plasma (and presumably muscle) ammonia accumulation during exercise, as compared to sea level or initially at high altitude (148). Ammonia stimulates glycogenolysis via phosphofructokinase activation, thus a reduction in muscle ammonia with acclimatization could contribute to a glycogen sparing effect. Furthermore, ammonia inhibits pyruvate dehydrogenase and pyruvate carboxylase leading to lactate accumulation from pyruvate. Decreased muscle ammonia accumulation would favor decreased glycogenolysis and lactate accumulation. High circulating lactate levels inhibit lypolysis during exercise (9). Therefore, reduced plasma lactate accumulation with acclimatization whether by decreased production, increased clearance or both would favor increased mobilization of fatty acids.

To summarize, energy metabolism is altered in the lowlander sojourning at altitude for prolonged periods. There is an apparent increase in the fat oxidation during exercise and probably also at rest. Muscle glycogen utilization and plasma lactate accumulation during exercise are reduced with acclimatization. An adaptation resulting in sparing of muscle glycogen stores during exercise may contribute to the increase in endurance experienced by the acclimatizing lowlander. Whether or not those adaptations persist upon return to sea level is an interesting question not yet addressed. Likewise, the metabolic responses of high-altitude residents to exercise remain to be studied.

NEUROHUMORAL RESPONSES

Changes in autonomic nervous system function may be the common mediator of cardiovascular and metabolic adaptations associated with altitude acclimatization (47). Most research on this topic has focused on the sympathetic branch of the autonomic nervous system. It is clear that sympathetic activity is increased during altitude acclimatization. As shown in Figure 13-13, there is a progressive rise in urinary norepinephrine excretion of lowlanders sojourning at high altitude (10,24,125). Plasma norepinephrine concentration also increases during prolonged altitude exposure (24). It is generally considered that the adrenal medulla is not involved in this response, because plasma (24) and urinary (10,125) epinephrine levels at high altitude are relatively unchanged from sea-level values. It may, however, be noteworthy that subjects exposed to simulated altitude in hypobaric chamber experiments have shown an increase in plasma (70) and urinary (5,70) epinephrine, which is

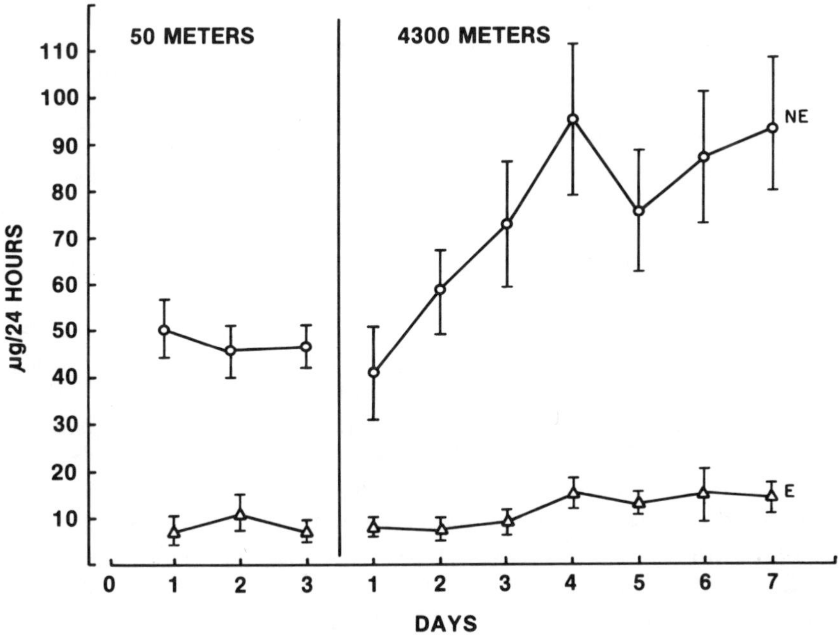

Figure 13-13. *Twenty-four hour urinary excretion (mean ± SE) of norepinephrine (NE) and epinephrine (E) in eight male sea-level residents before and during a seven-day sojourn at 4300 m. There were no significant changes in epinephrine excretion, but norepinephrine excretion was increased ($P < 0.05$) by the fourth day at high altitude. From Burse* et al. *(10).*

probably a general stress response to chamber confinement, not hypoxia. Norepinephrine levels rise during the first three to five days at altitude and appear to have plateaued by about the seventh day. Norepinephrine levels remain elevated as long as four days after descent to sea level (125). Norepinephrine binds to adenergic receptors in tissue. At least for beta-adenergic effects, the increased level of sympathetic stimulation during the first month or so at high altitude is modulated by a concomitant decrease in the number (133) and/or sensitivity (83) of recepters. Sympathetic activity in highland natives and acclimatized lowlanders who have remained three months or more at high altitude, appears to be similar to that in sea-level residents (115).

The parallel time course for changes in ventilation and changes in sympathetic activity during short- and long-term altitude acclimatization imply some relationship. Subjects exposed to hypobaric hypoxia with CO_2 added to the inspired air to prevent alkalosis and hypocapnia experience an immediate increase in ventilation (21) and urinary catecholamine excretion (20). Subjects exposed to equivalent alveolar hypoxia without CO_2 supplementation of inspired air show a "blunted" ventilatory response (21) and little or no change in urinary catecholamine excretion (20). Those observations provide support for the suggestion that increases in ventilation during altitude acclimatization "activate" the sympathetic nervous system.

Relatively little information is available regarding the effects of altitude acclimatization on parasympathetic activity. Subjects exercising submaximally at sea level and high altitude (4350 m) after a week of acclimatization experienced similar increments in heart rate following parasympathetic blockade with atropine (47). This suggests acclimatization had not altered parasympathetic activity. On the other hand, parasympathetic blockade with atropine produces an increase in the maximal heart rate of acclimatized subjects at 4600 m, whereas no effect is seen at sea level (57). No studies have investigated whether or not parasympathetic activity varies with duration of altitude residence. However, Vogel *et al.* (135) observed that maximal heart rate of highlanders was the same at 4350 m as at sea level after eight to 13 days, suggesting that parasympathetic activity is not increased in permanent or long-term residents at high altitude. Perhaps an increase in parasympathetic activity during early acclimatization is a transient adaptation that has the effect of protecting the lowlander's heart from excessive strain until normal autonomic regulation is re-established.

On arrival at high altitude, basal metabolic rate increases (42,125) and weight loss has occurred even though sufficient calories have been provided in the diet (107). Consistent with these observations, the levels of circulating thyroid hormones, thyroxine (T_4) and triiodothyronine (T_3), increase during initial exposure to actual or simulated altitudes (71,93,123). In five lowland residents, the turnover rate of T_4 was increased during the first three days of residence at 4300 m altitude, after which T_4 turnover appeared to decline back toward control values (125). The authors noted that there was also a concomitant increase in basal oxygen uptake that correlated with an increased rate of T_4 degradation (125). Therefore, a transient increase in levels of thyroid hormones could contribute to the increased metabolic activity that is observed during early altitude exposure. These increases in circulating thyroid hormones occurred when plasma thyroid stimulating hormone (TSH) levels were unchanged (93,97,123) or elevated (93) compared to sea-level values. TSH levels were reported increased at 5400 m and 6300 m altitude during an ascent of Mt. Everest (93). Even without an increase in TSH, large vascular fluid shifts at high altitude could result in displacement of thyroid hormones from tissue stores (123).

As discussed previously, altitude acclimatization is associated with sparing of muscle glycogen stores during exercise, which may be due to an increased mobilization and oxidation of free fatty acids. Another possible mechanism allowing conservation of endogenous glycogen stores could be an increased uptake and oxidation of plasma glucose, a process controlled largely by the pancreatic hormone insulin. Although acute high-altitude exposure has been reported to result in transient alterations in plasma insulin regulation (127), acclimatization to high altitude appears to restore insulin values to those seen prior to ascent (148) or slightly lower (123). With exercise at altitude, plasma insulin levels in acclimatized lowlanders have been reported to remain constant (148) or decrease slightly (123), similar to exercise at sea level. Thus, an increase in exogenous glucose uptake is not likely to contribute to the glycogen sparing adaptation. It appears that acclimatization to high altitude restores normal hormonal regulation of glucose in lowlanders who ascend to altitude. For high-altitude residents, little information is available concerning hormonal regulation of plasma glucose. It has been reported that high-altitude natives maintain lower fasting glucose levels (101,102). Furthermore, results of glucose tolerance tests indicate that altitude residents clear plasma glucose more rapidly than sea-level residents (100,101). This increase in glucose clearance seen in high-altitude natives appears to be a long-term adaptation to high-altitude residence, rather than due to dietary or genetic factors (102). These findings suggest that permanent residents of high altitude may maintain a greater level of plasma insulin and/or have an enhanced insulin sensitivity than lowlanders.

Cortisol, the major glucocorticoid produced by the human adrenal cortex, mediates adaptive responses to stress and fasting. Cortisol maintains blood glucose levels by stimulating hepatic gluconeogenesis, exerting catabolic effects on extrahepatic tissues to provide amino acids and other gluconeogenic precursors, and, in concert with catecholamines, enhancing lipolysis to provide free fatty acids. During early altitude acclimatization, resting plasma cortisol concentrations and excretion of cortisol metabolites are reported to increase when compared to pre-ascent values (89,117). When sea-level natives resided two weeks at 4300 m altitude, urinary cortisol metabolite concentrations, an index of cortisol synthesis and turnover, were increased over sea-level values only during the first week of exposure (89). After seven days at 4300 m, urinary metabolites of cortisol returned to pre-ascent values. Another study reported an increased rate of cortisol secretion during the first three days of exposure to 3500 m altitude that was followed by an immediate decline upon return to sea level (117). Most reports indicate that the plasma cortisol response to exercise is not affected by acclimatization to altitude (8,85). The temporarily increased cortisol concentrations may augment the action of catecholamines and other hormones leading to alterations in protein, lipid and carbohydrate metabolism. The return to normal resting plasma cortisol levels after about a week at altitude probably reflects an overall lessening of the effects of environmental stress due to acclimatization.

Resting growth hormone levels are unchanged in lowlanders at high altitude (99). Among its acute effects, growth hormone modulates (increases) the metabolic activity of cells. Circulating growth hormone levels have been observed to increase more rapidly during exercise at altitude (2850 m) in acclimatized lowlanders, as compared to exercise at sea level or under acute hypoxic conditions (99,127). High-altitude natives have elevated resting growth hormone levels compared to sea-level sojourners at altitude (99,128). Growth hormone levels increase more rapidly during exercise in high-altitude natives but appear to peak at the same level as in sea-level residents (99). It has been proposed that high-altitude natives may have a higher set point for hypothalmic release of growth hormone (128) concurrent with varying rates of hepatic clearance due to alterations in hepatic blood flow (99). A more rapid release of growth hormone, following initiation of exercise, may be a compensatory response at altitude enhanced by acclimatization. This may enable aerobic muscle metabolism to more rapidly meet energy requirements and limit development of oxygen deficit. This speculation awaits investigation.

A major hormonal controller of body fluid homeostasis is aldosterone, a mineralocorticoid steroid hormone synthesized by the adrenal medulla. Aldosterone acts on the kidney to increase sodium and thus water reabsorbtion. Aldosterone levels are changed substantially in unacclimatized lowlanders sojourning at high altitude. Aldosterone secretion is under the direct control of the renin-angiotensin mechanism, which can be summarized as follows: in response to decreased blood pressure or blood volume, the juxtamedullary apparatus of the kidney secretes renin, which catalyzes conversion of angiotensinogen in the blood to angiotensin I; angiotensin-converting enzyme in the vascular endothelium (especially of the lung) rapidly transforms angiotensin I to angiotensin II, which stimulates both synthesis and release of aldosterone as well as exerting a potent vasoconstrictor effect raising blood pressure.

During the early days of altitude acclimatization, aldosterone secretion at rest and during exercise is diminished when compared to pre-ascent values. An acute fall in aldosterone levels could contribute to a decrease in plasma volume. Aldosterone release has been reported to remain under the control of the renin-angiotensin mechanism during altitude acclimatization (59,82). During the early states of acclimatization, a generalized peripheral venoconstriction occurs resulting in a decrease in venous vascular capacity. A decrease in venous vascular capacity would translocate blood to the central circulation, and could be perceived by the kidney as an increase in total blood volume (136). This would result in a decrease in aldosterone secretion. Decreased aldosterone levels have also been observed with increased renin activity and decreased angiotensin converting enzyme activity, suggesting a blunting or uncoupling of the control mechanism during altitude exposure (97,88). Following acclimatization, resting and exercise levels of renin, angiotensin II, and aldosterone returned to sea-level values (82). Thus, a return to normal sea-level parameters of the renin-angiotensin-aldosterone axis may be a useful marker of the lowlander's acclimatization to high altitude.

Another important hormonal controller of body fluid homeostasis at sea level is arginine vasopressin. Arginine vasopressin is secreted in minute quantities by the posterior pituitary gland and increases water re-absorption from the renal tubules. Arginine vasopressin levels are unchanged at high altitude (55). Therefore, arginine vasopressin does not appear to play a primary role in the redistribution of body fluid contents during altitude acclimatization.

In summary, there are a number of significant neurohumoral responses experienced by the lowlander acclimatizing at high altitude. Some may mediate the acclimatization effects such as in the case of norepinephrine. Others, such as the thyroid hormones, cortisol and insulin, may mediate transient effects upon initial altitude exposure, and acclimatization restores these hormones to normal levels. In the case of aldosterone, it is not clear whether changes in the hormone level are in response to or mediate acclimatization. The effect of altitude acclimatization on many hormones, such as atrial natriuretic factor, remains to be clarified. Neurohumoral responses and adaptations are a particularly fertile area for further research.

PERCEPTUAL AND PSYCHOMOTOR FUNCTIONS

The detrimental effects of acute hypoxia on higher level mental functions, psychomotor performance and sensory perception are well documented and described in the previous chapter. Very little research addresses the effects of altitude acclimatization on these functions. A decrement in simple mental operations such as in coding tasks has been observed on arrival at high altitude (4300 m) and appears to be alleviated after about four days of exposure (36). This may reflect recovery from acute mountain sickness rather than a direct effect of altitude acclimatization. Changes in cerebral cortical electrical activity of subjects sojourning 12 days at 4300 m showed no tendency to return to normal and were interpreted as reflecting altered CNS function (40). Evaluation of mountaineers upon return to sea level from Himalayan climbing trips (>5300 m) indicated no permanent cerebral dysfunction (16). Sensory decrements (night vision) at altitude are still measureable after five weeks of acclimatization (112). The difficulties in objectively assessing these functions, undoubtedly, deters study of these functions, but this aspect of acclimatization remains open for much research.

The perception of effort appears to be altered with altitude acclimatization. Young *et al.* (145) had eight subjects rate their perceptions of "local" (muscular), "central" (cardiorespiratory) and "overall" (integrated) exertion during exercise (85% $\dot{V}O_2$ max) at sea level and at 4500 m two hours after arrival and again on the 18th day of residence. As shown in Figure 13-14, the local rating was predominant during exercise at sea level and under acute high altitude conditions. Following acclimatization, the subjects' local rating was reduced, and the central rating was now the dominant factor. The authors suggested that the increase in plasma (and muscle) lactate provided a significant cue for effort perception during exercise at sea level and under acute altitude conditions. With acclimatization to altitude, the plasma lactate did not increase nearly as much, therefore, the subject focused on other physiological cues for effort sensation.

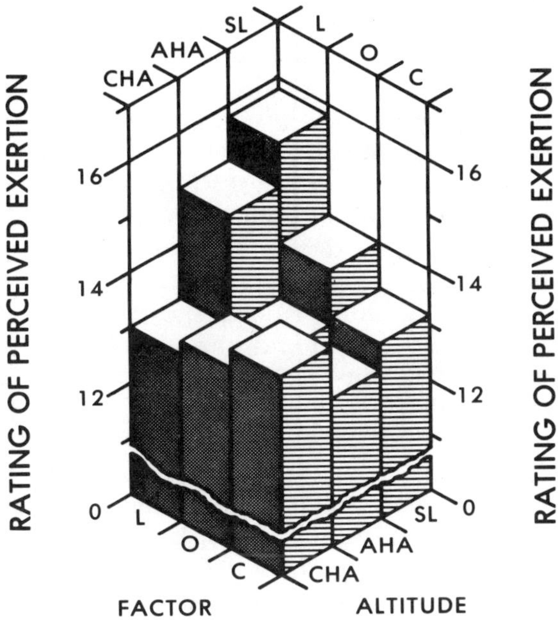

Figure 13-14. *Ratings of perceived exertion (RPE, Borg Scale) provided by subjects (N = 8) exercising (85% $\dot{V}O_2$ max) at sea level (SL), acute high altitude (AHA; 2 hours at 4300m) and chronic high altitude (CHA; day 18 of residence at 4300 m). Subjects differentiated their RPE into local (L; muscular), central (C; cardiorespiratory), and overall (O; integrated) ratings. From Young* et al. *(145).*

SPECIAL CONSIDERATIONS REGARDING ALTITUDE ACCLIMATIZATION

The similarity between the adaptations produced by physical training and altitude acclimatization raises the question of whether training at high altitude is of benefit to competitive athletes. The increased endurance, glycogen "sparing," reduction in blood lactate accumulation and changes in perception of effort during exercise all develop more rapidly during altitude acclimatization than during physical training programs at sea level. It is obvious that altitude acclimatization can improve the lowlander's overall performance at high altitude, particularly in sustained aerobic-type activities. However, because of the reduction in $\dot{V}O_2$ max at high altitude, athletes generally reduce the absolute intensity and/or duration of training sessions (3,13). Even after acclimatization, the $\dot{V}O_2$ max remains reduced at altitude (108,146,148) so that training intensity remains substantially lower than at sea level. An individual whose race "pace" is developed at altitude will find that pace too slow during sea-level competition. Furthermore, there is some evidence that detraining (i.e., effects of reduced physical activity) may be more rapid at altitude than at sea level (148). Therefore, for competition at sea level, programs emphasizing endurance training are found to be no more effective at altitude than at sea level (1), and may be less effective (13). The training effects of "interval" type training programs (repeated bouts of short-duration, high intensity exercise) are not detrimentally affected by hypoxia and may even be enhanced (4). It has been suggested that increasing emphasis on this type of training during early altitude acclimatization might best maintain muscle power and offset reductions in endurance training (3).

Because of the variability in the decrement in $\dot{V}O_2$ max experienced at high altitude (see Figure 13-7), not all persons experience the same performance handicap, and the best athlete at sea level may not retain the advantage at altitude. For competition at high altitude, elite athletes may require more than three weeks of acclimatization for optimal performance (3) though, this time requirement probably varies between individuals and elevations. When there is sufficient time available for acclimatization before competition, the lowlanders' training program must be planned in anticipation of the inability to sustain a sea-level endurance training program at high altitude. Athletes who are permanent high-altitude residents may have an advantage over the acclimatizing lowland athletes for quite some time at high altitude, but any physiological advantage appears to be minimized when the highlander comes down to sea level. Moderate-altitude (3100 m) residents and lowlanders experience a similar change in $\dot{V}O_2$ max when descending from high to low altitude (40,104). Even though an increased erythrocyte volume allows the highlander to have a greater arterial oxygen content than lowlanders, Figure 13-9 shows that maximal cardiac output of highlanders descending to sea level is not increased as high as that of lowlanders (2,134,135). Furthermore, the erythrocyte volume of highlanders sojourning at sea level decreases considerably in as little as ten days (16).

Compared with the number of investigations reporting altitude effects on males, studies of females at altitude are relatively few. Women appear to experience the same decrement in $\dot{V}O_2$ max at high altitude as do men (34). The effect of acclimatization on $\dot{V}O_2$ max, however, could be different for females than males since females reportedly show smaller hematocrit and hemoglobin increases compared to males (51). However, in another study, when females received dietary iron supplements ($FeSO_4$), their increase in hematocrit and hemoglobin was quantitatively similar to males, but the time course was slower (53). Thus, whether or not adaptations of the systemic oxygen transport mechanisms of acclimatizing females are the same as in males remains open for question. Upon return to sea level, the hematocrit of females decreased to pre-ascent values within two weeks (53), while that of males has been reported to return to normal after four to five weeks (86). Males, however, may regain plasma volume more rapidly upon return to sea level when compared to females (53); therefore, it is possible that there is a more pronounced or sustained erythropoetic response to high altitude in men. There is one investigation that compares ventilatory acclimatization of male and female lowlanders sojourning at 3600 m for three weeks (23). Resting ventilation and alveolar gas composition were reported to be similar. Thus, while more comparison studies are warranted, there appears to be no indication of major gender differences in the acclimatization process.

Females residing at high altitude undergo pregnancy in the face of the additional physiological challenge of environmental hypoxia. The maternal oxygen transport system must meet the needs of the mother as well as the developing fetus. Studies of pregnant women residing in Leadville, Colorado (3100 m), provide information about the combined adaptations to altitude and pregnancy (91,92). The blunted hypoxic ventilatory response normally seen in adult male altitude natives has been discussed earlier in this chapter. This blunted response is relieved in pregnant high-altitude residents so that ventilation is increased similarly to that of pregnant sea-level residents. Significantly increased ventilation during early pregnancy increases S_aO_2 compared to non-pregnant altitude residents (91). This increased S_aO_2 combined with her normally higher hemoglobin concentration allows the pregnant highlander to effectively maintain a C_aO_2 comparable to pregnant sea-level residents (91). In the third trimester, pregnant females at sea level and altitude both experience comparable decreases in hemoglobin concentration leading to decreased C_aO_2. The increased hypoxic ventilatory drive is an important response to pregnancy in high-altitude residents, since inadequate maternal oxygenation has been linked to low birth weight in infants (92). Mothers of smaller infants were characterized by hypoventilation, lower S_aO_2, and hemoglobin concentrations, when compared to mothers of larger babies at altitude (92).

SUMMARY

Altitude acclimatization allows humans to tolerate prolonged and even permanent exposure to environments having considerably less oxygen than that to which they were accustomed. Humans have permanently settled as high as 5300 m and have ascended, however briefly, to the highest point on earth (8870 m) without supplemental oxygen. Key adaptations experienced by the lowlander during altitude acclimatization include an increased ventilation, increased arterial oxygen content and, at least at moderate altitude, decreased cardiac output. Alterations in metabolism, body fluid regulation and neurohumoral responses also occur during acclimatization, but additional research is needed to elucidate factors mediating these changes. Endurance for submaximal exercise is increased compared to upon arrival at altitude, but maximal oxygen uptake changes little with acclimatization; therefore, overall physical work capacity remains less than at sea level. The effects of acclimatization, like the acute hypoxic responses, limit performance decrements at high altitude. With acclimatization, however, the strain, particularly on the lowlander's cardiovascular system, is reduced.

REFERENCES

1. Adams, W.C., E.M. Bernauer, D.B. Dill and J.B. Bomar. Effects of equivalent sea level and altitude training on $\dot{V}O_2$ max and running performance. *J. Appl. Physiol.* 39:262–266, 1975.
2. Alexander, J.K., L.H. Hartley, M. Modelski and R.F. Grover. Reduction of stroke volume in man following ascent to 3,100 m altitude. *J. Appl. Physiol.* 23:849–858, 1967.
3. Balke, B., J.T. Daniels and J.A. Faulkner. Training for maximum performance at high altitude. In: *Exercise at Altitude,* R. Margaria (ed.), New York: Excerpta Medica Found., 1967, p. 179–186.
4. Bannister, E.W. and W. Woo. Effects of simulated altitude training on aerobic and anaerobic power. *Eur. J. Appl. Physiol.* 38:55–69, 1978.
5. Becker, E.J. and F. Kreuzer. Sympathoadrenal response to hypoxia. In: *Biochemistry of Exercise,* Medicine and Sport, New York: Karger, 1969, Vol. 13, p. 188–191.
6. Bercowitz, H.Z., P.D. Wagner and J.B. West. Effect of change in P_{50} on exercise tolerance at high altitude: a theoretical study. *J. Appl. Physiol.* 53:1487–1495, 1982.
7. Bisgard, G.E., M.A. Busch and H.V. Forster. Ventilatory acclimatization to hypoxia is not dependent on cerebral hypocapnic alkalosis. *J. Appl. Physiol.* 60:1011–1015, 1986.
8. Bouissou, P., F. Peronnet, G. Brisson, R. He'lie and M. Ledoux. Metabolic and endocrine responses to graded exercise under acute hypoxia. *Eur. J. Appl. Physiol.* 55:290–294, 1986.
9. Boyd, A.E., S.R. Giamber, M. Mager and H.E. Lebovitz. Lactate inhibition of lipolysis in exercising man. *Metabolism* 23:531–542, 1974.
10. Burse, R.L., A. Cymerman and A. Young. Respiratory functions and muscle function during isometric handgrip exercise at high altitude. *Aviat. Space Environ. Med.* 58:39–46, 1987.
11. Busch, M.A., G.E. Bisgard and H.V. Forster. Ventilatory acclimatization to hypoxia is not dependent on arterial hypoxemia. *J. Appl. Physiol.* 58:1874–1880, 1985.
12. Buskirk, E.R. Decrease in physical working capacity at high altitude. In: *Biomedicine Problems of High Terrestrial Elevations,* Hegnauer, A.H., (ed.), Natick, MA.: U.S. Army Research Institute of Environmental Medicine, 1969, p. 204–222.
13. Buskirk, E.R., J. Kollas, R.F. Akers, E.K. Prokop and E.P. Reategui. Maximal performance at altitude and on return from altitude in conditioned runners. *J. Appl. Physiol.* 23:259–266, 1967.
14. Byrne-Quinn, E. Hypoxic and hypercapnic ventilatory drives in children native to high altitude. *J. Appl. Physiol.* 32:44–46, 1972.
15. Chiodi, H. Respiratory adaptations to chronic high altitude hypoxia. *J. Appl. Physiol.* 10:81–87, 1957.
16. Clark, C.F., R.K. Heaton and A.N. Wrens. Neuropsychological functioning after prolonged high altitude exposure in mountaineering. *Aviat. Space Environ. Med.* 54:202–207, 1983.
17. Comroe, J.H. *Physiology of Respiration,* Chicago, IL: Year Book Medical Publishers, p. 11, 1974.
18. Conlee, R.K. Muscle glycogen and exercise endurance: a twenty-year perspective. *Exer. Sport Sci. Rev.* 15:1–28, 1987.

19. Consolazio, C.F., L.O. Matoush, N.L. Johnson and T.A. Daws. Protein and water balances of young adults during prolonged exposure to high altitude (4300 m). *Am. J. Clin. Nutr.* 21:154–161, 1968.
20. Cruz, J.C., R.F. Grover, J.T. Maher, A. Cymerman and J.C. Denniston. Sustained venoconstriction in man supplemented with CO_2 at high altitude. *J. Appl. Physiol.* 40:96–100, 1976.
21. Cruz, J.C., J.T. Reeves, R.F. Grover, J.T. Maher, R.E. McCullough, A. Cymerman and J.C. Denniston. Ventilatory acclimatization to high altitude is prevented by CO_2 breathing. *Respiration* 39:121–130, 1980.
22. Cruz, J.C. and R.J. Zeballos. Influencia racial sobre la respuesta ventilatoria a la hipoxia e hipercapnia. *Acta Physiol. Lat. Am.* 25:23–32, 1975.
23. Cudkowicz, L., H. Spielvogel and G. Zubieta. Respiratory studies in women at high altitudes. *Respiration* 29:393–426, 1972.
24. Cunningham, W.L., E.J. Becker and F. Kreuzer. Catecholamines in plasma and urine at high altitude. *J. Appl. Physiol.* 20:607–610, 1965.
25. Cymerman, A., J.T. Maher, J.C. Cruz, J.T. Reeves, J.C. Denniston and R.F. Grover. Increased 2,3-diphosphoglycerate during normocapnic hypobaric hypoxia. *Aviat. Space Environ. Med.* 47:1069–1072, 1971.
26. DeGraff, A.C., R.F. Grover, R.L. Johnson, J.W. Hammond and J.M. Miller. Diffusing capacity of the lung in Caucasians native to 3100 m. *J. Appl Physiol.* 29:71–76, 1970.
27. Dejours, P., R.N. Kellogg and N. Pace. Regulation of respiration and heart rate response in exercise during altitude acclimatization. *J. Appl. Physiol.* 18:10–18, 1963.
28. Demendoza, S., E. Ineichen, E. Salazar, A. Zerpa, H. Nucele and C.J. Glueck. Lipids and lipoproteins in subjects at 1000 and 3500 m altitudes. *Arch. Environ. Health* 34:308–310, 1979.
29. Dempsey, J.A. and H.V. Forster. Mediation of ventilatory adaptations. *Physiol. Rev.* 62:262–346, 1982.
30. Dempsey, J.A., H.V. Forster and G.A. DoPico. Ventilatory acclimatization to moderate hypoxemia in man. *J. Clin. Invest.* 53:1091–1100, 1974.
31. Dempsey, J.A., H.V. Forster, N. Gledhill and G.A. DoPico. Effects of moderate hypoxemia and hypocapnia on CSF [H^+] and ventilation in man. *J. Appl. Physiol.* 38:665–674, 1975.
32. Dempsey, J.A., J.M. Thomson, H.V. Forster, F.C. Cerny and L.W. Chosy. HbO_2 dissociation in man during prolonged work in chronic hypoxia. *J. Appl. Physiol.* 38:1022–1029, 1975.
33. Dill, D.B., S.M. Horvath, T.E. Dahms, R.E. Parker and J.R. Lynch. Hemoconcentration at altitude. *J. Appl. Physiol.* 27:514–518, 1968.
34. Drinkwater, B.L., L.J. Folinsbee, J.F. Bedi, A.A. Plowman, A.B. Loucks and S.M. Horvath. Response of women mountaineers to maximal exercise during hypoxia. *Aviat. Space Environ. Med.* 50:657–662, 1979.
35. Eger, E.I., R.H. Kellogg, A.H. Mines, M. Lima-Ostos, C.G. Merrill and D.W. Kent. Influence of CO_2 on ventilatory acclimatization to altitude. *J. Appl. Physiol.* 24:607–615, 1968.
36. Evans, W.O., R.P. Carson and J.L. Shields. The effect of high terrestial environment on two different types of intellectual functioning. In: *Biomedicine of High Terrestrial Elevations*, A.H. Hegnauer (ed.), Natick, MA: U.S. Army Research Institute of Environmental Medicine, p. 291–294, 1969.
37. Fencl, V., R.A. Gabel and D. Wolfe. Composition of cerebral fluids in goats adapted to high altitude. *J. Appl. Physiol.* 47:508–513, 1979.
38. Forster, H.V., J.A. Dempsey, M.L. Birnbaum, W.G. Redden, J. Thodden, R.F. Grover and J. Rankin. Effect of chronic exposure to hypoxia on ventilatory response to CO_2 and hypoxia. *J. Appl. Physiol.* 31:586–592, 1971.
39. Forster, H.V., J.A. Dempsey and L.W. Chosy. Incomplete compensation of CSF [H^+] in man during acclimatization to high altitude (4,300 m). *J. Appl. Physiol.* 38:1067–1072, 1975.
40. Forster, H.V., R.J. Soto, J.A. Dempsey and M.J. Hoska. Effect of sojourn at 4300 m altitude on electroencephalogram and visual evoked response. *J. Appl. Physiol.* 39:109–113, 1975.
41. Gleser, M.A. and J.A. Vogel. Effects of acute alterations of $\dot{V}O_2$ max on endurance capacity of men. *J. Appl. Physiol.* 34:443–447, 1973.
42. Grover, R.F. Basal oxygen uptake of man at high altitude. *J. Appl. Physiol.* 18: 902–912, 1963.
43. Grover, R.F., J.C. Cruz, G. Jaimieson and E.B. Grover. Hypoxic ventilatory drive during exercise in man at 4300 m altitude. *J. Lab. Clin. Med.* (Abstract), 72:879–880, 1968.
44. Grover, R.F., R. Lufschanowski and J.K. Alexander. Alterations in coronary circulation of man following ascent to 3100 m. *J. Appl. Physiol.* 41:832–838, 1976.
45. Grover, R.F., J.T. Reeves, E.B. Grover and J.E. Leathers. Muscular exercise in young men

native to 3100 m altitude. *J. Appl. Physiol.* 22:555–564, 1967.

46. Grover, R.F., J.T. Reeves, J.T. Maher, R.E. McCullough, J.C. Cruz, J.C. Denniston and A. Cymerman. Maintained stroke volume but impaired arterial oxygenation in man at high altitude with supplemental CO_2. *Circ. Res.* 38:391–396, 1976.
47. Grover, R.F., J.V. Weil and J.T. Reeves. Cardiovascular adaptation to exercise at high altitude. *Exer. Sport Sci. Rev.* 14:269–302, 1986.
48. Guleria, J.S., J.N. Pande, P.K. Sethi and S.B. Roy. Pulmonary diffusing capacity at high altitude. *J. Appl. Physiol.* 31:536–543, 1971.
49. Hackett, P.H., J.T. Reeves, R.F. Grover and J.V. Weil. Ventilation in human populations native to high altitude. In: *High Altitude and Man,* J.B. West and S.K. Lahiri (eds.), Bethesda, MD: Amer. Physiol. Soc., 1984, p. 179–191.
50. Hackett, P.H., J.T. Reeves, C.D. Reeves, R.F. Grover and D. Rennie. Control of breathing in Sherpas at low and high altitude. *J. Appl. Physiol.* 49:374–379, 1980.
51. Hannon, J.P. High altitude acclimatization in women. In: *The Effects of Altitude on Physical Performance,* R.F. Goddard (ed.), Chicago, IL: The Athletics Institute, 1967, p. 37–44.
52. Hannon, J.P., K.S.K. Chinn and J.H. Shields. Effects of acute high altitude exposure on body fluids. *Fed. Proc.* 28:1178–1184, 1969.
53. Hannon, J.P., J.L. Shields and C.W. Harris. Effects of altitude acclimatization on blood composition of women. *J. Appl. Physiol.* 26:540–547, 1969.
54. Hansen, J.E., J.A. Vogel, G.P. Stelter and C.F. Consolazio. Oxygen uptake in man during exhaustive work at sea level and high altitude. *J. Appl. Physiol.* 23:511–522, 1967.
55. Harber, M.J., J.D. Williams and J.J. Morton. Anti-diuretic hormone excretion at high altitude. *Aviat. Space Environ. Med.* 52:38–40, 1981.
56. Hartley, L.H., J.K. Alexander, M. Modelski and R.F. Grover. Subnormal cardiac output at rest and during exercise in residents at 3100 m altitude. *J. Appl. Physiol.* 23:839–848, 1967.
57. Hartley, L.H., J.A. Vogel and J.C. Cruz. Reduction of maximal heart rate at altitude and its reversal with atropine. *J. Appl. Physiol.* 36: 362–365, 1974.
58. Havel, R.J. Autonomic nervous system and adipose tissue. Am. Physiol. Soc. In: *Handbook of Physiology. Adipose Tissue.* Washington, D.C.: 1965, Sect. 5, Chapt. 58, p. 575–582.
59. Hogan, R.P., T.A. Kotchen, A.E. Boyd and L.H. Hartley. Effect of altitude on renin-aldosterone system and metabolism of water and electrolytes. *J. Appl. Physiol.* 35:385–390, 1973.
60. Horstman, D., R. Weiskopf and R.E. Jackson. Work capacity during 3-week sojourn at 4300 m; effects of relative polycythemia. *J. Appl. Physiol.* 49:311–318, 1980.
61. Huang, S.Y., J.K. Alexander, R.F. Grover, J.T. Maher, R.E. McCullough, R.G. McCullough, L.G. Moore, J.B. Sampson, J.V. Weil and J.T. Reeves. Hypocapnia and sustained hypoxia blunt ventilation on arrival at high altitude. *J. Appl. Physiol.* 56:602–606, 1984.
62. Huang, S.Y., X.H. Ning, Z.N. Zhou, Z.Z. Gu and S.T. Hu. Ventilatory function in adaptation to high altitude: Studies in Tibet. In: *High Altitude and Man,* J.B. West and S.K. Lahiri (eds.), Bethesda, MD: Amer. Physiol. Soc. 1984, p. 173–177.
63. Jain, S.C., J. Bardhan, Y.V. Swamy, B. Krahna and H.S. Nyar. Body fluid compartments in humans during acute high-altitude exposure. *Aviat. Space Environ. Med.* 51:234–236, 1980.
64. Johnson, R.L., S.S. Cassidy, R.F. Grover, J.E. Schutte and R.H. Epstein. Functional capacities of lungs and thorax of beagles after prolonged residence at 3100 m. *J. Appl. Physiol.* 59:1773–1782, 1985.
65. Jung, R.C., D.B. Dill, R. Horton and S.M. Horvath. Effects of age on plasma aldosterone levels and hemoconcentration at altitude. *J. Appl. Physiol.* 31:593–597, 1971.
66. Klain, G.J. and J.P. Hannon. Effects of high altitude on lipid components of human serum. *Proc. Soc. Exp. Biol. Med.* 129:646–649, 1969.
67. Klausen, K. Cardiac output in man in rest and work during and after acclimatization to 3,800 m. *J. Appl. Physiol.* 21:609–616, 1966.
68. Klausen, K., S. Robinson, E.D. Micahel and L.G. Myhre. Effect of high altitude on maximal working capacity. *J. Appl. Physiol.* 21:1191–1194, 1966.
69. Knuttgen, H.G. and B. Saltin. Oxygen uptake, muscle high-energy phosphates and lactate in exercise under acute hypoxic conditions in man. *Acta Physiol. Scand.* 87:368–376, 1973.
70. Kotchen, T.A., R.P. Hogan, A.E. Boyd, T.K. Li, H.C. Sing and J.W. Mason. Renin, noradrenaline and adrenaline responses to simulated altitude. *Clin. Sci.* 44:243–251, 1973.
71. Kotchen, T.A., E.H. Mongey, R.P. Hogan, A.E. Boyd, L.L. Pennington and J.W. Mason. Thyroid responses to simulated altitude. *J. Appl. Physiol.* 34:165–168, 1973.
72. Kreuzer, F. and P. Van L. Campagne. Resting pulmonary diffusing capacity for CO and O_2 at high altitude. *J. Appl. Physiol.* 20:519–524, 1965.
73. Kronenberg, R.S., P. Safar, J. Lee, F. Wright, W. Noble, E. Wahrenbrock, R. Hickey, E.

Nemoto and J.W. Severinghaus. Pulmonary artery pressure and alveolar gas exchange in man during acclimatization to 12,470 ft. *J. Clin. Invest.* 5O:827–837, 1971.

74. Krzywicki, H.J., C.F. Consolazio, H.L. Jackson, W.C. Nielson and R.A. Barnhart. Water metabolism in humans during acute high-altitude exposure. *J. Appl. Physiol.* 30:806–809, 1971.
75. Kryzwicki, H.C., C.F. Consolazio, L.O. Matoush, H.L. Johnson and R.A. Barnhart. Body composition changes during exposure to altitude. *Fed. Proc.* 28:1190–1194, 1969.
76. Lahiri, S., J.S. Brody, E.K. Motoyama and T.M. Velasquez. Regulation of breathing in newborns at high altitude. *J. Appl. Physiol.* 44:673–678, 1978.
77. Lahiri, S., F.F. Kass, T. Velasquez, C. Martinez and W. Pezzia. Irreversible blunted respiratory sensitivity to hypoxia in high altitude natives. *Resp. Physiol.* 6:360–374, 1969.
78. Lahiri, S., J.S. Milledge, H.P. Chattopadhyay, A.K. Bhattacharyya and A.K. Sinha. Respiration and heart rate of Sherpa highlanders during exercise. *J. Appl. Physiol.* 23:545–554, 1967.
79. Lenfant, C. Effect of chronic hypoxic hypoxia on the O_2-Hb dissociation curve and respiratory gas transport in man. *Resp. Physiol.* 7:7–29, 1969.
80. Lenfant, C., J.D. Torrance and C. Reynafarje. Shift in the O_2-Hb dissociation curve at altitude: mechanism and effect. *J. Appl. Physiol.* 30:625–631, 1971.
81. Maher, J.T., L.G. Jones and L.H. Hartley. Effects of high altitude exposure on submaximal endurance capacity of man. *J. Appl. Physiol.* 37:895–898, 1974.
82. Maher, J.T., L.G. Jones, L.H. Hartley, G.H. Williams and L.I. Rose. Aldosterone dynamics during graded exercise at sea level and high altitude. *J. Appl. Physiol.* 39:18–22, 1975.
83. Maher, J.T., S.C. Manchanda, A. Cymerman, D.L. Wolfe and L.H. Hartley. Cardiovascular responsiveness to beta-adrenergic stimulation and blockade in chronic hypoxia. *J. Appl. Physiol.* 228:477–481, 1975.
84. Malconian, M.K., P.B. Rock, J.T. Reeves, A. Cymerman, J.D. Sutton, B.M. Groves, P.D. Wagner, H. Donner, L. Trad, V. Forte, P.M. Young and C.S. Houston. Operation Everest II: Alveolar and arterial blood gases at extreme altitude. *J. Appl. Physiol.* (In Press), 1988.
85. Maresh, C.M., B.J. Noble, K.L. Robertson and R.L. Seip. Adrenocortical responses to maximal exercise in moderate-altitude natives at 447 torr. *J. Appl. Physiol.* 56:482–488, 1984.
86. Merino, C.F. Studies on blood formation and destruction in the polycythemia of high altitude. Blood: 5:1–32, 1950.
87. Milledge, J.S. and D.M. Catley. Renin, aldosterone and converting enzyme during exercise and acute hypoxia in humans. *J. Appl. Physiol.* 52:320–323, 1982.
88. Milledge, J.S., D.M. Catley, M.P. Ward, E.S. Williams and C.R.A. Clarke. Renin-aldosterone and angiotensin converting enzyme during prolonged altitude exposure. *J. Appl. Physiol.* 55:699–702, 1983.
89. Moncloa, F., J. Donayre and R. Guerra-Garcia. Endocrine function studies at high altitude II. Adrenal cortical function in sea level natives exposed to high altitudes. *J. Clin. Endocrinol. Metab.* 25:1640–1642, 1965.
90. Moore, L.G., A. Cymerman, S.Y. Huang, R.E. McCullough, R.G. McCullough, P.B. Rock, A.J. Young, P.M. Young, D. Bloedow, J.V. Weil and J.T. Reeves. Propranolol does not impair exercise oxygen uptake in normal man at high altitude. *J. Appl. Physiol.* 61:1935–1941, 1986.
91. Moore, L.G., D. Jahnigen, S.S. Rounds, J.T. Reeves and R.F. Grover. Maternal hyperventilation helps preserve arterial oxygenation during high altitude pregnancy. *J. Appl. Physiol.* 52:690–694, 1982.
92. Moore, L.G., S.S. Rounds, D. Jahnigen, J.T. Reeves and R.F. Grover. Infant birth weight is related to maternal arterial oxygenation at high altitude. *J. Appl. Physiol.* 52:695–699, 1982.
93. Mordes, J.P., F.D. Blume, S. Boyer, M.R. Zheng and L.E. Braverman. High-altitude pituitary-thyroid function on Mount Everest, *N. Eng. J. Med.* 308:1135–1138, 1983.
94. Nestel, P.J., M. Podlewlinski and N.H. Fidge. Marked increase in HDL in mountaineers. *Atherosclerosis* 34: 193–196, 1979.
95. Pugh, L.G.C.E., M.B. Gill, S. Lahiri, J.S. Milledge, M.P. Ward and J.B. West. Muscular exercise at great altitudes. *J. Appl. Physiol.* 19:431–440, 1964.
96. Rahn, H. and A.B. Otis. Man's respiratory response during and after acclimatization to high altitude. *Am. J. Physiol.* 157:445–462, 1949.
97. Rastogi, G.K., M.S. Malhotra, M.C. Srirastava. Study of the pituitary-thyroid functions at high altitude in man. *J. Clin. Endocrinol. Metab.* 44:447–452, 1977.
98. Raynaud, J., J.P. Matineaud, M.C. Tillous and J. Durand. Oxygen deficit and debt in submaximal exercise at sea level and high altitude. *J. Appl. Physiol.* 37:43–48, 1974.
99. Raynaud, J., L. Drouet, J.P. Martineand, J. Bordachar, J. Coudert and J. Durand. Time course of plasma growth hormone during exercise in humans at altitude. *J. Appl. Physiol.* 50:229–233, 1981.

100. Reategui, E.P. Intravenous glucose tolerance test at sea level and at high altitudes. *J. Clin. Endocrinol. Metab.* 23:1256–1261, 1963.
101. Reategui, E.P. Studies on the metabolism of carbohydrates at sea level and at high altitudes. *Metabolism* 11:1148–1154, 1962.
102. Reategui, E.P., E.R. Buskirk and P.T. Baker. Blood glucose in high altitude natives and during acclimatization to altitude. *J. Appl. Physiol.* 29:560–563, 1970.
103. Reategui, E.P., R. Lozano and J. Valdirieso. Body composition at sea level and high altitudes. *J. Appl. Physiol.* 16:589–592, 1961.
104. Reeves, J.T., R.F. Grover and J.E. Cohn. Regulation of ventilation during exercise at 10,200 ft. in athletes born at low altitude. *J. Appl. Physiol.* 22:546–554, 1967.
105. Reeves, J.T., B.M. Groves, J.R. Sutton, P.D. Wagner, A. Cymerman, M.K. Malconian, P.B. Rock, P.M. Young and C.S. Houston. Operation Everest II: preservation of cardiac function at extreme altitude. *J. Appl. Physiol.* 63:531–539, 1987.
106. Reeves, J.T., J. Halpin, J.E. Cohn and F. Daoud. Increased alveolar-arterial oxygen difference during simulated high altitude. *J. Appl. Physiol.* 27:658–661, 1969.
107. Rose, M.S., C.S. Houston, C.S. Fulco, G. Coates, D. Carlson, J.R. Sutton and A. Cymerman. Operation Everest II: Effects of a simulated ascent to 29,000 feet on nutrition and body composition. Technical Report T15-87, U.S. Army Research Institute of Environmental Medicine, Natick, MA, 1987.
108. Saltin, B., R.F. Grover, C.G. Blomquist, L.H. Hartley and R.L. Johnson. Maximal oxygen uptake and cardiac output after 2 weeks at 4,300 m. *J. Appl. Physiol.* 25:400–409, 1968.
109. Samaja, M., A. Veicsteinas and P. Cerretelli. Oxygen affinity of blood in altitude Sherpas. *J. Appl. Physiol.* 47:337–341, 1979.
110. Sanchez, C., C. Merino and M. Figallo. Simultaneous measurement of plasma volume and cell mass in polycythemia of high altitude. *J. Appl. Physiol.* 28:775–778, 1970.
111. Schilling, J.A., R.B. Harvey, E.L. Becker, T. Velasquez, G. Wells and B. Balke. Work performance at altitude in man and dog. *J. Appl. Physiol.* 8:381–387, 1956.
112. Schull, W.J., R.I. Goldsmith, J. Clench, R.E. Ferrell, S.A. Barton and F. Rothhammer. 2,3-DPG and night vision. *Aviat. Space Environ. Med.* 52:41–44, 1981.
113. Severinghaus, J.W., C.R. Bainton and A. Carceben. Respiratory insensitivity to hypoxia in chronically hypoxic man. *Resp. Physiol.* 1:308-334, 1966.
114. Severinghaus, J.W., R.A. Mitchell, B.W. Richardson and M.M. Singer. Respiratory control at high altitude suggesting active transport regulation of CSF pH. *J. Appl. Physiol.* 18:1155–1166, 1963.
115. Sharma, S.C., R.S. Hoon, V. Balasubranian and K.J. Chadha. Urinary catecholamine excretion in temporary residents at high altitude. *J. Appl. Physiol.* 44:725–727, 1978.
116. Siri, W.E., C. Reynafarje, N.I. Berlin and J.H. Lawrence. Body water at sea level and at altitude. *J. Appl. Physiol.* 7:333–334, 1954.
117. Slater, J.D.H., R.E. Tuffley, E.S. Williams, C.H. Beresford, P.H. Sonkson, R.H.T. Edwards, R.P. Ekina and M. McLaughlin. Control of aldosterone secretion during acclimatization to hypoxia in man. *Clin. Sci.* 37:327–341, 1969.
118. Sorensen, S.C. Ventilatory acclimatization to hypoxia in rabbits after denervation of peripheral chemoreceptors. *J. Appl. Physiol.* 28:836–839, 1970.
119. Sorenson, S.C. and A.H. Mines. Ventilatory responses to acute and chronic hypoxia in goats after sinus nerve section. *J. Appl Physiol.* 28:832–835, 1970.
120. Sorensen, S.C. and J.W. Severinghaus. Respiratory sensitivity to acute hypoxia in man born at sea level living at high altitude. *J. Appl. Physiol.* 25:211–216, 1968.
121. Sorensen, S.C. and J.W. Severinghaus. Irreversible respiratory insensitivity in man born at high altitude. *J. Appl. Physiol.* 25:217–220, 1968.
122. Stenberg, J., B. Ekblom and R. Messin. Hemodynamic response to work at simulated altitude, 4000 m. *J. Appl. Physiol.* 21:1589–1594, 1966.
123. Stock, M.J., C. Chapman, J.L. Stirling and I.T. Campbell. Effects of exercise, altitude and food on blood hormone and metabolite levels. *J. Appl. Physiol.* 45:350–354, 1978.
124. Surks, M.I. Metabolism of human serum albumin in man during acute exposure to high altitudes (14,100 ft). *J. Clin. Invest.* 45:1442–1451, 1966.
125. Surks, M.I. H.J. Beckwitt, and C.A. Chidsey. Changes in plasma thyroxine concentration and metabolism, catecholamine excretion and basal oxygen uptake during acute exposure to high altitude. *J. Clin. Endocrinol. Metab.* 27:789–799, 1967.
126. Surks, M.I., K.S. Chinn and L.R.O. Matoush. Alterations in body composition in man after acute exposure to high altitude. *J. Appl. Physiol.* 21:1741–1746, 1966.
127. Sutton, J.R. Effect of acute hypoxia on the hormonal responses to exercise. *J. Appl. Physiol.* 42:587–592, 1977.
128. Sutton, J.R. and F. Garmedia. Hormonal responses to exercise at altitude in sea level and mountain man. In: *High Altitude Physiology and Medicine.* W. Brendel and R.A. Zink (eds.). New York: Springer-Verlag, 1980. p. 165–171.

129. Sutton, J.R., N.L. Jones and L.G.C.E. Pugh. Exercise at altitude. *Ann. Rev. Physiol* 45: 427–437, 1983.
130. Thomson, J.M., J.A. Dempsey, L.W. Chosy, N.T. Shahidi and W.G. Reddan. Oxygen transport and oxyhemoglobin dissociation during prolonged muscular work. *J. Appl. Physiol.* 37:658–664, 1974.
131. Torrance, J.D., C. Lenfant, J. Cruz and E. Marticorena. Oxygen transport mechanisms in residents at high altitude. *Resp. Physiol.* 11:1–15, 1970/71.
132. Vincent, J., M.F. Hellot, E. Vargas, H. Gautier, P. Pasquis and R. LaFrancois. Pulmonary gas exchange, diffusing capacity in natives and newcomers at high altitude. *Resp. Physiol.* 34:219–231, 1978.
133. Voelkel, N.F., L. Hegstrand, J.T. Reeves, I.F. McMurty and P.B. Moninoff. Effects of hypoxia on density of B-adrenergic receptors. *J. Appl. Physiol.* 50:363–366, 1981.
134. Vogel, J.A., L.H. Hartley, J.C. Cruz and R.P. Hogan. Cardiac output during exercise in sea level residents at sea level and high altitude. *J. Appl. Physiol.* 36:169–172, 1974.
135. Vogel, J.A., L.H. Hartley and J.C. Cruz. Cardiac output during exercise in altitude natives at sea level and high altitude. *J. Appl. Physiol.* 36:173–176, 1974.
136. Weil, J.V., E. Bryne-Quinn, D.J. Battock, R.F Grover and C.A. Chidsey. Forearm circulation in man at high altitude. *Clin. Sci.* 40:235–246, 1971.
137. Weil, J.V., E. Bryne-Quinn, I.E. Sodal, G.F. Filley. Acquired attenuation of chemoreceptor function in chronically hypoxic man at high altitude. *J. Clin. Invest.* 50:186–195, 1971.
138. Weil, J.V., G. Jamieson, D.W. Brown and R.F. Grover. The red cell mass-arterial oxygen relationship in normal man. *J. Clin. Invest.* 47:1627–1639, 1968.
139. Weiskopf, R.B. and J.W. Severinghaus. Diffusing capacity of the lung for CO in man during acute acclimatization to 14,246 ft. *J. Appl. Physiol.* 32:285–289, 1972.
140. West, J.B. Diffusing capacity of the lung for carbon monoxide at high altitude. *J. Appl Physiol.* 17:421–426, 1962.
141. West, J.B., P.H. Hackett, K.H. Maret, J.S. Milledge, R.M. Peters, C.J. Pizzo and R.M. Winslow. Pulmonary gas exchange on the summit of Mt. Everest. *J. Appl. Physiol.* 55:678–687, 1983.
142. Winslow, R.M. High altitude polycythemia. In: *High Altitude and Man,* J.B. West and S. Lahiri (eds.). Bethesda, MD: American Physiological Society, 1984, p. 163–172.
143. Winslow, R.M., C.G. Monge, N.J. Statham, C.G. Gibson, S. Charache, J. Whittenburg, O. Moran and R.V. Berger. Variability of oxygen affinity of blood: human subjects native to high altitude. *J. Appl. Physiol.* 51:1411–1416, 1981.
144. Young, A.J., A. Cymerman and R. Burse. The influence of cardiorespiratory fitness on the decrement in maximal aerobic power at high altitude. *Eur. J. Appl. Physiol.* 54:12–15, 1985.
145. Young, A.J., A. Cymerman and K.B. Pandolf. Differentiated ratings of perceived exertion are influenced by high altitude exposure. *Med. Sci. Sports Exer.* 14:223–228, 1982.
146. Young, A.J., W.J. Evans, A. Cymerman, K.B. Pandolf, J.J. Knapik and J.T. Maher. Sparing effect of chronic high-altitude exposure on muscle glycogen utilization. *J. Appl. Physiol.* 52:857–862, 1982.
147. Young, A.J., W.J. Evans, E.C. Fisher, R.L. Sharp, D.L. Costill and J.T. Maher. Skeletal muscle metabolism of sea-level natives following short-term high-altitude residence. *Eur. J. Appl. Physiol.* 52:463–466, 1984.
148. Young, P.M., P.B. Rock, C.S. Fulco, L.A. Trad, V.A. Forte and A. Cymerman. Altitude acclimatization attenuates plasma ammonia accumulation during submaximal exercise. *J. Appl Physiol.* 63:758–764, 1987.

14

Medical Problems Related To Altitude

Mark K. Malconian, D.O.

Paul B. Rock, D.O., Ph.D.

OUTLINE

INTRODUCTION

Each year large numbers of people from low altitude regions travel to areas of high terrestrial altitude for recreational activities including hiking, climbing, skiing and general sightseeing. Others such as miners, scientists and military personnel go to high altitude to work. In some areas of the world including the Tibetan Plateau and the Andean Altiplano, these high altitude sojourners are greeted by members of large indigenous populations which have been residing at lofty elevations for generations. All these people, both sojourners and permanent residents, are exposed to a variety of environmental conditions that have an adverse effect on their performance, health and well being.

Many of the environmental factors which cause medical problems in high altitude environments are not necessarily unique to high altitude, and most of them can be found in other environments. The particular combination of conditions and the nature of exposure to them is unique, however, and results in a distinct set of medical problems. The majority of these medical problems are, in some way, influenced by the progressively lower partial pressure of oxygen in the atmosphere which is a ubiquitous feature of all high altitude regions.

In this chapter we will review the present understanding of high altitude medical problems along with their treatment and prevention. The bulk of the chapter will concentrate on problems related to hypobaric hypoxia, for most of those problems are unique to the high altitude environment. Many of the other problems are well discussed in other chapters of this book. In our description of hypoxia-related problems, we will attempt to point out which information is based on the results of directed investigation and which is anecdotal. In spite of an increasing interest in high altitude medical problems, many are not well researched, and what is known of them relies heavily on scanty clinical reports, a fact that has not always been acknowledged in previous reviews.

PROBLEMS ASSOCIATED WITH ACUTE HYPOXIA

Hypoxia at high altitude is a direct consequence of decreased barometric pressure of the atmosphere, which causes a decreased partial pressure of oxygen. The partial pressure of oxygen in mountain areas ranges from 110mmg at 3,000m (10,000 ft) to 50mmg (1/3 of sea level) at the summit of Mt. Everest (29,000 feet) (65). Alveolar oxygen pressure at those altitudes ranges from approximately 58mmhg to 30mmhg (36,65). At the lower altitudes the decrease in barometric pressure and partial pressure of inspired oxygen (P_IO_2) results in little physiologic compromise, for the hemoglobin saturation changes little. At altitudes equivalent to the summit of Mt. Everest, the hemoglobin saturation has been measured as low as 40% (62). The physiologic consequences of this profound hypoxia gives rise to many medical problems (62).

Acclimation to high altitude is the process by which an individual adapts to low oxygen (Chapter 13). The physiologic changes which constitute acclimation are reversible and function primarily to increase oxygen availability at the tissue level. The most significant changes occur in the cardiovascular and pulmonary systems, but there may be adaptation at the cellular level as well. In a sense, hypoxia-induced medical problems at altitude can be viewed as related to acclimation. If acclimation is complete and the body successfully compensates for the hypoxia, no medical problems develop. Both incomplete acclimation and overcompensation result in characteristic physiologic states we recognize as altitude induced pathology.

Acute Mountain Sickness

Acute Mountain Sickness (AMS) is a symptom complex occurring in unacclimatized individuals who ascend rapidly to high altitude. It is self-limiting, but can be very debilitating. With appropriate measures it is preventable (7,9,11,14,15,47,56).

The symptoms of AMS include severe headache, lassitude, irritability, nausea, vomiting, anorexia, indigestion, flatus, constipation and sleep disturbances characterized by periodic breathing. In afflicted individuals 35% may have localized rales and many have decreased urine output despite adequate hydration (12,14,19,50). Onset of symptoms usually begins 6–12 hours after ascent and peak in intensity in 24–48 hours, resolving in 3–7 days as acclimatization takes place. A small number of individuals may have symptoms longer (50).

The reported incidence of AMS varies from 8–100% of exposed individuals and is directly dependent on both the rate of ascent and the final altitude reached (10,50). The highest incidence occurs in those sea level residents who fly into high mountain area, because flying is the fastest means of ascent and allows no acclimatization. A few individuals experience symptoms as low as 2,500m (8,000 ft), but it is much more common over 3,000m (10,000 ft). Although there is some individual variation in susceptibility to AMS, virtually every one will experience some symptoms of AMS if they go rapidly over 4,200m (14,000 ft) (50). Acclimatization at intermediate altitude decreases the incidence of AMS, but even well acclimatized individuals seem to get a headache upon reaching 5,550m (18,000 ft).

The pathophysiology of AMS is not entirely clear (10,18,20,50,61). Although it is understood that hypoxia is the triggering factor in AMS, the myriad of effects of hypoxia on the human physiology make it difficult to delineate the exact pathophysiology. The symptoms and associated physiological changes resemble closely subclinical cerebral edema. It is known that hypoxia causes a marked increase in cerebral blood flow, that may lead to increased intracranial pressure (ICP). People with AMS show a decreased hypoxic ventilatory response (1,10,14,30,41,50,59). This may further increase cerebral blood flow by a more profound hypoxia and a relative hypercarbia when compared to controls. Hypoxia causes pulmonary hypertension, which can lead to right ventricular overload and increased central venous pressure that may increase intracranial pressure leading to some cerebral edema. Hypoxia may cause cytotoxic injury to brain cells and the blood brain barrier leading to intracellular and extracellular edema causing an increase in ICP. Thus, there are several possible mechanisms to explain the occurence of mild cerebral edema in AMS (1,5,10,11,18,50,61). Research is being conducted in this area to define this pathophysiology.

Diagnosis of AMS is based upon the presence of symptoms in conjunction with a rapid ascent to altitudes over 3,000m (10,000 ft). The differential diagnosis includes such factors as dehydration, hangover, exhaustion, hypothermia, migraine and hysteria (10,50). It is important to rule out full blown High Altitude Cerebral Edema and High Altitude Pulmonary Edema (see below) both of which can be rapidly fatal if left untreated.

Although AMS is self-limiting the symptoms are debilitating enough often to require treatment. Treatment consists of descent, oxygen, acetazolamide and in limited cases, furosemide (50). Descent is the treatment of choice and should continue until symptomatic improvement occurs. Often a descent as little as 300m will be all that is required (50). Oxygen administration for obvious reasons is equally effective, however, often impractical in mountain environments. Anecdotal reports of acetazolamide 250mg every six hours will relieve symptoms, although it is more effective if used prophylactically (2,3,8,10,13,50). Furosemide may be useful if antidiuresis is present, however care must be taken not to give furosemide to individuals who are dehydrated (a common condition in people at altitude). For mild cases symptomatic treatment may be all that is required. Analgesics have varying success in treating cephalgia of AMS. Any respiratory depressant is contraindicated. Nausea and vomiting may be abated with compazine (10,50).

Many potential prophylactic measures have been formally or informally investigated for AMS. These include varying the rate of ascent, pharmacologic agents, dietary manipulations and various folk remedies. The ideal agent would prevent symptoms caused by a rapid ascent while allowing the body to acclimatize so that the prophylactic measure could be discontinued.

The use of gradual ascent, called staging allows the body to acclimatize and thus prevent AMS symptoms (13,21,50). The most widely recommended profile is to take one day of rest at 2,500m (8,000 ft) and one additional day at rest for every 600m (2,000 ft) gain in altitude above that (13,50). Because the exaggerated desaturation and hypoxia occurring during sleep appears to contribute to symptoms, it is advisable to sleep as low as possible (10,50,60,61). This is reflected in the moutaineers's axiom, "climb high, sleep low."

Pharmacologic agents are useful in prophylaxis of AMS in individuals who do not have time for gradual ascent. Military personnel often fall into this category. Acetazolamide 250 mg every six hours is effective in preventing or reducing AMS symptoms in many people at altitude (2,3,8,9,13,33,56). Currently, the Army recommends that acetazolamide be continued for 48 hours after initial ascent (50). It is common practice in mountaineering expeditions to continue acetazolamide throughout the ascent phase which may last many weeks. Acetazolamide does not prevent symptoms in all people and side effects such as paraesthesias are common. It is contraindicated in people with sulfa allergies.

A number of other prophylactic agents have been recommended but none are in widespread use. Dexamethasone 4 mg every six hours has been shown recently to be effective in preventing AMS symptoms (29). Spironolactone has also been used experimentally with varying success (50). A number of other agents including naproxen, dilantin, ergotamine and furosemide have been shown not to be effective. The chewing of cocoa leaves is a folk remedy of the South American Indians, but can not be recommended for many reasons. Finally, there is some evidence that a diet high in carbohydrates may be helpful in preventing AMS (22,50).

High Altitude Cerebral Edema

High Altitude Cerebral Edema (HACE) is a clinically apparent cerebral edema seen in individuals who ascend rapidly to high altitude. Although it has a low incidence of occurrence, HACE is potentially fatal if left untreated (5,10,18,25,50,56).

The early symptoms of HACE resemble those of AMS (5,10,18,50). In fact HACE may be a severe form of AMS (5,25,50). Early symptoms include severe headache, nausea, vomiting and extreme lassitude. Truncal ataxia, and change of mental status help differentiate early HACE from AMS (5,9,18,50). Left untreated a variety of focal and generalized symptoms will manifest including visual changes, anesthesias, paresthesias, rigidity, hemiparesis, clonus, pathological reflexes, hyperreflexia, bladder and bowel dysfunction, hallucinations and seizures (5,10,15,18,25,50,56). Other clinical manifestations of increased intracranial pressure will be seen if one has the appropriate diagnostic tools. Left untreated HACE will progress to obtundation, coma and death.

The incidence of HACE is low, occurring in about 1% of individuals exposed to altitude (5,10,15,18,50,56). Rapid ascent and lack of acclimatization are predisposing factors. It has been reported as low as 2,700m (9,000 ft) (18,50), but the majority of cases occur above 3,600m (12,000 ft). High altitude pulmonary edema (see below) often occurs with HACE.

Pathological examination of brain tissue from HACE patients reveals gross cerebral edema (10,18,50). The mechanism of edema formation may be cytotoxic (due to hypoxia), vasogenic (due to increased cerebral blood flow and increased blood pressure), or most likely a combination of the above (5,10,11,18,25,50).

The diagnosis of HACE is based upon the presence of symptoms in conjunction with rapid exposure to high altitude (10,18,50). Differential diagnosis includes, cerebral thrombosis or hemorrhage, infections, migranous encephalopathy and severe cerebral hypoxia from high altitude pulmonary edema. Because it is difficult to make a definitive diagnosis of HACE in mountainous conditions, all suspected cases should be evacuated to lower altitudes without delay (10,18,50).

The definitive treatment of HACE is to descend. In general, the greater the descent the better, preferably to an altitude of less than 2,500m (8,000 ft) (18,50). Descent of more than 300m (1,000 ft) may be required to cause any clinical improvement. Supplemental oxygen can be a useful adjunct, but should not be used as a substitute for descent (9,10,18,50). Dexamethasone 4–6 mg q.i.d. is a recognized treatment (10,18,50). Furosemide and mannitol has also been used with some success (50). Their use must be tempered with the awareness that many HACE victims are dehydrated despite the presence of HACE.

While there is no definitive evidence for effective prophylactic measures for HACE, the close relation between HACE and AMS would suggest that the measures previously discussed for the prevention of AMS may also be effective for HACE (particularly gradual ascent) (18,50).

High Altitude Pulmonary Edema

High Altitude Pulmonary Edema (HAPE) is a noncardiogenic pulmonary edema caused by altitude exposure in unacclimitized, but otherwise healthy individuals (10,24,26,38,50). Young active males seem particularly susceptible (10). Like HACE, HAPE can be rapidly fatal if left untreated, but usually responds to descent in altitude.

HAPE usually manifests itself 12–96 hours after a rapid ascent to high altitude (10,24,26,38,50). The onset is often subtle, manifested by fatigue, dyspnea on exertion and nonproductive cough (10,24,26,38,50). Frequently, AMS symptoms (nausea and headache) are also present (10,24,26,38,50). Early signs of HAPE include tachypnea, tachycardia and rales (10,24,50). As HAPE progresses, dyspnea at rest and orthopnea occur. The cough often becomes productive with blood tinged and eventually pink frothy sputum (10,50). Cyanosis becomes apparent. Mental status changes may occur indicative of severe hypoxia or HACE (10,18,50). Chest x-ray in the HAPE patient reveals multiple small patchy infiltrates, which are asymmetric and enlarged pulmonary vasculature with a normal size heart (Figure 14-1) (10,24,26,50). EKG may show tachycardia, right ventricular overload, right axis deviation, T wave inversions and other nonspecific ST-T wave changes (26,50). Without a previous EKG for comparison, an EKG may be of little useful information. If left untreated HAPE can run a fulminant course progressing to coma and death in less than 12 hours (10,50).

HAPE occurs in unacclimatized individuals who ascend rapidly to high altitude (10,24,26,38,50). The incidence is higher than that of HACE and appears to vary with age and activity. Children and

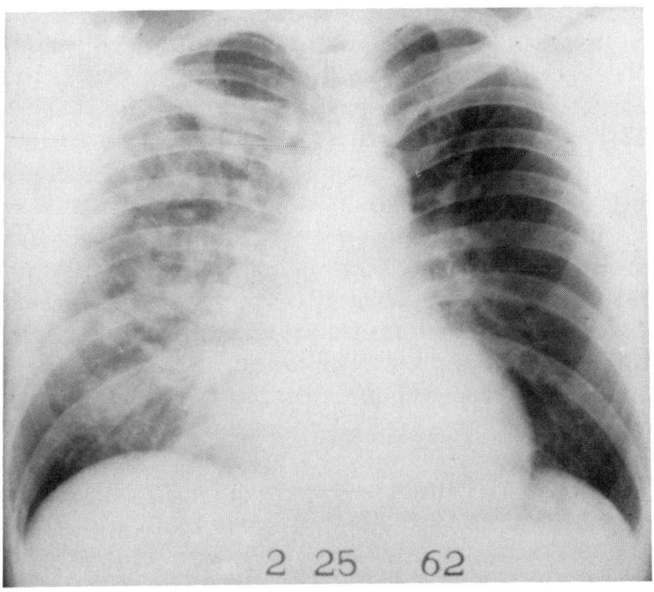

Figure 14-1. *High Altitude Pulmonary Edema (HAPE). (Courtesy of Charles S.Houston, M.D.)*

young adults have a much higher incidence than older individuals (50). Exercise at altitude, by further increasing an already increased pulmonary artery pressure may increase susceptibility to HAPE or more likely exacerbate the condition (24,50,61). HAPE is also more likely to occur in individuals who have previously experienced HAPE (50). Subclinical HAPE evidenced by rales, may occur much more frequently than previously thought (12,50,56). Reports have described rales in one-third to one-half of persons at altitude higher than 3,500m (11,500 ft) (10,24,38,61).

The pathophysiology of HAPE has recently become better understood. Recent findings in HAPE include high pulmonary artery pressure (compared to assymptomatic controls at altitude), normal left atrial filling pressure (4,10,48,50,51), normal cardiac function, blunted chemosensitivity and hypoxic ventilatory depression (17,28) and high protein pulmonary lavage samples with predominance of macrophages and complement activators (52). These findings suggest a high pressure, high permeability edema similar to neurogenic pulmonary edema (50). Other evidence indicates that HAPE may be a pure pressure edema due to redistributed blood flow away from microembolized vessels caused by hypoxia induced coagulopathy.

The diagnosis of HAPE is based upon the signs and symptoms of pulmonary edema in, otherwise, healthy individuals exposed to high altitude. The differential diagnosis includes pneumonia, congestive heart failure, pulmonary embolus and in the military setting, exposure to chemical warfare agents.

Treatment of HAPE depends on its severity, but virtually all cases will respond to descent (10,50). In severe cases descent is mandatory (10,50). Because exercise and hypoxia increases pulmonary artery pressure, descent should be via litter using supplemental oxygen if possible (50). Mild cases may be treated in place with bed rest and oxygen, but descent is mandatory for all but mild cases (10,24,38,50). Pharmacologic agents which have been useful in treating severe HAPE include furosemide and morphine sulphate (10,50). While morphine can depress respiration, its cautious use is justified similarly to its use in cardiogenic pulmonary edema (50). While using either of these agents it must be remembered that despite HAPE or HACE the person may be volume depleted and the addition of a diuretic and vasodilator may aggravate the presence of volume depletion (10,50). Aminophylline, isoproterenol and digoxin are not useful (10,50). Dexamethasone is not useful unless HACE accompanies HAPE (10,50).

Prophylactic measures for HAPE involve avoiding the risk factors for all altitude illnesses, a gradual ascent to allow for acclimatization, limited physical activity until acclimatization occurs, and because arterial desaturation is greatest during sleep, sleep at as low an altitude as possible. Avoiding cold exposure is also helpful (4,10,50). Acetazolamide may also be useful in preventing HAPE especially in persons with a prior history of HAPE (10,14,50).

High Altitude Retinal Hemorrhages (HARH)

High Altitude Retinal Hemorrhages (HARH) are the result of an altitude induced retinopathy. Although common, they are self limited and are of little consequence unless they occur in the macula (6,10,37,58). Most likely they have a similar pathophysiologic mechanism as AMS and HACE (6,10,37,58).

HARH are normally asymptomatic unless they occur in the macula. Macular hemorrhage will result in blurred vision and scotomata (10). Fundoscopic exam will reveal disc hyperemia and engorgement and increased tortousity of vessels in virtually all personnel exposed to high altitude without supplemental oxygen. Typically retinal hemorrhages appear as flame type hemorrhages in the superficial layers of the retina, but hemorrhages in the deeper layers are possible. Multiple hemorrhages are common (Figure 14-2a, 14-2b, 14-3a, 14-3b).

The incidence of HARH varies directly with altitude. They are unknown below 3,000m (10,000 ft), but the incidence approaches 100% above 6,800m (22,000 ft) (6,10,37,50,58). Strenuous exercise may increase the risk of retinal hemorrhage by increased systolic blood pressure, as well as a forced valsalva maneuver during a technical climb or while defacating. Unlike other altitude illnesses HARH are not related to the state of altitude acclimatization, and multiple occurrences are possible throughout any sojourn.

The exact mechanism of HARH is unknown. Most believe it results from pressure surges caused by exercise or other activity in retinal vessels, which are maximally dilated due to hypoxia induced increases in cerebral blood flow (6,10,37,50,58).

HARH usually go unrecognized, but may be easily diagnosed by fundoscopic exam. The differential diagnosis includes hemorrhage from vascular disease, diabetes mellitis, septic infarcts and organic hypoxia from cardiac and respiratory disease.

HARH are self limited and resolve one to two weeks after descent. There is no treatment other than descent and there is no prophylaxis. Because non-macular hemorrhages are of little consequence, descent is not necessary. When a macular hemorrhage is diagnosed, descent is imperative in order to promote healing, prevent further hemorrhages and prevent visual deficits (6,37,58).

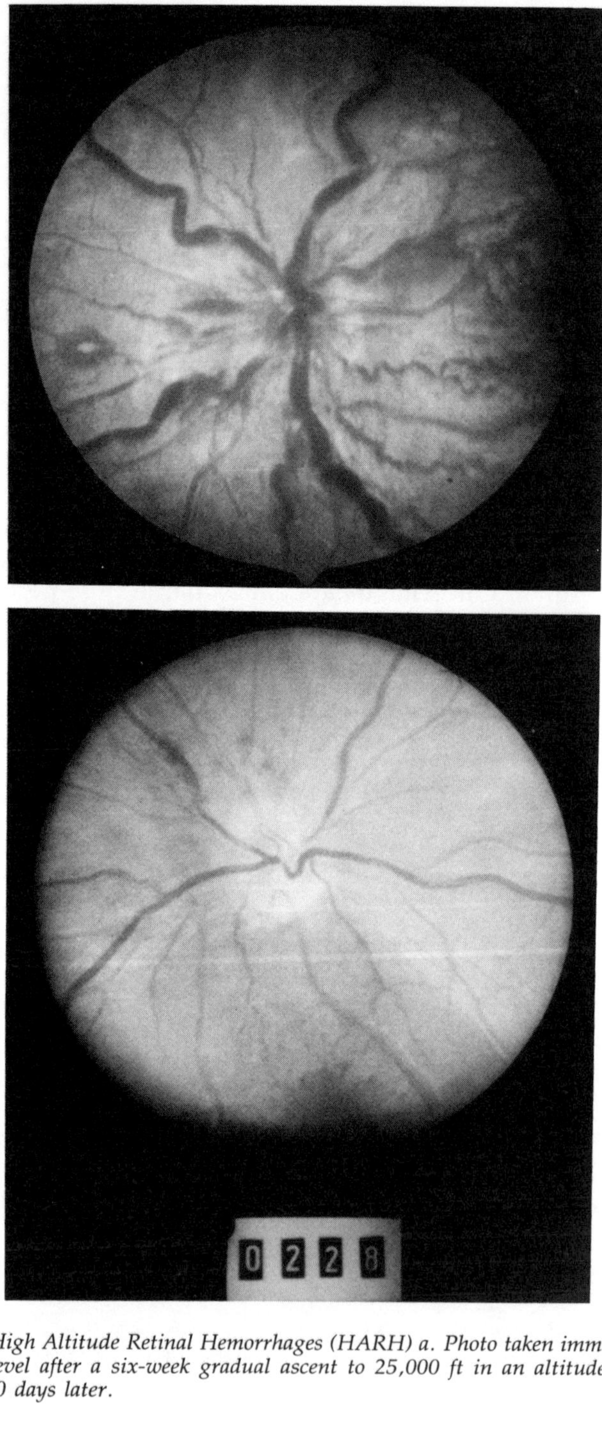

Figure 14-2. *High Altitude Retinal Hemorrhages (HARH) a. Photo taken immediately upon return to sea level after a six-week gradual ascent to 25,000 ft in an altitude chamber. b. Same subject 10 days later.*

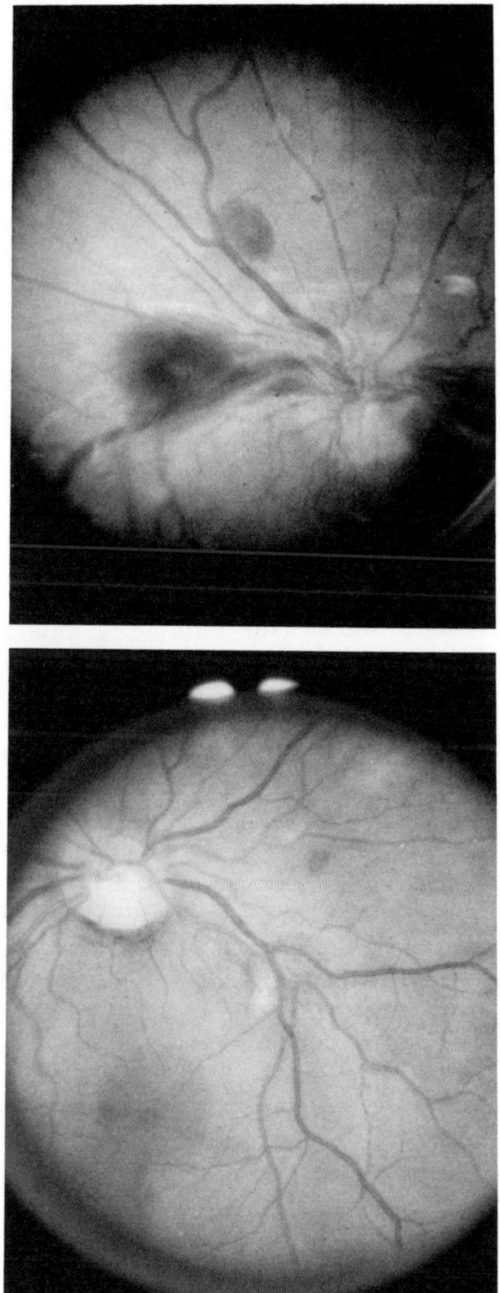

Figure 14-3. *High Altitude Retinal Hemorrhages (HARH) (Courtesy of Charles S.Houston, M.D.) a. HARH after several days at 18,500 ft. b. Same subject two weeks later.*

Generalized Peripheral Edema

Generalized edema occurs in some individuals when initially exposed to high altitude and with repeated exposure to high altitude. It is characterized by a pronounced edema of the face and upper extremities, decreased urine output and weight gain (50). Although uncomfortable, it is a benign condition. The edema resolves with descent but will reoccur with subsequent ascent. It is more common in females but is not related to menses or birth control pills (50). Furthermore it does not appear to have any relation with HACE or HAPE. The diagnosis is based on history and physical findings. The differential diagnosis includes cardiogenic edema and allergic reactions. Edema can be treated successfully with diuretics and salt restriction (50). Prophylaxis with the same regime is usually successful in preventing edema in susceptible individuals who must travel to high altitude.

Disorders Of Coagulation

People who ascend to high altitude appear to be at increased risk for venous thrombosis, pulmonary embolus and stroke (10,35,44,50,55). The possible causes for these phenomena include hypoxia induced clotting abnormalities, polycythemia, dehydration, cold and venous stasis during prolonged periods of inactivity during inclement weather (10). Clotting abnormalities observed include thrombocytopenia, decreased fibrinogen, decreased factor VIII and increased 2,3-diphosphoglycerate (DPG) and fibrin degradation products (FDP) (10,35,44,50,55). What is puzzling is that people have been observed with abnormal clotting parameters with no clinical manifestations and also with clinical manifestations from a coagulation disorder with normal laboratory clotting parameters (50). There are no general trends in the clotting parameters upon ascent to high altitude; however, there are numerous case reports of markedly abnormal clotting parameters in some individuals. More studies are needed to delineate this problem.

PROBLEMS ASSOCIATED WITH CHRONIC HYPOXIA

Reentry Pulmonary Edema

Reentry Pulmonary Edema is HAPE that occurs in high altitude natives upon returning to high altitude after traveling to low altitude areas. The occurence of Reentry Pulmonary Edema may be related to the rate of ascent upon return to high altitude. Case reports are more frequent in high altitude areas where rapid ascent is possible and absent from areas where rapid ascent is not possible

(10,27,53). It has been postulated that this occurs due to increased smooth muscle component of the chronically hypoxic pulmonary arteries that, when exposed to hypoxia, develop abnormally high pulmonary artery pressure (10,27,50,53).

Chronic Mountain Sickness

Chronic Mountain Sickness (CMS) is an ailment of people who reside at high altitude for a prolonged period of time. It is characterized by an erythrocythemia in high altitude residents (hemoglobin > 20gm/dl) (10,31,39,40,50,66). It occurs most commonly in people born at low altitude who move to high altitude, but also occurs in people who have always resided at high altitude. It was first described by Monge in 1928 when studying inhabitants of the Andes, and is often referred to as Monge's disease (40). Incidence of the disease is low in women and some populations living at high altitude have marked differences in incidence (10,42,50).

Symptoms are those of primary polycythemia: lethargy, dizziness, headache, sleep disturbances and mental status changes. Diagnosis is made by the presence of symptoms in high altitude residents with marked polycythemia (10).

Studies of CMS patients have shown excessive hypoxia for a given altitude when compared to other residents, blunted hypoxic ventilatory response, hypoventilation, widened A-a gradient and severe oxygen desaturation during sleep (10,16,31,32,39,46,50,64).

The surest form of treatment is relocation to lower altitude. When not practical, phlebotomy may provide subjective improvement. Low flow oxygen during sleep may also provide some improvement. Medroxyprogesterone acetate and acetazolamide have been shown to be successful, presumably due to the respiratory stimulant effect of these drugs (10,31,39,50).

PRE-EXISTING MEDICAL PROBLEMS AGGRAVATED BY HIGH ALTITUDE

There is little data available about the effect of altitude on pre-existing medical problems. Much of the following discussion is based on speculation. Pre-existing medical problems are, however, of great concern to many who journey to altitude.

Pulmonary Disease

The results of pulmonary disease at sea level may be pulmonary hypertension, V/Q mismatch, impaired diffusion capacity and decreased ventilatory muscle capacity. These impairments may cause illness at even modest altitudes.

People with pulmonary hypertension at sea level will experience further increases in pulmonary artery pressure at altitude from hypoxia, and thus worsening their condition; perhaps making them more susceptible to other altitude illnesses.

It seems obvious that persons who are hypoxic from chronic obstructive pulmonary disease (COPD) at sea level will have greater hypoxia and may be more symptomatic at high altitude than at sea level. These patients may want to avoid going to altitude. Further, people with COPD who are compensated or assymptomatic at sea level may experience symptoms upon ascending to altitude (50). An individual with COPD and modest sea level hypoxia, may experience a substantial decrease in oxygen saturation at a modest altitude, due to hemoglobin oxygen affinity at lower plasma oxygen partial pressures (10,50). Also, individuals with COPD may not be able to increase minute ventilation in response to the hypoxia of altitude the way a healthy individual would, thus causing them more profound hypoxia. In patients with COPD, who might be able to increase ventilation upon ascending to altitude, and who have decreased respiratory muscle reserve might conceivably develop respiratory muscle fatigue and failure. Thus, it would seem prudent that in people with COPD with sea level hypoxia or carbon dioxide retention that sojourns to high altitude be done with great care and possible supplemental oxygen. As stated previously much of the above is speculation and based on anecdotal reports.

It would seem likely that patients with obstructive sleep-apnea at sea level, would be more symptomatic at altitude. Hypoxia during obstructive apneic periods would likely be more profound than at sea level. It is difficult to predict in central sleep-apnea the effect of sleeping at high altitude due to the changes in arterial pH, cerebral spinal fluid pH and chemoreceptor sensitivities that may take place at altitude.

Studies from Colorado have shown lung disease to be more common at high altitude (42,43), and that elderly people with lung disease at altitude improve upon relocating to lower altitude (49).

Coronary Artery Disease

The question often arises as to the effect of going to altitude in patients with Coronary Artery Disease (CAD). It has been shown in healthy individuals performing maximal exercise at 29,000 ft in an altitude chamber (after a six-week gradual ascent), that there was no evidence of ischemia by EKG (51). Further, during near maximal exercise there was no evidence of ischemia by echocardiogram (57) and cardiac output was maintained (when compared to sea level values) (48). It appears in the healthy heart there is not an increased risk of ischemia under severe hypoxia at altitude. One report of patients with CAD studied 11 patients with CAD at altitudes up to 3,170m (45). In this study all patients who had normal exercise EKG's at sea level also did at 3,170m. Only patients symptomatic at sea level were also symptomatic at altitude. Further studies need to be done. It is also worth considering patients with CAD who have a degree of coronary artery spasm. It is known that hypoxia can induce spasm in these patients. Therefore, they may be at risk for coronary artery spasm upon ascent to altitude. At the present time it is unclear as to the risk of patients with CAD who ascend to high altitude.

Congestive Heart Failure

Heart failure both right sided and left would be expected to decompensate at altitude due to fluid retention, hypoxia and its effect on the pulmonary vasculature. Hackett has observed repeatedly older people with CHF who are brought to 2,000m, decompensate (10). Thus it would seem prudent for patients with CHF to avoid going to altitude.

PROBLEMS NOT ASSOCIATED WITH HYPOXIA

Cold Injuries

Cold injuries in high altitude environments are caused by a combination of environmental and physiological factors. In general, the temperature decreases 2° C for every 300m elevation (50). In addition, the climate of mountains causes a high windchill factor and much lower effective temperatures. The low temperatures, combined with peripheral vasoconstriction, dehydration and hemoconcentration at high altitude causes the full spectrum of cold injury including, hypothermia, frost nip, frost bite and immersion foot (50). The diagnosis and treatment of cold injury is discussed in chapter 11. Prevention of cold injury is based upon keeping personnel adequately hydrated, nourished and protected from cold exposure.

Dehydration

Dehydration at high altitudes results from the combination of low environmental humidity with a number of physiological and behavioral factors. There is an increased loss of fluids at altitude due to hypoxia induced fluid shifts, diuresis and hyperventilation (50). At the same time, personnel often have a decreased fluid intake due to blunted thirst sensation from cold, hypoxia, nausea from AMS and the lack of potable water. Dehydration is a serious problem in and of itself, but also greatly increases susceptibility to cold injury. Dehydration can be prevented by a conscious effort to imbibe fluids in the absence of thirst (50).

Solar Radiation Injuries

Ultra-violet (UV) radiation injury at high altitude is the result of decreased UV wave filtering capacity of the thinner atmosphere and increased reflection of light from snow and rock surfaces. Injuries include sunburn and snowblindness. Sunburn can occur rapidly and can be severe. Treatment is the same as at low altitude and can be prevented with limiting skin exposure with clothes and sunscreens. Snow blindness (photophthalmia) can develop within 12 hours of sun exposure (50). It results from UV absorption by the eye. Although self-limiting it can be extremely painful and debilitating. Opthalmic steroid preparations may provide symptomatic relief (50). Proper eye protection will prevent its occurrence (50).

Nutrition

Like dehydration, several physiological and behavioral factors contribute to poor nutrition at high altitude. Weight loss occurs due to decreased caloric intake secondary to AMS, and lack of energy to prepare and consume food. Evidence also exists for fat malabsorbtion at high altitude presumably due to hypoxia. Anecdotal and recent investigations suggest diets high in carbohydrates may be helpful due to ease of digestion and increased ventilation (50).

CONCLUSIONS

Exposure to high terrestrial altitude is associated with a distinct set of medical problems caused by a unique combination of environmental stressors including hypobaric hypoxia, cold, low humidity, increased solar radiation and rugged terrain. Many of the problems related to prolonged hypoxic exposure are unique to high altitude and can be viewed as resulting from inappropriate acclimation to the hypoxic stress. Little is known of the precise pathophysiologic mechanisms involved, but some guidelines for treatment and prevention exist none-the-less. The most effective treatments seem to involve decreasing the level of hypoxemia by descending to lower altitudes. Much research needs to be done in this area. The medical problems which are not primarily related to hypoxia are better understood in terms of their pathophysiology and are treated as they would be in any other environment.

REFERENCES

1. Anholm, J.D., C.S. Houston and T.M. Hyers. The relationship between acute mountain sickness and pulmonary ventilation at 2,835 meters (9.300 feet). *Chest* 75:33–36, 1979.
2. Birmingham Medical Research Expeditionary Society Mountain Sickness Study Group: Acetazolamide in control of acute mountain sickness. *Lancet* 1:180–183, 1981.
3. Cain, S.M. and J.E. Dunn II. Low doses of acetazolamide to aid accomodation of men to altitude. *J. Appl. Physiol.* 21:1195–1200, 1965.
4. Chauca, D. and J. Bligh. An additive effect of cold exposure and hypoxia on pulmonary artery pressure in sheep. *Res. Vet. Sci.* 21:123–124, 1976.
5. Dickinson, J.G. High altitude cerebral edema: Cerebral acute mountain sickness. *Sem. Resp. Med.* 5:151–158, 1983.
6. Frayser, R., C.S. Houston, A.C. Bryan, I.D. Rennie and G. Gray. Retinal hemorrhage at high altitude. *N. Engl. J. Med.* 282:1183–1184, 1970.
7. Gray, G.W., A.C. Bryan, R. Frayser, C.S. Houston and I.D.B. Rennie. Control of acute mountain sickness. *Aerospace Med.* 42:81–84, 1971.
8. Greene, M.K., A.M. Kerr, I.B. Mcintosh and R.J. Prescott. Acetazolamide in prevention of acute mountain sickness: a double blind controlled cross-over study. *Br. Med. J.* 233:811–813, 1981.
9. Hackett, P.H. *Mountain Sickness: Prevention, Recognition and Treatment.* New York: American Alpine Club, 1980.
10. Hackett, P.H. and T.F. Hornbein. Disorders of high altitude. In Murray, J.F. and J.A. Nadel (eds.), *Textbook of Resiratory Medicine.* Philadelphia: Saunders, 1988.
11. Hackett, P.H. and D. Rennie. Acute mountain sickness. *Sem. Resp. Med.* 5:132–140, 1983.
12. Hackett, P.H. and D. Rennie. Rales, peripheral edema, retinal hemorrhage and acute mountain sickness. *Am. J. Med.* 67:214–218, 1979.
13. Hackett, P.H. and D. Rennie. Avoiding mountain sickness. *Lancet* 2:938, 1978.
14. Hackett, P.H., D. Rennie, S.E. Hofmeister, R.F. Grover, E.B. Grover and J.T. Reeves. Fluid retention and relative hypoventilation in acute mountain sickness. *Respiration* 43:321–329, 1982.
15. Hackett, P.H., D. Rennie and H.D. Levine. The incidence, importance, and prophylaxis of acute mountain sickness. *Lancet* 2:1149–1154, 1976.
16. Hackett, P.H., J.T. Reeves, R.F. Grover and J.V. Weil. Ventilation in human populations native to high altitude. In West, J.B. and Lahiri, S. (eds.), *High Altitude and Man,* Washington, D.C.: American Physiological Society, 1984, pp. 179–191.
17. Hackett, P.H., R. Roach, R.B. Schoene, G. Harrison and W. Mills. Blunted chemosensitivity and hypoxic ventilatory depression in high altitude pulmonary edema. *Fed. Proc.* 44:1563, 1985.
18. Hamilton, A.J., A. Cymerman and P. Black. High altitude cerebral edema. *Neurosurgery* 19:841–849, 1986.
19. Hannon, J.P., K.S.K. Chinn and J.L. Shields. Effects of acute high altitude exposure on body fluids. *Fed. Proc.* 283:1178–1184, 1969.
20. Hansen, J.E. and W.O. Evans. A hypothesis regarding the pathophysiology of acute mountain sickness. *Arch. Environ. Health* 21:666–669, 1970.
21. Hansen, J.E., C.W. Harris and W.O. Evans. Influence of elevation of origin, rate of ascent and a physical conditioning program on symptoms of acute mountain sickness. *Military Med.* 132:585–592, 1967.
22. Hansen, J.E., L.H. Hartley and R.P. Hogan. Arterial oxygen increase by high carbohydrate diet at altitude. *J. Appl. Physiol.* 33:441–445, 1972.
23. Heath, D. and D.R. Williams. *Man at High Altitude.* Edinburgh: Churchill Livingstone, 1981.
24. Houston, C.S. Acute pulmonary edema of high altitude. *N. Engl. J. Med.* 263:478–480, 1960.
25. Houston, C.S. and J.D. Dickinson. Cerebral form of high altitude illness. *Lancet* 2:758–761, 1975.
26. Hultgren, H.N. *High Altitude Pulmonary Edema. Biomedical Problems of High Terrestrial Altitudes.* Springfield, VA.: Federal Scientific and Technical Information, 1967, pp. 131–141.
27. Hultgren, H.N. and E.A. Marticorena. High altitude pulmonary edema:Epidemiologic observations in Peru. *Chest* 74:372–376, 1978.
28. Hyers, T.M., C.H. Scoggin, D.H. Will, R.F. Grover and J.T. Reeves. Accentuated hypoxia at high altitude in subjects susceptible to high altitude pulmonary edema. *J. Appl. Physiol.* 46:41–46, 1979.
29. Johnson, T.S., P.B. Rock, C.S. Fulco, L.A. Trad, R.F. Spark and J.T. Maher. Prevention of acute mountain sickness by dexamethasone. *N. Engl. J. Med.* 310:683–686, 1984.
30. King, A.B. and S.M. Robinson. Ventilation response to hypoxia and acute mountain sickness. *Aerospace Med.* 43:419–421, 1972.
31. Kryger, M.H. and R.F. Grover. Chronic mountain sickness. *Sem. Resp. Med.* 5:164–168, 1983.

32. Kryger, M., R. McCullough, R. Doekel, D. Collins, J.V. Weil and R.F. Grover. Excessive polycythemia of high altitude: Role of ventilatory drive and lung disease. *Am. Rev. Resp. Dis.* 118:659–665, 1978.
33. Larson, E.B., R.C. Roach, R.B. Schoene and T.F. Hornbein. Acute mountain sickness and acetaazolamide: Clinical efficacy and effect on ventilation: *J. Am. Med. Assoc.* 248:328–332, 1982.
34. Lassen, N.A. and J.W. Severinghaus. Acute mountain sickness and acetazolamide. In Sutton, J.R., C.S. Houston and G. Coates (eds.), *Hypoxia and Cold*. New York: Praeger, 1987, pp. 493–504.
35. Maher, J.T., P.H. Levine and A. Cymermen. Human coagulation abnormalities during acute exposure to hypobaric hypoxia. *J. Appl. Physiol.* 41:702–707, 1976.
36. Malconian, M.K., P.B. Rock, J.T. Reeves, A. Cymerman, J.R. Sutton, B.M. Groves, P.D. Wagner, H. Donner, L. Trad, V. Forte, P.M. Young and C.S. Houston. Operation Everest II: Alveolar and arterial gases at extreme altitude. *J. Appl. Physiol.* (In review).
37. McFadden, D.M., C.S. Houston, J.R. Sutton, A.C.P. Powles, G.W. Gray and R.S. Roberts. High altitude retinopathy. *J. Am. Med. Assoc.* 245:581–586, 1981.
38. Menon, N.D. High altitude pulmonary edema. *N. Engl. J. Med.* 273:66–72, 1965.
39. Monge, C.C. and M.C. Monge. Natural acclimatization to high altitudes: Clinical conditions in life at high altitudes. *Scientific Publ. No. 140,* Washington, D.C.: Pan American Health Organization, 1966, pp. 46–52.
40. Monge, M.C. La Enfermedad de Los Andes. Sindromes Eritremicos. *Ann. Fac. Med.* (Peru) 11:1, 1928.
41. Moore, L.G., G.L. Harrison, R.E. McCullough, R.G. McCullough, A.J. Micco, A. Tucker, J.V. Weil and J.T. Reeves. Low acute hypoxic ventilatory response and hypoxic depression in acute altitude sickness. *J. Appl. Physiol.* 60:1401–1406, 1986.
42. Moore, L.G. and J.G. Regensteiner. Adaptation to high altitude. *Ann. Rev. Anthropol.* 12:285–304, 1983.
43. Moore, L.G., A.L. Rohr, J.K. Maisenbach and J.T. Reeves. Emphysema mortality is increased in Colorado residents of high altitude. *Am. Rev. Resp. Dis.* 126:225–228, 1982.
44. O'Bradovich, H., M. Andrew, G.W. Gray and G. Coates. Hypoxia alters blood coagulation during acute decompression in humans. *Physiologist* 25:276, 1982.
45. Okin, J.T. Response of patients with coronary heart disease to exercise at varying altitudes. In Vogel, J.H.K. (ed.), *Hypoxia, High Altitude and the Heart*. A monograph in the series: "Advances in Cardiology," Basel: S. Karger., 1970.
46. Penaloza, D. and F. Sime. Chronic cor pulmonale due to loss of altitude acclimatization (chronic mountain sickness). *Am. J. Med.* 50:728–743, 1971.
47. Ravenhill, T.H. Some experiences of mountain sickness in the Andes. *J. Trop. Med. Hygiene* 20:313–320, 1913.
48. Reeves, J.T., B.M. Groves, J.R. Sutton, P.D. Wagner, A. Cymerman, M.K. Malconian, P.B. Rock, P.M. Young and C.S. Houston. Operation Everest II: Preservation of cardiac function at extreme altitude. *J. Appl. Physiol.* 63:531–539, 1987.
49. Regensteiner, J.G. and L.G. Moore. Migration of the elderly from high altitudes in Colorado. *J. Am. Med. Assoc.* 253:3124–3128, 1985.
50. Rock, P.B. Medical problems of high terrestrial altitudes. *U.S. Army Flight Surgeons Manual,* 1986.
51. Rock P, M. Malconian, H. Donner, B. Groves, J. Reeves, J. Sutton, J. Alexander, H. Hultgren, A. Cymerman and C. Houston. Operation Everest II: Electrocardiography During Maximal Exercise At Extreme Altitude. *Med. Sci. Sports Exerc.* 18: S74, 1986.
52. Schoene, R.B., P.B. Hackett, W.R. Henderson, E.H. Sage, M. Chow, R.C. Roach, W.J. Mills and T.R. Martin. High altitude pulmonary edema; characteristics of lung lavage fluid. *JAMA* 256:1, 1986.
53. Scoggin, C.E., T.M. Hyers, J.T. Reeves and R.F. Grover. High altitude pulmonary edema in the children and young adults of Leadville, Colorado. *N. Engl. J. Med.* 297:1269–1273, 1977.
54. Severinghaus, J.W., H. Chiodi, E.I. Eger II, B. Brandstater and T.F.Hornbein. Cerebral blood flow in man at high altitude. *Circ. Res.* 19:274–282,1966.
55. Singh, I., I.S. Chohan and N.T. Matthew. Fibrinolytic activity in high altitude pulmonary edema. *Indian J. Med. Res.* 57:210, 1969.
56. Singh, I., P.K. Khanna, M.C. Srivastava, M. Lal, S.B. Roy and C.S.V. Subramanyam. Acute mountain sickness. *N. Engl. J. Med.* 280:175–184, 1969.
57. Suarez, J.M., J.K. Alexander and C.S. Houston. Operation Everest II: Left ventricular systolic function in man at high altitude assessed by two dimensional echocardiography. *Am. J. Cardiol.* 60:137–142, 1987.
58. Sutton, J.R. High altitude retinal hemorrhage. *Sem. Resp. Med.* 5:159–168, 1983.
59. Sutton, J.R., A.C. Bryan, G.W. Gray, E.S. Horton, A.S. Rebuck, W. Woodley, I.D. Rennie

and C.S. Houston. Pulmonary gas exchange in acute mountain sickness. *Aviat. Space Environ. Med.* 47:1032–1037, 1976.
60. Sutton, J.R., C.S. Houston, A.L. Marnsell, M.D. McFadden, P.H. Hackett, J.R.A. Rigg and A.C.P. Powles. Effect of acetozolmide on hypoxia during sleep at high altitude. *N. Engl. J. Med.* 301:1329–1331, 1979.
61. Sutton, J.R. and N. Lassen. Pathophysiology of acute mountain sickness and high altitude pulmonary edema. *Bull. Europ. Physiopath. Resp.* 15:1045–1052,1979.
62. Sutton J, J. Reeves, P. Wagner, B. Groves, A. Cymerman, P. Young, M. Malconian and C. Houston. Oxygen uptake during exercise at extreme simulated altitude maintained by marked reduction in mixed venous oxygen tension. *Fed. Proc.* 45: 882, 1986.
63. Wagner, P., J. Reeves, J. Sutton, A. Cymerman, B. Groves, M. Malconian and P. Young. Possible limitation of maximal O_2 uptake by peripheral diffusion. *Am. Rev. Resp. Dis.* 131: A207, 1986.
64. Weil, J.V., M.H. Kryger and C.H. Scoggin. Sleep and breathing at high altitude. In *Sleep Apnea Syndromes.* Guilleminault, C., and W.C. Dement (eds): New York; Liss, 1978, pp. 119–136.
65. West, J.B. Man on the summit of Mount Everest. In West, J.B. and S. Lahiri (eds.), *High Altitude and Man.* Bethesda: American Physiological Society, 1984, pp. 5–17.
66. Winslow, R.M. High altitude polycythemia. In West, J.B. and S. Lahiri, (eds.), *High Altitude and Man.* Bethesda: American Physiological Society, 1984, pp. 163–172.

15

Hyperbaric Physiology and Human Performance

STEPHEN R. MUZA, Ph.D.

OUTLINE

INTRODUCTION

For eons, man has been exposed to the terrestrial extremes of heat, cold and altitude, but only since the 16th century has man been able to venture for extended durations into hyperbaric environments. A natural high pressure environment, significantly greater than 760 torr, only exists underwater. The ambient pressure increases by 1 atmosphere (ATM) for every 10.0 and 10.4 m of depth in sea and fresh water, respectively. By comparison, in a mine shaft a depth of about 6,000 m below sea level (SL) would be required to increase ambient pressure by 1 atmosphere. Modern engineering technologies have exposed man to hyperbaria in deep mines, high pressure caissons, where construction work is done underneath bodies of water, and chambers used for therapeutic and experimental purposes. The earliest excursions into the underwater environment were accomplished by breath-hold diving (Table 15-1). Extension of submersion time had to wait for the development of a means to carry or deliver an adequate air supply to the diver. Breathing through hollow tubes (snorkel) extending above the water's surface limited the diving depth to less than 0.5 m. The development of the diving bell permitted extended duration underwater operations. However, dive duration was limited by the supply of air in the bell, its rate of use, and the accumulation of carbon dioxide in the bell. The first practical diving suit was of the "hard hat" style using surface-supplied air. This style of deep-sea diving outfit has remained virtually unaltered since its inception. Self Contained Underwater Breathing Apparatus (SCUBA) was developed along two lines: closed-circuit and open-circuit systems. The first practical SCUBA was a closed-circuit system supplying 100% oxygen from a tank and utilizing a carbon dioxide scrubber in the rebreathing circuit. This system could not be safely used at depths below 8 m because of the toxic effects of oxygen under pressure. The first successful and safe open-circuit SCUBA was developed by two Frenchmen, Jacques-Yves Cousteau and Emile Gagnan, during World War II. The "Aqua-Lung" as it was called has become the most widely used diving equipment available to sport or professional divers. More recently, development of mixed gas and saturation diving techniques using surface-supplied or SCUBA diving systems has greatly extended the depths and durations to which humans can safely dive. In addition to man's entry into the underwater environment, starting midway through the 19th century increasing numbers of men were exposed to high pressures while working in cassions. Unlike the relatively short duration diving operations, these construction crews worked daily 6-12 hr at moderately high pressures (2-5 ATM).

TABLE 15-1. *Historical Development of Diving.*

YEAR	DEVELOPMENT
ancient	breath-hold diving
4th century BC	primitive diving bell
1691	modern diving bell
1774	surface-supplied helmet diving apparatus
1837	Siebe's improved diving (hard hat) dress
1839	pressurized caisson
1878	Fleuss closed-circuit oxygen-rebreather SCUBA
1943	open-circuit compressed air SCUBA
1962	saturation diving techniques

The rapid and expanded entry of humans into the hyperbaric environment during the last century has brought with it a recognition of the medical problems and risks inherent to the high pressure environment. Understanding the physiological basis of these clinical observations has guided the development of diving technologies, by which divers can avoid or are protected from the adverse effects of hyperbaria on human physiological function and performance. In this chapter only the effect of increased ambient pressure per se on human physiological function will be reviewed. For a review of the thermal problems associated with water immersion, the reader is directed to chapters dealing with cold stress and injury.

GAS LAWS

Basic to the study of hyperbaric physiology is an understanding of the physical laws describing the behavior of gases. The kinetic theory of gases establishes the foundation upon which the gas laws were formulated. Simplified, the kinetic theory of gases describes ideal gas molecules which are in continuous random motion elastically colliding with each other and the walls of their container. The summation of the gas molecules' impact on the container's walls produces a force (pressure), which is directly proportional to the average kinetic energy of the gas. At a given temperature, the kinetic energies of all gases are identical. Thus, the pressure produced by the kinetic activity of any gas is affected by the same factors. The following gas laws describe the behavior of gases as a function of the temperature, pressure and volume of the gas.

Boyle's Law states that the pressure and volume of a gas are inversely related. For a given number of moles of gas, the product of pressure (P) and volume (V) is constant (K) at a given temperature:

$$PV = K$$

Boyle's Law is important for understanding the effect of ambient pressure changes on the volumes and pressures in body cavities.

Charle's Law states that the volume of a gas is directly proportional to the absolute temperature (T) assuming the pressure remains constant. Thus:

$$V_1 \cdot V_2^{-1} = T_1 \cdot T_2^{-1}$$

where the subscript 1 indicates initial conditions and the subscript 2 the final conditions. If the gas volume remains constant, then the pressure is directly proportional to the absolute temperature. Since Boyle's and Charle's laws describe interrelationships between the temperature, pressure and volume of any gas, these laws have been combined to yield the general gas law:

$$PV = nRT$$

where n is the number of gas molecules and R the gas constant whose numerical value depends on the units of the variables used. The general gas law can also be written to permit the calculation of new conditions:

$$(P_1 \cdot V_1) \cdot T_1^{-1} = (P_2 \cdot V_2) \cdot T_2^{-1}$$

Another important aspect of the kinetic theory of gases is that in a mixture of gases, each gas molecule behaves as if no other gases were present. Dalton's Law of partial pressures states that the total pressure exerted by a mixture of gases is equal to the sum of the individual pressures exerted by each of the gases comprising the gas mixture. The pressure exerted by each component gas is called its partial pressure (P_x). For atmospheric air containing nitrogen (N_2), oxygen (O_2), carbon dioxide (CO_2) and water vapor (H_2O) the total barometric or ambient pressure (P_B) would be:

$$P_B = PN_2 + PO_2 + PCO_2 + PH_2O$$

The partial pressure of a gas in a mixture can be calculated if the fractional concentration (F_x) of the gas and the total gas pressure are known. For example, the PO_2 in dry atmospheric air at standard SL P_B is:

$$PO_2 = FO_2 \cdot P_B = 0.2095 \cdot 760 \text{ torr} = 159 \text{ torr}$$

If the P_B is doubled by decending 10 m in sea water, than the PO_2 will increase to 318 torr since the P_B will be 1520 torr. It is important to recognize that the chemical and physiological actions of a gas are directly dependent upon the magnitude of the partial pressure of that gas and that only free gas molecules (nonchemically bound) contribute to the gas's partial pressure.

In a liquid, the partial pressure of the gas will be equal to its partial pressure in the gas phase if the gas is in free contact with the liquid. The liquid is saturated with the gas when its partial pressure is equal to the gas phase partial pressure. However, the concentration of a gas in a liquid depends not only upon its partial pressure but also on its solubility in the liquid:

$$\text{vol }\% = P_x \cdot C_s$$

where C_s is the coefficient of solubility commonly expressed as volume percent per torr of pressure (vol % $\cdot$ torr^{-1}). Henry's Law states that for gas of low solubility, at a given temperature the concentration in the liquid will be nearly proportional to the partial pressure of the gas in the gas phase. The lower the temperature of the liquid, the higher the solubility of the gas. The quantity of gas which dissolves in a body during a hyperbaric exposure is not only dependent upon the partial pressure and solubility of the gas, but also the duration of the exposure. Under identical conditions of temperature and pressure, gases with low solubility will require less time to saturate a liquid than gases of higher solubility. In diving, the longer and deeper the dive, the greater the saturation of the body's tissue.

BREATH-HOLD DIVING

Diving without underwater breathing apparatus is the most widely used technique for entering the underwater environment. Needing no special equipment, humans can pursue sport or professional tasks submerging for 1-3 min. The addition of simple equipment (mask, fins) can enhance a diver's performance by improving visual acuity and mobility. The use of a weight during descent and assistance during ascent extends the depth to which an exceptional breath-hold diver can reach to about 70 m. By comparison, diving mammals can reach depths in excess of 900 m and remain submerged for over 2 hr (28). Although the physiological events occurring during a breath-hold dive are complex, the simplistic nature of breath-hold diving provides an excellent model in which to examine the effects of the hyperbaric environment on human physiology and performance.

Pressure

Since the majority of the body is composed of water and other relatively incompressible materials, as one descends the increased ambient pressure is evenly transmitted throughout the body's tissues. However, per Boyle's law, during a breath-hold dive the increased ambient pressure compresses the gas containing body cavities. For a given pressure change, the magnitude of the volume change in these cavities is dependent upon the compliance of the cavities' walls. In the human, air is contained in the airways, lungs, sinuses, middle ears and gastrointestinal tract. Since the gastrointestinal tract is very compliant, gas trapped within it easily equalizes with the external pressure by compression. During normal breathing, the internal pressure in the airways, lungs, sinuses and middle ears equalize with the external pressure by venting air between the cavities and external environment. However, during a breath-hold dive the internal pressure of these cavities equalizes with the ambient pressure by the venting of air between the compressible (lungs) and noncompressible cavities. If a passage between these cavities is obstructed (e.g., eustachian tube), then an unequalized pressure difference will exist, if the ambient pressure changes. Since the vascular system is in equilibrium with the external pressure, increased transmural pressure of vessels transiting air cavities with lower internal pressure may cause the vessels to exude fluids and ultimately rupture causing hemorrhage. Injury resulting from pressure differences across the walls of gas filled cavities is called barotrauma. In diving, a general term applied to the pain and tissue damage associated with unequalized pressure differences is the "squeeze."

The air in the lung during a breath-hold dive can be compressed until the lung volume equals the noncompressible, residual volume (RV) of the lungs (44). As the breath-hold diver descends, the increasing ambient pressure compresses the chest wall as if the diver were exhaling. In Fig. 15-1, lung volume has been plotted as a function of immersion depth. The lung volumes were calculated assuming the last inspiration before submerging was 85% of vital capacity (25) with a face mask volume of 200 ml. In accordance with Boyle's law, the proportionally greater change in lung volume occurs close to the surface since pressure doubles during the first 10 m but not again until 30 m. The total incompressible volume equals the lung RV plus the face mask volume. From this model, it can be estimated that the depth at which the lung volume will equal the incompressible volume is 20 m. Should the diver descend beyond this depth, the pulmonary vessels may rupture producing pulmonary congestion, edema and hemorrhage. Therefore, respiratory mechanics limit the depth of a breath-hold diver to the ratio between total lung capacity (TLC) and RV (1,44). In the average adult male and female, the $TLC \cdot RV^{-1}$ ratio is 4:1–5:1. This corresponds to a maximum safe depth of 30–40 m (4–5 atmospheres absolute, ATA, where $ATA = 1 + Depth\ (m) \cdot 10^{-1}$). The maximal depth at which professional breath-hold divers work is generally near the limit predicted by their TLC:RV ratio. However, many are known to briefly dive several meters deeper. The diver who set the world breath-hold record of 73 m descended 12 m below his depth limit based on his TLC:RV ratio (47). Greater depths could be achieved if the RV were reduced. During deep breath-hold dives, a shift of approximately 1 liter (L) of blood into the thoracic cavity has been observed (47) which may displace air and decrease RV, thus extending the depth limit. Also, it has been proposed (1) that at depths below which RV is achieved, displacement of the diaphragm towards the head may cause partial collapse of the lung allowing further compression of the gas, thus postponing rupture of the pulmonary vessels.

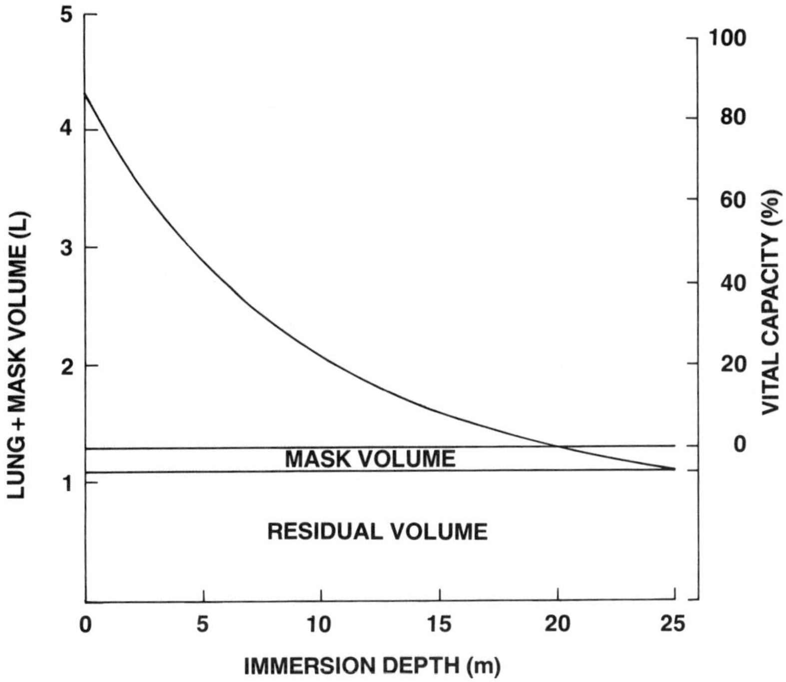

Figure 15-1. *The volume of the lung plus face mask plotted against the immersion depth. Lung volume was calculated according to Boyle's Law correcting for water vapor. (modified from Agostoni, (1)).*

Alveolar Gas Exchange

During a breath-hold dive, alveolar gas exchange is complicated by the flux of O_2, CO_2 and N_2 between the lung and blood and the change of pressure due to compression and expansion of the lung. In the absence of ambient pressure changes the alveolar gas exchange and composition during breath-holding is relatively uncomplicated. Assuming that the metabolic rate remains constant, the alveolar PO_2 falls approximately linearly during the first 2 min of breath-holding at 1 ATA then declines at a slower rate as hypoxia develops (31). Arterial saturation declines and venous blood O_2 content progressively decreases. At the onset of a breath-hold, the alveolar PCO_2 rises rapidly as CO_2 is transferred at a high rate from the mixed venous blood to the lungs. The progressive depletion of O_2 reduces the lung volume causing further increases of alveolar PCO_2. The arterial PCO_2 increases as a result of the increased alveolar PCO_2 and reduced capacity of hemoglobin for CO_2 (Haldane Effect). The rate of transfer of CO_2 from the blood into the lung progressively declines and may stop or reverse direction (39, 41).

During a breath-hold dive, in addition to the aforementioned fluxes, alveolar gas exchange is affected by the compression and decompression occurring during the descent, bottom and ascent phases of an actual dive. In Fig. 15-2 the alveolar partial pressures of O_2 and CO_2 during a hypothetical breath-hold dive are schematically illustrated. For comparison, the alveolar PO_2 and PCO_2 changes occurring during breath-hold at 1 ATA are included in Fig. 15-2.

Studies by Lanphier and Rahn (30,31) in which alveolar gas samples were obtained during breath-hold dives showed that during the descent phase of the dive, alveolar PO_2 and PCO_2 increases according to Boyle's Law as the ambient pressure rises. The maximum alveolar PO_2 attained is a function of its initial surface value,

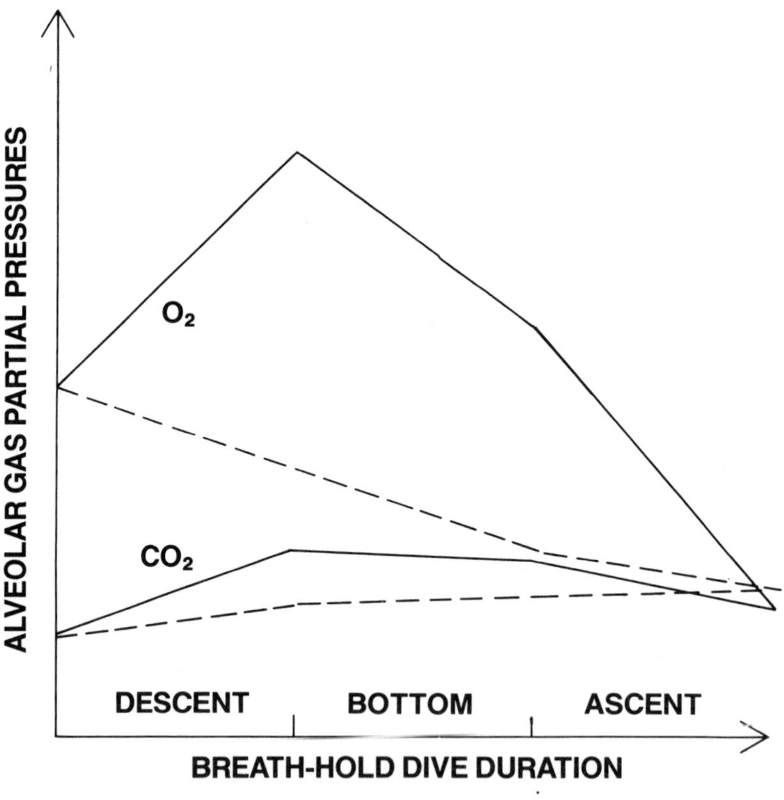

Figure 15-2. *Time courses of alveolar PO_2 and PCO_2 during breath holding at constant ambient pressure (dashed line) and 3 phases (descent, bottom, ascent) of a hypothetical dive (solid line) (adapted from Lanphier and Rahn, (30)).*

the metabolic rate, the depth achieved and the duration of the descent phase (30). Assuming a constant metabolic rate and surface alveolar PO_2, a rapid descent to a given depth will achieve a higher alveolar PO_2 by minimizing oxygen depletion. During the dive's bottom phase, oxygen moves out of the lung at a rate dictated by the metabolic demand.

However, due to the high ambient pressure, alveolar PO_2 is maintained above its comparable level during breath-hold at 1 ATA. Consequently, arterial saturation remains high and oxygen flux from lung to blood continues during the bottom phase. The dive's ascent phase produces a rapid reduction of the alveolar PO_2. At low alveolar PO_2 (<80 torr), the rate of oxygen transfer to the blood usually decreases and may stop (30).

Like the alveolar PO_2, during the breath-hold dive's descent phase, the alveolar PCO_2 rises as ambient pressure increases (Fig. 15-2). As alveolar PCO_2 increases, the gradient between the lung and mixed venous blood reverses and CO_2 is transferred from the lung into the arterial blood. At the bottom, the transfer of carbon dioxide into the blood continues at a progressively lower rate until the alveolar and mixed venous CO_2 contents become equal. As the lung volume decreases due to depletion of oxygen and flux of nitrogen into the blood, the alveolar PCO_2 may increase slightly (30). During ascent, the fall in alveolar PCO_2 causes the reestablishment of carbon dioxide transfer from venous blood to lung. This flux of CO_2 into the lung will attenuate the rate of alveolar PCO_2 decrease during the ascent. However, at the end of the breath-hold dive, at the surface the alveolar PCO_2 is typically lower than at the end of a similar 1 ATA breath-hold due, to storage of CO_2 in the body's tissues and the finite rate at which CO_2 can be transported to the lung via the blood (30).

The maximum duration to the end (break-point) of a voluntary breath-hold is dependent upon many factors including: psychological factors, alveolar and arterial gas tensions, oxygen consumption rate and ambient pressures. The psychological factors probably account for the great variability observed in maximum breath-hold times in humans. Lin (36) reported that a review of the literature yield breath-hold times ranging from 20–270 seconds (s) in air with no hyperventilation prior to a single maximal inspiration. Breath-hold duration is directly proportional to the initial lung volume (40). Conversely, unpleasant respiratory sensations arising from the respiratory muscles or lung, due to the absence of normal cyclic movement, may stimulate the desire to breathe (36). High alveolar and arterial PO_2 prolong breath-hold time, while increasing arterial PCO_2 is associated with the onset of involuntary respiratory muscle contraction and an increased desire to breathe (35). As expected, breath-hold duration is inversely related to metabolic rate but physical movements and conscious attention to performing tasks may delay the breaking point (36). Increased ambient pressure prolongs breath-hold duration (23, 36). As a result of the high ambient pressure, alveolar PO_2 is increased and the arterial blood saturation remains high longer than at 1 ATA. However, the likewise increased alveolar PCO_2 would oppose delaying the break-point. But high arterial PO_2 increases the tolerance to hypercapnia. This mechanism may operate during dives to attenuate the effect of the increased arterial PCO_2 on the onset of the breath-hold break-point.

Breath-hold duration can be increased by hyperventilation and or inhalation of oxygen prior to the apnea. The hyperventilation reduces the CO_2 content in the body and, consequently, the arterial PCO_2. Inhalation of pure oxygen obviously increases the body's O_2 stores. Many authors have written of the dangers associated with excessive hyperventilation prior to submergence. Each year numerous drownings occur as a result of the victim losing consciousness while submerged. Consciousness is lost due to the development of hypoxia (18). Recall that while on the bottom, the high ambient pressure keeps the alveolar PO_2 elevated and arterial blood saturated. Thus the diver maintains normal function. However, during ascent the alveolar PO_2 drops as the ambient pressure decreases (Fig. 15-2). If the diver initiates the ascent with a low alveolar PO_2, the alveolar PO_2 reduction during ascent may drop the arterial PO_2 below 40 torr at which central nervous system function is impaired and loss of consciousness may occur. Hyperventilation prior to diving decreases the body's carbon dioxide content more than it increases the oxygen content due to the lower solubility of oxygen. The lower alveolar PCO_2 delays the onset of the break-point allowing the alveolar PO_2 to decline to hypoxic levels.

Cardiovascular Responses

During breath-hold diving, the voluntary apnea, facial submergence, water temperature and increased ambient pressure along with other factors generate the physiological events traditionally described as the "diving response" (42). Compared to diving mammals, humans demonstrate a relatively mild diving response. In man, the principal cardiovascular responses to diving include: bradycardia, peripheral vasoconstriction and hypertension (36). The bradycardia is mediated by vagal parasympathetic efferents and can be abolished by administration of atropine (20). During a breath-hold the total peripheral resistance increases by 26–53%, blood pressure is usually increased and cardiac output generally unchanged (36). Sympathetic neural and humoral stimulation are probably responsible for the peripheral vasoconstriction and hypertension. These diving responses can be elicited by breath-holding without submergence. However, facial immersion in cold water potentiates the bradycardia (diving reflex). In humans, apnea and stimulation of cold receptors on the face accounts for the majority of the decrease in heart rate. Other factors which potentiate the diving response in man are high initial heart rate and metabolic rate, low alveolar PO_2 and large lung volume (36). Concerning the latter, if the subject breath-holds by closing his glottis and relaxing his respiratory muscles, the inward recoil of the chest wall will produce large intrathoracic pressures which impede venous return. Consequently, cardiac output may decrease and tachycardia may occur (17). This impediment of venous return can be reversed as the subject dives. Ferrigno *et al.* (19) observed decreasing intrathoracic pressure as subjects made simulated dives to 20 m. The authors suggested that at 3 ATA the thoracic volume was smaller than the subjects' normal relaxation volume. Assuming the respiratory muscles were relaxed, at this lung volume the chest wall should recoil outwards reducing intrathoracic pressure, thus enhancing venous return and cardiac output. Consequently, compared to the surface control (8 s end-expiratory pause during head-out immersion) stroke volume, heart rate and cardiac output were not altered during a simulated, breath-hold dive to 20 m (19). Two previous reports indicated a tendency for attenuation of bradycardia with increased depth (26, 42). It was suggested that the elevation of alveolar PO_2 at increased ambient pressure may be responsible for the slight attenuation of the bradycardia during breath-hold dives (42). In man, as a mechanism for reducing oxygen consumption and prolonging breath-holding the diving response is not very effective. For example, it has been demonstrated (50) that compared to breath holding in air, cold water immersion enhanced the bradycardia but decreased breath-hold duration due to increased oxygen consumption associated with shivering.

Acclimation

In this book, numerous examples of human acclimation to the environmental extremes of heat, cold and altitude are presented. Generally, repeated exposure to the given environmental extreme over a prolonged period of time is required to induce acclimation. Likewise, repeated exposure to the high pressure, hypoxia and hypercapnia associated with breath-hold diving does produce adaptations in these divers which may reduce stress and enhance diving performance. In humans, acclimation to breath-hold diving manifests itself by alterations in pulmonary function, chemosensitivity and cardiovascular control.

Several studies have demonstrated increased vital capacity (15, 49), decreased residual volume (46) and increased inspiratory muscle strength (49) in breath-hold divers compared to non-divers. A principal limit to the maximum depth a breath-hold diver can safely attain is his TLC:RV ratio. Greater depths could safely be achieved if the ratio is increased. Total lung capacity is greater in active breath-hold divers compared to non-diver controls (15, 49). Furthermore, longitudinal studies of US Navy breath-hold divers at the beginning of their diving duties and after one year of duty showed about a 4% increase of total lung capacity and a 15% decrease of residual volume (46). In these divers, one year of repeated breath-hold diving increased their TLC:RV ratio from 3.65 ± 0.57 to 4.38 ± 0.97. Thus, their depth limit was increased from 26 to 34 m. Increased vital capacities are predominately due to an increased inspiratory reserve volume (49). Increased inspiratory muscle strength has also been reported in Korean women breath-hold divers (49). At the surface, the divers remain submerged below the neck and inspiratory efforts are opposed by hydrostatic pressures on the chest wall. This may provide the training stimulus for increased inspiratory muscle strength development. Appropriately, the divers' expiratory muscle strength is not significantly different from non-divers (49).

Breath holding causes elevation of arterial PCO_2 and reduction of arterial PO_2 which are potent stimuli to the respiratory centers. Increasing one's tolerance to high PCO_2 and low PO_2 could reduce the stress of breath holding and delay the break-point of the apnea thus prolonging the dive. Again, both in the Ama (Japanese and Korean women breath-hold divers) and US Navy breath-hold divers, lower ventilatory hypercapnic responsiveness has been reported (46, 49). In the US Navy divers, the decreased ventilatory response to hypercapnia occurred over the first year of breath-hold diving and was lost after a 3-month period of no diving (46). It has been suggested (46) that the decreased ventilatory sensitivity to hypercapnia may be related to the increased tidal volumes and decreased breathing frequency observed in these divers. It is not clear whether the ventilatory hypoxic responsiveness is attenuated in breath-hold divers. Studies of breath holding divers have reported the ventilatory hypoxic sensitivity to: 1) be unchanged (49); 2) be decreased (46), or 3) possibly be attenuated only in severe hypoxic (alveolar $PO_2 < 50$ torr) situations (37). Further investigation is needed before hypoxic chemosensitivity in breath-hold divers is clearly understood.

As previously discussed, apnea, especially with facial immersion, produces bradycardia. Whereas in diving mammals the bradycardia is part of a set of circulatory adjustments operating to conserve oxygen, thus prolonging submergence; in man oxygen consumption is not significantly reduced. However, Hong *et al.* (27) have reported that divers (breath-hold and SCUBA) demonstrate a greater bradycardia than non-divers. No systemic physiological differences were found which accounted for the difference in the degree of bradycardia developed. But, since the divers were more accustomed to breath holding, the familiarity with the procedure may have lessened the stress and contributed to a lower heart rate.

PROBLEMS OF DEEP AND LONG DIVES

The development of underwater breathing apparatus expanded both the depth and duration to which humans could enter and remain in a hyperbaric environment. Underwater breathing apparatus, whether surface-supplied (hard-hat) or self-contained (SCUBA), extended the depth-limit by supplying air to the lungs at a pressure equivalent to the ambient pressure. This eliminated the "squeeze" on the lungs, airways and ears present during breath-hold diving. However, breathing compressed air at depth is not without its problems.

As a diver descends, the density of the gas in the breathing circuit and lungs increases since the delivery pressure increases to balance the ambient pressure. For example, if a diver requires a 1 L tidal volume both at the surface and 10 m depth, then twice as many air molecules must enter the breathing circuit and lungs at 2 ATA to provide the same tidal volume. Increasing gas density with depth will increase the work of breathing per liter respired gas by elevating airflow resistance (24). The increased gas density will reduce maximal ventilatory flows ultimately producing ventilatory limitations on aerobic performance (51). Six ATA air pressure appears to be the threshold at which ventilatory flow limitations may hinder maximal work performance (51). However, substituting helium for the nitrogen in the breathing gas will reduce the gas density and forestall these ventilatory problems as well as others described later. As the gas density increases with depth, divers may hypoventilate to minimize the increase in the work of breathing. Furthermore, since open circuit SCUBA gas supplies are depleted in direct proportion to the ambient pressure, divers may consciously hypoventilate to conserve their air supply. A potentially hazardous outcome of hypoventilation is CO_2 retention with subsequent carbon dioxide intoxication. An arterial PCO_2 greater than 70 torr depresses central nervous system cellular metabolism producing lethargy, narcosis, unconsciousness and ultimately death (21). Carbon dioxide narcosis is not a common occurrence, because the stimulating effects of CO_2 in the 40–70 torr range alert the diver that CO_2 retention is occurring. However, increased oxygen and nitrogen partial pressures at depth have deleterious effects on human physiological function and performance. The various pathophysiological processes arising from high pressure exposure are: 1) oxygen poisoning; 2) inert gas narcosis; 3) high pressure nervous syndrome; and 4) decompression sickness.

Oxygen Poisoning

The study of the noxious influence of oxygen on biological organisms received little attention until the work of Paul Bert in the late 19th century (10). He observed that birds and other small animals quickly developed convulsions of the central nervous system at oxygen tensions above 3 ATA. Prolonged exposure to the high pressure oxygen environment resulted in continued convulsions of decreasing severity at increasing intervals until death ensued (10). Later, Smith (48) showed that an oxygen tension of approximately 600 torr has an irritating effect on the lung and produces inflammation. Subsequent studies have confirmed that during exposure to inspired PO_2 ranging from about 570–1,500 torr, the lung is the first vital organ to be severely affected by oxygen poisoning (16). At higher inspired partial pressures of oxygen, the central nervous system (CNS) appears to be initially affected (2, 10, 29), but pulmonary oxygen toxicity transpires concurrently with the development of the central nervous system pathophysiology. This is illustrated in Fig. 15-3 where the tolerance limits of resting, normal men to pulmonary and CNS oxygen poisoning from inspired PO_2 is plotted as a function of exposure duration.

The pathology of pulmonary oxygen toxicity includes: edema, congestion, inflammation, atelectasis, fibrin formation and lung consolidation, pneumonia, bronchitis with bronchiectasis, hyperplasia, hypertrophy, desquamation and degenerative changes in alveolar cells, and sclerotic changes with narrowing, thickening and hyalinization of pulmonary arterioles (16). In humans, these pathological developments are accompanied by a reduced vital capacity and substernal distress aggravated by deep breathing. The mechanisms responsible for the entire syndrome of pulmonary oxygen toxicity contain multiple interacting factors, which include direct toxic effects upon the lung tissue and indirect toxic actions of oxygen at extrapulmonary sites (16). Oxygen at high pressure exerts direct toxic effects on the cell by the combined inactivation of enzymatic reac-

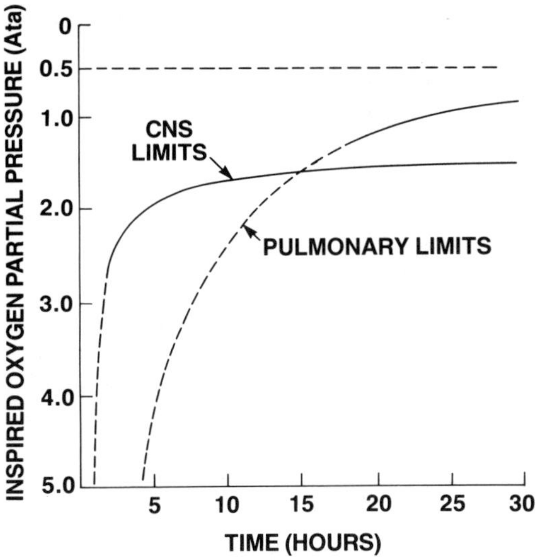

Figure 15-3. *Tolerance limits of resting, normal men to high inspired PO_2 as a function of exposure duration. The minimum inspired PO_2 at which symptoms of pulmonary and CNS oxygen poisoning will emerge is assumed to be 0.5 and 2.0 ATA respectively. At an infinitely high inspired PO_2, the rectangular hyperbolas defining pulmonary and CNS tolerance limits are assumed to converge near zero. Dashed lines indicate predicted limits. (Modified from USN Diving Manual, 1975.)*

tions, membrane lipids and proteins and the oxidation of nonprotein cellular constituents to inactive forms (22). Consequently, the essential metabolic cellular processes are disrupted. By indirect mechanisms, raised cerebral oxygen tensions increase neural excitability, inducing generalized sympathetic discharge which enhances pulmonary capillary permeability and development of pulmonary edema (16, 29, 52). Hyperoxic environments also produce a decrease in lung surfactant activity due to elevated intra-alveolar cholesterol levels, also attributed to increased sympathetic activity (9). The alterations in pulmonary mechanics resulting from decreased surfactant activity could precipitate the development of pulmonary edema and contribute to the pathogenesis of pulmonary oxygen poisoning (9, 16).

As described above, central nervous system dysfunction occurs at oxygen partial pressures well above 1 ATM (2). In humans, oxygen toxicity of the CNS generally manifests itself first with localized muscle twitching, particularly in small muscles of the face and occasionally of the hand. The muscle twitching increases in severity until a rigid tonic phase involving the body's major muscle groups occurs. Consciousness is lost and vigorous clonic contractions occur.

The pattern of brain electrical activity is similar to a grand mal epileptic seizure (29). The susceptibility to CNS oxygen toxicity is extremely variable between subjects and even varies daily within subjects (53). Also, performing exercise in the hyperoxic environment increases the susceptibility to oxygen poisoning. The mechanisms involved in the pathogenesis of CNS oxygen toxicity are not well established. It is believed that disruption of normal neural membrane excitation and transmission occurs, possibly resulting from: 1) membrane damage; 2) sodium/potassium pump dysfunction; 3) altered neurotransmitter metabolism or function (53). It has been proposed that oxygen toxicity may be potentiated by elevated arterial PCO_2 (29). The increased arterial PCO_2 may cause cerebral vasodilation elevating the cerebral tissue PO_2 and thus the direct toxic effects of oxygen on the neural cellular processes.

Inert Gas Narcosis

Exposure to compressed air produces symptoms of narcosis and intoxication. Generally, these symptoms are pleasant and resemble alcohol intoxication including: euphoria; hyperexcitability; impaired intellectual function, perception, memory; and impaired neuromuscular function (5). Early observations of these symptoms were made in the middle of the 19th century, but extensive research into this impairment was not made until the 1930's. Behnke, Thompson and Motley (3) were the first to suggest that high partial pressure of nitrogen in the compressed air environment was responsible for the narcosis. Subsequent studies confirmed the narcotic effect of nitrogen and other inert gases at high pressure. Consequently, the syndrome has been given the name nitrogen narcosis or inert gas narcosis.

The decrements in performance caused by inert gas narcosis increase with increasing pressure. Breathing compressed air, some subjects will experience symptoms at depths of about 30 m (4 ATA). Unconsciousness occurs when the PN_2 approaches 23,000 torr (38 ATA). The narcosis usually appears immediately upon reaching an individual's critical depth and disappears rapidly during decompression. The symptoms increase in severity with increasing pressure but not duration of exposure (5). The susceptibility to inert gas narcosis is extremely variable between subjects. Generally, experienced divers are less affected. In fact, it has been shown that repeated exposure to high pressure nitrogen results in acclimation to nitrogen narcosis (12). The U.S. Navy has established 90 m as the practical limit for compressed air diving operations due to the high incidence and severity of nitrogen narcosis at this and greater depths. Additionally, hypoventilation resulting from breathing gases of increased density causes carbon dioxide retention which potentiates the nitrogen narcosis (4). The narcotic potency of the inert gases varies. Xenon will produce anesthesia at atmospheric pressure (32), whereas performance decrements characteristic of narcosis have not been found when breathing helium and oxygen (heliox) at 46 ATA (8). The narcotic potencies of the inert gases are strongly correlated to their lipid solubility. Therefore, the order of the inert gases from least to most lipid soluble and narcotic is: helium, neon, hydrogen, nitrogen, argon, krypton and xenon (5). The advantage of breathing heliox mixtures was recognized early, and first put to practical use by the U.S. Navy in 1939.

Inert gas narcosis appears to be the result of physical actions of the inert gases on neural tissue. Since the narcotic potency of the inert gases correlates very well with their lipid solubility, it is believed that the site of action is the lipid phase of the neural membrane. At high pressures the inert gases may physically act on central nervous system post-synaptic membranes altering their permeability to cations and effectively depressing synaptic transmission (5). The membrane permeability may be altered by decreasing the stability of the lipid phase of the membrane. It has been suggested that pharmacological agents, which enhance membrane stability, may ultimately provide divers protection from compressed air narcosis (5).

High Pressure Nervous Syndrome

The first recognition that high pressure itself could pose a problem in diving occurred during a series of deep dives in the mid 1960's using heliox to counter the narcotic effects of nitrogen. Generally, humans compressed beyond 18 ATA show decrements in motor and intellectual performance. Further symptoms of the High Pressure Nervous Syndrome (HPNS) include: dizziness, nausea, vomiting, tremor of the hands, arms and torso, and modification of electroencephalogram (EEG) activity (7). The EEG changes include appearance of slow waves in the theta band (4-6 Hz) and depression of alpha activity (8-13 Hz). The EEG changes are accompanied by behavioral changes including intermittent bouts of daytime somnolence termed microsleep. These signs and symptoms of HPNS are very different from those that could be expected from helium narcosis, since no tremors, nausea, vomiting or EEG changes are present in nitrogen narcosis. Also, with narcosis the symptoms do not disappear with time at depth (6). Usually the symptoms of HPNS do not occur simultaneously in a given individual, nor for a given dive do all the symptoms occur at the same depth. Generally, the tremors appear first, followed by the EEG changes and daytime drowsiness (43). Furthermore, there is apparently no correlation between the EEG changes and psychomotor performance, or occurrence of tremor (43).

The symptoms of HPNS increase in severity with depth. Therefore, like inert gas narcosis, it was believed that HPNS may limit the depth to which man could dive. However, in experimental dives using slower rates of compression the symptoms of HPNS are attenuated (6, 7). Depending on the compression speed, the clinical and EEG symptoms appear at different depths. The onset of tremors varies by as much as 10 ATM and the EEG changes by 20 ATM (43). Using compression curves with intermediate stages also attenuates the severity of the HPNS and increases the depth at which the symptoms appear (6). During the halts in the compression, the intensity of the symptoms dissipate suggesting that adaptation to the high pressure environment is occurring (7). With slow compression schedules and breathing heliox, dives to 52 ATA have been accomplished with generally good recovery of performance after reaching depth. However, from a practical point of view, the use of slow (2.5 $m \cdot h^{-1}$) or interrupted compression curves is not economical in terms of time or money.

In the early 1970's, studies demonstrated that there is an antagonism between the narcotic affects of high pressure nitrogen and the CNS hyperexcitability underlying the signs and symptoms of HPNS (33). The addition of nitrogen into the heliox breathing mixture was proposed to suppress the HPNS. It has been shown that using trimix (nitrogen-helium-oxygen) breathing gases with various concentrations of nitrogen does not abolish HPNS but does attenuate the symptomatology (43). Generally, approximately 4–5% nitrogen (by volume) has been found to minimize the symptoms of HPNS without producing narcotic induced performance decrements. Furthermore, the HPNS symptomatology is generally more transitory with symptoms usually disappearing within 24 hr at constant depth (45).

HPNS is believed to be a manifestation of general neuronal hyperexcitability in response to hydrostatic pressure. This hyperexcitability may result from the effects of pressure per se on nerve cell membranes causing alterations in permeability (7). Both neuroanatomical and pharmacological studies have demonstrated in animals that the pressure is acting at a subcortical site and that the cortex provides a substantial degree of inhibition to counter the convulsions which arise from the subcortical sites (11). The findings that strychine and picrotoxin potentiate HPNS, suggest that high pressure may cause the convulsive effects by blocking the inhibitory process mediated by the neurotransmitters glycine and γ-aminobutyric acid (GABA) (11). However, the authors point out that these findings have not ruled out the hypothesis that HPNS is the outcome of a generalized increased excitation of the CNS and not removal of inhibitory mechanisms (11). Finally, whether the narcotic inert gases antagonize the effects of high pressure at the molecular level or simply neutralize the effects exerted on discrete sites in the central nervous system remains unresolved (13).

Decompression Sickness

When air is breathed under pressure, the nitrogen diffuses into the various tissues of the body. As long as the nitrogen pressure gradient is maintained from the inspired air to the tissues, nitrogen uptake continues by the body. Consequently, in accordance with Henry's Law, the amount of nitrogen absorbed increases with the partial pressure of the inspired nitrogen (depth) and the duration of the exposure (time).

Different tissues of the body become saturated at different rates. Nitrogen uptake for various parts of the body is related to the different tissue perfusion rates with blood and solubility of the tissue for nitrogen (54). Since nitrogen is more soluble in lipid than in water, at similar perfusion fatty tissues will absorb about five times as much nitrogen as will watery tissues. However, the fatty tissues will require more time to become saturated. At a given partial pressure of nitrogen, an exposure of about 24 hr is required to completely saturate all tissues of the body. This length of time is independent of the partial pressure of the inspired nitrogen. For other inert gases, the time required to reach saturation will be different. For example, since helium is less soluble than nitrogen, saturation of the body with helium will require less time. Also, since exercise increases tissue perfusion, the time required to reach saturation will decrease.

The reverse of saturation is the process of desaturation. When the partial pressure of the inert gas in the lungs is reduced by changing the gas's inspired partial pressure, the inert gas will diffuse down its pressure gradient from the tissues to the blood, from the blood to the alveolar gas, and then to the expired air. Some parts of the body will desaturate more slowly due to reduced perfusion or greater content of the inert gas. The major difference between saturation and desaturation is that the body will accommodate a large and relatively sudden increase in the inspired partial pressure of a gas, but not a sudden and large decrease. In the latter process the tissues will be supersaturated and the gas will come out of solution faster than it can be transported by the circulation to and through the lungs. The bubbles of gas which form in the tissues and blood result in a condition known as decompression sickness or the bends. The gas bubbles probably form in the tissue and microcirculation. Bubbles in the venous circulation are termed "silent bubbles" and are not considered to pose any pathological risk. However, arterial bubbles are believed to be the cause of the neurological symptoms seen in severe decompression sickness (14). Decompression sickness produces a wide variety of symptoms ranging from skin rash to mild discomfort and pain in the joints and muscles, paralysis, numbness, hearing loss, vertigo, the chokes (shortness of breath, chest pain), unconsciousness and death. The inert gases are primarily responsible for bubble formation since oxygen is removed by metabolism and carbon dioxide is very soluble and chemically converted to bicarbonate.

The rate at which a diver can be brought to the surface depends on two factors: the depth of the dive and the amount of time (bottom time) spent there. This relationship is illustrated in Fig. 15-4. At modest increased pressure (2 ATA), a diver can remain sub-

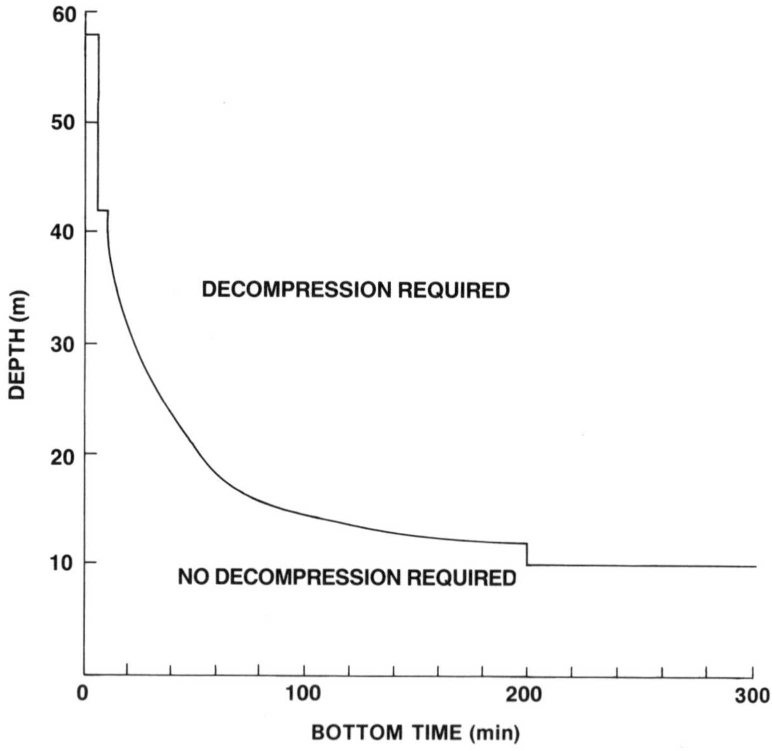

Figure 15-4. *No decompression required diving limit plotted as a function of diving depth versus bottom time. Assumes single dive with minimum of 12 hr between dives. (Adapted from USN Standard Air Decompression Table (38).)*

merged for relatively long durations and ascend directly to the surface without stopping for decompression. Although the diver's tissue nitrogen partial pressure will be higher than sea level pressure, bubble formation will be slight and probably cause no signs of decompression sickness. Dives conducted at depth and durations likely to cause decompression sickness on ascent, require the use of staged decompression. This is a procedure in which the diver ascends at a specified rate to various depths in a stepwise manner. At each depth, the diver remains for a specified period before continuing the ascent. This procedure controls the supersaturation of inert gas in tissue fluids so that symptoms of decompression sickness related to bubble formation do not occur (54). The draw back to staged decompression is the prolonged time required to accomplish the process. For example, a diver working for 1 h at a depth of 45 m will require a total staged ascent time of nearly 2 hr (U.S. Navy Standard Air Decompression Table, 38). The decompression tables in use for sport and professional divers do not insure complete desaturation of the body's tissues upon surfacing. Consequently, repeated dives conducted within 12 hr of each other must use repetitive dive decompression schedules which account for the residual nitrogen remaining in the body at the start of the subsequent dive.

The development of decompression schedules has generally followed the principles known as the "Haldane Method" (54). These principles are: 1) the estimation of the percent of tissue saturation or desaturation as a function of pressure-time exposures, and 2) the determination of the maximum inert gas pressure difference permissible without the development of decompression sickness. For each 1 ATM increase of ambient pressure, the whole human body can store about 1 L of nitrogen. The blood can only carry about one twenty-sixth of this nitrogen. Therefore, each time a unit of blood passes through the lungs it would pick up about 4% of the excess nitrogen and transport it to the body's tissues (54). Consequently, the rate of saturation follows an exponential curve for each tissue site. Desaturation of the tissue would follow a similar exponential function again, peculiar to each tissue. In order to maximize inert gas removal from the body, the largest possible pressure difference between the inspired and tissue inert gas partial pressures is desired. To avoid bubble formation, a maximum pressure difference of twice the ambient pressure has been determined to be acceptable. Therefore, decompression schedules, involving stages or stops at lesser depths (lower ambient pressure) for specified periods, were developed based on the estimated percent of inert gas tissue saturation and depth of dive (54). Continued research has strived to increase our knowledge of inert gas transport and storage. Mathematical models have been devised to develop and test new decompression schedules. Generally, dives of extended duration require longer decompression schedules while use of inert gases with lower lipid solubility decrease decompression duration. Research efforts are evaluating the effectiveness of using multiple inert gases to lower the partial pressure of each inert gas, to reduce decompression time or minimize decompression sickness (34).

SATURATION-EXCURSION DIVING

As previously described, if the inspired partial pressure of an inert gas is maintained constant for a sufficient duration, the body's tissues will become saturated with this gas and no net flux of gas will occur between the tissues and inspired gas. Consequently, once saturation is achieved, continued exposure to the elevated pressure will not increase the time required for desaturation to occur. If divers are maintained at the increased pressure for the entire period required to perform their mission, they only have to undergo decompression once. Using underwater habitats or surface compression chambers with a submersible pressurized personnel transfer capsule, divers can be maintained at high pressure for extensive periods of time (38). This permits the divers to spend the majority of their underwater time performing productive work.

Excursion tables have been developed which permit the divers to ascend and descend specified depths from their saturation depth. These excursion tables are based on the same principals as the previously reviewed decompression schedules. Many experimental saturation dives have been conducted to depths beyond 650 m and many military and commercial operational dives use saturation diving techniques in the open sea.

As with all diving operations, saturation divers are prone to the varied pathophysiological processes caused by hyperbaria. In particular, due to the generally long duration (days) spent at high pressure, pulmonary oxygen poisoning is a greater possibility if the inspired PO_2 exceeds 0.5 ATA (Fig. 15-3). Generally, use of compressed air in saturation diving habitats is limited to depths less than 15 m, because of oxygen poisoning. At greater depths, nitrox, heliox or trimix breathing gases are generally used with the PO_2 kept between 0.2–0.5 ATA (38).

SUMMARY

Modern technological developments have enhanced human performance in the underwater hyperbaric environment. Breath-hold diving still remains the most widely practiced method of entry into the natural hyperbaric environment underwater. Although simple in nature, breath-hold diving causes significant alterations in physiological function. The increased hydrostatic pressure squeezes the chest wall to small volumes. Alveolar gas exchange is disrupted as carbon dioxide flows from alveolar gas to arterial blood due to increased alveolar PCO_2. Arterial PO_2 is likewise elevated but can fall to severe hypoxic levels on ascent. The diving response in man principally consists of bradycardia, peripheral vasoconstriction and hypertension. Unlike in diving mammals, in humans the diving response does not appear to reduce metabolic rate. Professional breath-hold divers demonstrated adaptations which improve their ability to dive deeper and remain submerged longer than nondivers. The development of underwater breathing apparatus permitted humans to dive deeper and longer. This exposed humans to a number of stresses associated with increased partial pressures of oxygen, inert gases and pressure per se. Oxygen poisoning, inert gas narcosis and high pressure nervous syndrome all decrease human performance in hyperbaric environments and threaten life itself. Understanding of the mechanisms involved in these syndromes has led to the development of new breathing mixtures and procedures, in order to alleviate or prevent performance decrements and injury. Further research will elucidate the cellular mechanisms involved in these pathophysiological processes. Saturation and excursion diving techniques have evolved, based on the principles of gas diffusion, transport and solubility in body tissues. These techniques have not only increased the depth and duration to which humans can work in the underwater environment, but also their productivity.

REFERENCES

1. Agostoni, E. Limitation to depths of diving mechanics of chest wall. In: *Physiology of breath-hold diving and the Ama of Japan*. H. Rahn and T. Yokoyama (eds.), Washington, D.C.: National Academy of Sciences National Research Council Publication 1341, 1965,pp 139–145.
2. Bean, J.W. Effects of oxygen at increased pressure. *Physiol. Rev.* 25: 1–147, 1945.
3. Behnke, A.R., R.M. Thompson, and E.P. Motley. The psychological effects from breathing air at 4 atmospheres pressure. *Am. J. Physiol.* 112: 554–558, 1935.
4. Bennett, P.B. Cortical CO_2 and O_2 at high pressures of argon, nitrogen, helium, and oxygen. *J. Appl. Physiol.* 20: 1249–1252, 1965.
5. Bennett, P.B. Inert gas narcosis. In: *The Physiology and Medicine of Diving and Compressed Air Work*. P.B. Bennett and D.H. Elliott, (eds.), London: Bailliere Tindall, 1975, pp. 207–230.
6. Bennett, P.B. The high pressure nervous syndrome: Man. In: *The Physiology and Medicine of Diving and Compressed Air Work*. P.B. Bennett and D.H. Elliott, (eds.), London: Bailliere Tindall, 1975, pp. 249–263.
7. Bennett, P.B., and M. McLeod. Probing the limits of human deep diving. *Phil. Trans. R. Soc. Lond.* B304: 105–117, 1984.
8. Bennett, P.B., and J.E. Towse. Performance efficiency of men breathing oxygen-helium at great depths between 100 ft. to 1500 ft. *Aerospace Med.* 42: 1147–1156, 1971.
9. Bergren, D.R., and D.L. Beckman. Hyperbaric oxygen and pulmonary surface tension. *Aviat. Space Environ. Med.* 46: 994–995, 1975.
10. Bert, P. *Barometric Pressure: Researches in Experimental Physiology*. translated by M.A. Hitchcock and F.A. Hitchcock, Columbus, OH: College Book Co., 1878, pp. 715–775.

11. Bowser-Riley, F. Mechanistic studies on the high pressure neurological syndrome. *Phil. Trans. R. Soc. Lond.* B304: 31–41, 1984.
12. Brauer, R.W. Acclimation to nitrogen narcosis. In: *Nitrogen Narcosis.* R.W. Hamilton and K.W. Kizer, (eds.), Bethesda, MD: Undersea Med. Soc., vol VII, 1985, pp. 1–17.
13. Brauer, R.W., J.A. Dutcher, W. Hinson, and W.S. Vorus. Effect of habituation to subanesthetic N_2 or N_2O levels on pressure and anesthesia tolerance. *J. Appl. Physiol.* 62: 421–428, 1987.
14. Brubakk, A.O., R. Peterson, A. Grip, B. Holand, J. Onarheim, K. Segadal, T.D. Kunkle, and S. Tonjum. Gas bubbles in the circulation of divers after ascending excursions from 300 to 250 msw. *J. Appl. Physiol.* 60: 45–51, 1986.
15. Carey, C.R., K.E. Schaefer, and H. Alvis. Effect of skin diving on lung volumes. *J. Appl. Physiol.* 8: 519–523, 1956.
16. Clark, J.M, and C.J. Lambertsen. Pulmonary oxygen toxicity: A review. *Pharmacol. Rev.* 23: 37–133, 1971.
17. Craig, A.B., Jr. Effects of submersion and pulmonary mechanics on cardiovascular function in man. In: *Physiology of breath-hold diving and the Ama of Japan.* H. Rahn and T. Yokoyama, (eds.), Washington, D.C.: National Academy of Sciences National Research Council Publication 1341, 1965, pp. 295–302.
18. Dejours, P. Hazards of hypoxia during diving. In: *Physiology of breath-hold diving and the Ama of Japan.* H. Rahn and T. Yokoyama, (eds.), Washington, D.C.: National Academy of Sciences National Research Council Publication 1341, 1965, pp. 183–195.
19. Ferrigno, M., D.D. Hickey, M.H. Liner, and C.E.G. Lundgren. Simulated breath-hold diving to 20 meters: cardiac performance in humans. *J. Appl. Physiol.* 62: 2160–2167, 1987.
20. Finley, J.P., J.F. Bonet, and M.B. Waxman. Autonomic pathways responsible for bradycardia on facial immersion. *J. Appl. Physiol.* 47: 1218–1222, 1979.
21. Folk, G.E. *Textbook of Environmental Physiology.* Philadelphia: Lea & Febiger, 1974, pp. 352–371.
22. Haugaard, N. Cellular mechanisms of oxygen toxicity. *Physiol. Rev.* 48: 311–373, 1968.
23. Hesser, C.M. Breath holding under high pressure. In: *Physiology of breath-hold diving and the Ama of Japan.* H. Rahn and T. Yokoyama, (eds.), Washington, D.C.: National Academy of Sciences National Research Council Publication 1341, 1965, pp. 165–181.
24. Hesser, C.M., D. Linnarsson, and L. Fagraeus. Pulmonary mechanics of breathing at maximal ventilation and raised air pressure. *J. Appl. Physiol.* 50: 747–753, 1981.
25. Hong, S.K., H. Rahn, D.K. Kang, S.H. Song, and B.S. Kang. Diving patterns, lung volumes, and alveolar gas of the Korean diving women (Ama). *J. Appl. Physiol.* 18: 457–465, 1963.
26. Hong, S.K., S.H. Song, P.K. Kim, and C.S. Suh. Seasonal observations on the cardiac rhythm during diving in the Korean Ama. *J. Appl. Physiol.* 23: 18–22, 1967.
27. Hong, S.K., T.O. Moore, G. Seto, H.K. Park, W.R. Hiatt, and E.M. Bernauer. Lung volumes and apneic bradycardia in divers. *J. Appl. Physiol.* 29: 172–176, 1970.
28. Kooyman, G.L. How marine mammals dive. In: *A companion to animal physiology.* Cambridge, UK: Cambridge University Press, 1982, pp. 151–160.
29. Lambertsen, C.J. Effects of oxygen at high partial pressure. In: *Handbook of Physiology, Respiration.* Vol II. W.O. Fenn and H. Rahn, (eds.), Washington, D.C.: American Physiol. Soc., 1965, pp. 1027–1046.
30. Lanphier, E.H. and H. Rahn. Alveolar gas exchange during breath-hold diving. *J. Appl. Physiol.* 18: 471–477, 1963.
31. Lanphier, E.H., and H. Rahn. Alveolar gas exchange during breath holding with air. *J. Appl. Physiol.* 18: 478–482, 1963.
32. Lawrence, J.H., W.F. Loomis, C.A. Tobias, and F.H. Turpin. Preliminary observations on the narcotic effect of xenon with a review of values for solubilities of gases in water and oils. *J. Physiol.* 105: 197–204, 1946.
33. Lever, M.J., K.W. Miller, W.D.M. Paton, and E.B. Smith. Pressure reversal of anaesthesia. *Nature, Lond.* 231: 368–371, 1971.
34. Lillo, R.S., E.T. Flynn, and L.D. Homer. Decompression outcome following saturation dives with multiple inert gases in rats. *J. Appl. Physiol.* 59: 1503–1514, 1985.
35. Lin, D.A., Lally, T.O. Moore, and S.K. Hong. Physiological and conventional breath-hold breaking points. *J. Appl. Physiol.* 37: 291–296, 1974.
36. Lin, Y.C. Breath-hold diving in terrestrial mammals. *Exercise and Sport Sciences Reviews.* R.L. Terjung (ed.), Philadelphia: The Franklin Institute Press 10: 270–307, 1982.
37. Masuda, Y., A. Yoshida, F. Hayashi, K. Sasaki, and Y. Honda. Attenuated ventilatory responses to hypercapnia and hypoxia in assisted breath-hold divers (Funado). *Jpn. J. Physiol.* 32: 327–336, 1982.
38. Miller, J.W., ed. *NOAA Diving Manual.* Washington, D.C.: U.S. Department of Commerce, U.S. Government Printing Office, 1979.

39. Mithoefer, J.C. Mechanism of pulmonary gas exchange and CO_2 transport during breath holding. *J. Appl. Physiol.* 14: 706–710, 1959.
40. Mithoefer, J.C. The breaking point of breath holding. In: *Physiology of breath-hold diving and the Ama of Japan.* H. Rahn and T. Yokoyama (eds.), Washington, D.C.: National Academy of Sciences National Research Council Publication 1341, 1965, pp. 195–205.
41. Mithoefer, J.C. Breath-holding. In: *Handbook of Physiology, Respiration.* Vol II. Washington, D.C.: American Physiological Society, 1965, pp. 1011–1025.
42. Moore, T.O., Y.C. Lin, D.A. Lally, and S.K. Hong. Effects of temperature, immersion, and ambient pressure on human apneic bradycardia. *J. Appl. Physiol.* 33: 36–41, 1972.
43. Naquet, R., C. Lemaire, and J.C. Rostain. High pressure nervous syndrome: psychometric and clinico-electrophysiological correlations. *Phil. Trans. R. Soc. Lond.* B 304: 95–102, 1984.
44. Rahn, H. The physiological stresses of the Ama. In: *Physiology of Breath-Hold Diving and the Ama of Japan.* H. Rahn and T. Yokoyama, (eds.), Washington, D.C.: National Academy of Sciences National Research Council Publication 1341, 1965, pp. 113–137.
45. Rostain, J.C., M.C. Gardette-Chauffour, and R. Naquet. HPNS during rapid compression of men breathing $He\text{-}O_2$ and $He\text{-}N_2\text{-}O_2$ at 300 m and 180 m. *Undersea Biomed. Res.* 7: 77–94, 1980.
46. Schaefer, K.E. Adaptation to breath-hold diving. In: *Physiology of Breath-Hold Diving and the Ama of Japan.* H. Rahn and T. Yokoyama, (eds.), Washington, D.C.: National Academy of Sciences National Research Council Publication 1341, 1965, pp. 237–252.
47. Schaefer, K.E., R.D. Allison, J.H. Dougherty, Jr., C.R. Carey, R. Walker, F. Yost and D. Parker. Pulmonary and circulatory adjustments determining the limits of depth in breath-hold diving. *Science* 162: 1020–1023, 1968.
48. Smith, J.L. The pathological effects due to increase of oxygen tension in the air breathed. *J. Physiol. (Lond)* 24: 19–35, 1899.
49. Song, S.H., D.H. Kang, B.S. Kang, and S.K. Hong. Lung volumes and ventilatory responses to high CO_2 and low O_2 in the Ama. *J. Appl. Physiol.* 18: 466–470, 1963.
50. Sterba, J.A., and C.E.G. Lundgren. Diving bradycardia and breath-holding time in man. *Undersea Biomed. Res.* 12: 139–150, 1985.
51. Van Liew, H.D. Mechanical and physical factors in lung function during work in dense environments. *Undersea Biomed. Res.* 10: 255–264, 1983.
52. Wood, J.D., G.F. Perkins, A.E. Smith, and J.M. Reaux. Response of the cardiovascular system in oxygen toxicity. *Aerospace Med.* 43: 162–167, 1972.
53. Wood, J.D. Oxygen toxicity. In: *The Physiology and Medicine of Diving and Compressed Air Work.* P.B. Bennett and D.H. Elliott, (eds.), London: Bailliere Tindell, 1975, pp. 166–184.
54. Workman, R.D., and R.C. Bornmann. Decompression Theory: American Practice. In: *The Physiology and Medicine of Diving and Compressed Air Work.* P.B. Bennett and D.H. Elliot, (eds.), London: Bailliere Tindell, 1975, pp. 308–330.

16

Air Quality and Human Performance

KENT B. PANDOLF, Ph.D.

OUTLINE

INTRODUCTION

In addition to the environmental extremes of heat, cold, altitude and hyperbaria already discussed in this book, poor air quality or air pollution is another environmental stress or known to affect human physiological performance. The individual air pollutants have been classified as primary or secondary pollutants (81,88). Primary pollutants are those emitted directly to the environment from sources such as gasoline powered vehicles or industrial plants and exert their effects with little or no chemical change. These pollutants include carbon monoxide, sulfur oxides, nitrogen oxides and primary particulates. Secondary pollutants are those which develop from the interaction of primary pollutants and include ozone, peroxyacetyl nitrate and certain aerosols. Four population groups have been shown to be especially susceptible to air pollution effects: asthmatics and others with respiratory disorders, athletes, children and the elderly (81,94).

Three historic air pollution episodes (the Meuse Valley, Belgium, 1930; Donora, Pennsylvania, 1948; London, 1952) and the associated medical consequences sensitized the public to the health problems of poor air quality, while stimulating research leading to the development of control measures and standards. These three air pollution episodes were associated with greater than 4000 excess deaths resulting from the interactive effects of poor air quality and existing disease states such as heart and lung disorders (81,93). In part, the end result of these and other pollution episodes has led to the development of public guidelines to help assess the potential health problems associated with poor air quality as illustrated in Tables 16-1 and 16-2. Table 16-1 presents the possible adverse health effects of four common air pollutants (carbon monoxide, ozone, sulfur dioxide, total suspended particulates) as a function of a specific averaging time in hours and a standard described as "unhealthy." Table 16-2 displays the pollution standards index (PSI) ranging from numerical values of 0 to 500 for five common pollutants (total suspended particulates, sulfur dioxide, carbon monoxide, ozone, nitrogen dioxide) and the associated health effects descriptor.

The early research studies concerning the physiological effects of the various air pollutants on man were conducted at rest while more recent research has evaluated submaximal and maximal exercise responses. The adverse physiological effects of air pollutant inhalation appear related to the intensity of exercise with generally greater adverse effects at higher intensities. The physiological effects of the various air pollutants primarily involve the respiratory tract with the site of the effect decided by the pollutants' solubility and/or size (81). The nose is very effective in removing large particulates and highly soluble gases, but is poor in the filtering of small particulates and low soluble gases. During nasal breathing, it is estimated that 99.9 percent of sulfur dioxide, which is a highly-soluble gas is removed; however, oral breathing which is generally employed during exercise bypasses this major defense mechanism (81,89). Some pollutants, such as ozone which is a low soluble gas, may act at different sites along the respiratory tract. During exercise, ozone, nitrogen dioxide and sulfur dioxide generally exert their effect on the small airways and proximal alveoli resulting in alterations in mucous secretion and mucociliary clearance (44,81). Carbon monoxide penetrates to the alveoli where it binds with hemoglobin

TABLE 16-1. *Possible Adverse Effects of Air Pollution on Health With Air Pollution Index Greater than 100 ('Unhealthy').*

Pollutant	Averaging Time (hr)	Primary NAAQS	Explanation Given
CO	8	9 ppm	Impaired exercise tolerance in persons with cardiovascular disease
	1	35 ppm	Decreased physical performance in normal adults
O_3	1	0.12 ppm	Aggravation of chronic lung disease and asthma Irritation of the respiratory tract in healthy adults Decreased visual acuity; eye irritation Decreased cardiopulmonary reserve in healthy subjects
SO_2	24	0.14 ppm	Increased hospital admissions for respiratory illness in elderly patients with related illness
TSP	24	150 $\mu g/m^3$	Aggravation of chronic lung disease and asthma Aggravation of cardiorespiratory disease symptoms in elderly patients with heart or chronic lung disease Increased cough, chest discomfort and restricted activity

CO = carbon monoxide, O_3 = ozone, SO_2 = sulfur dioxide, TSP = total suspended particulates.

Averaging time defines the specific time in hours which the specific pollutant being monitored is averaged. For example, the level of CO fluctuates over an 8-hour period from 0 ppm to above 100 ppm. If the 8-hour average is greater than 9 ppm, then the level has exceeded the primary National Average Air Quality Standards (NAAQS).

Source: Environmental Protection Agency.

in the blood to impede oxygen transport. Thus, carbon monoxide affects exercise performance through impaired transport of oxygen in the blood while ozone, sulfur dioxide and nitrogen dioxide possibly cause an impairment in the ability to ventilate maximally (44).

Pulmonary function tests have been used over the years to evaluate the effects of the various air pollutants. These tests can be divided into five categories: (a) lung volume measurements, (b) lung capacity measurements, (c) forced spirometry measurements, (d) airway resistance measurements and (e) measurements of pulmonary diffusing capacity (81). Several of these measurements are illustrated schematically in Figure 16-1.

There are three static measurements of lung volume which are performed with a spirometer (21,117). *Tidal Volume* (TV) is the volume of air inhaled or exhaled during each breathing cycle. *Inspiratory Reserve Volume* (IRV) is the maximum amount of air that can be inhaled after a normal inhalation. *Expiratory Reserve Volume* (ERV) is

TABLE 16-2. *Comparison of PSI Values with Pollutant Concentrations and Descriptor Words.*

Index Value	Air Quality Level	TSP (24-hr), $\mu g/m^3$	SO_2 (24-hr), $\mu g/m^3$	CO (8-hr), $\mu g/m^3$	O_3 (1-hr), $\mu g/m^3$	NO_2 (1-hr), $\mu g/m^3$	Health Effect Descriptor
		Pollutant Levels					
500	Significant Harm	600	2620	57.5	1200	3750	
400	Emergency	500	2100	46.0	1000	3000	Hazardous
300	Warning	420	1600	34.0	800	2260	
							Very Unhealthful
200	Alert	350	800	17.0	400	1130	
							Unhealthful
100	NAAQS	150	365	10.0	235	*	
							Moderate
50	50% of NAAQS	50	80	5.0	118	*	
							Good
0		0	0	0.0	0	*	

*No index values reported at concentration levels below those specified by 'alert level' criteria.

$\mu g/m^3$ = ppm × molecular weight/0.024.

Source: Environmental Protection Agency.

the maximum amount of air that can be exhaled after a normal exhalation. A fourth lung volume, *Residual Volume* (RV), is the volume of air left in the lungs after a maximal exhalation.

There are four static lung capacity measurements (21,117). *Total Lung Capacity* (TLC) is the amount of air in the lungs after a maximal

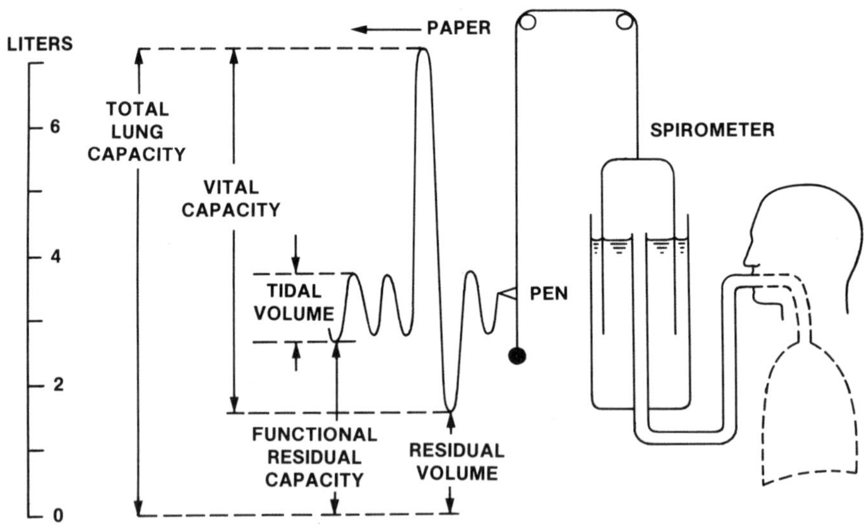

Figure 16-1. *Lung volumes and lung capacities. Reprinted with permission from West (117).*

inhalation. *Vital Capacity* (VC) is the volume of air that can be expelled from the lungs through a maximal effort after a maximal inhalation. *Inspiratory Capacity* (IC) is the maximal volume of air that can be inhaled from the resting expiratory level. *Functional Residual Capacity* (FRC) is the volume of air left in the lungs at the resting expiratory level.

Three forced expiratory measurements have been used to evaluate the effects of the various air pollutants (81). *Forced Expiratory Volume* (FEV_t) measures the volume of air exhaled by a maximal effort at a specific time. In practice, this measurement is made during the first ($FEV_{1.0}$) or third ($FEV_{3.0}$) second. In some instances, FEV_t is expressed as a percentage of the forced vital capacity (FVC) as $FEV_{1.0}$/FVC% or $FEV_{3.0}$/FVC%. *Forced Midexpiratory Flow* (FEF) measures the average rate of airflow over the middle half of the FVC or $FEF_{25\%-75\%}$. *Maximal Voluntary Ventilation* (MVV, normally expressed as $L \cdot min^{-1}$) is the maximum volume of air that can be breathed through voluntary effort over a 12-second period when the individual is instructed to breath as hard and fast as possible.

Three measurements are commonly used to evaluate airflow obstruction (81,117). *Airway Resistance* (R_{aw}) represents the difference between the alveolar and mouth pressures divided by the flow rate. The R_{aw} is primarily a measurement of large airway resistance. *Static and Dynamic Lung Compliance* (C_{st}, C_{dyn}) are measurements of the detensibility or "elastic resistance" of the lungs (21). *Closing Volume* (CV) is a measure of the lung volume at which peripheral airways begin to close. It is thought to be sensitive to alterations in the small airways.

The *Diffusing Capacity* of the lung (D_L) is a measure of the lung's ability to move a gas from the alveoli to the capillary blood. Carbon monoxide is the gas most suitable for measuring the D_L.

The potential interactive effects of various air pollutants may pose a greater threat to human physiological performance than that found with each single pollutant. Pollutants may interact in three ways: (a) additively, (b) synergistically or (c) antagonistically (81). An *additive* interaction implies that the total physiological effect is simply the sum of the individual pollutant effects which is exemplified by ozone and nitrogen dioxide (57). In contrast, ozone and peroxyacetyl nitrate seem to display a *synergistic* effect (27), where the combined effect of the pollutants is greater than the sum of the individual pollutant effects. An *antagonistic* interaction, which has not been observed to date, results when the combined pollutant effects are less than the sum of the individual effects.

More recently, research interest has focused on human "adaptation" to the various pollutants, most noticeably ozone. Repeated daily exposure to ozone has been associated with an improved exercise performance by the third or fourth exposure day (46,59). However, several authors (20,46) caution that the acquired insensitivity to ozone or "adaptation" may not be protective, but rather repeated exposure to low levels of this pollutant may be undesirable because of the potential for developing changes or injury to the lung. This same caution should be considered when studying "adaptation" to the other pollutants.

The rate and severity of air pollution episodes is known to be influenced by environmental and meteorological factors, not to mention the time of day. Figure 16-2 displays the daily and seasonal fluctuations for ozone and carbon monoxide. As shown in this figure, primary pollutants such as carbon monoxide and also the nitrogen oxides manifest daily peaks in clear association with peak traffic conditions and display their highest levels in mid-winter (81). Secondary pollutants such as ozone show a distinctive pattern related to the sunlight hours with peak daily values in the afternoon, and peak seasonal values in the summer or early autumn (81). In addition to sunlight, other meteorological factors known to influ-

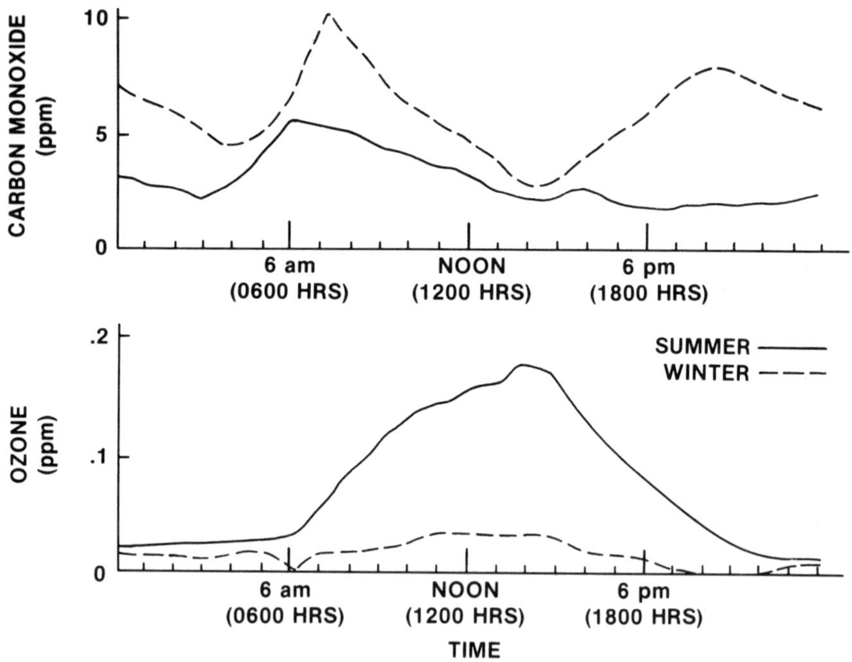

Figure 16-2. *Carbon monoxide (ppm) and ozone (ppm) observations from the Los Angeles area which illustrate the daily and seasonal fluctuations. Modified from a version originally published by McCafferty (81).*

ence the severity of air pollution episodes are the wind speed and the vertical temperature gradient.

The combination of certain pollutants with the environmental extremes of heat, cold or altitude may result in additive and/or synergistic effects. In addition, extremes in the percent relative humidity are known to be associated with adverse health effects. For instance, the adverse effects of sulfur dioxide become more pronounced at high humidity where ozone levels are enhanced by low humidity (12,81). Some authors (12) suggest an optimal range of 40 to 60% relative humidity to help minimize the performance degradation of these pollutants.

When the relative effects of environmental heat and adverse air quality are compared, heat would appear to be the more important stressor in terms of human performance. Several authors (28,49,95,96) report that heat stress was more effective in reducing human exercise performance than carbon monoxide and/or peroxyacetyl nitrate. In contrast, the combination of heat stress and ozone has been shown to have an additive effect in reducing exercise performance (41,47).

Little direct experimental evidence exists concerning the combined effects of environmental cold and air pollution. However, breathing cold air during exercise has been shown to enhance bronchoconstriction (112). Whether the degree of bronchoconstriction would be further enhanced by certain air pollutants in combination with cold air exposure is worthy of investigation.

Carbon monoxide exposure is known to impair maximal exercise performance at sea level in terms of a reduced maximal aerobic power. At elevations greater than about 1500 meters, maximal aerobic power is known to decrease linearly (87). The hypoxic hypoxia which occurs during exercise at altitude is similar to that induced by carbon monoxide. One might expect the effects of carbon monoxide to be enhanced at altitude due to the reduction in mean capillary oxygen pressure (51).

THE PRIMARY POLLUTANTS

Carbon Monoxide

Carbon monoxide (CO) is reported to be the most commonly occurring of all the air pollutants in urban environments, with the total emissions from this pollutant being greater than all other pollutants combined (44). Carbon monoxide is thought to impair cardiorespiratory function, particularly during physical exercise, by binding with hemoglobin in the blood (COHb) to impede oxygen transport. In fact, CO has a 200 times greater affinity for hemoglobin than that of oxygen, and is known to shift the oxygen dissociation curve to the left, resulting in more difficult tissue oxygen extraction. In general, CO is associated primarily with motor vehicle emissions and the outdoor environment; however, CO toxicity has been reported for young hockey players skating in an indoor hockey rink and attributed to a gasoline-powered resurfacing machine (7).

Submaximal Exercise Performance. Little impairment of cardiorespiratory function and no major physical performance decrements have been reported in healthy individuals at COHb levels of less than 20% throughout a wide range of submaximal exercise intensities (30–75% $\dot{V}O_2$max) of both short and prolonged duration (31,49,90,113). Submaximal heart rate (HR) was found to be significantly increased with CO administration at these same exercise intensities (31,49,90,113), and added respiratory distress, in terms of increased pulmonary ventilation ($\dot{V}_E$), was reported (31,113) at the higher submaximal intensities (70–75% $\dot{V}O_2$max). It should be remembered that the COHb levels reported in these studies (range, 10.7-20.1% COHb) are well above those normally associated with poor air quality (81).

In contrast to healthy individuals, cardiovascularly-impaired individuals are at significant risk during submaximal exercise even at low COHb levels (~2.5 – 3.0% COHb). In an investigation of 10 male patients with documented coronary artery disease, exposure for 90 min to heavy freeway traffic increased the average COHb level to 5.08% causing a decreased exercise time to angina onset with significant reductions in systolic blood pressure and HR at angina (10). In this same study, ischemic ST-segmental depressions were observed in three of these 10 men while breathing freeway air in contrast to no such abnormalities while breathing compressed-purified air during freeway traffic. In two investigations each studying 10 male patients with documented angina, exercise time for onset of angina was evaluated while breathing either 50 ppm CO (COHb = 2.7%) for two hr (11), or 50 ppm CO (COHb = 2.9%) and 100 ppm CO (COHb = 4.5%) for four hr (4). During both studies at either CO concentration, average exercise angina onset times were reduced when compared to values while breathing compressed-purified air. In addition, the duration of angina was significantly prolonged after breathing 100 ppm CO, but not after breathing 50 ppm CO (4). In general, deeper and more prolonged ischemic ST-segmental depressions were noted in these patients after breathing CO (4).

Maximal Exercise Performance. In contrast to human performance during submaximal exercise, maximal exercise performance for healthy individuals, in terms of maximal exercise time and/or

$\dot{V}O_2max$, appears to be inversely related to the CO concentration (31,61,90,113). The critical level at which COHb *significantly* influences $\dot{V}O_2max$ has been reported as 4.3% (61). However, even lower COHb levels have been associated with significant decrements in maximal exercise time (28,95,96).

After maximal treadmill exercise in five young men at an average COHb level of 15.4%, Pirnay *et al.* (90) observed a 15.1% reduction in $\dot{V}O_2max$. These authors also noted a small ($\bar{X} = 2.4\ b \cdot min^{-1}$), but significant increase in HR_{max} and a relative hyperventilation with CO exposure. After maximal upright cycle ergometer exercise involving eight young male volunteers at an average COHb level of 20.5%, Vogel and Gleser (113) report a 22.6% reduction in $\dot{V}O_2max$ which was cited as proportional to the arterial desaturation. While maximal cardiac output ($\dot{Q}_{max}$), stroke volume (SV), HR_{max} and peak lactate concentrations were not different, peak $\dot{V}_E$ was lower during maximal exercise with CO (113). These same authors suggest that the leftward shift in the oxyhemoglobin dissociation curve with CO exposure and the associated decrease in oxygen carrying capacity can account for the entire decrement in $\dot{V}O_2max$. Ekblom and Huot (31) evaluated maximal cycle and treadmill exercise responses in ten well-trained subjects at COHb levels ranging from 4.8 to 21.2%. During either cycle (mean COHb = 7.5 and 20.7%) or treadmill (mean COHb = 7.1 and 19.3%) maximal exercise, average maximal exercise time and $\dot{V}O_2max$ were significantly reduced at all of these CO concentrations. For both types of maximal exercise, peak $\dot{V}_E$ and peak blood lactate concentrations remained unchanged, while HR_{max} generally displayed a slightly lower value after CO exposure. During additional maximal treadmill exercise experiments over a wider range of COHb levels, the $\dot{V}O_2max$ ($r = 0.85$) and maximal exercise time ($r = 0.79$) predictively decreased with increasing COHb level (31). These same authors conclude from an extrapolation of their findings that exercise time would be reduced to zero at 46% COHb.

Reports by Drinkwater *et al.* (28) and Raven *et al.* (96) appear to eminate from a common database involving 20 young men (10 smokers and 10 non-smokers) who exercised maximally by performing a modified Balke treadmill test. The $\dot{V}O_2max$ was unchanged for these subjects with COHb levels ranging from 2.5–4.5%. The higher COHb levels were associated with the smokers. In one of these studies (28), the maximal exercise time was reduced for non-smokers during CO exposure. In a study of 16 older men (nine non-smokers and seven smokers) who performed maximal treadmill exercise, Raven et al. (95) report no significant reductions in $\dot{V}O_2max$ with CO exposure (non-smokers COHb = 2.3%, smokers COHb = 4.5%). However, the older male smokers had a $\dot{V}O_2max$ that was 27% less than the non-smokers of comparable age. In a study by Aronow and Cassidy (9), ischemic ST-segmental depressions of greater than 1.0 mm were noted in 10% of clinically normal individuals during maximal treadmill exercise following inhalation of 100 ppm CO (COHb, 3.95%). In this same study, maximal exercise time was significantly reduced with CO.

In four male volunteers who exercised maximally on a treadmill, Horvath and colleagues (61) noted a 4.9 and 7.0% reduction in maximal exercise time when the levels of COHb were 3.3 and 4.3%, respectively. However, these same authors report that $\dot{V}O_2max$ is not significantly decreased until COHb levels exceed 4.3%. Figure 16–3 illustrates the relationship between the percent increase in COHb and the associated percent decrement in $\dot{V}O_2max$ (58). When COHb levels range from about 4 to 33%, there is a linear decline in $\dot{V}O_2max$ expressed by the following formula: percent decrease in $\dot{V}O_2max$ = 0.91 (% COHb) + 2.2 (58). No studies have been reported to our knowledge on human adaptation to CO during either submaximal or maximal exercise.

Sulfur Oxides

Sulfur oxides (SO_x) result from fossil fuel combustion and include primarily sulfur dioxide (SO_2), sulfuric acid and sulfate. It is estimated that about 98% of the atmospheric sulfur from the burn-

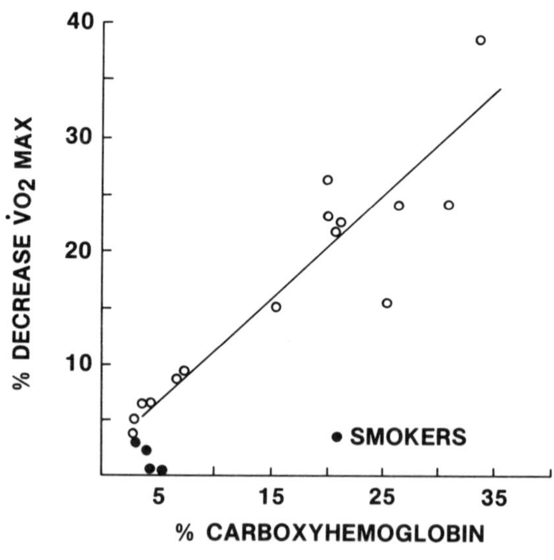

Figure 16-3. *Relationship between the percent carboxyhemoglobin and the percent decrement in maximal oxygen uptake ($\dot{V}O_2max$) for smokers and non-smokers. Reprinted with permission from Horvath (58).*

ing of fossil fuels is initially in the form of SO_2 which may convert to sulfuric acid and sulfate (81). Sulfur dioxide is a highly soluble gas which exerts its main influence as an upper respiratory tract irritant and can cause a reflex bronchoconstriction and increased airway resistance. As indicated earlier, the shift from nose to mouth breathing, known to occur during physical exercise, will result in less efficient absorption of SO_2 and consequently greater pulmonary flow resistance as illustrated in Figure 16-4. While submaximal exercise performance has been evaluated with this pollutant, no studies have been reported concerning maximal exercise performance.

Submaximal Exercise Performance. Eight healthy young males performed intermittent cycling exercise for two hrs (15 min exercise, 15 min rest) at an exercise intensity sufficient to double their pul-

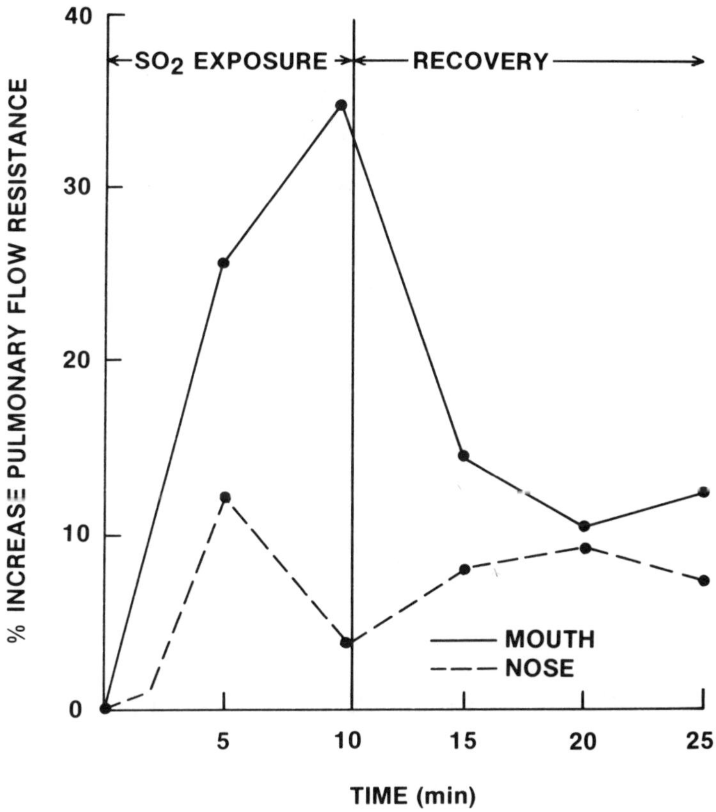

Figure 16-4. *Differences in the percent increase in pulmonary flow resistance over time for healthy individuals breathing sulfur dioxide (SO_2) orally or nasally. Modified from a version originally published by Speizer and Frank (109).*

monary ventilation while breathing 0.37 ppm SO_2 (57). These authors observed no significant effect on ventilatory function as exemplified by the maximum mid-expiratory flow rate (MMFR) at this SO_2 concentration. In a study of four sensitive subjects (respiratory hyperractivity to inhaled irritants) using the same experimental protocol and SO_2 concentration as Hazucha and Bates (57), Bell *et al.* (19) confirmed that 0.37 ppm SO_2 had no effect on FVC, $FEV_{1.0}$ and delta N_2. At 0.40 ppm SO_2, Bedi *et al.* (17) had nine adult men perform intermittent treadmill exercise (15 min walk, 15 min rest) for two hrs which elevated $\dot{V}_E$ during exercise to 30 $\ell \cdot min^{-1}$. This SO_2 concentration was insufficient to produce significant pulmonary function differences (i.e., FVC, $FEV_{1.0}$, IC, FRC, RV, TLC, $FEF_{25-75\%}$, $FEF_{50\%}$, $FEF_{75\%}$, MVV, R_{aw} and CV). While exposed to 0.75 ppm SO_2 for four hrs, 11 healthy men performed two 15-min bouts of treadmill walking (1.78 $m \cdot s^{-1}$, 10% grade). Evaluation from a battery of 19 measurements of pulmonary function indicated that the airway resistance, lung volume and air flow responses to 0.75 ppm SO_2 were not significantly altered (110). At a concentration of 1.0 ppm SO_2, Folinsbee et al. (38) had 22 healthy young men exercise intermittently on a treadmill (walking, 1.56 $m \cdot s^{-1}$) for two hrs (30 min exercise, 10 min rest) with the exercise adequate to elevate the $\dot{V}_E$ to 38 $\ell \cdot min^{-1}$. However, this SO_2 concentration was reported to be below the threshold for significant pulmonary effects as shown from measurements of FVC, $FEV_{1.0}$, $FEF_{25-75\%}$, and specific airway conductance (38). In contrast, Snell and Luchsinger (108) report a small but significant decrease in the maximum expiratory flow from one half vital capacity ($MEF_{50\%VC}$) following 15 min inhalation of 1.0 ppm SO_2 by mouth in nine healthy adult volunteers. However, Kreisman *et al.* (71) evaluated eight trained subjects during light cycling exercise (50 W) at 1.0 ppm SO_2 and reported no significant changes in maximal expiratory flow rate as an index of respiratory function.

At a concentration of 3.0 ppm SO_2, Kreisman *et al.* (71) demonstrate a significant decrease in the maximal respiratory flow rate in nine subjects during light cycling exercise (~50 W). In additional experiments utilizing the same experimental protocol and test subjects mentioned, Snell and Luchsinger (108) show more pronounced decreases in $MEF_{50\%VC}$ at 5.0 ppm SO_2. At 5.0 ppm SO_2, Wolff et al. (119) evaluated tracheobronchial clearance in 10 healthy adults during cycling exercise for 30 min at an intensity estimated to require 70–75% of HR_{max}. These authors report that the speed of clearance was enhanced during exercise at 5.0 ppm SO_2, but suggest that this should not be interpreted as a beneficial effect of SO_2. However, Andersen *et al.* (3) showed a slowing of clearance during exposure to 5 ppm SO_2. In conclusion, it would appear that the threshold level of SO_2 which affects human physiological performance in healthy individuals during submaximal exercise is between 1.0 and 3.0 ppm SO_2.

Asthmatic individuals and possibly others with pulmonary hyperractivity appear to have a lower threshold and greater bronchomotor responsiveness during SO_2 exposure at rest (65,79,104,105). In seven asthmatic adult volunteers performing moderate cycle exercise at 67 W (400 $kpm \cdot min^{-1}$) for 10 min, Sheppard *et al.* (104) report a significant increase in specific R_{aw} at 0.50 ppm of SO_2. In their two most responsive asthmatic subjects, inhalation of 0.10 ppm SO_2 at this same exercise intensity also significantly increased the specific R_{aw}. During exposure to 0.50 ppm SO_2 while performing cycle exercise for 5 min at 92 W (550 $kpm \cdot min^{-1}$), Kirkpatrick et al. (65) evaluated six adult asthmatic subjects during either oral, nasal or oronasal breathing. For these asthmatic individuals, nasal breathing provided slightly more protection against increases in specific R_{aw} during exercise, but both oral and oronasal breathing of 0.50 ppm SO_2 resulted in significant bronchoconstriction in these people. In 23 asthmatic volunteers who performed 5 min of heavy cycling exercise at a mean power output of 122 W (730 $kpm \cdot min^{-1}$), Linn *et al.* (79) evaluated SO_2 concentrations of 0.0, 0.2, 0.4 and 0.6 ppm. Pulmonary measurements included thoracic gas volume (V_{tg}), specific R_{aw}, FVC, $FEV_{1.0}$, $FEV_{2.0}$, $FEV_{3.0}$, peak expiratory flow rate (PEFR) and flow rates with 75, 50 and 25% FVC remaining. These authors report highly significant changes in most of these pulmonary measures at 0.60 ppm SO_2, few changes at 0.40 ppm and no significant changes at 0.20 ppm. Thus, the threshold concentration of SO_2 for asthmatics during submaximal exercise would appear to be between 0.20 and 0.50 ppm of SO_2.

Adaptation. Several studies suggest that healthy and asthmatic individuals may adapt to SO_2, but unfortunately none of this research investigated adaptation to SO_2 during physical exercise. In an investigation of industrial workers who were exposed on a regular basis to concentrations of 10 ppm SO_2, Amdur *et al.* (5) report that these individuals were not responsive to levels of 5 ppm SO_2 as shown in non-exposed individuals. Anderson *et al.* (3) conducted experiments involving 15 young men exposed for six hrs to levels of 1, 5 or 25 ppm SO_2 where the SO_2 level was slowly increased during the initial 1 to 1 1/2 hrs of exposure to reach the target concentration. These same authors state that the test subjects tolerated all levels including 25 ppm "very well," possibly due to the very slow rise to the target level, but the experimenters, who had to occasionally enter the chamber, were quite disturbed by the seemingly abrupt changes in SO_2 level and found the 25 ppm level to be an "almost intolerable discomfort." In eight asthmatic subjects performing 3 min of voluntary eucapnic hyperpnea at 0.50 ppm SO_2 repeated three times at 30-min intervals, Sheppard et al. (103) show significantly greater tolerance in terms of specific R_{aw} with each repeated exposure.

Nitrogen Oxides

The nitrogen oxides (NO_x) develop from high temperature combustive processes involving nitrogen and oxygen with elevated levels being prevalent during heavy motor vehicle use, with aircraft at airports; and, the smoke associated with cigarettes or fire fighting. Motor vehicles have been suggested to account for about 40% of the NO_x emissions (81). The nitrogen oxides include nitrous oxide, nitric oxide, nitrogen dioxide, dinitrogen trioxide, dinitrogen pentoxide and nitrate ions. Of these, nitrogen dioxide (NO_2) is known to be potentially harmful to health and is the only NO_x studied to any extent in man. Acute exposure to high concentrations of NO_2 (estimated range, 200–4000 ppm) in farmers filling silos (silo-filler's disease) was associated with early severe pulmonary edema, subsequent bronchiolitis fibrosa obliterans, and ultimately death in two of the four exposed individuals (80). Some suggest that those individuals with chronic bronchitis, chronic obstructive pulmonary disease and possibly other respiratory disorders may be highly susceptible to the adverse effects of NO_2 exposure.

Submaximal Exercise Performance. With a shift from nasal to oral breathing, which occurs during physical exercise, less NO_2 is absorbed while more of this pollutant reaches the small airways and alveoli resulting in increased airway resistance. Only a few studies involving intermittent light to moderate physical exercise have evaluated the singular effects of NO_2. Maximal exercise responses to this pollutant have not been reported to date.

Sixteen healthy male volunteers performed intermittent light cycling exercise for two hrs (15 min exercise, 5 min rest) while exposed to 1.0 ppm NO_2 (56). The exercise intensity was sufficient to double the resting ventilation. Most of the pulmonary function tests and associated measurements, which were described earlier in the Introduction to this chapter, were evaluated in this study. These authors report no statistically significant pulmonary function changes associated with NO_2 exposure except for a marginal loss in FVC after NO_2 exposure for two consecutive days. However, these same authors suggest that the long term effects of this pollutant need to be evaluated as well as its effects on individuals with pulmonary hyperractivity.

Posin *et al.* (92) evaluated the blood biochemical changes of ten healthy young men, who were exposed to concentrations of either 1 or 2 ppm NO_2, while performing light exercise for 2.5 hrs employing a previously described experimental protocol (56). At both NO_2 concentrations, decreases were reported for hemoglobin, hematocrit and erythrocyte membrane enzyme acetylcholinesterase. Levels of peroxidized red blood cell lipids were increased only at 2 ppm NO_2 while glucose-6-phosphate dehydrogenase was elevated only after two exposures to the higher NO_2 concentration. These authors conclude that minimal but significant blood biochemical changes were associated with NO_2 inhalation; however, the experimental procedures appeared to contribute to some of these changes.

Fifteen healthy young men were exposed for two hrs to 0.62 ppm NO_2 while exercising at 45% $\dot{V}O_2$max for various time periods (40). During the two hr exposure period, exercise involved 15, 30 or 60 min. Pulmonary function measurements included: TV, FVC, $FEV_{1.0}$, $FEF_{25-75\%}$, $FEF_{50\%}$, $FEF_{75\%}$, IC, ERV, FRC, MVV and R_{aw}. There appear to be no profound alterations in these pulmonary functions and other cardiorespiratory responses during intermittent exercise at concentrations of 0.62 ppm NO_2 (40). While inhaling 0.50 ppm NO_2 for four hrs, 10 healthy men performed two 15-min bouts of treadmill walking (1.78 $m \cdot s^{-1}$, 10% grade), which resulted in no significant effects on a battery of 19 measurements of pulmonary function (110).

From these limited number of studies, NO_2 exposure does not appear to adversely affect human physiological performance during submaximal exercise in healthy individuals. No studies have been reported to date on human adaptation to NO_2 during physical exercise.

Primary Particulates

Primary particulate matter mainly includes dust, soot and smoke (81,88). Soot emanates primarily from the incomplete combustion of fossil fuels and environmental erosion, while the major sources of dust and smoke include dust storms, forest fires, wind storms and volcanos. The receptor site reached by particulate matter is determined by the particle size with the frequency of breathing, tidal volume and other factors, such as nasal versus oral breathing influencing total particle disposition (22). Thus, elevated levels of particulate matter may increase the effective dose of this type of pollutant for individuals performing physical exercise. However, the physiological performance effects of particulates have not been directly evaluated using exercise in man.

In man, inhalation of fine charcoal dust (118), inert dust particles (30) or cigarette smoke particles (86) results in a decreased airway conductance and an increased airway resistance. After inhalation of fine charcoal dust in nine healthy men, Widdicombe *et al.* (118) report an average decrease of 41% for airway conductance (increased resistance) and an associated bronchoconstriction. After breathing inert dust particles in five healthy middle-age men, Dubois and Dautrebande (30) observed a significant increase in airway resistance and pulmonary resistance with often a slight decrease in lung compliance. In 36 healthy subjects (21 smokers and 15 nonsmokers) and 22 patients with diagnosed cardiopulmonary disease (all smokers) after inhaling 15 puffs of cigarette smoke in a five-min period, Nadel and Comroe (86) show a significant decrease in the airway conductance/thoracic gas volume ratio. These same authors suggest that these changes are not dependent on nicotine or the nitrogen oxides in cigarette smoke, but are probably related to the submicronic particle matter (86).

In general, particles greater in size than 10 microns are not suitable for respiration while particles less than 5 microns can cause adverse health effects such as inflammation, congestion and/or ulceration (58,81). Particles between 3 and 5 microns usually settle in the upper respiratory tract while particles between 0.5 and 3 microns generally reach the alveoli (81). Larger particles are generally discharged from the respiratory tract faster than smaller particles (72). Phagocytosis and mucous transport are the usual processes employed for lung clearance after particle inhalation (64).

The question of adaptation to atmospheric particles in man is limited to the inferential observations from one study. In three healthy human subjects, who repeatedly inhaled aluminum dust (mean size = 0.68 microns) three times between short intervals of normal air breathing, Dautrebande *et al.* (23) concludes that the pneumoconstriction became greater and respiration more frequent with each dose of dust. Thus, at least for dust, adaptation in man appears questionable.

THE SECONDARY POLLUTANTS

Ozone

Ozone (O_3) is produced in oxygen containing atmospheres primarily from the interaction of hydrocarbons and nitrogen dioxide in the presence of solar ultraviolet radiation. As illustrated earlier in Figure 16-2, the O_3 level is related to the sunlight hours and usually reaches peak concentrations near midday. Whereas, the current U.S. air quality standard for O_3 is 0.12 ppm (one hour averaging time), Kleinfeld and Giel (66) report three case studies of ozone poisoning in male welders exposed to 9.2 ppm O_3 while Kelly and Gill (63) describe O_3 poisoning in a male industrial crane operator with all cases leading to severe and prolonged illness. Kelly and Gill (63) further state that inhalation of 50 ppm O_3 for 30 min could be fatal.

Submaximal Exercise Performance. Ozone is a potent airway irritant capable of causing a reflex bronchoconstriction in the upper airways at sufficient concentrations. In 12 healthy volunteer subjects exposed to 0.20 ppm O_3 for two hrs with intermittent light cycling exercise (33–50 W), capable of doubling the resting pulmonary ventilation, Linn et al. (75) report nosignificant adverse effects on arterial blood oxygenation or lung mechanics. After two hrs of intermittent light cycling exercise (15 min exercise, 15 min rest) adequate to double the pulmonary ventilation of eight healthy young men breathing 0.37 ppm O_3, Hazucha and Bates (57) report a "just significant decrease" in maximal mid-expiratory flow rate (MMFR). In four normal and four sensitive volunteer subjects, Bell et al. (19) attempted to reproduce the observations of Hazucha and Bates (57) in Los Angeles residents. While exposed to the same O_3 level (0.37 ppm) and experimental methodology as the earlier study (57), these authors (19) did not observe significant changes in mean pulmonary functions (FVC, $FEV_{1.0}$, TLC, RV, ΔN_2, closing capacity, and flow rates with 50% and 25% FVC remaining) at this O_3 concentration. After two hrs of intermittent treadmill walking and rest (15 min walk, 15 min rest) which was sufficient to produce a $\dot{V}_E$ of 30 $\ell \cdot min^{-1}$ in nine young adult men breathing 0.40 ppm O_3, Bedi *et al.* (17) report significant decreases in maximum expiratory flow ($FEV_{1.0}$, $FEF_{25-75\%}$ and $FEF_{50\%}$), FVC and IC. In six healthy young adults who inhaled 0.45 ppm O_3 for two hrs while performing intermittent light cycle exercise (33–75 W) yielding a $\dot{V}_E$ of 27 $\ell \cdot min^{-1}$, Bedi *et al.* (16) display significant decrements in FVC, $FEV_{1.0}$, $FEV_{3.0}$, $FEF_{25-75\%}$, and TLC; however, $\dot{V}_E$, $\dot{V}O_2$, and $\dot{V}CO_2$ were not significantly altered at this O_3 level.

During moderate treadmill exercise (1.56 $m \cdot s^{-1}$, up an incline) for two hrs (30 min exercise, 10 min rest), producing a $\dot{V}_E$ of 38 $\ell \cdot min^{-1}$ in 22 young adult men while breathing 0.30 ppm O_3, Folinsbee et al. (38) report significant reductions in forced expiratory measurements (FVC, $FEV_{1.0}$ and $FEF_{25-75\%}$). While inhaling 0.40 ppm O_3 for four hrs, 12 healthy men performed two 15-min bouts of treadmill walking (1.78 $m \cdot s^{-1}$, 10% grade) which resulted in significant decreases in specific R_{aw}, FVC and $FEF_{50\%}$ (110). In a comparison of six female and six male subjects who performed three intensities of cycling exercise (~30–60% $\dot{V}O_2max$) for one hr each at 0.0, 0.20, 0.30 or 0.40 ppm O_3, Lauritzen and Adams (73) show greater decrements in FVC, $FEV_{1.0}$ and respiratory rate for the females at the same total O_3 effective dose as the males. These same females displayed an increased $\dot{V}_E$ with O_3 exposure, but no effect on $\dot{V}O_2$ and HR. A sample of 135 healthy male volunteers, divided into six groups, was exposed to intermittent moderately heavy treadmill exercise (15 min rest, 15 min exercise sufficient to increase $\dot{V}_E$ to 65 $\ell \cdot min^{-1}$) either at 0.0, 0.12, 0.18, 0.24, 0.30 or 0.40 ppm O_3 (83). While coughing occurred at all levels of O_3 exposure, McDonnell *et al.* (83) report small changes in FVC, $FEV_{1.0}$ and $FEF_{25-75\%}$ at 0.12 and 0.18 ppm O_3 with larger changes at O_3 concentration equal to or greater than 0.24 ppm. During exercise, these same authors note alterations in TV, specific R_{aw} and respiratory rate accompanied with increased subjective distress at O_3 levels greater than or equal to 0.24 ppm (83). While breathing 0.15 or 0.30 ppm O_3 for one hr, DeLucia and Adams (25) evaluated the performance of six healthy males during cycle exercise at 25, 45 or 65% $\dot{V}O_2$ max, and reported no changes in $\dot{V}_E$, $\dot{V}O_2$ or blood biochemistry at any exercise intensity. However, all subjects showed some signs of toxicity at 45 and 65% $\dot{V}O_2$max with significant decrements in VC, $FEV_{1.0}$ and MMFR observed at 65% $\dot{V}O_2$max while inhaling 0.30 ppm O_3. During moderately heavy cycling exercise (45–75% $\dot{V}O_2$max) which immediately followed two hrs of exposure to 0.37, 0.50 or 0.75 ppm O_3, Folinsbee et al. (43) show no significant alterations in submaximal $\dot{V}O_2$, HR or $\dot{V}_E$ at any of these O_3 levels. However, other measurements of pulmonary function such as TV and respiratory rate appeared to be somewhat disturbed by O_3 exposure. In an evaluation of the threshold for O_3 toxicity during cycling exercise, Adams et al. (1) had eight trained male volunteers complete 18 different protocols involving three O_3 levels (0.20, 0.30 and 0.40 ppm), two exercise intensities ($\dot{V}_E$ = 33 and 66 $\ell \cdot min^{-1}$), and different exercise durations between 30 and 80 min. These same authors report the threshold for O_3 toxicity to involve moderately heavy exercise (65% $\dot{V}O_2$max) at O_3 levels between 0.20 and 0.30 ppm.

During heavy cycling exercise (75% $\dot{V}O_2$max) for one hr involving seven trained athletes exposed to 0.21 ppm O_3, Folinsbee *et al.* (37) show significant decreases in FVC, $FEV_{1.0}$, $FEF_{25-75\%}$ and MVV with the magnitude of these alterations similar to those seen during moderate intermittent exercise at 0.24 ppm O_3 for two hrs. Symptoms of distress were also quite severe (37). While breathing 0.0, 0.20 or 0.35 ppm O_3 during one hr of cycle exercise which simulated training or competition (30 min at ~85% $\dot{V}O_2$max), Adams and Schelegle (2) evaluated the performance of 10 well-trained runners and reported no significant changes in HR, $\dot{V}O_2$, $\dot{V}_E$ and alveolar ventilation with O_3 exposure, but showed significant decrements in FVC and $FEV_{1.0}$ with enhanced unpleasant subjective symptoms during O_3 exposure. Three of these 10 subjects were not able to complete the training and competitive simulations at 0.35 ppm O_3 because of respiratory discomfort (2). Employing a similar experimental design to that described above (2), Schelegle and Adams (100) exposed 10 competitive endurance athletes while breathing either filtered air, 0.12, 0.18 or 0.24 ppm O_3 to a one hr competitive simulation cycling protocol (30 min at 86% $\dot{V}O_2max$). While no difference in HR, $\dot{V}O_2$, $\dot{V}_E$, respiratory rate or TV with O_3 were found, these authors show a general increase in the inability to complete these simulations with increasing O_3 concentration (significant at 0.24 ppm O_3), which appeared to be primarily associated with enhanced subjective symptoms of distress. During two hrs of heavy intermittent treadmill exercise (1.30 $m \cdot s^{-1}$, 15% grade; 15 min rest, 15 min exercise) involving 23 boys (8–11 yr) exposed to 0.12 ppm O_3, McDonnell et al. (82) report small decrements in $FEV_{1.0}$ without induction of cough or any changes in exercise $\dot{V}_E$ or HR.

In summary, submaximal exercise performance at light to moderate exercise intensities during O_3 exposure does not appear to be limited in terms of the cardiorespiratory system; however, decrements in pulmonary functions and enhanced subject discomfort become apparent. At heavy intensities of submaximal exercise, O_3 exposure can limit performance, primarily from severe respiration discomfort and associated changes in pulmonary functions without altering cardiorespiratory responses.

Maximal Exercise Performance. In contrast to submaximal exercise performance, the literature concerning maximal exercise performance and O_3 exposure in man is quite limited. After two hrs of intermittent light cycling exercise involving 13 adult men breathing either filtered air (FA) or 0.75 ppm O_3, Folinsbee *et al.* (42) measured cycling $\dot{V}O_2max$ and reported a 10% reduction with 0.75 ppm O_3 compared to FA. Reductions were also observed for maximal exercise intensity (10%), maximal $\dot{V}_E$ (16%) and HR_{max} (6%) during O_3 exposure. In a study of nine healthy men following 30 min of breathing either 0.00, 0.15 or 0.30 ppm O_3, Savin and Adams (99) show no change in anaerobic threshold, exercise capacity or cycling $\dot{V}O_2max$ associated with O_3 concentration. After two hrs of resting exposure to 0.00, 0.25, 0.50 or 0.75 ppm O_3 in eight men and five women, Horvath et al. (60) report no significant changes in treadmill $\dot{V}O_2max$, HR_{max} or maximal performance time associated with O_3 exposure. Thus, maximal exercise performance may be lowered at relatively high O_3 levels, but further research is necessary in order to support this claim.

Adaptation. Human adaptation to O_3 exposure was initially suggested from experimental observations involving residents of

Southern California who were chronically exposed to elevated levels of O_3 and were less sensitive to this pollutant than Canadian residents who are far less often exposed to O_3 (19,52). After exposure for four consecutive days (two hrs/day) to 0.50 ppm O_3 in six male volunteers with respiratory hyperractivity, Hackney *et al.* (54) reports pulmonary function decrements during the first three days of exposure, which were mostly reversed by the fourth day. This reversal suggests adaptation.

In a study of 14 healthy subjects (10 men and 4 women) exposed for five consecutive days (three hrs/day) to 0.40 ppm O_3, Farrell *et al.* (33) suggests that the maximal response to O_3 in terms of impaired pulmonary functions occurred for most subjects on the second exposure day with adaptation thereafter. A series of experiments (39,48,59) concerning adaptation to O_3, which typically employed a two hr exposure with intermittent exercise at various O_3 concentrations (0.20, 0.35, 0.42 or 0.50 ppm) and involved 75 volunteer subjects (62 men and 13 women), has been summarized by Folinsbee *et al.* (35). It can generally be concluded from these experiments that adaptation to consecutive daily exposures of O_3 varies from two to five days, lasts for less than two weeks (range, 7–20 days) and does not differ between genders (34,35,39,48,59). Figure 16-5 illustrates the number of consecutive days required to produce adaptation to 0.42 ppm O_3 for 24 of these subjects (35). After exposure to 0.20 or 0.40 ppm O_3 for two hrs on three to five consecutive days in two groups of healthy adult volunteers (n = 7 each group), Dimeo *et al.* (26) shows peak changes in bronchial reactivity to occur after the first or second exposure. Adaptation occurs after the third, fourth or fifth exposure. In a group of 11 generally healthy volunteers exposed initially to 0.47 ppm O_3 for four consecutive days (two hrs per day with intermittent exercise) and then after a four-day interval without exposure followed by seven-day intervals for four weeks, Linn et al. (78) report peak changes in $FEV_{1.0}$ and subjective symptoms on the second consecutive exposure day. Adaptation generally appears on the fourth exposure day. These authors also state that adaptation is partly lost after a four-day interval without O_3 exposure and seems to be completely lost after a seven-day interval.

In a recent study of eight trained males exposed to 0.35 ppm O_3 for four consecutive days (one hr per day), Foxcroft and Adams (46) report significant decrements in $\dot{V}O_2$max, maximal exercise performance time and pulmonary functions on the first exposure day with significant improvement in $\dot{V}O_2$max and maximal exercise time, but not impaired pulmonary functions after the fourth consecutive exposure day. While adaptation is seen to occur after two to five days of consecutive exposure to O_3, it should be remembered that this adaptation may be eventually harmful to man because of the associated suppression of normal defense mechanisms (81).

Peroxyacetyl Nitrate

Peroxyacetyl nitrate (PAN), regarded as a secondary pollutant, was first detected in the exhaust of motor vehicles in the early 1950s (81). This pollutant is known to cause eye irritation and ocular function disturbances while producing some pulmonary function alterations in man. Relatively few studies have investigated the effects of PAN during submaximal and/or maximal exercise, while no research has been reported concerning adaptation to this pollutant.

Submaximal Exercise Performance. During submaximal treadmill exercise (35% $\dot{V}O_2$max) of 210 min duration while breathing 0.24 ppm PAN, no remarkable changes in cardiorespiratory, metabolic or thermoregulatory responses were observed in either younger (22–26 yr, n = 10) or older (45–55 yr, n = 9) healthy male volunteers (49,97). These responses included $\dot{V}O_2$, $\dot{V}_E$, $\dot{V}_E/\dot{V}O_2$, HR, cardiac index, rectal temperature (T_{re}) and mean skin temperature ($\bar{T}_{sk}$). There were no differences in these responses related to age (49). In these

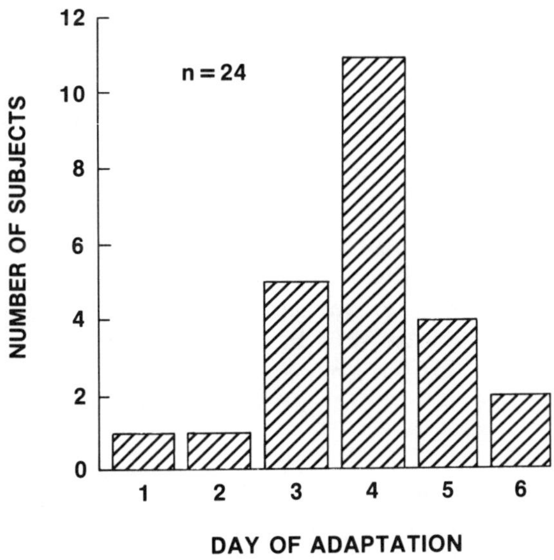

Figure 16-5. *Number of days necessary to produce adaptation to ozone (O_3) for 24 individuals at 0.42 ppm O_3 for two hr exposures during five consecutive days. The two subjects indicated on day 6 never demonstrated adaptation. Reprinted with permission from Folinsbee* et al. *(35).*

same experiments, FVC was reduced 4–7% in the younger subjects with PAN, but the significance is questionable (97).

Other pulmonary function measures ($FEV_{1.0}$, $FEV_{2.0}$, $FEV_{3.0}$, $FEV_{1.0}$/FVC%, MMFR, IC and ERV) showed no significant differences due to this pollutant (97). Subjective complaints of eye irritation, blurred vision and eye fatigue were inhanced in the presence of PAN (49). The concentration of PAN (0.24 ppm) in these few experiments may be at or slightly below the threshold level needed for demonstrable physiological effects which might suggest additional studies during submaximal exercise at slightly higher concentrations. Nevertheless, some minor changes in pulmonary function and subjective discomfort have been observed to date during prolonged exposure to PAN, with little change in cardiorespiratory, metabolic or thermoregulatory responses during light submaximal exercise.

Maximal Exercise Performance. During maximal exercise involving a modified treadmill Balke test while breathing 0.27 ppm PAN, Raven *et al.* (96) report no significant decrement in $\dot{V}O_2max$ for 20 healthy men (range, 21–30 yr), who were divided in two groups (10 non-smokers and 10 smokers). In another study from the above mentioned laboratory, utilizing similar test methodology but evaluating 16 somewhat older men (range, 40–57 yr) divided into two groups (9 non-smokers and 7 smokers), Raven et al. (95) show no significant change in $\dot{V}O_2max$ while breathing 0.27 ppm PAN. In addition, these same authors report no alterations in maximal $\dot{V}_E$, HR_{max}, maximal respiration rate, maximal time walked, post-exercise lactate and oxygen debt associated with breathing 0.27 ppm PAN. While the concentration of 0.27 ppm PAN is seemingly below the threshold to become physiologically meaningful for healthy individuals in a thermally comfortable environment, these authors caution that this PAN concentration might be a threshold level for other groups such as those with pulmonary impairments (95). Therefore, it would seem premature, particularly regarding maximal exercise performance, to conclude that PAN has no adverse effects on exercise performance, until higher concentrations of this pollutant are evaluated, and at risk individuals are studied.

Aerosols

Of those investigations concerning the effects of aerosol pollutants in man, the majority have involved the study of sulfate aerosols, and sulfuric acid (H_2SO_4) or nitrate aerosols (110). The sulfate aerosols which are ammonia neutralization products of (H_2SO_4) are ammonium sulfate [$(NH_4)_2SO_4$] and ammonium bisulfate (NH_4HSO_4), while the nitrate aerosol most studied is ammonium nitrate (NH_4NO_3).

Other reports have investigated SO_2 combined with sodium chloride (NaCl) aerosol; and, saturated and unsaturated aldehydes. Not all of these reports have included the effects of physical exercise in their evaluation of these aerosol pollutants.

Sulfate Aerosols. In a study involving two hr exposures with light intermittent cycle exercise (25–50 W; 15 min exercise, 15 min rest) to either $(NH_4)_2SO_4$ or NH_4HSO_4 at a concentration of 100 $\mu g/m^3$, Avol *et al.* (13) evaluated both healthy (n=12) and asthmatic (n=12) adult men, and reported no significant adverse changes in pulmonary functions or recorded clinical symptoms. After four hr exposures including two, 15 min treadmill exercise sessions (1.78 $m \cdot s^{-1}$, 10% grade) involving healthy male volunteers in either 133 $\mu g/m^3$ $(NH_4)_2SO_4$ (n=13) or 116 $\mu g/m^3$ NH_4HSO_4 (n=15), Stacy *et al.* (110) also observed no significant changes in a battery of 19 measurements of pulmonary function. Thus, the sulfate aerosols at "worst case" ambient concentrations elicit minimal adverse effects relative to some of the other pollutants.

Sulfuric Acid. After 5–15 min resting exposures of 15 normal human subjects to a concentration range of 350 to 5000 $\mu g/m^3$ H_2SO_4, Amdur *et al.* (6) observed shallower and more rapid breathing as a function of increasing concentration with the retention of inhaled H_2SO_4 averaging 77%. In a study of normal (n=17) and asthmatic (n=17) adults breathing H_2SO_4 aerosol in concentrations up to 1000 $\mu g/m^3$ for 10 min, Sackner *et al.* (98) report no changes in lung volumes, distribution of ventilation, ear oximetry, dynamic breathing mechanics, oscillation mechanics of the chest-lung wall, pulmonary capillary blood flow, D_L, $\dot{V}O_2$ and pulmonary tissue volume. While breathing 100 $\mu g/m^3$ H_2SO_4 in six normal and six asthmatic subjects utilizing the experimental methodology described in the previous paragraph, Avol *et al.* (13) showed no significant changes in pulmonary functions and subjective symptoms. At a concentration of 100 $\mu g/m^3$ H_2SO_4 in 11 healthy males also employing the same test methodology described in the previous paragraph, Stacy *et al.* (110) demonstrated no significant alterations in a battery of 19 tests for pulmonary function. In experiments involving 10 healthy nonsmokers exposed for one hr to concentrations of 110, 330 or 980 $\mu g/m^3$ H_2SO_4 during rest, Leikauf *et al.* (74) observed marked alterations in bronchial mucociliary clearance, displaying a dose dependent pattern for these subjects. Thus, H_2SO_4 does not appear to be harmful to cardiorespiratory function for single brief exposures within the concentration range discussed; however, prolonged exposures, multiple exposures, larger aerosol particles and high relative humidity could cause this pollutant to have adverse effects on man.

Nitrate Aerosols. In a study involving two hr exposures with light intermittent cycle exercise (25–50 W; 15 min exercise, 15 min rest) at a concentration of 200 $\mu g/m^3$ NH_4NO_3, Kleinman *et al.* (68) examined 20 normal and 19 asthmatic adult volunteers and reported no substantial changes in pulmonary functions or overall reported symptoms associated with this pollutant. At a concentration of 80 $\mu g/m^3$ NH_4NO_3 in 12 healthy male volunteers using the same methodology previously described, Stacy et al. (110) showed no adverse effects on a battery of 19 pulmonary function measurements with this pollutant. While somewhat limited, the experimental evidence does not support major alterations in respiratory responses associated with this pollutant aerosol.

Sulfur Dioxide and Sodium Chloride Droplet Aerosol. In resting experiments of 60 min duration in nine adolescent asthmatic subjects exposed to 1 ppm SO_2 plus 100 $\mu g/m^3$ NaCl droplet aerosol, Koenig *et al.* (69) report significant decreases in maximal flow at 50 and 75% of expired vital capacity ($\dot{V}_{max50}$ and $\dot{V}_{max75}$) with no changes in FRC, FEV, and total respiratory resistance (R_T). After 30 min rest followed by 10 min of intermittent treadmill exercise (1.07 $m \cdot s^{-1}$, 12% grade) involving eight adolescent asthmatic volunteers exposed to 1 ppm SO_2 plus 1000 $\mu g/m^3$ NaCl, Koenig *et al.* (70) observed significant changes in $\dot{V}_{max50}$, $\dot{V}_{max75}$, $FEV_{1.0}$ and R_T with no change in FRC. However, these same authors could not certify that these exposure effects were due to the mixture (SO_2+NaCl) rather than SO_2 alone.

Saturated and Unsaturated Aldehydes. During 5–30 min exposures to formaldehyde (12 males, 13.8 ppm), acrolein (12 males each, 0.80 ppm and 1.22 ppm), crotonaldehyde (12 males, 4.1 ppm), acetaldehyde (14 males, 134 ppm), propionaldehyde (12 males, 134 ppm), butyraldehyde (15 males, 230 ppm) and isobutyraldehyde (15 males, 207 ppm), Sim and Pattle (106) observed that acrolein and crotonaldehyde were highly irritant while acetaldehyde, proprionaldehyde, butyraldehyde and isobutyraldehyde were essentially non-irritant. Formaldehyde was intermediate in terms of irritability compared to these two groups.

A summary of the effects of all of the air pollutants, discussed thus far on submaximal or maximal exercise performance in normal individuals, is presented in Table 16-3. This table is an updated version of a previously published table by the author (87).

INTERACTIONS BETWEEN POLLUTANTS

As discussed in the Introduction to the chapter, the various air pollutants may interact in three ways: (a) additively, (b) synergistically, or (c) antagonistically. The differentiation between additive

TABLE 16-3. *Effects of Air Pollutants on Exercise Performance of Normal Individuals.*

Air Pollutant Exercise Intensity	Performance Decrement	No Effect	Selected References
CO			
—Submaximal Exercise		X	Drinkwater *et al.* 1974; Ekblom and Huot, 1972; Gliner *et al.* 1975; Horvath *et al.* 1975; Pirnay *et al.* 1971; Raven *et al.* 1974; Vogel and Gleser, 1972.
—Maximal Exercise	X		
SO_2			
—Submaximal Exercise	?		Bedi *et al.* 1979; Bell *et al.* 1977; Folinsbee *et al.* 1985; Huzacha and Bates, 1975; Kreisman *et al.* 1976; Snell and Luchsinger, 1969; Stacy *et al.* 1983; Raven *et al.* 1979; Wolff *et al.* 1975.
—Maximal Exercise	?		
NO_2			
—Submaximal Exercise		X	Folinsbee *et al.* 1978; Hackney *et al.* 1978; Horvath and Folinsbee, 1978; Posin *et al.* 1978; Raven, 1979; Stacy *et al.* 1983.
—Maximal Exercise	?		
Particulates			
—Submaximal Exercise	?		DuBois and Dautrebande, 1958; Nadel and Comroe, 1961; Widdicombe *et al.* 1962.
—Maximal Exercise	?		
O_3			
—Submaximal Exercise		X	Adams *et al.* 1981; Adams and Schelegle, 1983; Bedi *et al.* 1979; Bedi et al. 1982; Bell *et al.* 1977; DeLucia and Adams, 1977; Folinsbee *et al.* 1975; Folinsbee *et al.* 1977; Folinsbee *et al.* 1984; Folinsbee *et al.* 1985; Hazucha and Bates, 1975; Horvath *et al.* 1979; Lauritzen and Adams, 1985; Linn *et al.* 1979; McDonnell *et al.* 1983; McDonnell *et al.* 1985; Savin and Adams, 1979; Schelegle and Adams, 1986; Stacy *et al.* 1983.
—Maximal Exercise	?		
PAN			
—Submaximal Exercise		X	Drinkwater *et al.* 1974; Gliner *et al.* 1975; Raven *et al.* 1974; Raven *et al.* 1976; Raven, 1979.
—Maximal Exercise	?		
Aerosols			
—Submaximal Exercise	?		Amdur *et al.* 1952; Avol *et al.* 1979; Kleinman *et al.* 1980; Koenig *et al.* 1980; Koenig *et al.* 1981; Leikauf *et al.* 1981; Sackner *et al.* 1983; Sim and Pattle, 1957; Stacy *et al.* 1983.
—Maximal Exercise	?		

and synergistic effects is frequently difficult and usually involves the specific pollutant concentration (81). No pollutant interactions have been reported that are antagonistic.

Additive Interactions

Ozone in combination with nitrogen dioxide has been studied at various concentrations (O_3=0.25–0.50 ppm, NO_2=0.30–0.50 ppm) during light to moderate intermittent exercise of two to four hr duration (36,53,55). In general, the effects attributed to O_3 alone were not increased with O_3 and NO_2 in combination. Exposure to NO_2 has been shown earlier in this chapter not to produce significant effects during submaximal exercise. Thus, this interaction is defined as additive.

In two of the above mentioned studies (53,55), CO at a concentration of 30 ppm was evaluated in combination with O_3 (0.25–0.50 ppm) and NO_2 (0.30 ppm) during light intermittent exercise of two to four hr duration in healthy and sensitive individuals. It has been demonstrated that CO does not produce significant effects during submaximal exercise. The addition of CO to the pollutant mixture (O_3 + NO_2) failed to produce detectable effects and, therefore, this three-pollutant interaction is deemed additive.

Ozone in combination with total suspended particulates (TSP) has been evaluated at various concentrations (O_3=0.15–0.22 ppm, TSP=200–295 $\mu g/m^3$) during light to heavy intermittent or continuous exercise of one to two hr duration in healthy and/or asthmatic individuals (14,15,76). The combined exposure to O_3 and TSP produced, generally, the same response as O_3 separately at the same concentration. Thus, this interaction is defined as additive by inference.

In a series of experiments involving a total of 137 healthy male volunteers during moderate intermittent exercise of four hr duration, Stacy *et al.* (110) evaluated the effects of O_3, NO_2 and SO_2 each in combination with H_2SO_4, $(NH_4)_2SO_4$, NH_4HSO_4, or $NO_2H_2SO_4$. The concentrations of these pollutants were 0.40 ppm O_3, 0.75 ppm SO_2, 0.50 ppm NO_2, 100 $\mu g/m^3$ H_2SO_4, 133 $\mu g/m^3$ $(NH_4)_2SO_4$, 116 $\mu g/m^3$ NH_4HSO_4 and 80 $\mu g/m^3$ NH_4NO_3. None of the four aerosols separately, NO_2 or SO_2 separately, or combinations of NO_2 or SO_2 with any of the four aerosols caused significant effects. Therefore, these interactions would be termed additive. The response of O_3 with some of the aerosols, particularly O_3 and H_2SO_4, produced effects greater than those for O_3 separately; however, the statistical criteria for this apparent synergistic response was not quite significant.

The oxides of nitrogen and sulfur when combined have been cited by McCafferty (81) to demonstrate additive effects with SO_2

acting at once, but for a shorter period of time, and NO_2 acting later but for a longer period of time. However, Linn et al. (77) evaluated 24 normal and 19 asthmatic volunteers during intermittent light exercise for two hrs at 0.50 ppm NO_2 combined with 0.50 ppm SO_2 (normals) or 0.30 ppm SO_2 (asthmatics). These same authors conclude that there was no marked synergistic interaction between NO_2 and SO_2 as should be expected from these pollutants given the general negative finding in studies of each pollutant separately at these concentrations.

Epidemiological observations over a 13-year period involving adult mortality of two communities with widely different air pollution levels resulted in the hypothesis by Morris *et al.* (85) that smoking and air pollution are additive. However, Shephard (101) proposes that exposure to tobacco smoke could increase the toxicity of some air pollutants through a variety of mechanisms, which ultimately could lead to synergistic effects. More recently, Shephard *et al.* (102) studied the interaction of O_3 and cigarette smoke in 32 volunteers (26 men, 6 women) during two hr exposures to intermittent light exercise. These same authors report little evidence that cigarette smoking effects the acute toxicity of O_3.

The interactive effects of PAN and CO (PANCO) have been studied during both submaximal and maximal exercise (49,95,96,97). At the low concentrations studied during either submaximal (50 ppm CO, 0.24 ppm PAN) or maximal exercise (50 ppm CO, 0.27 ppm PAN), these authors report that exposure to these two pollutants either singly or in combination, did not cause adverse physiological effects during submaximal or maximal exercise. Thus, the interactive effects of PAN and CO would be considered additive.

Synergistic Interactions

In a study involving eight healthy men from Montreal during light intermittent cycling exercise while breathing 0.37 ppm O_3 combined with 0.37 ppm SO_2, Hazucha and Bates (57) report that the combined effect of these two pollutants on pulmonary functions (FVC, $FEF_{50\%}$, $FEF_{25-75\%}$, $FEV_{1.0}$) was "much greater" than when either pollutant was breathed separately which suggests a synergistic effect. However, Bell et al. (19) were unable to confirm the synergistic effects of O_3 combined with SO_2 reported earlier by Hazucha and Bates (57), while utilizing a similar experimental protocol and evaluating nine volunteers (four Montreal subjects and five Los Angeles subjects). These same authors suggest three possible explanations for the differences between these two studies: (a) enhanced sensitivity of the Montreal volunteers, (b) adaptation of the Los Angeles residents particularly to O_3 and (c) greater effects due to the chemical formation of sulfur containing aerosols in the Montreal study. Additional studies of the combined exposure to O_3 and SO_2 at different concentrations (0.40 ppm O_3 + 0.40 ppm SO_2; 0.30 ppm O_3 + 1.0 ppm SO_2) during intermittent exercise also have not supported the suggested synergistic effect of these two pollutants (17,18,38). In a study involving 19 healthy human volunteers performing light intermittent cycling exercise for two hrs while breathing a mixture containing 0.37 ppm O_3, 0.37 ppm SO_2 and 100 $\mu g/m3$ H_2SO_4, Kleinman *et al.* (67) indicated a 3.7% decline in $FEV_{1.0}$. These same authors suggest that the mixture produced a slight but not substantial change as O_3 alone should be expected to decrease $FEV_{1.0}$ by 2.8% under similar exposure conditions.

The interactive effects of O_3 (O.45 ppm) in combination with PAN (0.30 ppm) have been investigated during light intermittent cycling exercise in 10 healthy adult men (27). PAN exposure alone did not induce any changes in pulmonary functions. Decrements in pulmonary functions (FVC, $FEV_{1.0}$, $FEV_{2.0}$, $FEV_{3.0}$, $FEF_{25-75\%}$, IC, ERV and TLC) were 10% greater after combined exposure to O_3 and PAN than after O_3 exposure alone. While these findings suggest a synergistic effect between O_3 and PAN, additional studies are required before this hypothesis can be accepted.

Stacy *et al.* (110) indicate that sulfuric acid, sulfate aerosols $(NH_4)_2SO_4$, NH_4HSO_4 or nitrate aerosols (NH_4NO_3) do not interact synergistically with SO_2 or NO_2 during intermittent light exercise in man. However, McCafferty (81), after a careful review of the related literature, suggests that the interactions between SO_2 and aerosols; and, certain other pollutants with sulfuric acid aerosols, the particulate sulfates and the particulate nitrates could be synergistic. The degree of synergism is thought to depend on the size of the aerosol with greater synergistic effects associated with breathing smaller particle aerosols. Unfortunately, the majority of studies that suggest a synergistic effect for the above pollutants involve small animal experimentation with even fewer studies evaluating human performance during submaximal exercise. Nevertheless, high concentrations of small particle size aerosols are known to exist in urban areas suggesting the need for further research on the interactive effects of these aerosols with certain other pollutants during exercise in man.

The interactive effects between the various pollutants during human performance of muscular exercise needs to be addressed in greater detail. Additional experiments should consider a wider range of concentrations between interactive pollutants, a variety of subject populations (normal vs. hyperractive, older vs. younger, etc.) at low, moderate and high exercise intensities.

POLLUTANT INTERACTIONS WITH THE ENVIRONMENT

As mentioned in the Introduction to this chapter, the combined effects of certain pollutants with the environmental extremes of heat, cold or altitude during physical exercise may result in additive and/or synergistic outcomes. While a number of investigations have been reported concerning the relative effects of environmental heat and adverse air quality, far fewer reports have been published on the interaction of high terrestrial altitude and air pollution. No studies have been reported on the effects of environmental cold and poor air quality.

Environmental Heat

Elevated concentrations of air pollution are often associated with excessive heat and humidity (32). Since excessive heat and humidity alone are related to increases in morbidity and mortality, it is not unreasonable to expect that with the addition of poor air quality morbidity and mortality rates could further increase (29). Certainly, human performance of submaximal or maximal exercise could be expected to suffer under the combined stresses of excessive heat, humidity and poor air quality.

Carbon Monoxide and/or Peroxyacetyl Nitrate. While the atmospheric pollutants of CO and PAN, either singly or in combination (PANCO), have been evaluated during exercise-heat stress, the environmental conditions were limited to 30% rh at 25° C and 35° C (28,49,95,96,97). However, during the experiments both maximal and submaximal (35% $\dot{V}O_2$max) exercise intensities were studied. Pollutant concentrations were limited to 50 ppm CO for all experiments with 0.24 ppm PAN for submaximal and 0.27 ppm PAN for maximal exercise. All subjects were described as healthy young or healthy middle-aged adult men. During submaximal exercise, no significant changes in physiological responses were reported while breathing CO, PAN or PANCO at 35° C (49,97). Nevertheless, subjective complaints, particularly for PAN and PANCO, were greater at 35° C. Gliner et al. (49) conclude that the pollutants CO, PAN or PANCO have relatively little effect on the physiological responses to submaximal exercise, compared to increased ambient temperature. The $\dot{V}O_2$max was not altered during exposure to either CO, PAN or PANCO at 35° C (28,95). While breathing filtered air, exposure to this level of heat stress (35° C) was more effective in lowering $\dot{V}O_2$max (~4%) than exposure to either single pollutant or the two pollutants in combination at 25° C. Drinkwater *et al.* (28) conclude that heat stress was more effective in decreasing exercise capacity than any of these pollutant conditions. However, these same authors speculate that the combination of CO and heat stress may be important in the more pronounced respiratory disturbances seen at this elevated ambient temperature compared to 25° C during maximal exercise. In conclusion, these authors suggest that heat stress was generally more effective in reducing human submaximal or maximal exercise performance than carbon monoxide and/or peroxyacetyl nitrate.

Ozone. Submaximal exercise performance (~40% $\dot{V}O_2max$) during O_3 exposure (0.50 ppm) has been evaluated at four different environmental conditions (25° C, 45% rh; 31° C, 85% rh; 35° C, 40% rh; 40° C, 50% rh) for a two hr period in 14 healthy young men (41). A trend was found that indicated a greater impairment in pulmonary function during combined exposure to O_3 and heat stress. Decrements in pulmonary function after exposure to ozone and heat were greatest immediately following exercise. Reductions in VC and MVV were significant during the most extreme heat exposure (40° C, 50% rh), while exercise $\dot{V}_E$ was highest at this ambient temperature (plus O_3). Since heat and O_3 were not related to additional reductions in any other pulmonary flow variables compared to O_3 alone, these same authors suggest that some other mechanism(s) besides bronchoconstriction is related to the reported decrements in pulmonary function during exposure to these combined stresses. Folinsbee *et al.* (41) conclude that environmental heat stress may increase the response to O_3 suggesting a possible additive effect.

In a study involving 10 aerobically trained young women performing moderately heavy exercise (66% $\dot{V}O_2max$) for one hr while breathing either filtered air, 0.15 or 0.30 ppm O_3 at 24 and 35° C, Gibbons and Adams (47) found that O_3 exposure, combined with heat stress (35° C), produced an interactive effect on alveolar volume and respiratory rate with near significant interactions for FVC and $FEV_{1.0}$. In addition, there were more subjective complaints during O_3 exposure in the heat, and these combined stressors were more likely to prompt premature cessation of exercise. In a recent study involving 17 top caliber endurance cyclists simulating competition conditions including submaximal cycling exercise (70% $\dot{V}O_2max$) for one hr followed by incremented exercise to exhaustion while breathing filtered air, 0.12 or 0.20 ppm O_3 in a warm environment (31° C), Gong *et al.* (50) report the threshold for significant impairment of exercise performance in this warm environment to be between 0.12 and 0.20 ppm O_3.

In addition to the adverse effects associated with poor air quality combined with elevated ambient temperature, extremes in the percent relative humidity are also thought to precipitate problems, when associated with certain air pollutants (12,81). Low relative humidity is suggested to enhance the adverse health effects of O_3, while high relative humidity is thought to intensify the adverse effects of SO_2 and also probably NO_2 (12,81). It has been suggested that for SO_2 the high humidity accelerates catalytic oxidation of this pollutant to form aerosols, which were more irritating than SO_2 itself (12,81). Arundel *et al.* (12) recommends an optimal range of 40 to 60% relative humidity to minimize the adverse effects of these pollutants. Nevertheless, the effects of extremes in percent relative humidity have not been systematically evaluated during human performance of submaximal or maximal exercise.

Future research is needed to more fully appraise the detrimental effects of heat stress and/or extremes in relative humidity for all of the air pollutants discussed during performance of muscular exercise. Encompassed in the research should be a range of pollutant concentrations, various subject populations and a wide range of exercise intensities.

Environmental Cold

In addition to many of the air pollutants, breathing cold air during exercise has been shown to induce a reflex bronchoconstriction, particularly prevalent in asthmatic individuals. The temperature and humidity of the inhaled cold air during exercise are critical in the resultant degree of bronchoconstriction. Unfortunately, no studies have directly evaluated the interactive effects of breathing cold, polluted air and the subsequent degree of airway obstruction during exercise. By inference, Pierson *et al.* (89) recently imply that breathing cold, polluted air could adversely affect respiratory function and consequently athletic performance.

Exercise-induced asthma or bronchoconstriction has been reported to take place in about 12% of the population (62). Approximately 11% of the 1984 Olympic athletes were observed to have exercise-induced asthma or bronchoconstriction (114). Strenuous physical exercise of 6–12 min duration can result in airway obstruction with the symptoms being most pronounced 5–10 min after cessation of exercise (107). Running is most likely to cause bronchoconstriction, while swimming is least likely with cycling and walking being intermediate (107). Nasal breathing should help warm and humidify the air and lessen bronchoconstriction as will wearing a cold weather mask during exercise (107). However, cold air breathing during exercise should enhance or accentuate the bronchoconstriction response.

In a study of eight young asthmatic adults who performed heavy cycling exercise of short duration ($\bar{X} = 160.5 \pm 57.5$(S.D.) W; 3.41

± 1.97 (S.D.) min) while breathing air at either ambient (23 to 26° C) or sub-freezing (−11 to −15° C) temperatures, Strauss *et al.* (112) report a marked enhancement in the post-exercise bronchospastic response in these subjects after cold air inhalation relative to ambient air breathing. Ambient air inhalation also produced significant bronchoconstriction. Further, the effects of breathing cold air at rest were quite small as demonstrated from some of these experiments. Additional studies by this group suggest that the potentiating effects of cold air inhalation are local with the proximate stimulus related to cooling of the intrathoracic airways (24).

Increasing the humidity of ambient (22–25° C) or warm (36° C) air temperatures, while performing heavy physical exercise of short duration, has been shown to blunt the degree of airway obstruction in asthmatic children, adolescents and young adult volunteers (8,111). Whether increasing the humidity of cold air during exercise would result in a significant lessening of the severe bronchoconstrictor response is open to question.

Physical exercise and cold air inhalation are thought to be synergistic in terms of the degree of exercise-induced bronchospasm. Strauss *et al.* (112) states that "less well known, but equally important, is the fact that short term exposure to low levels of atmospheric pollutants, like nitrogen dioxide, can increase the sensitivity of asthmatic patients to other bronchoconstrictor substances. . . ." Might not these atmospheric pollutants also act synergistically with exercise and cold air inhalation to produce an even greater degree of exercise-induced bronchoconstriction?

High Terrestrial Altitude

As stated in the Introduction to this chapter, the effects of carbon monoxide at high altitude may be enhanced due to a greater degree of hypoxia. The hypoxemia associated with exercise at altitude is known to shift the oxygen dissociation curve to the right, which permits more O_2 to be released to the tissues at a given blood PO_2 (115). However, CO shifts the oxygen dissociation curve to the left which makes tissue oxygen extraction more difficult. Collectively, the elevated COHb levels resulting from CO exposure and the lower O_2 tension of high altitude may produce additive effects (84).

Only a few studies have evaluated human physiological performance during exercise at high altitude while exposed to CO. In an early study involving three male subjects performing light cycle exercise at a simulated altitude of 4877 m (16,000 ft) while breathing relatively low concentrations of CO, Forbes *et al.* (45) reported an increased CO uptake associated with the observed hyperventilation at high altitude. However, when the values at sea level and simulated altitude were corrected to a standard ventilation rate of 10 L per min (ambient pressure), the differences in CO uptake became insignificant. In another early investigation involving 10 adult men exposed for 25 min (5 min treadmill walking; 20 min rest) to three different simulated altitudes (2134, 3048, 4724 m) at COHb levels of 6 and 13%, Pitts and Pace (91) suggest a 1% increase in blood COHb was equivalent to a 102 m increase in altitude. These same authors indicate that this relationship was established for the altitude range of 2134 to 3048 m (7,000-10,000 ft) and for COHb levels up to 13%.

In more recent experiments involving 12 male adults (six smokers and six non-smokers) performing moderate intensity cycling exercise (53% $\dot{V}O_2max$) for 30 min at 3048 m (10,000 ft) while breathing CO (COHb = 4.2%), Wagner *et al.* (115) state that this COHb concentration and level of altitude represented only marginal additional stressors to respiratory function. When compared to observations at either sea level, sea level while breathing CO, or altitude, the combination of hypoxic stressors (CO plus altitude) during exercise produced slightly greater increases in $\dot{V}_E$ and respiratory rate, but no difference in $\dot{V}O_2$ and TV. During low level CO exposure (COHb=5.1%) involving nine healthy male volunteers at an altitude of 1610 m (5282 ft), Weiser *et al.* (116) show that the impairments in maximal exercise capacity and exercise performance are of the same magnitude as observed at sea level. Since major reductions in $\dot{V}O_2max$ occur at altitudes greater than about 1500 m with an approximate 3% decrement in $\dot{V}O_2max$ per 300 m of additional ascent (87), it is not unreasonable to propose that low-level CO exposure may have a detrimental effect at higher altitudes (≥2000 m).

TABLE 16-4. *Effect of High Altitude on Carboxyhemoglobin (COHb) Concentrations after Carbon Monoxide (CO) Exposure for Eight to Twelve Hours.*

	CO Exposure (ppm)	
COHb(%)	Sea Level	1500 m
1.0	9	6
1.5	12	8
2.0	15	10

After an extensive review of the literature, Mitchell *et al.* (84) conclude that the current National Ambient Air Quality Standards derived at sea level for CO are probably to lenient for altitudes of 1500 m or greater. Table 16-4 shows the CO exposure level (ppm) at sea level and 1500 m necessary to produce equal COHb concentrations of 1, 2 and 3% after 8–12 hrs. These same authors state that California is the only state to use 6 as compared to 9 ppm as an eight hr CO exposure standard for areas greater than 1500 m. In conclusion, additional research is necessary to fully understand the interactive effects of CO during altitude exposure. No definitive research has been reported concerning the potential adverse effects of the other pollutants during exercise performance at altitude.

SUMMARY

The various air pollutants have been classified as primary or secondary pollutants. Primary pollutants are emitted directly to the environment from their source and include carbon monoxide, sulfur oxides, nitrogen oxides and primary particulates. Secondary pollutants develop from interactions of primary pollutants and include ozone, peroxyacetyl nitrate and certain aerosols.

Carbon monoxide does not appear to cause decrements in submaximal exercise performance in healthy individuals; however, cardiovascularly-impaired individuals appear to be at significant risk during submaximal exercise even at low carboxyhemoglobin levels. Maximal exercise performance for healthy individuals seems to be altered by breathing carbon monoxide with the critical concentration being 4.3% carboxyhemoglobin. The threshold level of sulfur dixoxide, which effects submaximal exercise performance in healthy individuals, is between 1.0 and 3.0 ppm while asthmatic individuals and possibly others with pulmonary hyperractivity are affected at a lower threshold concentration between 0.20 and 0.50 ppm. No studies have been reported concerning maximal exercise performance during sulfur dioxide exposure. Several studies suggest that healthy and asthmatic individuals may adapt to sulfur dioxide, but, unfortunately, no research has investigated adaptation to this pollutant during physical exercise. While no studies have been reported which evaluate maximal exercise performance, nitrogen dioxide exposure does not appear to adversely affect submaximal exercise performance in healthy individuals. The physiological performance effects of breathing primary particulates have not been directly evaluated during exercise in man.

Ozone exposure does not appear to limit submaximal exercise performance at light to moderate exercise intensities. At heavy exercise intensities, ozone exposure can limit performance primarily due to severe respiratory discomfort and changes in pulmonary functions. While adaptation occurs after two to five consecutive days of exposure to ozone, this adaptation could eventually be harmful because of the associated suppression of the normal defense mechanisms. Submaximal and maximal exercise performance have not been altered dramatically during peroxyacetyl nitrate exposure at the concentrations tested. The sulfate aerosols, sulfuric acid and the nitrate aerosols elicit minimal adverse effects relative to some of the other pollutants when tested singly.

The various pollutants may interact in three ways: (a) additively, (b) synergistically, or (c) antagonistically. Clear distinction between additive and synergistic interactions is frequently difficult and usually is dependent upon the specific pollutant concentrations. In general, ozone in combination with nitrogen dioxide represent an additive interaction, while ozone combined with peroxyacetyl nitrate seem to suggest synergistic effects. No pollutant interactions have been reported that are antagonistic. Human performance of submaximal or maximal exercise can be expected to suffer under the combined stresses of excessive heat, humidity and poor air quality. The interactive effects of breathing cold polluted air should increase the degree of exercise-induced bronchospasm and adversely effect exercise performance in susceptible individuals. The adverse effects of certain pollutants such as carbon monoxide may be enhanced at high altitude due to a greater degree of hypoxemia.

REFERENCES

1. Adams, W.C., W.M. Savin and A.E. Christo. Detection of ozone toxicity during continuous exercise via the effective dose concept. *J. Appl. Physiol.* 51:415–422, 1981.
2. Adams, W.C. and E.S. Schelegle. Ozone and high ventilation effects on pulmonary function and endurance performance. *J. Appl. Physiol.* 55:805–812, 1983.
3. Andersen, I., G.R. Lundqvist, P.L. Jensen and D.F. Proctor. Human response to controlled levels of sulfur dioxide. *Arch. Environ. Health* 28:31–39, 1974.
4. Anderson, E.W., R.J. Andelman, J.M. Strauch, N.J. Fortuin and J.H. Knelson. Effect of low-level carbon monoxide exposure on onset and duration of angina pectoris: a study in ten patients with ischemic heart disease. *Ann. Intern. Med.* 79:46–50, 1973.
5. Amdur, M.O., W.W. Melvin, Jr. and P. Drinker. Effects of inhalation of sulfur dioxide in man. *Lancet* 1:758–759, 1953.
6. Amdur, M.O., L. Silverman and P. Drinker. Inhalation of sulfuric acid mist by human subjects. *Arch. Ind. Hyg.* 6:305–313, 1952.
7. American Medical Association. Carbon monoxide intoxication associated with use of a resurfacing machine at an ice-skating rink. *JAMA* 251:1016, 1984.
8. Amirav, I., R.J. Dowdeswell and M. Plit. Respiratory heat loss in exercise-induced asthma: measurement and clinical application. *S. Afr. Med. J.* 69:227–232, 1986.
9. Aronow, W.S. and J. Cassidy. Effect of carbon monoxide on maximal treadmill exercise: a study in normal persons. *Ann. Intern. Med.* 83:496–499, 1975.
10. Aronow, W.S., C.N. Harris, M.W. Isbell, S.N. Rokaw and B. Imparato. Effect of freeway travel on angina pectoris. *Ann. Intern. Med.* 77:669–676, 1972.
11. Aronow, W.S. and M.W. Isbell. Carbon monoxide effect on exercise-induced angina pectoris. *Ann. Intern. Med.* 79:392–395, 1973.
12. Arundel, R.A.V., E.M. Sterling, J.H. Biggin and T.D. Sterling. Indirect health effects of relative humidity in indoor environments. *Environ. Health Perspect.* 65:351–361, 1986.
13. Avol, E.L., M.P. Jones, R.M. Bailey, N.N. Chang, M.T. Kleinman, W.S. Linn, K.A. Bell and J.D. Hackney. Controlled exposures of human volunteers to sulfate aerosols. *Am. Rev. Respir. Dis.* 120:319–327, 1979.
14. Avol, E.L., W.S. Linn, D.A. Shamoo, T.G. Venet and J.D. Hackney. Acute respiratory effects of Los Angeles smog in continuously exercising adults. *J. Air Pollut. Control Assoc.* 33:1055–1060, 1983.
15. Avol, E.L., W.S. Linn, T.G. Venet, D.A. Shamoo and J.D. Hackney. Comparative respi-

ratory effects of ozone and ambient oxidant pollution exposure during heavy exercise. *J. Air Pollut. Control Assoc.* 34:804–809, 1984.
16. Bedi, J.F., D.M. Drechsler-Parks and S.M. Horvath. Duration of increased pulmonary function sensitivity to an initial ozone exposure. *Am. Ind. Hyg. Assoc. J.* 46:731–734, 1985.
17. Bedi, J.F., L.J. Folinsbee, S.M. Horvath and R.S. Ebsenstein. Human exposure to sulfur dioxide and ozone: absence of a synergistic effect. *Arch. Environ. Health* 34:233–239, 1979.
18. Bedi, J.F., S.M. Horvath and L.J. Folinsbee. Human exposure to sulfur dioxide and ozone in a high temperature-humidity environment. *Am. Ind. Hyg. Assoc. J.* 43:26–30, 1982.
19. Bell, K.A., W.S. Linn, M. Hazucha, J.D. Hackney and D.V. Bates. Respiratory effects of exposure to ozone plus sulfur dioxide in Southern Californians and Eastern Canadians. *Am. Ind. Hyg. Assoc. J.* 38:696–706, 1977.
20. Bromberg, P.A. and M.J. Hazucha. Is "adaptation" to ozone protective? *Am. Rev. Respir. Dis.* 125:489–490, 1982.
21. Comroe, J., R. Forster, A. Dubois, W. Briscoe and E. Carlsen. *The Lung: Clinical Physiology and Pulmonary Function Tests.* Chicago: Yearbook Medical Publishers, 1962.
22. Corn, M. and G. Burton. The irritant potential of pollutants in the atmosphere. *Arch. Environ. Health* 14:54–61, 1967.
23. Dautrebande, L., W.C. Alford, B. Highman, R. Downing and F.L. Weaver. Studies on aerosols V.: effect of dust and pneumodilating aerosols on lung volume and type of respiration in man. *J. Appl. Physiol.* 1:339–349, 1948.
24. Deal Jr., E.C., E.R. McFadden Jr., R.H. Ingram Jr. and J.J. Jaeger. Effects of atropine on potentiation of exercise-induced bronchospasm by cold air. *J. Appl. Physiol.* 45:238–243, 1978.
25. DeLucia, A.J. and W.C. Adams. Effects of O_3 inhalation during exercise on pulmonary function and blood biochemistry. *J. Appl. Physiol.* 43:75–81, 1977.
26. Dimeo, M.J., M.G. Glenn, M.J. Holtzman, J.R. Sheller, J.A. Nadel and H.A. Boushey. Threshold concentration of ozone causing an increase in bronchial reactivity in humans and adaptation with repeated exposures. *Am. Rev. Respir. Dis.* 124:245–248, 1981.
27. Drechsler-Parks, D.M., J.F. Bedi and S.M. Horvath. Interaction of peroxyacetyl nitrate and ozone on pulmonary functions. *Am. Rev. Respir. Dis.* 130:1033–1037, 1984.
28. Drinkwater, B.L., P.B. Raven, S.M. Horvath, J.A. Gliner, R.O. Ruhling, N.W. Bolduan and S. Taguchi. Air pollution, exercise and heat stress. *Arch. Environ. Health* 28:177–181, 1974.
29. Driscoll, D.M. Base lines for measuring adverse effects of air pollution: some evidence for weather effects on mortality. *Environ. Res.* 4:233–242, 1971.
30. Dubois, A.B. and L. Dautrebande. Acute effects of breathing inert dust particles and of carbachol aerosol on the mechanical characteristics of the lungs in man: changes in response after inhaling sympathomimetic aerosols. *J. Clin. Invest.* 37:1746–1755, 1958.
31. Ekblom, B. and R. Huot. Response to submaximal and maximal exercise at different levels of carboxyhemoglobin. *Acta Physiol. Scand.* 86:474–482, 1972.
32. Ellis, F.P. Mortality from heat illness and heat-aggravated illness in the United States. *Environ. Res.* 5:1–58, 1972.
33. Farrell, B.P., H.D. Kerr, T.J. Kulle, L.R. Sauder and J.L. Young. Adaptation in human subjects to the effects of inhaled ozone after repeated exposure. *Am. Rev. Respir. Dis.* 119:725–730, 1979.
34. Folinsbee, L.J. Effects of ozone exposure on lung function in man: a review. *Rev. Environ. Health* 3:211–240, 1981.
35. Folinsbee, L.J., J.F. Bedi, J.A. Gliner and S.M. Horvath. Concentration dependence of pulmonary function adaptation to ozone. In *The Biomedical Effects of Ozone and Related Photochemical Oxidants.* Edited by M.A. Mehlman, S.D. Lee and M.G. Mustafa. Princeton Junction N.J.: Princeton Scientific Publishers, Inc., pp. 175–187, 1983.
36. Folinsbee, L.J., J.F. Bedi and S.M. Horvath. Combined effects of ozone and nitrogen dioxide on respiratory function in man. *Am. Ind. Hyg. Assoc. J.* 42:534–541, 1981.
37. Folinsbee, L.J., J.F. Bedi and S.M. Horvath. Pulmonary function after 1 h continuous heavy exercise in 0.21 ppm ozone. *J. Appl. Physiol.* 57:984–988, 1984.
38. Folinsbee, L.J., J.F. Bedi and S.M. Horvath. Pulmonary response to threshold levels of sulfur dioxide (1.0 ppm) and ozone (0.3 ppm). *J. Appl. Physiol.* 58:1783–1787, 1985.
39. Folinsbee, L.J., J.F. Bedi and S.M. Horvath. Respiratory responses in humans repeatedly exposed to low concentrations of ozone. *Am. Rev. Respir. Dis.* 121:431–439, 1980.
40. Folinsbee, L.J., S.M. Horvath, J.F. Bedi and J.C. Delehunt. Effect of 0.62 ppm NO_2 on cardiopulmonary function in young male nonsmokers. *Environ. Res.* 15:199–205, 1978.
41. Folinsbee, L.J., S.M. Horvath, P.B. Raven, J.F. Bedi, A.R. Morton, B.L. Drinkwater, N.W. Bolduan and J.A. Gliner. Influence of exercise and heat stress on pulmonary function during ozone exposure. *J. Appl. Physiol.* 43:409–413, 1977.
42. Folinsbee, L.J., F. Silverman and R.J. Shephard. Decrease of maximum work performance following ozone exposure. *J. Appl. Physiol.* 42:531–536, 1977.

43. Folinsbee, L.J., F. Silverman and R.J. Shephard. Exercise responses following ozone exposure. *J. Appl. Physiol.* 38:996–1001, 1975.
44. Folinsbee, L.J. and P.B. Raven. Exercise and air pollution. *J. Sports Sci.* 2:57–75, 1984.
45. Forbes, W.H., F. Sargent and F.J.W. Roughton. The rate of carbon monoxide uptake by normal men. *Am. J. Physiol.* 143:594–608, 1945.
46. Foxcroft, W.J. and W.C. Adams. Effects of ozone exposure on four consecutive days on work performance and $\dot{V}O_2$max. *J. Appl. Physiol.* 61:960–966, 1986.
47. Gibbons, S.I. and W.C. Adams. Combined effects of ozone exposure and ambient heat on exercising females. *J. Appl. Physiol.* 57:450–456, 1984.
48. Gliner, J.A., S.M. Horvath and L.J. Folinsbee. Preexposure to low ozone concentrations does not diminish the pulmonary function response on exposure to higher ozone concentrations. *Am. Rev. Respir. Dis.* 127:51–55, 1983.
49. Gliner, J.A., P.B. Raven, S.M. Horvath, B.L. Drinkwater and J.C. Sutton. Man's physiologic response to long-term work during thermal and pollutant stress. *J. Appl. Physiol.* 39:628–632, 1975.
50. Gong Jr., H., P.W. Bradley, M.S. Simmons and D.P. Tashkin. Impaired exercise performance and pulmonary function in elite cyclists during low-level ozone exposure in a hot environment. *Am. Rev. Respir. Dis.* 134:726– 733, 1986.
51. Hackney, J.D. Relationship between air pollution and cardiovascular disease; a review. In *Clinical Implications of Air Pollution Research.* Edited by A.J. Finkel and W.C. Duel. Acton, MA: Publishing Sciences Group, Inc., pp. 89–106, 1976.
52. Hackney, J.D., W.S. Linn, S.K. Karuza, R.D. Buckley, D.C. Law, D.V. Bates, M. Hazucha, L.D. Pengelly and F. Silverman. Effects of ozone exposure in Canadians and Southern Californians: evidence for adaptation? *Arch. Environ. Health* 32:110–116, 1977.
53. Hackney, J.D., W.S. Linn, D.C. Law, S.K. Karuza, H. Greenberg, R.D. Buckley and E.E. Pedersen. Experimental studies on human health effects of air pollutants III: two-hour exposure to ozone alone and in combination with other pollutant gases. *Arch. Environ. Health* 30:385–390, 1975.
54. Hackney, J.D., W.S. Linn, J.G. Mohler and C.R. Collier. Adaptation to short-term respiratory effects of ozone in men exposed repeatedly. *J. Appl. Physiol.* 43:82–85, 1977.
55. Hackney, J.D., W.S. Linn, J.G. Mohler, E.E. Pedersen, P. Breisacher and A. Russo. Experimental studies on human health effects of air pollutants II: four-hour exposure to ozone alone and in combination with other pollutant gases. *Arch. Environ. Health* 30:379–384, 1975.
56. Hackney, J.D., F.C. Thiede, W.S. Linn, E.E. Pedersen, C.E. Spier, D.C. Law and D.A. Fischer. Experimental studies on human health effects of air pollutants IV: short-term physiological and clinical effects of nitrogen dioxide exposure. *Arch. Environ. Health* 33:176–181, 1978.
57. Hazucha, M. and D.V. Bates. Combined effect of ozone and sulphur dioxide on human pulmonary function. *Nature* 257:50–51, 1975.
58. Horvath, S.M. Impact of air quality in exercise performance. In *Exercise and Sport Sciences Reviews.* Edited by D.I. Miller. Philadelphia: The Franklin Institute Press, pp. 265–296, 1981.
59. Horvath, S.M., J.A. Gliner and L.J. Folinsbee. Adaptation to ozone: duration of effect. *Am. Rev. Respir.* Dis. 123:496–499, 1981.
60. Horvath, S.M., J.A. Gliner and J.A. Matsen-Twisdale. Pulmonary function and maximum exercise responses following acute ozone exposure. *Aviat. Space Environ. Med.* 50:901–905, 1979.
61. Horvath, S.M., P.B. Raven, T.E. Dahms and D.J. Gray. Maximal aerobic capacity at different levels of carboxyhemoglobin. *J. Appl. Physiol.* 38:300–303, 1975.
62. Katz, R.M. Prevention with and without the use of medications for exercise-induced asthma. *Med. Sci. Sports Exerc.* 18:331–333, 1986.
63. Kelly, F.J. and W.E. Gill. Ozone poisoning. *Arch. Environ. Health* 10:517–519, 1965.
64. Kilburn, K. Cilia and mucous transport as determinants of the response of lung to air pollutants. *Arch. Environ. Health* 14:77–91, 1967.
65. Kirkpatrick, M.B., D. Sheppard, J.A. Nadel and H.A. Boushey. Effect of the oronasal breathing route on sulfur dioxide-induced bronchoconstriction in exercising asthmatic subjects. *Am. Rev. Respir. Dis.* 125:627–631, 1982.
66. Kleinfeld, M. and C.P. Giel. Clinical manifestations of ozone poisoning: report of a new source of exposure. *Am. J. Med. Sci.* 231:638–643, 1956.
67. Kleinman, M.T., R.M. Bailey, Y.C. Chang, K.W. Clark, M.P. Jones, W.S. Linn and J.D. Hackney. Exposures of human volunteers to a controlled atmospheric mixture of ozone, sulfur dioxide and sulfuric acid. Am. Ind. Hyg. Assoc. J. 42:61–69, 1981.
68. Kleinman, M.T., W.S. Linn, R.M. Bailey, M.P. Jones and J.D. Hackney. Effect of ammonium nitrate aerosol on human respiratory function and symptoms. *Environ. Res.* 21:317–326, 1980.

69. Koenig, J.Q., W.E. Pierson and R. Frank. Acute effects of inhaled SO_2 plus NaCl droplet aerosol on pulmonary function in asthmatic adolescents. *Environ. Res.* 22:145–153, 1980.
70. Koenig, J.Q., W.E. Pierson, M. Horike and R. Frank. Effects of SO_2 plus NaCl aerosol combined with moderate exercise on pulmonary function in asthmatic adolescents. *Environ. Res.* 25:340–348, 1981.
71. Kreisman, H., C.A. Mitchell, H.R. Hosein and A. Bouhuys. Effect of low concentrations of sulfur dioxide on respiratory function in man. *Lung* 154:25–34, 1976.
72. LaBelle, C.W. and H. Brieger. Synergistic effects of aerosols II: effects on rate of clearance from the lung. *Arch. Indus. Health* 20:100–105, 1959.
73. Lauritzen, S.K. and W.C. Adams. Ozone inhalation effects consequent to continuous exercise in females: comparison to males. *J. Appl. Physiol.* 59:1601–1606, 1985.
74. Leikauf, G., D.B. Yeates, K.A. Wales, D. Spektor, R.E. Albert and M. Lippmann. Effects of sulfuric acid aerosol on respiratory mechanics and mucociliary particle clearance in healthy nonsmoking adults. *Am. Ind. Hyg. Assoc. J.* 42:273–282, 1981.
75. Linn, W.S., M.P. Jones, E.A. Bachmayer, K.W. Clark, S.K. Karuza and J.D. Hackney. Effect of low-level exposure to ozone on arterial oxygenation in humans. *Am. Rev. Respir. Dis.* 119:731–740, 1979.
76. Linn, W.S., M.P. Jones, E.A. Bachmayer, C.E. Spier, S.F. Mazur, E.L. Avol and J.D. Hackney. Short-term respiratory effects of polluted ambient air: a laboratory study of volunteers in a high-oxidant community. *Am. Rev. Respir. Dis.* 121:243–252, 1980.
77. Linn, W.S., M.P. Jones, R.M. Bailey, M.T. Kleinman, C.E. Spier, D.A. Fischer and J.D. Hackney. Respiratory effects of mixed nitrogen dioxide and sulfur dioxide in human volunteers under simulated ambient exposure conditions. *Environ. Res.* 22:431–438, 1980.
78. Linn, W.S., D.A. Medway, U.T. Anzar, L.M. Valencia, C.E. Spier, F.S.D. Tsao, D.A. Fischer and J.D. Hackney. Persistence of adaptation to ozone in volunteers exposed repeatedly for six weeks. *Am. Rev. Respir. Dis.* 125:491–495, 1982.
79. Linn, W.S., T.G. Venet, D.A. Shamoo, L.M. Valencia, U.T. Anzar, C.E. Spier and J.D. Hackney. Respiratory effects of sulfur dioxide in heavily exercising asthmatics. *Am. Rev. Respir. Dis.* 127:278–283, 1983.
80. Lowry, T. and L.M. Schuman. "Silo-filler's disease"-a syndrome caused by nitrogen dioxide. *JAMA* 162:153–160, 1956.
81. McCafferty, W.B. *Air Pollution and Athletic Performance.* Springfield, IL: Charles C. Thomas Publisher, 1981.
82. McDonnell, W.F., R.S. Chapman, M.W. Leigh, G.L. Strope and A.M. Collier. Respiratory responses of vigorously exercising children to 0.12 ppm ozone exposure. *Am. Rev. Respir. Dis.* 132:875–879, 1985.
83. McDonnell, W.F., D.H. Horstman, M.J. Hazucha, E. Seal, Jr., E.D. Haak, S.A. Salaam and D.E. House. Pulmonary effects of ozone exposure during exercise: dose-response characteristics. *J. Appl. Physiol.* 54:1345–1352, 1983.
84. Mitchell, R.S., F.N. Judson, T.S. Moulding, P. Weiser, L.L. Brock, D.L. Keible and J. Pollard. Health effects of urban air pollution: special consideration of areas at 1,500 m and above. *JAMA* 242:1163–1168, 1979.
85. Morris, S.C., M.A. Shapiro and J.H. Waller. Adult mortality in two communities with widely different air pollution levels. *Arch. Environ. Health* 31:248–254, 1976.
86. Nadel, J.A. and J.H. Comroe, Jr. Acute effects of inhalation of cigarette smoke on airway conductance. *J. Appl. Physiol.* 16:713–716, 1961.
87. Pandolf, K.B. Importance of environmental factors for exercise testing and exercise prescription. In *Exercise Testing and Exercise Prescription for Special Cases—Theoretical Basis and Clinical Application.* Edited by J.S. Skinner. Philiadelphia, PA: Lea & Febiger, pp. 77–98, 1987.
88. Pierson, W.E., D.S. Covert and J.Q. Koenig. Air pollutants, bronchial hyperreactivity, and exercise. *J. Allergy Clin. Immunol.* 73:717–721, 1984.
89. Pierson, W.E., D.S. Covert, J.Q. Koenig, T. Namekata and Y.S. Kim. Implications of air pollution effects on athletic performance. *Med. Sci. Sports Exerc.* 18:322–327, 1986.
90. Pirnay, F., J. Dujardin, R. Deroanne and J.M. Petit. Muscular exercise during intoxication by carbon monoxide. *J. Appl. Physiol.* 31:573–575, 1971.
91. Pitts, G.C. and N. Pace. The effect of blood carboxyhemoglobin concentration on hypoxia tolerance. *Am. J. Physiol.* 148:139–151, 1947.
92. Posin, C., K. Clark, M.P. Jones, J.V. Patterson, R.D. Buckley and J.D. Hackney. Nitrogen dioxide inhalation and human blood biochemistry. *Arch. Environ. Health* 33:318–324, 1978.
93. Raven, P.B. Air pollution and physical activity. *Sports Med. Dig.* 6:1–3, 1984.
94. Raven, P.B. Heat and air pollution: the cardiac patient. In *Heart Disease and Rehabilitation.* Edited by M.L. Pollock and D.H. Schmidt. Boston: Houghton and Mifflin Book Company, pp. 563–586, 1979.
95. Raven, P.B., B.L. Drinkwater, S.M. Horvath, R.O. Ruhling, J.A. Gliner, J.C. Sutton and N.W. Bolduan. Age, smoking habits, heat stress, and their interactive effects with carbon

monoxide and peroxyacetylnitrate on man's aerobic power. *Int. J. Biometeor.* 18:222–232, 1974.

96. Raven, P.B., B.L. Drinkwater, R.O. Ruhling, N. Bolduan, S. Taguchi, J. Gliner and S.M. Horvath. Effect of carbon monoxide and peroxyacetyl nitrate on man's maximal aerobic capacity. *J. Appl. Physiol.* 36:288–293, 1974.
97. Raven, P.B., J.A. Gliner and J.C. Sutton. Dynamic lung function changes following long-term work in polluted environments. *Environ. Res.* 12:18–25, 1976.
98. Sackner, M.A., D. Ford, R. Fernandez, J. Cipley, D. Perez, M. Kwoka, M. Reinhart, E.D. Michaelson, R. Schreck and A. Wanner. Effects of sulfuric acid aerosol on cardiopulmonary function of dogs, sheep and humans. *Am. Rev. Respir. Dis.* 118:497–510, 1978.
99. Savin, W.M. and W.C. Adams. Effects of ozone inhalation on work performance and $\dot{V}O_2$max. *J. Appl. Physiol.* 46:309–314, 1979.
100. Schelegle, E.S. and W.C. Adams. Reduced exercise time in competitive simulations consequent to low level ozone exposure. *Med. Sci. Sports* Exerc. 18:408–414, 1986.
101. Shephard, R. Cigarette smoking and reactions to air pollutants. *Can. Med. Assoc. J.* 118:379–381, 1978.
102. Shephard, R.J., B. Urch, F. Silverman and P.N.J. Corey. Interaction of ozone and cigarette smoke exposure. *Environ. Res.* 31:125–137, 1983.
103. Sheppard, D., J. Epstein, R.A. Bethel, J.A. Nadel and H.A. Boushey. Tolerance to sulfur dioxide-induced bronchoconstriction in subjects with asthma. *Environ. Res.* 30:412–419, 1983.
104. Sheppard, D., A. Saisho, J.A. Nadel and H.A. Boushey. Exercise increases sulfur dioxide-induced bronchoconstriction in asthmatic subjects. *Am. Rev. Respir. Dis.* 123:486–491, 1981.
105. Sheppard, D., W.S. Wong, C.F. Uehara, J.A. Nadel and H.A. Boushey. Lower threshold and greater bronchomotor responsiveness of asthmatic subjects to sulfur dioxide. *Am. Rev. Respir. Dis.* 122:873–878, 1980.
106. Sim, V.M. and R.E. Pattle. Effect of possible smog irritants on human subjects. *JAMA* 165:1908–1913, 1957.
107. Sly, R.M. History of exercise-induced asthma. *Med. Sci. Sports Exerc.* 18:314–317, 1986.
108. Snell, R.E. and P.C. Luchsinger. Effects of sulfur dioxide on expiratory flow rates and total respiratory resistance in normal human subjects. *Arch. Environ. Health* 18:693–698, 1969.
109. Speizer, F. and N. Frank. A comparison of changes in pulmonary flow resistance in healthy volunteers acutely exposed to SO_2 by mouth and nose. *Br. J. Ind. Med.* 23:73–79, 1966.
110. Stacy, R.W., E. Seal, Jr., D.E. House, J. Green, L.J. Roger and L. Raggio. A survey of effects of gaseous and aerosol pollutants on pulmonary function of normal males. *Arch. Environ. Health* 38:104–115, 1983.
111. Strauss, R.H., E.R. McFadden, Jr., R.H. Ingram, Jr., E.C. Deal, Jr. and J.J. Jaeger. Influence of heat and humidity on the airway obstruction induced by exercise in asthma. *J. Clin. Invest.* 61:433–440, 1978.
112. Strauss, R.H., E.R. McFadden, Jr., R.H. Ingram, Jr. and J.J. Jaeger. Enhancement of exercise-induced asthma by cold air. *N. Engl. J. Med.* 297:743–747, 1977.
113. Vogel, J.A. and M.A. Gleser. Effect of carbon monoxide on oxygen transport during exercise. *J. Appl. Physiol.* 32:234–239, 1972.
114. Voy, R.O. The U.S. Olympic Committee experience with exercise-induced bronchospasm, 1984. *Med. Sci. Sports Exerc.* 18:328–330, 1986.
115. Wagner, J.A., S.M. Horvath, G.M. Andrew, W.H. Cottle and J.F. Bedi. Hypoxia, smoking history, and exercise. *Aviat. Space Environ. Med.* 49:785–791, 1978.
116. Weiser, P.C., C.G. Morrill, D.W. Dickey, T.L. Kurt and G.J.A. Cropp. Effects of low-level carbon monoxide exposure on the adaptation of healthy young men to aerobic work at an altitude of 1,610 meters. In *Environmental Stress: Individual Human Adaptations.* Edited by L.J. Folinsbee, J.A. Wagner, J.F. Borgia, B.L. Drinkwater, J.A. Gliner and J.F. Bedi. New York: Academic Press, Inc., pp. 101–110, 1978.
117. West, J.B. *Respiratory Physiology—the Essentials.* 3rd Edition, Baltimore: Williams and Wilkins, 1985.
118. Widdicombe, J.G., D.C. Kent and J.A. Nadel. Mechanism of bronchoconstriction during inhalation of dust. *J. Appl. Physiol.* 17:613–616, 1962.
119. Wolff, R.K., M. Dolovich, G. Obminski and M. Newhouse. Effect of sulfur dioxide on tracheobronchial clearance at rest and during exercise. In *Inhaled Particles.* Edited by W.H. Walton and B. McGovern. New York: Pergamon Press, pp. 321–332, 1977.

ACKNOWLEDGEMENT

The author gratefully acknowledges the expert technical assistance of Ms. Edna R. Safran in the preparation of the chapter.

Index